Ocean Variability & Acoustic Propagation

Ocean Variability
& Acoustic Propagation

edited by

John Potter

and

Alex Warn-Varnas
SACLANT Undersea Research Centre,
La Spezia, Italy

sponsored by

North Atlantic Treaty Organization
US Navy Office of Naval Research
Centro Ricerche Energia Ambiente (ENEA)

SPRINGER SCIENCE+BUSINESS MEDIA, B.V.

Proceedings of the Workshop on
Ocean Variability & Acoustic Propagation
La Spezia, Italy
June 4–8, 1990

Library of Congress Cataloging-in-Publication Data

Ocean variability & acoustic propagation / edited by John Potter and
  Alex Warn-Varnas ; sponsored by North Atlantic Treaty Organization,
  US Navy Office of Naval Research, Centro ricerche energia ambiente
  (ENEA).
       p.   cm.
  Papers from a workshop.
  Includes indexes.
  ISBN 978-94-010-5462-1   ISBN 978-94-011-3312-8 (eBook)
  DOI 10.1007/978-94-011-3312-8
   1. Underwater acoustics. 2. Oceanography.   I. Potter, John.
  II. Warn-Varnas, Alex. III. North Atlantic Treaty Organization.
  IV. United States. Office of Naval Research. V. Centro ricerche
  energia ambiente Santa Teresa. VI. Title: Ocean variabiligy and
  acoustic propagation.
  QC244.O34   1991
  534'.23--dc20   90-25672

ISBN 978-94-010-5462-1

TABLE OF CONTENTS

Ocean Variability & Acoustic Propagation
Lerici 4–8 June 1990

PREFACE

Fifteen years ago NATO organised a conference entitled 'Ocean Acoustic Modelling'. Many of its participants were again present at this variability workshop. One such participant, in concluding his 1975 paper, quoted the following from a 1972 literature survey:

'... history presents a sad lack of communications between acousticians and oceanographers'

Have we done any better in the last 15 years? We believe so, but only moderately. There is still a massive underdeveloped potential for acousticians and oceanographers to make significant progress together. Currently, the two camps talk together insufficiently even to avoid simple misunderstandings, such as those in Table 1.

Table 1   Oceanographic and acoustic jargon (from an idea by Pollard)[1]

Jargon	Oceanographic use	Acoustic use
db or dB	decibar (depth in m)	decibel (energy level)
PE	primitive equations	parabolic equations
convergence zone	converging currents (downwelling water)	converging rays (high energy density)
front	thermohaline front	wave, ray or time front
speed	water current speed	sound propagation speed

[1]   The list goes on.

The time-varying aspect of the ocean makes descriptive oceanography a 4-dimensional problem, with spatial scales from 10^7 to 10^{-3} m, energy cascading ever-downward over 10 orders of magnitude. In-situ sensing alone is ultimately unable to provide a sufficient description. The ocean is fundamentally opaque to electromagnetic radiation, but this limitation on information transfer is largely compensated by its ability to propagate compressional energy over thousands of kilometers. Sound, then, is the only natural communication mechanism in the ocean, be it for geophysical surveying, whale song or remote sensing of ocean structures. Conversely, the acoustician needs to know the ocean structure if he wishes to correctly predict an acoustic field, a 7-dimensional problem. The futures of oceanography and acoustics are thus inextricably bound together. It is surprising, then, to find that acousticians and oceanographers have not felt each other's interests more pertinent. The confusion in terminology is a trivial problem to resolve, the lack of mutual interest that gives rise to it remains a serious obstacle.

Consider the most basic acoustic problem, that of predicting an acoustic field. To what spatial resolution do the acousticians need to know the ocean for modelling? This is a question often asked by oceanographers of acousticians, who can rarely agree on an answer. We believe that the question is fundamentally unanswerable by acousticians. Acoustic energy within the body of the ocean is influenced only by gradients in the local refractive index. The acoustician should give an answer which is limited to the acoustic parameter space. He might say, for example, that he needs an ocean data point every time the refractive index changes by 0.02%. How this is mapped onto the physical space of the oceanographer depends on whether we are mapping a thermocline, abyssal water, front or whatever. This question can only be answered by the oceanographer. Table 2 gives our idea of order-of-magnitude values for the acoustic space, oceanographic space and the resulting required oceanographic sampling resolution (in metres) in the vertical and horizontal directions.

Table 2   *Required oceanographic measurement resolution for acoustic modelling*

Acoustic scale		Oceanographic variability scale					
		mixed region $[O(10^2)$ km]		eddy $[O(10^1)$ km]		front $[O(10^0)$ km]	
		surface	deep	surface	deep	surface	deep
long-range [$O(10^3)$ km range, $O(10^1)$ Hz freq.]	hor.	10^4	10^5	10^3	10^3	200	200
	vert.	50	500	15	15	15	25
medium-range [$O(10^2)$ km range, $O(10^2)$ Hz freq.]	hor.	10^3	10^4	400	400	100	100
	vert.	5	100	2	5	2	5
short-range [$O(10^1)$ km range, $O(10^3)$ Hz freq.]	hor.	250	10^3	200	200	50	50
	vert.	1	25	1	2	0.5	0.5

Once we decide how to sample our acoustic and oceanographic fields, how are we to connect them? Recently, some success has been achieved in using oceanographic forecasting and nowcasting as input to acoustic models. This is very fine, but not a true collaboration because the process is serial. First the oceanographers do their work, independently of the acousticians. Then the acousticians do theirs, with no reference to how the oceanographers were able to provide the input, only that it exists. The reverse serial process would be that in which acoustic observations are used to calculate the oceanographic physical structure (tomography). The oceanographers are then let loose on the dynamic problems to explain these structures. A more satisfactory unification would be to work acoustically and oceanographically in parallel. This may be achieved in a number of ways, perhaps through simulated annealing; slowly 'cooling' the freedom of parameter changes in the oceanographic and acoustic parameter space simultaneously. The solution provides not only the source location but also the range-dependence of the intervening medium.

If there is one issue to pound home the need to work properly together, it is the treatment of de-terministic versus stochastic descriptions. In the ocean, larger scale features tend to determine the evolution of smaller scales, by entropic disordering of the available energy. In general, small scales do not play a strong role in governing the behaviour of larger features. This assumption is also commonly applied to relate the ocean to acoustics. The ocean environment is split into a deterministic (large-scale) component plus a stochastic (small-scale) residue. Firstly, the classifi-cation of deterministic versus stochastic may have nothing to do with scale, so we are in immediate trouble. Secondly, the line (in the spatial frequency domain) is drawn arbitrarily; usually a choice is forced onto the acoustician by the limited resolution of the ocean data. Acoustic models are run on the 'deterministic' component to obtain the mean acoustic field. The 'stochastic' component is then thought of as super-imposing a variability. This is equivalent to assuming a monotonic mapping of scales from the ocean to the acoustic field, with no cross-coupling. Clearly this is technically wrong but it has often been hopefully assumed that, apart from some special cases, the approximation is not too bad, i.e. the cross-coupling is weak and local. One of the products of this workshop was that where ocean variability is strong (almost everywhere), the coupling extends from the small (ocean scale) to the large (mean acoustic field). The mapping of oceanography into acoustics is strongly non-linear. It is time to abandon the general use of the separable assumption. Further, we must come to terms with what is truly deterministic in the ocean and what must be treated statistically, or in some other (as yet) undeveloped way.

The five-day workshop comprised one session per day, with the intention of rubbing a roughly equal number of oceanographers and acousticians together over a mixed bag of presentations. A prerequisite was that all participants should also be authors of a paper. This helped to engender an active atmosphere, in which everyone felt truly involved. The allocation of papers to sessions was kept rather arbitrary, to inhibit participants from favouring particular sessions over others of equal importance. Thus they were unable to exercise preconceptions about which presentations were most pertinent to their particular interest. Over half each working day was devoted to discussion, either immediately following each presentation, or during the end-of-day 'brain-storming' period, steered by the session chairmen. Their impressions are summarised in the closing articles to be found in this book at the end of each session. For this reason the sessions and papers are published in the order in which they were presented at the workshop.

To help the reader find his way around the book, both author and subject indices are included at the back. We hope that the reader will find the contents of this book informative, interesting and fun. For whatever good things come of this, the participants are to be credited; not only for their written contributions but also for their enthusiasm and creativity which made it all work. The session chairmen deserve special recognition, for theirs was the hardest job. We are most grateful to the Director and staff of ENEA, who allowed us to use their laboratory and facilities for the workshop. Finally, to our workshop secretaries – Anna Bizzarri, Caroline Durville and Jeanne van den Beuken – go our respect and thanks for having taken such good care of us throughout the week, always with suave efficiency and a smile.

John Potter &
Alex Warn-Varnas.

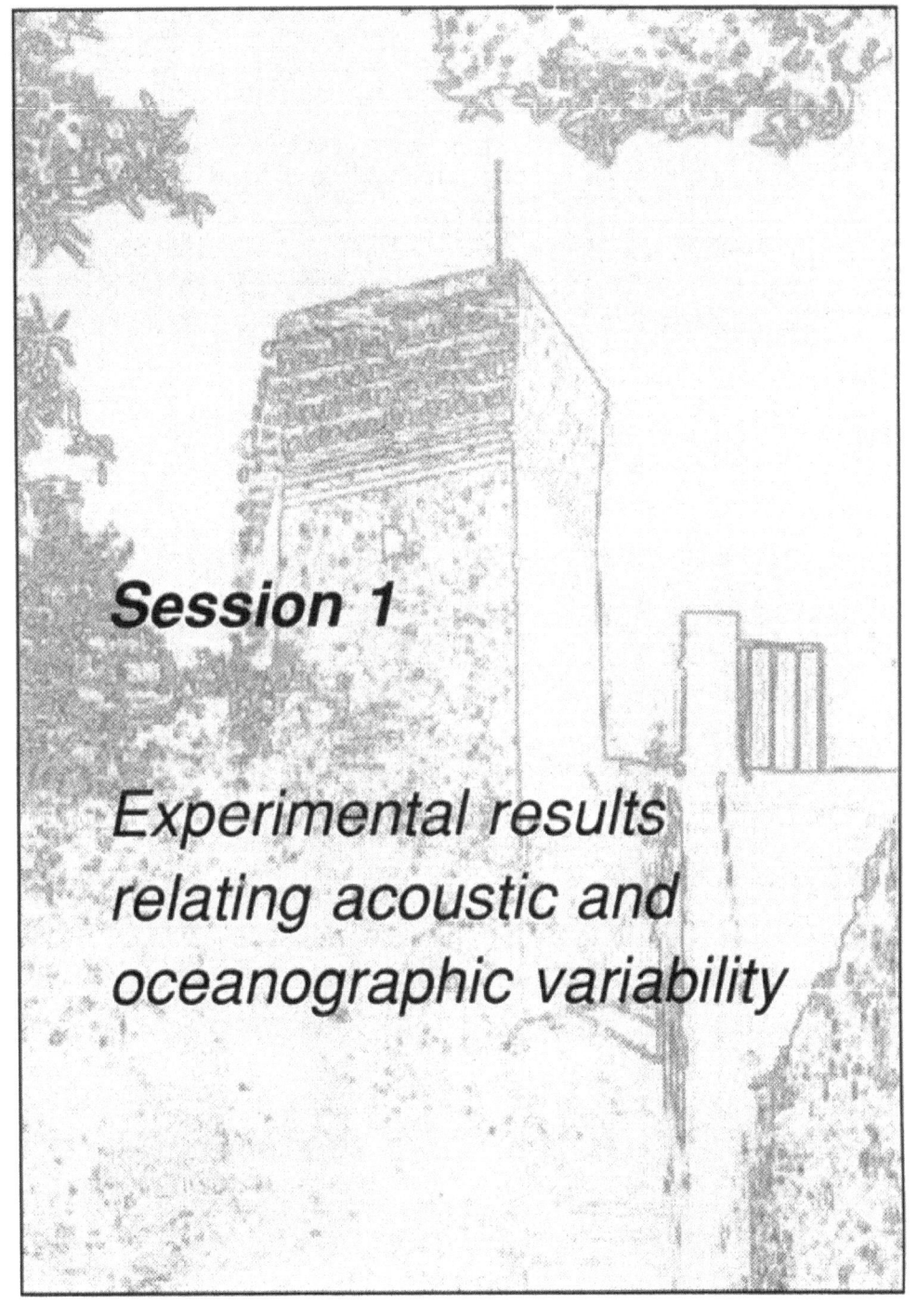

Session 1

*Experimental results
relating acoustic and
oceanographic variability*

DETERMINISTIC-STOCHASTIC OCEANOGRAPHIC DESCRIPTIONS FOR OCEAN-ACOUSTIC EXPERIMENTS

M. G. BRISCOE
Office of Naval Research
800 North Quincy Street
Arlington, Virginia 22217-5000
U.S.A.

ABSTRACT. The SACLANT Research Centre led the way twenty years ago in the design and performance of oceanographic-acoustic experiments that took into account the inevitable variability of the ocean, and hence of the acoustic propagation through it. One of these early experiments will be described from the perspective of modern knowledge of the ocean. The principal conclusion is that ocean acoustic experiments should assume ocean variability and be designed accordingly. The corollary conclusion is that knowledge of the oceanic processes causing the variability can greatly simplify the measurement and description problem. The blending of statistical and deterministic ocean descriptions, and the effect of sparse knowledge, will briefly be discussed.

1. Introduction

1.1 HISTORICAL BACKGROUND

Twenty years ago at the (then) NATO SACLANT ASW Research Centre (SACLANTCEN), I was part of an experiment [1] to combine oceanographic measurements and acoustic measurements so as to try and answer a very straightforward question that had been asked by Director I.R. Van Batenburg:

"Given as much oceanographic data as possible, how well would predictions of propagation loss based on this oceanographic data compare with what is observed acoustically?"

We designed and performed an experiment called the Joint Oceanic Underwater Sound Trial (JOUST) to address this question. The significant aspect of the experiment was an overt attempt to take into account the variability of the oceanography and of the acoustic propagation by careful sampling of both fields and by statistical analyses based on the physics of both processes.

Looking back on that experiment from the perspective of twenty years of hindsight, I can see that we were far ahead of our time and impressively insightful about the problem, and yet most aspects of the experiment were, never the less, rather naïve. The next part of this paper (Section 2) is a brief description of the oceanographic side of that experiment and its major results. The ocean analysis described here was never incorporated into the ocean-acoustic joint analysis; we didn't really know how to do the analysis then. We are smarter

3

J. Potter and A. Warn-Varnas (eds.), Ocean Variability & Acoustic Propagation, 3–22.
© 1991 *Kluwer Academic Publishers.*

4

now; it could be fun and useful to return to the data set.

The short conclusion of that 1970 SACLANTCEN experiment was that the ability to describe the ocean exceeded the ability to model the acoustic propagation through it. At the time, the acoustic propagation model was based on ray-tracing; today, the propagation model would probably be based on the parabolic equation, which has never been verified to the level of detail of that old SACLANTCEN experiment, except for some specialized, high-frequency, short-range experiments like MATE [2].

The final part of this paper (Section 3) is based on a brief comparison of the scales of variability of the ocean and the consequent smoothing provided by long wavelength sound propagating through short-scale variability. It suggests some caution to those who wish to go into the variable ocean and make comparisons between oceanographic data, acoustic propagation models, and acoustic measurements.

2. The 1970 JOUST Experiment at SACLANTCEN

The only available report on JOUST is classified [1] because of its acoustic content and operational conclusions. The description here addresses the oceanographic part of JOUST and how one can describe oceanographic variability so as to be useful acoustically. This statistical oceanographic description was begun but not completed at the time of JOUST, and was not incorporated into the acoustic analysis of JOUST.

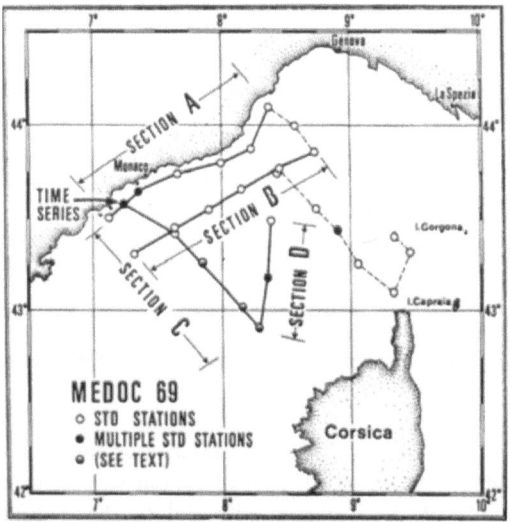

Figure 1. SACLANTCEN portion of the MEDOC [3] project in 1969. Salinity-Temperature-Depth (STD) stations are shown. The figure is from the study in [1]; the half-black circles are not relevant here.

In February 1969 an international "Mediterranean Oceanography" expedition called MEDOC took place in the vicinity of the Gulf of Lyon [3]; the objective was to examine the formation of cooling and sinking of surface water when the cold, dry Mistral winds blow down the Rhone valley and out over the Mediterranean Sea. This "deep water formation" process, specifically its prediction and modeling, remains today an important problem in oceanography and central to the ocean's response to global climate change.

SACLANTCEN took part in MEDOC-69 with oceanographic surveys of the eastern end of the study region. Figure 1 shows the SACLANTCEN survey patterns. The general result of the survey was nearly isothermal, isohaline water to at least 500 m depth in the far offshore region (e.g., Section D), and increasingly complicated water masses nearer the French coast. Figure 2 shows the spatial complexity along Section B, for example, and Figure 3 displays a time series of profiles at the indicated point southwest of Monaco. The situation of colder water overlying warmer water is not unusual in the winter northwest Mediterranean near the coast where river, ice and snow runoff

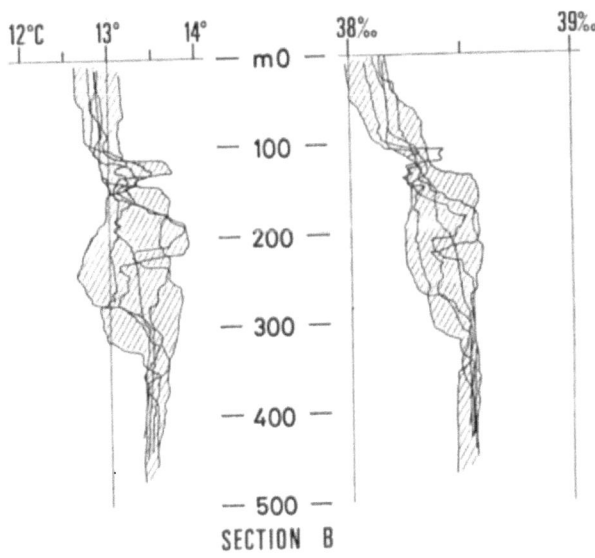

Figure 2. Overplot of temperature (left) and salinity (right) profiles from Section B in Figure 1.

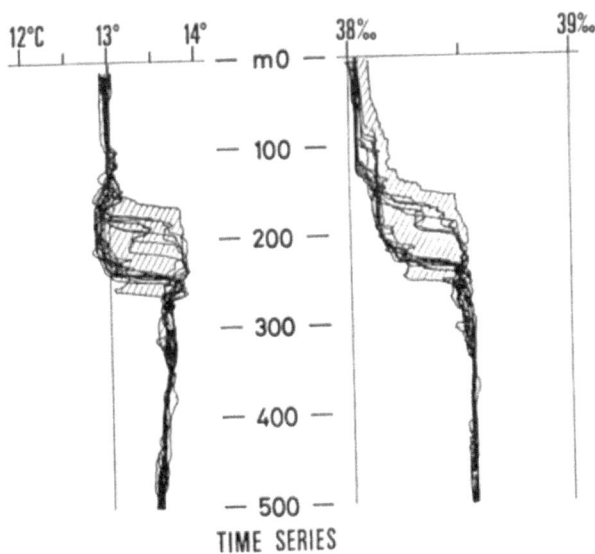

Figure 3. Overplot of temperature (left) and salinity (right) profiles from Time Series station on Section A in Figure 1.

provides fresher, lighter water near the surface.

The plan for the Joint Oceanographic Underwater Sound Trial (JOUST) was to return to the eastern MEDOC region in February 1970; the goal was to work in both the complicated region near the coast and to exploit the simplicity of the ocean in the far offshore region. The critical element of the JOUST planning was the attempt to take into account the spatial and temporal variability of the ocean, as observed in 1969. The principal descriptor of the variability was the structure function, defined for the process h on the variable x with lag ξ:

$$D_h(x;\xi) = <[h(x) - h(x+\xi)]^2>$$ (1)

where $< >$ signifies the expected value over the ensemble. For stationary processes, the structure function reduces to:

$$D_h(\xi) = 2\sigma_h^2[1 - C(\xi)]$$ (2)

where σ_h^2 is the variance of the process h and $C(\xi)$ is the normalized correlation function. Essentially all theoretical treatments (e.g., [4-6] over a 40-year period) of acoustic propagation through a variable ocean use the correlation function as the basic descriptor, hence its importance in this context.

The modeling of acoustic propagation demands the purely spatial structure function in Equation (1) but since the measurements along a propagation path can only be made during a time very long compared to the time it takes for the acoustics to "sample" the propagation path, actual *measurements* produce the mixed space-time structure function:

$$D_h(x,t;\xi,\tau) = <[h(x,t) - h(x+\xi,t+\tau)]^2>$$ (3)

where now there are lags both in space (ξ) and time (τ). It is also possible to measure purely temporal structure functions at a point:

$$D_h(x,t;\tau) = <[h(x,t) - h(x,t+\tau)]^2>$$ (4)

2.0.1. *Hypothesis.* The essential hypothesis is that one can estimate the desired purely spatial structure function (1) from a combination of the mixed space-time (3) and purely temporal (4) structure functions; this is analogous to the definition of a transfer function. Dropping physical variables and showing only lag variables, the hypotheses is:

$$D_h(\xi) \propto \frac{D_h(\xi,\tau)}{D_h(\tau)}$$ (5)

The validity of this hypothesis will be discussed later.

2.1 SAMPLING STRATEGY

In JOUST, the oceanographic sampling plan was designed to provide the data to give the mixed space-time and the purely temporal structure functions; it was to be worked out later how they might be combined to provide the medium description to enable theoretical analysis of the acoustic propagation. Figure 4 (from [1]) shows what actually took place: on 12 Feb 70 there were oceanographic and acoustic measurements near the French coast, and on 17 Feb 70 there were measurements in the simpler offshore waters. In both cases, the SACLANTCEN ship Maria Paolina G. (MPG) sat stationary in the water making shipborne expendable bathythermograph (SBT) profiles every 5 or 10 minutes, while 2 or 3 minesweepers (FILICUDI and PIOPPO from Italy and CHRYSANTHEME from France) steamed along the test section making SBT drops at the same moment as did the MPG. The

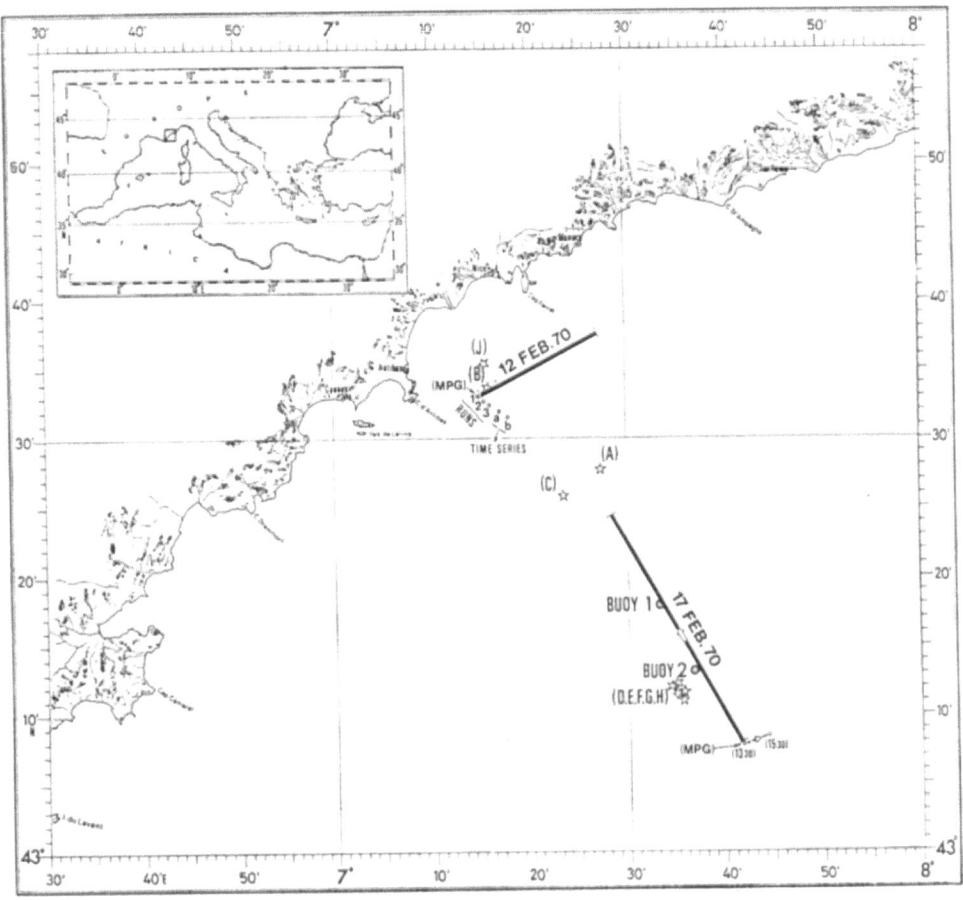

Figure 4. Measurements sites during JOUST [1], in the northwestern portion of the surveys shown in Figure 1. The inset shows the location of the detailed chart. The stars are locations of Salinity-Temperature-Depth (STD) stations A through J. Two moored thermistor buoys were located along the 17 Feb test section; their data are not discussed here. The position of the Maria Paolina G. (MPG) for the start of each of three runs (1-2-3) on 12 Feb is shown, as is the start position for two time-series stations (a-b). On 17 Feb the drift of the MPG over the run is shown.

minesweepers kept a fixed distance from each other so a structure function could be calculated from their data that would be based on paired time series, moving together through the sampling space, at a fixed spacing Δx:

$$D_h(x,t;\Delta x,\tau) = <[h(x,t)-h(x+\Delta x,t+\tau)]^2>$$

(6)

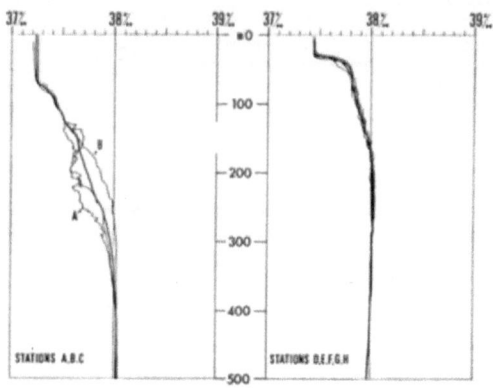

Figure 5. Salinity profiles at stations A-B-C (left) and D-E-F-G (right) of Figure 4. At the offshore position (right) there is a shallow and strong halocline, and little variability at depth.

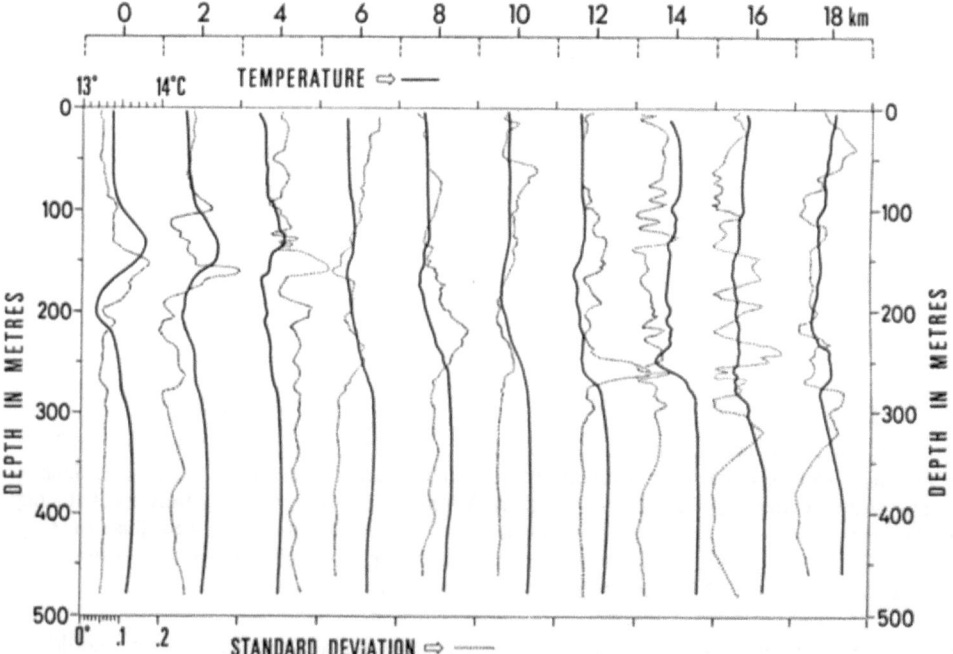

Figure 6. Temperature data along the 18 km test section of 12 Feb in Figure 4. The temperatures are means over all the data within 2 km range bins. The standard deviations are over the same data in the same bins.

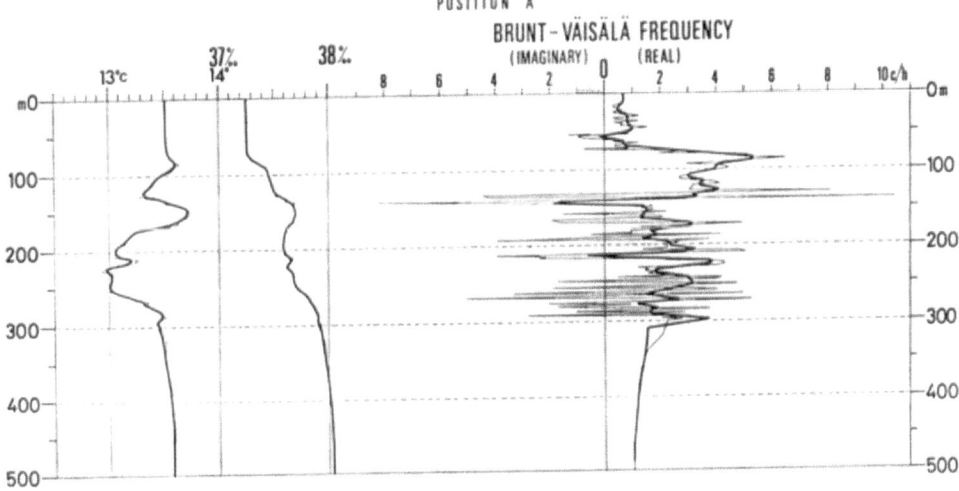

Figure 7. Temperature, salinity, and Brunt-Väisälä (buoyancy) frequency profiles for station A in Figure 4. The heavy lines are smoothed versions of the raw data (light line).

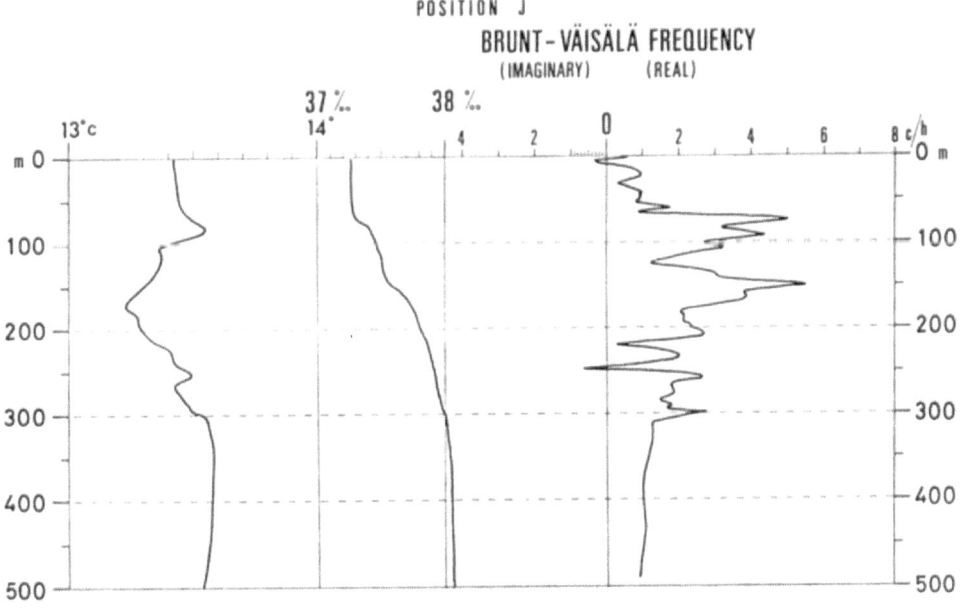

Figure 8. Same as Figure 7, but for station J (no raw data are shown).

2.2 RESULTS FOR OCEANOGRAPHIC VARIABILITY

Time series stations of salinity near the 12 Feb runs are shown in Figure 5 as stations A,B, and C; the quite different and more stable structure near the 17 Feb runs are also in Figure 5 as stations D,E, F, and G.

The spatial variability of temperature along the 12 Feb test section is shown in Figure 6, with the mean temperature profile in each 2-km range bin shown as a solid line (these were the profiles used in the range-dependent ray-tracing for the acoustic analysis in [1]) and the standard deviation given as a dotted line. The end of the test section nearest the MPG appears to have its greatest variability in the 100-200 m depth interval, whereas the other end of the test section is most variable in a deeper range. Note that, for constant salinity, the variance of sound speed in (m/s)^2 is approximately 20 times the variance of temperature in (deg C)^2.

Smoothed Brunt-Väisälä frequencies are shown in Figures 7 (station A) and 8 (station J); the corresponding temperature and salinity profiles are on the left sides of each figure. Figure 7 also shows the unsmoothed data as a light line to illustrate the smoothing that has been applied. The significance of these plots is that one can expect internal wave fluctuations of several cycles per hour (10-30 min periods) in the test areas.

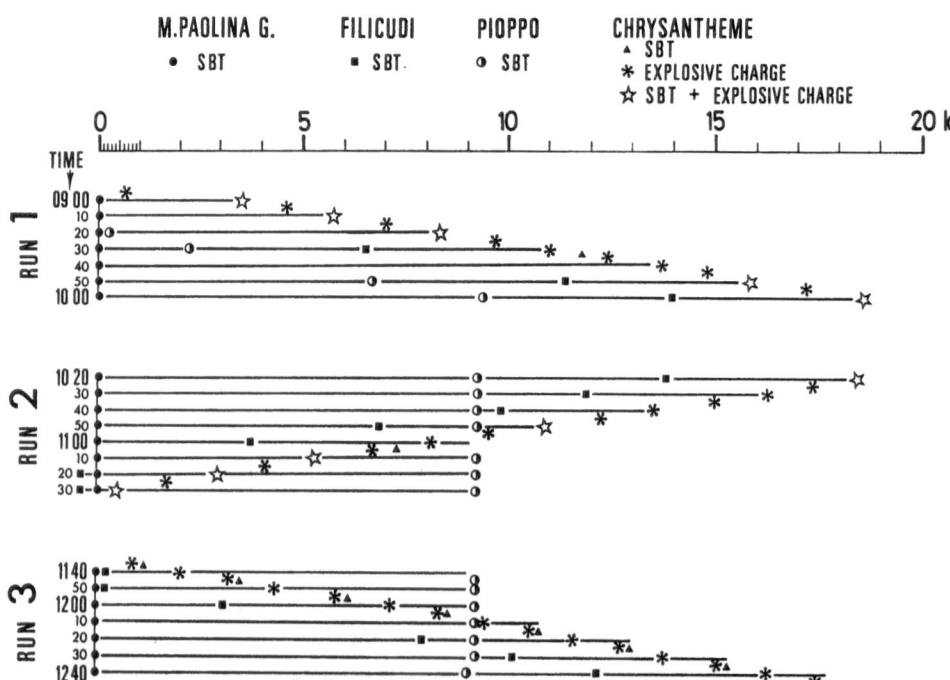

Figure 9. Feb 12 test runs along track shown in Figure 4. The shipboard expendable bathythermograph (SBT) positions are shown as circles, squares, or triangles for the four ships involved. Positions of explosive charges for the acoustic side of the experiment are shown as asterisks or stars. The Figure shows that the MPG remained stationary while the Chrysantheme and the Filicudi ran out to the end of the range, came back, and went out again. The Pioppo ran out half way, then held position. All four ships took simultaneous SBT profiles nominally every ten minutes.

2.2.1. *Sampling for Structure functions.* The 12 Feb data (Figure 9) is comprised of three runs (out, back and out again) over the 19 km test section, and had the additional feature of PIOPPO stopping at about 9.5 km range to maintain a time series at that point. The ship spacing for the paired, moving time series was about 5 km. All sampling was every 10 minutes over the 3h 40min total experiment.

The 17 Feb data (Figure 10) ran out once to 35 km range, with 10 km ship spacing, and 5 minute drop rates for the SBTs. The whole test was just 2h long.

(For reference, the weather in between 12 and 17 Feb, and later on the 17th, was too rough for safe working. Also, the asterisks and stars in Figures 5 and 6 show when and where explosive charges were dropped; see [1] for the acoustic results.)

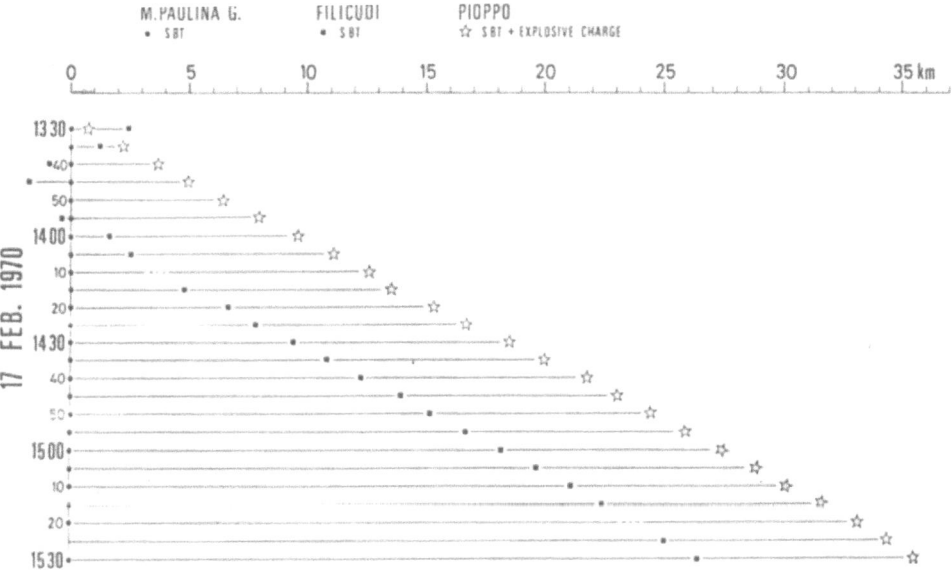

Figure 10. Same as Figure 9 but for 17 Feb. Only three ships are involved, the test section is longer, and the SBT profile interval is every 5 minutes.

2.2.2. *Analysis.* The principal analysis of the oceanographic data is contained in Figures 11 through 14. Figures 11 and 12 show the average temperature structure function (i.e., mean-squared change in temperature) versus depth for the stationary ship (MPG) and a moving ship (CHRYSANTHEME on 12 Feb and PIOPPO on 17 Feb); the average is over all data from the test section on that day, for a fixed time lag of 10 or 50 minutes, as indicated. In Figure 11 for 12 Feb, the essential features are:

• the moving ship sees more mean-squared change in temperature during a 10 minute interval than does the stationary ship, over most but not all depth ranges;
• the moving ship generally sees more mean-squared change in temperature during a 50 minute interval than it does during a 10 minute interval;
• the additional mean-squared temperature change accumulated during a 50 minute lag versus a 10 minute lag is greatest near the surface and least at depth;
• the moving ship appears to encounter variability in the 300 m depth vicinity that is not seen at all by the stationary ship.

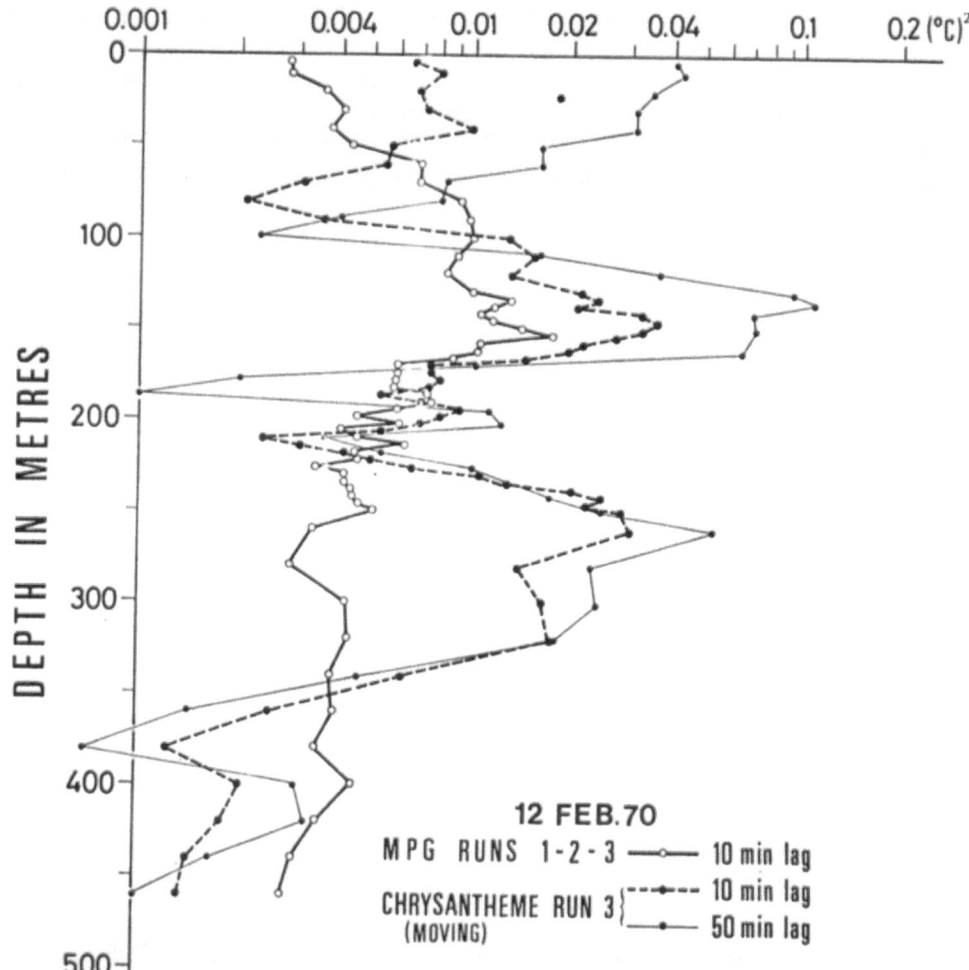

Figure 11. For the 12 Feb runs, the plot shows the mean-square change in the measured temperature, versus depth, for a ten-minute interval between profiles (heavy solid and heavy dashed line) and for a 50-minute internal between profiles (light line). Data from the stationary MPG and the moving Chrysantheme are shown.

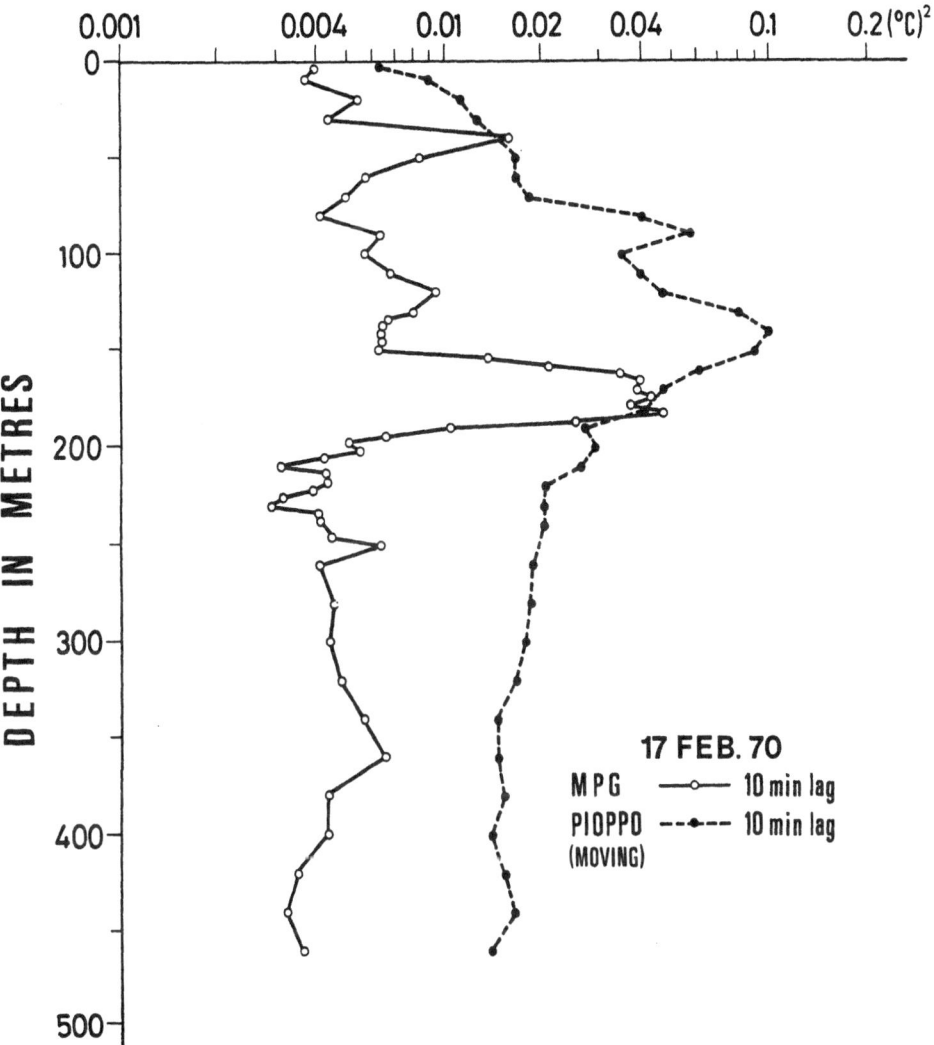

Figure 12. Same as Figure 11, but for 17 Feb. Only the mean-square changes over 10-minute intervals are shown.

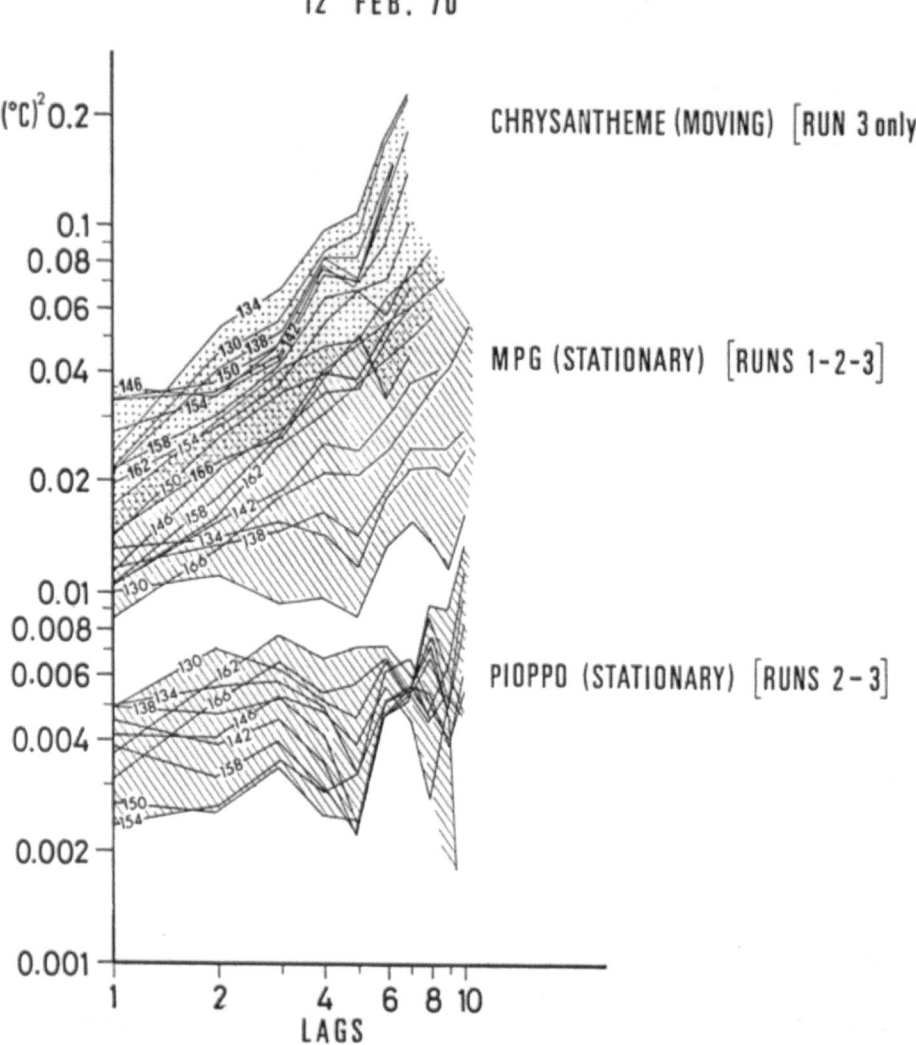

Figure 13. Structure functions of temperature for 12 Feb. The plot shows the mean-square change in temperature versus the number of lags (one lag is ten minutes) for two stationary ships and one moving ship, with depth as a parameter. That is, the temperatures along a given depth (between 130 and 166 m) are used to calculate the structure functions. The two stationary ships were separated about 9 km.

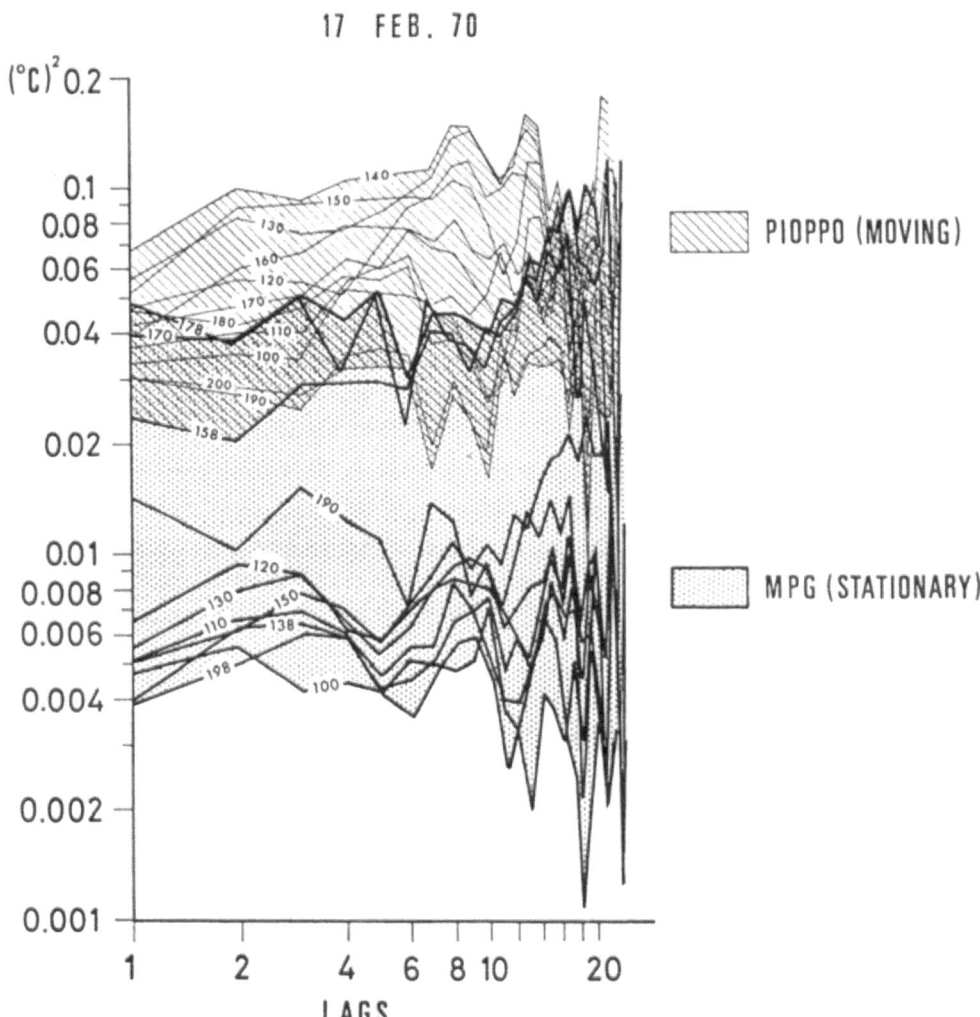

Figure 14. Same as Figure 13 but for 17 Feb. One lag equals five minutes, and only two ships are shown.

Figure 12 for 17 Feb shows similar features:

- the moving ship always sees more mean-squared temperature change over a ten minute lag than does the stationary ship;
- the main variability at the stationary ship is in the 150 to 200 m depth range; at the moving ship it is from 75 to 225 m.

Note that the levels of the mean-square temperature variability are essentially identical for 12 and 17 Feb even though (Figure 5) the general water mass structure is quite different.

Figures 11 and 12 showed the structure functions versus depth for only a 10 or 50 minute lag, averaged over the entire test section. If there had been no spatial variability, we might have expected the structure functions from the moving ship to have been the same as those from the stationary ship; that is, both would have shown temporal variability only. Figures 13 and 14 show the complementary pictures: the structure functions are plotted versus all lags, for each of several depths within the most variable depth regions from Figures 11 and 12, for each of the ships. (In terms of sound speed variability in $(m/s)^2$, multiply the vertical axis by 20.) The results are:

- the general shape of each structure function is an increasing value for an increasing lag, especially on 12 Feb; note that one lag on 12 Feb is 10 minutes, but is only 5 minutes on 17 Feb;
- the two stationary ships on 12 Feb show up to an order of magnitude different variability;
- the moving ships show more structure function than the stationary ships, and the slope of the structure function is steeper for the moving ships;
- all the plots become statistically less reliable at the longer lags where fewer data are available;
- the structure functions do not flatten out at the largest lags, so all the variability of the system has not yet been encountered over the total duration or length of the runs.

2.3 COMPARISON WITH PREVIOUS RESULTS

Earlier measurements of temperature structure functions are available, for example, from the Western North Atlantic [7] and the Eastern North Pacific [8], both from early or late summer data. No such analyses seem to be available for the Mediterranean Sea in any season. It should be noted that such analyses have not been popular since the advent of the Garrett-Munk spectrum for internal waves [9]; once the fashion became to display frequency or wavenumber spectra of the data, lower-order physical domain displays are quite rare.

It should also be noted that spectra and structure functions are related for stationary processes that have power-law descriptions. That is, data yielding a structure function

$$D_h(\tau) = A \ \tau^m \qquad (7)$$

will have a spectrum asymptotically proportional to

$$\int_{-\infty}^{+\infty} \tau^m e^{i\omega\tau} d\tau \ \to \omega^{-(m+1)} \qquad (8)$$

where τ is the lag variable for the structure function and ω is the corresponding Fourier variable for the spectrum.

Reference [7] gives a space-time temperature structure function for a sensor towed at about 100 m depth; over a range of 0.1 to 20 km the slope is close to 2/3. At 10 km lag, the value is about 0.03 (deg C)². In JOUST it took about 32.4 minutes to go 10 km (corresponding to a nominal ship speed of 10 knots). On Feb 17, which had 5 minute sampling, the structure function for 6 to 7 lags is about 0.03 to 0.1 (deg C)² for the moving ship. On Feb 12, with 10 minute lags, the 3 to 4 lag structure function was about 0.03 to 0.09 (deg C)². The slopes for the moving ships in JOUST vary from 2/3 to 1 on 12 Feb, and from about 1/4 to 1/2 on 17 Feb. This suggests that the horizontal correlation scales were relatively short on 17 Feb, but otherwise results were similar to those from the Atlantic.

Reference [8] shows space-time structure functions at 30 and 60 m depths that change slope from about 1.1 at smaller lags to about 2/3 at longer lags; the change occurs at about 220 m lags. In JOUST there were no lags any shorter than about 1500 m (5 minutes at 10 knots) so no corresponding data can be shown. At 90 m in [8] the only slope observed is 1.1. Thus, there is fair correspondence with the moving-ship JOUST values for 12 Feb. The temporal structure functions in [8] change from 2/3 slope at shorter lags to almost 2 at longer lags; the change occurs at about 1.5 minute lags. Again, the shortest lags in JOUST were 5 minutes so the change in slope cannot be compared. The stationary ship on 12 Feb saw slopes from almost flat to a little stronger than 2/3; on 17 Feb the stationary-ship structure functions were almost flat, but were highly variable and suggested 25 to 30 minute fluctuations rather than a broad spectrum.

2.4 INTERNAL WAVES

Modern internal wave theory and observations, which began with [9] (see also [10]-[13]), suggest that internal wave motions should produce spectra with frequency or wavenumber slopes between -2 and -3/2, depending on the detailed situation. This corresponds to structure functions of slopes 1/2 to 1, which in fact encompasses the entire range of moving-ship JOUST observations; the stationary ship sometimes saw flatter slopes. Slopes corresponding to internal wave theory are no proof of the existence of internal waves, but detailed reviews (e.g., [11]) have concluded that whenever internal waves can exist, they do exist.

If the JOUST variability is indeed caused by internal waves, what does this tell us about the estimation of the purely spatial structure function from the measured space-time and temporal structure functions, as in the initial hypothesis of Equation (5)?

2.4.1. *Spectral Forms.* The class of internal wave wavenumber-frequency spectra initiated by [9] and updated recently in [13] is based on the conclusion that observed two-point coherences cannot be explained by a separable spectrum in which a wavenumber function simply multiplies a frequency function. Rather, the form used in [13] is:

$$E(\omega,j) = B(\omega)H(j)E \qquad (9)$$

where the j are vertical mode numbers, E is an energy parameter [13],

$$H(j) = \frac{(j^2+j_*^2)^{-1}}{\sum_{1}^{\infty}(j^2+j_*)^{-1}} \qquad (10)$$

and j. corresponds to the effective mode number bandwidth (i.e., how many modes are *really* important?). In this form of the spectrum, the coherences depend upon the bandwidth in mode number space: larger bandwidths give decreased coherences. The $B(\omega)$ function is just the observed frequency spectrum and contains the scaling information relative to energy levels and dependence on Brunt-Väisälä frequency; typically it falls off as ω^{-2} or $\omega^{-3/2}$, for example:

$$B(\omega) = 2\pi^{-1}\omega^{-1}f(\omega^2-f^2)^{-1/2} \tag{11}$$

where f is the Coriolis frequency for the given latitude.

The temporal structure function suggested by Equation (9) is therefore proportional to $\tau^{1/2}$ to τ^1, but multiplied by a complicated mode number function. The space-time structure function is obscured by the form of Equation (9).

The form in Equation (9) was motivated [9] by there being very good measurements available of the frequency spectrum; however, there is nothing that requires this form. One could just as well say:

$$F(\alpha,j) = A(\alpha)G(j)E \tag{12}$$

where α is the modulus of the horizontal wavenumber. Observations [14] tend to give $A(\alpha)$ as going like α^{-2} to $\alpha^{-3/2}$. In this case, the spatial structure function would be proportional to $\xi^{1/2}$ to ξ^1, multiplied by a possibly different mode number function than in Equation (9).

The conversion from $E(\omega,j)$ to $F(\alpha,j)$ depends upon the dispersion relation that relates α and ω; for simple oceans and internal wave frequencies that are not too high this is

$$\alpha = \frac{\pi j}{N_0 b} (\omega^2-f^2)^{1/2} \tag{13}$$

where N_0 is a scale Brunt-Väisälä frequency for the selected ocean regime, and b is an e-folding scale depth for the Brunt-Väisälä profile; typically, $b \sim O(1\ km)$. Some algebra gives:

$$F(\alpha,j) = \frac{2\pi^{-1}(\frac{\pi f j}{N_0 b})}{\alpha^2+(\frac{\pi f j}{N_0 b})^2} H(j)E \tag{14}$$

For low horizontal wavenumbers, $\alpha << \pi f j N_0^{-1}b^{-1}$, the F-spectrum is not dependent on α and falls off as $1/j$ times the $H(j)$ similarity function. For high horizontal wavenumbers, $\alpha >> \pi f j N_0^{-1}b^{-1}$, the F-spectrum goes as α^{-2} times $jH(j)$. This suggests the high mode numbers are emphasized relative to the lower modenumbers; that is, at short horizontal scales, the equivalent mode number bandwidth is large, hence the correlation scales are small. The cross-over from low to high wavenumbers occurs at $\alpha=\pi f j N_0^{-1}b^{-1}$, which corresponds for the first mode to horizontal scales of about 30 km in the Atlantic and somewhat less in the Mediterranean; the scales for the j-th mode are j-times smaller.

2.4.2. *Structure Functions.* In terms of structure functions, Equation (14) says the high horizontal wavenumber part of the internal wave spectrum should produce a spatial structure function that goes like ξ^1 times a complicated mode number function. The implied structure functions for time and space are therefore

$$\begin{array}{ll} E(\omega,j) \Rightarrow & D(\tau) \propto \tau^1 H(j) \\ F(\alpha,j) \Rightarrow & D(\xi) \propto \xi^1 jH(j) \end{array} \tag{15}$$

from which the modenumber bandwidth function can be eliminated to express the purely spatial structure function in terms of the purely temporal structure function, the lag variables, and a mode number:

$$D(\xi) \propto \frac{\xi^1}{\tau^1} j D(\tau) \tag{16}$$

Note this says that the spatial structure function is different for each mode of the internal wave field.

In Equation (16) the ratio ξ/τ is a speed defined by the lags in time and space. No lags at all in space forces a zero spatial structure function. The product of the structure functions in Equation (15) gives the implied space-time structure function

$$D(\xi,\tau) \propto \xi^1 \tau^1 j [H(j)]^2 \tag{17}$$

and substituting for the $H(j)$ using the first of Equations (15) gives

$$D(\xi,\tau) \propto \frac{\xi^1}{\tau^1} j [D(\tau)]^2 = V_s j [D(\tau)]^2 \tag{18}$$

where V_s is just the speed of the platform used to obtain the space-time data. Since the entire approach is predicated on internal wave theory, the $D(\tau)$ in Equation (18) really means a temporal structure function that goes as $\tau^{1/2}$ to τ^1, and thus a space-time structure function that goes as τ^1 to τ^2.

Equation (18) allows the stationary and moving results in Figures 13 and 14 to be compared. In Figure 13, the moving CHRYSANTHEME in fact does have structure functions that are about twice the slope of the MPG stationary structure functions, but the stationary PIOPPO results are too flat to fit into the internal wave-based theory here. Figure 14 also shows the moving ship to have a steeper slope, as predicted by Equation (18), but again the slopes are too flat to fit into the internal wave theory as developed here.

I do not believe that internal wave theory, *per se*, is inappropriate for the oceanic description needed here; rather, I believe the existing internal wave models are based on observations in the major ocean basins and need some modification to be appropriate to the Mediterranean. There has been *no* Mediterranean data ever used in any internal wave model. Some possible differences between the Mediterranean and the major ocean basins are (1) a very different density, hence Brunt-Väisälä profile, (2) extremely weak surface tides hence even weaker internal tides, and (3) little wind or surface wave forcing over most of the Mediterranean, most of the time.

2.5 CONCLUSIONS ABOUT JOUST OCEANOGRAPHIC VARIABILITY

The original idea of trying to estimate purely spatial structure functions from a combination of temporal and mixed space-time structure functions was a good idea. As internal wave theory has advanced over the years, however, it has shown that one can replace the complicated oceanographic sampling of JOUST with simpler sampling plus the knowledge of the process that causes the variability. Equation (16) illustrates this simplification: all that is needed is the temporal structure function from a fixed location, plus internal wave theory, to estimate the purely spatial structure function. This is another good hypothesis, and needs to be tested!

However, since in JOUST the mixed space-time structure function was measured, the internal consistency of the data sets can be checked. It seems that the mixed space-time structure function is approximately what one would expect if the internal wave theory were strictly correct and rigorously developed, but the fundamental assumption of variability driven by internal waves *as described in existing models* is, in fact, questionable. Further study, as they say, is warranted.

The lessons learned are:

- knowledge of the underlying processes driving the ocean variability can greatly simplify the ocean sampling problem;
- the heuristic derivations in this paper suggest useful relationships can be obtained between temporal, spatial, and mixed space-time ocean measurement programs; the relationships depend strongly on the assumptions about the processes involved;
- it would be difficult to measure too much oceanography; the need for consistency checks is paramount.

3. Scales of Oceanographic Variability

Oceanic inhomogeneities of scales large compared to acoustic wavelengths could be treated deterministically if it were possible to measure, map, model, or otherwise describe them. That is, the propagation problem is one of refraction rather than scattering. For example, for 100 kHz sound, essentially all of oceanography has larger scales; for 1 kHz sound, oceanic microstructure is of smaller scale; for 100 Hz sound, all the fine and microstructure and some of the internal waves have smaller scales; for 10 Hz sound, much of the rest of the internal wave field has smaller (vertical) scales; finally, for 1 Hz sound, even some of the oceanic frontal structures begin to have smaller scales. But for no acoustic frequencies above the mHz range are the eddies and most of the frontal structures in the ocean of smaller scales.

Putting this another way, for the acoustic frequency range between 10 Hz and 1 kHz, one can anticipate that internal waves and fine-and-microstructure will have comparable or somewhat smaller scale sizes than the acoustic wavelengths, but fronts and eddies will be larger in scale.

The consequence of the overlapping scales of acoustic wavelengths and oceanographic processes is a mixed deterministic-stochastic propagation problem, exacerbated by there often being no good description available of the deterministic part of the field.

In fact, I submit that the central problem in ocean acoustics is not knowing what the ocean is; even though the scale sizes may permit a deterministic, refractive calculation to describe the sound propagation, the data available rarely allow that calculation to be made. Thus, we are faced with:

- (i)   deterministic and known ocean structure
- (ii) deterministic but unknown
- (iii) statistical and known
- (iv) statistical and unknown.

The third and fourth categories are the least of the problems: the small-scale statistical structure is often approximately known, and in any case has effects small compared with the first two categories. The first category is the goal of the Navy Ocean Modeling and Prediction Program (NOMP): using external inputs, model the large-scale, slowly changing ocean structure. The serious problems come with category two, especially if the sound wavelength is short enough that many ocean processes are involved in the unknown determinism; this is the situation for frequencies higher than O(100 Hz).

A concrete example of this mixed deterministic-stochastic problem in the face of sparse data is propagation of 500 Hz sound through an oceanic frontal region. The sound wavelength is just 3 m so most of the oceanic processes need to be included in the propagation calculation. At best, we can anticipate that the scales of 3 m to perhaps 10 km can be described statistically for the internal wave field, and the scales of 20 km and up may be able to be modeled numerically for the eddy field, but what about the front itself? It has enormous variability on the scales of a few hundred meters to tens of km, with little or no statistical description possible (it is variable but not stochastic); it is deterministic on a scale smaller than we can model, and it is unknown. The kind of front that appears when one smooths the output of airdropped bathythermographs taken every 10 or 20 km is quite unlike what is present in the ocean; examine any towed thermistor chain section across a frontal zone to test this assertion. Also, the front of the North wall of the Gulf Stream is unique; it is unwise to transport ideas from there to other fronts driven by other, quite different, dynamics or thermodynamics, such as the fronts in the Greenland-Iceland-Norwegian Sea or in the Mediterranean.

3.1 THE CHALLENGE

I submit that the most important progress we can make in acoustic propagation modeling is to learn how to take into account mostly unknown -- but presumably bounded -- fields. We may be nearing the point that we need a "kinetic theory" of propagation that will give us the "average propagation" of the sound. That is, we should develop statistical mechanical approaches instead of calculating the interaction of every molecule with every other molecule to estimate the temperature of the gas. Small steps have been made in this direction; large steps are needed.

As an oceanographer I should be gratified whenever I hear an acoustician say that all he or she needs to know is the sound speed field everywhere, and every aspect of the shape of the surface and bottom, because my career is assured. But this will not yield progress. For progress we need to accept our lack of knowledge and to develop intellectual approaches that are matched to the inherently unknown nature of the problem; this is the challenge to both the oceanographers and the acousticians, and I believe success will depend on both communities working together to develop the methods.

4. Acknowledgements

The work described here has mostly been carried around in my file cabinets and brief cases for nearly two decades. Figures 1 through 14 were all prepared at and by SACLANTCEN. I wish to thank my colleagues at SACLANTCEN, Woods Hole, and the Office of Naval Research for listening over the years to evolving versions of this presentation.

5. References

1. Allan, T.D., Briscoe, M.G., Clarke, R.H., Gerrebout, J.R., Krol, H.R., Özturgut, E., Padley, J.B.W., and Thompson, M. (1972) 'Comparison of measurements and ray tracing prediction of total propagation loss for the Western Ligurian Sea in February 1970 (JOUST)', Technical Report TR-215, NATO SACLANT ASW Research Centre, La Spezia, Italy (NATO CONFIDENTIAL). DTIC: AD 596 183.

2.   Ewart, T.E., and Reynolds, S.A. (1984) 'The mid-ocean acoustic transmission experiment, MATE', J. Acoustical Society America 75, 785-802.

3.   MEDOC Group (1970) 'Observation of formation of deep water in the Mediterranean Sea, 1969', Nature 227, 1037-1040.

4.   Liebermann, L. (1951) 'The effect of temperature inhomogeneities in the ocean on the propagation of sound', J. Acoustical Society of America 23, 563-570.

5.   Watson, J.G., Siegmann, W.L., and Jacobson, M.J. (1977) 'Acoustically relevant statistics for stochastic internal-wave models', J. Acoustical Society America 61, 716-726.

6.   Uscinski, B.J., Potter, J.R., and Akal, T. (1989) 'Broadband acoustic transmission fluctuations during NAPOLI 85, an experiment in the Tyrrhenian Sea: preliminary results and an arrival-time analysis', J. Acoustical Society America 86, 706-715.

7.   Voorhis, A.D. and Perkins, H.T. (1966) 'The spatial spectrum of short-wave temperature fluctuations in the near-surface thermocline', Deep-Sea Research 13, 641-654.

8.   Williams, R.B. (1968) 'Horizontal temperature variations in the upper water of the open ocean', J. Geophysical Research 73, 7127-7132.

9.   Garrett, C., and Munk, W. (1972) 'Space-time scales of internal waves', Geophysical Fluid Dynamics 2, 225-264.

10. Garrett, C., and Munk, W. (1975) 'Space-time scales of internal waves: a progress report', J. Geophysical Research 80, 291-297.

11. Briscoe, M.G. (1975) 'Internal waves in the ocean', Rev. Geophysics and Space Physics 13, 591-598 and 636-645.

12. Garrett, C., and Munk, W. (1979) 'Internal waves in the ocean', Ann. Rev. Fluid Mech. 11, 339-369.

13. Munk, W. (1981) 'Internal waves and small-scale processes', in B.A. Warren and C. Wunsch (eds.), Evolution of Physical Oceanography, MIT Press, Cambridge, Massachusetts, pp. 264-291.

14. Katz, E.J., and Briscoe, M.G. (1979) 'Vertical coherence of the internal wave field from towed sensors', J. Physical Oceanography 9, 518-530.

EXPERIMENTAL OCEAN ACOUSTIC FIELD MOMENTS VERSUS PREDICTIONS

T.E. Ewart and S.A. Reynolds
Applied Physics Laboratory
University of Washington
1013 N.E. 40th Street
Seattle, WA 98105
United States

ABSTRACT. Results for various moments of a propagating ocean acoustic complex wavefield will be presented from field and numerical experiments. They will be compared with theoretical predictions. It will be shown that the ocean environment must be well understood to achieve substantive agreement between theory and measurement. The comparisons will include complex field correlations, intensity correlations, phase correlations and intensity probability distributions. An attempt will be made to summarize the significance of these results within our current understanding, and to identify areas of research that will extend our understanding of volume scattering.

1. Introduction

In this paper we will focus on coupling our understanding of ocean processes having horizontal scales tens of kilometers and smaller, with predictions of the fluctuations of sound propagated through the ocean. We will discuss the formalism that allows us to put oceanic variability into the acoustic context; we will then discuss predictions of the results of ocean and numerical experiments; finally, we will summarize the level of our understanding and our view of what needs to be done next.

The oceanic sound speed variability is usually described by two-point statistics. As the underlying oceanographic processes are Gaussian or close to Gaussian, this would appear to be a reasonable assumption. Two-point statistics are also used to describe the acoustic phase, complex field, or intensity fluctuations. The intensity of a propagating wave begins with a log normal distribution, builds to distributions with high skewness and kurtosis, and finally, at saturation of the field, approaches exponential. Because the process is not Gaussian, two-point statistics are not adequate to describe the intensity fluctuations. The pdf's of intensity are discussed in Ewart and Percival (1986) and Ewart (1989). The implications of such distributions will be discussed.

1.1 THE STOCHASTIC OCEAN

The coordinate system we use to discuss stochastic ocean behavior is shown in Figure 1. The two-point separation coordinates are $\xi = x_1 - x_2$, $\eta = y_1 - y_2$, $\zeta = z_1 - z_2$, $\tau = t_1 - t_2$, and the Fourier conjugate variables in the wavenumber/frequency domain are α_1, α_2, β, and ω.

J. Potter and A. Warn-Varnas (eds.), Ocean Variability & Acoustic Propagation, 23–40.

24

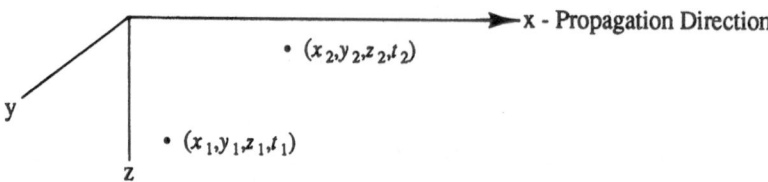

Figure 1. Coordinate system.

1.1.1 *Representing the Sound Velocity Fluctuations.* The root mean square index of refraction fluctuations are from the sound speed fluctuations, δC (vertical displacements or velocities in the propagation direction),

$$\langle\mu^2\rangle^{\frac{1}{2}} = \frac{\langle\delta C^2\rangle^{\frac{1}{2}}}{C_o},$$

where C_0 is the reference sound speed. We represent the two-point statistics of μ with the power spectrum

$$S(\alpha_1,\alpha_2,\beta,\omega).$$

If we assume horizontal isotropy, the spectrum may be written

$$S(\alpha,\beta,\omega), \quad \text{where } \alpha^2 = \alpha_1^2 + \alpha_2^2.$$

The integral of the spectrum over the positive frequencies is $\langle\mu^2\rangle$. The horizontally isotropic medium correlation function is

$$R(\xi,\eta,\zeta,\tau) = F\left\{S(\alpha,\beta,\omega)\right\},$$

where $F\{..\}$ indicates a Fourier transform. In what follows, variability in the transverse separation coordinate, η, will not be considered. That is, we treat $R(\xi, \zeta, \tau)$ and $R(0,0,0) = \langle\mu^2\rangle$. In virtually all theoretical treatments of scattering, the medium is represented by a projection of the medium correlation function in the direction of wave propagation. This projected correlation function is called the transverse correlation function, abbreviated here as TCF. We can think of the medium as consisting of δ-correlated phase changing screens that are statistically described by the TCF. The TCF is obtained from the correlation function by

$$R_\perp(\zeta,\tau) = \int_{-\infty}^{\infty} \frac{R(\xi,\zeta,\tau)\,d\xi}{R(0,0,0)} = \int_{-\infty}^{\infty} \frac{R(\xi,\zeta,\tau)\,d\xi}{\langle\mu^2\rangle}.$$

$R_\perp$ is the function we will discuss in the stochastic ocean context.

The ocean processes we will consider in this work are the tides, internal waves and finestructure. Finestructure is the name given to the poorly understood portion of the oceanic fluctuations in space/time that do not possess a wave-like dispersion relation, but give appreciable variance in

δC. We will discuss material from three sources: numerical experiments using parabolic equation wave propagation methods, the Mid-Ocean Acoustic Transmission Experiment (MATE) and the AIWEX Acoustic Transmission Experiment (AATE).

1.1.2 The MATE Ocean Displacement Spectra. The MATE oceanographic setting (1000 m depth, Lat. 46°46'N., Long. 130°47'W) is characteristic of open ocean conditions. The only exceptions are strong baroclinic tides caused by the presence of several seamounts. The oceanographic measurements during MATE were sufficient to overdetermine $S(\alpha,\beta,\omega)$. They included three velocity/temperature/salinity moorings (30 days extent) and depth profiles of temperature and salinity taken throughout the experiment (one set of yo-yo profiles was taken every 20 minutes for 24 hours). In addition, three runs (one 20 km isobaric and two 22 km depth cycling) of the Self-Propelled Underwater Research Vehicle (SPURV) measured temperature and salinity with sensor packages separated vertically by 1 m. Analysis of the SPURV records supports the assumption of horizontal isotropy. The various projections of $S(\alpha,\beta,\omega)$ are the one-dimensional spectra or coherences measured by one or more of the sensors. Details of the data analysis for finestructure and internal waves, that we only briefly describe are found in Levine and Irish (1981) and Levine et al. (1986).

The temperature spectrum is shown in Figure 2a and the velocity spectrum for the 30 day moorings is illustrated in Figure 2b. The tidal and the inertial/buoyancy frequencies are depicted, as well as the internal wave model. These are representative of the temporal measurements at various spatial separations.

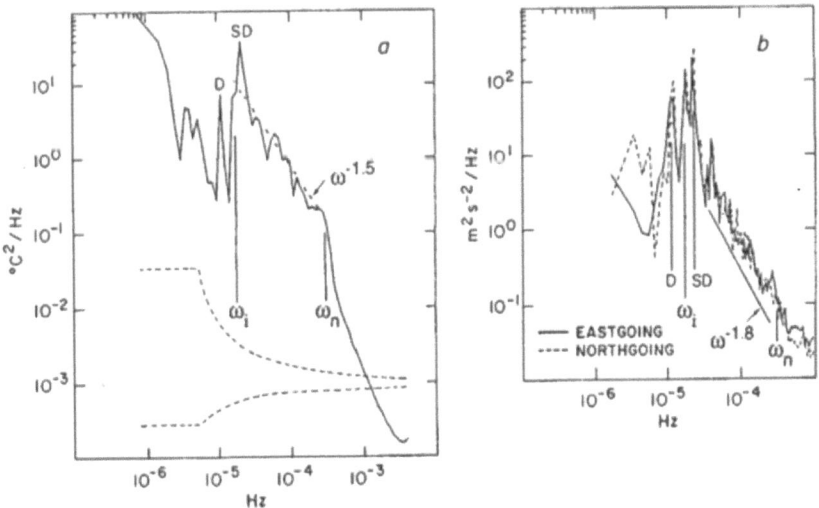

Figure 2. Temperature and velocity spectra from MATE. Frequencies are expressed as Hz.

Figure 3 shows the spatial spectra. Results from the vertical (yo-yo) profiling and from the SPURV isobaric profiling is illustrated in Figure 3a and Figure 3b, respectively. From these, the MATE model of internal waves and finestructure developed as a fit to the projections of $S(\alpha,\beta,\omega)$. Subsequently, we will show that insight into the models can also be obtained from the MATE acoustic measurements.

1.1.3 *The MATE TCF.* The general form of the TCF is written for the case of separable vertical and time correlations as

$$R_{\perp,IW,FS}(\zeta,\tau) = \int_{-\infty}^{\infty} \frac{R_{IW,FS}(\xi,\zeta,\tau)\,d\xi}{R_{IW,FS}(0,0,0)} = L_{p;IW,FS} \frac{\sigma_{IW,FS}(\zeta)}{\sigma_{IW,FS}(0)} \frac{\psi_{IW,FS}(\tau)}{\psi_{IW,FS}(0)}, \text{ where} \tag{1}$$

$$L_{p;IW,FS} = R_{\perp,IW,FS}(0,0).$$

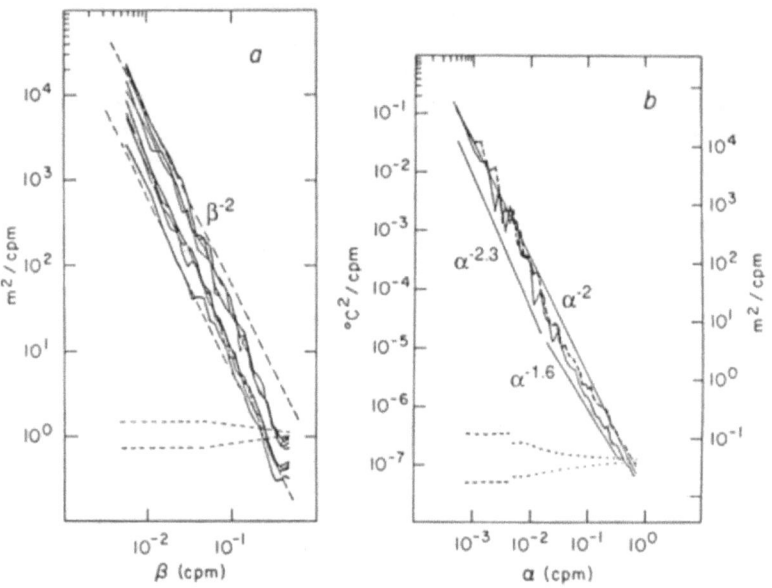

Figure 3. Vertical and horizontal spatial spectra of MATE. The wave numbers are expressed as cycles per meter.

For internal waves, following Uscinski (1980) we write

$$R_{\perp,IW}(\zeta,\tau) = 2\,G_o^{-1}H_o^{-1} \int_{\omega_i}^{\omega_n} \frac{G(\omega)}{r_\omega} \left[\int_{\beta_c}^{\infty} \frac{H(\beta)}{\beta} \cos(\beta\zeta)d\beta \right] \cos(\omega,\tau)\,d\omega \tag{2}$$

where $G(\omega)$ and $H(\omega)$ are obtained from the model presented in Levine et al. (1986); r_ω arises from the internal wave dispersion relationship

$$r_\omega^2 = \frac{\alpha^2}{\beta^2} = \frac{\omega^2 - \omega_i^2}{\omega_n^2 - \omega^2}, \text{ and} \qquad G_o = \int_{\omega_i}^{\omega_n} G(\omega)d\omega, \quad H_o = \int_{\beta_c}^{\infty} H(\beta)\,d\beta. \tag{3}$$

ω_i and ω_n are the inertial and buoyancy frequencies, respectively. β_c is the vertical wave number corresponding to the lowest internal wave mode. The model in this form is written

$$G(\omega) = \frac{(\omega^2 - \omega_i^2)^{1/2}}{\omega^P}, \quad H(\beta) = \frac{\beta_*}{\beta_*^2 + \beta^2}, \quad \beta_* = t\,(\omega_n^2 - \omega^2)^{1/2}. \tag{4}$$

Except for the variable spectral slope, p, this model is the same as that of Desaubies (1976). This model uses a continuous vertical wavenumber representation rather than the modal representation used by Garrett and Munk (e.g. Munk 1981). The spectral slope, $p = 3$ is used by Desaubies and Garrett and Munk. The parameter t is the bandwidth parameter from Desaubies. The ζ-τ or β-ω functions are assumed separable.

For finestructure, we use a modified form of the Levine and Irish (1981) and Ewart et al. (1986) representation. In addition to internal waves, two processes are postulated: a low wavenumber and low frequency process that is characterized by a slow decay compared to the inertial period, and another process that is characterized by high wavenumbers and frequencies, that interacts with the acoustic field through modulation by the internal wave field. The second process has very low variance, and thus little effect on the acoustic propagation. It has been discussed extensively in Ewart et al. (1983), and will not be discussed here except in the context of the acoustic phase correlations. The low frequency finestructure model has high-wavenumber/high-frequency asymptotic dependence in the spectral domain α^{-2}, β^{-2}, and ω^{-2}. The forms of R_{FS} and $R_{\perp FS}$ are

$$R_{FS}(\xi,\zeta,\tau) = \exp\left\{-\left[\left[\frac{\xi}{L_H}\right]^2 + \left[\frac{\zeta}{L_v}\right]^2\right]^{1/2}\right\} e^{-|\tau/\tau_0|}, \text{ and} \tag{5}$$

$$R_{\perp FS}(\zeta,\tau) = L_{p;FS}\frac{\sigma_{FS}(\zeta)}{\sigma_{FS}(0)}\frac{\psi_{FS}(\tau)}{\psi_{FS}(0)} = L_{p,IW}\frac{\zeta}{L_v}K_1(\zeta/L_v)\, e^{-|\tau/\tau_0|}.$$

We have taken $L_{p;FS} = 2 \cdot L_{H,IW} = L_{p,IW}$. K_1 is the K_1 Bessel function.

1.2 THE MOMENTS OF THE ACOUSTIC FIELD

The complex acoustic field, E, is conveniently thought of as a phasor

$$E(x,z,t) = A(x,z,t)\, e^{i\,\phi(x,z,t)}.$$

We will include a discussion of the three moments of E; in what follows, numbered subscripts refer to lags in space and/or time. The phase correlations are

$$R_\phi(\zeta,\tau) = <\phi_1\phi_2>. \tag{6}$$

The complex field correlations are

$$M_{II}(\zeta,\tau) = <E_1E_2^*>, \tag{7}$$

where * indicates complex conjugate. They are also known as the mutual coherence functions, and include the angular spectrum. The fourth moment of the field is written as

$$M_{IV}(\zeta,\tau) = <E_1E_2^*E_3E_4^*>.$$

Generally, fourth moment results are presented as intensity correlations. This is accomplished by collapsing the subscripts $2 \rightarrow 1$ and $4 \rightarrow 3$, so that M_{IV} is the 2-point intensity correlation

$$M_{IV}(\zeta,\tau) = <I_1 I_2>.$$ (8)

The scintillation index, or normalized intensity variance is written

$$SI = \frac{M_{IV}(0,0) - M_{II}^2(0,0)}{M_{II}^2(0,0)} = \frac{<I^2> - <I>^2}{<I>^2}.$$

1.3 THE SCATTERING PARAMETERS

We have introduced the representations of the ocean sound velocity and the acoustic field moments. We will now review the scattering parameters that have arisen naturally from theoretical formulations. We will need these in the discussions of scattering regimes, and in comparing theory with experiment. There are many different parameters in the literature. Two formulations are common today. The scattering parameters of the path integral formulation Λ, Φ are defined in Dashen (1979), or Esswein and Flatte (1980). Those of the moment equation formulation γ, X are discussed in Uscinski (1977). γ is the local scattering strength equal to the Fresnel length divided by the scattering length. The scattering length is defined as the length of medium required for the wave to pick up a mean squared phase deviation of 1 rad^2 (single scatter). The Fresnel length is that range required for the wave to acquire diffraction effects. $\gamma > 1$ for most ocean acoustic regimes; hence, acoustic wave propagation in the ocean is virtually always characterized by weak multiple scattering with small diffractive effects. X is the propagation range scaled by the Fresnel length. $X < 1$ for most ocean acoustic propagation. They are defined as

$$\gamma = k^3 <\mu^2> L_p L_v^2, \qquad X = \frac{range}{k L_v^2},$$ (9)

where k is the acoustic wavenumber, L_p is the integral range scale (see Equation 1) and L_v is the vertical correlation scale for δC. For constant or linear sound velocity profiles, $\Phi^2 = \gamma X$ and $X = 6\Lambda$. Otherwise, the parameters are functions of the range along characteristics.

1.4 STATUS OF THE THEORETICAL PREDICTIONS

Moment theory solutions are available as full range predictions of M_{II} and M_{IV}. The first full range solution for M_{IV} for plane wave propagation was reported by Uscinski (1982). The theory given in multiple integral form must be evaluated numerically for arbitrary TCF's. Point source propagation, sound velocity profiles, differing TCF's and more precise evaluations for M_{IV} have become available.

Evaluations of the formal solutions for M_{IV} are available in the literature as a zero order approximation and a first order correction. They have been called the zeroth and first order predictions of M_{IV}. The theory and its evaluation is discussed extensively in Ballard and Uscinski (1990). We will assume here that $M_{II}(\zeta,\tau)$, and $M_{IV}(\zeta,\tau)$ can be predicted accurately if ocean parameters are known. Such theories will be discussed in other papers during these proceedings.

2. The Phase Correlations

During the time that the MATE oceanographic measurements were conducted, acoustic pulses having center frequencies near 2, 4, 8, and 13 kHz were transmitted along a shallow angle (3°) path (18.1 km long, and 1000 m deep) between a fixed set of co-located transmitters and fixed, spatially separated receivers (four at the corners of a rectangle 3 m high by 253 m transverse to the propagation path) for 15 days (Ewart and Reynolds, 1984). We will present a sketch of the results, and indicate their relevance to the theme of this paper concerning the importance of detailed oceanographic understanding when attempting to explain acoustic measurements.

We can learn a great deal about oceanographic questions from the temporal phase correlations (Equation 6). The tidal, internal wave and finestructure processes are clearly seen in the phase spectrum from the 15 day 2 kHz travel time record. This is plotted in Figure 4. The data spectra for the other frequencies of MATE are virtually identical out to temporal frequencies well above the buoyancy frequency, indicating the geometric nature of the phase fluctuations. The data has been expressed in $<\mu^2>^{1/2}$ units by using the product γX from Equation 9, with $k = 2 \cdot \pi \cdot 2083/1480$, and L_p from the MATE model. Thus, the integral of the spectrum is $<\mu^2>$. The diurnal, semidiurnal, and quarter-diurnal (overtone of the semidiurnal) tidal lines are evident, as is the sharp fall-off at ω_i and ω_n. The dashed lines indicate a fit of the data using a simultaneous deterministic/stochastic inverse to a model that includes a trend function, tides, finestructure and internal waves. In the inverse processing, the finestructure and internal waves were modeled as

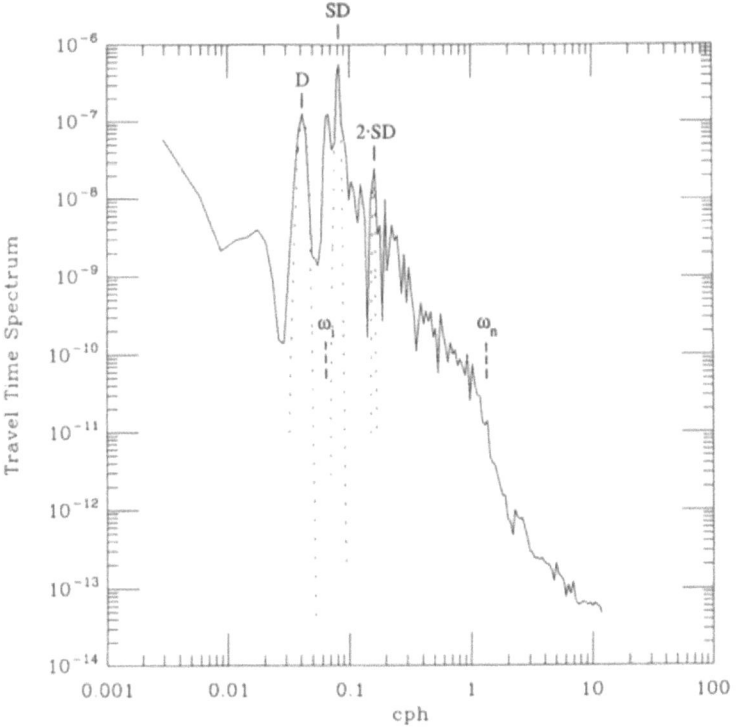

Figure 4. MATE 2 kHz travel time spectrum in $<\mu^2>$ units per cycle per hour.

the Fourier transforms of Equations 5 and 2, respectively. [We can do this because the phase spectrum in the geometric acoustics regime is related to the Fourier transform of the TCF by γX. See Uscinski (1986)]. The tides were modeled as a sum of sine and cosine terms with independent coefficients. The spectral estimates were computed using the DPSS method described by Slepian (1978). Four discrete prolate spheroidal windows were used; the windowing is evident in the tide lines. This stochastic inverse method is the frequency domain equivalent of a similar, inverse computation using correlation functions published by Ewart (1986). For our purposes, we use the inverse to remove the deterministic tides from the data spectra. The determinism of the tides will be assumed. However, the large baroclinic tides are almost certainly time dependent over the 15 days.

We can apply the same technique to a mooring measurement of temperature. The 30 day temperature record was resampled over the 15 days of the acoustics experiment and interpolated to the same time grid. The modeled form of the temperature record is a different integral of $S(\alpha,\beta,\omega)$, i.e., the moored spectrum is

$$\int_{-\infty}^{\infty}\int_{-\infty}^{\infty} S(\alpha,\beta,\omega)\,d\alpha\,d\beta.$$

We use the linear T-S relation of MATE and the conversion from pressure, temperature, and salinity to sound velocity to get the moored spectrum at the ray depth in $<\mu^2>$ units. Plots of both the travel time and temperature spectra in those units are illustrated in Figure 5. The plots include only the stochastic components, with the tidal components removed as above.

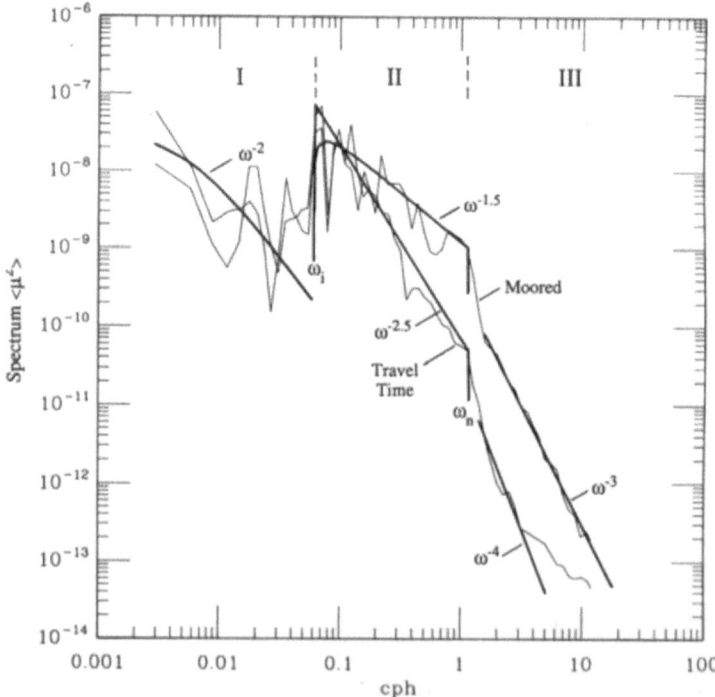

Figure 5. Temperature and travel time spectra with tide lines removed are plotted in $<\mu^2>$ units per cycle per hour.

Three distinct regions, separated by ω_i and ω_n are indicated. The proposed theoretical model for each of the regions is shown. In region I, we see that the moored and the travel time spectrum are identical within statistical limits. Those limits are large, due to the short (15 day) time record. This supports the model having no distinct dispersion relation, but the short record will not allow determination of τ_0. The 500 hour value of τ_0 used in the model arises from the constraint of equal variance of finestructure and internal waves found in the β domain (Levine and Irish, 1981). In region II, we see strong evidence that the internal wave model is correct. The differing spectral forms of $\omega^{-1.5}$ for the moored spectrum and $\omega^{-2.5}$ for the travel time spectrum support the dispersion relation, the spectral cutoffs, the value of p and the normalization of the model. The ω^{-3} spectral slope of the region III moored spectrum demonstrates the correctness of the ω^{-2} high wavenumber finestructure model. The ω^{-4} spectral slope of the travel time spectrum provides strong evidence that the high wavenumber finestructure is advected by internal waves (hence the effect of the internal wave dispersion relation).

We have attempted to demonstrate both the complexity of the ocean TCF as well as the large diversity of oceanographic data needed to confirm ocean spectral models. The ability of the acoustic field to give us an integral constraint on the model through the phase correlations must be emphasized. The clear need for acoustic measurements that include vertical measurements of E to allow for spectral testing in the vertical wavenumber domain is obvious. This will be discussed subsequently.

3. Numerical Simulations and Theory

We can learn a great deal about our understanding of volume fluctuations from numerical simulations of the propagation of E. The technique of using parabolic equation propagation to test moment theoretical predictions was initiated by Macaskill and Ewart (1984). Since then, the technique has been modified to include a point source initial condition and other important physics. We will discuss numerical simulations in the context of the validity of the moment equation. The importance to ocean acoustics is that the moment theories are full range theories, and not asymptotic at short or long range.

In the introduction, we noted that two-point statistics provide an insufficient basis for understanding the intensity fluctuations. Ewart (1989) used parabolic equation simulations of propagating fields with plane wave initial condition to demonstrate that the Generalized Gamma distribution function is a candidate model of the intensity distribution for wave propagation in random media. Figure 6 shows a contour plot from that paper of the scintillation index as a function of acoustic frequency and range. Similar plots for the skewness and kurtosis were also presented in that paper. The input ocean model used for the TCF was $R_\perp(\zeta) = (1 + |\zeta|/L_v)\exp(-|\zeta|/L_v)$ which is asymptotically β^{-4} in the vertical wavenumber domain. The normalization was typical of mid-ocean internal waves with $\langle\mu^2\rangle = 3.0 \cdot 10^{-9}$, $L_p = 4600\,\text{m}$, and $L_v = 150\,\text{m}$. For frequencies above 1 kHz, the SI rises to a maximum, called the focus of the medium, and decays to the exponential distribution value of one. The noise cutoff demonstrates that true saturation is difficult to achieve for the mid-ocean case. (Variance, skewness and kurtosis plots for differing values of $\langle\mu^2\rangle$, L_p, and L_v can be easily produced.) For our purposes, it is important to note that Uscinski's full range theory for plane wave propagation is in excellent agreement with the SI as plotted, and with the vertical wavenumber decomposition of the intensity variance.

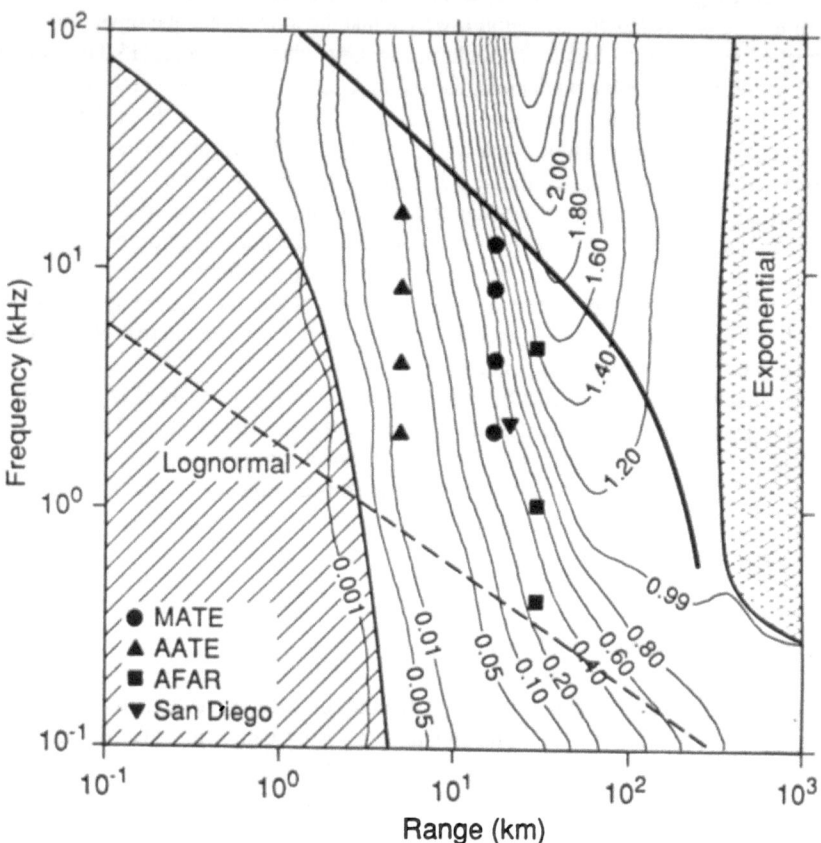

Figure 6. Contours of SI vs frequency and range. Crosshatched areas are the asymptotic regions for the log-normal and exponential distributions. The solid line shows the range limit for a 200 dB source (for Sea State IV noise levels), and the dashed line is the multiple scattering limit $\gamma X = 1$.

Figure 7 shows the same plot for the case of point source initial condition and the Levine and Irish (1981) model of linear internal waves. This figure was obtained using zero-order evaluation of the full-range theory. The normalizations of the theory are the same as above for the mid-ocean case. The scintillation index peak is a little higher, and occurs somewhat earlier in range at higher acoustic frequencies. Other than that, the contours are very similar. This has been included to demonstrate the robust nature of the full range theory. Uscinski (1990) has demonstrated that when using parabolic equation propagation in polar coordinates (for that case the natural initial condition is point source), the predictions of M_{II} and M_{IV} of moment theory with point source initial condition agree to within statistics with the simulations. However, he used the Gaussian TCF, and more work remains to be done for power law media like internal waves.

The results presented in Ewart (1989) demonstrate that two-point statistics are quite incapable of modeling the region of the medium focus. This has profound implications in signal processing, where the theories are always based on Gaussian quadrature components of the signal. For our purposes, we have shown the ability of the full range moment theories to provide accurate predictions - when the ocean statistics are known. Thus, we can predict at least two-point statistics of intensity. We turn our attention now to field measurements of E.

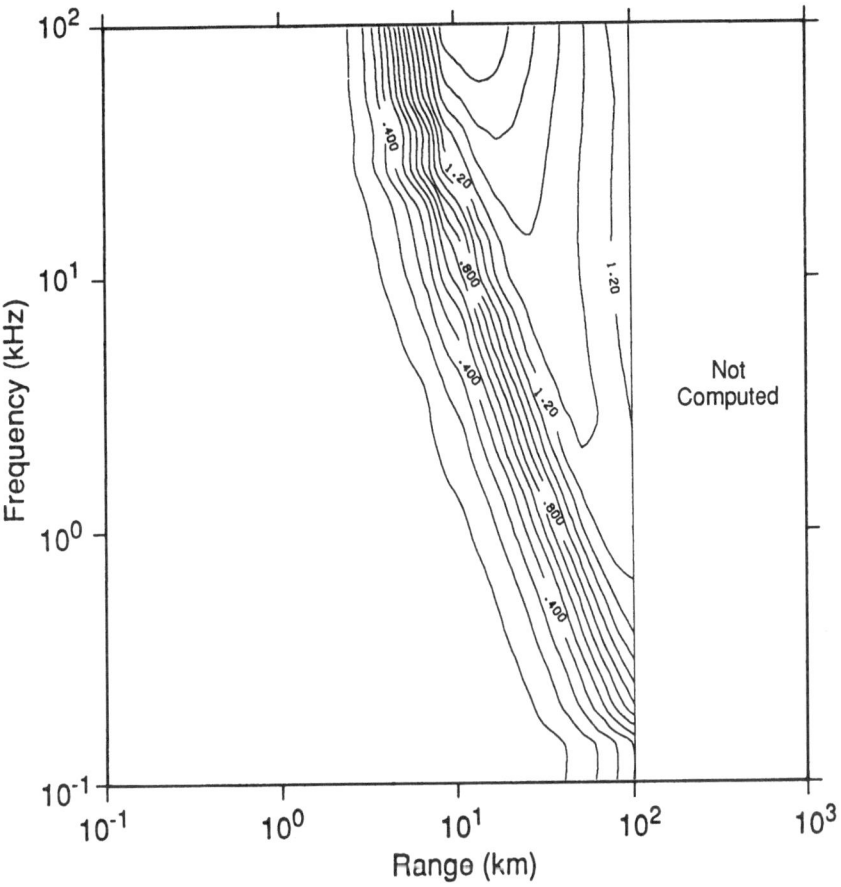

Figure 7. This illustration is similar to Figure 6 for point source propagation and internal waves.

4. MATE Lower Path Results

State-of-the-art ability to predict 4th-moments of the wave field is evidenced by the comparisons made between theory and the MATE acoustic measurements. The zeroth and first order correction evaluations of M_{IV} are discussed in Ballard and Uscinski (1990) in the MATE context. Predicted and observed levels of the 8 kHz SI are 1.50 and 1.47 respectively. Comparison with other frequencies are also quite good and are described in Ewart and Reynolds (1990). At this writing, evaluation of the first order correction for the temporal intensity spectrum has not been completed. The zeroth order prediction for the 8 kHz intensity auto-spectrum is shown in Figure 8 where the internal wave and finestructure medium correlation functions described above have been used; the zeroth order evaluation of SI is 1.19. The first order correction to the spectrum should improve the fit to the observations. Measurements from a shallow (up-ray) that encountered a more energetic scattering regime are also described in that paper.

34

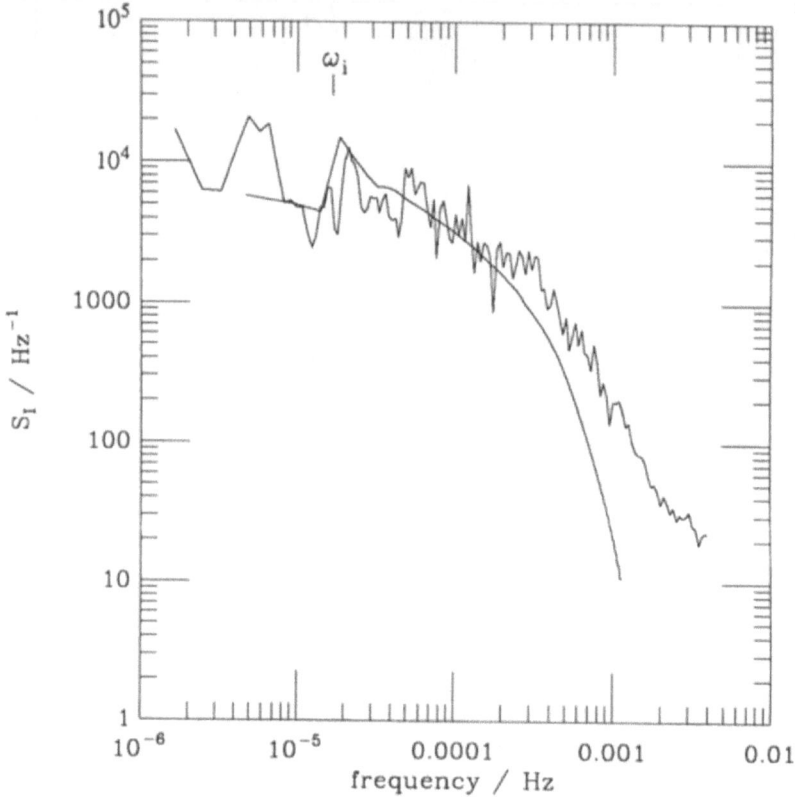

Figure 8. MATE 8 kHz intensity auto-spectrum compared with zeroth order prediction. The internal wave model parameters are $t = 3.1 \ m^{-1}$ for the Desaubies bandwidth parameter (corresponding to a GM modal bandwidth, $j_* = 6$), a buoyancy frequency at the turning depth of $\omega_n = 1.85 \cdot 10^{-3} \ s^{-1}$, and a finestructure time parameter of $\tau_0 = 500h$. The inertial frequency, $\omega_i = 1.06 \cdot 10^{-4} \ s^{-1}$.

The spectral enhancement near the inertial frequency is a feature not seen previously. This is because earlier evaluations truncated the medium correlation functions at the first zero-crossing. Evaluations of the temporal intensity spectra (zeroth-order) are shown in Figure 9 for a few ranges up to and beyond 18.1 km for the 8 kHz case. This figure shows the behavior of the scattered intensity with range, and demonstrates the predictive capability for varying scattering regimes.

Hence, the prediction of the SI and its spectral decomposition has been validated by measurement. Improvement in the spectrum is expected when the first order corrections are completed. We again emphasize that the medium statistics are known significantly better at the MATE site than is typical of most ocean acoustic experiments. Predictions for the intensity cross-spectra do not compare as favorably with the MATE observations (see Ewart et al., 1985). Both Uscinski and Beran (1990) believe that this is the result of a serious failure in moment theory for the cross-frequency case.

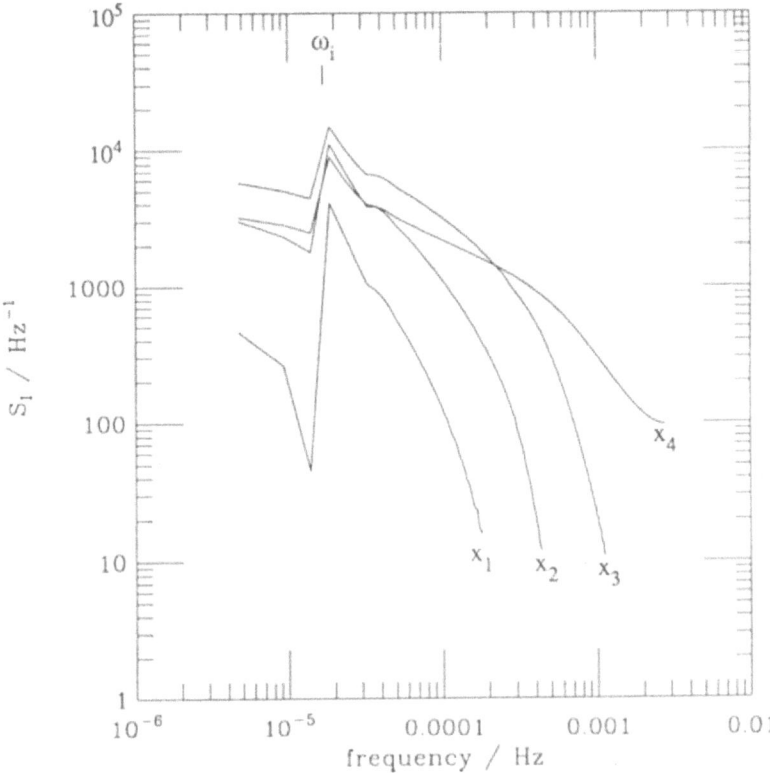

Figure 9. Zeroth-order evaluations for the 8 kHz intensity auto-spectra as a function of propagation range through the MATE scattering conditions. [x_1, x_2, x_3, x_4 = 4.5, 9.0, 18.1 (the MATE range), and 36.2 km, respectively.] The prediction at 4.5 km mimics the TCF, as expected for weak scattering.

5. AATE Measurements of E(z,t)

MATE demonstrated that extensive environmental measurements must be made simultaneously with acoustic field measurements so that scattering conditions can be determined. Although a long series of temporal acoustic measurements were made, only a few spatially separated measurements of E are available. For better understanding of the spatial characteristics of the scattered field, AATE (conducted under multi-year arctic ice in the Beaufort Sea, 225 miles north of Prudhoe Bay, Alaska) was designed to make both vertical and temporal measurements of E. The transmission experiment consisted of four co-located transmitters (2,4,8,16 kHz) suspended beneath the ice at 153 m depth. These were positioned 6.43 km from a depth cycling array of 3 receivers, separated by 51 m (the depth cycle was 51 m providing a 153 m vertical aperture). Simultaneous environmental measurements were made by several investigators. All measurements were part of AIWEX (Arctic Internal Wave Experiment). A compilation of AIWEX results is continuing, however, it is known that the scattering conditions differ significantly from open ocean conditions.

36

Figure 10 displays the travel time and log-intensity spectra measured over two time periods during AATE. (A strong wind event occurred between the two time periods making interpretation of the travel time measurements across the event difficult if not impossible.) The spectra in Figure 10 may be compared to those taken during MATE. Predictions from the weak-scattering theory (Rytov) of Desaubies (1978) are shown with the observations from before the wind-event. The acoustic fluctuations are significantly less energetic (≈ 1/50th) than those expected under canonical GM, open ocean conditions (the GM-parameters have been adjusted for the AIWEX buoyancy frequency profile). In addition, the spectral slope of the travel time spectra is a power less than the prediction. These results are similar to the results obtained using measurements made at the AIWEX environmental moorings (Levine, 1990). Note that the travel time spectra observed after the wind event display a peak at the local inertial frequency. This feature is also seen in the two dimensional travel time spectra shown in Figure 11 calculated from the 2 kHz data set. The presence of the inertial peak after the wind event shows a serious lack of stationarity. This complicates modeling, but also indicates that additional interesting oceanographic

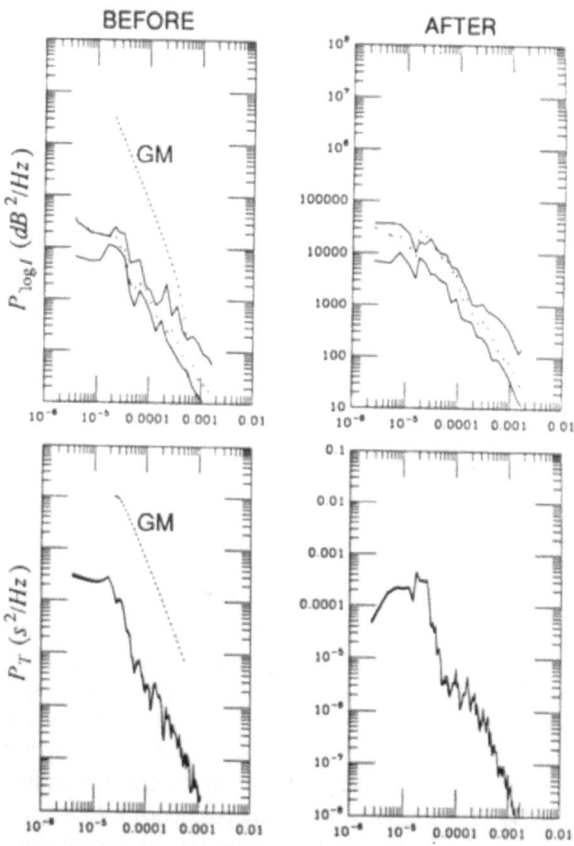

Figure 10. AATE observed log-intensity and travel time spectra for a 3 day time period before a wind event and a 4.3 day time period after the same event. To show the temporal behavior, a fixed depth data series was obtained. Spectra obtained from the 2, 4 (dashed) and 8 kHz data sets are shown. Predictions obtained using the canonical GM-model are overplotted on the left-hand graphs.

processes were present after the wind event. Sorting out the mechanisms requires close examination of the environmental data taken during these two time periods.

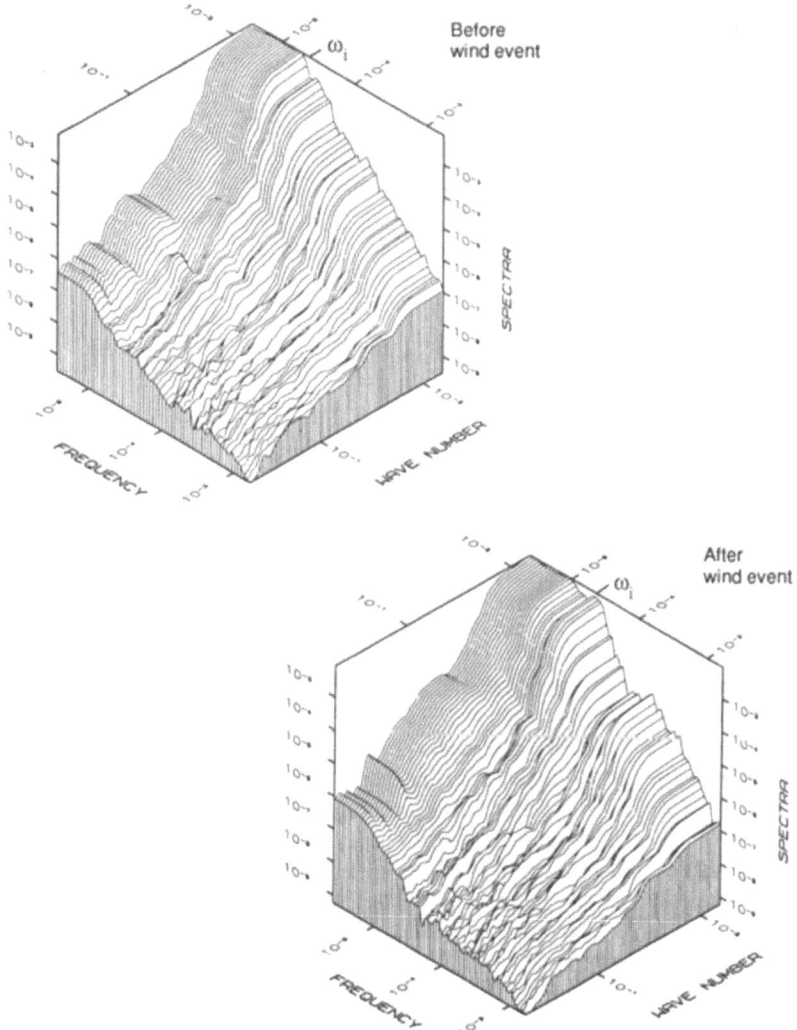

Figure 11. AATE travel time spectrum as a function of β and ω estimated from the 2 KHz measurements. A peak is observed near the inertial frequency in the spectrum from after the wind event.

A measurement of the angular spectrum is shown in Figure 12a as estimated from the 16 kHz data set. The peak narrows significantly (Figure 12b), when the time averaged arrival time is removed from the data. Removal of the deterministic arrival time structure reduces the angular spread. Further gains in the array performance are possible when scattering effects are included in the processing (Uscinski and Reeve, 1990).

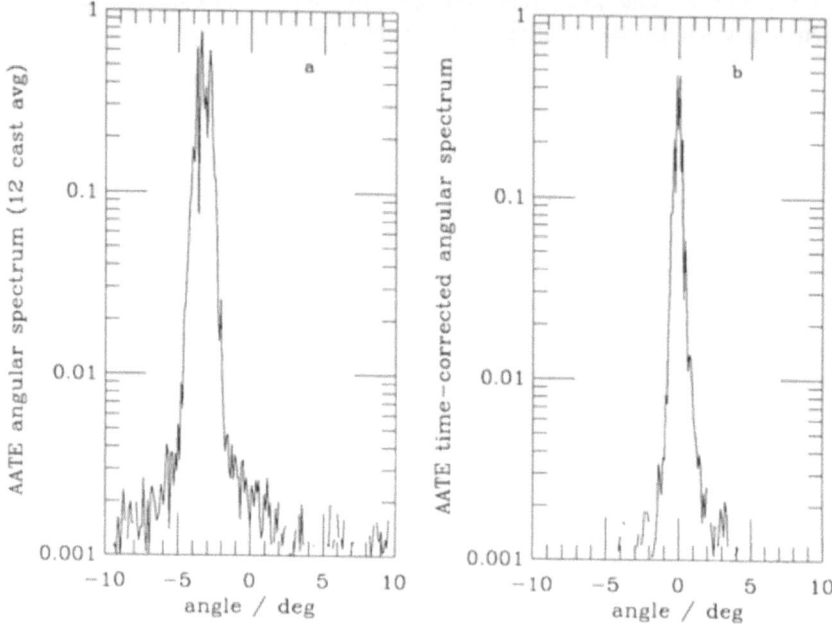

Figure 12. (a) AATE 16 kHz angular spectrum, (b) AATE 16 kHz angular spectrum (corrected for average travel time).

Intensity measurements taken during AATE also exhibit unexpected trends. Weak scattering theory predicts that the log-intensity spectra should have a spectral dependence similar to the travel times and the observed variance should scale with acoustic frequency. The measurements in Figure 10 display variances that scale as expected, however, the spectral behavior differs from that observed in the travel times. We also see that for the travel times shown in Figure 11 over the internal wave frequency bandwidth, the β-dependence is not uniform, contrary to theory. Understanding these observations will require close interaction between the environmental and acoustic investigations. We can say that without the environmental observations, connecting the acoustic measurements to the scattering processes would be speculative at best.

6. Discussion

Our understanding of the fluctuations of sound propagation in the ocean appears to be limited not by our ability to predict those fluctuations, but by our understanding of the oceanic processes. For some time, full range theories have been available for the phase correlations and the second moment of the field for known TCF's. Full range fourth moment predictions are now available, rather than approximations asymptotic to short or long range, for many important acoustic range-frequency regimes. The predictions are limited by our knowledge of the TCF. We have demonstrated this with examples from field and numerical experiments. Prediction of the MATE intensity spectra as we have outlined here, shows us that extensive oceanic measurements and thoughtful interpretation of the results in terms of models are required to achieve success in these predictions.

On the other hand, the results of AATE demonstrate that our understanding is limited when oceanic processes are not statistically stationary. The possibility of intermittency must be included in newer theories. In MATE and AATE, we are dealing with carefully isolated Fermat paths, where the definitions of phase and amplitude are unambiguous. Long range propagation can make path identification either impossible or meaningless. Variance plots like Figure 6 provide guidance on how the propagation in range-frequency behaves, but leaves out important physics. For long range propagation models that include ocean statistics, it will be necessary to model internal waves and finestructure at the same time as the larger scale dynamics, e.g., boundary currents and eddies.

Evaluations of M_{IV} are numerically tedious for realistic ocean TCF's. It is our intent to make some of the evaluation software available through the Office of Naval Research Code 1125OA.

6.1 WHAT SHOULD WE DO NEXT?

Clearly, much has been learned, but the problems outlined above will require advances in theory, numerical modeling, and field experiments. We mentioned the failure of the cross frequency fourth moment. This is a critical area if we are to understand the performance of large aperture wide-band arrays. More work remains to be done on theory that includes arbitrary sound speed profiles. It is evident that the issues of intermittency and statistical stationarity must be included in future research. However, there is a critical lack of solid field experiments where numerical and theoretical predictions can be tested with "real ocean" results. We need to conduct careful ocean/acoustic experiments like MATE. Such experiments require the collaboration of scientists from many disciplines. To carry out ocean and acoustic measurements at the edge of our understanding requires scientists that are working at the edge in each discipline. It will simply not be good enough for acousticians to throw in a few CTD's, or for oceanographers to forge ahead without considering acoustic issues.

The following experiment in volume scattering should be done next. Relatively short range, i.e., 0.5-2.0 convergence zone propagation at frequencies ranging from above 100 Hz to about 10 kHz should be done with the following attributes:
- Include "fixed" vertical acoustic array measurements in two-way transmission.
- Include multi-range transmission.
- Do at least a 30 day time series.
- Design the experiment collaboratively with Oceanographers, Theorists, and Modelers.
- Oceanography must include space/time velocity measurements for dynamics.
- Exploit stochastic inverse from M_{II} and phase.
- Carry out the operation at several statistically differing sites.

7. Acknowledgements

The authors would like to thank Dr. John Ballard for carrying out the evaluations of theory. Nina Triffleman edited our efforts and produced it in camera ready finished form. This work was carried out as part of the collaborative research efforts between our Group and the Ocean Acoustics Group at DAMTP, University of Cambridge directed by Dr. Barry Uscinski. This research was sponsored by the Department of the Navy, Office of the Chief of Naval Research, under Grant N00014-90-J-1260. This article does not necessarily reflect the position or the policy of the Government, and no official endorsement should be inferred.

40

8. References

J. Ballard and B.J. Uscinski, "Large intensity fluctuations in a randomly varying ocean: Part I. Theory and its evaluation," Unpublished manuscript, (1990).

R. Dashen, "Path integrals for waves in random media," *J. Math. Phys.*, **20**(5), pp. 894-920 (1979).

Y.J.F. Desaubies, "Acoustic-phase fluctuations induced by internal waves in the ocean," *J. Acoust. Soc. Am.*, **60**(4), pp. 795-799 (1976).

Y.J.F. Desaubies, "On the scattering of sound by internal waves in the ocean", *J. Acoust. Soc. Am.*, **64**(5), pp. 1460 -1469 (1978).

R. Esswein and S.M. Flatte, "Calculation of the strength and diffraction parameters in oceanic sound transmission," *J. Acoust. Soc. Am.*, **67**(5), pp. 1523-1531 (1980).

T.E. Ewart, C. Macaskill and B.J. Uscinski, "Intensity fluctuations. Part II: Comparison with the Cobb experiment," *J. Acoust. Soc. Am.*, **74**(5), pp. 1484-1499 (1983).

T.E. Ewart, "Acoustic propagation, internal waves, and finestructure," *Proc. Instit. Acoust.*, **8** (Part 5), pp. 106-122 (1986).

T.E. Ewart, "A model of the intensity probability distribution for wave propagation in random media," *J. Acoust. Soc. Am.*, **86**(4), pp. 1490-1498 (1989).

T.E. Ewart and D.B. Percival, "Forward scattered waves in random media - the probability distribution of intensity," *J. Acoust. Soc. Am.*, **80**(6), pp. 1745-1753 (1986).

T.E. Ewart and S.A. Reynolds, "The mid-ocean acoustic transmission experiment, MATE," *J. Acoust. Soc. Am.*, **75**(3), pp. 785-802 (1984).

T.E. Ewart and S.A. Reynolds, "Instrumentation to measure the depth/time fluctuations in acoustic pulses propagated through Arctic internal waves," *J. Atmosph. Ocean. Tech.*, **7**(1), pp. 129-139 (1990).

T.E. Ewart and S.A. Reynolds, "Large intensity fluctuations in a randomly varying ocean: Part II. MATE results," Unpublished manuscript (1990).

M.D. Levine, "Internal waves under the Arctic pack ice during the Arctic internal wave experiment: The coherence structure," *J. Geophy. Res.*, **95**(C5), pp. 7347-7357 (1990).

M.D. Levine, J.D. Irish and T.E. Ewart, "A statistical description of temperature finestructure in the presence of internal waves," *J. Phys. Ocean.*, **11**(5), pp. 676-691 (1981).

M.D. Levine, J.D. Irish, T.E. Ewart and S.A. Reynolds, "Simultaneous spatial and temporal measurements of the internal wave field during MATE," *J. Geophys. Res.*, **91**(C8), pp. 9709-9719 (1986).

C. Macaskill and T.E. Ewart, "Computer simulation of two-dimensional random wave propagation," *I.M.A. J. Appl. Math.*, **33**, pp. 1-15 (1984).

D. Slepian, "Prolate-spheroidal wave functions, Fourier analysis and uncertainty - V: the discrete case," *Bell Syst. Tech. J.* **75**(5), pp.1371-1430 (1978).

B.J. Uscinski, *The Elements of Wave Propagation in Random Media*, McGraw-Hill, New York, 153 pp., (1977).

B.J. Uscinski, "Parabolic moment equations and acoustic propagation through internal waves," *Proc. R. Soc. Lond.*, **A**(372), pp. 117-148 (1980).

B.J. Uscinski, "Intensity fluctuations in a multiple scattering medium - solution of the fourth moment equation," *Proc. R. Soc. Lond.* **A**(380), pp. 137-169 (1982).

B.J. Uscinski, "Acoustic scattering by ocean irregularities: Aspects of the inverse problem," *J. Acoust. Soc. Am.*, **79**(2), pp. 347-356 (1986).

B.J. Uscinski, "Numerical simulations and moments of the field from a point source in a random medium," *J. Mod. Optics* (in press).

B.J. Uscinski and M.J. Beran, Personal Communications (1990).

B.J. Uscinski and D.E. Reeve, "The effect of ocean inhomogeneities on array output," (submitted to) *J. Acoust. Soc. Am.* (1990).

TWO DIMENSIONAL ACOUSTICAL PROPAGATION IN A STRATIFIED SHEAR FLOW

DAVID M. FARMER[1,2] and DANIELA DiIORIO[1,2]
[1]Institute of Ocean Sciences
P.O. Box 6000
9860 West Saanich Road
Sidney, B.C. V8L 4B2, Canada
[2]Department of Physics
University of Victoria
P.O. Box 1700
Victoria, B.C. V8W 2Y2, Canada

ABSTRACT. Acoustical scintillation measurements provide a basis for determining some of the properties of a turbulent flow in the coastal environment. We discuss an experiment in which two-dimensional arrays of projectors and hydrophones are used to examine the anisotropy of the refractive index field in a high Froude number tidal flow through a channel. Both slowly varying and rapidly fluctuating signals are examined. Analysis of the rapidly fluctuating component using the two-dimensional log-amplitude correlation function, shows the flow to be strongly anisotropic, with maximum variability in the horizontal over scales of <10m. The two-dimensional angle of arrival distribution is correlated during strongly sheared flow, a result that can be explained in terms of the passage of inclined planes of higher density fluid associated with shear flow instabilities passing through the acoustical path.

Introduction And Experimental Approach

The coastal environment is often characterised by enhanced mixing, increased variability in water properties and strong tidal effects. Acoustical propagation measurements have the potential for serving as a sensitive probe of coastal oceanographic processes, which can differ markedly from those prevalent in the less active waters of the open ocean. Here we describe preliminary results of a high frequency forward scatter experiment specifically designed to shed light on some 2-dimensional features of the turbulent environment of a tidal channel.

The experiment was conducted in Cordova Channel, British Columbia, a 3 km long, 1 km wide, 30m deep channel between James Island and Vancouver Island (Figure 1). This is part of a larger area in which fresh water, primarily from the Fraser River, mixes with water of higher salinity before reaching the Pacific Ocean through the Strait of Juan de Fuca. Except for brief periods during slack water, the flow in Cordova Channel is characterised by a turbulent bottom boundary layer that can

41

J. Potter and A. Warn-Varnas (eds.), Ocean Variability & Acoustic Propagation, 41–55.

Cordova Channel

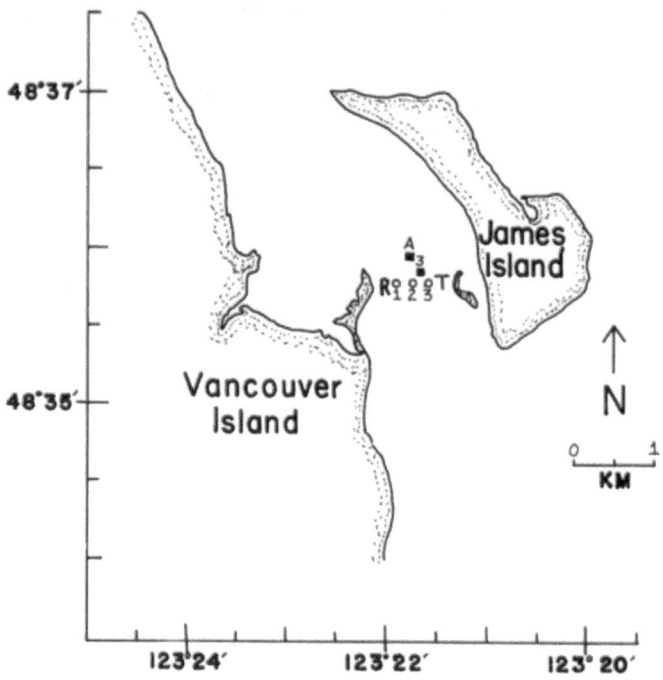

Figure 1. Cordova Channel showing instrument locations, together with the transmitter array (T) and hydrophone array (R). Recording current meters are at (o) 1, 2, and 3; the CTD profiles and time-series were obtained at stations (■), A, and 3.

spread upwards throughout the water column and is responsible for mixing the variable properties of the inflow.

Typical density profiles, which are dominated by salinity, are shown for different times in the tidal cycle in Figure 2. These may be referenced to the corresponding time in the tidal cycle, which is shown in Figure 5. In profiles (a) and (b) the flow is to the south (ebb) and there is significant stratification. In (b) the current is weaker than (a) and the mixing is incomplete in the lower layer. In (c) the flow is also to the south and stronger; both layers are well mixed. In (d) the ebb flow reaches 1 ms^{-1} and the water column is very thoroughly mixed. Profile (e) is close to slack water; a small but significant pycnocline appears at depth. Finally, profile (f), taken during strong northerly flow (flood tide), shows the entire water column overturning. The profiler descends at 1 ms^{-1}, which slightly exceeds the flow speed. Thus, even if the profile is changing rapidly during the measurement time (30s), this profile must have been taken at a time of local static instability. It turns out that this portion of the tidal cycle is a period of strong vertical shear.

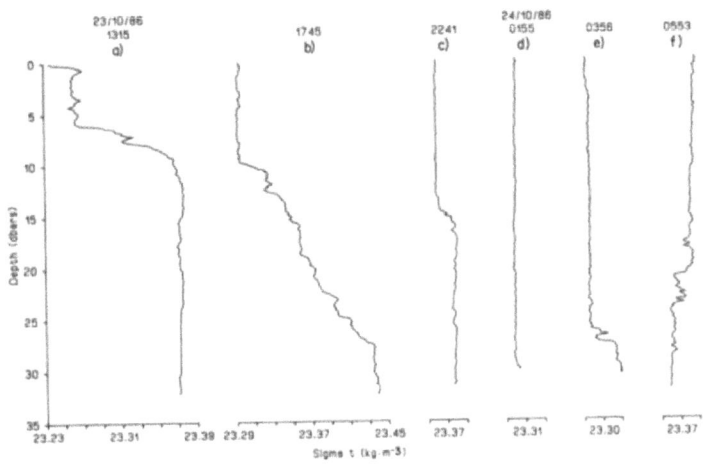

Figure 2. Density (σ_t) profiles at stations A and 3, October 23 - 24, 1986.

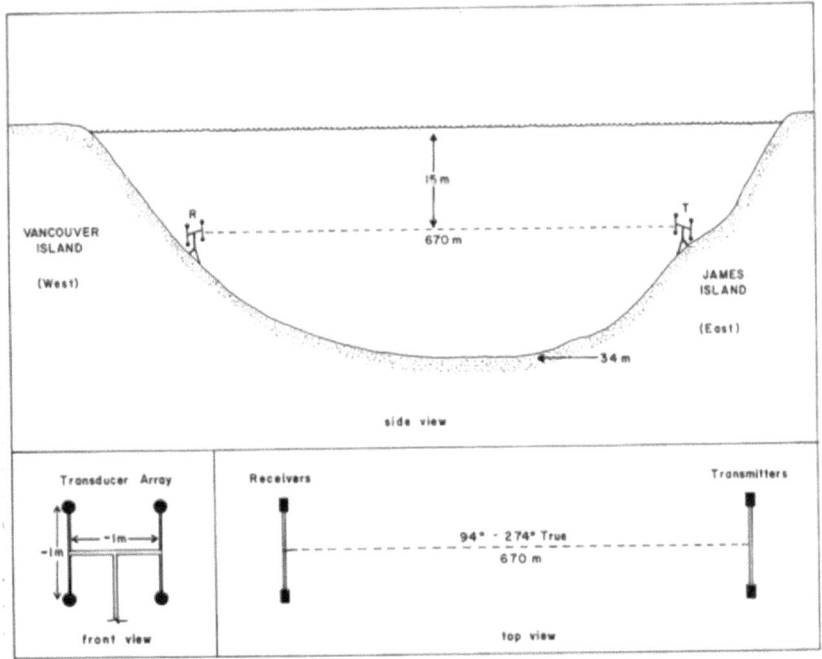

Figure 3. Deployment scheme for acoustic array.

44

Our experiment made use of a set of 4 projectors operating at a frequency of 67 kHz, and 4 hydrophones, rigidly mounted in square arrays of side 1m on firmly secured tripods, facing each other across the channel (Figure 3). Both the location and some of the measurement concepts followed earlier studies in Cordova Channel using a single source and two receivers (Farmer, Clifford and Verrall, 1987), but the technical approach was new. In addition to the use of 2-dimensional arrays, timing and control was achieved by radio rather than by cable. Identical coded sequences (63 bit, phase encoded m-sequences) were transmitted from each of the four sources in succession, so that transmissions cycled through the entire array 17 times a second. For a portion of the time series two closely spaced alternating frequencies (69 and 65 kHz) were used at a slightly lower repetition rate. The signal from each source was detected simultaneously at each of the four hydrophones, following which a specially built computer was used to decode the quadrature demodulated and detected signals in real time. The correlation peaks were subsequently extracted from each decoded pseudo random noise sequence by interpolation across the five highest points. The procedure leads to a time series of phase and amplitude measurements for each hydrophone, corresponding to the transmission from each source. Independently computed time of arrival based on the interpolated peak location allows resolution of the phase ambiguity.

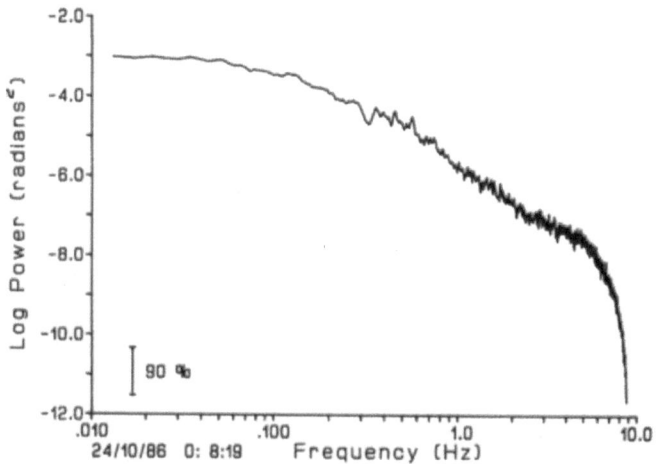

Figure 4. Sample acoustical phase difference spectrum. The data have been low pass filtered at 5.0 Hz.

Measurement related noise provides a limit to the useful resolution of phase at the highest sampling frequencies. We therefore low pass filter the data at 5 Hz using a 2nd order Butterworth filter (Figure 4). Most of the temporal variability in the signal occurs at much lower frequencies. The small delay of 14 ms between transmission of successive sources within a cycle is therefore negligible, so that following the low pass filter the signals from different sources in any given cycle can for most purposes be treated as though they are instantaneous.

The transmission path was at a nominal depth of 15m. Pulse discrimination was sufficient to separate the direct path from surface and bottom reflections. Measurements of current were obtained using Aanderaa recording current meters as indicated in Figure 1, with the instruments moored at depths of 15 and 16m so as to include the propagation path.

Low Frequency Fluctuations

Sound travelling across the channel can be modified in various ways by the intervening water. Fine scale variability in sound speed results in fine scale structure in the space- and time-evolving acoustic field detected at the hydrophones. The horizontal translation of this pattern, which is readily determined from the cross-correlation between horizontally spaced hydrophones, is related to the path weighted perpendicular component of the flow field (c.f. Farmer, Clifford and Verrall, 1987).

However, we can also look at the longer scale fluctuations over scales of minutes to hours. The amplitude, or log-amplitude, remains essentially unchanged, but significant fluctuations occur in phase. These exist because of gross changes in the mean sound speed over the path through the tidal cycle. More interesting, however, are the fluctuations in angle-of-arrival. Since the signal is detected with a 2-dimensional array, we determine the horizontal and vertical components of arrival angle θ_y, θ_z

$$\sin\theta_y = \Delta\phi_y \bar{c}/\omega d_y \ ,$$

$$\sin\theta_z = \Delta\phi_z \bar{c}/\omega d_z \ ,$$

(1)

where $\Delta\phi_y$, $\Delta\phi_z$ are the observed horizontal and vertical phase differences, from which the phase ambiguity has been removed as discussed above, $\bar{c}$ the mean sound speed, ω the acoustical frequency and d_y, d_z the horizontal and vertical transducer spacings. (The vertical angle measured at the receiver is referenced to the path connecting the receiver and transmitter and increases downwards.)

The calculations were carried out both to examine the slow variations associated with 'deterministic' deflections of the acoustical path, and also the pattern of rapid fluctuations. Figure 5 shows low pass (5.0 Hz) filtered time series of θ_y and θ_z. The horizontal deflections are much less than the vertical.

Slowly varying deflections of this sort must come about from changes in the overall current and sound speed structure through the tidal

46

cycle. Since the current is directed primarily along the channel, we should not expect any large influence from this effect. Changes in the vertical stratification on the other hand, can result in significant deflections in the vertical arrival angle. For a mean vertical sound speed gradient C_z and path length L, the vertical deflection is

$$\theta_z = \tan^{-1}\left[\frac{L}{2C_o}\, C_z\right] .$$ (2)

The deflection increases with gradient. In Figure 5 we have included mean gradients deduced from CTD profiles scaled by (2) for comparison. The CTD calculations are noisy, since they are based on a local measurement, whereas the path deflections represent an integration

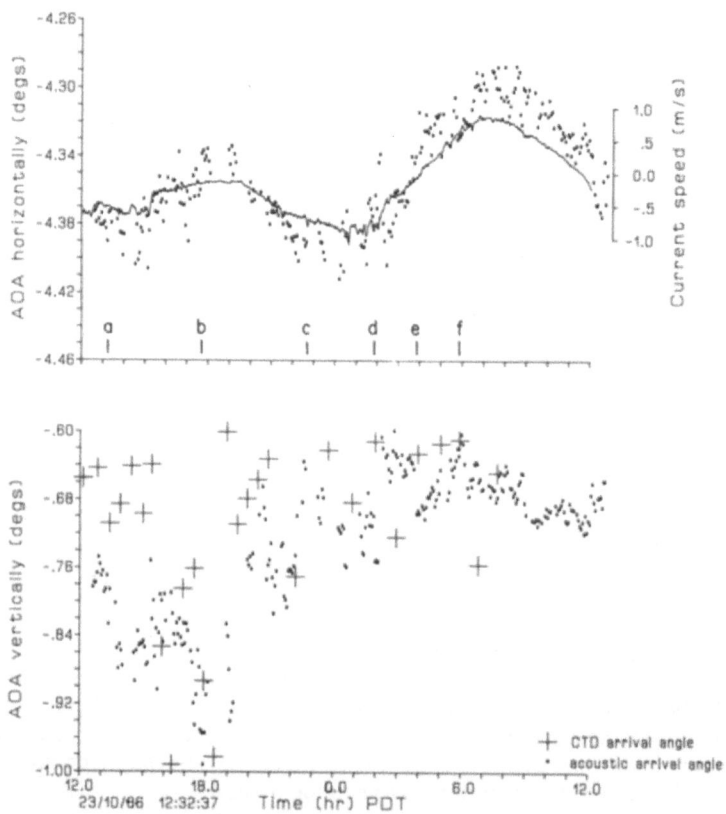

Figure 5. (Above) Horizontal angle of arrival — dots. Measured current speed scaled by equation 5 — solid line. (Below) Vertical arrival angle — dots. Measured vertical sound speed gradient scaled by equation 2 — +.

across the channel; nevertheless the large vertical deflection near the beginning of the time series is consistent with the observed increase in gradient.

Horizontal deflections are associated with horizontal (along-channel) sound speed gradients and horizontal currents. Over longer time scales (i.e. >100s) mean horizontal sound speed gradients tend to be very small. On the other hand, the tidal current can advect downstream the acoustic pulse during its passage, producing a significant deflection. For a mean flow speed $\bar{U}$, the pulse is advected $\bar{U}\tau$ downstream, where

$$\tau = (L/c)(1 - \bar{U}^2/c^2)^{\frac{1}{2}} + \delta L/c \tag{3}$$

and $\delta L = (L^2 + [\bar{U}\tau]^2)^{\frac{1}{2}} - L$ is the incremental distance the sound must travel to reach the hydrophone, which is negligibly small compared to the first term. The deflection angle is therefore

$$\theta_y = \sin^{-1}[(\bar{U}/c)/(1 - \bar{U}^2/c^2)^{\frac{1}{2}}], \tag{4}$$

or, to first order for $\bar{U} \ll c$,

$$\theta_y = \bar{U}/c, \tag{5}$$

which is just the Mach number for the flow. Note that this result is independent of the path length. Measured current speed is shown in Figure 5, scaled according to (5). Clearly (5) explains the gross features of the horizontal deflection, but not the details. Departure of the flow from a horizontally uniform current together with transverse sound speed gradients will result in differences between the speed measured by the current meter and that predicted by (5).

Short Period Fluctuations

Rapid fluctuations are associated with smaller scale variability in the water properties advected through the acoustical path. For a fully developed and uniform turbulent flow this variability is expressed in terms of a 'structure constant' C_n^2 of the refractive index field which can be determined, for example, from the variance of the acoustical log-amplitude χ (c.f. Farmer, Clifford and Verrall, 1987):

$$\langle C_n^2 \rangle_\chi = 8.07 k^{-7/6} L^{-11/6} \sigma_\chi^2, \tag{6}$$

where k is the acoustical wavenumber, L the path length and σ_χ^2 the log-amplitude variance.

Expression (6) is derived on the basis of assumptions relating to the similarity structure of turbulence and its presumed isotropy (Tatarski, 1971). Figure 6 shows calculations of C_n^2 for the period of the experiment. Very low values occur near, but not quite at, slack water. These low values represent times at which very well mixed water is moving relatively slowly through the channel. Large values are associated not just with high flow speeds, but with the advection of imperfectly mixed water.

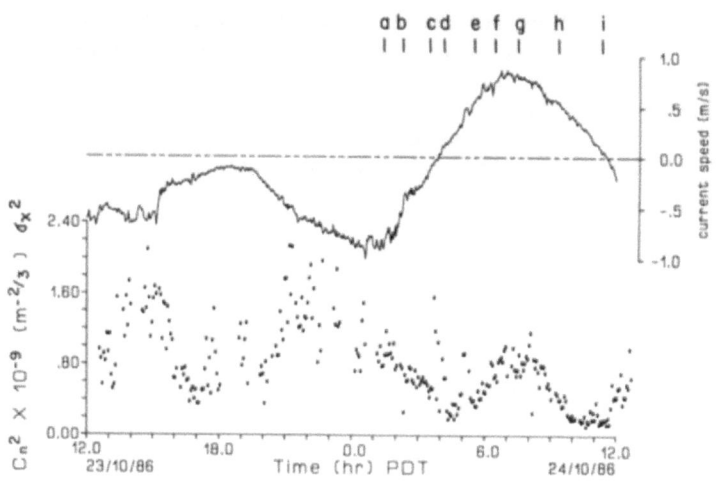

Figure 6. Structure parameter C_n^2 (dots) together with observed current speed (solid line). Letters correspond to Figure 7.

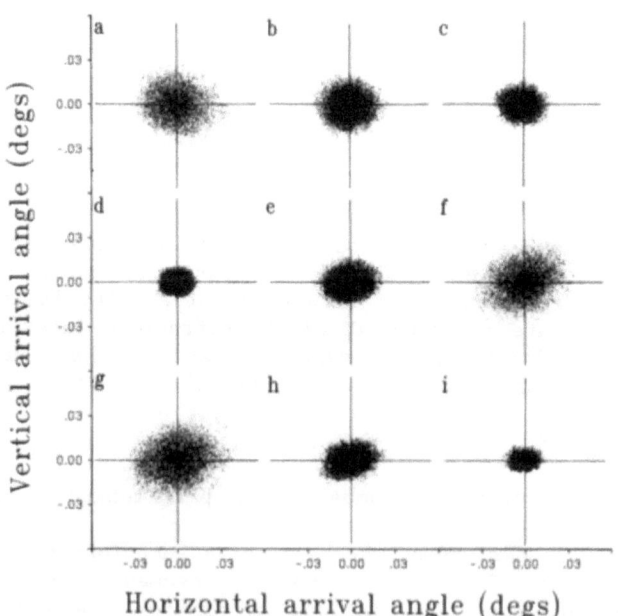

Figure 7. Two dimensional angle of arrival distribution through the tidal cycle (see Figure 6). Each plot is based on 30 minutes of data.

How closely does Nature approach the idealised conditions of isotropic turbulence in this well stirred flow? The fluctuations on which, for example, the calculation of Figure 6 are based, are sensitive only to smaller scale refractive index structures, for example Fresnel scale variability ($\sqrt{\lambda L}$ = 3m) in the case of log-amplitude variance. Even for the idealised and unachievable concept of homogeneous, isotropic turbulence, there must always be an outer scale for which it cannot approximate real conditions. This will be imposed by either the flow boundaries, or by the effects of stratification. We can explore the degree of symmetry in the turbulent structure over these scales, through 2-dimensional calculations of the rapid fluctuations observed with our 2-dimensional array.

Consider first the 2-dimensional angle-of-arrival distribution. Figure 7 shows distributions of the band-pass filtered data (0.1 Hz to 5 Hz) based on 30 minute data sets through the tidal period. If the turbulence were isotropic we would expect an isotropic distribution, but this is not observed. Moreover there is a systematic distortion, implying a correlation between vertical and horizontal deflections which is most pronounced towards the end of the flood tide (Frames f to h in Figure 7). We will return to this feature subsequently.

Theories have been developed for sound scattering from anisotropic turbulence under the assumption that the refractivity spectrum $\Phi(k,\theta)$ is separable in wavenumber k and orientation θ in the plane perpendicular to the path (Lee and Harp, 1969; Gorelov and Dotsenko, 1989). For the simplest case of elliptical anisotropy, it can be shown that the covariance over the horizontal hydrophone separation d_y and time lag τ is

$$C(d_y,\tau) = 4\pi^2 k^2 \int_0^k d\kappa \int_0^L ds \; \kappa\Phi(\kappa) \; \{J_0[\kappa|d_y-V\tau|] +$$

$$aJ_2[\kappa|d_y-V\tau|]\} \; \frac{\sin^2}{\cos^2}\left[\frac{\kappa^2 s}{2k}\right] , \tag{7}$$

where a defines the degree of anisotropy, J_n is the nth order Bessel function and V is the current velocity. The diffraction term $\sin^2(\;)$, $\cos^2(\;)$, applies to the log-amplitude and phase respectively. This expression applies to plane waves, which approximates the present experiment where we use the signal from multiple projectors in the measurement (Lee and Harp, 1969).

The elliptical anisotropy coefficient a is determined by a least squares fit of (7) to the measured time-lagged covariance between a pair of horizontal hydrophones. Figure 8 shows the result for the period of the experiment. A value of a = 0 corresponds to perfect symmetry, -ve values to a vertically oriented, +ve values to a horizontally oriented distribution. It is apparent from this figure that the data are consistent with horizontally oriented anisotropy during strong flows and vertically oriented anisotropy for brief periods during weak flows. At no time is there a strong vertical anisotropy.

50

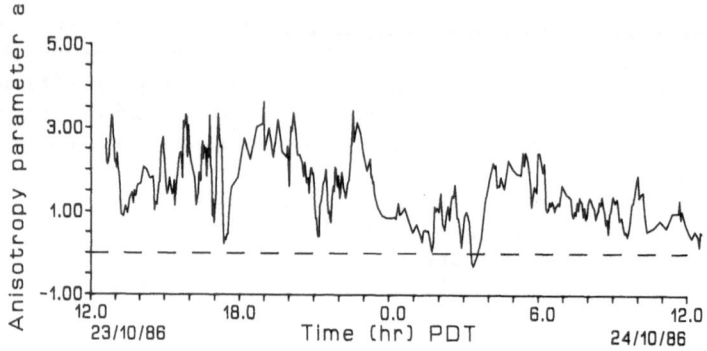

Figure 8. The anisotropy parameter *a* for the period corresponding to Figure 6.

Discussion

Although Cordova Channel is an energetic environment, its properties (salinity, temperature, sound speed, density) are far from being isotropically distributed. This is apparent both from the angle-of-arrival distribution and from a fit to a stochastic model of the anisotropic correlation function. Both of these results pose interesting questions.

Considering first the angle-of-arrival distribution, we note (Figure 7) that for a significant part of the tidal cycle the horizontal deflections exceed the vertical deflections. Moreover, for a small portion of the measurement period (Figure 7 f to h), vertical deflections are positively correlated with horizontal deflections: the resulting scatter distribution has an inclination with respect to the horizontal of about 35°. The period during which this inclination occurs coincides with a marked vertical shear in the measured current (Figure 9). We identify two possible explanations. One possibility is that fluctuations in vertical sound speed gradient result in fluctuations in acoustical path depth, with corresponding horizontal deflections due to differential horizontal pulse advection in the vertically sheared flow. A second possibility is that the stratification in the sheared flow consists in part of inclined planes of sound speed structure, which would result in correlated vertical and horizontal angle-of-arrival fluctuations.

Taking the first possibility, we calculate the horizontal deflection $\Delta\theta_y$ associated with a change in vertical sound speed gradient C_z, for given vertical shear U_z:

$$\tan\theta_y = \frac{1}{L} \int \frac{U(z)}{C(z)} \, dS = \frac{1}{L} \int \frac{U_0 + z\, U_z}{C_0 + z\, C_z} \, dS, \tag{8}$$

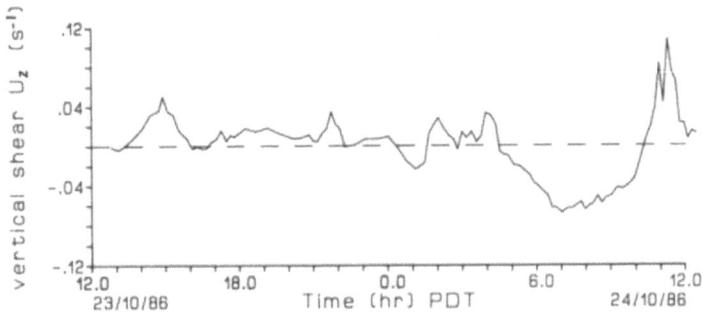

Figure 9. The vertical shear U_z for the experimental period, determined acoustically (Farmer, Clifford and Verrall, 1987).

where U_o, C_o are the flow speed and sound speed at the projector-hydrophone depth and integration is carried out along the refracted path. For small deflections we assume a uniform vertical shear U_z and sound speed gradient C_z. Integration of (8) yields

$$\tan \theta_y = L^{-1}C_z^{-1} \left\{ (U_o - U_z C_z^{-1} C_o) \; \ell n \left| \frac{1 + \sin \theta_z}{1 - \sin \theta_z} \right| + \right.$$

$$\left. 2U_z C_z^{-1} \; \frac{\theta_z C_o}{\cos \theta_z} \right\} \; . \tag{9}$$

We are concerned here not with the gross horizontal deflection, which is dominated by the slowly varying tidal current $U_o(t)$, but with the rapid deflections $\Delta\theta_y$ about the mean, associated with fluctuations in vertical shear ΔU_z caused by eddying motions. Similarly, small vertical deflections $\Delta\theta_z$ occur as a result of fluctuations in the vertical sound speed gradient ΔC_z (see equation 2). From (9),

$$\tan \Delta\theta_y = \Delta U_z \Delta C_z^{-1} \left[\frac{\Delta\theta_z}{\sin \Delta\theta_z} - \cos \Delta\theta_z \right] \; . \tag{10}$$

Thus small deflections in the horizontal arrival angle will depend upon small deflections in the vertical arrival angle in a way that is determined by the vertical shear and sound speed gradient. With the small angle approximation (10) can be expressed as

$$\frac{\Delta\theta_y}{\Delta\theta_z} = \Delta U_z \; \frac{L}{3C_o} \; . \tag{11}$$

Thus the orientation depends only on the vertical shear. For a positive shear the slope of the angle-of-arrival distribution is positive. This is opposite to the measured results. Figure 9 shows a strong negative shear whereas Figure 7 f to h shows positive slopes. While the shear effect is undoubtedly present, it cannot explain the observed results.

Suppose instead that angle-of-arrival deflections are caused by the passage of inclined planes of higher density, and thus higher sound speed fluid through the acoustical path. The presence of such coherent structures is a common feature of high Froude number, high Reynolds number flows and is associated with the braids that connect successive Kelvin-Helmholtz like instabilities in the case of density stratification (c.f. Koop and Browand, 1979). Figure 10 shows an example of such an instability; the figure is an echo-sounding of such a feature in the stratified shear flow of Knight Inlet (c.f. Farmer and Freeland, 1983). No such data are available from Cordova Channel, but the environment has some similarities. Even if the Froude number is sufficiently high that the dynamic significance of the stratifica-tion is negligible, turbulent boundary layers are still characterised by plumes that are inclined with the shear (Head and Bandyopadhyay, 1981). Either braids, or inclined plumes will consist of filaments of higher sound speed water drawn upwards, which will cause a deflection first in one direction and then in the other as they pass through the acoustical path.

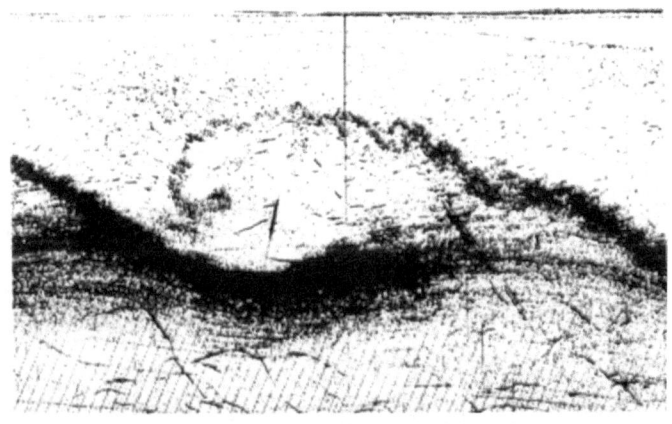

Figure 10. Echo-sounding of shear flow instability typical of tidally forced flow over topography. (Vertical scale 70m, horizontal scale approximately 10 minutes). Although this image was not taken in Cordova Channel, it is an example of the kind of structure that can lead to strong acoustical anisotropy in mixing layers.

Consider the effect of such a filament in a sheared flow. As the filament, which is of higher sound speed than the surrounding water, passes through the acoustical path, it first deflects the path upwards at right angles to the filament's orientation, then downwards. In the (unlikely) event that the filament stretched evenly and uniformly across the channel, the deflections would be

$$\Delta\theta_s = \frac{L}{2C_o}\frac{dC}{ds} ,$$ (12)

where the deflection $\Delta\theta_s$ is normal to the filament. We might expect that in general filaments will be less than the channel width, but the basic concept is nevertheless illustrated by (13). For a filament with orientation ϕ to the vertical, the orientation of the angle-of-arrival distribution is

$$\frac{\Delta\theta_z}{\Delta\theta_y} = \pi/2 - \phi .$$ (13)

If the shear is negative (corresponding to z positive downwards), ϕ is positive and thus $\Delta\theta_z/\Delta\theta_y$ is positive. This contrasts with (11), for which a negative shear leads to a negative value of $\Delta\theta_z/\Delta\theta_y$

In Figure 7 f to h, deflections towards positive $\Delta\theta_z$ are correlated with deflections towards positive $\Delta\theta_y$. Thus the orientation of the distribution is consistent with the passage of inclined braids or filaments in the shear flow, but is inconsistent with the shear effect defined by (11). The fact that a pronounced correlation occurs simultaneously with the occurrence of vertical shear supports the interpretation in terms of shear flow instabilities. In contrast, the shear effect not only predicts a distribution with an orientation opposite to that observed, but is also rather small for the present example. Vertical deflection magnitudes are consistent with a vertical sound speed gradient of 0.01 s^{-1}, which is not atypical for the flow, but horizontal deflections would require a vertical shear of .45 s^{-1} which is much greater than observed. Clearly the shear effect must occur for a stratified shear flow, but is too small to dominate the observed deflections in the present example.

The vertical component of sound speed variability along the path exceeds the horizontal variability for fluctuation periods of a few 10s of seconds and longer. In fact, most of the horizontal angle-of-arrival deflections in the low-passed time series (Figure 5) can be accounted for by horizontal advection of the sound path by the tidal current. The low frequency horizontal deflections are just 0.25 the amplitude of the low frequency vertical deflections. On the other hand the high passed data, which is sensitive to features with a horizontal scale less than 10-15m, is horizontally anisotropic. This is true even when the current is not strongly sheared and thus cannot be expected to give rise to shear flow instabilities as discussed above. The angle-of-arrival distributions in Figure 7 exhibit in general greater horizontal than vertical deflections.

Except close to slack water, the stochastic model of log-amplitude correlations persistently exhibits a smaller horizontal than vertical correlation length corresponding to a greater horizontal refractive index variability at time scales of less than 10s. These results might seem counterintuitive, since we normally anticipate stratification to be vertical rather than horizontal. However, it must be emphasised that this is a high Froude number mixing flow and we are describing the variability associated with rapidly fluctuating components (<10s). Mixing of different water masses entering the channel will occur intermittently and will proceed rapidly throughout the depth of this shallow channel. This will lead to significant horizontal variability, which is indeed apparent in the time series measurements from the CTD at fixed depth (Figure 11). The vertical profile (Figure 2f) shows overturning on the scale of the full water column depth. Such an effect must be associated with rapid and localised vertical exchanges, resulting in significant horizontal variability.

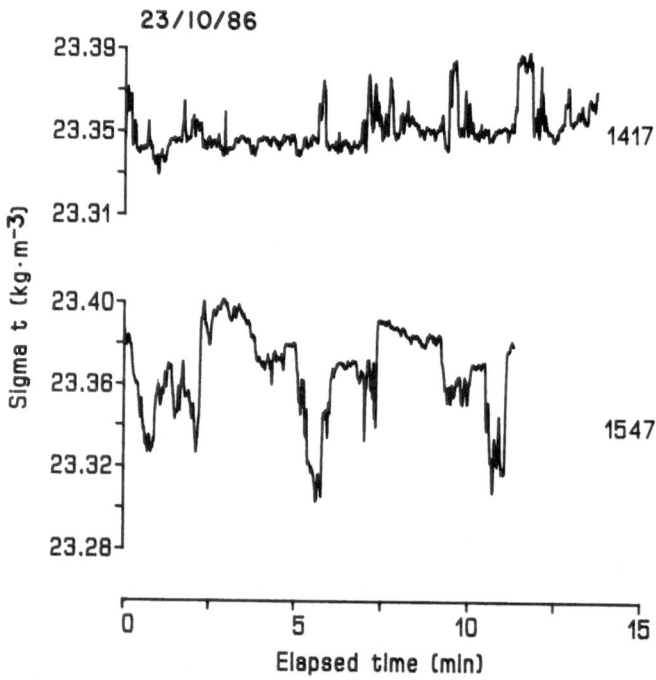

Figure 11. Density (σ_t) time series from a CTD suspended at a depth of 15 meters at station A.

Concluding Comment

The results discussed here illustrate the need for consideration of fine scale features in the flow dynamics when interpreting variability in acoustical propagation, even for the well developed turbulent flow of the present example. The two-dimensional angle-of-arrival appears to be a sensitive signal for examining this variability over a wide range of scales. In principle, the perpendicular flow speed, vertical shear, mean sound speed and mean sound speed gradient can all be recovered from a simple two-dimensional array of the sort discussed here. These properties, together with analysis of the two-dimensional features apparent in the angle-of-arrival distribution and the log-amplitude time series, provide an effective tool for studying stratified shear flows in the coastal environment.

Achnowledgements

This work received support from the Canadian Panel on Energy Research and Development, and the U.S. Office of Naval Research. We are grateful to David Lemon and his colleagues at Arctic Sciences Ltd., for their assistance with the experiment.

References

Farmer, D.M., Clifford, S.F. and Verrall, J.A. (1987) 'Scintillation structure of a turbulent tidal flow', *J. of Geophys. Res.*, **92**(C5), 5369-5382.

Farmer, D.M. and Freeland, H.J. (1983) 'The Physical Oceanography of Fjords', In: Progress in Oceanography, M.V. Angel and J.J. O'Brien eds., Pergamon Press, **12**(2), 147-220.

Gorelov, V.N. and Dotsenko, S.V. (1989) 'Correlation description of two-dimensional anisotropic sound fields', *Sov. Phys. Acoust.*, **35**(3), 255-258.

Head, M.R. and Bandyopadhyay, P. (1981) 'New aspects of turbulent boundary-layer structure', *J. Fluid Mech.*, **107**, 297-338.

Koop, C.G. and Browand, F.K. (1979) 'Instability and turbulence in a stratified fluid with shear', *J. Fluid Mech.*, **93**, 135-159.

Lee, R.W. and Harp, J.C. (1969) 'Weak scattering in random media, with applications to remote probing', *Proc. IEEE*, **57**(4), 375-406.

Tatarski, V.I. (1971) 'The effects of the turbulent atmosphere on wave propagation', translated from Russian, Israeli Program for Scientific Translation, Jerusalem.

THE EFFECTS OF SOUND SPEED ON THE SHAPE OF THE OCEAN IMPULSE RESPONSE

R. L. FIELD and M. K. BROADHEAD
Naval Oceanographic and Atmospheric Research Laboratory
Stennis Space Center, Mississippi 39529-5004
USA

ABSTRACT. The effects of sound speed on the shape of the ocean impulse response measured within the shadow zone are investigated at a range of 12 km and depths of 489, 677, and 841 m. Measured impulse responses are computed by cross-correlating received signals at these depths with a source replica of 35-55 Hz bandwidth. Time waveform envelopes of the measured responses are compared with envelopes computed by a time-domain parabolic equation model for three sound speed cases. It is found that the envelopes of these responses are predictable enough to remove propagational distortion by deconvolving the data envelope with the model envelope and thus extract useful source signal characteristics.

Introduction

This paper investigates the effects of ocean sound speed on the shape of ocean impulse responses near the ocean surface. The near surface is defined to be the shadow zone which lies between the direct arriving wavefront and the ocean surface. In this zone, the acoustic pressure decays rapidly away from the direct wavefront. This low pressure region is found to be sensitive to the details of the ocean sound speed structure near the ocean surface[1,2]. Because diffraction effects dominate in this zone, acoustic wave equation models must be used to predict the acoustic pressure.

One of the objectives of the experiment discussed here was to characterize the distortion of transient signals due to propagation effects. If the ocean impulse response can be predicted then, in principle, the inverse of this predicted response could be used to remove the distortion from the signal. The purpose of this study is three fold: (1) to show that transient distortion within this low intensity zone is dependent on the ocean sound speed structure, (2) to show that very subtle features of the ocean impulse response within the shadow zone can be predicted and (3) that knowledge of these sound speed dependent features allows useful signal parameters to be restored through deconvolution. Three sound speed cases are constructed based on bathythermograph measurements taken over the acoustic track. Acoustic propagation for each sound speed case is simulated with a time-domain parabolic equation model. The model results are compared with the acoustic data at 12 km in range and 489, 677, and 841 m in depth. Deconvolution techniques are used to extract the source waveform envelope for each sound speed case. The results of the deconvolution determine the best sound speed case.

J. Potter and A. Warn-Varnas (eds.), Ocean Variability & Acoustic Propagation, 57–68.

58

Experiment

An acoustic transient experiment, TRANSECT-D, (Transient Source Extraction and Classification Test – Deep water) was performed in the Pacific Ocean at 34°N, 140°W, in July of 1989. Figure 1 shows the acoustic track over which a linear frequency modulated (LFM) signal was transmitted by an HLF-2AH source and received by a 128 element vertical line array. The array had a 23 m sensor separation with a top hydrophone depth of 489 m. The source transmitted transient signals in a 26 to 140 Hz frequency band for ranges out to two convergence zones. Here we are concerned with receiver depths of 489, 677, and 841 m, a range of 12 km, and a signal frequency band of 35 to 55 Hz. Bathythermographs were dropped by aircraft (AXBTS) and the source ship (XBTS) along the entire acoustic track over a five day period.

ACOUSTIC DATA

The transmitted LFM signal was recorded by an accelerometer mounted on the source. The recorded signature is cross-correlated with the LFM signal received on the array. This matched filter result is a bandlimited impulse response from source to receiver.

The correlation process can be represented as follows. The received signal, $r(t)$, for fixed source-receiver positions, is given by the convolution integral

$$r(t) = \int_{-\infty}^{+\infty} s(u)\,g(t-u)\,du + n(t) \qquad (1)$$

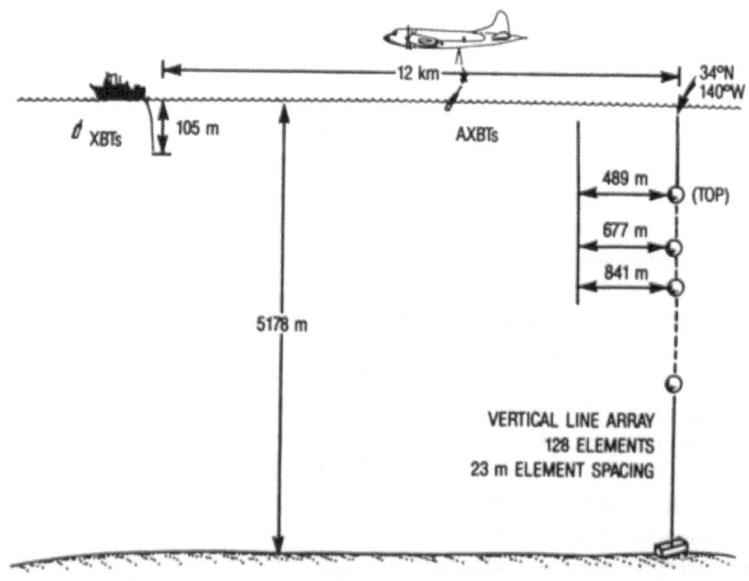

Figure 1. Transient experiment: TRANSECT – D.

where, $s(t)$, is the transmitted LFM signal, $g(t)$, is the ocean impulse response or Green's function and $n(t)$ is noise. Replacing the integral by asterisk notation

$$r(t) = g(t) * s(t) + n(t) \tag{2}$$

Equation 2 assumes the ocean is time and spatially invariant. (The term spatially invariant should not be confused with range independence). These assumptions should be valid over signal transmission times and for source-receiver locations that are almost fixed. In the experiment, the source was towed at approximately 5 kts.

If $s(t)$ is cross-correlated with $r(t)$ then Equation 2 becomes

$$\begin{aligned} h(t) &= [(g(t) * s(t)) \ x \ s(t)] + [n(t) \ x \ s(t)] \\ &= [g(t) * s(t) \ x \ s(t))] + [(n(t) \ x \ s(t)] \\ &= [g(t) * a(t)] + [n(t) \ x \ s(t)] \end{aligned} \tag{3}$$

where $a(t)$ is the autocorrelation of the source signal, $s(t)$, and x denotes cross-correlation. If $n(t)$ and $s(t)$ are uncorrelated, $h(t)$ has optimal signal-to-noise properties[3].

Referring to Equation 2, the output of the matched filter, Equation 3, can be interpreted as the propagation of the source autocorrelation function, $a(t)$. The normalized autocorrelation is shown in Figure 2 along with its envelope. The waveform was filtered to a 35-55 Hz passband to achieve maximum signal-to-noise. The autocorrelation function shown in Figure 2 will be the transmitted transient signal.

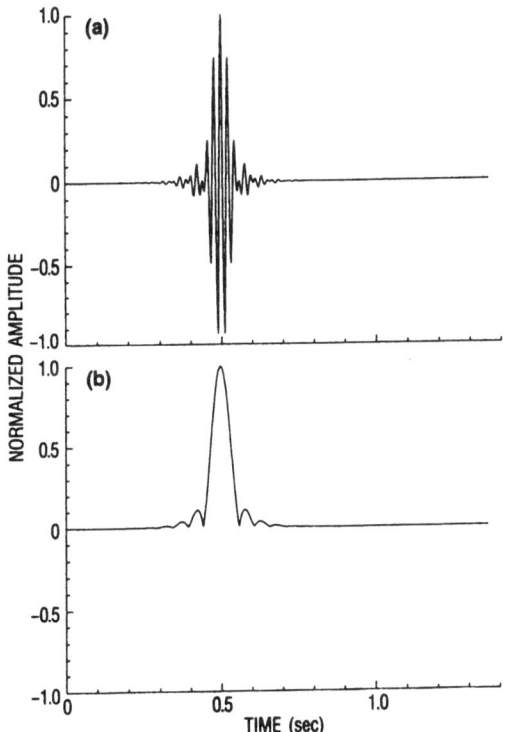

Figure 2. Source autocorrelation (a) and its envelope (b), range = 0, source depth =105 m.

Since we are studying the effects of sound speed in the shadow zone, the beginning of this zone at the 12 km range will be located. Figure 3 shows the matched filter output at depths of 489, 1241, and 1335 m. The direct and bottom-interacting paths are labeled in the figure. As receiver depth decreases, the direct path diminishes in amplitude relative to the bottom-interacting paths. Figure 3 shows that, at the 12 km range, 1335 m is close to the ray theory limit for direct path propagation. A ray theory model predicted the depth limit of direct path propagation to be 1400 m at this range. Twenty-five meters shallower, the ray model predicted the bottom reflected path as the first arrival. The 1335 m depth at 12 km range marks the beginning of the shadow zone and where the fine structure of the ocean sound speed begins to influence the acoustic propagation.

The direct path arrival will be windowed from the matched filtered data, $h(t)$, at receiver depths of 489, 677, and 841 m. The windowed results are renormalized and their envelopes computed at each depth using the Hilbert transform. The data envelopes displayed in Figure 4 will be modeled for three sound speed cases.

SOUND SPEED DATA

AXBTS were dropped on July 6 and 8. Each drop covered different parts of the track. XBTS were dropped from the source ship on July 8 and 9. XBTS were also taken at the array site on July 7 and 10. Signal transmission occurred on July 9. Sound speed profiles were computed from the AXBT and XBT data. Figure 5 shows all profiles computed from A and XBTs gathered at

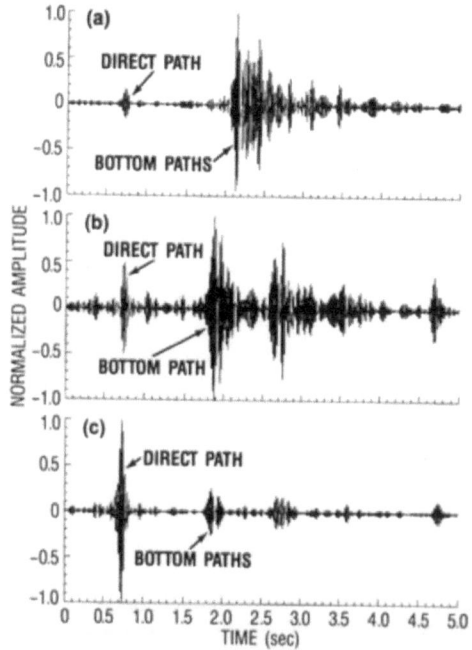

Figure 3. Matched filter outputs, range = 12 km, receiver depth = (a) 489 m, (b) 1241 m, and (c) 1335 m.

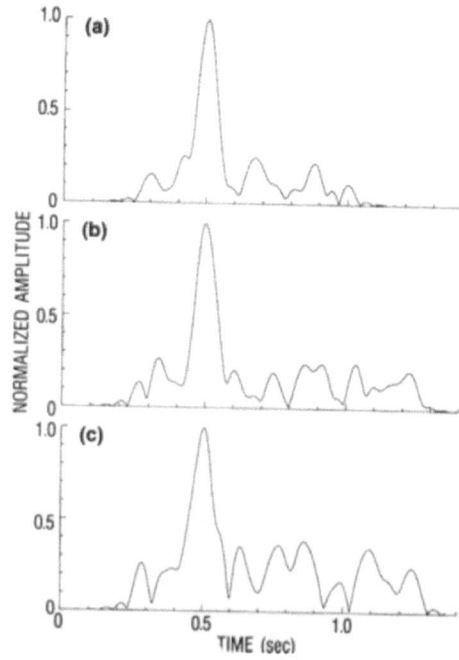

Figure 4. Direct path envelopes, range = 12 km, receiver depth = (a) 489 m, (b) 677 m, and (c) 841 m.

their respective ranges. The array is at zero range. The sound speed profiles displayed in Figure 5 show a consistent trend over the track during the period of July 6 through 9. Because of this consistency, a sound speed grid is formed using the surface fitting program ZGRID[4]. Sound speed profiles are computed every 2 km using all AXBTS and XBTS gathered over the track. Figure 6 shows the gridded profiles out to two convergence zones. Figure 7 shows the gridded sound speed data 12 km from the array where model and data envelopes will be compared. Figure 8 shows the profiles computed from XBTS taken at the array site on the 7th and 10th along with the profiles computed from the AXBTs dropped within the 12 km of the array. The three AXBTS shown in Figure 8 were taken at approximately 10 km from the array but on different days and at different times. The profiles of Figure 8 show the near surface (0-200 m) sound speed variability that has been removed in the gridding process (Fig. 7).

From Figures 7 and 8, three sound speed cases are made. The gridded sound speeds shown in Figure 7 form the first case. The second and third cases are the single profiles taken at the array site on the 7th and 10th, respectively. These two cases were arbitrarily chosen from the set of profiles displayed in Figure 8. The gridded data forms a sound speed structure with slight range dependence. Here the near surface sound speed variations observed within the first 200 m have been smoothed by the ZGRID process. The two range independent cases retain these near surface velocity variations. Below 200 m there is little difference between the three sound speed cases.

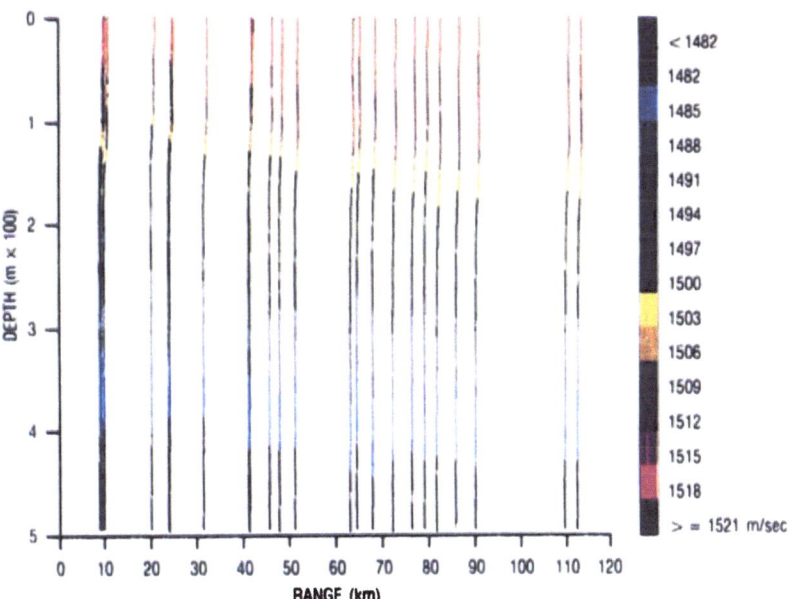

Figure 5. Sound speed profiles computed from AXBTs and XBTs gathered over two convergence zones.

62

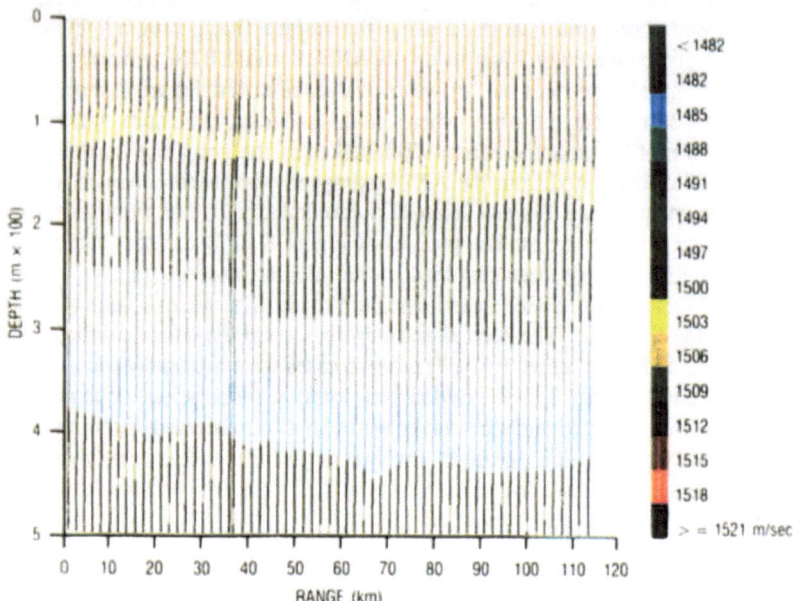

Figure 6. Gridded sound speed profiles computed from AXBTs and XBTs gathered over two convergence zones.

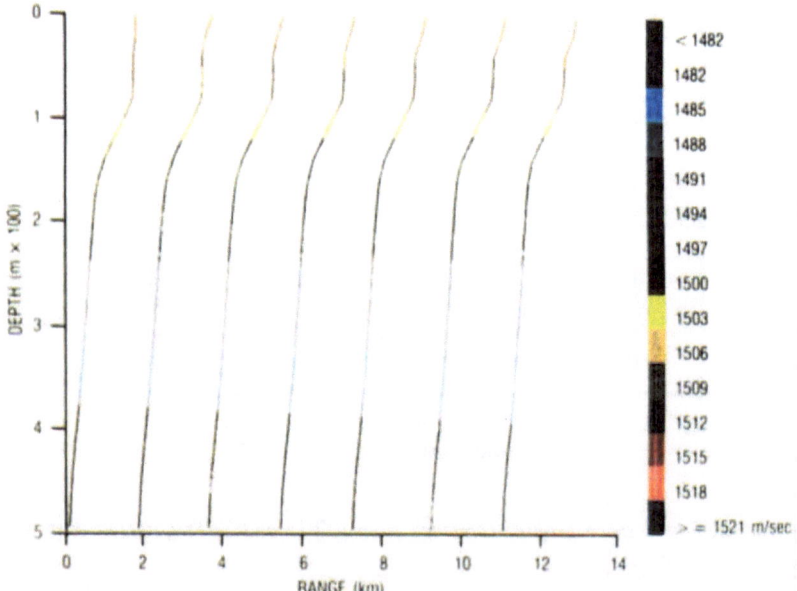

Figure 7. Gridded sound speed profiles computed from AXBTs and XBTs gathered within 12 km of the array.

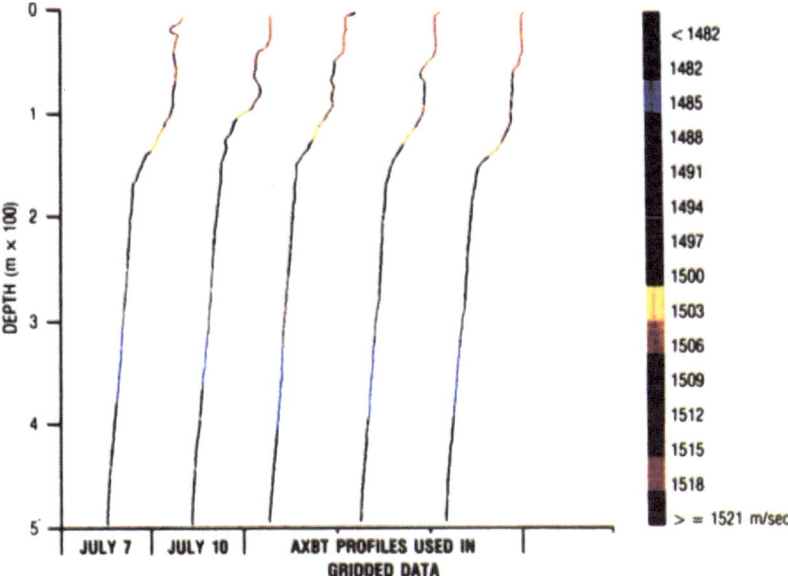

Figure 8. Sound speed profiles computed from AXBTs and XBTs gathered within 12 km of the array.

Propagation Model

A time-domain parabolic equation (TDPE) model[5] is used to predict the shape of the impulse response as a function of range and depth. The autocorrelation of the LFM signal, Figure 2b, is propagated by the model to the 12 km range for all three sound speed cases. The time waveform computed by the TDPE model is given by

$$h_m(t) = g_m(t) * a(t) \tag{4}$$

where $h_m(t)$ is the modeled matched filter output and $g_m(t)$ is the modeled impulse response. Normalized envelopes are computed for each $h_m(t)$ within the upper 1200 m of the water column. Figure 9 displays the contours of these normalized envelopes in depth and relative time for each sound speed case. The depths that will be compared with the data are labeled on the figure. The color scale is in dBs. The -30 dB color is approximately 0.03 amplitude units on a normalized linear scale.

The three cases show a marked difference from each other within the upper 1200 m of the water column. The gridded sound speed case shown in Figure 9a is characterized by a single main peak (red) consisting of direct path energy extending from 1200 m up to 250 m. At this depth, diffraction energy emanating from the near surface begins to interfere with the direct path energy. From just above 489 m to just above 677 m the direct path broadens due to the inter-fering diffractions.

The July 7th case shown in Figure 9b shows the near surface diffractions extending deeper into the water column. At 489 m, there are two main peaks. The second peak is the diffraction front from the near surface. At 677 m, the interference of diffraction and direct path energy results in three main peaks.

64

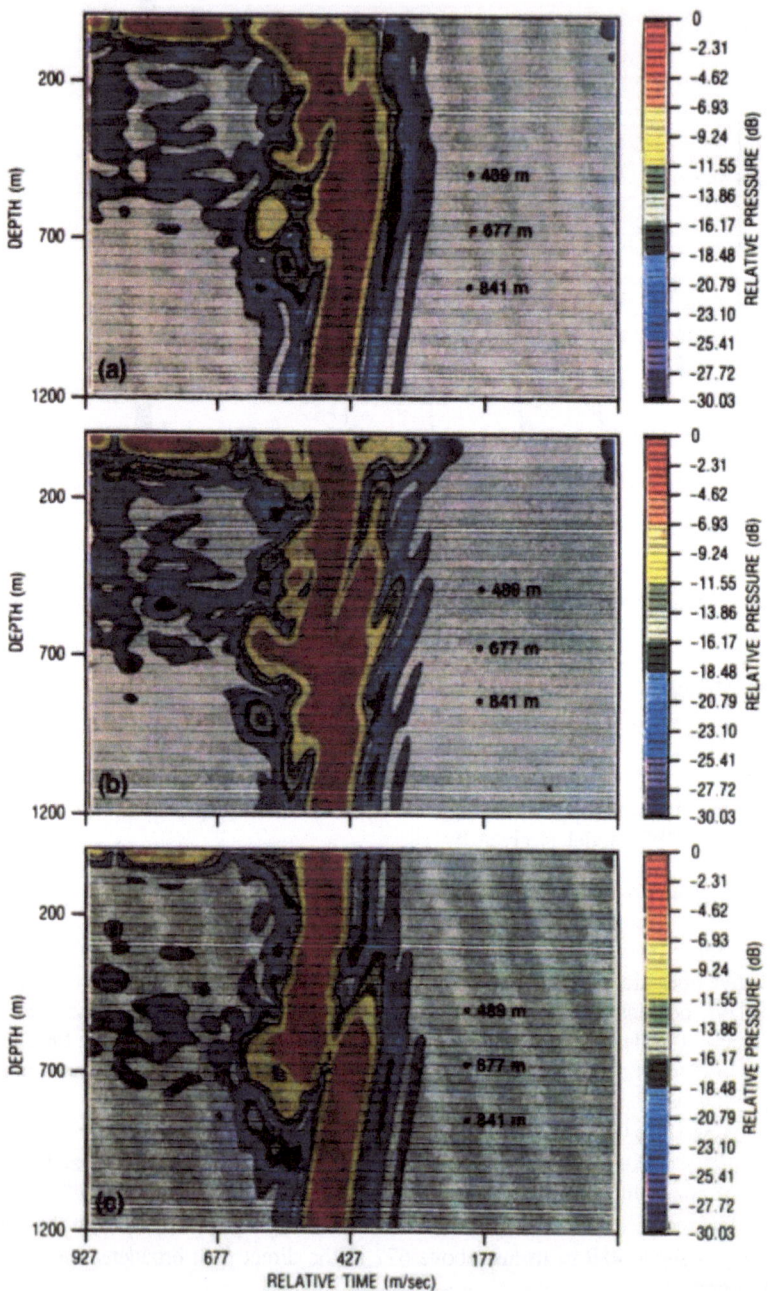

Figure 9. Contours of model envelopes, range = 12 kim, (a) gridded, (b) July 7, and (c) July 10.

The July 10th case (Figure 9c) predicts primarily diffraction energy and little direct path energy at the 489 m depth. The diffractions and direct path remain separated with little mixing over most of the water column.

Results

Figures 10, 12 and 14 compare model and data envelopes at the 489, 677, and 841 m depths, respectively. The model envelopes are time slices of Figure 9 at each depth and for each sound speed case. In each figure the leading edge of the envelopes are aligned to compare waveforms. Model envelopes are in red. The data (in black) is repeated for each sound speed case.

In order to assess the best sound speed case, an inverse filter is designed by an iterative deconvolution algorithm[6] for each model envelope. When this inverse is applied to the model envelope from which it was designed and the result filtered with the transmitted envelope, a perfect extraction of the source autocorrelation envelope is obtained. These inverse and filtering operations are applied to the data for each model sound speed case. The results are rectified to allow better comparison with the all-positive data envelopes. Figures 11, 13 and 15 are the deconvolution results associated with Figures 10, 12 and 14, respectively. The results are graded on the ability of each sound speed case to extract the envelope of the source autocorrelation shown in black.

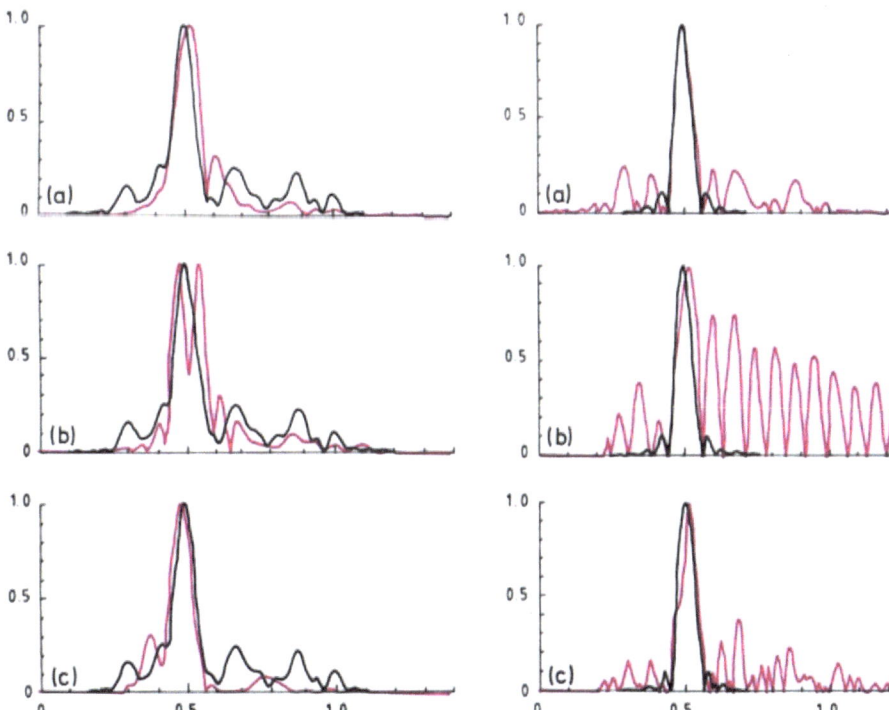

Figure 10. Direct path envelopes, receiver depth = 489 m, black = data, red = model, (a) gridded, (b) July 7, and (c) July 10.

Figure 11. Deconvolved envelopes, receiver depth = 489 m, black = source, red = deconvolved data, (a) gridded, (b) July 7, and (c) July 10.

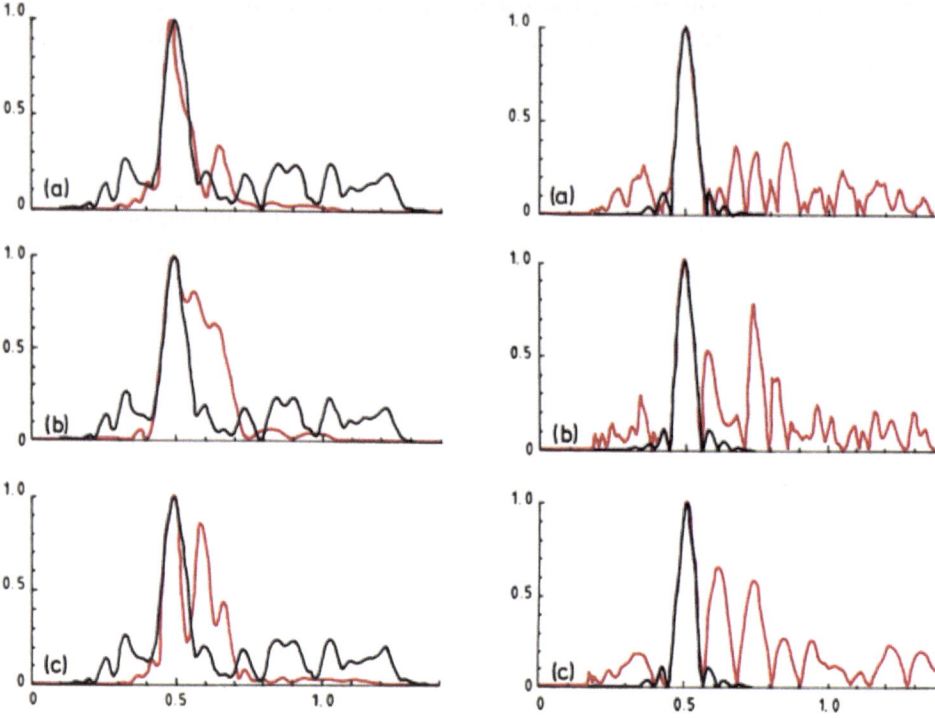

Figure 12. Direct path envelopes, receiver depth = 677 m, black = data, red = model, (a) gridded, (b) July 7, and (c) July 10.

Figure 13. Deconvolved envelopes, receiver depth = 677 m, black = source, red = deconvolved data, (a) gridded, (b) July 7, and (c) July 10.

Figure 10 compares the three sound speed cases for the 489 m depth. Figure 10a is the gridded sound speed case. It predicts the envelope of the main peak reasonably well except for a shift in the arrival time of the main peak relative to the leading edge. The arrival time and amplitude of the secondary peak at 0.6 s are off with respect to the data peak at 0.675 s. However, their shapes are similar. Note that both data and model exhibit the same knee. The minor peaks from 0.8 to 1.0 s are in good agreement in overall shape and arrival time. The model predicts a lower amplitude for these events than those observed in the measurements.

Figure 10b shows the twin peaks alluded to before in the July 7th case (Fig. 9b). The main peak is not well modeled by this sound speed case. However, the minor peaks around 0.8 to 1.0 s exhibit the same characteristic shape as the data.

July 10th is shown in Figure 10c. The main peak agrees well with the data, however, the minor peaks from 0.6 to 1.0 s are not very close. The main peak predicted here is the diffraction front emanating from the surface (Fig. 9c).

The gridded sound speed case is the best overall result at the 489 m depth as evidenced by the deconvolution results of Figure 11. The width of the main peak of the deconvolved data (red) is restored to its original width (black) along with the arrival time of the peak relative to the leading edge. The sidelobes are lower and the first zero-crossings of the main peak are almost restored. None of these attributes are recovered from the range independent cases.

Figure 12 compares the three sound speed cases at the 677 m depth. The gridded case appears to give a better fit to the data with respect to the main peak. One would expect the deconvolution

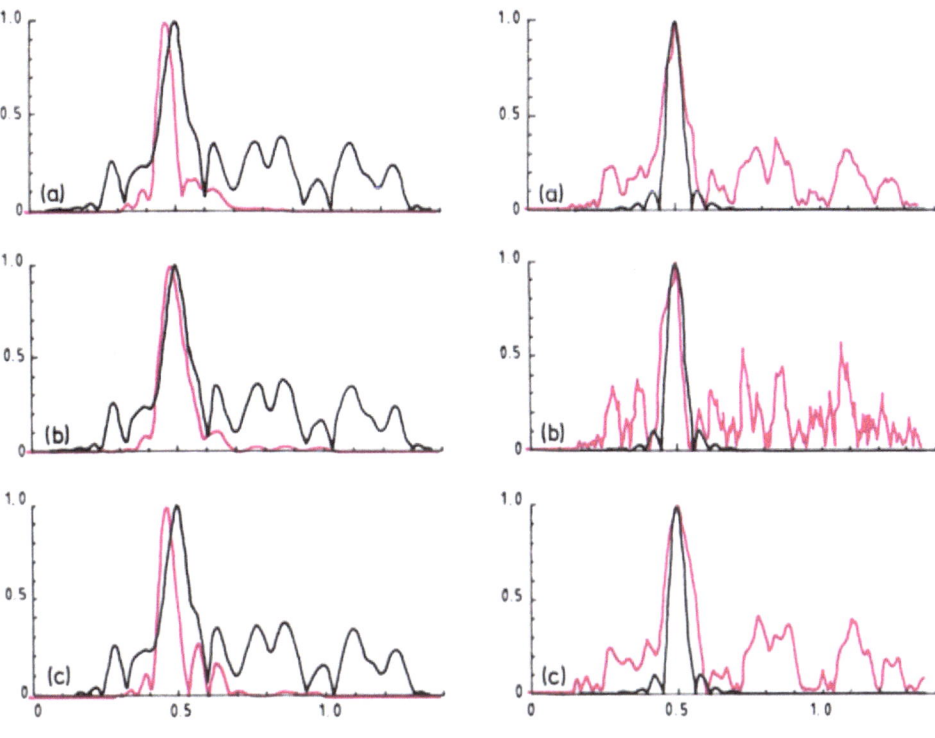

Figure 14. Direct path envelopes, receiver depth = 841 m, black = data, red = model, (a) gridded, (b) July 7, and (c) July 10.

Figure 15. Deconvolved envelopes, receiver depth = 841 m, black = source, red = deconvolved data, (a) gridded, (b) July 7, and (c) July 10.

results to be considerably better for the gridded case. However, Figure 13 shows that the gridded and July 10th cases restore the main peak attributes previously discussed about the same. The main difference between Figures 13a and 13c is in the sidelobe levels. Based on these sidelobe levels, the gridded case is superior.

The reason for the main peak restoration for the July 10th sound speed case can be seen by examining Figure 12c more closely. Here the arrival time of the main peak and all relative maxima following the main peak (up to 1.0 s) of both data and model are in close agreement. The main discrepancy is in the predicted amplitudes of the relative maxima. For example, the events predicted by the model (red) between 0.8 and 1.0 s are approximately an order of magnitude lower than the data.

Figures 14 and 15 show the results at the 841 m depth. Here the July 7th case is the better fit. Figure 14b shows that the main peak parameters of the data are well predicted and the arrival times and shapes of the model peaks from 0.5 to 1.0 s also agree with the data. The amplitudes of the events from 0.7 to 1.0 s are, again, an order of magnitude lower for the model results. The deconvolution result, Figure 15b, would be improved if these amplitude discrepancies could be resolved.

Conclusions

We have shown that transient signal distortion within the shadow zone is dependent on the ocean sound speed structure.The gridded sound speed case which smoothed out near surface sound speed anomalies gave better overall agreement at 489 m. The range independent sound speed profile taken on July 7th gave better agreement at 841 m. At 677 m, both the gridded and July 10th sound speed cases restored the main peak attributes equally. Additional model results and data must be analyzed to determine the best sound speed case for all depths.

We have shown that very subtle features of the ocean impulse response can be predicted and that these predictions can be used to restore certain attributes of the transmitted signal envelope. The main discrepancy between the data and model predictions is in the relative amplitude of the later arriving diffractions. Model amplitude predictions are about an order of magnitude lower than the measurements.

These results suggest the desirability of reconstructing the near surface sound speed structure by iterative modeling. In this approach the gridded data would be the starting sound speed profile. This profile would be perturbed until the least mean square difference between model and data is found.

Acknowledgments

This work has been approved for public release and was supported by the Office of Naval Research and the Naval Oceanographic and Atmospheric Research Laboratory (Contribution No. 90:051:244). The authors would like to thank Dr. Janice Boyd for the sound speed data without which this work would not have been possible.

References

1. Officer, C. B. (1958) Introduction to the Sound Transmission, McGraw-Hill Book Company, Inc., New York, pp. 172-178.
2. Morse, R. W. (1950) 'The dependence of shadow zone on the surface velocity gradient', Technical Report, Research Analysis Group, Department of Physics, Brown University, (Contract N7 onr-291, Task Order III, Subcontract I).
3. Robinson, E. A. and Treitel, S. (1980) Geophysical Signal Analysis, Prentice-Hall, Inc., Englewood Cliffs, New Jersey, pp. 333-338.
4. Taylor, J., Richards, P. and Halstead, R. (1971) 'Computer routines for surfaces generation and display', Manuscript Report Series 16, Department of Energy, Mines and Resources, Marine Science Branch, Ottawa, Canada.
5. Collins, M. D. (1988) 'The time-domain solution of the wide-angle parabolic equation including the effects of sediment dispersion', J. Acoust. Soc. Am. 84, 2114-2125.
6. Ioup, G. E. (1981) 'Always-convergent iterative noise removal and deconvolution', Bull. Am. Phys. Soc. 26, 1213.

THE EFFECT OF SEASONAL TEMPERATURE FLUCTUATIONS IN THE WATER COLUMN ON SEDIMENT COMPRESSIONAL WAVE SPEED PROFILES IN SHALLOW WATER

SUBRAMANIAM D. RAJAN and GEORGE V. FRISK
Woods Hole Oceanographic Institution
Woods Hole, MA 02543

ABSTRACT. In shallow water areas, the temperature of the water column undergoes large fluctuations during the course of a year. In the Gulf of Mexico, for example, temperature fluctuations of as much as twelve degrees Centigrade have been observed. These seasonal variations in water column temperature affect the pore water temperature of the bottom sediments which, in turn, affects the compressional wave speed profile. Using Biot theory, it is shown that the sediment compressional wave speed varies approximately linearly with pore water temperature and this effect is, to first order, independent of the porosity and sediment type. It is also shown that the velocity ratio (ratio of sound speeds in the water and the sediment at the sediment/water interface) is independent of the temperature but dependent on sediment type. These effects are demonstrated using two experimental measurements made at the same site in the Gulf of Mexico but during different seasons. In both cases, perturbative inversion techniques were used to infer the sound velocity profiles in the bottom. The differences between the two profiles fall within the error bounds predicted by linear inverse theory everywhere except the top 10 m of sediment, where the differences are attributed to the seasonal temperature fluctuation phenomenon. The experimental results suggest that the influence of the water column extends to greater depths than those predicted by theory.

1. Introduction

The propagation of acoustic energy in shallow water is strongly influenced by the geoacoustic properties of the sediment layers. These properties depend on the physical properties of the sediment material, the porosity of the sediments, the overburden pressure, and sediment temperature. In this paper, we study the flux of heat from the water column across the water/sediment interface, the changes in pore water temperature which result from this heat flux, and the consequent change in the compressional wave speed profile of the sediment.

It has been observed that in some shallow water regions the water column temperature experiences fluctuations as large as 12 degrees Centigrade over a one year period. This seasonal fluctuation causes a temperature disturbance to propagate into the sediment which is superimposed on the mean temperature structure of the sediment. The variation of water column temperature, therefore, results in seasonal fluctuation of the sediment temperature which, in turn, influences the acoustic properties of the sediments. The magnitude of the temperature fluctuations and the depth to which such changes are felt depend on the magnitude of the temperature change at the water/sediment interface and the sediment

69

J. Potter and A. Warn-Varnas (eds.), Ocean Variability & Acoustic Propagation, 69–80.

material. Though the changes in sediment properties are confined to a region close to the water/sediment interface, their impact on the propagation of sound in the water column is significant [1].

This paper is organized as follows. In Sec. 2, we present the theory of heat flux across the water/sediment interface by considering a one-dimensional heat equation. The effect of changes in the pore water temperature on the compressional wave speed of the sediment is then studied using Biot's theory. In Sec. 3, we briefly describe the details of experiments conducted by Rubano [2] and WHOI [3] at a site in the Gulf of Mexico and present the results of inversions performed on data collected in these experiments. In the next section, we compare the two inferred profiles and investigate the reasons for the observed differences in the top 10 m of sediment. We conclude that a primary cause for this difference is the difference in sediment temperature profiles during the two experiments and compare the experimentally observed values with those predicted by theory. In the concluding section, we summarise our results and propose directions for further research.

2. Theory

To study the flux of heat into the sediment layers, we model the heat flow as a one-dimensional problem. The heat flow then satisfies the differential equation [4]

$$\frac{d^2 T}{dz^2} = \frac{1}{\alpha^2} \frac{dT}{dt} \tag{1}$$

where z represents the depth dimension, T is the temperature, α^2 is the thermal diffusivity of the material and t is the time. The thermal diffusivity is related to the thermal conductivity of the material by the equation [4]

$$\alpha^2 = \frac{K}{S\rho} \tag{2}$$

where K, S and ρ are the thermal conductivity, specific heat and density, respectively. Consider a sinusoidal thermal fluctuation $T_0 \exp(i\omega t)$ applied at the interface $z = 0$. Let the solution $T(z,t)$ to Eq. (1) be of the form $U(z) \exp(i\omega t)$. Substituting this in Eq. (1), we obtain

$$\frac{U''}{U} = i\omega \frac{1}{\alpha^2} \tag{3}$$

where the primes indicate differentiation with respect to depth. The solution has to satisfy the initial condition $T(0,t) = T_0 \exp(i\omega t)$. The solution is therefore

$$T(z,t) = T_0 \exp(-kz) \exp\{i(\omega t - kz)\} \tag{4}$$

where $k = (\frac{\omega}{2\alpha^2})^{1/2}$. The temperature therefore decays with depth. Note that there exists a phase delay as well. From Eq. (4) we observe that the depth to which changes in water column temperature will affect pore water temperature is dependent on the diffusivity of the sediment material.

Measurements of thermal conductivity of marine sediments have been reported in the literature [5,6]. The thermal conductivity of the sediments is dependent on the porosity of the material; the higher the porosity (or the water content) the lower the thermal conductivity. The thermal diffusivity of the sediments is obtained from the thermal conductivity using the empirical relation [7]

$$\alpha^2 = 1.53K - 0.7 \times 10^{-3} \tag{5}$$

where K is expressed as calories/deg C/cm/sec and α^2 is expressed as cm^2/sec. For example, the thermal conductivity of silty sediments has an average value of 0.0019 calories/deg C /cm/sec [3], yielding a value of 0.0022 cm^2/sec for thermal diffusivity. Using this value of diffusivity in the expression for k in Eq. (4) we note that the temperature effects are observable only up to a depth of 4 m into the sediment.

We now study the variations of sediment compressional wave speed due to changes in the temperature of the pore water in the sediment. The sediment is modelled as a water-saturated porous medium and Biot's theory is applied to compute the compressional wave speed, which depends on the following parameters [8]:

1. Bulk modulus of the sediment grain material (K_r)

2. Bulk modulus of the fluid (K_w)

3. Bulk modulus of the frame (K_f)

4. Porosity of the sediment (n)

5. Shear modulus of the frame (μ)

6. Density of sediment material (ρ)

The compressional wave speed is given by $\sqrt{H/\rho}$, where H is [8]

$$H = \frac{(K_r - K_f)^2}{\{K_r[1 + n(K_r/K_w - 1)]\} - K_f} + K_f + \frac{4\mu}{3} \tag{6}$$

Temperature has no effect on the compressibility of the grain material [9]. The bulk modulus of the frame (K_f) and of the pore fluid (K_w) are both dependent on temperature. However the effect of temperature on K_f is much smaller than its effect on K_w and can be neglected. The bulk modulus K_w at different temperatures T is obtained from the relation $K_w(T) = C_p^2(T) \times$ density of the pore fluid. The compressional wave speed $C_p(T)$ of the fluid is computed using the empirical equation [10]

$$C_p(T) = 1449 + 4.6T - 0.055T^2 + 0.0003T^3 + (1.39 - 0.012T)(S - 35) + 0.017d \tag{7}$$

where T is the temperature in degrees Centigrade, S is the salinity in parts per thousand and d is the depth in meters. The density ρ of the sediment, which varies with the porosity n, is given by $\rho = n\rho_w + (1 - n)\rho_r$, where ρ_w is the density of the pore fluid and ρ_r is the density of the sediment material. The variation of compressional wave speed with temperature for silty sediments computed using Eq. (6) is presented in Fig. 1. The physical properties of the sediment material used in computing the compressional wave speeds are:

72

- $$K_r = 3.6{\times}10^{10} \text{ N/m}^2$$

- $$K_f = 5.2{\times}10^7 \text{ N/m}^2$$

- $$\mu = 5.7{\times}10^7 \text{ N/m}^2$$

For calculating C_p, a pore water salinity of 35 parts per thousand and a depth of 30 m were assumed. We observe that the sediment compressional wave speed varies approximately linearly with respect to temperature at a rate $(2.5 \text{ m/s/deg}\,C)$ which is nearly independent of the porosity. The velocity ratio between the sound speeds in the water and in the sediment at the water/sediment interface is a parameter that is commonly used to characterize the sediments. In Fig. 2, the velocity ratio for the sediment is plotted as a function of temperature. It is seen that for a given porosity, the velocity ratio is independent of temperature. In [1], we show that the rate of variation of the compressional wave speed with temperature is to first order independent of sediment type as well, whereas the velocity ratio, though independent of temperature, is dependent on sediment type.

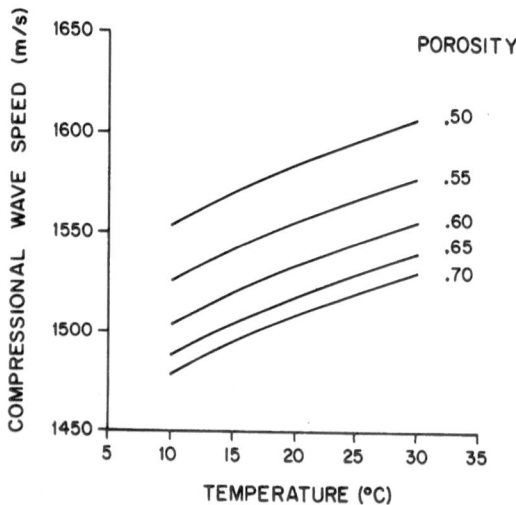

Figure 1: Variation of compressional wave speed with temperature for silt

3. Experiment

On April 6, 1975, L. Rubano [2] performed a broadband experiment in the Gulf of Mexico using explosive charges dropped along a track by ship and aircraft and received by a bottom moored vertical array spanning the water column. From the measured field, the group velocity dispersion curve of the first normal mode was obtained (Fig. 3). Rubano used this to obtain a geoacoustic model for the area by forward modelling.

On September 9, 1985, WHOI conducted a synthetic aperture array experiment at the site previously studied by Rubano. The experimental configuration and details are described in [3]. The complex pressure field was measured as function of range at two frequencies (140 Hz and 50 Hz), and the depth-dependent Green's function was obtained from the

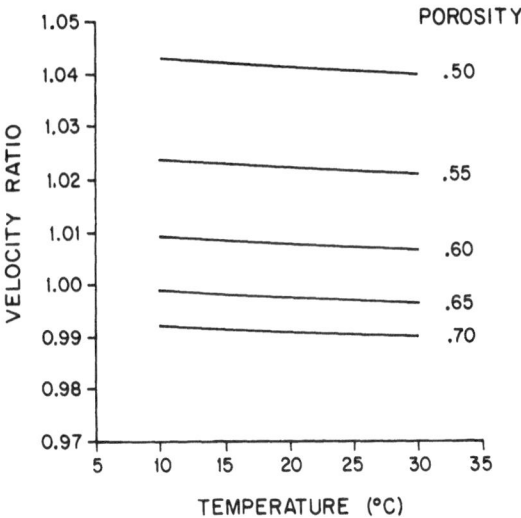

Figure 2: Variation of velocity ratio with temperature for silt

measured field via the Hankel transform [3]. The normal modes of the waveguide are the poles of the depth-dependent Green's function and appear as strong resonance peaks when one transforms the pressure field data. The modal eigenvalues are experimentally measured by picking the peak locations of the Green's function. The modal eigenvalues thus obtained are the input data for an inverse algorithm to estimate the compressional wave speed in the sediment layers.

3.1 INVERSION OF NARROWBAND DATA

In a 1987 paper, Rajan et al [11] described how one can invert for the compressional wave speed profile from modal eigenvalue data, giving examples from data taken in Nantucket Sound, Massachusetts. We will apply that same technique here. As the details of that inversion scheme are discussed extensively in that article, we will just briefly outline the method.

From first order perturbation theory, one can write an integral equation for the sound speed profile perturbation, $\Delta c(z)$, from some background profile, $c(z)$, which relates these quantities to the modal eigenvalues. It is:

$$\Delta k_m = \frac{1}{k_m} \int_0^\infty \rho^{-1}(z) \mid \varphi_m(z) \mid^2 k^2(z) \frac{\Delta c(z)}{c(z)} dz \qquad (8)$$

where

- k_m is the mth eigenvalue for some assumed background geoacoustic model,

- Δk_m is the difference between the measured eigenvalue and the background model eigenvalue,

- $\rho(z)$ is the density profile for the background model,

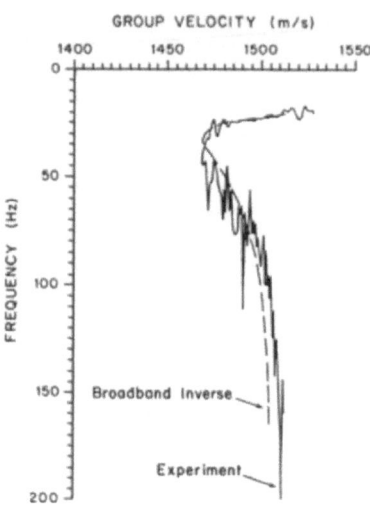

Figure 3: Group velocity dispersion curve (experiment and theory).

- $\varphi_m(z)$ is the normalized mode function for the mth mode of the background model,

- $k(z)$ is $\omega/c(z)$, where ω is the (angular) frequency and $c(z)$ is the background sound speed profile, and

- $\Delta c(z)$ is the perturbation in sound speed from the background profile, which is the answer we desire.

Equation (8) is a Fredholm integral equation of the first kind, and generally represents an underdetermined problem for the bottom sound speed profile. To make the problem tractable, it is standard practice to constrain the possible solutions. The method we will use is one which picks out from all the possible solutions the one that is the smoothest, the so called regularization method.

Using the modal eigenvalues obtained from field measurements, the perturbative inverse was performed, initially using a constant gradient background profile going from 1500 m/sec to 2000 m/sec in the bottom over 200 m, and with an assumed constant density of $\rho = 1.56$ g/cm^3. After iteration of the inverse, which combined the set of eigenvalues from the two frequencies used, 50 and 140 Hz, into an overall data vector, the technique converged to the results shown in Figure 4. The $c(z)$ model predicted is close to the water sound speed (1545 m/sec) near the water-sediment interface and approaches the bottom background model deeper in the sediments.

To evaluate the accuracy of the compressional wave speed profile obtained by inversion, we compare in Fig. 5 the experimentally determined depth-dependent Green's function and the Green's function computed using the bottom model in Fig. 4. A good agreement between the two is observed. Similar results are obtained if one compares the measured pressure field with the field predicted by the model (Fig. 6).

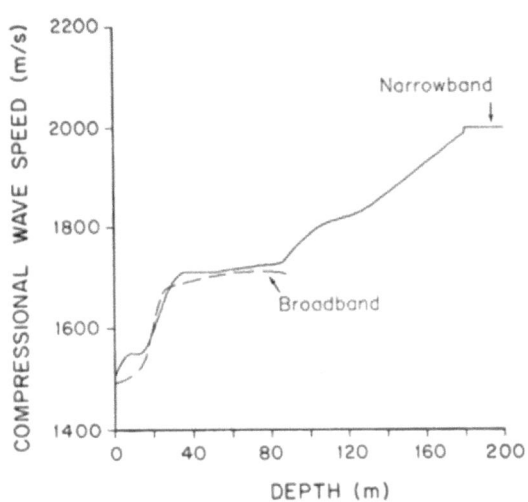

Figure 4: Profiles from narrowband and broadband inversions

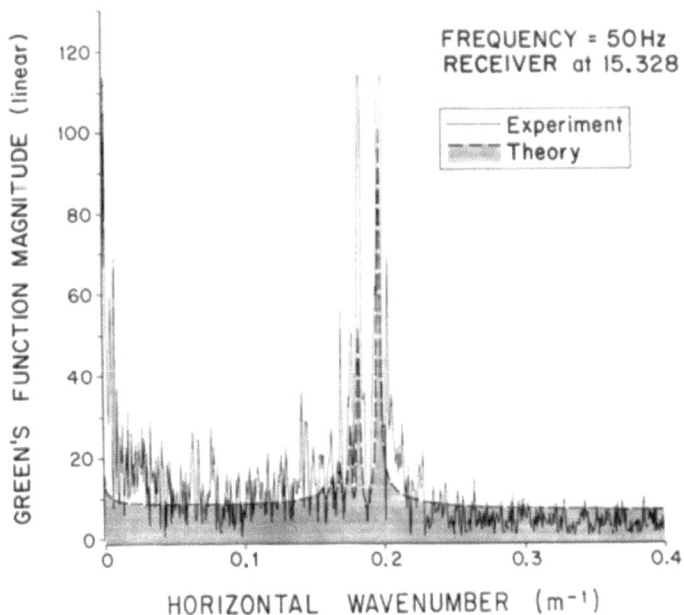

Figure 5: The depth-dependent Green's function (theory and experiment).

76

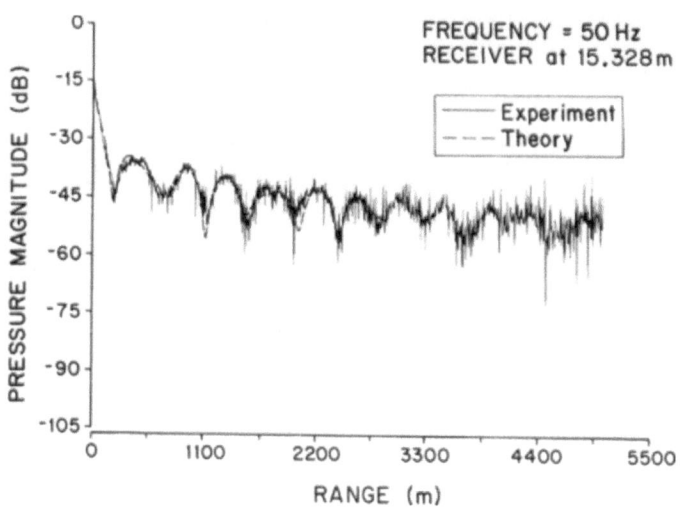

Figure 6: The pressure field (theory and experiment).

3.2 INVERSION OF BROADBAND DATA

In order to process broadband shallow water data using linear inverse theory in a manner analogous to the treatment of narrowband data, Rajan et al [11] derived an expression which relates the measured group velocity dispersion curve to the unknown sound speed profile through an integral equation. The expression which relates the group velocity dispersion curve to the bottom sound speed is:

$$
\left(\frac{1}{v_m^g} - \frac{1}{v_m^{g(0)}}\right)\Big|_{\omega_0} = \frac{1}{k_m}\int_0^\infty dz \left\{\left[2\omega_0^{-1}\rho^{-1}(z)\mid\varphi_m(\omega_0,z)\mid^2 k^2(z)\frac{\Delta c(z)}{c(z)}\right]\right.
$$

$$
+ \left[\frac{1}{k_m v_m^{g(0)}}\rho^{-1}(z)\mid\varphi_m(\omega_0,z)\mid^2 k^2(z)\frac{\Delta c(z)}{c(z)}\right]
$$

$$
+ \left.\left[\rho^{-1}(z)\left\{\frac{d}{d\omega}\mid\varphi_m^2(\omega_0,z)\mid\right\}k^2(z)\frac{\Delta c(z)}{c(z)}\right]\right\}, \tag{9}
$$

where all the quantities have the same meaning as in Eq. (8) and

- v_m^g is the experimentally measured group velocity for the mth mode at ω_0, and

- $v_m^{g(0)}$ is the group velocity of the mth mode at ω_0 for the background model.

In order to use Eq. (9), we took Rubano's group velocity data and fit a smooth curve to it in a best least squares sense. This gave us $v_1^g(\omega)$ over an acoustic bandwidth from about 20 to 120 Hz. The background model was taken to be a model close to the narrowband inverse result. As the group velocity dispersion curve is often rather smooth, only a few frequencies can suffice for sampling. We used 13 frequencies in our inversion to obtain the bottom model result shown in Fig. 4. We note that the profile obtained in this inversion is terminated by a half-space at 88 m whereas the profile obtained from narrowband inversion extends to a depth of 182 m. This is due to the fact that the mode function of the first normal mode does not have much energy at depths greater than 80 m for the frequencies considered, and therefore the inversion algorithm cannot reconstruct the compressional wave speed at these depths; the solution merges to the starting model, i.e., a terminating half-space at 88 m. In the discussion that follows we will therefore compare the two profiles only up to a depth of 88 m. To evaluate the accuracy of the inversion we compare in Fig. 3 the group velocity dispersion curve experimentally obtained with the curve predicted for the reconstructed model. The agreement once again is good.

4. Discussion

The method for obtaining the total variance of the estimates obtained by inversion is described in [3]. The upper and lower limits of the profiles obtained by putting error bars equal to one standard deviation are shown in Fig. 7. We note that in the top 10 m of sediment the two profiles are well outside the tolerance limits for one standard deviation. Another parameter that is used to quantify the reliability of the estimate is the resolution length. It has been shown in [3] that the resolution lengths near the sediment/water interface for both the profiles are small. We therefore conclude that the differences between the two profiles cannot be explained as due to the variance of the estimates or inadequate resolution. These differences can only be explained by invoking other causes.

We have seen that the temperature of the pore water affects the compressional wave speed profile in the sediment layers. The two experiments that we consider here were performed at two different seasons with quite different water column temperature structures. Therefore the temperature differences in the water column and the resulting differences in the pore water temperatures are likely candidates for explaining the observed differences between the two profiles.

We now examine the seasonal fluctuation in water column temperature at the experimental site and its effect on the compressional wave speed. The monthly variations of the surface and bottom temperatures for a station located at 28° 07′ N and 96° 13′ 30″ W was measured by Temple et al [12] during the years 1963 and 1964. This station is close to the site where the Rubano and WHOI experiments were conducted. Since the Rubano and WHOI experiments were conducted during 1975 and 1985, respectively, we compare the water column temperatures during the two experiments with the temperatures during corresponding periods in the data of [12] to establish that these data are representative of conditions during 1975 and 1985. Unfortunately, the water column temperature during Rubano's experiment is not available to us. An alternative approach is to obtain the water column temperature during the experiment from the water column sound speed using Eq. (7). For the bottom sound speed of 1518 m/s obtained in Rubano's experiment and an assumed salinity of 35 parts per thousand, the estimated temperature is 18.6 deg C. During the WHOI experiment, the water column temperature was measured using a CTD and had a value of 29.05 deg C. The water column temperatures in [12] for corresponding days in 1963 and 1964 (obtained by linear interpolation) are: April 6, 1963 - 19.8 deg C; April 6, 1964 - 17.7 deg C; September 9, 1963 - 23.8 deg C; September 9, 1964 - 28.1 deg C. We will

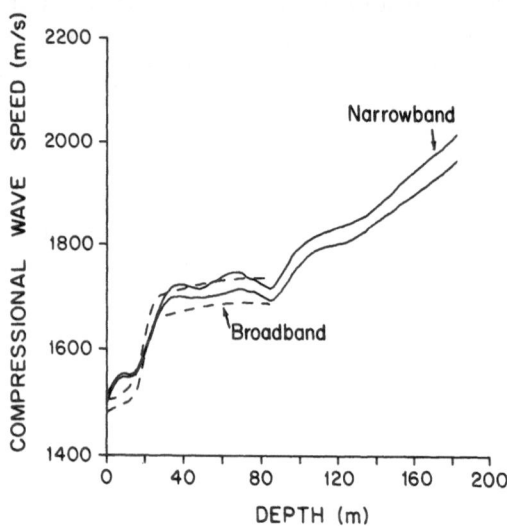

Figure 7: Error bars for reconstructed profiles

therefore assume that the data in [12] are representative of the temperature fluctuations during the years 1975 and 1985 and model the temperature fluctuations at the experimental site as approximately sinusoidal with a peak-to-trough value of about 12 deg C. The maximum value of these fluctuations occur in the months of Aug./Sep., and the minimum value occurs in Jan./Feb..

The temperature fluctuation in the sediment layers as a result of the temperature variation in the bottom can be obtained using Eq. (4). Coring data and other information available for this area indicate that the top 10 m of the sediment is composed of silty clay and sand-silt-clay [13]. These sediments have porosity values ranging from 66 to 75%. Assuming an average value of 70% for the porosity, the thermal diffusivity for the sediment [14] is 0.234×10^{-6} m^2/sec. The resulting sediment temperature profiles for the months April and September are shown in Fig. 8, from which we note that the difference in temperature near the top of the sediment is about 10 deg C. We now check whether this difference in sediment temperature results in the observed sound speed difference. From Fig. 1, we note that for silty sediments the compressional wave speed varies approximately linearly with temperature (2.5m/s/deg C). A ten degree difference in sediment temperature therefore corresponds to a sound speed difference of about 25 m/s. The sound speed differences near the water sediment interface for the two profiles in Fig. 4 lies between 20 to 30 m/s in the top 3 m of the sediments. Therefore the sound speed differences predicted by theory compare favourably with the experimentally observed differences in this region. However, the major anomaly between theory and experiment is the significant difference that is observed in the experimentally determined estimates of the compressional wave speed profiles at depths greater than 3 m. At the present time, we cannot give definite reasons for this difference. A probable cause is movement of pore water into and out of the sediment layers. Equation (4) for the temperature in the sediment layers assumes that the water/sediment layer is static, and there is no relative motion between the frame material and the fluid in

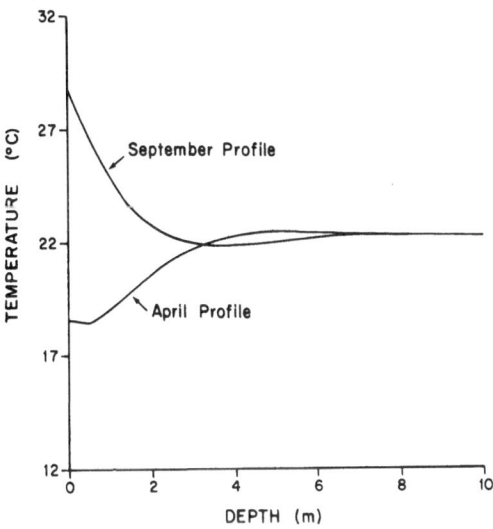

Figure 8: Temperature profiles in the sediment

the pores. This may not be strictly true, especially in shallow water areas where wave and tidal action can cause the pore fluid to be pumped in and out of the sediments. If there is considerable pore water movement, then the temperature effects may be felt in deeper layers. This is only speculation and more experiments are required to investigate pore water movement and its impact on the thermal structure in the sediments.

5. Conclusions

The variation of the sediment compressional wave speed profile with season due to temperature fluctuations in the water column was studied and it was shown that the effect can be substantial, especially in the top few meters of sediment in shallow water areas. The variability was computed using a one-dimensional theory for heat transfer across the water-sediment interface combined with a Biot model for the sediments. The effect was demonstrated experimentally by the inversion of two data sets acquired at the same site in the Gulf of Mexico but at different seasons. The experimental results indicate that the influence of the water column is felt to substantially greater depths in the sediment than those predicted by theory, an effect which requires further investigation.

6. Acknowledgements

The authors gratefully acknowledge the support of ONR contract N00014-89-k-0055. This paper is WHOI contribution No. 7405.

References

1. Rajan, S. D. and Frisk, G. V. ' Seasonal variations of the compressional wave speed profile in the Gulf of Mexico ', submitted to J. Acoust. Soc. Am.

2. Rubano, L. A. (1980) ' Acoustic propagation in shallow water over a low-velocity bottom ', J. Acoust. Soc. Am. 67, 1608-1613.

3. Lynch, J. F., Rajan, S. D., and Frisk, G. V. (1990) ' A comparison of broadband and narrowband modal inversions for bottom geoacoustic properties at a site near Corpus Christi, Texas ', to be published in J. Acoust. Soc. Am.

4. Carslaw, H. S. and Jaeger, J. C. (1959) Conduction of Heat in Solids, Oxford University Press, London.

5. Ratcliffe, E. H. (1960) ' The thermal conductivity of ocean sediments ', J. Geophys. Res. 65, 1535-1541.

6. Sclater, J. G., Corry, C. E., and Vacquer, V. (1969) ' In situ measurement of the thermal conductivity of ocean floor sediments ', J. Geophys. Res. 74, 1070-1081.

7. Von Herzen, R. and Maxwell, A. E. (1959) ' The measurement of thermal conductivity of deep sea sediments by a needle probe method ', J. Geophys. Res. 64, 1557-1563.

8. Stoll, R. D. (1974) ' Acoustic waves in saturated sediments ', in L. D. Hampton (ed.), Physics of Sound In Marine Sediments, Plenum, New York, pp. 19-38.

9. Shumway, G. (1958) ' Sound velocity vs temperature in water saturated sediments ', Geophysics XXIII, 494-505.

10. Kinsler, L. E. and Frey, A. R. (1962) Fundamentals of Acoustics, John Wiley, New York.

11. Rajan, S. D., Lynch, J. F., and Frisk, G. V. (1987) ' Perturbative inversion methods for obtaining bottom geoacoustic parameters in shallow water ', J. Acoust. Soc. Am. 82, 998-1017.

12. Temple, R. F., Harrington, D. L., and Martin, J. A. (1970) ' Monthly temperature and salinity measurements of continental shelf waters of the Northwestern Gulf of Mexico ', National Oceanic and Atmospheric Administration, NOAA Technical Report NMFS SSRF - 707, pp. 1 -20.

13. Matthews, J. E., Bucca, P. J., and Geddes, W. H. (1985) ' Preliminary environmental assessment of the project Gemini site - Corpus Christi, Texas ', Naval Ocean Research and Developement Activity, Report 120.

14. Hamilton, E. L. (1979) ' Sound velocity gradients in marine sediments ', J. Acoust. Soc. Am. 65, 909-922.

SLICE89: A SINGLE SLICE TOMOGRAPHY EXPERIMENT

BRUCE M. HOWE, JAMES A. MERCER, ROBERT C. SPINDEL
Applied Physics Laboratory
University of Washington
Seattle, Washington 98105

PETER F. WORCESTER, JOHN A. HILDEBRAND, WILLIAM S. HODGKISS, Jr.
Scripps Institution of Oceanography
University of California, San Diego
La Jolla, California 92093

TIMOTHY F. DUDA and STANLEY M. FLATTÉ
Institute of Marine Sciences
University of California, Santa Cruz
Santa Cruz, California 95064

ABSTRACT. Acoustic travel times in the ocean are sensitive both to the mean sound speed field and to high-wavenumber ocean features with wavelengths corresponding to ray loop "wavelengths" and their harmonics (70 km and smaller). The quasi-sinusoidal rays can be thought of as taking a partial Fourier transform of the sound speed field (Cornuelle and Howe, 1987). The SLICE89 experiment was conducted in the northeast Pacific during summer 1989 to determine how well high-wavenumber information can be resolved in practice. A broadband acoustic source near the sound channel axis transmitted to hydrophones 1000 km away. Preliminary travel time data from multiple receivers distributed through the water column are combined to construct time fronts. The measured time fronts are compared with predictions based on the sound speed field as calculated from CTD and XBT data taken concurrently. Although inversions of the measured travel time data are not yet complete, simulations are presented to show the horizontal resolution that can be attained.

Introduction

Early in the history of ocean acoustic tomography it was recognized that more receivers widely spaced in the vertical could provide additional information. Initially it was thought that the gain would be primarily due to having additional measurements with independent internal wave noise. Cornuelle and Howe (1987) showed that the travel times gave information about horizontal ocean scales comparable to and smaller than a convergence zone. They estimated how more hydrophones in the vertical would improve the resolution of these ocean scales.

SLICE89 was designed to test these ideas. We describe the experiment, time front predictions and recordings, and simulated inversions.

J. Potter and A. Warn-Varnas (eds.), Ocean Variability & Acoustic Propagation, 81–86.
© 1991 *Kluwer Academic Publishers.*

82

1. The Experiment

The location was approximately 600 n.mi. north of Hawaii. The center frequency and bandwidth of the Hydroacoustics HLF-5 source were 250 Hz and 83 Hz, respectively. Phase-coded linear maximal sequence transmissions lasting 135 seconds were made on a schedule comprising both 10 minute and 1 hour intervals. The frequent sampling was intended to eliminate aliasing of the fluctuating pulses. There were 400 transmissions over 14 days.

The receiving array, 1000 km distant, is described in detail by Sotirin and Hildebrand (1988). The array used here had 50 hydrophones spaced at 60 meter intervals starting near the surface. The array was suspended from the three-point-moored research platform FLIP. Both the source assembly and the receiving hydrophones were acoustically navigated relative to a GPS-located network of bottom-mounted transponders.

During the experiment 10 CTD casts were made along the slice. To obtain high horizontal resolution, three AXBT flights were also made, each dropping approximately 60 probes.

2. Time Fronts

Predicted ray arrival times for each hydrophone are shown in Fig. 1, together with a measured pulse. The similarity is striking. We believe this is the first measurement to show graphically

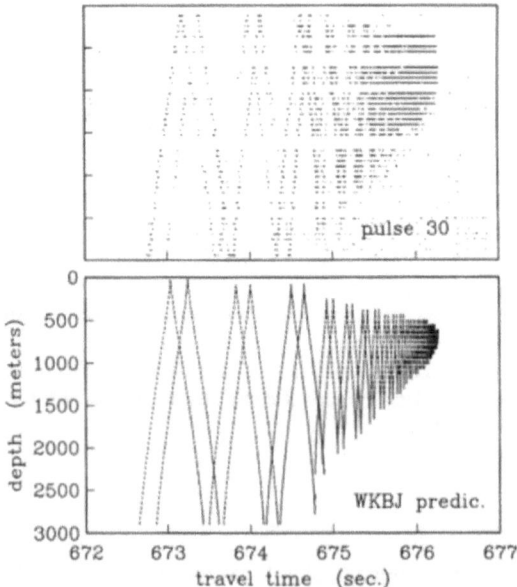

Figure 1. One of the processed pulses is shown in comparison with a range-independent ray arrival pattern. The prediction was made using range-averaged CTD data. The processed pulse, sampled at 1 kHz, is displayed using dots at intensity levels exceeding a threshold. The recorded pulse has been offset in time to reflect range mismatch with the arrival predictions, with the final arrivals appearing coincident. Differences between measured and predicted arrivals for identifiable pulse segments are input data for environmental mapping.

the multiple time fronts of a pulse sweeping by a vertical array. Although quantitative comparison is not possible until measured travel times are corrected for mooring motion and clock drift, we can address the implications of the recordings to beamforming gain with a fluctuating field.

Simple plane-wave beamforming is not possible when nonplanar pulse distortions exceed one-quarter wavelength. One can predict acoustic distortions induced by synoptic ocean scales (Fig. 1), but internal waves additionally perturb the pulse stochastically. The perturbations reduce the vertical extent over which beamforming provides gain, determining how much coherent versus incoherent gain can be attained from a long vertical array. The uncorrected pulses appear stable over the 4 days of processed data, and after correction will allow quantification of both internal wave and subinertial frequency wavefront distortions.

A not insignificant data processing problem has been to "track" ray arrivals over time; a specific peak in the arrival pattern must be associated with a specific ray, and it must be tracked in succeeding arrival patterns. This has been done using a pattern recognition algorithm based on the travel time (and vertical arrival angle if available) pattern. The tracking should be significantly easier because of the strong pattern shown in Fig. 1.

While the earlier, off-axial arrivals show a distinctive pattern, the more axial, late arriving energy does not. This phenomenon has been observed in previous measurements of this kind, and no convincing explanation has been offered. We expect the detailed analysis of these data will shed some light on this mystery.

3. Inverse Simulations

A sound speed field $C(x,z)$ was derived from the AXBT temperature field using the T-S relationship obtained from CTD measurements. The corresponding perturbation sound speed field $\delta C(x,z) = C(x,z) - C_0(z)$ is shown in Fig. 2. There are distinct steplike features apparent at 600 km and 900 km. For the purpose of simulation, an idealized step change in $\delta C(x)$ was synthesized to represent one of these features.

3.1. OCEAN MODEL

The perturbation sound speed field is represented by $\delta C(x,z) = \sum a_{ij} X_i(x) Z_j(z)$. The set of horizontal basis functions $X_i(x)$ consists of a truncated Fourier series, and the set of vertical basis functions $Z_j(z)$ consists of three triangle functions (linear splines) over the depth ranges 0–100 m, 0–100–400 m, and 100–400–1000 m.

The damped, weighted least-squares procedure used for fitting the acoustic travel time data to the model requires the specification of the data and model error covariances. Travel time errors were taken to be 1 ms; this assumes that internal wave variance and mooring motion travel time perturbations can be successfully averaged over time. The a priori model variances used for the three vertical modes were 10, 3, and 1 m/s rms, respectively. The horizontal structure (sines and cosines) was specified using a wavenumber spectrum that is flat for wavelengths greater than 200 km, and follows a k^{-2} shape for smaller wavenumbers, k being wavenumber. One-hundred harmonics in a 2000 km domain were modeled, so the shortest wavelength was 20 km.

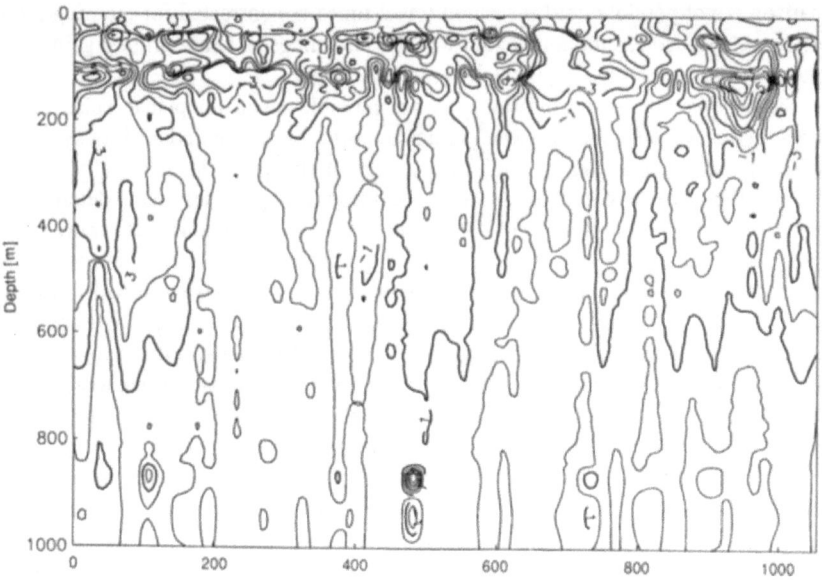

Figure 2. The perturbation sound speed field (meters per second) from 60 AXBT profiles.

3.2. INVERSE PROBLEM

The perturbation travel time for a particular ray Γ_i is given by

$$\delta T_i = T_i - T_{0_i} = -\int_{\Gamma_i} \frac{\delta C(x,z)}{C_0^2} \, dx$$

or

$$d = Gm,$$

where d contains the data δT_i, $G = \partial d/\partial m$, and m contains the a_{ij} parameters which model the $\delta C(x,z)$ field. The travel time in the unperturbed ocean is given by T_0. Only 11 widely spaced receivers out of the possible 50 were used; 200 rays with the corresponding travels times were used. The m corresponding to the step function in δC was used with $d = Gm$ to calculate simulated travel times.

A useful byproduct of an inversion is the resolution matrix. In this case, the diagonal elements of the matrix show how well each wavenumber is resolved by the measurements; the diagonal term for the 100 m mode is plotted in Fig. 3. The resolution of the mean is high, as is the resolution for wavenumbers corresponding to ray wavenumbers and their harmonics. There is a low-wavenumber gap in the resolution; eddies in this region affect all rays roughly the same.

Figure 4 shows the perturbation sound speed at 100 m for both the true field and the tomographic inverse result. The acoustic tomographic reconstruction of the total field is lacking the low-wavenumber information because of the above mentioned gap in the resolution spectrum. The tomographic reconstruction of the high-pass field (wavenumbers 35–49 and 70–98) captures the small-scale features. This is quantified in the plot of the total and high-pass errors.

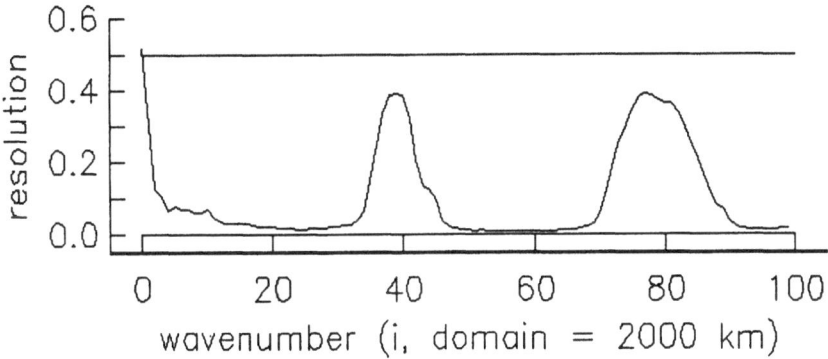

Figure 3. The resolution spectrum for the 100 m mode. The maximum resolution is approximately 0.5, because the expansion domain for the Fourier components was 2000 km, twice the domain in which measurements were available.

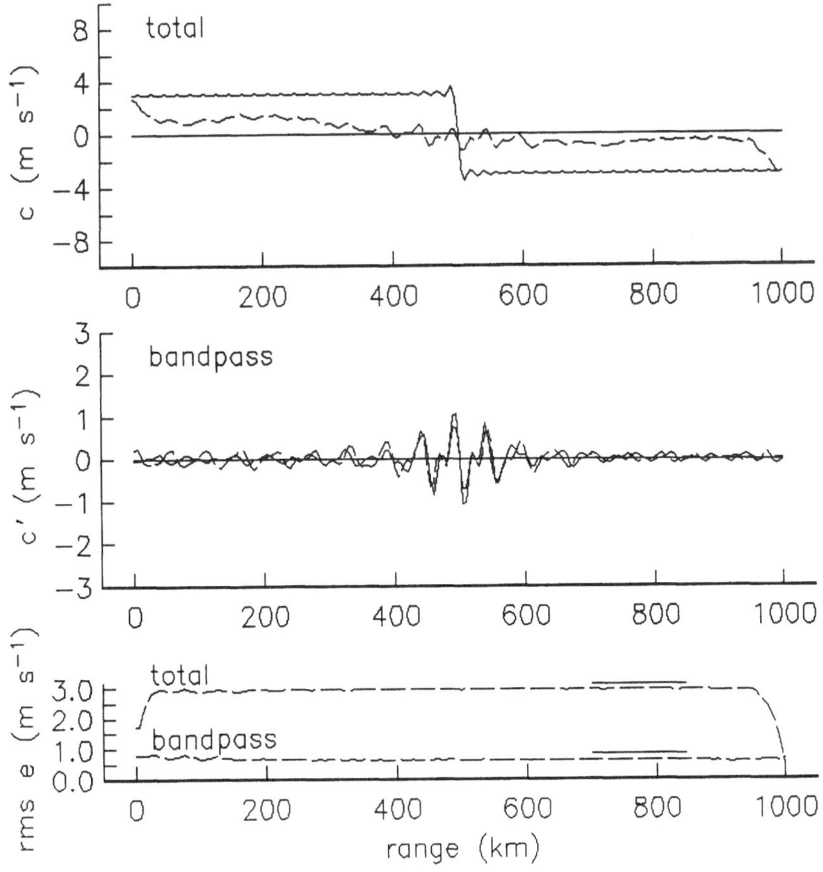

Figure 4. The sound speed at 100 m, from AXBTs (solid) and from the tomographic reconstruction (dashed). a) Total sound speed, b) the high-pass filtered sound speed, and c) rms errors. The horizontal lines at 800 km show the a priori errors.

4. Conclusions

The measurements discussed here show graphically the multiple time fronts of a pulse sweeping by a vertical array. The agreement between the measurements and ray theory is not unexpected; it is clear, though, that the later, more axial arrival pattern (or lack thereof), needs to be better understood.

The inversion simulations are not unrealistic and should be a reasonable indicator of what will be possible with the actual data.

Acknowledgements. Our thanks to the people and organizations who assisted in this experiment. K. Metzger of the University of Michigan kindly provided software. J. Boyd of NOARL was responsible for the environmental data. This work was supported by the Office of Naval Technology, contract N00014-87-K-0760.

References

Cornuelle, B. and Howe, B. (1987) 'High spatial resolution in vertical slice ocean acoustic tomography', J. Geophys. Res., 92, 11,680–11,692.

Sotirin, B. and Hildebrand, J. (1988) 'Large aperture digital acoustic array', IEEE J. Oceanic Eng., 13, 271–281.

MARGINAL ICE ZONE OCEANOGRAPHIC VARIABILITY AND ITS EFFECTS ON ACOUSTIC PROPAGATION

R. W. MEREDITH and P. M. JACKSON
Naval Oceanographic and Atmospheric Research Laboratory
Arctic Acoustics Branch
Stennis Space Center, MS 39529
USA

ABSTRACT. The Marginal Ice Zone (MIZ) is a complex region both acoustically and oceanographically. The presence of the Polar Front associated with the East Greenland Current causes substantial temporal and geographic variability. In addition, changing combinations of ice cover and open water produces complex range-dependent environments for the propagation of sound at all but the very lowest frequencies. In April and May of 1988, personnel from the Naval Oceanographic and Atmospheric Research Laboratory (NOARL) used the USCGC NORTHWIND to establish an ice camp in the MIZ between Greenland and Svalbard in the Fram Strait, and conduct an environmental acoustic experiment. A comprehensive set of environmental measurements were made that included expendable bathythermographs, conductivity-temperature depth profiles, both vertical and time series, current meter casts, meteorological and navigational measurements, and satellite imagery. Presented here are statistical analyses of oceanographic temporal and spatial variations associated with the Marginal Ice Zone. Additionally, acoustic modeling was used with these data inputs to make propagation predictions at 24, 115, 273, and 2000 Hz. The effects of these oceanographic variations on predicted transmission loss are discussed.

1. Introduction

The propagation of sound in the Marginal Ice Zone (MIZ) off the eastern coast of Greenland is among the least understood and most complex in the entire Arctic[1,2]. The presence of the Polar Front associated with the East Greenland Current causes substantial variability in sound speed profiles and therefore, in acoustic propagation. In addition, changing combinations of ice cover and open water presents complex environments for acoustic propagation.

The Office of Naval Technology (ONT) sponsored and NOARL conducted environmental acoustics measurements in the MIZ off the northeast coast of Greenland, in the Fram Strait, during the Spring, 1988 from an ice camp supported by the USCGC NORTHWIND. Other participants included the Naval Ocean Systems Center, the Applied Physics Laboratory, University of Washington, Ocean Sensors, Inc., the Navy Polar Oceanographic Center (NPOC), and Patrol Wing Five Detachment. This study focuses on one effect of the physical oceanographic environment in acoustic propagation; namely, changing sound speed profiles. This paper examines

J. Potter and A. Warn-Varnas (eds.), Ocean Variability & Acoustic Propagation, 87–101.
© 1991 *Kluwer Academic Publishers.*

acoustic propagation along a 70 nmi path (from open water, into the icepack), transverse to the MIZ. Along this path, the depth of the Polar Front increases; nominally at approximately 1 m of depth per kilometer of range. At the Polar Front, the intermingling of the colder, less saline Polar Water and warmer, more saline Atlantic Intermediate Water causes mixing and layering resulting in temporal and spatial fluctuations in the sound speed gradient. These two components create an interesting and unique oceanographic medium in which the acoustic sound speed gradient can change rapidly over short distances and in short time periods. This is the environment we wish to study acoustic transmission loss.

2. Experimental Overview

In April and May of 1988 personnel from the NOARL established an ice camp in the MIZ between Greenland and Svalbard in the Fram Strait, to conduct environmental acoustic propagation experiment. Comprehensive and intense environmental measurements were gathered. These data include expendable bathythermographs, conductivity-temperature-depth (CTD) profiles, both vertical and time series current meter casts, meteorological and navigational measurements, and satellite imagery. Oceanographic data was collected aboard the USCGC NORTHWIND, from NORTHWIND based HH-52 helicopters, and from the ice camp. Navigational fixes and meteorological data collected at the ice camp include wind speed and direction, peak gusts, air temperature, barometric pressure, and solar radiation. In addition, Advanced Very High Resolution Radiometer (AVHRR) satellite imagery was collected to find the extent of regional ice coverage during the exercise. Additional details of the environmental measurements, the sensor calibrations, and preliminary analysis are available in Reference 3.

3. Oceanography

Figure 1 shows the oceanographic environment in the marginal ice zone and is taken from Reference 2. Dominating the circulation in the upper 500 m of the exercise area is the southward flowing East Greenland Current that circulates both Polar Water and below that, water of Atlantic origin termed Atlantic Intermediate Water (AIW). AIW composes most the East Greenland Current transport, and is characterized by temperature and salinity ranges of 0° to 3°C and 34 to 35 ppt and acoustic sound velocities >1445 m/s. It has its origin in the Fram Strait where North flowing Atlantic Water entrained in the West Spitzbergen Current branches to the west, mixes with Polar Water, and returns southward as a subsurface water mass in the East Greenland Current. This water mass has a width of approximately 100 km or less and occurs at depths between 50 and 300 meters.

A surface layer of Polar Water overrides the AIW and originates in the Arctic and flows through Fram Strait along the eastern coast of Greenland. Polar water is characterized by temperatures less than 0°C, salinities between 30 and 34 ppt and acoustic sound speed <1445 m/s. A major oceanographic feature, the Polar Front, lies between the cold, low saline Polar Water and the warm, high salinity Atlantic Water. This front is not vertical, but slopes toward the west as a function of depth. The position of front snakes and curves both horizontally and with depth and eddies are easily formed. This area is very dynamic. Historically, the mean position of this front lies near the 1000 m depth contour and like most fronts, is characterized by the appearance of fine structure—here caused by double diffusion processes and inter-fingering of the two water masses. This frontal variability and the dissimilar water masses on either side of the Polar Front caused large temporal and spatial variability in the acoustic sound

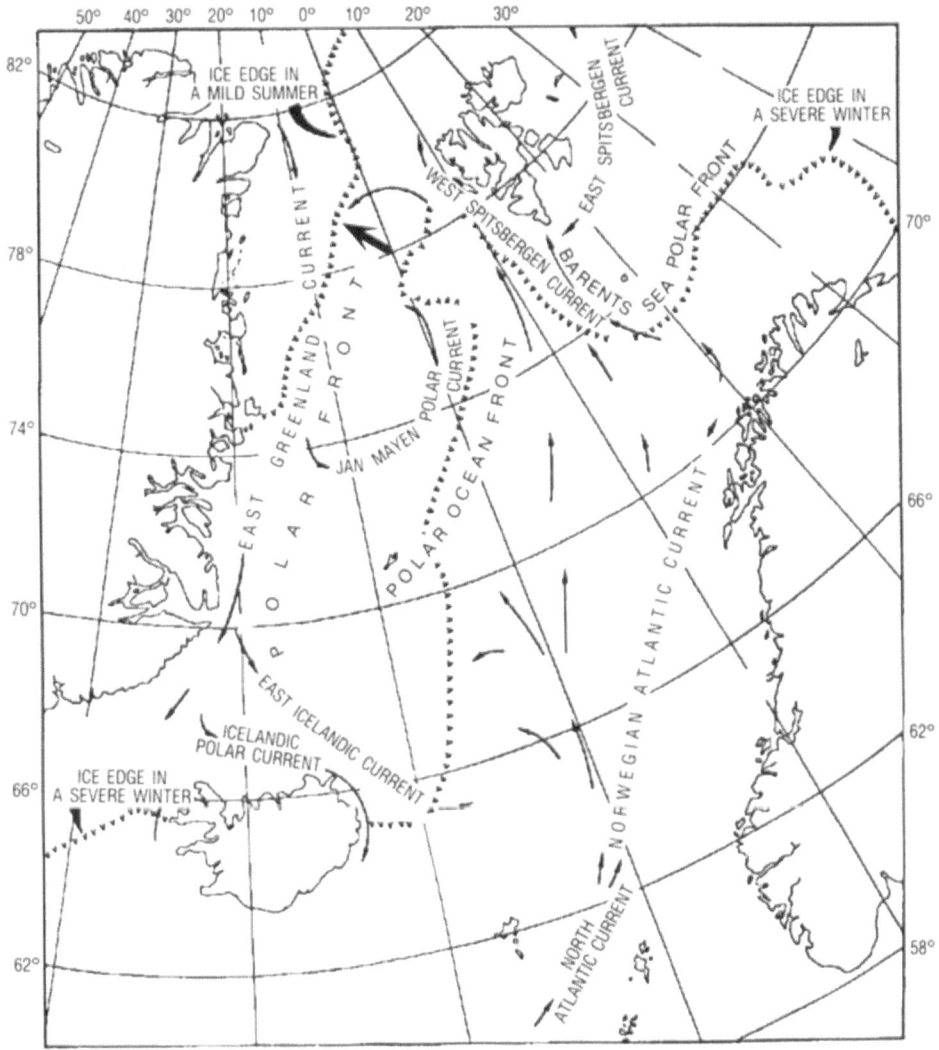

Figure 1. Major currents and oceanographic feature of the East Greenland marginal ice zone, taken from Reference 2.

speed profile during the exercise. The slope of the polar front is not as great as observed elsewhere in the literature since the Polar Front was traversed at an oblique angle.

Figure 2 shows a composite of the six sound speed profiles used for acoustic modeling and the approximate depth of the Polar Front. The range span for each profile is given in Table I, measured from the eastern most profile.

These sound speed profiles are "extended" linearly from 500 m to the bottom, based on water depth. Temperature, and hence sound speed, inversion layers tens of meters thick, are common just below the Polar Front, in the depth range 100-500 m. The nominal sound speed excursion

90

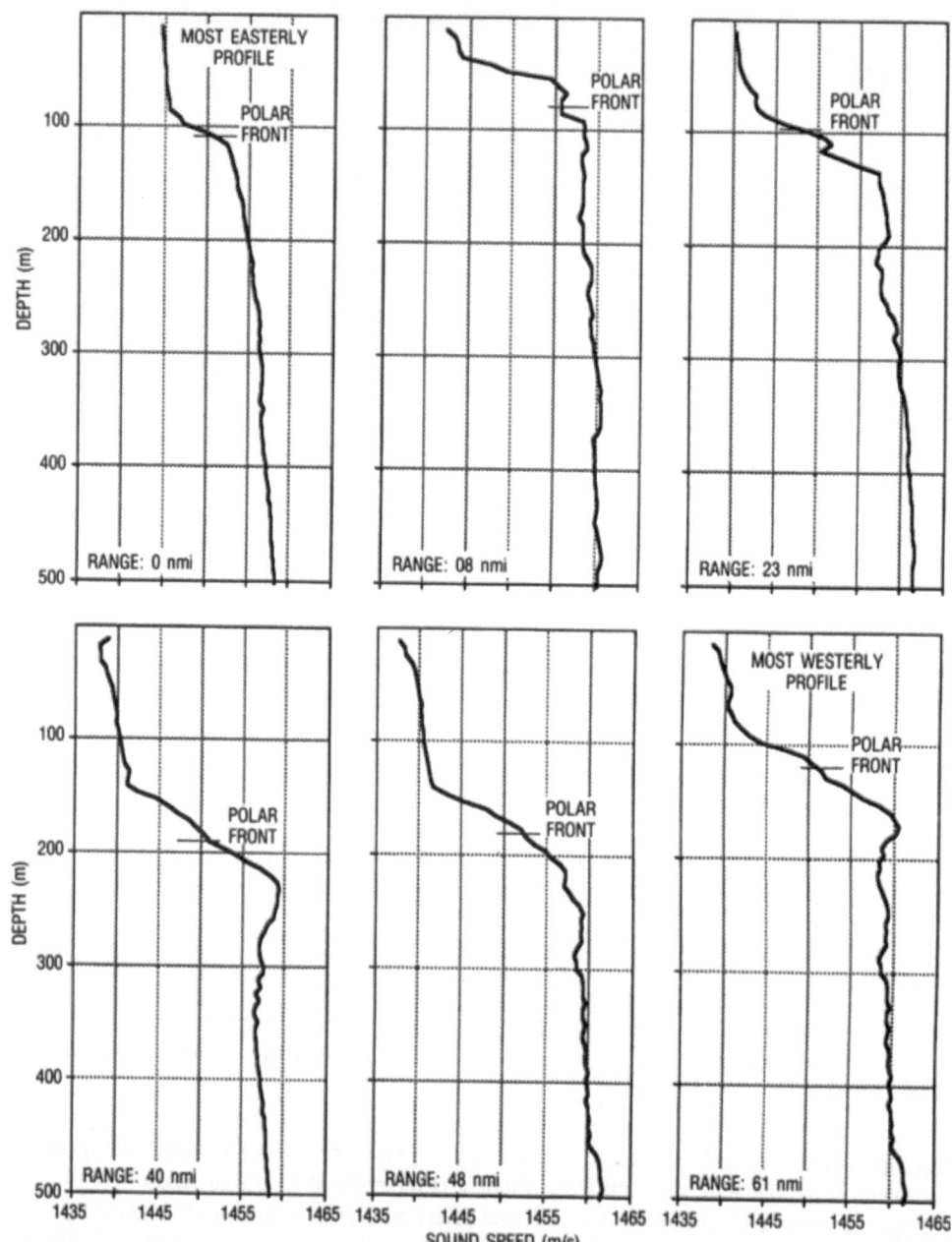

Figure 2. Six sound speed profiles measured transverse to the Polar Front over a range span of 70 nmi. All six are used for the range-dependent acoustic models and only the most easterly is used for the range-independent modeling.

Table I. Range span for each profile used in range-dependent acoustic modeling and the approximate depth of the center of the Polar Front.

OBSERVATION	RANGE FROM EASTERN END OF TRACK (nmi)	POLAR FRONT DEPTH (m)
# 82	0	104
# 83	8	80
# 89	23	104
#109	40	189
#105	48	176
#101	61	115

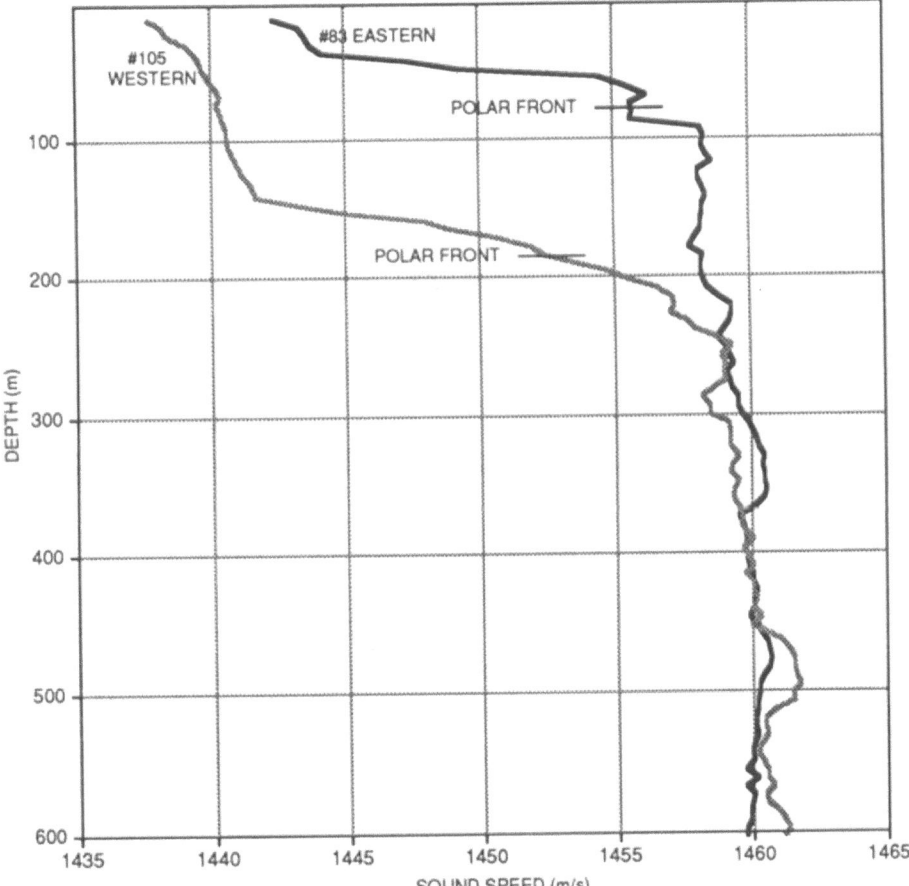

Figure 3. Comparison of the two sound speed profiles taken 40 nmi and 2 days apart to demonstrate sound speed variability.

within any layer is 1.5 m/s. Profile #83 and #105 from Table I are plotted together in Figure 3 to demonstrate the variability in sound speed profiles. These two profiles are separated by 40 nmi and were measured 2 days apart. Depending on depth and range and the position of the Polar

92

Front, sound speed varies by as much as 15 m/s. Above the front differences of 5 m/s are common and below the front differences average 2.5 m/s. The sound speed gradient, due primarily to temperature, at the interface of the two water masses shows significant spatial variability. Fluctuation of the gradient within each profile increases as the Polar Front approaches the surface and decreases as the front deepens. Significant temporal variability was observed in the sound speed structure and although upcasts are not presented here, marked differences exist between the down and upcasts of the CTDs (a time span of about 30 minutes). Figure 4 shows two pairs of CTDs, all downcasts, and all taken within 24 hours. The first pair, taken 30 minutes apart, and the second pair, taken 60 minutes apart show markedly different sound speeds. Above the Polar Front the sound speed difference averages 0.5 m/s for the 30 minute time span and

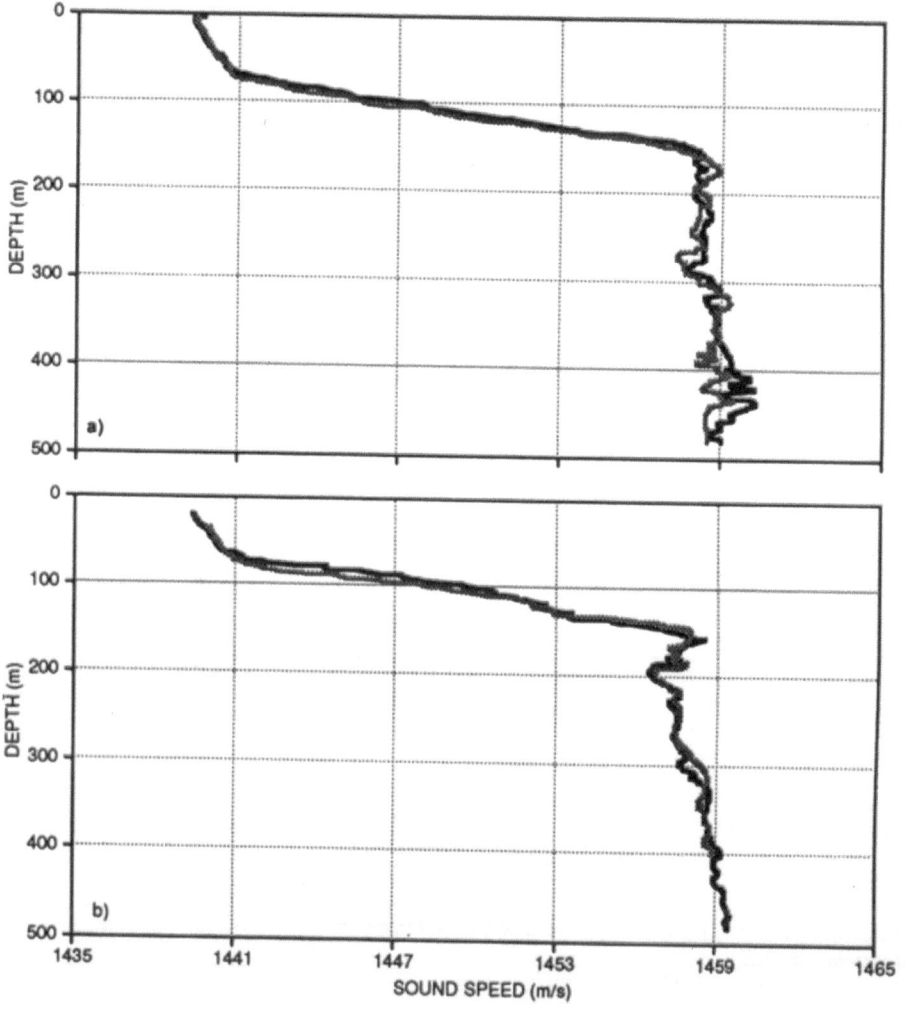

Figure 4. Comparison of sound speed profiles taken a) 30 minutes apart, and b) 60 minutes apart.

5.0 m/s for the 60 minute span. Below the front the differences are 1.5 m/s for the 30 minute span and 2.0 m/s for the 60 minute span. The 30 minute pair seem to zig-zag across each other while the 60 minute pair seem to be offset one from another by about 2 m/s. This tendency causes average differences to be misleading. The 30 minute pair has more zero crossing and thus the average contains more low values near the crossings. The 60 minute pair has less zero crossings and the average represents more of an offset between the two profiles. The 30 minute pair shows more variance below the front than the 60 minute pair, and the 60 minute pair shows a larger difference near the front.

4. Acoustic Modeling

Presented here are comparisons and analysis of modeled acoustic transmission loss for two environmental scenarios; one range dependent; the other range independent. The range-dependent scenario employs six sound speed gradients computed from sound speed profiles in Figure 2. The range-independent environmental scenario uses a single sound speed gradient, the most easterly one used in the range-dependent scenario. For each scenario the acoustic transmission loss is modeled for two source depths, one near and one below the Polar Front. This modeling is needed to support and aid in future processing and interpretation of the experimental acoustic data and is used here to evaluate the effects of changing sound speed on acoustic propagation.

4.1 ACOUSTIC MODELING OVERVIEW

The same bathymetry is used for both scenarios and was obtained from a standard US Navy data base. Acoustic effects due to the spatial distribution of ice coverage were also kept the same for both environmental scenarios. Ice loss was determined from guidelines established in Reference 4 and is the minimum of either a free surface perturbation model or the Gordon-Bucker empirical surface loss model. The ice was modeled as a Gaussian surface with a 4 m RMS roughness and a constant keel spacing of 100 m. Geoacoustic bottom loss parameters were calculated based on the geographic region using a thin sandy-silt layer 50 m thick, a thick layer sediment from 50 – 500 m, and a fully absorbing basement below 500 m.

The two acoustic models used here are from Reference 5. The Split-Step Parabolic Equation (PE) acoustic model is used to make acoustic transmission loss predictions at 24, 115, and 273 Hz. The SSPE uses the Tappert-Hardin split step algorithm and solves the integral solution to the wave equation via a FFT. Time and transform size restrict the use of the SSPE model at frequencies above 300 Hz. ASTRAL, a hybrid acoustic model, incorporates elements of both normal modes and geometric ray acoustics. ASTRAL models propagation at 2000 Hz and assumes adiabatic invariance to propagate mode-like envelopes and obtain acoustic pressure. Comparisons of predicted transmission loss computed from these acoustic propagation models for the two environmental scenarios follows.

4.2 ACOUSTIC MODELING RESULTS

Figure 5 shows acoustic transmission loss predictions vs. horizontal range, at 24 Hz using both a range-dependent and a range-independent environment. Range is measured from the eastern most sound speed profile, westward through the Polar Front. At 24 Hz, there can be significant differences in transmission loss depending on the choice of environmental scenarios. Because both environmental scenarios use the same beginning profile, there are no differences out to a

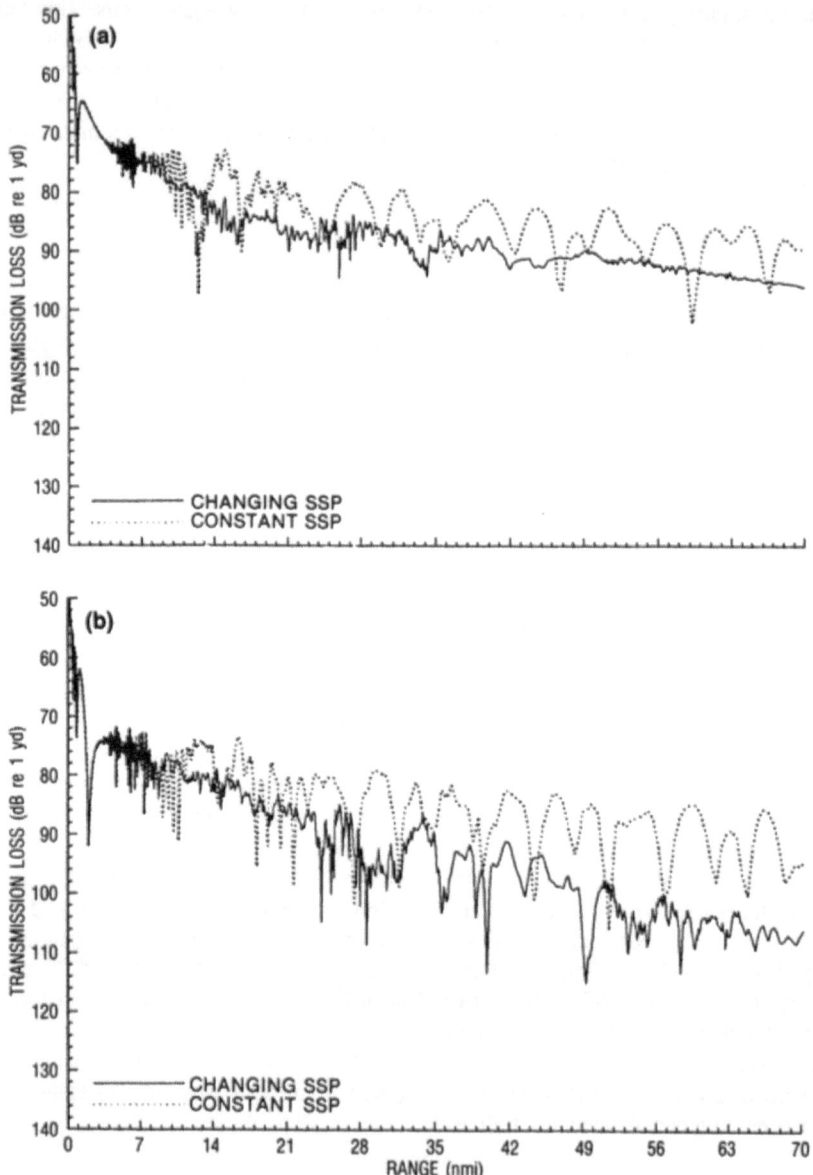

Figure 5. Acoustic transmission loss vs horizontal range, at 24 Hz using both a range-dependent and range-independent environment. Range is measured from the eastern most sound speed profile through the Polar Front. a) Source depth: 130 m, b) source depth: 300 m.

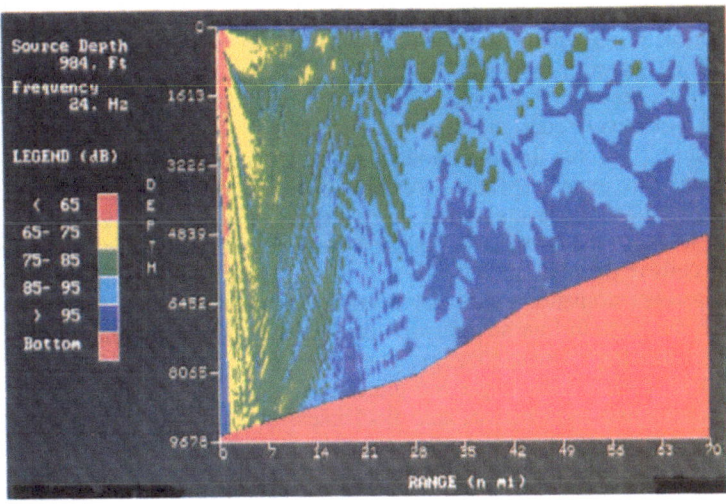

Figure 6. 24 Hz acoustic field intensity vs range for the range-independent scenario with a source depth of 300 m.

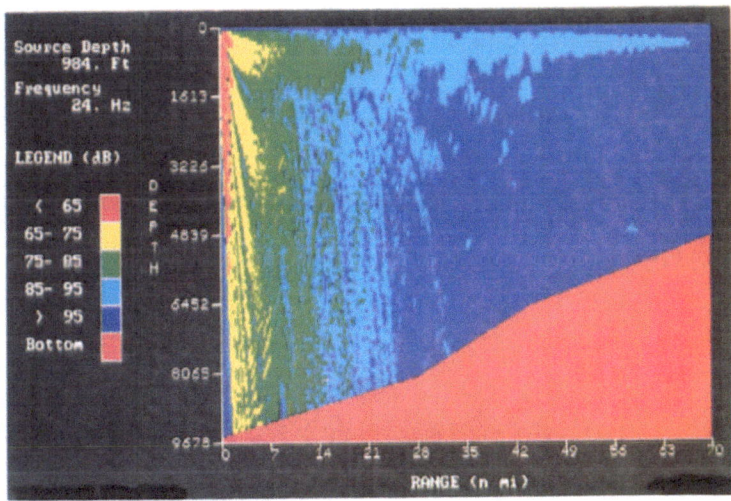

Figure 7. 24 Hz acoustic field intensity vs range for the range-dependent scenario with a source depth of 300 m.

range of 8 nmi, then as the environment changes the differences between the two scenarios become apparent. TL differences between the two scenarios reach 6 dB by 20 nmi and beyond 50 nmi, 15 dB differences appear constant. When compared to the range-dependent scenario, the range-independent scenario shows more structure in transmission loss levels and characteristic deep nulls at a relatively constant range spacing. This is due to the smooth structure of the easterly sound speed profile that has a rounded knee and no relative minimum below the knee. At shorter ranges, the range-dependent scenario shows less of the characteristic deep nulls and their range spacing and the levels exhibit fewer severe fluctuations. For 24 Hz, using the range-independent scenario, the overall transmission loss level is insensitive to source depth, and although the nulls in the transmission loss shift in range, the null spacings are not affected. For the range-dependent scenario, source depth influences transmission loss level, but only after 30 nmi. Here, the western four sound speed profiles have a relative minimum below the knee creating a duct that traps acoustic energy. The deeper source displays 10 – 12 dB more loss at 70 nmi and loss levels exhibit larger and more frequent level fluctuations.

Figure 6 shows the modeled acoustic field intensity vs. range, at 24 Hz, for the range-independent environmental scenario. Figure 7 is the same for the range-dependent scenario. These two figures illustrate how much the range-independent scenario overestimates the acoustic propagation. The differences at the greater ranges and depths emphasize the effects of the range-dependent sound speed profiles.

Figure 8 is similar to Figure 5 for 115 Hz. This and the remaining higher frequencies, show significantly increased TL structure and variability, up to 20 dB. Because of the increased

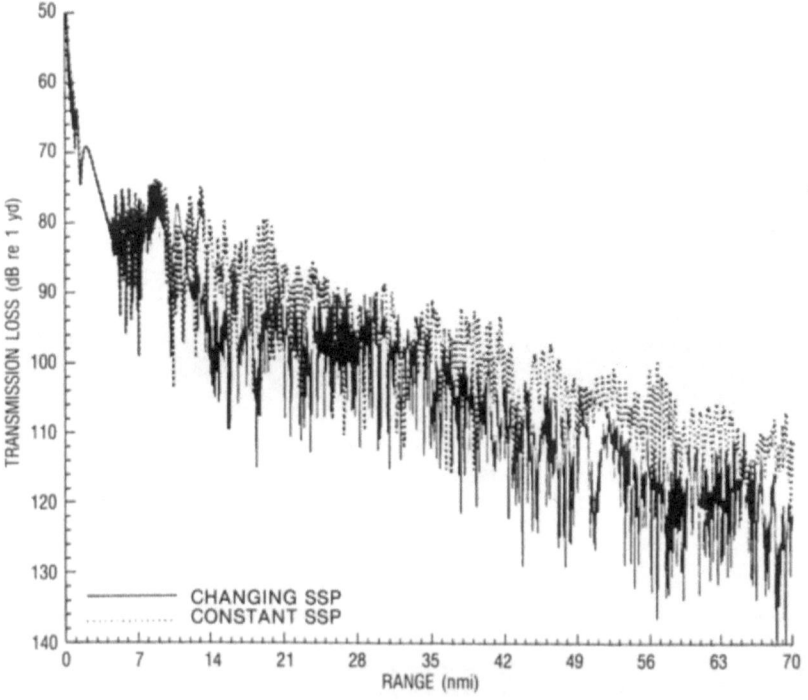

Figure 8. Acoustic transmission loss vs range at 115 Hz showing increased TL level structure and variability, up to 20 dB. Because of this variability subsequent TL levels are averaged over a 0.5 nmi range span to estimate TL differences between the environmental scenarios.

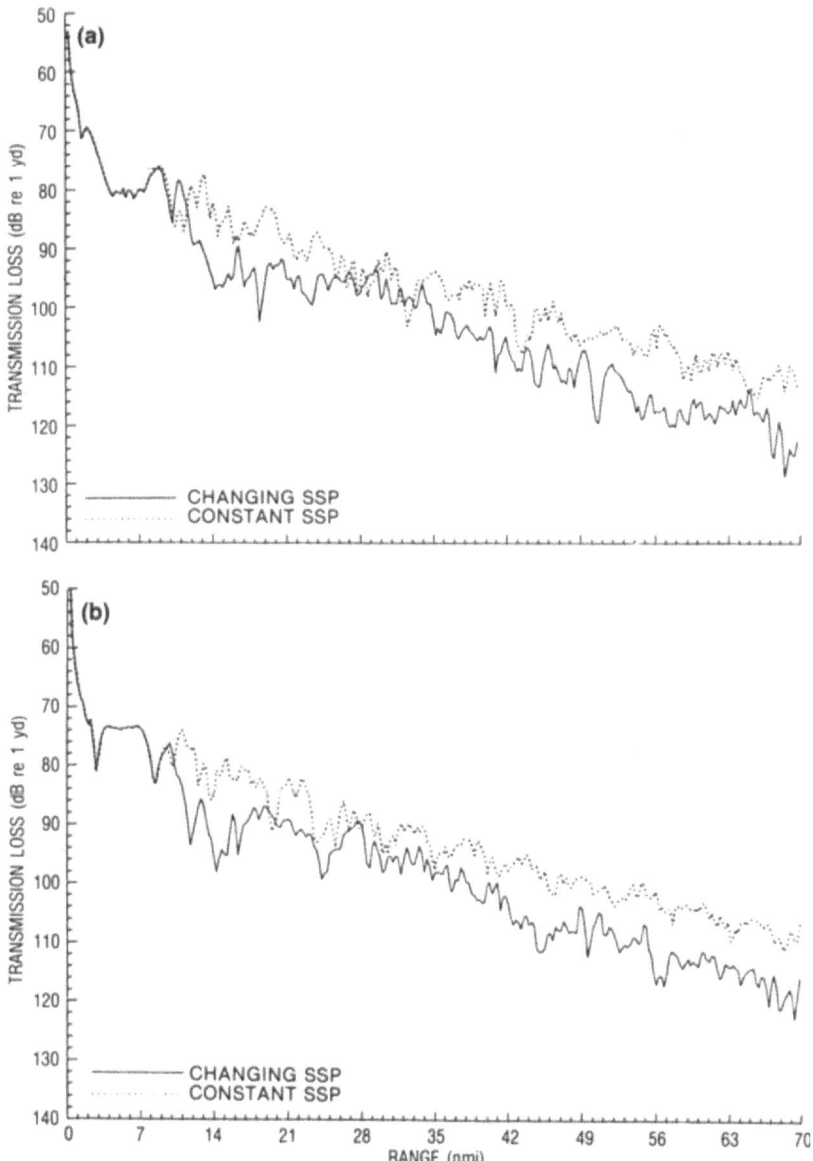

Figure 9. Averaged acoustic transmission loss vs horizontal range, at 115 Hz using both a range-dependent and range-independent environment. a) Source depth: 130 m, b) source depth: 300 m.

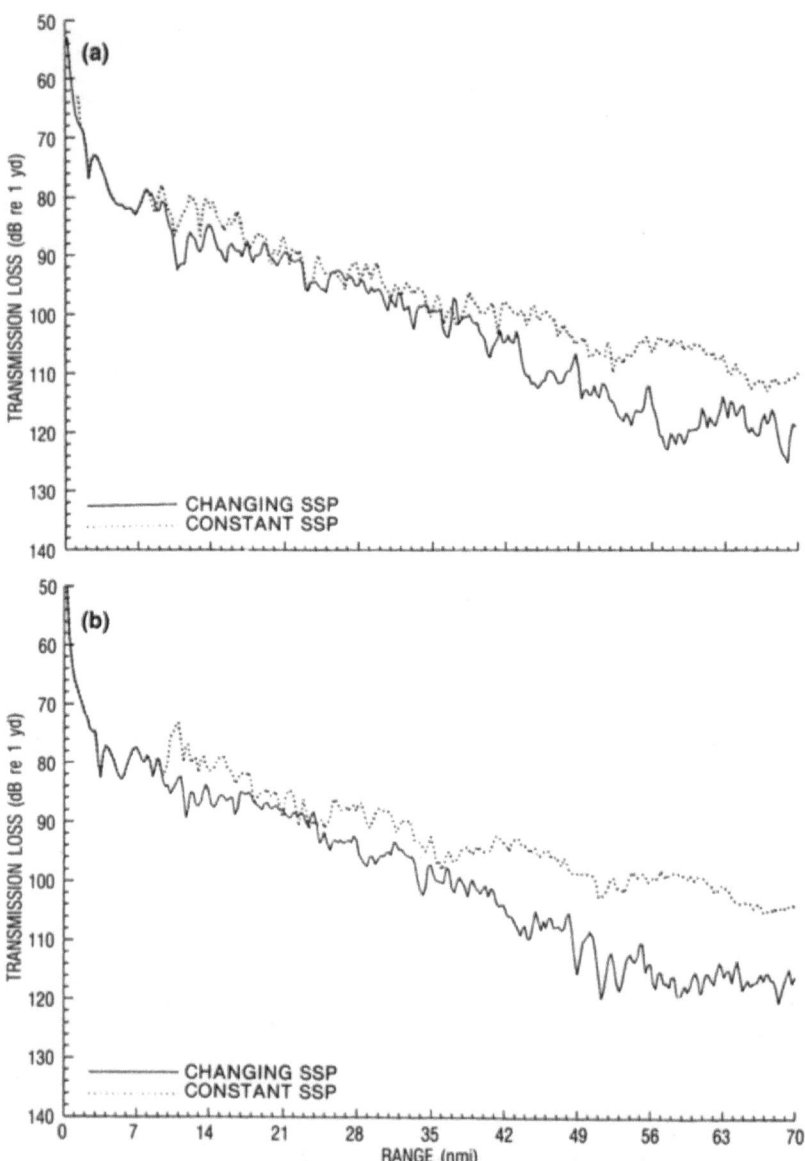

Figure 10. Averaged acoustic transmission loss vs horizontal range, at 273 Hz using both a range-dependent and range-independent environment. a) Source depth: 130 m, b) source depth: 300 m.

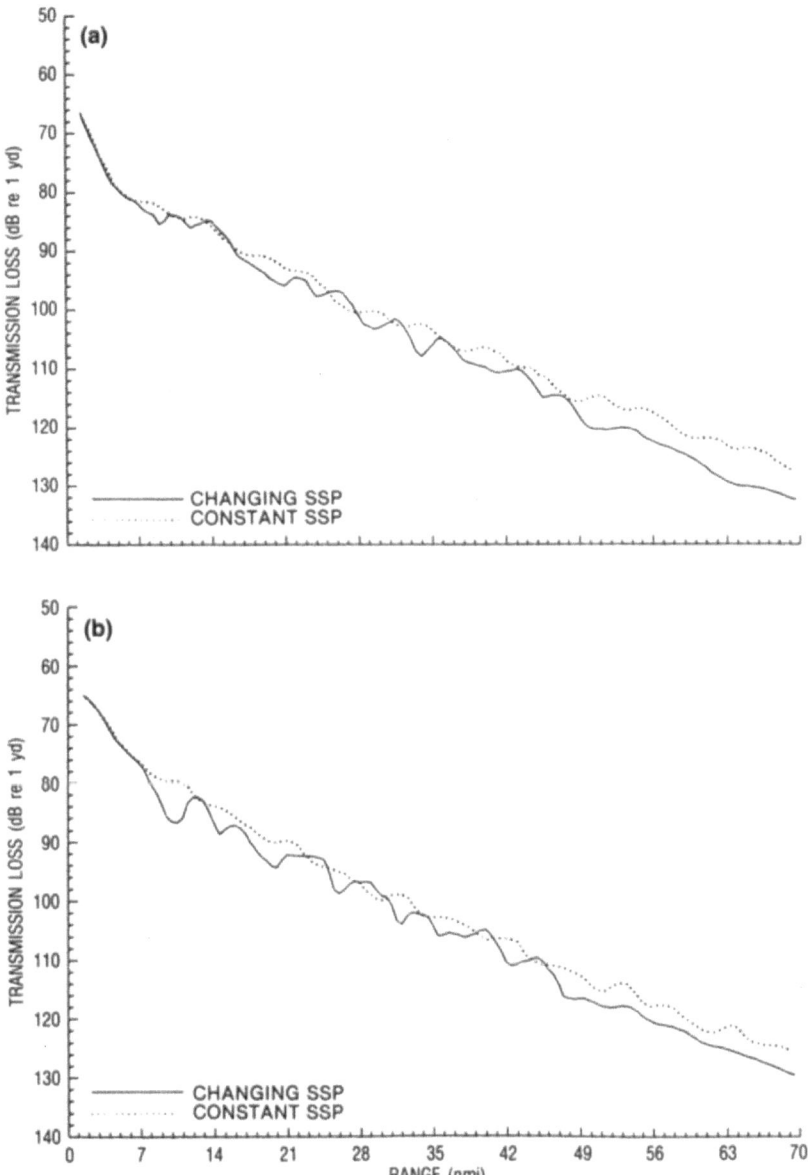

Figure 11. Averaged acoustic transmission loss vs horizontal range, at 2000 Hz using both a range-dependent and range-independent environment. a) Source depth: 130 m, b) source depth: 300 m.

structure in TL levels for these frequencies, levels are range averaged over 0.5 nmi to better estimate TL differences between the environmental scenarios. Figures 9-11 are modeled TL results for 115, 273, and 2000 Hz. At these higher frequencies, no source depth effects were observed for either environmental scenario. Table II summarizes typical differences in the two scenarios for frequency, source depth, and range.

Table II. Transmission loss summary. TL difference is the range-dependent modeled TL minus the range-independent modeled TL.

FREQUENCY/ WAVELENGTH (Hz) (m)	SOURCE DEPTH (m)	TRANSMISSION LOSS DIFFERENCE (dB) HORIZONTAL RANGE (nmi)		
		14	28	48
24 (60.5)	130	4	5	6
	300	4	7	11
115 (12.6)	130	6	5	8
	300	6	2	8
273 (5.3)	130	2	1	2
	300	2	6	6
2000 (0.7)	130	2	2	2
	300	2	2	2

These results show that the environmental scenario used to analyze the experimental data is important and that a constant sound speed should not be assumed over a 70 nmi range. The range-dependent scenario always gives a higher loss and this additional loss increases with horizontal range and frequency up through 273 Hz. At 2000 Hz there is an average 2 dB difference in TL for the two scenarios at both source depths. At these frequencies, over the 70 nmi range span, the acoustic propagation is dominated more by surface and bottom interaction than by changing sound speed profiles.

Summary and Conclusions

Acoustic modeling results have been presented for both range-independent and range-dependent sound speed scenarios. These results prove that the environmental modeling is important to analyze the MIZ experimental data and that a constant sound speed cannot be assumed over a 70 nmi range transverse to the Polar Front. Sound speed profiles change quickly and significantly over short time periods as well and these changes impact on acoustic transmission loss at low frequencies and over long ranges. Transmission loss (TL) differences between the two environmental scenarios varies from 2 – 12 dB. The range-dependent model always gives higher TL than range-independent model. At 2000 Hz, the difference between the range-dependent model and the range-independent model is a constant 2 dB because sound speed fluctuations are negligible compared to the losses due to the effects of bottom and surface interactions.

Acknowledgments

This research was supported by the Office of Naval Technology with technical management provided by the Naval Oceanographic and Atmospheric Research Laboratory. The authors are grateful and indebted to Mr. Paul Bucca for assistance with the oceanographic data.

References

1. D. A. Horn and G. L. Johnson (1985) 'MIZEX east, past operations and future plans,' Proceedings of the Arctic Oceanography Conference and Workshop, Stennis Space Center, MS, June 11-14.
2. B. G. Hurdle, editor (1986) The Nordic Seas, Springer-Verlag, New York.
3. R. W. Meredith, P. J. Bucca, and K. McCoy (1988) 'Icecamp environmental measurements and analysis in support of Arctic acoustics experiments in the marginal ice zone during May 1988,' NOARL Report 210, Naval Oceanographic and Atmospheric Research Laboratory, Stennis Space Center, MS.
4. R. E. Keenan (1989) 'Ice surface loss models evaluation and recommendations,' SAIC-89/1456, Science Applications International Corporation, Mashpee, MA. (June).
5. L. G. Miller, I. J. Lucas, and L. Gainey (1989) 'ICECAP Arctic modeling system,' PSI Technical Report No. 406430, PSI Inc., McLean, VA.

INTERNAL WAVE INDUCED FLUCTUATIONS IN THE OCEANIC DENSITY AND SOUND SPEED FIELDS

ROBERT PINKEL AND JEFFREY T. SHERMAN
Marine Physical Laboratory, A-013
Scripps Institution of Oceanography
La Jolla, California 92093
USA

ABSTRACT. Internal waves distort the sound speed field in the sea. The time variation associated with this distortion results in fluctuations in acoustic intensity and phase in forward transmission experiments. The effort to predict the spectrum of internal wave induced phase fluctuations experienced initial success. Corresponding success was not achieved in the prediction of acoustic intensity, which is sensitive to much smaller vertical scales than phase.

The actual sound speed field which causes the scattering is seldom monitored. Rather, it is estimated using a combination of direct measurements and existing internal wave and fine structure models.

Here a set of ten thousand profiles of sound speed is considered. The profiles extend from the surface to 550 m. They were obtained over a 20 day period in the eastern North Pacific Ocean. A vertical wavenumber-frequency spectrum of sound speed fluctuations is estimated and compared with a corresponding spectrum of isopycnal displacement.

Of particular relevance is the frequency dependence of the sound speed spectrum at the vertical wavenumber corresponding to the Fresnel scale of a given acoustic propagation experiment. Most internal wave spectral models assume a frequency dependence which is independent of vertical wavenumber. The isopycnal displacement spectrum is generally consistent with this assumption. However, the sound speed fluctuation spectrum displays a frequency dependence which changes significantly with vertical wavenumber. At vertical scales greater than 100 m and frequencies below .1 cph, the spectrum is independent of frequency. It then decreases with increasing frequency. At 10 m scale, the spectrum is white in frequency to 1 cph, decreasing with further increasing frequency.

Discrepancies between spectra observed in fixed and isopycnal following reference frames are often discussed in terms of the "fine-structure" contamination of the fixed depth measurements. Here, it is suggested that the non-linear deformation of the sound speed profile by the internal wavefield is responsible for the difference in spectral forms. A simple numerical simulation is used to illustrate the effect.

Introduction

Oceanic acoustic transmissions are influenced by internal wave induced fluctuations in the

103

J. Potter and A. Warn-Varnas (eds.), Ocean Variability & Acoustic Propagation, 103–118.
© 1991 *Kluwer Academic Publishers.*

background sound speed field. These modulate both the travel-time and intensity of acoustic signals transmitted over ranges of a few kilometers or more. The effect is of sufficient strength to constitute a major constraint in the design of acoustic transmission and detection systems. Key to the design of acoustic systems which either maximize or minimize the influence of the internal wavefield is an accurate scattering theory which can predict the influence of the wavefield on acoustic transmission. The evolution of a class of such theories (Desaubies, 1976, 1978, Munk and Zachariasen, 1976, Flatté et. al., 1979, Ucinski, 1977) has been one of the more significant developments of the past 15 years.

The experimental verification of these theories has proved challenging. It is necessary to conduct an ambitious acoustic transmission experiment while simultaneously measuring the vertical wavenumber-frequency spectrum of sound-speed fluctuations along the acoustic path. Both classes of measurement require formidable experimental effort (Ewart 1976, Ewart and Reynolds, 1984). The two dimensional spectrum of sound speed fluctuations has not been measured directly in conjunction with any acoustics experiment. One dimensional measurements, such as time series or horizontal/vertical profiles of temperature, have been obtained. A variety of internal wave spectral models (Garrett and Munk, 1972, 1975, Munk, 1981, Levine and Irish, 1981) has been used to infer the relevant two-dimensional spectrum from the one dimensional observations. When theoretical predictions and experimental results disagree, it is difficult to establish whether the problem is with the scattering theory or with the method used to infer the background scattering field.

To illustrate the level of uncertainty, one can recount the history of the classic transmission experiments conducted by Ewart and co-workers (Ewart, 1977, Ewart and Reynolds, 1984) in the vicinity of Cobb Seamount, in the northeastern Pacific Ocean. The initial Cobb experiment, conducted in 1971, produced measurements of travel time fluctuation which were in excellent agreement with the weak scattering theory of Desaubies (1976). Observations of intensity fluctuation, however, disagreed (Figure 1). More variance in intensity fluctuation was observed at high frequency (near Vaisala frequency and above) than was predicted by theory. The initial tendency was to hypothesize a field of "frozen fine-structure" (Ewart, 1977, Unni and Kaufman, 1981) which existed along with the internal wave induced fluctuations. Alternatively, Flatté, Leung and Lee (1980) suggested that the vertical self-advection of the wavefield should be considered in applying existing internal wave models to the acoustic transmission problem. These changes to the modeling approach significantly reduced the discrepancy between theory and observation, but did not eliminate it.

Subsequently, a more complete experiment, MATE, was conducted at Cobb Seamount. A variety of one dimensional measurements of the background temperature field were obtained. This time the intensity fluctuation data were interpreted in terms of multiple scattering theory rather than weak, Rytov scattering (Ewart and Reynolds, 1984). The discrepancy between theory and experiment was greatly reduced relative to that of the previous single scattering model. Yet a significant discrepancy remained, again at high frequency. Are modifications to the scattering theory called for? Should the sound speed fluctuation field be modeled differently?

In an effort to assist in the resolution of this issue, we present estimates of the vertical wavenumber-frequency spectrum of sound speed obtained from a profiling CTD system mounted on the Research Platform FLIP. The CTD data is also used to track the vertical displacement of isopycnals associated with internal wave motion. A vertical wavenumber-frequency spectrum of isopycnal displacement is also produced. The comparison between the spectra of sound speed and vertical displacement can be used to verify the adequacy of

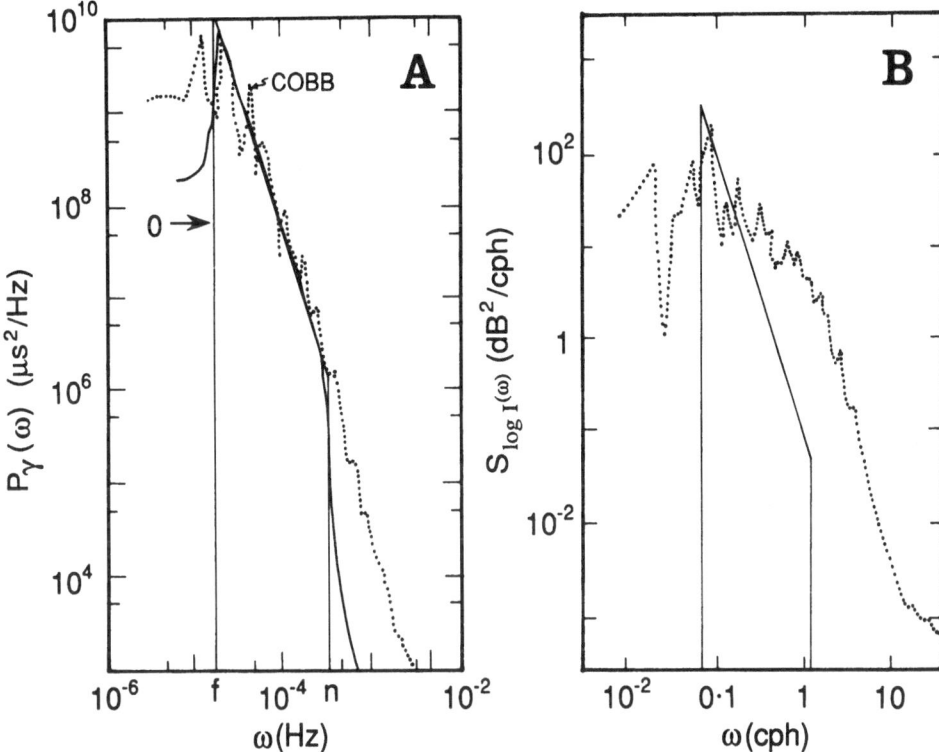

Figure 1. Power spectra of acoustic phase fluctuation (A) and log-intensity (B) from the Cobb Seamount Experiment (Ewart 1976). The theoretical prediction of Desaubies (1976,8) fits the phase fluctuation spectrum well (solid line). The log-intensity agreement is poor

existing environmental models and to suggest improved modeling approaches for use in acoustic fluctuation studies.

Experiment Description

The data are derived from conductivity, temperature and pressure profiles obtained from a profiling CTD system on the Research Platform FLIP. Ten thousand profiles, from the surface to 560 m, were collected in the 1986 PATCHEX Experiment. The PATCHEX site is 34°N, 127°W, about 500 km west of Pt. Conception, California. Water depth was 4 km. FLIP was placed in a two point taut moor, restricting lateral motion to an area of several hundred meters square.

Seabird Electronics model SBE-9 CTD systems were used. These were profiled every three minutes, at a drop rate of 3.8 m/s. The instruments have been packaged in an unconventional manner to accommodate the high fall rate. The modifications include the construction of an open cylindrical frame to protect and support the instrument and the addition of a weighted (25 kg lead shot) nose piece to increase the fall rate. Also, a static pressure port is interfaced to the Digiquartz pressure sensor to reduce the magnitude of turbulent pressure fluctuations associated with the high fall rate.

In PATCHEX, one CTD profiled from 1-300 m while a second covered the depth range 260-560 m. The instruments are sampled at 12 Hz, corresponding to a sampling distance of 32 cm. Given the high fall rate, it is not necessary to pump the conductivity cell to achieve reasonable spatial resolution. The relative phase and amplitude response of the temperature and conductivity sensors is estimated by comparison of temperature and conductivity gradient cross spectra using data from nearly isohaline regions of the water column. The processing method is discussed in greater detail in Sherman (1989). Potential temperature, salinity, potential density, and sound speed are obtained from the response corrected CTD data. The Chen and Millero (1977) equation of state is used in the sound speed computation. A series of representative sound speed profiles is presented in Figure 2.

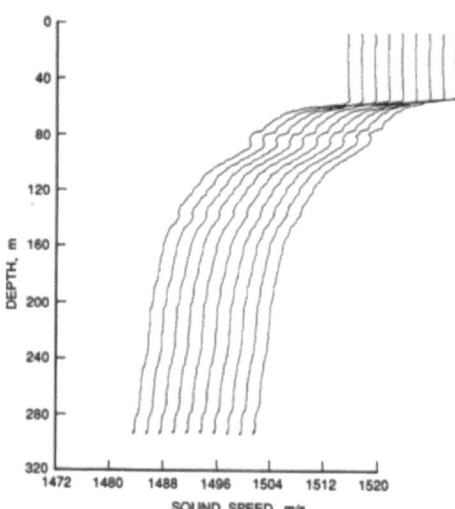

Figure 2. Representative profiles of sound speed from the PATCHEX experiment. The scale is appropriate for the left-most profile. Successive profiles are offset by 3 min. in time and 2 m/s in velocity.

Distinctly separate approaches are used in the preparation of the sound speed and density profiles for subsequent statistical analysis. To gain maximal knowledge of the internal wavefield, the vertical displacement of a set of constant density (isopycnal) surfaces is tracked. A set of 560 isopycnals, of mean separation 1 m, is followed for the duration of the experiment. The isopycnal following measurements are unaffected by so called fine-structure contamination and provide a graphic view of the internal wavefield (Figures 3, 4).

To a first approximation, acoustic ray paths are not advected by the internal wavefield. Thus, the relevant environmental factor in acoustic propagation studies is the fluctuation in sound speed at fixed depth. Time changes in sound speed at fixed depth result from the vertical (and horizontal) advection of the instantaneous sound speed field.

$$\frac{Dc}{Dt} = \gamma \frac{Dp}{Dt}$$

$$\frac{\partial c}{\partial t} = -\bar{u} \cdot \nabla c + \gamma \frac{Dp}{Dt}$$

(1)

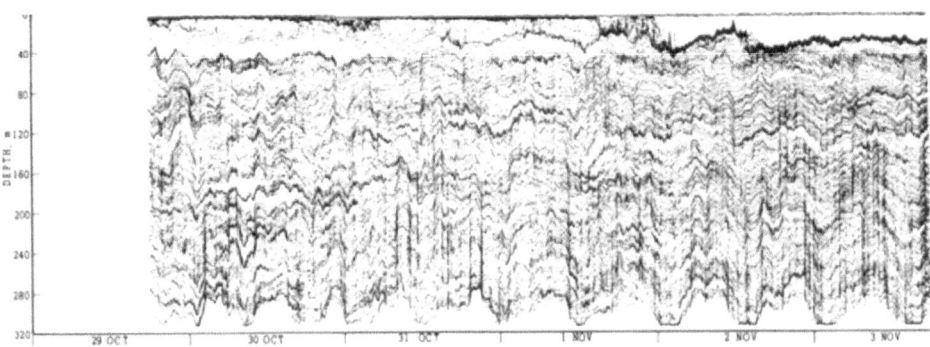

Figure 3. Isopycnal depth fluctuations resulting from internal wave vertical displacement. Data were obtained using the upper CTD on FLIP.

Here c is the speed of sound, p is the pressure and γ relates the change in sound speed to changing pressure. In the special case that the sound speed gradient is constant, vertically directed, and of sufficient magnitude that the pressure effect can be neglected, sound speed fluctuations can be related to vertical motion.

$$\frac{\partial c}{\partial t} = -w\frac{\partial c}{\partial z} \tag{2}$$

This special case is seldom realized in the ocean. Herein lies the uncertainty in applying existing models of the internal wavefield to the acoustic fluctuation problem.

A twelve hour record of sound speed fluctuation, at depths of 220-235 m is presented in Figure 5. Data from adjacent 1 m depths are offset by 1 m/s in sound speed, so that details of the fluctuations can be easily seen. The top four time series (220-224 m) in Figure 5 are replotted in Figure 6, without this offset. This gives a short realization of the actual field described by the wavenumber frequency spectrum to be presented.

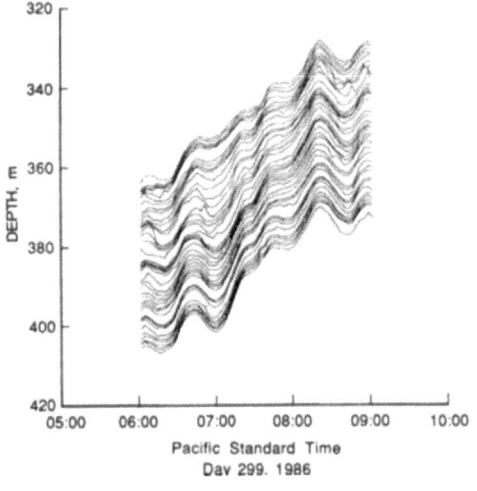

Figure 4. A close-up view of the PATCHEX isopycnal displacement data set.

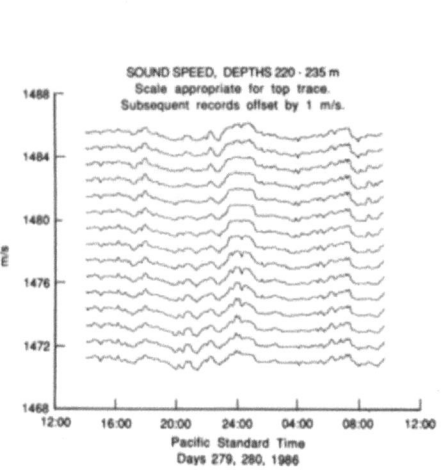

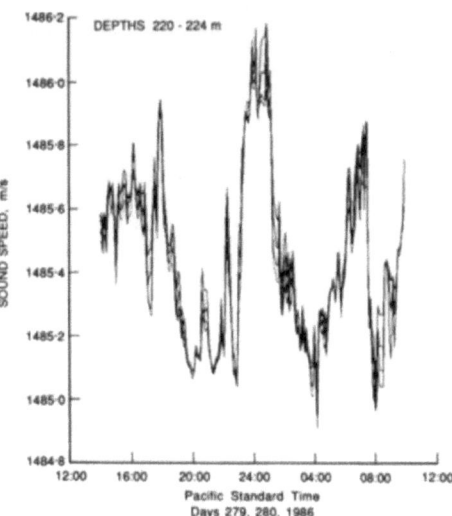

Figure 5. Representative time series of sound speed from PATCHEX, for the depth range 220-235 m. The scale is appropriate for the top series. Subsequent records are offset by 1 m/s.

Figure 6. The top four traces in Figure 5 are here replotted on a common velocity scale. Note the non-sinusoidal aspect of these fluctuations.

The non-sinusoidal aspect of the fixed depth signals is apparent in Figures 5 and 6. Extended periods of limited variation are seen, (Figure 5) when a layer of nearly uniform sound speed surrounds the measurement depth. These periods are interspersed with intervals of rapid change, as high gradient regions are advected past the reference depth. Note that the variability of the sound speed gradient does not necessarily render the adjacent time series dissimilar. Indeed many features of these records are recognizable over the 15 m span. However, the non-constancy of the sound speed gradient induces extremely non-sinusoidal behavior in the fixed depth measurements. One would hesitate to characterize this field as the sum of a smoothly oscillating variable (such as in Figure 4) plus an unrelated field of additive fine-structure.

Spectral Analysis

Successive three minute values of isopycnal displacement were averaged together, to reduce the number of data to process. The resulting six-minute sampled time series were divided into three 10-day records, overlapped by 50% in time. These were then first differenced in time. A triangular data window was applied prior to Fourier transformation. Fourier coefficients were produced in 1200 frequency bands, from .1 to 120 cpd.

The depth-frequency array of Fourier coefficients was then transposed, to produce depth profiles of complex Fourier coefficients in each of the frequency bands. Isopycnals whose mean depths were between 150 and 500 m were selected for subsequent processing. This avoided the zone of maximum variability just below the base of the mixed layer.

Traditionally (Pinkel 1975, 1984) some attempt is made to remove deterministically non-stationary aspects of internal wave signal, resulting from the depth variability of the Vaisala frequency, N. Here, the function $W(z) = K(N^2(z) - \sigma^2)^{1/4}$ was used to weight the time Fourier coefficients, prior to Fourier transformation in depth. The constant K was chosen such that variance of the depth-series was unaltered by this weighting. The internal wave frequency is given by σ. An analytic approximation to the observed Vaisala profile,

$$N(z) = N_o e^{-z/1000} \tag{3}$$

was used in the weighting process.

It is also common to stretch the depth scale as suggested by WKB theory, to further reduce the signature of wave refraction. This was not done in the present case. Experience has shown that the stretching process has limited effect on the gross spectral form. The amplitude weighted Fourier coefficients in each frequency band were first differenced in depth, and multiplied by a triangle window over the 150-500 m observational range. The 350 m complex depth series was zero filled to 512 m (points) and then Fourier transformed in depth. Wavenumber-frequency spectral estimates were produced in 512 wavenumber bands, 1200 frequency bands, averaged over the three overlapping data blocks, and recolored to form the spectral estimates. The spectral estimates were then smoothed to varying degrees in both wave number and frequency to increase statistical stability. The details of the smoothing process are discussed in conjunction with the figures presented below.

The Eulerian sound speed data are processed in a generally similar manner. However, no attempt has been made to weight or stretch the data, to make it appear as if the fluctuations are homogeneous in the vertical. To the extent that the non-homogeneous nature of the field influences the acoustic scattering problem, it is appropriate that this be reflected in the spectral estimates.

The depth region 150-500 m was again selected for vertical Fourier analysis. As before, the sound speed data were first differenced in depth and multiplied by a triangular data window. The 350 m depth series was augmented to 512 m total length (by "zero filling") and was Fourier transformed. Spectral estimates were formed from the squared vertical wavenumber frequency Fourier coefficients.

Wavenumber-Frequency Spectra

The resulting wavenumber-frequency spectra of sound speed and isopycnal displacement are presented in log (frequency) linear (wavenumber) format in Figures 7 and 8. These spectral estimates are averaged over four adjacent Fourier bands in wavenumber. The frequency averaging is logarithmic. There is no averaging at low frequency, extreme averaging at high frequency. Thus the statistical significance of the spectral estimates varies within the plots, from 16 degrees of freedom at low frequency to several hundred at high frequency.

Statistical confidence limits are not an issue in the present study. The spectra to be compared represent data extracted from the same volume of the ocean at the same time. The observed differences are a consequence of the differing observational reference frames.

The isopycnal displacement spectrum displays an σ^{-2} frequency dependence at low wavenumber. A Vaisala cut-off is seen at frequencies above 2 cph. The sound speed spectrum has an $\sigma^{-1.5}$ frequency dependence at low wavenumber, with no apparent Vaisala

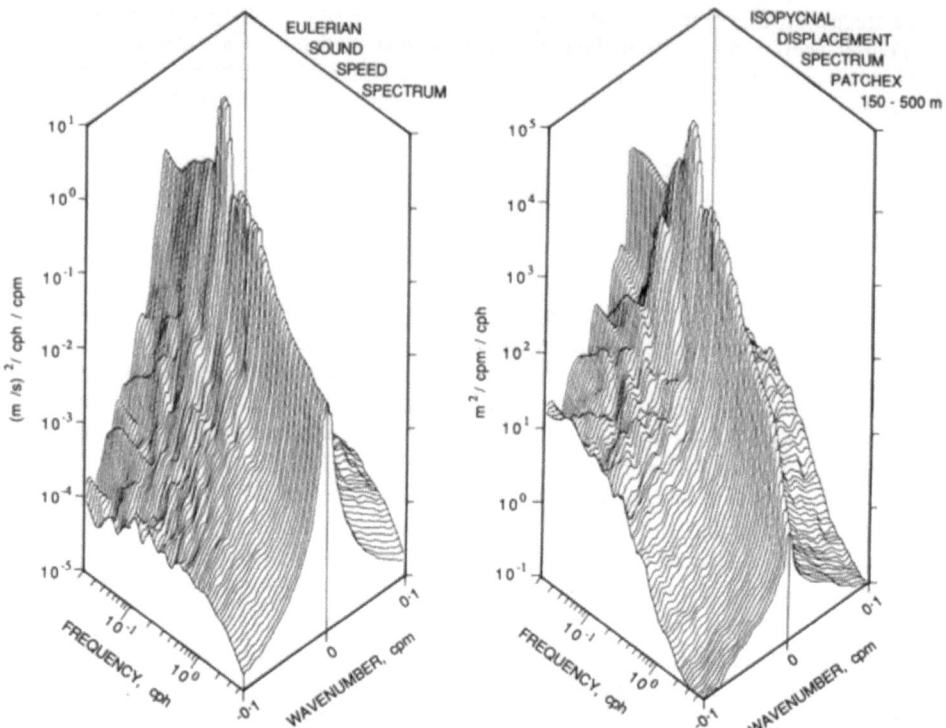

Figure 7. The PATCHEX wavenumber-frequency spectrum of sound speed fluctuation. The vertical wavenumber axis is linear. Variance at positive wavenumbers is associated with fluctuations with downward phase propagation.

Figure 8. The PATCHEX wavenumber-frequency spectrum of isopycnal displacement. The vertical wavenumber axis is linear. Variance at positive wavenumbers is associated with downward phase propagation.

cut-off. In part, the absence of a cut-off is a result of the wavenumber smoothing employed. The zero wavenumber band, corresponding to the 150 - 500 m mean sound speed fluctuation, indeed shows a cut-off, (Figure 11). However, when averaged with the adjacent wavenumber bands this cut-off is obscured.

The zero wavenumber band of the displacement spectrum is associated with a variance of 21.4 m^2. The total variance in the wavenumber bands corresponding to upward (downward) phase propagation is 8.47 m^2 (10.82 m^2). The up-down assymmetry is due to the internal tide, which has predominately downward phase propagation at the PATCHEX site.

The zero wavenumber band of the sound speed spectrum corresponds to a variance of .093 m^2/s^2. The total variance in the bands corresponding to upward (downward) phase propagation is here .04821 m^2/s^2 (.04759 m^2/s^2). There is no significant vertical assymmetry in this spectrum.

The frequency dependence of the sound speed spectrum near the Fresnel wavenumber, $k_f = 1/(\lambda R)^{1/2}$ is central to the prediction of acoustic amplitude fluctuation statistics (Flatté,

1983). Here λ is the acoustic wavelength and R is the transmission path length. The spectra of Figures 7 and 8 are terminated at .1 cpm. From the exposed edges, one can see that the frequency slope of the isopycnal-following spectrum is steeper than that of sound speed, at 10 m scale. Successive cuts of these spectra at smaller wavenumbers (Figure 9 A - D) demonstrate that the frequency dependence differs at 20 and 40 m scales, as well. The 40 m scale is within the range of Fresnel scales spanned in the MATE experiment (Ewart and Reynolds, 1984).

In Figures 10 and 11, it is seen that these differences persist at all vertical scales. Here, cross sections of the spectra are presented as a function of frequency at a variety of fixed wavenumber bands. The zero wavenumber (150-500 m mean) fluctuations show a Vaisala cutoff in both figures. However, the cross-sections are significantly different at all non-zero wavenumbers, even at vertical scales of 320 m!

Discussion

The differences between the spectrum of isopycnal displacement and that of sound speed fluctuation are not unexpected. Traditionally, such differences have been discussed in term of the "contamination" of fixed depth measurements by the vertical advection of "fine-structure" past the sensing instruments. This notion is quantified by the equation

$$\frac{\partial\theta}{\partial t}(z,t) = -w(z,t)\frac{\partial\theta}{\partial z}(z,t) \tag{4}$$

which is derived from the statement

$$\frac{D\theta}{Dt} = 0 \tag{5}$$

under the assumption that time changes due to horizontal advection, $\bar{u}\cdot\nabla_H$, are negligible relative to those due to vertical advection. The disparity between fixed depth and isopycnal following measurements has been attributed to small scale detail in the vertical gradient field, $\partial\theta/\partial z$ (Phillips 1971, Garrett and Munk, 1971). The detail required to be consistent with observations is typically in excess of that predicted by linear internal wave models. An additive field of "fine-structure contamination" is frequently hypothesized (Muller et. al., 1978, McKean, 1974) to account for observations. Such a field can be incorporated into acoustic propagation models (Unni and Kaufman, 1981) and adjusted so as to provide excellent agreement between acoustic experiment observations and theory (Ewart, 1990 -- this conference). Parameters necessary to describe the horizontal and vertical wave number dependence of the fine-structure spectrum, as well as its gross variance, must be added to propagation models.

While excellent agreement between model and experiment is possible, understanding of the overall problem is less than satisfactory because the essential physics of this hypothesized "fine structure" field is unknown. The results of pioneering experiments such as MATE (Ewart and Reynolds, 1984) can not be used to predict acoustic propagation parameters at other depths or locations unless the geographic variation of this "fine-structure field" is predictable.

112

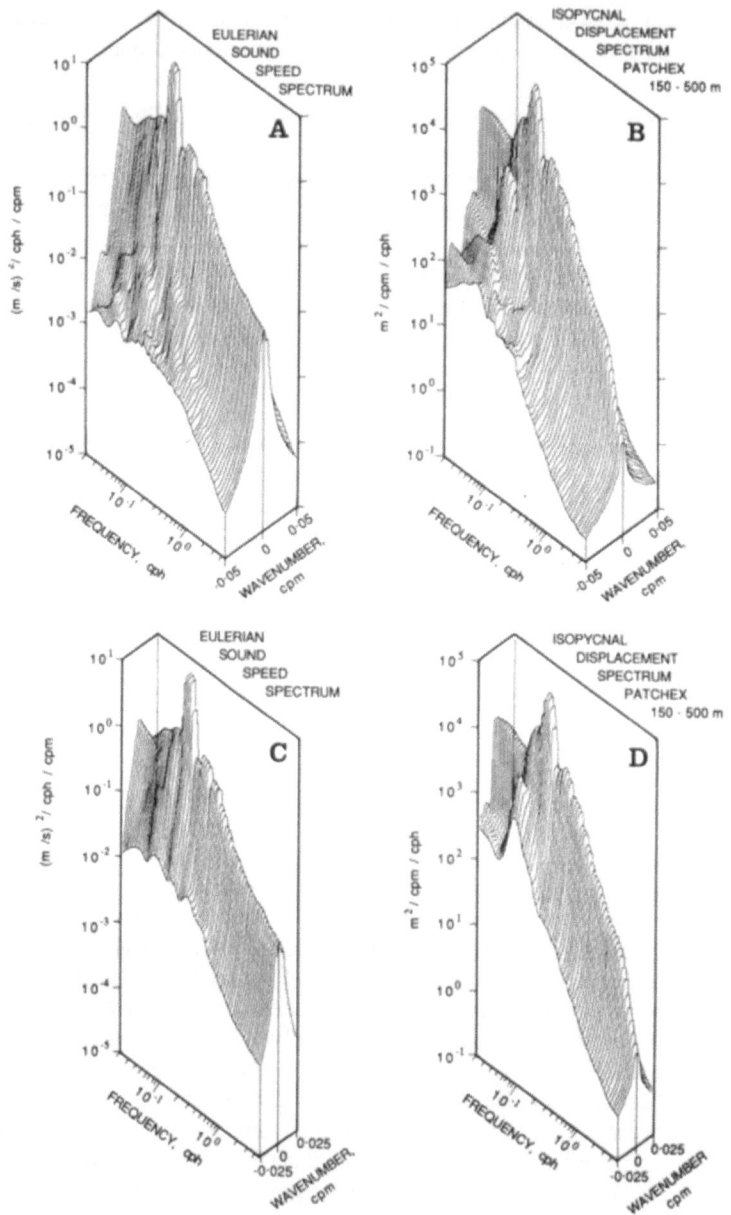

Figure 9. The sound speed spectrum of Figure 7 is here truncated at 20 m (A) and 40 m (C) inverse wavenumber, to emphasize the frequency dependence at these vertical scales. The contrast with the displacement spectrum, similarly truncated (B, D) is apparent.

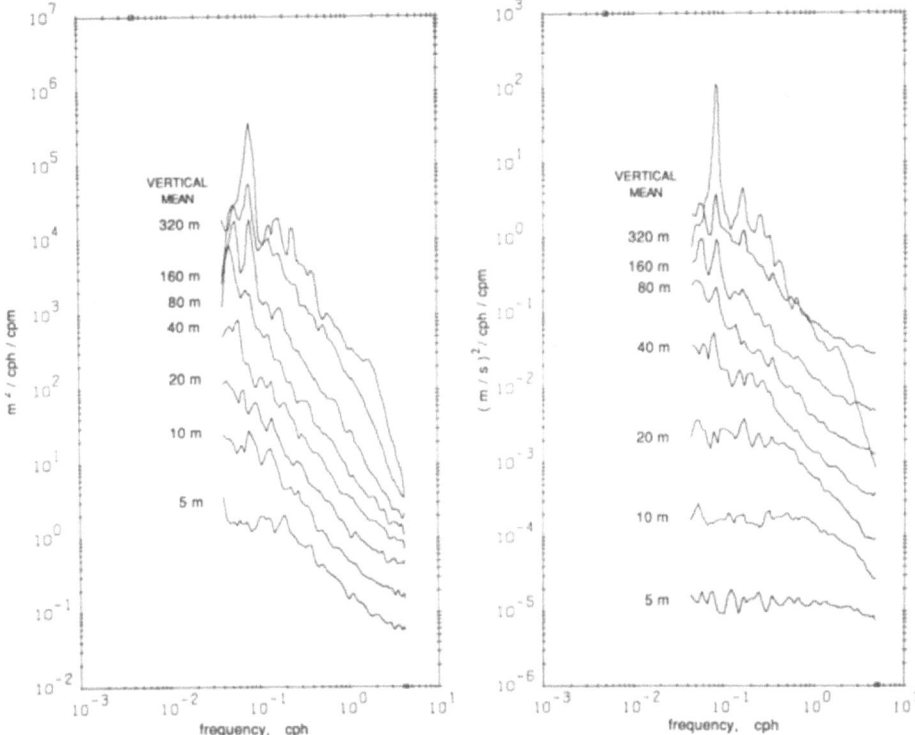

Figure 10. Cross sections of the isopycnal displacement spectrum at 5 - 320 m inverse wavenumber. The frequency spectrum of the 150 -500 m vertical mean displacement is also shown.

Figure 11. Cross sections of the sound speed fluctuation spectrum at 5 - 320 m inverse wavenumber. The frequency spectrum of the 150 -500 m vertical mean sound speed fluctuation is also shown. This has a steeper spectral slope than the spectrum at any non-zero wavenumber. A pronounced Vaisala cut-off is also seen. These details are obscured in Figures 7 and 9 by the cross wavenumber smoothing employed to increase statistical confidence.

From the perspective of physical oceanography, the fine-structure issue is also of concern. Approximately two-thirds of the typical variance of strain and shear in the thermocline (at vertical scales greater than 1 m) is found at scales less than 10 m. It is disheartening that the physics governing motions on this scale remains unknown, given the many years of effort.

The challenge in the present case is to explain observed differences at large vertical scale as well as small. Large scale differences are presumably a result of the gross curvature of sound speed profile. This is seen by writing

$$c(t, z_o) = C(z_o - \eta(t, z_i(t))) \tag{6}$$

$$\approx C(z_o) + \eta(t, z_o)\frac{dC}{dz}\Big|_{z_o} + \eta^2(t, z_o)\frac{d^2C}{dz^2}\Big|_{z_o} \tag{7}$$

Here, $C(z)$ is the undistorted sound speed profile, $\eta(t, z)$ is the instantaneous excursion of an isopycnal whose mean depth is z from its mean depth, and z_i is the mean depth of the isopycnal which is at the depth of interest, z_o, at time t, i.e., $z_i = z_o - \eta(t, z_i)$.

Two levels of approximation are employed in progressing from equation (6) to (7). The Taylor series is truncated at second order. Also the vertical displacement at depth z_o is substituted for the exact quantity, the displacement at the time dependent z_i.

The power spectrum of sound speed at fixed depth, $c(t, z_o)$, can be expected to differ significantly from that of isopycnal displacement, $\eta(t, z_o)$, in situations where the higher order terms in the Taylor expansion are significant. Differences will be seen even at large vertical scales in upper thermocline measurements. The curvature, in particular, is a large scale phenomenon which indeed characterizes the thermocline.

In deep, upward refracting experiments such as MATE, large scale differences between fixed depth and isopycnal following measurements will be less significant. Below the sound channel axis, the curvature of the sound speed profile is smaller relative to the gradient. The linear approximation to (7) will be correspondingly more accurate.

Differences in spectral detail at small scale still remain, however. These must be explained through direct study of the fine scale field.

Yves Desaubies has been involved in the early study of the fine-structure problem (Joyce and Desaubies, 1977). After an extensive study of vertical profiles of temperature (Desaubies and Gregg, 1981), Desaubies suggested that an additive field of "fine-structure" might not be necessary to account for highly irregular temperature profiles. A vertical velocity field, perhaps due to internal waves, will deform the distribution of tracers such as temperature, sound velocity, or density. Extensive variance can be produced in the tracer field at scales smaller than any energetic scale in the velocity field. The magnitude of the velocity field must be sufficient that non-linear effects are important. Desaubies demonstrated that the typical oceanic internal wavefield has sufficient energy.

Closely related to Desaubies suggestion is the attempt by Flatté, et. al, (1980) to explain acoustic scattering results in terms of the "self-advection" of the internal wavefield. These authors noted that if the low frequency - small scale components of the internal wavefield are advected vertically by the high-frequency large scale components, the resulting modeled sound speed fluctuations more closely resembled the Cobb Seamount (Ewart, 1976) observations. Munk (1981) also noted the self-advective aspect of the internal wavefield. He suggested that small scale components of the wavefield be termed "compliant waves" if their phase velocity was less than the characteristic horizontal orbital velocity, typically 10 cm/s, associated with the larger scale waves. The Desaubies suggestion is more extreme than the Munk or Flatté et. al. views, in that small vertical scale variance in both passive scalars and perhaps shear, $\partial u/\partial z$, $\partial v/\partial z$, is created from an underlying velocity field which has no associated small scale strain rate $\partial w/\partial z$, $\partial u/\partial x$, $\partial v/\partial y$. (Specifically, the diagonal elements of the strain rate tensor are much smaller than the off-diagonal elements).

If this hypothesis is correct, one should be able to model the "fine scale" field associated with any given large scale internal wavefield with no need for adjustable free parameters. Since the depth dependence of the large scale wavefield is more or less known, a degree of predictability for the fine scale field is suggested. This would enable the a-priori prediction of acoustic transmission performance, in contrast to the present ability to justify the observations after-the-fact.

As a first step toward investigating this "self-distortion" hypothesis, a simple numerical simulation was performed. An internal wave - like vertical velocity field

$$w(z,t) = e_o \sum_k \sum_\sigma a(k)b(\sigma)e^{ikz - \sigma t + \phi_{k\sigma}} \qquad (8)$$

was generated numerically. Here k is vertical wavenumber, σ is frequency, and $\phi_{k\sigma}$ is a random phase. This format is chosen such that the associated power spectrum of vertical velocity is separable, consistent with the Garrett - Munk 1972 hypothesis:

$$E_w(k,\sigma) = E_o A(k)B(\sigma)$$
$$A(k) = a(k)^2 \qquad (9)$$
$$B(\sigma) = b(\sigma)^2$$

The velocity field is generated from a superposition of 356×64 sinusoidal constituents. Frequencies are selected in 356 bands corresponding to periods from inertial (21 hr) to Vaisala (1/3 hr). The spectral frequency dependence, $\beta(\sigma)$, is set to increase linearly, between inertial and semi-diurnal tidal frequencies. At higher frequency $\beta(\sigma)$ is held constant, in a rough approximation of observation. Constituents are evaluated at +/- 32 wavenumber bands in each frequency band. Wavenumbers correspond to vertical scales of 12 to 384 m. The wavenumber dependence of the input spectrum, $A(k)$, is given in Figure 12.

A set of reference surfaces $\{\eta_i\}$ is introduced into this velocity field. Initially the surfaces are separated by equal increments in depth. The time evolution of the position of surfaces is properly given by

$$\frac{\partial \eta_i}{\partial t} = w + u \cdot \nabla_H \eta_i \qquad (10)$$

In the general non-linear case, the terms on the right hand side of (10) are evaluated at $z = \eta_i$. The standard linear approximation is to evaluate the vertical velocity at $z = \langle \eta_i \rangle$ and neglect the non-linear term. In this simulation, an ad. hoc. "tunnel vision" approximation is made. The vertical velocity is still evaluated at $z = \eta_i$, yet the lateral advective term is neglected. In other words, the isopycnal is vertically advected by the velocity it experiences at its instantaneous depth, rather than its mean depth.

As time passes, the position of the tracer surfaces evolve. An interesting special case is seen to occur. As adjacent surfaces get close together, they experience nearly the same velocity field. As a result, they tend to move in tandem, remaining "stuck" together for a long time. The tracer field tends to assume the classical "sheets and layers" configuration. The

116

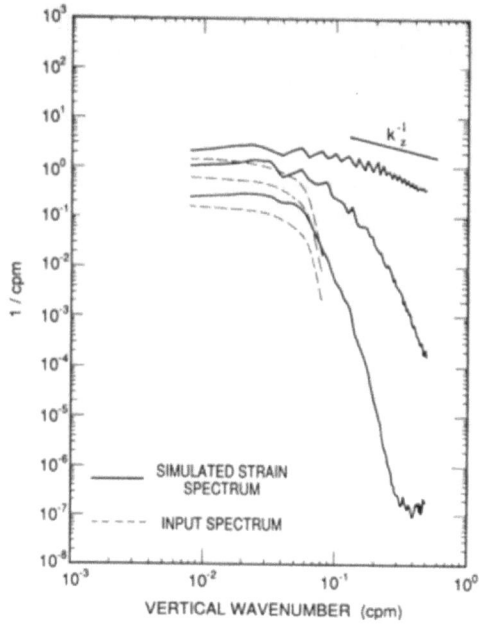

Figure 12. Simulated vertical wavenumber spectra of strain, $\partial\eta/\partial z$, generated from a motion field with a strain rate spectrum given by the dashed line. As the variance of the generating spectrum is increased, both the energy and the form of the resultant strain spectrum are altered.

tracers congregate into isolated high gradient sheets, separated by thicker, tracer-free, layers. This tendency of the tracers to gather together is greatly increased as the overall amplitude (energy) of the velocity field is increased.

Representative vertical wavenumber spectra of vertical "strain", $\partial\eta/\partial z$ are plotted in Figure 12. The corresponding wavenumber spectra of the vertical derivative of the generating field, $E_o k^2 A(k)$, are also presented. As the energy level of the generating field is increased, the strain spectral energy also increases. In addition, the form of the strain spectrum is altered, becoming progressively flatter with increasing input energy. In Figure 12, the highest energy level was arbitrarily selected to produce a k^{-1} slope at high wavenumber, consistent with present observations (Gregg, 1977).

We emphasize that this is an essentially kinematic exercise. To model the "true" internal wave spectrum dynamically, one needs to specify energy sources and sinks, and parameterize the relevant non-linear interactions. Even as a kinematic study, the present effort suffers in that lateral advective terms are not modeled. These terms are thought to be of the same order as the non-linear effects included here (Desaubies and Gregg, 1981).

The point here is that considerable "fine-structure" can be generated in a tracer field associated with a motion field which has "no" fine-structure. The task of properly modeling and predicting this effect is beyond the scope of this work. Whether an "added field" of motions is still required to explain observations once these non-linear kinematic effects are understood remains to be seen.

Summary

Given the differences between the spectra of sound speed at fixed depth, and isopycnal depth fluctuation (at fixed density), it is not surprising that there is difficulty in fitting acoustic experimental results using existing internal wave models. The reliance on an <u>additive field</u> of "fine-structure" to match experiment with model is perhaps unsatisfactory. In particular, one cannot yet predict the parameters of this fine-structure field, as it varies in depth, time, and geographic location. Until this problem is remedied, otherwise capable acoustic models will be unable to predict acoustic performance in regions where the "fine-structure" is not already known.

To the extent that "fine-structure" results from the self-advection of the wavefield, fine-structure effect might be predictable from simple internal wave models, with no added free-parameters. A more complete study of this issue is called for. If the kinematic approach fails to describe observations such as those presented here, an extended attack on the fine scale dynamics of the thermocline will be necessary to support the acoustic prediction effort.

Bibliography

Chen, C.T. and F.J. Millero, 1977: "Speed of Sound in Sea Water at High Pressures", J. Acoust. Soc. Am., 62, 1129-1135.

Desaubies, Y.J.F., 1976: "Acoustic Phase Fluctuations Induced by Internal Waves in the Ocean", J. Acoust. Soc. Am., 60, 795-800.

Desaubies, Y.J.F., 1978: "On the Scattering of Sound by Internal Waves in the Ocean", J. Acoust. Soc. Am., 64, 1460-1469.

Desaubies, Y.J.F. and M.C. Gregg, 1981: "Reversible and Irreversible Fine Structure", J. Phys. Oceanogr., 11, 541-556.

Ewart, T.E., 1976: "Acoustic Fluctuations in the Open Ocean - A Measurement Using a Fixed Refracted Path", J. Acoustic. Soc. Am., 60, 46-59.

Ewart, T.E. and S.A. Reynolds, 1984: "The Mid Ocean Acoustic Transmission Experiment, MATE", J. Acoust. Soc. Am., 75, 785-802.

Flatté, S.M., R. Dashen, W.H. Munk, K.M. Watson and F. Zachariasen, 1979: "Sound Transmission Through a Fluctuating Ocean", Cambridge University Press, New York.

Flatté, S.M., R. Leung, and S.Y. Lee, 1980: "Frequency Spectra of Acoustic Fluctuations Caused By Oceanic Internal Waves and Other Fine Structure", J. Acoust. Soc. Am., 68, 1773-1779.

Flatté, S.M. 1983: "Wave Propagation Through Random Media: Contributions from Ocean Acoustics", Proc. IEEE, 71, 1267-1294.

Garrett, C.J.R. and W.H. Munk, 1971: "Internal Wave Spectra in the Presence of Fine Structure", J. Phys. Oceanogr., 1, 196-202.

Garrett, C.J.R. and W.H. Munk, 1972: "Space-Time Scales of Internal Waves", Geophys. Fluid. Dyn., 3, 225-264, 1972.

Garrett, C.J.R. and W.H. Munk, 1975: "Space-Time Scales of Internal Waves - a Progress Report", J. Geophys. Res., 80, 291-299.

Gregg, M.C., 1977: "A Comparison of Fine Structure Spectra from the Main Thermocline", J. Phys. Oceanogr., 7, 33-40.

Joyce, T.M. and Y.J.F. Desaubies, 1977: "Discrimination Between Internal Waves and Temperature Fine Structure", J. Phys. Oceanogr., 7, 22-32.

Levine, M.D. and J.D. Irish, 1981: "A Statistical Description of Temperature Fine Structure in the Presence of Internal Waves", J. Phys. Oceanogr., 11, 676-691.

McKean, R.S., 1974: "Interpretation of Internal Wave Measurements in the Presence of Fine Structure", J. Phys. Oceanogr., **4**, 200-213.

Muller, P., D.J. Olbers and J. Willebrand, 1978: "The IWEX Spectrum", J. Geophys. Res., **83**, 479-499.

Munk, W.H., 1981: "Internal Waves and Small Scale Processes, in Evolution of Physical Oceanography", B.A. Warren and C. Wunsch, eds., MIT Press, Cambridge, Mass., 264-291.

Munk, W.H. and F. Zachariasen, 1976: "Sound Propagation through a Fluctuating Stratified Ocean: Theory and Observation", J. Acoust. Soc. Am., **59**, 818-838.

Phillips, O.M., 1971: "On Spectra Measured in an Undulating Layered Medium", J. Phys. Oceanogr., **1**, 1-6.

Pinkel, R., 1975: "Upper Ocean Internal Wave Observations from FLIP", J. Geophys. Res., **80**, 3892-3910.

Pinkel, R., 1984: "Doppler Sonar Observations of Internal Waves: The Wavenumber Frequency Spectrum", J. Phys. Oceanogr., **14**, 1249-1270.

Sherman, J.T., 1989: "Observations of Fine Scale Vertical Shear and Strain in the Upper Ocean", Ph.D. Thesis, U. of California, San Diego.

Ucinski, B.J., 1977: "The Elements of Wave Propagation in Random Media", McGraw-Hill, New York, 150 pp.

Unni, S. and C. Kaufman, 1981: "Acoustic Fluctuations due to the Temperature Fine Structure of the Ocean", J. Acoust. Soc. Am., 676-680.

GYRE-SCALE RECIPROCAL ACOUSTIC TRANSMISSIONS

PETER WORCESTER AND BRIAN DUSHAW
Scripps Institution of Oceanography
University of California, San Diego, CA 92093 USA

BRUCE HOWE
Applied Physics Laboratory
University of Washington
Seattle, Washington 98105

ABSTRACT. Three acoustic transceivers moored in a triangle approximately 1000 km on a side simultaneously transmitted broadband signals to one another for four months during summer 1987. The triangle extended in latitude from the Subtropical Front to the Subarctic Front in the North Central Pacific Ocean. Individual ray arrivals were resolved and their arrival times measured with a precision of a few milliseconds using phase-coded transmissions centered at 250 Hz. Using CTD data taken on the three legs, mean sound speed profiles were calculated using the Chen and Millero and Del Grosso sound speed algorithms. The measured arrival patterns agree best with arrival patterns predicted using the Del Grosso algorithm. Predicted and measured absolute travel times also agree best using the Del Grosso algorithm. All resolvable ray paths are surface-reflected, and the travel times therefore represent both range and depth averages. The sum of the travel times of oppositely traveling signals depends on the sound speed field. The sum travel times decreased as the summer thermocline formed during the experiment. The travel times suggest that the thermocline deepened episodically on the northern leg of the triangle. Historical data gives a low mesoscale energy level for the experimental area, with small temperature fluctuations in the main thermocline. The sum travel times correspondingly show little low-frequency variability except for that caused by the formation of the summer thermocline. The transmissions along the east and west legs of the triangle propagated through a frontal region, which was located at 34°–38° N. CTD and XBT data collected on the deployment and recovery cruises suggests that the frontal region was stable throughout the experiment. The difference in the travel times of oppositely traveling signals depends on the current field. Since the experimental area has both a low mesoscale energy level and low mean currents, the differential travel times are dominated by tidal currents. Tidal currents determined acoustically agree well with Schwiderski's tidal model and with barotropic tidal currents determined from a current meter mooring located approximately in the center of the northern leg of the triangle. At frequencies below one cycle per day the differential travel times give depth and range-averaged currents of a few mm/s, with an eastward current along the northern leg. The circulation around the entire triangle corresponds to relative vorticities of about $10^{-8} s^{-1}$, of the same magnitude as expected from Sverdrup dynamics.

1. Introduction

This experiment is the first to measure reciprocal acoustic transmissions over gyre scale ranges. Related experiments conducted in 1976 (Worcester, 1977) and 1983 (Howe et al., 1987) were at much shorter ranges. In 1976 the first midocean reciprocal transmissions at 25 km range were made with instruments suspended from ships. In 1983 two moored instruments separated by 300 km yielded a 21 day time series of reciprocal travel times. In the 1987 reciprocal acoustic transmission experiment three second generation moored transcievers were deployed for four months in a triangle at ranges of 745 km, 995 km and 1275 km. All three instruments transmitted

119

J. Potter and A. Warn-Varnas (eds.), Ocean Variability & Acoustic Propagation, 119–134.
© 1991 *Kluwer Academic Publishers.*

simultaneously, providing reciprocal transmissions along all three legs of the triangle. In addition, instrument locations were surveyed to within a few tens of meters using the NAVSTAR Global Positioning System, making possible comparison of the measured travel times with acoustic predictions. Preliminary results on the currents obtained from the differential travel times measured in this experiment were previously reported by Worcester et al. (in press).

Section 2 gives a brief description of the acoustic travel time forward problem. Section 3 is a description of the experiment. In section 4, the absolute travel times are compared with predictions based on sound speed fields calculated using the sound speed algorithms of Del Grosso (1974) and Chen and Millero (1977). CTD data obtained at the time of transceiver deployment provide the temperature and salinity fields. In section 5, tidal amplitudes and phases estimated from differential travel times are shown to agree well with estimates made from current meter data and from a numerical model of the tides. In section 6, variability of vertical arrival angle is shown. In sections 7 and 8 inversions of low frequency travel times give depth and range averaged sound speed changes and currents. Finally, section 9 shows the estimated gyre-scale areal averaged relative vorticity derived from the modeled currents to be of order $10^{-8} s^{-1}$.

2. Theory

Acoustic propagation varies due to perturbation of the sound speed and current fields by gyre or mesoscale processes, internal waves, inertial currents, tides, wind forcing, and other oceanographic processes. The travel time T_i for ray path Γ_i is given by

$$T_i(t) = \int_{\Gamma_i} \frac{ds}{c_0(\mathbf{x}) + \delta c(\mathbf{x},t) + \mathbf{u}(\mathbf{x},t)\tau},$$

where $c_0(\mathbf{x})$ is a reference sound speed profile, $\mathbf{u}(\mathbf{x},t)\tau$ is the component of current velocity in the direction of the ray path, $\delta c(\mathbf{x},t)$ is the sound speed perturbation, ds is an increment of arc length, and t is geophysical time. Fluctuations in travel time are due to sound speed perturbations $\delta c(\mathbf{x},t)$ or currents $\mathbf{u}(\mathbf{x},t)$.

Forming the sum and difference in travel time of oppositely traveling pulses shows that the sum travel time depends linearly on the sound speed perturbation field and the differential travel time depends linearly on the current field. Expanding the above equation in the small parameters $\delta c(\mathbf{x},t)$ and $\mathbf{u}(\mathbf{x},t)\tau$, and forming the sum of the travel times of oppositely traveling pulses gives

$$(T_i^+ + T_i^-) \approx 2T_0 - 2\int_{\Gamma_i} \frac{\delta c(\mathbf{x})}{c_0^2(\mathbf{x})} ds,$$

where T_0 is the travel time for the reference profile. This equation is approximate because the path Γ_i is the path determined by $c_0(\mathbf{x})$ only; it is assumed the perturbation $\delta c(\mathbf{x},t)$ and current $\mathbf{u}(\mathbf{x},t)\tau$ do not significantly change the ray path.

Forming the difference of the travel times of oppositely traveling pulses gives

$$(T_i^+ - T_i^-) \approx -2\int_{\Gamma_i} \frac{\mathbf{u}(\mathbf{x})\tau}{c_0^2(\mathbf{x})} ds.$$

Since sum and difference travel times can be approximated as linear functions of the unknown perturbations $\delta c(x,t)$ and current $u(x,t)$ along the ray path, all of the techniques of linear inverse theory are applicable to compute the sound speed perturbation field and current field from the measured sum and difference travel times. These inversions are usually made for the low frequency, or large scale, component of the travel times. Appendix 1 gives a brief review of the inverse method used here.

The structure of the ray paths determines their sensitivity to internal waves and the amount of spatial information available. For example (and unfortunately), in the 1987 experiment only the deep turning, surface reflected rays have been resolved, so all rays sample the thermocline (and internal wave field) about equally and only range and depth averaged estimates of the current and sound speed fields are possible. Sum travel times from rays that turn at a variety of depths are required to give information on the vertical structure. With an adequate array of transceivers the three dimensional field of sound speed (or currents) can be computed for the sampled region.

3. The Experiment

Three acoustic transciever moorings were deployed north of Hawaii in a triangle extending from the Subtropical Front to the Subarctic Front from May through September 1987 (Fig. 1). The experiment was located in this area because the mesoscale variability levels are relatively low, reducing mesoscale noise in measurements of gyre-scale averages. Two or three fronts were present in the experiment area. The experiment was co-located with the Barotropic ElectroMagnetic and Pressure EXperiment (BEMPEX), but only overlapped with it during May and June 1987 (Luther et al., 1987). A subsurface current meter mooring with Vector Measuring Current Meters (VMCM's) at about 73, 173, 943, 2498, 5650, and 5722 m was deployed at 40.647°N, 163.025°W, between the northern transciever moorings, from September 1986 through September 1987. As part of BEMPEX, a number of bottom mounted electrometers, magnetometers, and pressure sensors were deployed in the vicinity of the current meter mooring.

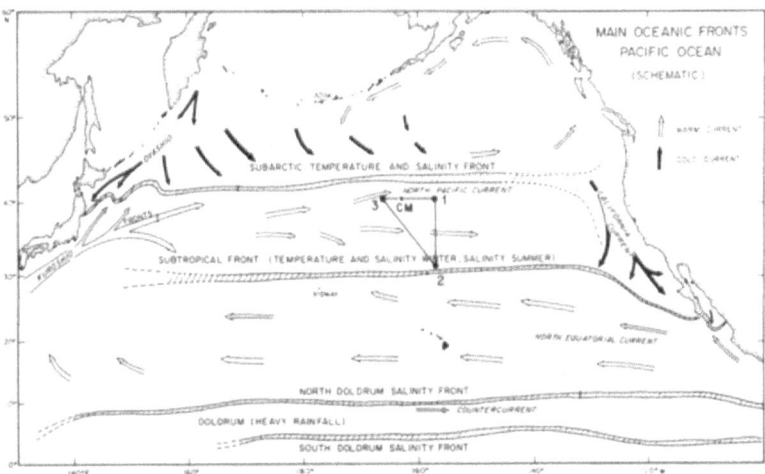

Fig. 1. The geometry of the 1987 gyre-scale reciprocal acoustic transmission experiment, with acoustic transceivers at locations 1, 2, and 3. A current meter mooring is on the northern leg of the acoustic triangle, at location CM. A bottom mounted electrometer was also located at CM as part of BEMPEX. (Reproduced from Worcester et al. (in press), as modified from Roden (1975).)

Each acoustic transciever consisted of an acoustic source with 100 Hz bandwidth at 250 Hz center frequency and a four-element receiving array with 3/2λ separation between elements (Worcester et al., 1985a). Pulse compression techniques were used to improve the signal-to-noise ratio without degrading the travel time resolution (Worcester et al., 1985b). The effective resolution between ray arrivals was that of a pulse containing four cycles at 250 Hz, i.e., 16 ms.

The transcievers were moored near the depth of the sound channel axis. The motion of each transciever was tracked using bottom-mounted acoustic transponders, allowing the pulse travel times to be corrected for transciever motion. Daily average arrival patterns for the north leg show a set of early arrivals that are stable and well-resolved (Fig. 2). The pattern is quite complex immediately preceding the final cut-off, however.

The transmissions in opposite directions were found to be quite reciprocal at all three ranges, except for changes in travel times due to currents. Daily average arrival patterns from oppositely traveling transmissions show similar structure at both receivers for all three legs (Fig. 3). The signal-to-noise ratio is relatively low for transmissions between Moorings 2 and 3 due to the long range (1275 km), making the low amplitude early arrivals difficult to discern.

The initial and final sound speed fields were determined from XBT and CTD measurements made during the deployment and recovery of the transceivers (Fig. 4). Using these sound speed fields acoustic propagation calculations can be used to predict the travel time arrival pattern, and so ray paths can be identified with pulse arrivals. Prior to the experiment, calculations based on climatological sound speed profiles suggested that the temporal resolution of the instruments would be adequate to resolve purely refracted rays that turned in the main thermocline. Only rays that turned at or near the surface have been resolved to date, however. The ray paths then cycle through almost the entire water column (Fig. 5), and the travel times average the sound speed and current field in both range and depth. Little information is therefore available from the acoustic travel times on the depth structure of the current field for this experiment.

The total number of acoustic transmissions is limited by the energy stored in the batteries that power the sources. The sources transmitted simultaneously at two-hour intervals every fourth day.

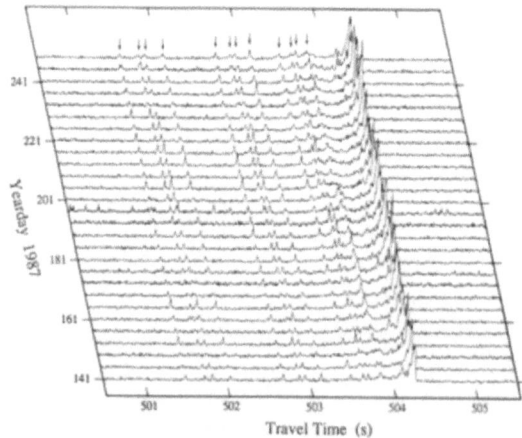

Fig. 2. Daily average amplitudes for transmissions from Mooring 3 to Mooring 1. Path 1 was used to align the arrivals prior to averaging.

4. Absolute Travel Times and a Comparison of Two Sound Speed Algorithms

Using the CTD data taken during experiment deployment, different sound speed fields can be calculated using the algorithms of Del Grosso and Chen and Millero. The predictions of absolute travel times and pulse arrival patterns using the Del Grosso sound speed fields agree best with the measurements.

The difference between the Del Grosso and the Chen and Millero sound speed profiles increases with depth, with Del Grosso slower (Fig. 6). This difference causes the predicted pulse arrival patterns for the Chen and Millero profile to be more dispersed in time. The range averaged sound speed profiles were used with an acoustic normal mode code to predict the arrival patterns for all three legs. The predicted pulse arrival patterns for the Del Grosso profile agree well with the measurements, while the Chen and Millero patterns disagree by a few tenths of a second (Fig. 7). No travel time adjustment has been made to align the patterns. The predicted pattern uses the measured sound speed field and range; the measured pattern uses the measured absolute travel times. The north leg agrees especially well since it is almost range-independent, consistent with the range-average sound speed profile used to make the prediction. The east and west legs do not agree quite as well with the Del Grosso prediction since they contain significant range-dependence, but Del Grosso still gives significantly better results than Chen and Millero.

Using the range dependent sound speed field, ray tracing was used to obtain predictions of absolute travel times for the resolved rays. The range-dependent prediction for the east leg using the Del Grosso algorithm agrees significantly better with the measured arrival pattern than range-independent predictions (Fig. 8). The predictions based on the Chen and Millero algorithm differ by 0.3 sec from the measured absolute travel times, while the predictions based on the Del Grosso algorithm differ from measurement by about 30 msec (Table I). The results in Table I are averages over adjacent rays that turn at approximately the same depths.

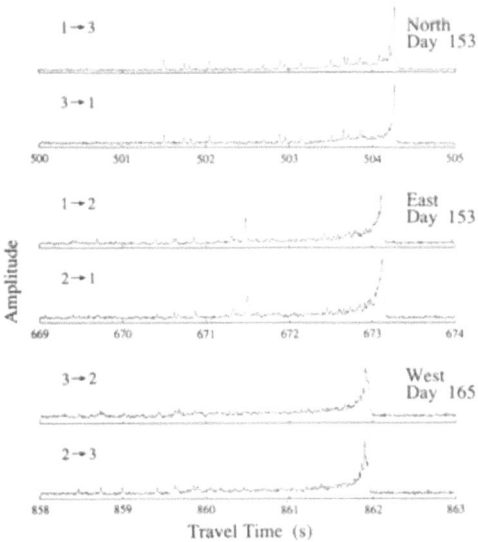

Fig. 3. Daily average reciprocal arrival patterns for the day indicated and for each path of the triangle. The day is chosen to have clear arrivals. The time scale is absolute measured travel time.

124

The error in predicted absolute travel times due to mooring positioning error is roughly 20 msec. This accuracy is achieved through the use of the Global Positioning System (GPS) to locate the mooring and transponder positions. The error of the CTD measurements is 0.003°C, 0.002 ‰ for salinity and 3 decibars for pressure. The temperature error dominates the other error sources in the calculation of sound speed, giving an error of 0.014 m/s in sound speed, which is an order of magnitude smaller than the difference between the two algorithms. The largest error is introduced by attempting to estimate the range dependent sound speed field from a discrete set of CTD and XBT profiles. Internal wave induced temperature fluctuations significantly exceed 0.003°C, for example.

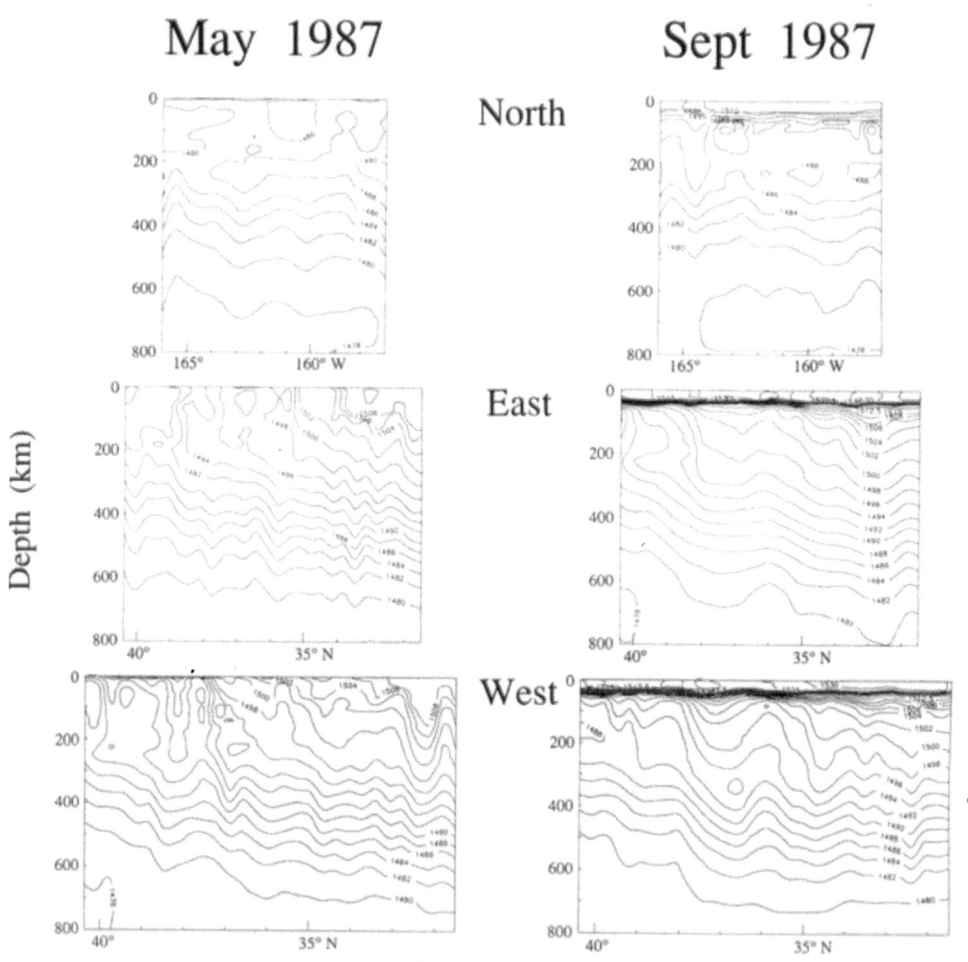

Fig. 4. Initial and final sound speed fields for all legs computed using the Chen and Millero algorithm on the XBT and CTD data. The CTD data was used to construct an average T-S relation for each leg, and the T-S relation was used to assign salinity to the XBT measurements. Three or four fronts are present on the east and west legs. The summer thermocline has formed by recovery.

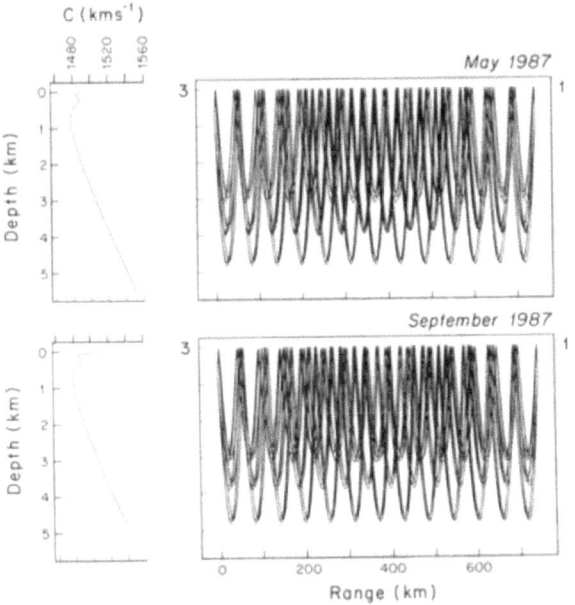

Fig. 5. Sound speed profiles and ray paths for the northern leg of the triangle. Only resolved ray paths are shown. The sound speed profile changes due to the formation of the summer thermocline. In May all of the ray paths are reflected from the surface, but in September some of the paths have turning points at the bottom of the summer thermocline. (Reproduced from Worcester et al. (in press).)

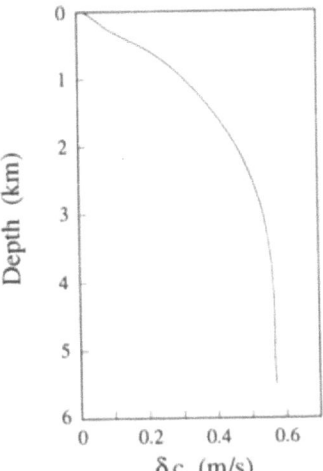

Fig. 6. The range-averaged sound speed profile for the North leg computed from CTD data using the Chen and Millero algorithm minus that computed used the Del Grosso algorithm. The Del Grosso algorithm gives lower velocity at depth.

126

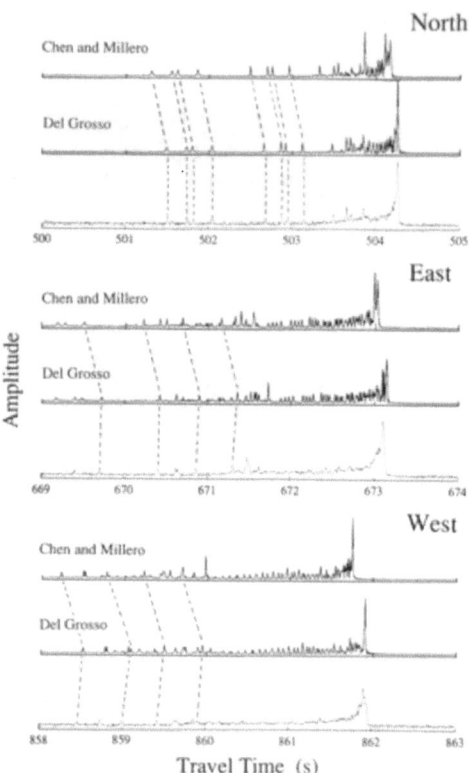

Fig. 7. Measured and predicted arrival patterns for all three legs using the sound speed algorithms of Chen and Millero and Del Grosso. CTD data obtained during deployment are used to compute range-averaged sound speed profiles. The predictions were obtained using a broadband, range-independent, normal mode code. Predicted arrival patterns using the Chen and Millero algorithm show greater dispersion and have shorter travel times than those predicted using the Del Grosso algorithm.

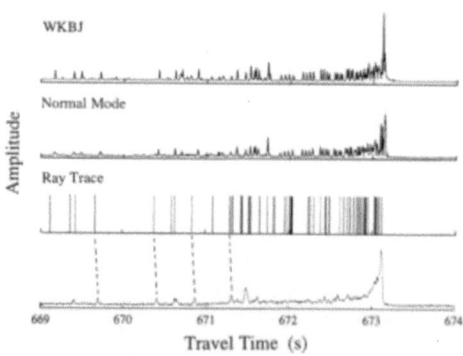

Fig. 8. Comparison of three acoustic propagation predictions with the measured arrival pattern for the east leg. Measurements and predictions are made for data taken at the time of deployment.

TABLE I.
Comparison of daily average sum travel times with predictions
made using the Del Grosso and Chen and Millero
sound speed algorithms.

		$\Delta T = T_{measured} - T_{predicted}$	
		Del Grosso (ms)	Chen and Millero (ms)
North leg	(Rays 1-4)	36	236
Day 141	(Rays 5-8)	46	249
East leg	(Rays 1,2,4)	22	231
Day 153	(Rays 5-8)	27	225
West leg	(Rays 1,2,4)	29	294
Day 165			

5. Measured Travel Times

The measured travel times were filtered into low (< 1 cy/day) (i.e., a daily average) and high frequencies. The changes in the low frequency sum travel times show the formation of the summer
thermocline (Fig. 9). In May, winter storms had deepened the mixed layer to the main thermocline. By September, the summer thermocline had formed giving a roughly 60 m deep surface
layer about 7 degrees warmer than at the start of the experiment. The change in travel times are
consistent with this temperature change in the surface layer.

Low frequency differential travel times show a variability of a few msec (Fig. 10).

6. Fluctuations in Pulse Arrival Angle

Fluctuations in pulse arrival angle, amplitude, and phase were also measured. Presented here are
the measurements of vertical arrival angle. Arrival angle is more sensitive to non-reciprocity than
are the travel times. The vertical arrival angle as a function of time for several rays shows little
variation over the duration of the experiment (Fig. 11). Table II compares measured and
predicted values for the north leg. The measured arrival angles agree well with predictions. The
arrival angle fluctuations are close to the theoretical precision with which arrival angle can be
measured with the simple vertical array used in this experiment (Worcester et al., 1985b),
although internal wave induced fluctuations are also certainly present. Similar results were
obtained for the other two legs.

7. Tidal Estimates

This experiment is the first to measure a significant tidal component in differential travel times.
Previous estimates of tidal components involved one-way travel times (Munk et al., 1981), while
the acoustic path in the 1983 reciprocal acoustic transmission experiment was oriented perpendicular to the principle tidal axis (Howe et al., 1987).

The amplitudes and phases of the tidal constituents estimated from the high frequency differential travel times agree well with estimates from the VMCM data and from a numerical model of the tides. Tidal amplitudes and phases were estimated separately for each ray path by least squares fitting the differential travel times to a sum of sines and cosines with unknown amplitudes at eight tidal frequencies. Table III shows the average of the ray path estimates for the north leg; the error bars are standard deviation of those estimates. These results agree well in both amplitude and phase with estimates made from the VMCM data obtained at the current meter mooring approximately in the center of the north leg (Chave et al., in press) and with Schwiderski's numerical model of the tides (Schwiderski, 1980a and 1980b).

Estimates of tidal amplitudes using the VMCM data are smaller than those using the differential travel times. One possible cause of this discrepancy is the frequent stalls occurring on the deep current meters. This impeded extraction of the barotropic tides from the VMCM data and reduced their amplitude. Another possible explanation is that the VMCM data is not corrected for mooring motion; this would reduce the estimated tidal amplitudes slightly. The acoustic tidal estimate has been corrected for mooring motion occurring during the transmission.

8. Inversions of Low Frequency Sum Travel Times

Inversions of the sum travel times using the technique described in Appendix 1 yield the depth and range average change in the sound speed field (Fig. 12). The modeled sound speed changes are consistent with the formation of the summer thermocline. A sudden change in the sound speed field around yearday 200 is observed on the north leg, but not on the other two legs. One possible explanation is the movement of a front across the north leg. Another is that the north leg may be subject to greater relative warming than the other two legs.

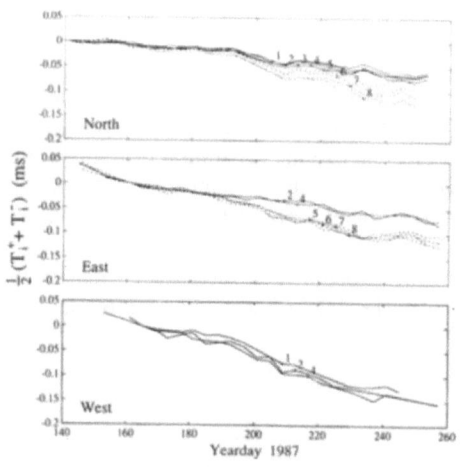 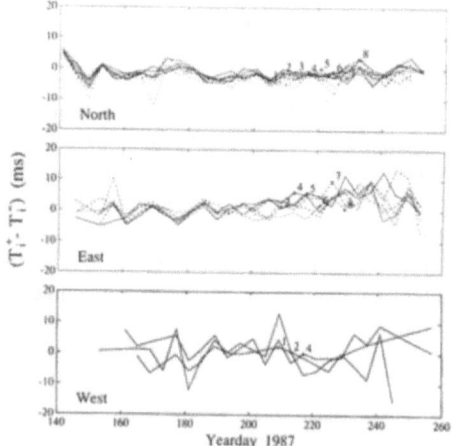

Fig. 9. Change in daily average sum travel times for each leg. Changes are relative to the day indicated. Only the best paths are shown.

Fig. 10. Daily average differential travel times for each leg. Only the best paths are shown.

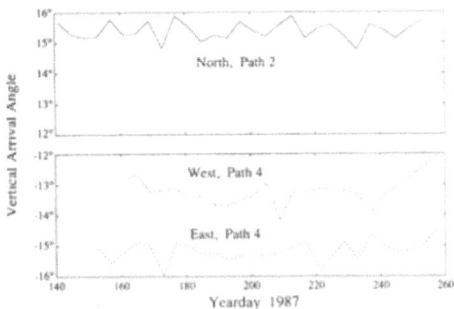

Fig. 11. Daily average vertical arrival angles for a typical ray path from each leg. Since all of the resolved ray paths are surface-reflected or nearly surface reflected, the arrival angles are relatively large.

TABLE II
Measured and predicted vertical arrival
angles θ for transmissions
from Mooring 1 to Mooring 3.

Ray Path	$\theta_{measured}$ (degrees)	$\theta_{predicted}$ (degrees)
1	15.6 ± 0.49	16.3
2	15.4 ± 0.55	16.1
3	-15.2 ± 0.50	-16.0
4	-15.0 ± 0.56	-15.8
5	13.3 ± 0.50	14.1
6	13.0 ± 0.52	13.6
7	-12.8 ± 0.55	-13.7
8	-12.6 ± 0.57	-13.4
9	10.8 ± 0.70	11.5
10	10.7 ± 1.01	11.2
11	-10.4 ± 0.77	-11.0
12	-10.0 ± 0.74	-10.6

9. Inversions of Low Frequency Differential Travel Times

Inversions of the differential travel times yield the depth and range averaged currents (Fig. 13). The error limit for these modeled currents are shown in the plots. The variations in currents are significant. There are no conventional methods which can easily measure these range averaged currents.

The acoustic derived currents from the north leg can be compared with some of the BEMPEX current measurements (Chave et al., in press). Fig. 14 shows the acoustic currents compared to the barotropic current estimated from the VMCM data and the inherently barotropic electromagnetic field measurements. The VMCM data is shown as a daily average value every fourth day and the electric field data has been low pass filtered by using a one day running mean filter. The agreement is encouraging in spite of the fact that range averaged current measurements are not expected to agree precisely with point current measurements.

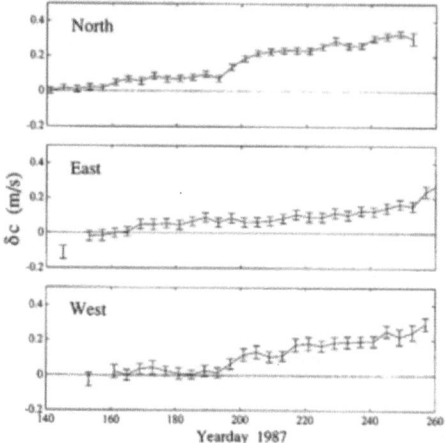

Fig. 12. Depth and range average change in sound speed computed from the sum travel times measured on each leg of the triangle in RTE87. The change in sound speed is relative to the day indicated. Points separated by the four days between transmissions are connected by lines. Gaps larger than four days are not connected.

TABLE III

Amplitude and Greenwich Epoch of the eastward barotropic
tidal current components for the north leg.

Tidal Component		Acoustic Tomography	Current Meter Mooring	Schwiderski Numerical Model
M_2	cm/s	1.23±0.13	0.93±0.02	1.28
	°G	221±5	217±1.4	222
S_2	cm/s	0.52±0.09	0.39±0.02	0.66
	°G	268±15	277±3.2	270
N_2	cm/s	0.17±0.06	0.11±0.02	0.16
	°G	194±37	213±12.1	184
K_2	cm/s	0.22±0.15	0.12±0.02	—
	°G	248±28	279±10.7	—
O_1	cm/s	0.54±0.06	0.39±0.02	0.33
	°G	101±3	119±3.2	99
K_1	cm/s	0.80±0.10	0.61±0.02	0.45
	°G	128±9	140±2.1	127
P_1	cm/s	0.24±0.08	0.19±0.02	—
	°G	135±36	132±6.8	—
Q_1	cm/s	0.19±0.16	0.05±0.02	—
	°G	45±37	18±25.3	—

10. Circulation

Reciprocal transmissions around a triangle or other closed figure give the circulation. By Stoke's theorem, this is equivalent to the areal averaged relative vorticity. Rossby made estimates of achievable precisions some years ago and concluded that oceanic vorticity meters were practical over a wide range of scales (Rossby, 1975). DeFerrari was the first to measure vorticity in this way, with triangular transciever arrays in the Straits of Florida (DeFerrari and Nguyen, 1986; Ko et al., 1989). His first array had approximately 20 km legs; the second 50 km legs. The ray paths in both experiments interacted heavily with the bottom, but it still proved possible to interpret the results.

This experiment is the first to measure circulation over gyre scales. The areal averaged relative vorticity was estimated by integrating the depth and range averaged currents around the triangle as in Stoke's theorem (Fig. 13). This estimate is robust to changes in model assumptions. The estimated error bars indicate that the observed fluctuations in relative vorticity are significant, although these results are preliminary. The Sverdrup balance between wind stress curl and vorticity gives estimates of vorticity of order $10^{-8}s^{-1}$, of the same order as these estimates.

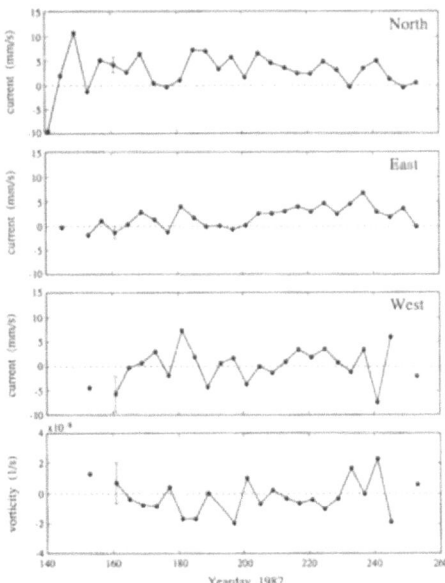

Fig. 13. Depth-averaged currents and areal-average relative vorticity computed from the differential travel times measured on each leg of the triangle in RTE87. Times at which measurements are available are marked by dots. Points separated by the four days between transmissions are connected by lines. Gaps larger than four days are not connected. (Reproduced from Worcester et al., in press).

132

11. Discussion

Gyre scale reciprocal transmissions produce estimates of sound speed and current fields. With travel times for resolved rays that turn at a variety of depths and an adequate array of transcievers, depth and range dependent sound speed and current fields can be calculated. The long range transmissions in this experiment are all surface-reflected, however, and therefore inherently average in both range and depth. The estimated sound speed fields can in turn be used to estimate average heat content over a large region.

The 1987 gyre scale reciprocal transmission experiment has measured the change in sound speed field over the duration of the experiment. This change is due to the formation of the summer thermocline. The change is surprisingly different for each leg of the triangle, indicating that for an accurate estimate of change in heat content measurements along more sections are needed.

The current measurements are dominated by the tidal signal. Estimates of tidal amplitude and phase agree well with independent estimates, thus confirming the utility of long-range differential travel times in estimating currents. The currents based on the low frequency differential travel times are a few mm/s in magnitude and show considerable variability. The estimates of gyre scale relative vorticity were made from these low frequency currents.

The long range transmissions allow one to distinguish between the Chen and Millero and Del Grosso sound speed algorithms. The Del Grosso algorithm has been shown to be more accurate. The predictions based on the Del Grosso algorithm agree well with the mostly range independent north leg transmissions, indicating that it is an accurate algorithm. The difference between the predicted and measured travel times can be used to make corrections to the sound speed algorithm.

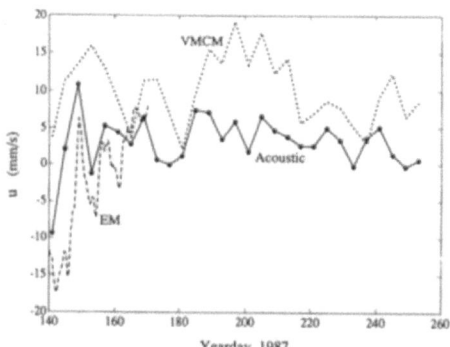

Fig. 14. Low frequency currents computed from the differential travel times measured along the northern leg of the triangle in RTE87, from the current meter mooring at location CM (Fig. 1), and from BEMPEX electrometer EB near the current meter mooring. The currents are approximately barotropic, although the vertical weighting differs slightly for each type of measurement. (Reproduced from Worcester et al., in press).

Appendix. A Brief Summary of the Inverse Method Used

Presented here is a brief summary of the inverse method used for the sum travel times. The formulation for the measured differential travel times is almost the same. Assume

$$\delta c(x,z) = \delta c(z) = \sum_k A_k \eta_k(z),$$

for example. This is a simple range independent model of the ocean. The $\eta_k(z)$ could be any or all of layers, quasi-geostrophic modes, etc. Then,

$$(T_i^+ + T_i^-) = 2T_0 - 2\sum_k A_k \int_{\Gamma_i} \frac{\eta_k(z)}{c_0^2(x)} ds, \text{ or}$$

$$t_i = \sum_k G_{ik} A_k + \varepsilon_i, \text{ where}$$

$$G_{ik} = -2 \int_{\Gamma_i} \frac{\eta_k(z)}{c_0^2(x)} ds,$$

$t_i = (T_i^+ + T_i^-) - 2T_0$ and ε_i is data noise.

Inverse theory provides a solution for the A_k from the t_i. In acoustic tomography this problem is most often underdetermined. We assume an a priori model covariance matrix, $(R_{mm})_{kk} = <(A_k)^2>$ diagonal, and an a priori data noise covariance matrix $(R_{nn})_{ii} = <(\varepsilon_i)^2>$, diagonal. With these define

$$R_{dd} = G R_{mm} G^T + R_{nn}, \text{ and}$$

$$G_p^{-1} = L = R_{mm} G^T R_{dd}^{-1}.$$

The G_p^{-1} is the inverse operator particular to the choice of ocean model and a priori variances. The model solution is

$$\hat{m} = L t, t \text{ a column vector.}$$

The model error covariance matrix is

$$E = R_{mm} - L G R_{mm}.$$

$L G$ is the resolution matrix: $\hat{m} = L G m$, where m is the "true" solution for the A_k's.
The model solution can be transformed to a depth average by an "integral" operator W.

$$\bar{m} = W^T \hat{m}, \text{ and}$$

$$\bar{E} = W^T E W.$$

Here $\bar{m}$ and $\bar{E}$ are scalars.
The major difficulty with the solution $\hat{m}$ is that it is often sensitive to the choice of ocean model and assumed a priori variances.

Acknowledgments

D. Behringer processed the CTD data obtained on the transceiver recovery cruise. A. Chave, J. Filloux, and D. Luther graciously gave permission to show the low frequency currents computed from the electrometer deployed at location EB in BEMPEX. A. Chave and D. Luther provided the tidal analysis of the current meter data. The current meter mooring was prepared by the Instrument Development Group at Scripps Institution of Oceanography. B. Cornuelle and W. Munk participated in many useful discussions during the analysis of the data. R. Spindel played a key role in the early design of the experiment. Credit for the success of the experiment belongs largely to the dedicated personnel who designed, fabricated, tested, and fielded the equipment: S. Abbott, K. Hardy, D. Horwitt, J. Kemp, S. Liberatore, D. Peckham, and R. Truesdale. D. Betts did the illustrations. B. Ma helped with the programming.

This work was supported by National Science Foundation Grants OCE-82-14918 and OCE-84-14978 and Office of Naval Research Contracts N00014-80-C-0217, N00014-84-G-0214, and N00014-87-K-0120.

References

Chave, A. D., Luther, D. S. and Filloux, J. H. (in press) 'Spatially-averaged velocity from the seafloor horizontal electric field', *Proceedings of the Fourth IEEE Working Conference on Current Measurement*, Clinton, MD, April 3–5, 1990.

Chen, Chen-Tung, and Millero, F. J. (1977) 'Speed of sound in seawater at high pressures', *J. Acoust. Soc. Am., 62*, 1129–1135.

DeFerrari, H.A. and Nguyen, H.B. (1986) 'Acoustic reciprocal transmission experiments, Florida Straits,' *J. Acoust. Soc. Am., Vol. 79*, 299–315.

Del Grosso, V. A. (1974) 'New equation for the speed of sound in natural waters (with comparisons to other equations)', *J. Acoust. Soc. Am., 56*, 1084–1091.

Howe, B. M., Worcester, P. F. and Spindel, R. C. (1987) 'Ocean acoustic tomography: Mesoscale velocity', *J. Geophys. Res., Vol. 92*, 3785–3805.

Ko, D.S., DeFerrari, H.A. and P. Malanotte-Rizzoli (1989) 'Acoustic tomography in the Florida Straits: Temperature, current, and vorticity measurements', *J. Geophys. Res., Vol. 94*, 6197–6211.

Luther, D.S., Chave, A.D. and Filloux, J.H. (1987) 'BEMPEX: A study of barotropic ocean currents and lithospheric electrical conductivity', *EOS, Vol. 68*, 618–619; 628–629.

Munk, W. H., Zetler, B., Clark, J. C., Porter, D., Spiesberger, J., and Spindel, R. (1981) 'Tidal effects on long-range sound transmission', *J. Geophys. Res., Vol. 86*, 6399–6410.

Roden, G.I. (1975) 'On North Pacific temperature, salinity, sound velocity and density fronts and their relation to the wind and energy flux fields', *J. Phys. Oceanogr., Vol. 5*, 557–571.

Rossby, T. (1975) 'An oceanic vorticity meter', *J. Marine Res., Vol. 33*, 213–222.

Schwiderski, E. W. (1980a) 'Ocean Tides, Part I — Global tidal equations', *Marine Geodesy, 3*, 161–217.

Schwiderski, E. W. (1980b) 'Ocean Tides, Part II — A hydrodynamical interpolation model', *Marine Geodesy, 3*, 219–255.

Worcester, P. F. (1977) 'Reciprocal acoustic transmission in a midocean environment', *J. Acoust. Soc. Am., 62*, 895–905.

Worcester, P. F., Dushaw, B. and Howe, B. M. (in press) 'Gyre-scale current measurements using reciprocal acoustic transmissions', *Proceedings of the Fourth IEEE Working Conference on Current Measurement*, Clinton, MD. April 3–5, 1990.

Worcester, P. F., Peckham, D. A., Hardy, K. R. and Dormer, F. O. (1985a) 'AVATAR: Second-generation transceiver electronics for ocean acoustic tomography', *OCEANS 85 Conference Record*, San Diego, CA, Nov. 12–14, 1985, 654–662.

Worcester, P. F., Spindel, R. C. and Howe, B. M. (1985b) 'Reciprocal acoustic transmissions: Instrumentation for mesoscale monitoring of ocean currents', *IEEE J. Oceanic Eng., Vol. OE-10*, 123–137.

Summary of Session 1

Briscoe and Ewart

The eight papers presented Monday covered topics including internal waves, acoustically-relevant descriptions of internal waves, the acoustic effects of internal waves and other processes, ocean acoustic experiments, and the use of acoustics to infer, or directly observe ocean processes (e.g. acoustic scintillation and tomography inverses). Appropriate partitions between deterministic processes (e.g. tides and fronts) and stochastic processes was an important theme. This theme and many others continued through the week, hence, it is not useful to summarize the discussions held only on Monday.

From the perspective of the week of discussions, but emphasizing the topics of Monday, we feel that the participants presented views of ocean systems that were either rather simple or rather complicated. Many of the latter did not appear to be systems describable as stochastically perturbed deterministic system, but rather as systems undescribable in any deterministic sense. They could also not be characterized by stationary random processes. We have attempted in Table 1 to summarize the extent of this parameter space.

Table 1 *The parameter space of oceanography from the perspective of deterministic-statistical descriptions*

	Deterministically describable	'Undescribable'	Statistically describable
Simple Ocean	OK (1)	(4)	OK (2)
Complicated Ocean	(3)	(4)	OK (2)

(1) *No problems. This is the regime desired, often hypothesized, usually modelled, and rarely observed.*

(2) *No first-order problems. The tools to make the oceanographic descriptions and to analyze the acoustic effects are mostly in hand. This is the regime most likely to allow the acoustics-to-oceanographic inverse results.*

(3) *'Tomographic Regime'. The kind of oceanography that acoustic tomography may best be suited for observing. This is the regime that numerical dynamical models are also trying to predict.*

(4) *Examples are observed in thermistor-chain temperature sections in the Baltic and in the Iceland-Faeroes area: the oceanographic structure is determined by density fields, but the temperature fields depend upon the intrusive interleaving of complicated, mixing water masses. One hopes with time that processes in (4) might move into the (1), (2) or (3) regimes as oceanographic and acoustic understanding improves.*

Our sense of the discussions is that the principal uses of the deterministic system approach will be for naval applications. The separable deterministic/stochastic system approach is being used and will be used for acoustical oceanography. This ranges from the stochastic inverse approaches, which can strongly supplement oceanographic measurements, to long range determination of the

J. Potter and A. Warn-Varnas (eds.), Ocean Variability & Acoustic Propagation, 135–136.
© 1991 *Kluwer Academic Publishers.*

energy levels of ocean variability. Acoustic tomography may be an exception if it can produce deterministic snapshots of the mesoscale.

An exciting point of discussion throughout the week was the idea of using ambient noise to image the large scale ocean structure using the ocean and its topography as a 'lens'. Another exciting issue was the introduction of non-spectral (fractal or chaotic) methods to describe ocean process, thus dragging us from a 2-point statistics world.

In summary, it appears that the most immediate exciting results will come from the 'statistically describable' regime of Table 1, but the long-range most important results will likely come from the 'presently undescribable' regime. We want to strongly recommend, based on the discussions, that combined ocean-acoustics experiments (carefully designed to be process oriented) be undertaken in the 'statistically describable' and 'undescribable' regimes. The purpose of the latter is to probe the limits of our lack of understanding and develop a high quality data base; purpose of the former to test ocean acoustic theoretical models and help confirm statistical oceanographic models.

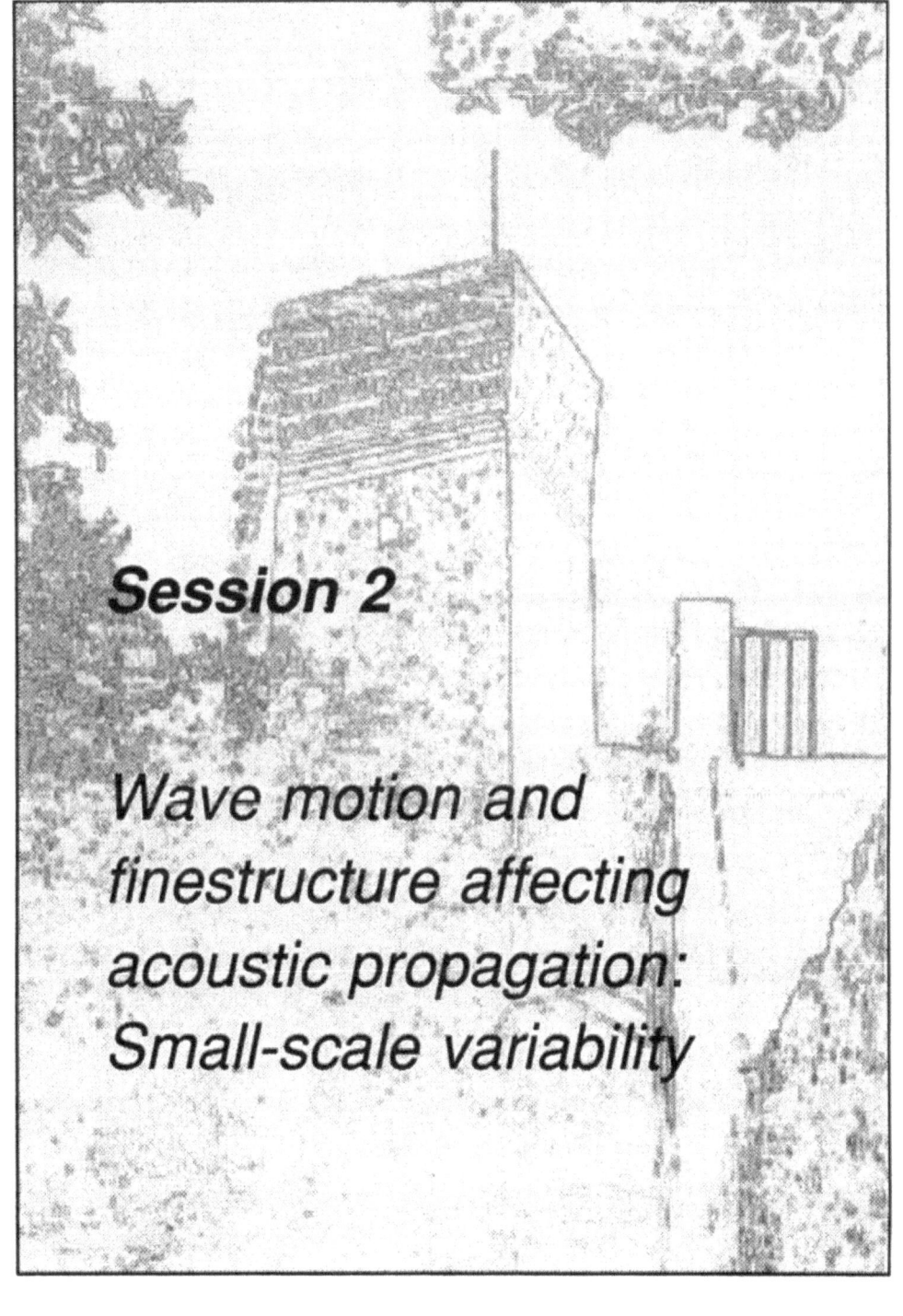

Session 2

Wave motion and finestructure affecting acoustic propagation: Small-scale variability

CHAOS IN UNDERWATER ACOUSTICS

MICHAEL G. BROWN, FREDERICK D. TAPPERT, GUSTAVO J. GONI, AND KEVIN B. SMITH

Rosenstiel School of Marine and Atmospheric Science
University of Miami, 4600 Rickenbacker Cswy.
Miami, FL   33149

ABSTRACT. The range–dependent sound propagation problem differs in a fundamental way from the range–independent problem, at least in the geometric limit. In the range–independent problem the divergence between neighboring ray trajectories is slow (power law behavior). In the generic range–dependent problem some ray trajectories exhibit extreme sensitivity to initial conditions (chaotic behavior) wherein neighboring trajectories diverge exponentially. Using simple range–dependent models of oceanic waveguides, involving both volume and boundary structure, the chaotic behavior of sound rays is discussed and illustrated via the construction of Poincare sections, the calculation of power spectra, and the calculation of Lyapunov exponents. It is argued that under chaotic conditions ray trajectories are not computable beyond some finite predictability horizon. Under conditions in which ray trajectories are predominently chaotic, the growth in range of the complexity of the wavefield is shown to be exponential: the number of eigenrays connecting a fixed source and receiver grows exponentially while the average intensity of ray arrivals decays exponentially. Some preliminary attempts to investigate the extent to which the chaotic behavior of ray trajectories carries over to finite frequency wavefields are discussed.

1.   INTRODUCTION

When it is assumed that the ocean's sound speed varies as a function of depth only and the ocean boundaries coincide with surfaces of constant depth, then the acoustic wave equation can be solved by a variety of techniques which depend on separating variables. This assumption is frequently unjustified. The solution to the range–dependent problem differs in a fundamental way from the solution to the range–independent problem, at least in the geometric limit. In the range–dependent problem there is no Snell invariant, and ray trajectories may exhibit chaotic motion, i.e., extreme sensitivity to initial conditions. This extreme sensitivity to initial conditions leads to a limited ability to predict acoustic fields.

In the following sections these ideas are illustrated by examining

139

J. Potter and A. Warn-Varnas (eds.), Ocean Variability & Acoustic Propagation, 139–160.
© 1991 *Kluwer Academic Publishers.*

sound ray propagation in two simple range–dependent oceanic waveguides. In the first, the ocean boundaries are flat but the sound speed in the ocean volume depends on both depth and range. The second waveguide consists of a range–independent downward refracting ocean volume overlying a bottom with range–dependent bathymetric variations. The latter problem is examined by introducing an area–preserving mapping to replace the ray equations. Both problems are discussed and analyzed in the context of recent developments in the studies of integrable and nonintegrable Hamiltonian systems. Our emphasis on Hamiltonian dynamics is natural inasmuch as the acoustic ray equations have Hamiltonian form.

Detailed modern discussions of Hamiltonian systems can be found in Henon (1983) and Lichtenberg and Lieberman (1982). Our presentation is largely pedagogical, in the hope that readers who are not familiar with these topics will be able to follow the presentation. We state without proof many results which are discussed in the aforementioned references. The novelty of our work is the application of these results to the underwater sound propagation problem. Results of similar studies have been presented by Palmer et al. (1988) and Abdullaev and Zaslavskii (1988).

2. RAY CHAOS IN THE OCEAN VOLUME

The ray equations consistent with the parabolic wave equation are (see Tappert, 1977)

$$\frac{dz}{dr} = \frac{\partial H}{\partial p}, \frac{dp}{dr} = -\frac{\partial H}{\partial z}, \tag{1a,b}$$

where

$$H(z,p,r) = \frac{1}{2} p^2 + V(z,r) . \tag{1c}$$

Here z and r are depth and range, respectively, and, by equations (1), $p = dz/dr = \tan \theta$ where θ is the ray angle with respect to the horizontal. The potential $V(z,r)$ may be thought of as either $\frac{1}{2} (1 - c_0^2/c^2(z,r))$ or $c(z,r)/c_0 - 1$ where $c(z,r)$ is the sound speed and c_0 is a reference value. These equations define a Hamiltonian system with one degree of freedom. Note that range is the time–like variable in these equations. If $V = V(z)$ (range–independent problem), the system is said to be autonomous. If $V = V(z,r)$ (range–dependent problem) the system is said to be nonautonomous. This distinction is crucial. It is most easily understood by examining the geometry of ray trajectories.

2.1 Ray Geometry

Solutions to equations (1) are trajectories $z(r)$, $p(r)$, which lie in the three–dimensional space (z,p,r). We shall confine our attention to the special case of sound speed structures which are periodic in r, $c(z,r) = c(z,r + \lambda)$. Then r can be defined modulo λ without loss of generality. Although this case is special, it suffices to illustrate all of the important ideas. Additionally, in the oceanic waveguides considered here, both ray depth and ray angle are bounded. With these assumptions ray trajectories lie in the bounded three dimensional space (z,p,r). There are two fundamentally different types of trajectory – those which fill volumes in (z,p,r) and those which lie on two–dimensional surfaces in this three dimensional space.

All trajectories are of the latter type (termed "regular") if the system is integrable, i.e., if there exists a constant of the motion I(z,p,r) for which

$$\frac{dI}{dr} = \frac{\partial I}{\partial z}\frac{dz}{dr} + \frac{\partial I}{\partial p}\frac{dp}{dr} + \frac{\partial I}{\partial r} = 0 .$$ (2)

If such a function exists each trajectory lies on a surface of constant I, thereby reducing the dimension of the accessible space from three to two. In a bounded phase space these surfaces are tori. Thus, in an integrable system with a bounded phase space, ray trajectories lie on a set of nested tori.

This is the situation encountered in a range–independent environment. Here H(z,p) is a constant of the motion – the first two terms on the rhs of (2) cancel by (1) and the third term is zero. In range–dependent environments it is extremely rare that there exists a constant of the motion: generically, in range–dependent environments, the acoustic ray equations are nonintegrable. Before discussing this situation further we introduce a simple technique – first used by Poincare – which is now commonly used to distinguish between area and volume filling trajectories. To distinguish between the two cases one need only examine a two–dimensional slice of the three–dimensional space. On this slice, volume and area filling trajectories in (z,p,r) will fill areas and lie on smooth curves, respectively.

For the periodically range dependent problems considered here, the simplest way to construct such a Poincare section is to view the ray trajectories, z(r), p(r), stroboscopically at integer multiples of λ, z(nλ), p(nλ), n = 0,1,2,... Some examples are shown in figure 1. Here, the trajectories, z(r), p(r), were numerically computed for many sets of initial conditions, z(o), p(o), using a perturbed Munk (1974) potential,

$$V(z,r) = \epsilon(e^{-\eta} + \eta - 1) + \delta \frac{2z}{B} e^{-2z/B}\cos (2\pi r/\lambda).$$ (3)

Here $\eta = 2(z - z_a)/B$ is a scaled depth coordinate, z_a = 1 km is the sound channel axis depth, the depth scale B = 1 km, and ϵ = 0.0057. In figure 1 the wavelength of the range dependent perturbation λ was taken to be 10 km while the perturbation strength δ was varied.

The first plot in figure 1 corresponds to the range–independent case, δ = 0. Here, for each set of initial conditions, the succession of points plotted all lie on smooth curves. (In some cases the curves appear to be broken because so few (500) points are plotted.) This is expected because in this case the ray equations are integrable. As the strength of the range–dependent perturbation δ increases, some of these closed curves break up into a series of closed curves ("islands") surrounded by speckled regions ("chaotic seas"). This phenomenon is due to a resonance between unperturbed (δ = 0) rays and the range dependent perturbation. The rays which form the five island structure, for instance, have a wavelength very close to 50 km. The 10 km wavelength perturbation induces a 5:1

142

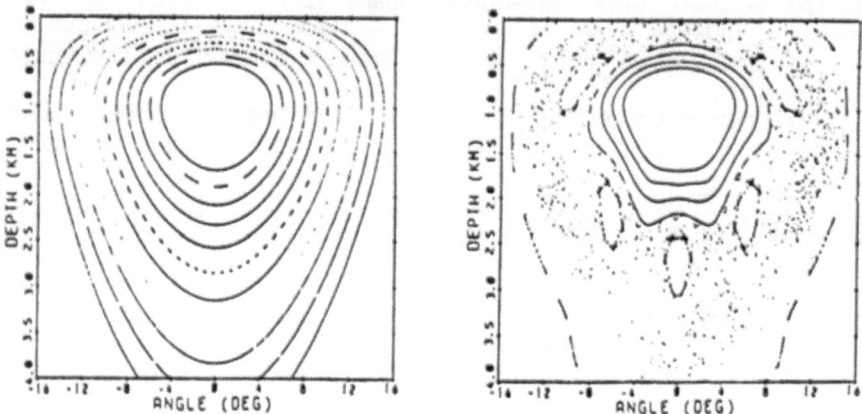

Figure 1. Poincaré sections computed using the potential $V(z,r)$ given in equation (3). In both cases the initial ray depth $z_0 = 1$ km and the initial ray angles are $\theta_0 = 5°, 6°, \ldots 15°$. 500 points are plotted for each ray trajectory. Left: $\delta = 0$. Right: $\delta = 0.01$.

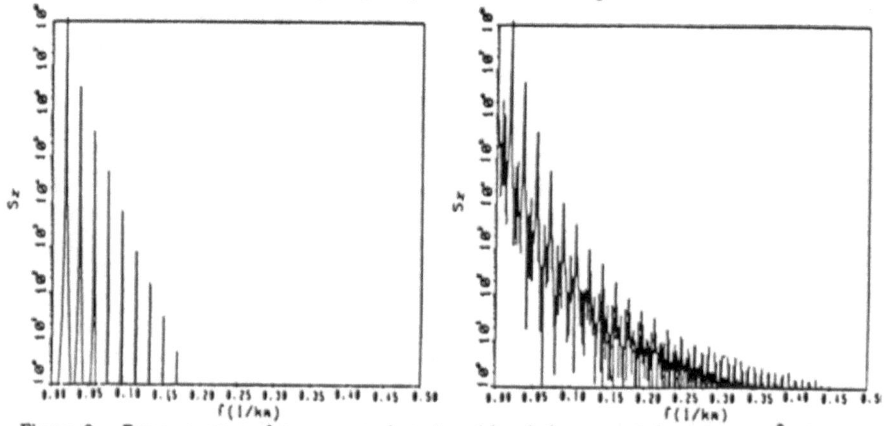

Figure 2. Power spectra of two ray trajectories $z(r)$. Left: $z_0 = 1$ km, $\theta_0 = 11°$, $\delta = 0$. Right: $z_0 = 1$ km, $\theta_0 = 12°$, $\delta = 0.01$.

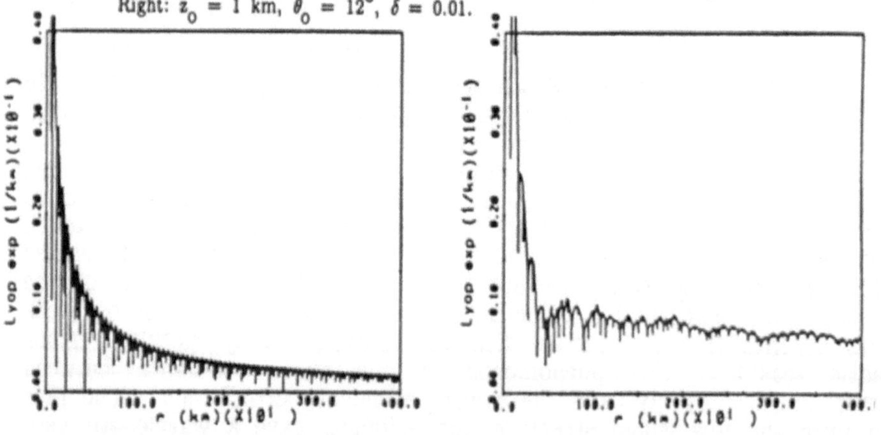

Figure 3. Plots of $(1/r) \ln |\lambda_1(r)|$ vs. r for the same two trajectories used in figure 2.

resonance and 5 islands are formed. The ray trajectories which form the speckled regions surrounding the islands (not seen in all cases) are termed "chaotic". They correspond to volume filling (in (z,p,r)) rays. A further increase in the strength of the range–dependent perturbation causes the regular barriers (termed "KAM invariant tori" for reasons discussed below) separating neighboring chaotic seas to break down, thereby forming larger connected chaotic regions.

All of this behavior is typical of integrable bounded Hamiltonian systems which are subjected to a nonintegrable perturbation. A central result to understanding this problem is the Kolmogorov–Arnol'd–Moser (KAM) theorem which states that when an integrable Hamiltonian system (range–independent sound speed structure) is subjected to a nonintegrable (range–dependent) perturbation, regular motion is preserved for most trajectories. Stated somewhat differently, the KAM theorem guarantees that most of the regular tori associated with an integrable Hamiltonian system survive a small perturbation to the Hamiltonian, although the tori become slightly distorted. When coupled with numerical simulations of the type shown in figure 1, this theorem provides important insight into the behavior of acoustic ray trajectories in weakly range–dependent oceans.

The Poincare sections shown in this figure give some geometric insight into the difference between regular and chaotic (a precise definition will be given later) ray trajectories. An important observation is that regular trajectories, because each such trajectory is constrained to lie on one of a set of smooth embedded surfaces, diverge from each other only very slowly. Chaotic trajectories, on the other hand, form a tangled web in phase space, becoming hopelessly intertwined with many other trajectories, possibly with very different initial conditions. Under such conditions the motion of individual ray trajectories cannot be predicted at long range and neighboring trajectories diverge very rapidly. These ideas will be discussed in more detail below.

2.2 Power Spectra

Power spectra of $z(r)$ provide additional insight into the difference between regular and chaotic trajectories. Spectra of regular trajectories consist of a finite number of isolated lines while spectra of chaotic trajectories are continuous and appear to be noisy. The connection between regular motion and discrete spectra can be seen geometrically. Motion on a torus is most naturally described using action–angle variables. In these coordinates the motion is periodic. Because the transformation between action–angle variable and $z(r)$, $p(r)$ is generally nonlinear, harmonics will be present in spectra of the latter. These spectra remain discrete, however. The connection between regular trajectories and discrete spectra can also be rationalized by noting that the long time evolution of such systems are predictable: all that is required is a knowledge of the wavenumbers, amplitudes, and phases of a finite number of sinusoids. The long time evolution of a chaotic system is, on the other hand, effectively unpredictable. This is consistent with a continuous, noisy spectrum. These ideas are illustrated in figure 2. Here, power spectra of $z(r)$ for two rays are shown. The first case corresponds to $\delta = 0$, the range–independent problem. Here, all trajectories are regular. The power spectrum of the trajectory shown consists of a small number of lines, as expected. The second spectrum was computed using one of the rays which fills the large chaotic sea in the $\delta = 0.01$ calculation shown in figure 1. The spectrum of this chaotic trajectory is broad band and noisy.

2.3 Lyapunov Exponents

We have seen that Poincare sections and power spectra are useful for addressing the question of whether a ray trajectory is chaotic. The Lyapunov exponent provides more information – it is a measure of how chaotic a trajectory is. The most important feature that distinguishes chaotic and regular motion is that under chaotic conditions neighboring ray trajectories diverge exponentially while under regular conditons the divergence is governed by a power law. The Lyapunov exponent is a quantitaive measure of this divergence. After describing how Lyapunov exponents are calculated, we consider some of the implications of chaotic motion.

The variational equations follow from the ray equations (1) upon setting $\xi = \delta z$, $\eta = \delta p$,

$$\frac{d}{dr}\begin{bmatrix} \xi \\ \eta \end{bmatrix} = \begin{bmatrix} \dfrac{\partial^2 H}{\partial p\, \partial z} & \dfrac{\partial^2 H}{\partial p^2} \\ -\dfrac{\partial^2 H}{\partial z^2} & -\dfrac{\partial^2 H}{\partial z\, \partial p} \end{bmatrix} \begin{bmatrix} \xi \\ \eta \end{bmatrix} = \begin{bmatrix} 0 & 1 \\ -\dfrac{\partial^2 V}{\partial z^2} & 0 \end{bmatrix} \begin{bmatrix} \xi \\ \eta \end{bmatrix} \quad (4)$$

These equations describe how small elements $\delta z\, \delta p$ of phase space are stretched along a ray trajectory. The variational equations (4) and the ray equations (1) define a system of four coupled equations which can be integrated to give $z(r)$, $p(r)$, $\xi(r)$, and $\eta(r)$ given a knowledge of their initial conditions $z(o)$, $p(o)$, $\xi(o)$, and $\eta(o)$. The variational equations can be combined to give

$$\frac{d^2 \xi}{dr^2} + \left[\frac{\partial^2 V}{\partial z^2} \right] \xi = 0. \quad (5)$$

Assume that $\xi_1(r)$ and $\xi_2(r)$, with

$$\xi_1(0) = 1 \quad , \quad \frac{d\xi_1}{dr}(0) = 0 , \quad (6a)$$

$$\xi_2(0) = 0 \quad , \quad \frac{d\xi_2}{dr}(0) = 1 , \quad (6b)$$

are two solutions to (5). They are linearly independent as their Wronskian is unity,

$$W(r) = W(o) = \xi_1(o)\frac{d\xi_2}{dr}(o) - \xi_2(o)\frac{d\xi_1}{dr}(o) = 1. \quad (7)$$

(The Wronskian is constant because the coefficient of $d\xi/dr$ in (5) is zero.) Note that from the first of the variational equations (4) $d\xi/dr = \eta$. It then follows from (7) that the Jacobi matrix

$$J(r) = \begin{bmatrix} \xi_1(r) & \xi_2(r) \\ \eta_1(r) & \eta_2(r) \end{bmatrix} \tag{8}$$

has determinant 1. The Jacobi matrix allows us to study sensitivity to initial conditions,

$$\begin{bmatrix} dz(r) \\ dp(r) \end{bmatrix} = \begin{bmatrix} \frac{\partial z}{\partial z_0}(r) & \frac{\partial z}{\partial p_0}(r) \\ \frac{\partial p}{\partial z_0}(r) & \frac{\partial p}{\partial p_0}(r) \end{bmatrix} \begin{bmatrix} dz_0 \\ dp_0 \end{bmatrix} = J(r) \begin{bmatrix} dz_0 \\ dp_0 \end{bmatrix}. \tag{9}$$

Let $\lambda_i(r)$ and $\bar{u}_i(r)$, $i = 1,2$, with $|\lambda_1(r)| \geq |\lambda_2(r)|$, denote the eigenvalues and eigenvectors of $J(r)$. Because $\det J(r) = 1$, $\lambda_1(r) \lambda_2(r) = 1$. The solution to (9), $[dz\ dp]^T$, is a linear combination of $\lambda_i(r)\ \bar{u}_i(r)$. Under chaotic conditions

$$|\lambda_1(r)| \sim e^{\nu r} \quad \text{and} \quad |\lambda_2(r)| \sim e^{-\nu r} \quad \text{as } r \to \infty . \tag{10}$$

The Lyapunov exponent is defined as

$$\nu \equiv \lim_{r \to \infty} \frac{1}{r} \ln |\lambda_1(r)| . \tag{11}$$

For a regular trajectory (10) is replaced by power law growth and decay of $\lambda_1(r)$ and $\lambda_2(r)$, respectively, and ν defined in (11) is zero.

Plots of $(1/r) \ln |\lambda_1(r)|$ vs r are shown in figure 3 for the same two trajectories that were used to produce figure 2. For the regular trajectory this curve appears to be approaching $\nu = 0$, while for the chaotic trajectory the curve appears to be approaching a value of ν close to $(200 \text{ km})^{-1}$. This behavior is consistent with the results shown in both figures 1 and 2.

We now consider one of the implications of exponential sensitivity to initial conditions associated with chaotic ray motion. Suppose we attempt to find the eigenrays connecting a fixed source and receiver separated by a distance r. This might be done by computing the trajectories of many rays, each leaving the source with a different angle θ, out to range r. If the ray depth at range r lies within some specified tolerance of the receiver depth we say that we have found an eigenray. Suppose we find by trial and error that in order to meet this tolerance criterion at range r, the ray launch angle must be specified with n_r bits of precision. It is natural to ask how n_r varies as a function of r. Under chaotic conditons n_r is proportional to r. The proportionality constant is the (scaled) Lyapunov exponent. If, for instance, θ must be specified with 12 bits of precision at $r = 120$ km and $\nu'(= \nu/\ln 2)$ is $(10 \text{ km})^{-1}$, then each 10 km increase in range would require another bit of precision in the specification of the launch angle. In this example 20 bits would be required to find eigenrays at r = 200 km, 30 bits would be required at 300 km, etc. One quickly runs into limitations imposed by finite precision computers. On a machine in which floating point numbers are stored with a 24 bit mantissa, eigenrays could be found only out to a range of somewhat less than 240 km in our example. At longer ranges, attempts to iteratively search for eigenrays are doomed to failure. Computed ray trajectories at these longer ranges will have effectively forgotten their initial conditions. With these ideas in mind, it is natural to refer to ν^{-1} as a "predictability horizon", an order of magnitude estimate of the range over which a ray trajectory can be predicted.

These ideas also provide some insight into a commonly given definition of chaos: chaos is unpredictable behavior in a low order dynamical system associated with extreme sensitivity to initial conditions.

3. RAY CHAOS INDUCED BY BOUNDARY INTERACTIONS

We now turn our attention to sound ray propagation in a simple ocean model in which ray trajectories interact with a range dependent boundary. The techniques used to study this problem differ somewhat from those used earlier. Here, the ray dynamics are studied using an area preserving mapping.

3.1 An Area Preserving Mapping

We consider sound ray propagation in a topless ocean in which the sound speed increases with height above the bottom so that rays are downward refracted. We assume that the bottom bathymetry $z_b(r)$ is known and that the bottom is rigid so that rays are specularly reflected. Let r_n and θ_n be the range and angle (in radians) at which a ray intersects the bottom. The situation is shown schematically in figure 4. If $z_b(r)$ is small compared to the turning height of a ray, then the bottom displacement can be neglected when determining the intersection of the ray with the bottom. This approximation allows the ray dynamics to be studied using the mapping

$$\theta_{n+1} = \theta_n + 2z_b'(r_n) , \tag{12a}$$
$$r_{n+1} = r_n + B(\theta_{n+1}) . \tag{12b}$$

Here $z_b'(r)$ is the bottom slope and $B(\theta)$ is the increment in range of the ray that leaves (and returns to) the bottom with angle θ. Equation (12a) says that upon reflection the outgoing ray angle is equal to the incoming ray angle plus twice the local bottom slope. Equation (12b) incrementally updates the total range traversed by a ray.

We shall assume that

$$z_b(r) = - h \cos (kr) \tag{13}$$

and

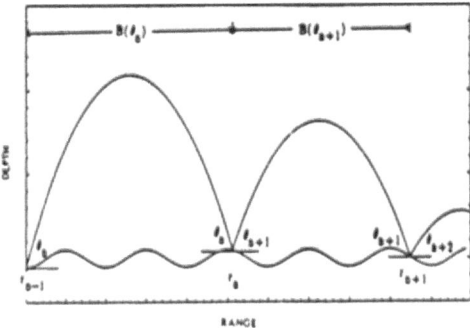

Figure 4. Segment of a ray in a downward refracting ocean overlying a sinusoidal bottom.

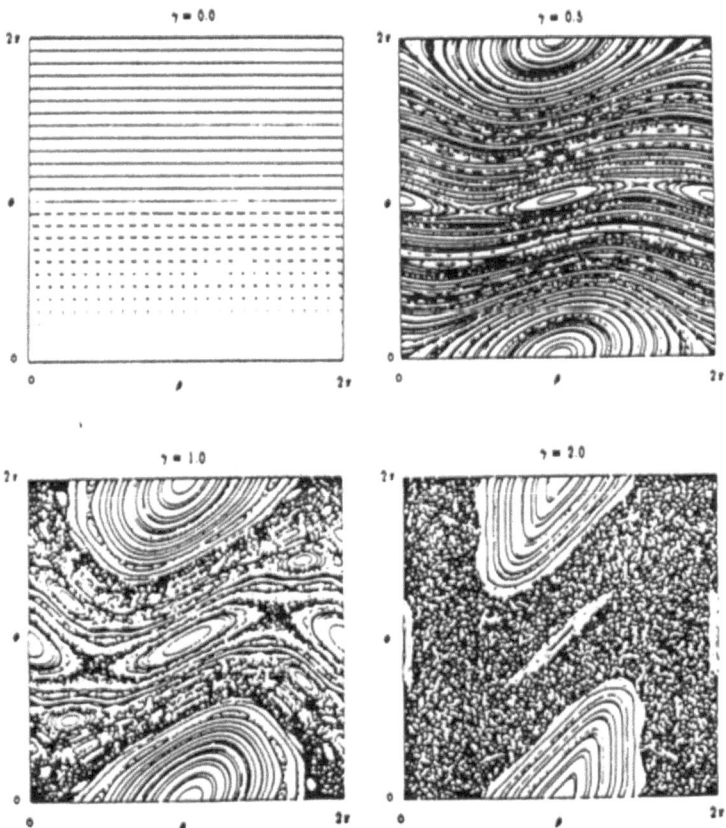

Figure 5. Iterates of the standard map (17) for 729 different initial contitions, ρ_0, ϕ_0. The mapping is iterated 300 times for each initial condition.

$$V(z) = \frac{c(z) - c_0}{c_0} = -gz \qquad (14)$$

with z positive down. The bottom slope $z_b'(r)$ is easily computed from (13). The range increment $B(\theta)$ can be computed using (14) and the ray equations (1),

$$B(\theta) = \frac{2}{g} \tan \theta \simeq \frac{2}{g} \theta . \qquad (15)$$

Inserting these expressions for the bottom slope and range increment into the mapping (12) gives

$$\theta_{n+1} = \theta_n + 2kh \sin kr_n , \qquad (16a)$$

and

$$r_{n+1} = r_n + \frac{2}{g} \theta_{n+1} . \qquad (16b)$$

In terms of the dimensionless variables $\rho_n = kr_n$ and $\phi_n = 2k\theta_n/g$ these equations become

$$\phi_{n+1} = \phi_n + \gamma \sin \rho_n , \qquad (17a)$$

and

$$\rho_{n+1} = \rho_n + \phi_{n+1} , \qquad (17b)$$

where the stochasticity parameter

$$\gamma = \frac{4k^2 h}{g} . \qquad (18)$$

The dimensionless mapping (17) has been extensively studied by others (see, e.g., Lichtenberg and Lieberman, 1982). It is referred to as the "standard map." The mapping depends on the single dimensionless parameter γ which, in our case, is four times the ratio of bottom curvature to ray curvature. It is easily verified that the mapping is area-preserving,

$$\det \begin{bmatrix} \dfrac{\partial \phi_{n+1}}{\partial \phi_n} & \dfrac{\partial \phi_{n+1}}{\partial \rho_n} \\[2ex] \dfrac{\partial \rho_{n+1}}{\partial \phi_n} & \dfrac{\partial \rho_{n+1}}{\partial \rho_n} \end{bmatrix} = 1 . \qquad (19)$$

This condition, which may be thought of as a discrete form of Liouvilles theorem, guarantees that the Hamiltonian character of the ray equations (1) is faithfully reproduced in the mapping (17). Stated somewhat differently, area-preservation (19) dictates that ϕ_n and ρ_n are canonically conjugate variables (like z and p in the ray equations (1)).

The simpliest way to study the mapping is to compute its iterates for a set of initial conditions (ϕ_0, ρ_0). Some examples are shown in figure 5. Note that ϕ_n and ρ_n are plotted modulo 2π. This is because the mapping (17) is 2π periodic in both variables. Results for several values of stochasticity parameter γ are given. If $\lambda = 2\pi/k = 100$ m and $g = 0.067$ km^{-1} then $\gamma = 0, 0.5, 1.0,$ and 2.0 correspond to $h = 0, 2.1, 4.2$ and 8.4

mm, respectively. This figure should be interpreted in precisely the same way that figure 1 was interpreted. When, for a given ray, successive iterates lie on smooth curves, the motion is regular. A succession of iterates that fills an area in an apparently random way signifies chaotic motion. This figure shows the same qualitative behavior that was seen in figure 1. The stochasticity parameter γ, like δ in figure 1, is a measure of the strength of a nonintegrable (range–dependent) perturbation to an integrable (range–independent) problem. As γ is increased chaotic seas surrounding regular islands are formed. A further increase in γ causes neighboring chaotic seas to merge. For values of γ less than the critical value $\gamma_c \approx 0.97$, the chaotic seas are all bounded. For $\gamma > \gamma_c$ a large chaotic sea exists wherein ray trajectories may wander without bound in the $\phi - \rho$ plane. This phenomenon is referred to as "global chaos".

3.2 Lyapunov Exponents

Lyapunov exponents for the mapping (17) (in base 2 these have units bits per bounce) are computed in much the same way that they were computed for the continuous ray trajectories. Differentiating the mapping (17) gives

$$\begin{bmatrix} d\phi_{n+1} \\ d\rho_{n+1} \end{bmatrix} = M_{n+1} \begin{bmatrix} d\phi_n \\ d\rho_n \end{bmatrix} \tag{20}$$

where, by area–preservation, the determinent of the real 2 x 2 matrix M_{n+1} is 1. Iterating (20) gives

$$\begin{bmatrix} d\phi_n \\ d\rho_n \end{bmatrix} = J_n \begin{bmatrix} d\phi_0 \\ d\rho_0 \end{bmatrix} \tag{21}$$

where the Jacobi matrix $J_n = M_n M_{n-1} \cdots M_1$ also has determinant 1. Like equation (9) this equation can be used to study sensitivity to initial conditions. The same argument leading to our earlier definition of the Lyapunov exponent (11) applies here. The Lyapunov exponent for the mapping (17) is then

$$\nu \equiv \lim_{n \to \infty} \frac{1}{n} \ln |\lambda_n^{(1)}| \tag{22}$$

where $|\lambda_n^{(1)}|$ is the modulus of the larger of the two eigenvalues of J_n. We defer until later giving an example of such a calculation.

3.3 Eigenrays

We turn our attention now to eigenrays. Because our analysis is based on the mapping (17) we restrict our attention to situations in which both source and receiver lie on the bottom. Specifically, we examine the behavior of eigenrays as the range r between source and receiver increases. Fans of rays for both a flat bottom and a sinusoidal bottom with amplitude h = 10 cm and wavelength $2\pi/k$ = 200 m are shown in figure

6. In both cases the sound speed gradient $c_0 g = 0.1$ sec^{-1}. These correspond to stochasticity parameters of $\gamma = 0$ and $\gamma = 5.9$, respectively. These models are used to produce all of the numerical results shown in this section. For $\gamma = 5.9$ almost all rays are chaotic.

Eigenrays for the mapping are easily found by constructing the curves $r_n(\theta_0)$, range vs. launch angle for rays which have $n - 1$ bottom bounces. Several examples of such curves are shown in figure 7. Eigenrays for a receiver at range r correspond to the solutions of $r_n(\theta_0) = r$, i.e., intersections of a curve $r_n(\theta_0)$ with a horizontal line. Of course, to find all of the eigenrays at range r this process must be repeated for many values of n. Before we show the results of carrying out this procedure, however, it is instructive to take a closer look at one of the $r_n(\theta_0)$ curves for the sinusoidal bottom.

Figure 8 shows a succession of blow ups of the $r_{13}(\theta_0)$ curve for $5.0°$ $< \theta_0 < 5.5°$. In the uppermost curve the sampling interval $\Delta\theta = 10^{-4}$ degrees. In the lowermost curve $\Delta\theta = 10^{-8}$ degrees. It would have been difficult to guess that such a small sampling interval is necessary after only 12 bounces. This number is not absolute. Our numerical simulations indicate that for each additional two bounces $\Delta\theta$ must be decreased by roughly a factor of ten.

Figure 9 shows the number of eigenrays N as a function of range r, computed using the procedure outlined above. Because of the huge difference in the number of eigenrays for the flat and sinusoidal bottom cases, different angular apertures were used: $4°$ for the flat bottom and 0.5 degrees for the sinusoidal bottom. This figure shows clearly that for the flat bottom N(r) grows linearly while for the sinusoidal bottom N(r) grows exponentially ($\sim \exp (r/1.9$ km$)$ in this figure).

Because the number of eigenrays is growing exponentially in the sinusoidal bottom case, energy conservation dictates that the corresponding average intensity

$$I_{av} = \frac{1}{N} \sum_{i=1}^{N} \left| \left[\frac{\partial r}{\partial \theta_0} \right]_i \theta_i \right|^{-1}, \tag{23}$$

where θ_i is the launch angle of an eigenray, should decay exponentially in range. This quantity is plotted as a function of range in figure 10. In this figure $I_{av} \sim \exp (-r/2.0$ km$)$. This is consistent with exponential growth of the number of eigenrays.

One would expect that the exponential growth of the number of eigenrays and the corresponding exponential decay of their average intensities is related to an average Lyapunov exponent for the rays under consideration (recall $5.0° < \theta_0 < 5.5°$). Because these rays are predominantly chaotic and most lie in the same chaotic sea the notion of an "average Lyapunov exponent" is a sensible one. For an individual ray

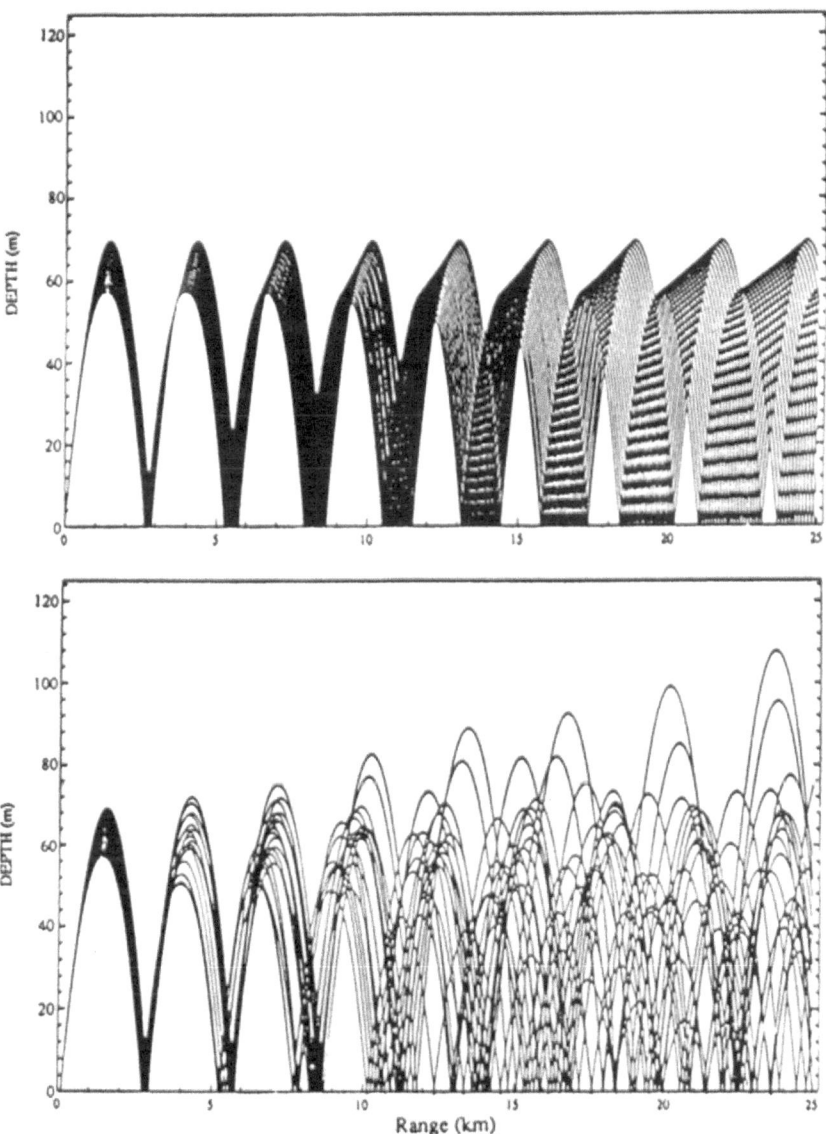

Figure 6. Fans of rays corresponding to launch angles $5.0° < \theta_0 < 5.5°$ over flat (upper panel) and sinusoidal (lower panel) bottoms.

152

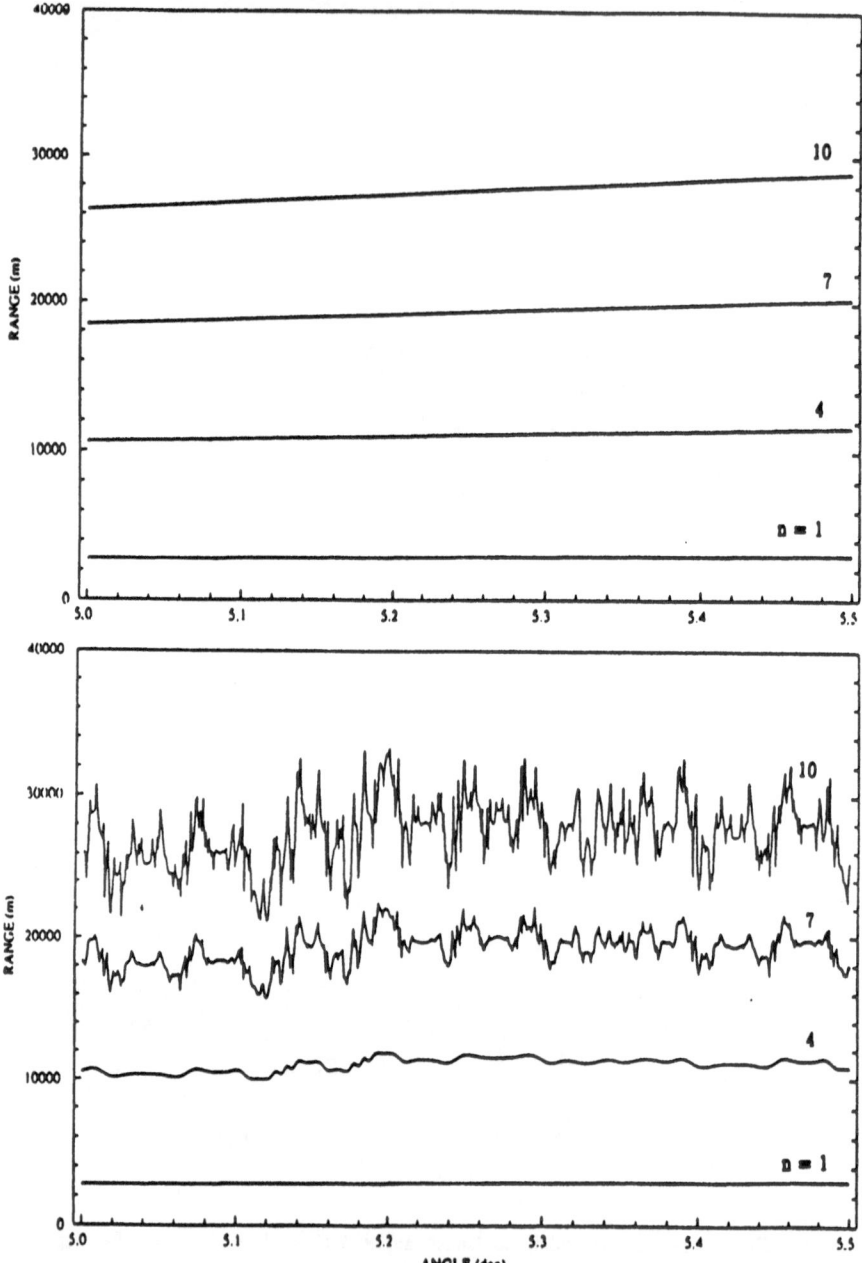

Figure 7. Plots of r_n vs. θ_o for flat (upper panel) and sinusoidal (lower panel) bottoms.

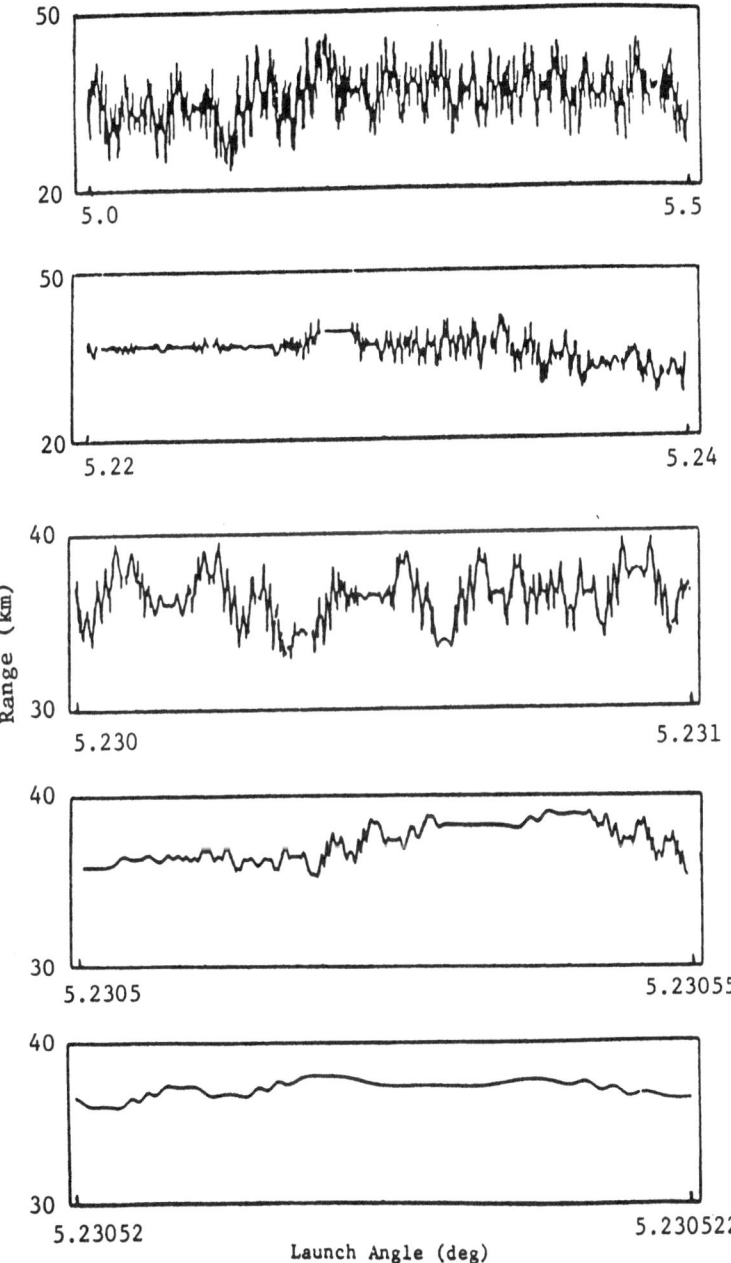

Figure 8. Plots of r_{13} vs θ_0 for a sinusoidal bottom. Each plot is a magnified version of some portion of the plot above.

154

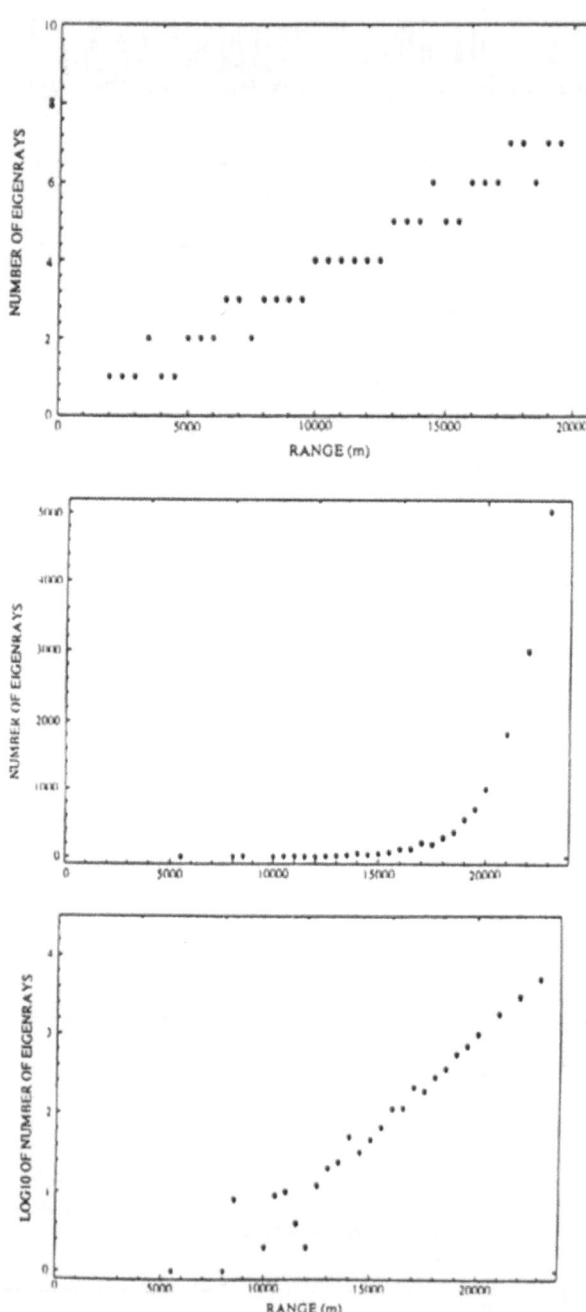

Figure 9. Plots of the number of eigenrays N as a function of range r. Upper panel: N vs r for a flat bottom. Center panel: N vs r for a sinusoidal bottom. Lower panel: log N vs r for a sinusoidal bottom.

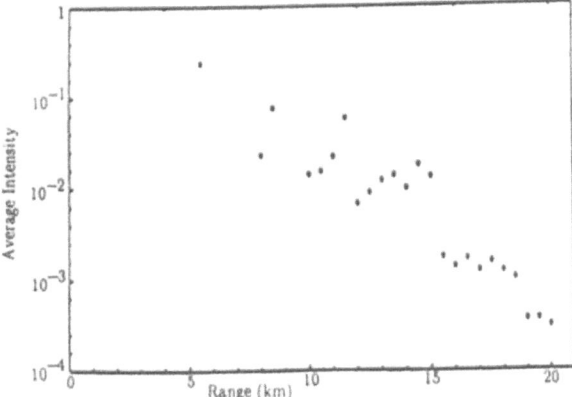

Figure 10. Plot of average eigenray intensity as a function of range for the sinusoidal bottom problem.

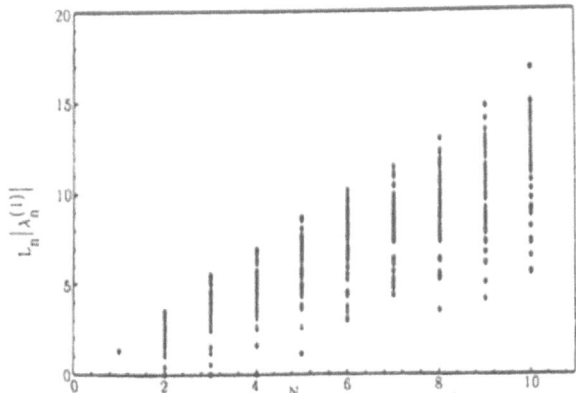

Figure 11. Plot of ln $|\lambda_n^{(1)}|$ vs. n for 51 rays whose lauch angle $5.0° \leq \theta_0 \leq 5.5°$. The average slope of the plotted points is an estimate of the average Lyapunov exponent.

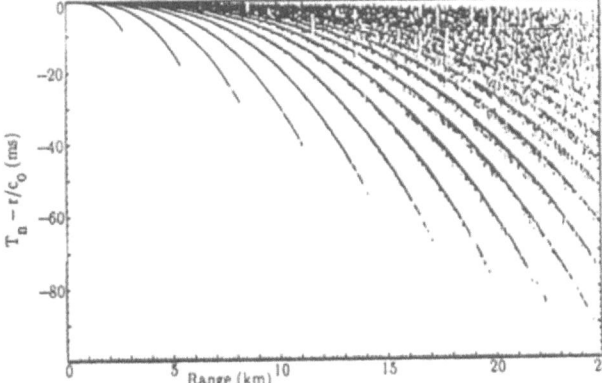

Figure 12. Plot of reduced travel time $T_N - r/c_0$ vs. range r for the sinusoidal bottom problem. Each cluster of points correspond to a fixed number of bottom bounces N, with N increasing upwards and to the right.

the Lyapunov exponent ν is defined by equation (22). To maintain consistency with the results shown in figures 6–10, however, we have plotted in figure 11 $\ln |\lambda_n^{(1)}|$ vs n, n = 0,1,2... 10, for 51 rays in the band $5.0° < \theta_0 < 5.5°$. The average slope of the points plotted is an estimate of the average Lyapunov exponent ν_{av}. From the figure $\nu_{av} \approx$ 1.3/bounce. Using 2.9 km/bounce as an average range increment – see figure 6 or 7 – this corresponds to $\nu_{av} \approx (2.2 \text{ km})^{-1}$. This is in rough agreement with our earlier results.

The distribution of travel times T as a function of range r gives additional insight into the complexity of the wavefield under chaotic conditions. In figure 12, T(r) is plotted for many (not all) rays in the band $1.0° < \theta_0 < 5.0°$. Each cluster of points in this figure corresponds to a different number of bottom bounces n. This figure shows that travel times for rays which have hit the bottom only a few times are tightly bound and clusters corresponding to different n are distinct. As n increases, travel time spreads grow and clusters corresponding to different n overlap with their neighbors. This figure shows that after a small number of bounces the wavefield is highly structured and predictable. After a large number of bounces, however, the wavefield is largely unstructured and appears to be unpredictable. This is chaos.

4. A SEARCH FOR WAVE CHAOS

Our discussion up until now has focused on chaotic behavior of ray trajectories in models of range–dependent oceanic waveguides. It is natural to ask, "To what extent does the unpredictability associated with chaotic ray trajectories carry over to finite frequency wavefields?" In this section we present some preliminary numerical results which address this question. It is natural to refer to chaotic behavior in finite frequency wavefields, if it exists, as "wave chaos."

In another context this phenomenon, a controversial one, is referred to as "quantum chaos." To date, most of the work on quantum chaos (see, e.g., Berry, 1987) has been concerned with the distribution of energy levels of quantum systems which exhibit chaotic motion in the classical limit. It is known that these energy levels obey different statistics in (a) systems which are classically chaotic and (b) systems which are classically regular. These important results provide little insight into our underwater acoustics problem, however. The reason is that these results apply to the solutions to boundary value problems whereas our underwater acoustic waveguide problems are initial value problems.

In this section, we confine our attention to one of many ideas which might be used to investigate wave chaos. The idea is reversibility or, more specifically, lack of reversibility in chaotic systems. The ray equations (1) can be integrated backwards in range by simply changing the sign of r. It is straightforward to verify that under regular conditions a ray trajectory can be computed forward, out to a very long range, and backwards, only to recover the initial conditions, $z(o)$, $p(o)$. For a chaotic trajectory this procedure does not work at ranges longer than a few e–folding distances, ν^{-1}. The reason is that at longer ranges the initial conditions of the ray will have been forgotten. One way to investigate wave chaos is to ask whether finite frequency wavefields "forget" their initial conditions in a similar fashion.

We have addressed this question by examining solutions to the parabolic wave equation,

$$i\ \frac{\partial\psi}{\partial r} + \frac{1}{2k_0} \frac{\partial^2\psi}{\partial z^2} - k_0\ V(z,r)\ \psi = 0\ , \qquad (24)$$

which are computed using the split–step Fourier algorithm (Tappert, 1977). In (24), $k_0 = \omega/c_0$ where ω is the angular frequency of the sound waves and the acoustic pressure

$$u(z,r) = \psi(z,r)\ (k_0 r)^{-1/2}\ e^{ik_0 r}. \qquad (25)$$

It should be noted that the wave equation (24) reduces to the ray equations (1) in the high frequency limit. The split step Fourier algorithm is well suited to looking at reversibility because depth dependent complex wavefields $\psi(z)$ can be stepped backwards as well as forwards in range using this technique.

Figure 13 shows examples of forward and back propagated fields out to a range of $r_{max} = 18.53$ km. These wavefields were computed using an acoustic frequency of 16 kHz. The azimuthal spreading factor in (25) was neglected in both the forward and back propagation problems. The initial $(r = 0)$ field for the forward propagation calculation was a narrow downward directed gaussian beam. The ocean structure was chosen to coincide with the sinusoidal bottom problem described in the previous sections. The potential $V(z,r)$ is given by (14) with $c_0 = 1510$ m s^{-1} and the bathymetry is given by (13) with $h = 61$ cm and $2\pi/k = 185$ m. This choice of parameters corresponds to a stochasticity parameter (eq. 17) $\gamma = 44.8$. The PE runs require a finite ocean depth and bottom sound speed. We used 98 m and 1740 m s^{-1}, respectively. In the back propagation calculation, the complex conjugate of the forward propagated field at r_{max} is used as the initial condition and the bathymetry is reversed.

In this environment ray trajectories are highly chaotic and one would expect that ray theory accurately describes the acoustic wavefield at 16 kHz. Figure 13 shows rather dramatically, however, that the wavefield at $r = r_{max}$ has not forgotten its initial conditions. Furthermore, our investigation of the dependence of back–propagated fields on r_{max} indicates that the results degrade slowly as r_{max} is increased. There is no evidence of exponential degradation of back–propagated fields. These and other numerical simulations that we have performed suggest that wave chaos does not exist.

158

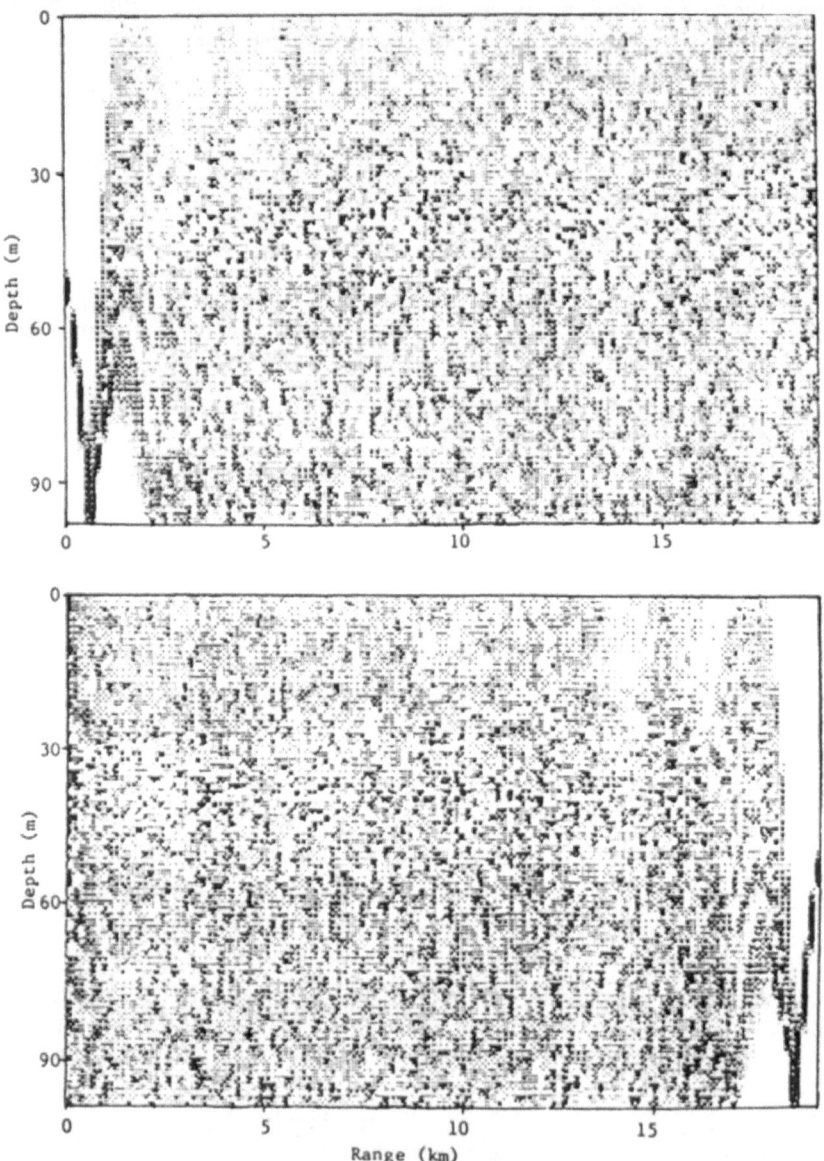

Figure 13. Gray scale plots of the logarithm of the intensity of forward (upper panel) and back (lower panel) propagated acoustic wavefields. The ocean model is described in the text.

5. DISCUSSION AND SUMMARY

We have argued that the generic range–dependent sound propagation problem differs in a fundamental way from the range–independent problem, at least in the geometric limit. In the range–dependent problem at least some ray trajectories exhibit chaotic motion wherein neighboring trajectories diverge from each other exponentially. Under conditions in which ray trajectories are predominantly chaotic we have shown that the complexity of the geometric wavefield grows exponentially in range. In the range independent problem, wavefield complexity also grows in range but at a much slower (power law) rate.

In this paper we have restricted our attention to periodically range–dependent models of ocean sound channels. It is natural to ask how this situation differs from the more realistic nonperiodic range–dependent problem. An important difference is that in the latter problem the phase space (z,p,r) is not bounded. As a result Poincare sections cannot be constructed. Power spectra and Lyapunov exponents are still useful diagnostic tools to identify chaotic motion. In all cases there is uncertainty associated with taking the limit $r \to \infty$ in estimating the Lyapunov exponent (11). The best one can do is to say that over some range of r ν appears to be approaching a well defined value. We are currently investigating the behavior of ray trajectories in numerically simulated ocean sound channels containing realistic mesoscale induced perturbations. The results are not yet complete.

Under chaotic conditions ray trajectories are not computable, due to extreme sensitivity to initial conditions, beyond some finite predictability horizon. This leads us to question whether there might be a fundamental limitation on our ability to compute finite frequency wavefields. The preliminary numerical experiments which we have performed suggest that no such limitation exists.

ACKNOWLEDGEMENTS

This work was supported by the Office of Naval Research and the National Science Foundation.

REFERENCES

Abdullaev, S.S., and G.M. Zaslavskii (1988) "Fractals and Ray Dynamics in a Longitudinally Inhomogeneous Medium," Sov. Phys. Acoust. <u>34</u>(4), 334–336.

Berry, M.V. (1987) "Quantum Chaology," Proc. Roy. Soc. Lon. A <u>413</u>, 183–198.

Henon, M. (1983) "Numerical Exploration of Hamiltonian Systems," in <u>Chaotic</u> <u>Behavior</u> <u>of</u> <u>Deterministic</u> <u>Systems</u> (Les Houches Lectrues 36), edited by G. Iooss, R.G.H. Helleman, and R. Stora, North Holland, Amsterdam, 171–271.

Lichtenberg, A.J., and M.A. Lieberman (1982) <u>Regular</u> <u>and</u> <u>Stochastic</u> <u>Motion</u>, Springer Verlag, New York, 499 pp.

Munk, W.H. (1974) "Sound Channel in an Exponentially Stratified Ocean with Application to SOFAR," J. Acoust. Soc. Am. <u>55</u>, 220–226.

Palmer, D.R., M.G. Brown, F.D. Tappert, and H.F. Bezdek (1988) "Classical Chaos in Nonseparable Wave Propagation Problems," Geophys. Res. Lett. 15(6), 569–572.

Tappert, F.D. (1977) "The Parabolic Approximation Method, in Lecture Notes in Physics, vol. 70, Wave Propagation and Underwater Acoustics, J.B. Keller and J.S. Papadakis (eds), Springer Verlag, New York, 224–287.

IMPULSE RESPONSE ANALYSIS OF OCEAN ACOUSTIC PROPAGATION

Stanley M. Flatté, John Colosi, Timothy Duda, and Galina Rovner
Physics Department and Institute of Marine Sciences
University of California at Santa Cruz
Santa Cruz, California 95064

ABSTRACT. An intuitive picture of ocean sound channel propagation from point sources is developed in terms of impulse response. The entire wavefront is described for fixed times that are appropriate for propagation to ≈ 1000 km. The wavefront characteristic that allows determination of the source depth is shown to be the distance between the two distinct wavefronts that result from the upgoing and down-going rays at the source. The effect of internal waves in terms of wavefront corrugations is described.

1. Introduction

The pattern of acoustic energy from a point source in the ocean has a complicated structure in space due to the ocean sound channel. The acoustic field from such a source also has a temporal structure. The temporal structure of the received field is determined by a combination of the temporal structure of the source and the propagation. If the temporal structure of the source is a narrow frequency line, then analysis of the problem in frequency space is appropriate. If the temporal structure of the source is impulse-like, then analysis in terms of the impulse response is appropriate. If the pulse is broad-band in a random or pseudorandom way, then either type of analysis may be useful.

It is the purpose of this article to point out two advantages to thinking about long-range ocean acoustic propagation in terms of impulse response, even at rather low frequencies (down to about 10 Hz).

The first reason involves the nature of propagation in the sound channel; this propagation becomes much simpler and more intuitive in terms of impulse response, because the resulting geometrical picture of propagating wavefronts is much easier to understand then the complicated pattern of phase and amplitude that results from propagation of a single frequency.

The second reason involves the addition of the effect of small-scale ocean structure such as internal waves to the propagation. The effect of internal waves on wavefronts can be expressed in terms of a small-scale corrugation added to the smoothly curved structure imposed by the sound channel. On the other hand, the effect of small-scale structure on a single-frequency field pattern is a confusing coherent superposition of many rays arriving at every point in space [1]. It is important to realize that the calculation of internal-wave effects

161

J. Potter and A. Warn-Varnas (eds.), Ocean Variability & Acoustic Propagation, 161–172.
© 1991 *Kluwer Academic Publishers.*

has been successful for arbitrary sound-speed profiles only by means of ray theory; that is, by identifying the energy arriving at a particular point on a wavefront as having moved through the ocean in a restricted region around the ray trajectory that connects the source with that point.

In Section 2, the impulse responses for a number of interesting ocean propagation cases are examined by calculating the propagation of wavefronts using ray theory. Simple properties of the wavefront pattern can be related to the geometry of the source and receiver depths and ranges, and to the character of the ocean sound channel. For example, the wavefront has a twinned or paired character, and the spacing between the pair of wavefronts increases with the depth of the source.

In Section 3, the effect of internal waves is described in terms of corrugations of the wave-fronts. The magnitude of the corrugation can be expressed in terms of a root-mean-square (rms) displacement, in meters, or in terms of a travel-time fluctuation: for example, in milliseconds. The correlation functions in the vertical and horizontal directions are also important in the characterization. These internal-wave effects can be calculated from ocean-acoustics theory developed over the past 15 years; the calculation consists of ordinary integrals along rays that can be done numerically [2-6].

A word about ray theory: We are interested in waves with frequency content confined to 10 Hz and above, so that the wavelength is less than 150 m. The ocean sound channel usually has characteristic scales that are considerably larger than that (averaged horizontally), so that ray theory is a valid and useful starting point. The effects of internal waves come from multiscale structures whose components range from a few meters to several kilometers. It has been shown that the calculation of internal-wave effects on forward propagation from a ray theory starting point is valid [7] down to frequencies in the neighborhood of 10-20 Hz.

2. Impulse Response: Wavefronts

The calculation of impulse response in a general range-dependent sound channel is a formidable task requiring massive computing resources. Calculations for a range-independent case are easier, but a full-wave solution is still formidable. One could at this point try several different routes; one is the use of full-wave techniques that depend on the energy in the sound channel being confined to a reasonably narrow cone of angles in one direction; another is the use of ray theory, with corrections. Methods that become increasingly valid at low frequency usually depend in some way on Fourier transforming in range: for example, the split-step Fourier algorithm for the parabolic wave equation, or the Fast Field Program [8]. Other methods that allow diffractive corrections to ray theory usually depend on the WKBJ method [9]. The low-frequency methods are computer intensive for impulse response. For example, determining impulse response using the parabolic wave equation requires many frequencies to synthesize a narrow pulse, and resolving a narrow pulse in space requires more closely spaced grid points than ordinarily used in single-frequency calculations. Furthermore, any wavefield method that does not identify rays as an integral part of the method will provide great difficulties to the addition of the effects of small-scale structure, such as internal waves.

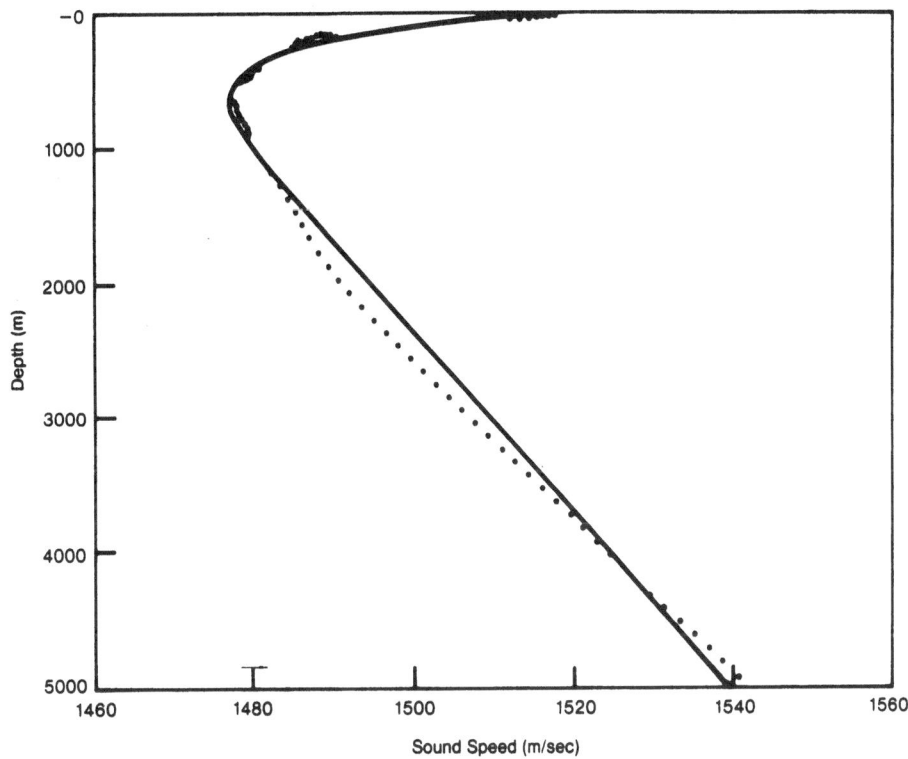

Figure 1. Sound-speed profile used in this study. The points represent an experimental profile in the Pacific, several hundred miles off the California coast [10]; the solid line is a canonical profile [11] with axis depth of 635 m, axis sound speed of 1477 m/s, sound-cannel width of 440 m, epsilon of 0.00223, and fractional slope in sound speed near the ocean bottom of 0.01012 km^{-1}. These parameters fit the experimental profile at the base of the mixed layer (40 m), the axis depth, and at 4700 m.

164

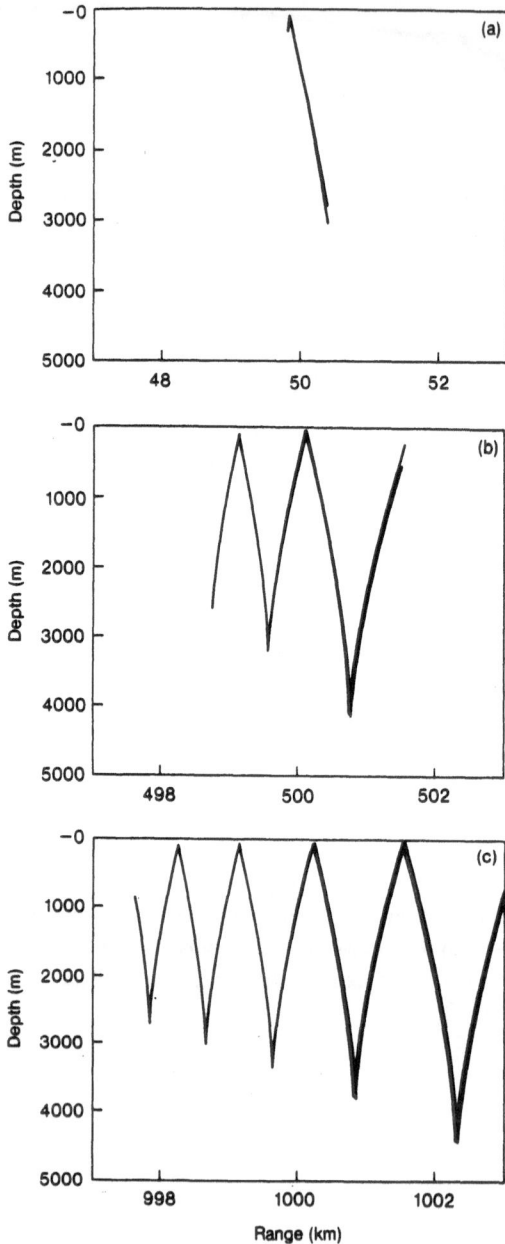

Figure 2. Wavefront reconstruction for propagation from a source at 100-m depth through a canonical profile. Note the twinned wavefronts separated by about 30 m at the leading edge.

When ray theory is valid to a first approximation, a simple and reasonably quick procedure can be used to generate wavefronts without finding the field at every point in space. One may simply trace a fan of rays from the source using the Hamiltonian form of ray optics, so that the step is in travel time rather than space. Then the raytrace is stopped on each ray when the appropriate time is reached, and the position (depth and range) of each ray is recorded. Finally, a depth-range plot of the points representing each and every ray will clearly show the wavefront at that particular time.

Figure 1 shows an example of a sound-speed profile from the Pacific Ocean a few hundred miles off the California coast [10]. The smooth solid curve is a canonical profile [11] with the parameters chosen to fit the observed sound speeds at the bottom of the mixed layer (40 m), the axis depth (635 m) and at 4700 m.

Figure 2 shows the wavefront structures at several ranges for a source at 100-m depth, and for the canonical profile of Figure 1. At 50-km range (Figure 2a) we see a relatively simple structure involving two wavefront segments. One of these wavefront segments comes from the rays that have a downward launch angle at the source, and the other comes from upward launch angles. (If there were no sound channel, one of these two wavefronts would correspond to the direct rays and the other would correspond to rays reflected from the ocean surface.) The upper edges of both wavefronts are in the process of touching the first convergence zone in this sound-channel propagation.

The wavefront structure of Figures 2b and 2c show a quasirepetitive sawtooth pattern with a repetition distance of about 1 km. As the wavefront traverses the ocean from 500 km to 1000 km, it snakes its way from about two of these patterns to about four of them, and the repetition distance changes in a slow, regular fashion. The distance between the closely spaced pair of wavefronts that results from upgoing and downgoing rays at the source undergoes very little change with propagation range. (This distance is measured near the leading edge of the wavefront structure, because the two wavefronts get gradually closer together as one moves toward the trailing edge of the structure. In fact the two wavefront structures connect at the trailing edge, since the wavefront must be one continuous line segment.)

Figure 3 is the result when the source is moved down to sound-channel axis (635 m). We see the same qualitative features, except that the separation between the two distinct wavefronts has changed to about 200 m.

The behavior of the temporal pattern of arrivals on a vertical array is qualitatively the mirror image of the wavefront pattern. The mirror image is not exact, because of the relative propagation of rays between a picture that represents a fixed time, and a picture that represents a fixed range. The time-depth pattern at fixed range could be called a 'timefront' rather than a 'wavefront'. Such timefronts were plotted by Munk and Wunsch in their original paper on acoustic tomography [12], and by Brown et al [13].

The separation between the twin wavefronts as a function of source depth is shown in Figure 4 for a range of 1000 km. The dependence is close to linear, with separation being the same order of magnitude as the source depth, at least for source depths below a few tens of meters.

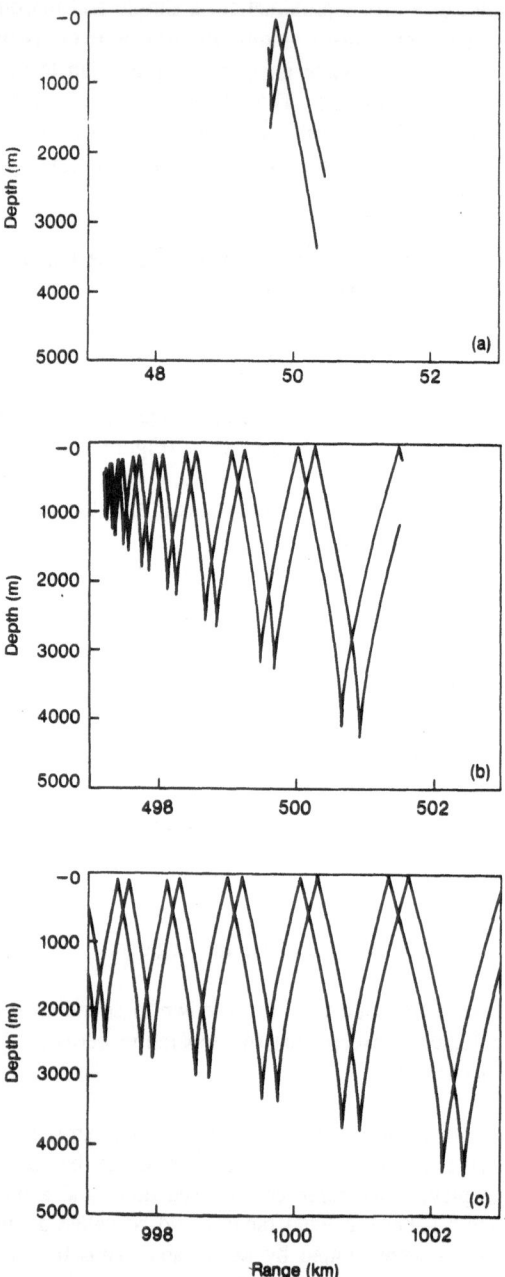

Figure 3. Wavefront reconstruction for propagation from a source at 635-m depth through a canonical profile. Note the twinned wavefronts separated by about 200 m at the leading edge.

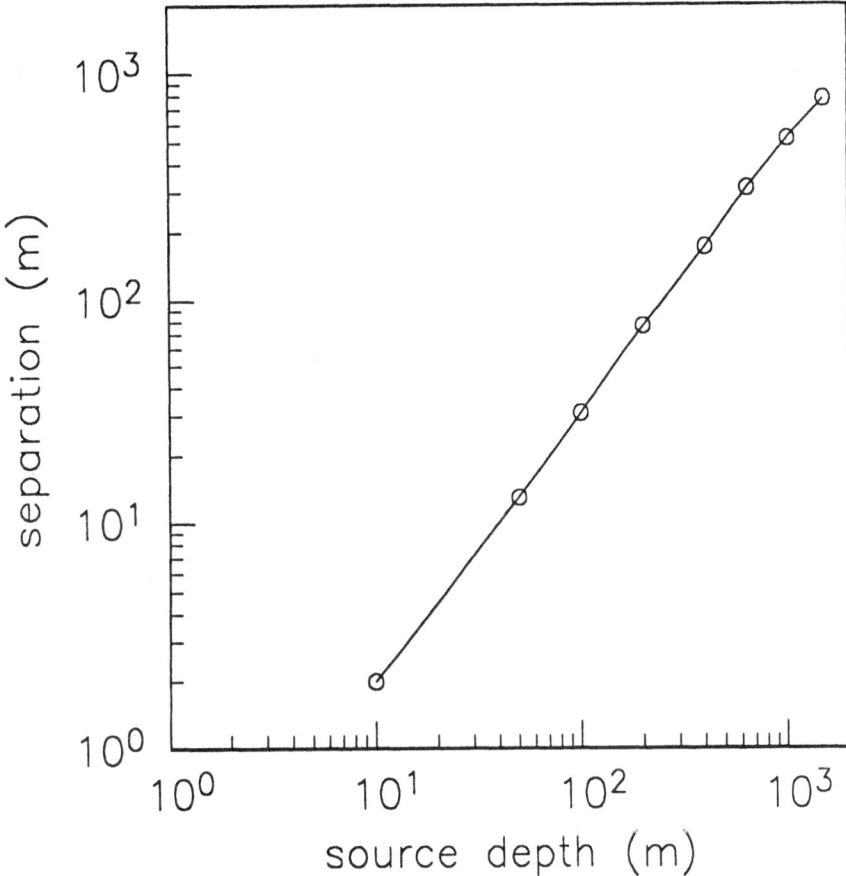

Figure 4. Separation between the twinned wavefronts as a function of source depth for travel time of 674.31 seconds (about 1000 km).

The results for a very smooth (canonical) profile were shown in Figure 2. If the experimentally measured profile with all its small-scale structure were to be used instead, each of the wavefront segments would fold back and forth onto itself, to a greater or lesser degree, such that a given receiver would see a microray pattern of arrivals very close (within a few tens of meters) to the arrival calculated from a smooth profile.

The effect of small-scale structure is much more rigorously evaluated by use of a statistical treatment of waves in random media [14 and 2]. Such a treatment is discussed in the next section.

3. The Effects of Internal Waves

Figure 5 shows a calculation of the rms travel-time fluctuation caused by a canonical field of internal waves for rays travelling in various sound-speed profiles. The typical value for propagating to 1000 km is between 5 and 10 ms. In other words, a given point on the wavefront structure resulting from an impulse will vary in range with an rms displacement of 8 to 15 m. The transverse structure of this wavefront corrugation can be calculated by use of the path-integral theory [7]. Two examples of such calculations are shown in Figure 6 for propagation to 270 km. The structure function being plotted is the expectation value of the square of the difference between the displacements at two points separated vertically on a given wavefront. Thus a value of 25 m² at a spacing of 100 m means that two receivers spaced vertically by 100 m will experience an rms difference in the arrival of a wavefront equivalent to a displacement difference of 5 m.

The conclusion from Figure 6, aside from the quantitative knowledge of the entire structure function, and equivalently, the correlation function or wavenumber spectrum of the corrugations, is that the vertical scale length of the wavefront corrugations will be in the range of several hundred meters when the range of propagation is many hundreds of kilometers.

The horizontal scale length of the wavefront corrugations is in the range of a few kilometers [7].

If the combination of pulse width, strength of internal-wave fluctuations, and propagation range create the conditions of strong fluctuations in the wavefield at the receiver, then the corrugated wavefront may develop a multiple-lamination character. These laminations, when observed with a single receiver, have been called microrays [2]. The transverse behavior of these laminations has yet to be investigated, but we expect that they will create a mini-sawtooth structure that is a small-scale version of the global sawtooth structure created by the large-scale structure of the sound channel itself. The vertical extent of these small-scale wavefront segments needs to be investigated both theoretically and experimentally.

The time of arrival obtained at a receiver by averaging over the microrays of one global ray should provide a measure of arrival-time fluctuations and structure functions that represent corrugations in the global wavefront, and that are primarily caused by the larger-scale internal waves. The spread and transverse coherence of the microray laminations should provide measures of the smaller-scale internal-wave behavior.

rms Travel Time

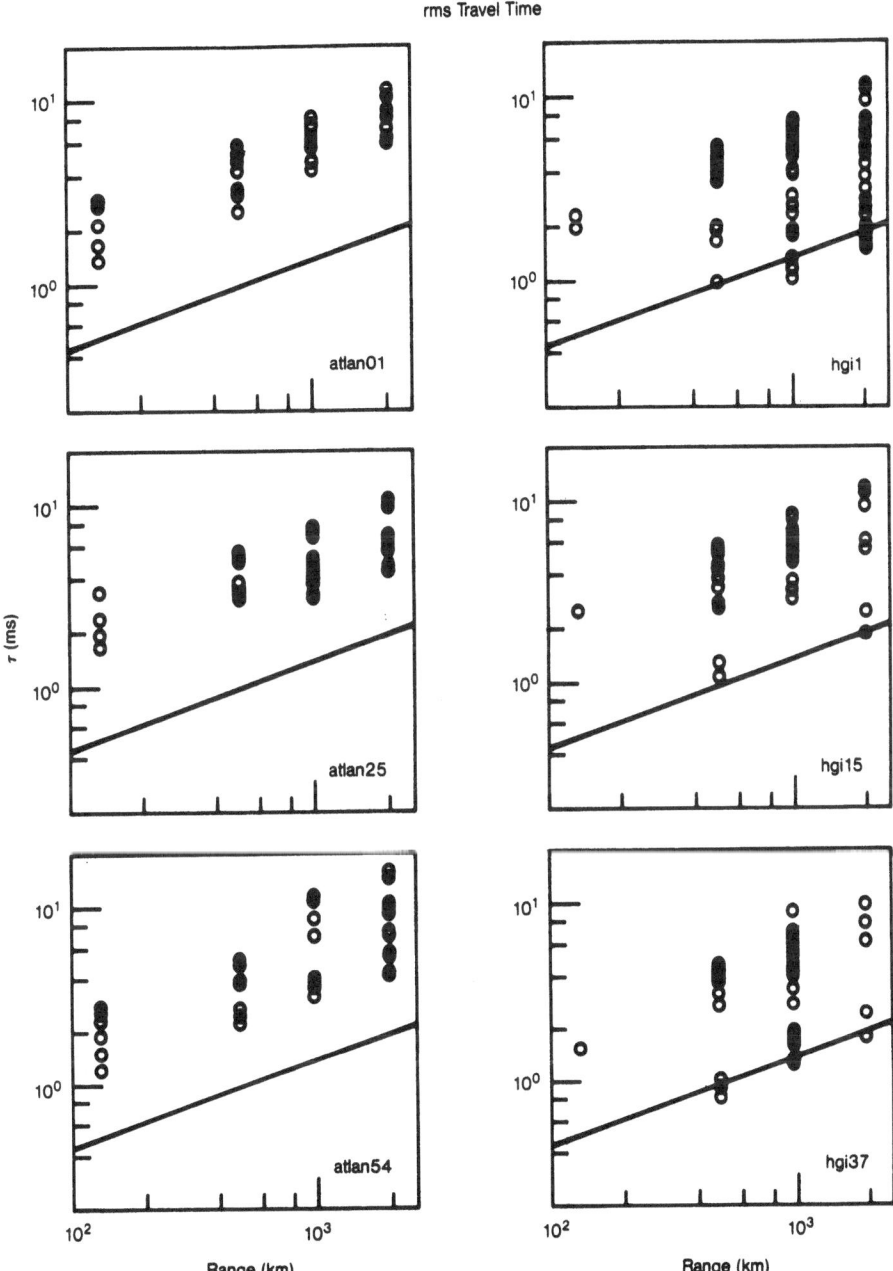

Figure 5. Travel-time fluctuation for six different sound-speed profiles. Each point represents a different launch angle from the source, which is at a depth of 400 m. The left (right) column uses profiles from the Atlantic (Pacific).

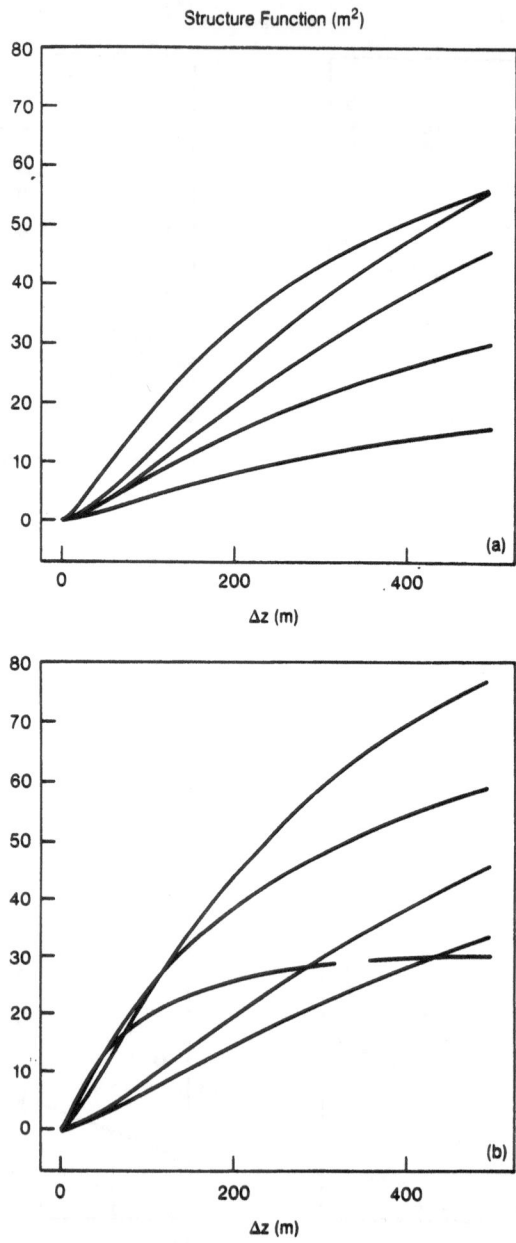

Figure 6. Structure functions for rays in an Atlantic profile that propagate to a range of 270 km. The different curves represent different launch angles from the source, which is at a depth of 1100 m.

4. Conclusions

An intuitive understanding of long-range ocean propagation has been presented in terms of impulse response: that is, the spatial structure of wavefronts propagating out from a point source. This spatial structure has been shown to have relatively simple qualitative features, such as a quasi-repetition distance of about 1 km, and a twinning of wavefronts into pairs whose separation is a good measure of the depth of the source.

The full wavefront structure can be thought of as being made up of a number of subwave-fronts, each of which spans a good fraction of the water column, and has a smooth spatial behavior.

The effect of internal waves is to corrugate these subwavefronts. The corrugations have rms displacements of 10-15 m for 1000-km propagation. The vertical correlation lengths of these corrugations are typically a few hundred meters, and the horizontal correlation lengths are typically a few kilometers.

In general it should be concluded that the signal processing that is developed for use of signals propagating over long ranges in the ocean should have an intimate connection to the physics of the propagation. It is suggested that the most intuitive way of understanding this physics is to think in terms of the wavefronts associated with the impulse response of this long-range ocean propagation.

The observation of the properties of long-range wavefronts is an exciting frontier in acoustical oceanography. Determination of the internal-wave spectrum should be possible from measurements of the corrugations and laminations of wavefront segments. This determination includes the strength of the spectrum, as well as the dissection of the spectrum into its contributions from various regions of vertical wavenumber and frequency. It should also be possible to monitor these internal-wave properties as a function of depth, geographical position, and geophysical time. Such a monitoring would be an illuminating constraint on models of ocean and atmosphere dynamical processes.

5. Acknowledgements

This work was partially supported by the Office of Naval Research (Code 1125OA) and the Office of Naval Technology. We gratefully acknowledge a grant from the W. M. Keck Foundation.

172

6. References

1. For example, Porter, M.B. et al (1987) 'Simulations of matched-field processing in a deep-water Pacific environment' IEEE J. Oceanic Eng. OE-12, 173-181, and Tolstoy, A. (1989) 'Sensitivity of matched-field processing to sound-speed profile mismatch for vertical arrays in a deep-water Pacific environment', J. Acoust. Soc. Am. 85, 2394-2404.

2. Flatté, S. M. (1983) 'Wave Propagation through Random Media: Contributions from Ocean Acoustics', PROC. IEEE, 71, 1267-1294.

3. Dashen, R., Flatté, S.M., and Reynolds, S.A. (1985) 'Path-integral treatment of acoustic mutual coherence functions for rays in a sound channel', J. Acoust. Soc. Am. 77, 1716-1722.

4. Reynolds, S.A., et al (1985) 'AFAR measurements of acoustic mutual coherence functions of time and frequency', J. Acoust. Soc. Am. 77, 1723-31.

5. Flatté, S.M., Reynolds, S.A., and Dashen, R. (1987) 'Path-integral treatment of intensity behavior for rays in a sound channel', J. Acoust. Soc. Am. 82, 967-972.

6. Flatté, S.M., et al (1987) 'AFAR measurements of intensity and intensity moments', J. Acoust. Soc. Am. 82, 973-980.

7. Flatté, S.M., and Stoughton, R. (1988) 'Predictions of internal-wave effects on ocean acoustic coherence, travel-time variance, and intensity moments for very long-range propagation', J. Acoust. Soc. Am. 84, 1414-1424.

8. Porter, M.B. (1990) 'The time-marched fast-field program (FFP) for modeling acoustic pulse propagation', J. Acoust. Soc. Am. 87, 2013-2023.

9. Brown, M. (1981) 'Application of the WKBJ Green's function to acoustic propagation in horizontally stratified oceans', J. Acoust. Soc. Am. 76, 1427-1432.

10. Freese, H. (1988) private communication.

11. Munk, W.H. (1974) 'Sound channel in an exponentially stratified ocean, with application to SOFAR', J. Acoust. Soc. Am. 55, 220-226.

12. Munk, W.H., and Wunsch, C. (1979) 'Ocean acoustic tomography: a scheme for large-scale monitoring', Deep-Sea Research 26A, 123-161.

13. Brown, M.G., Munk, W.H., Spiesberger, J.L., and P. Worcester (1980) 'Long-Range Acoustic Transmission in the Northwest Atlantic' J. Geophys. Res. 85, 2699-2703.

14. Flatté, S.M. et al (1979) Sound Transmission through a Fluctuating Ocean, Cambridge University Press, New York, Chapter 11.

DEPENDENCE OF NEAR - SURFACE
ACOUSTIC SCATTER ON WIND SPEED

B. EDWARD MCDONALD
Naval Research Laboratory
Washington DC USA 20375

ABSTRACT. Measurements of acoustic backscatter cross section per unit area of the ocean surface exceed Bragg scatter predictions. Volume scatter from bubble plumes accompanying breaking waves has recently been hypothesized as an explanation. This paper finds order - of - magnitude agreement between selected backscatter data and calculated volume scatter levels assuming acoustic inhomogeneities modeled after recent ocean measurements. Empirical data for whitecap coverage as a function of wind speed are combined with Born approximation backscatter cross section estimates for horizontally elliptical scattering patches whose strength decreases exponentially with depth. The model produces results in reasonable agreement with some of the data for backscatter as a function of wind speed without assuming bubble resonances. Estimates are cited on the effect of small numbers of resonant bubbles.

1. Introduction

Acoustic variability of the ocean's mixed layer is quite different from that of the deep ocean. This is due in part to the mixed layer's turbulence and short timescale response to winds and other local atmospheric forcing. The wind can alter acoustic propagation in the mixed layer by two means; by generating waves which cause rough surface scattering, and by altering the acoustic properties of the water itself. Changes in temperature, salinity, and microbubble content are examples of the latter. Scattering due to surface roughness has traditionally been estimated theoretically with a two - component model[1] of the roughness. On the small scales, capillary - gravity waves do the actual scattering. On a larger scale, swell contributes to a local average grazing angle for an incoming beam. In many instances, acoustic experiments conducted under various wind conditions have revealed higher backscatter than rough surface scatter theory predicts.[2] This is illustrated in Figure 1, where the discrepancy between representative data sets and rough surface scattering theory exceeds 10db for winds higher than about 10m/s. This level of unaccounted - for scatter motivates consideration of volume inhomogenieties in the acoustic index of refraction caused by vigorous churning of the upper mixed layer. A number of investigators have recently begun to consider the volume scatter hypothesis.

J. Potter and A. Warn-Varnas (eds.), Ocean Variability & Acoustic Propagation, 173–186.

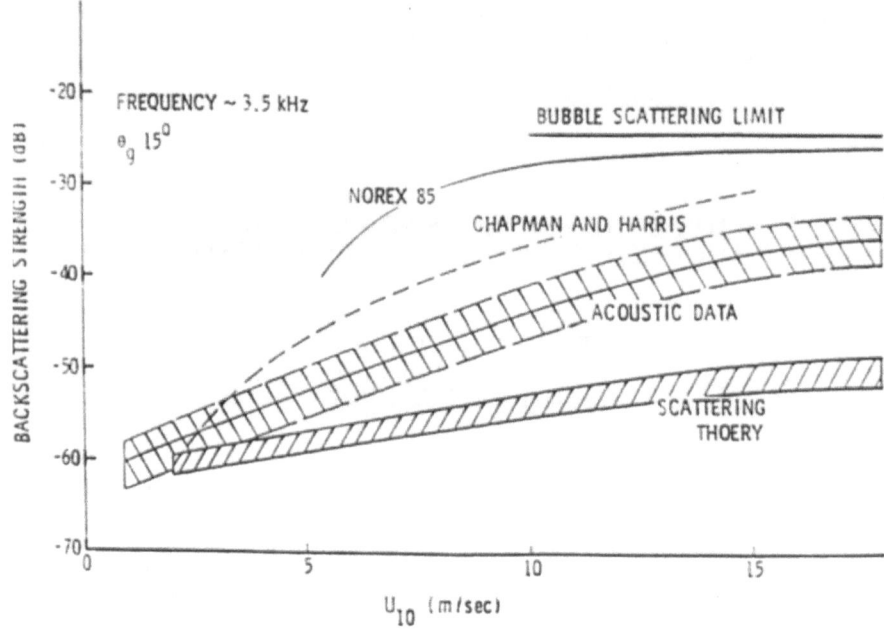

1 Comparison of results from rough surface scattering theory with relevant data sets (Figure courtesy S. McDaniel[2]). The lack of agreement at wind wind speeds above 5m/s motivates consideration of another scattering mechanism; *i.e.*, volume scatter.

2. Acoustic Alteration of the Mixed Layer: Fluid Mechanisms

2.1 Whitecapping

The onset of wind can lead to whitecapping, which causes intense local turbulence. Aside from noise generation, the dominant wind - related changes for acoustic propagation in the mixed layer are changes in termperature and air entrainment. A temperature drop from $20°C$ to $19°C$ lowers the sound speed c by approximately 2.7m/s. A much greater change in c results from entrainment of microbubbles by whitecaps. Figure 2 shows the dramatic change in c due to a small void fraction due to bubbles. A void fraction of 10^{-6} causes the sound speed to drop by about 12m/s. Surface values of the sound speed decrement δc_s in the range 10 - 20m/s have been measured in FASINEX.[3] The decrease in c due to near- surface cooling during a significant wind event is almost certainly dominated by the change due to air entrainment. In any case, the two effects reinforce each other and result in a layer of decreased sound speed. In the absence of deep convection, the sound speed decrement decreases approximately exponentially with depth, with an e- folding distance of 1 to 2 meters[3]. Horizontal structure within this near - surface layer contributes to acoustic scatter.

2.2 Langmuir Circulation

The material in this section is primarily descriptive and is intended to set the stage for future examination of low frequency backscatter ($f \lesssim$ 1kHz). Higher frequencies are considered in the results of section 3.

Deep convection patterns may be capable of pulling acoustically altered surface water down several meters into the water column. Evidence of this is seen in data presented by Farmer and Vagle[3] and by Zedel and Farmer.[4] The primary condidate for deep convection of acoustically altered surface water during wind events is Langmuir cell formation. Relevant features of this mechanism according to current understanding are as follows. Langmuir circulation patterns can be set in motion within approximately one - half hour after a wind commences.[5] The mechanism by which the cells sustain themselves is shown schematically in Figure 3. The energy source for the cells is the wind- driven vertical shear in the surface layer. Lines of vorticity $\zeta = \nabla \times \mathbf{v}$ near the surface are initially parallel to the ocean surface and perpindicular to the wind. Far below the surface, the flow is dominated by pairs of counterrotating rollers, whose vorticity is aligned alternately parallel and antiparallel to the wind.

Relevant properties of vortex dynamics which determine Langmuir cell development are: a) vortex lines cannot end in the fluid, but can be stretched indefinitely; b) they are advected with the fluid; and c) adjacent regions of co-aligned vorticity tend to wrap up together, yielding a larger region of the same net vorticity. When surface water is drawn down between two Langmuir cell rollers, a small depression developes in the vortex line denoted by $\zeta(t_1)$ in Figure 3 as part of the line is advected downward with the flow. Following the line's evolution from time t_1 to t_2, that portion of the

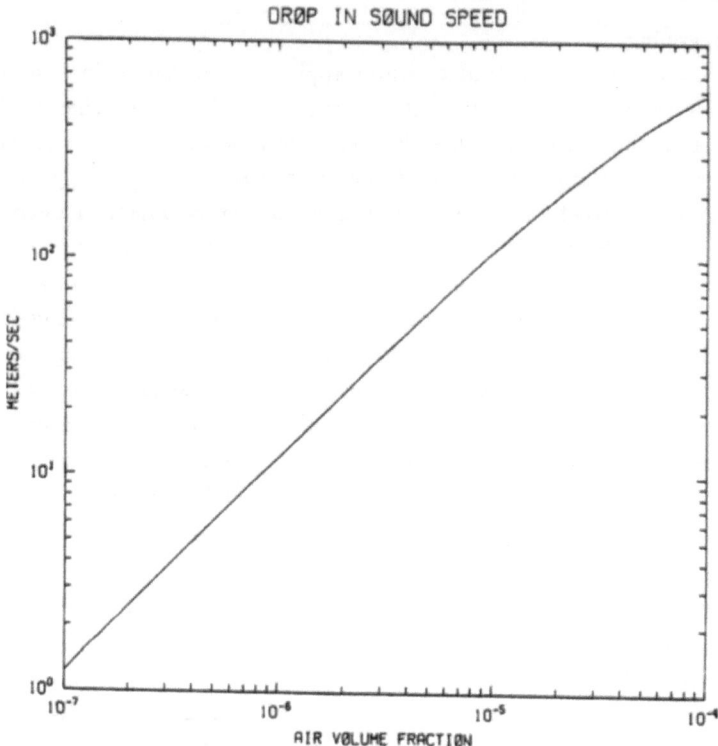

2 Effect of air entrainment on the sound speed of water.

LANGMUIR CIRCULATION DYNAMICS

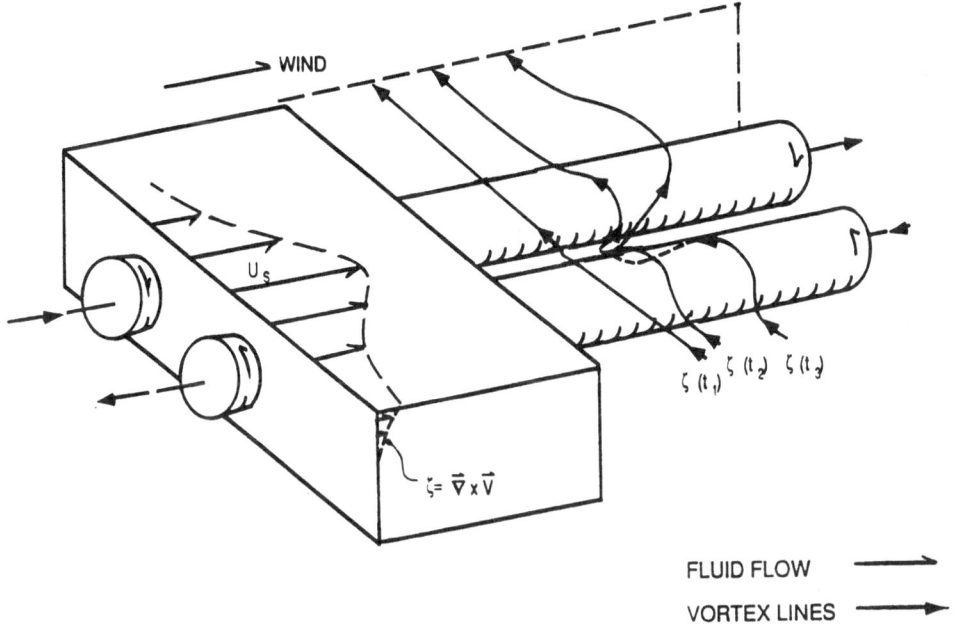

FLUID FLOW ⟶

VORTEX LINES ⟶

3 Schematic for Langmuir cell dynamics. The maximum surface drift u_s is observed to coincide with wind - aligned slicks and with downwelling. The latter is a manifestation of counter- rotating gyres beneath the surface.

vortex line which remains near the surface is dragged forward by the surface drift, while the portion caught between the rollers is pulled down into a region of low drift speed. Continuing on to time t_3, the portion of the line caught between the rollers has tilted forward to maintain contact with the near - surface portion. On each side of the symmetry plane between the rollers, $\zeta(t_3)$ has acquired a horizontal component parallel or antiparallel to the wind direction, and in agreement with the vorticity of the adjacent roller. The agreement of horizontal vorticities on a given side of the symmetry plane causes the vortex line to wrap up with the rollers in a manner that increases the net vorticity of each. The circulation pattern is then reinforced.

The velocity at which the water is pulled down between the rollers has been estimated empirically at 1 to 2 percent of the wind speed.[5] For a 15m/s wind, the sinking has been observed to be of order 15 to 30cm/s. The downwelling speed places constraints on the trajectories of entrained bubbles and their time- averaged scattering strength. If a downwelling flow is known or assumed, empirically derived relations between bubble radius and rise velocities can give upper bounds on the sizes of bubbles which may be carried to depth. Figure 4 illustrates bubble rise speed as a function of radius[6] as compared to theoretial estimates from classical drag laws.[7] The discrepancies between data and theory for bubbles of radius greater than about 5×10^{-3} cm is not well understood. One hypothesis under current consideration[8] has to do with vertical pressure gradient adjustment across the bubble surface. This hypothesis envokes the length scale $(\nu^2/g)^{1/3} \simeq 5 \times 10^{-3}$ cm (with ν the kinematic viscosity of water and g the acceleration of gravity), which is tantalizingly close to the radius at which the Stokes drag law begins to diverge from the data.

3. Acoustic Response: Estimates of Scattering Strength

We will consider here the level of scatter to be expected from nonresonant near- surface bubbles. Our desire is to consider scatter based on conservative assumptions before invoking resonance as a necessary ingredient (some doubt has been expressed that resonance has been seen in scatter experiments).

A simple scatter model is under development[8] for later comparison with representative data[2,9-11] on acoustic backscatter under the sea surface for frequencies between about 200Hz and 5kHz. Data are scarce below this range, and single- or multiple-bubble resonances may possibly become important above it. Basic features of the model at its current stage of development are as follows.

3.1 Geometry: Elliptical Patches in the Horizontal Plane

The model under consideration here assumes nonresonant bubble patches which are elliptically shaped in the horizontal, with the entrained air volume fraction horizontally uniform inside the patch, and decreasing with depth as $\exp(z/L)$ (see Figure 5). We approximate the surface to be a flat pressure release boundary (rough surface scatter

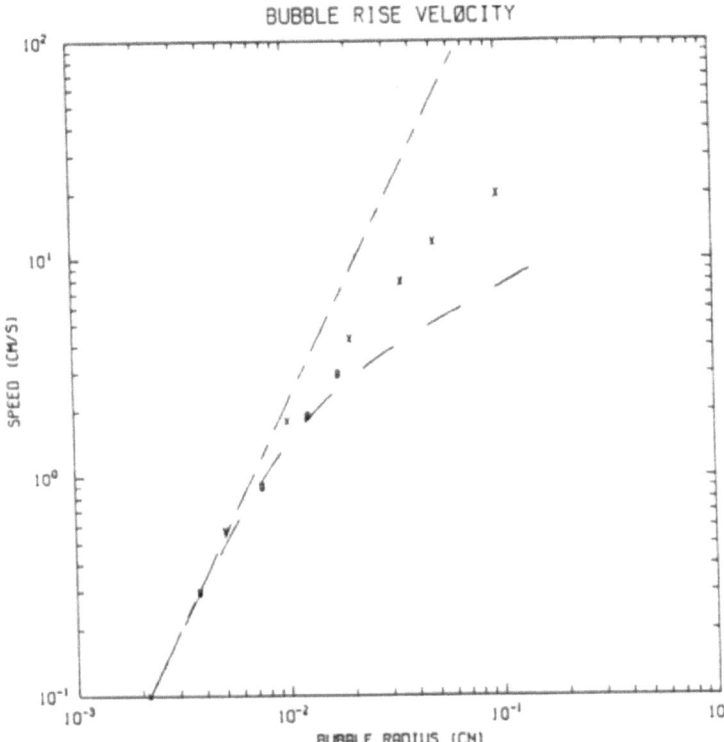

4 Bubble rise speed as a function of radius. Dot - dash line: Stokes drag result for a "hard sphere" bubble experiencing a force equal to the bubble's buoyancy. Dash line: Stokes result corrected for Reynolds stress. X and O points: data from ref. [6], fig. 2.6 and table 2.5, respectively.

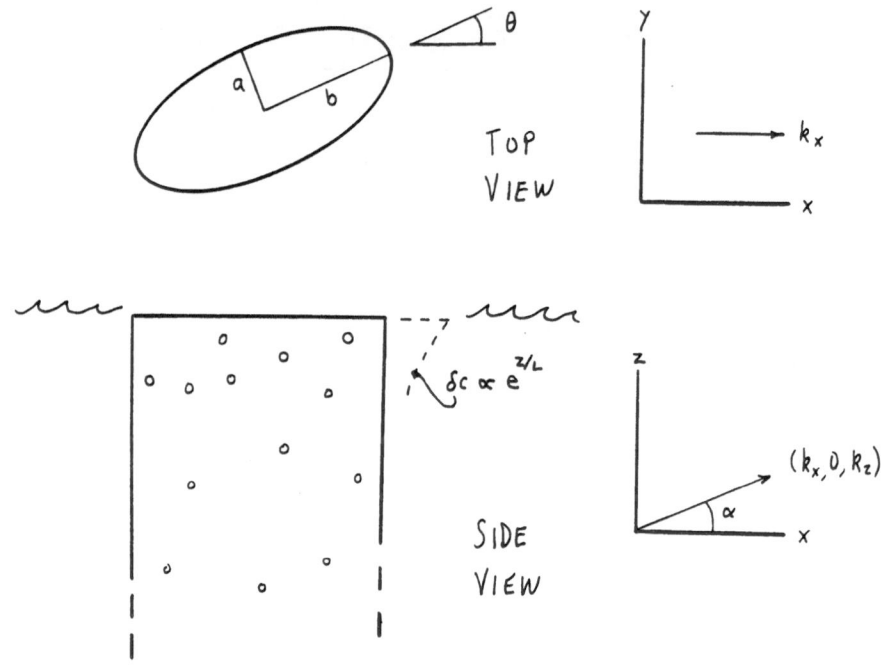

5 Geometry of hypothesized scattering patches: ellipse in horizontal with semimajor axis b oriented at an angle θ to the horizontal component of the incident acoustic wave. The sound speed defect δc is taken to be horizontally uniform in the ellipse, and to decrease exponentially with depth.

at significant wind speeds and small grazing angles has been estimated to be a small portion of the total backscatter as illustrated in Figure 1.)

3.2 Calculation of Backscatter Cross Section

Consider an incident beam which, in the absence of the surface, would be of the form $p_0 \exp i(\mathbf{k} \cdot \mathbf{r} - \omega t)$. When this beam encounters a single bubble of radius a at a depth z below a pressure release surface, and when the acoustic frequency is subresonant, the backscattered pressure field at a large distance r_0 back down the incoming beam direction is

$$p_{b\bullet} \simeq p_0 \frac{k^2 a^3}{r_0} \left(\frac{c_w^2 \rho_w}{3 c_a^2 \rho_a} \right) \exp 2i(\mathbf{k} \cdot \mathbf{r})(1 - e^{-2ik_\bullet z})^2 e^{i(2kr_0 - \omega t)}. \tag{1}$$

A term proportional to r_0^{-2} is neglected since it does not carry to great distances. Here c_w and ρ_w are the sound speed and density of water, while c_a and ρ_a are those for air. The origin of the vector $\mathbf{r}$ is taken to be within a small distance of the bubble (for purposes of later summing on bubbles in a local cloud), and r_0 is the distance from the field point to the local origin. Since (1) is proportional to the bubble volume, the scatter from a cloud of bubbles can be represented by an integral over the air volume in the cloud. The integral as opposed to a sum over individual bubbles is valid as long as the mean separation between bubbles is much smaller than the acoustic wavelength. The integral can also be cast in terms a continuum sound speed defect $\delta c(\mathbf{r})$, which is assumed to be a small fraction of the sound speed. The scattered power in the direction of the incident beam gives the backscatter cross section,

$$\frac{d\sigma}{d\Omega} = 4 \frac{k^4}{\pi^2} \left| \int \exp 2i(k_x x + k_y y) \sin^2 k_z z \left(\frac{\delta c(\mathbf{r})}{c} \right) dx \ dy \ dz, \right|^2 \tag{2}$$

where the integration is taken over a localized scattering volume. Equation (2) represents Born approximation scattering in the presence of a pressure release surface. Without loss of generality, take the incident wave vector to be $\mathbf{k} = (k \cos \alpha, 0, k \sin \alpha)$, where α is the grazing angle of the wave (see Figure 5). Take the elliptical scattering patch to have a semimajor axis oriented at an angle θ from the x direction. Carrying out the integral (2) over the ellipse and taking

$$\frac{\delta c}{c} = \frac{\delta c_\bullet}{c} \begin{cases} \exp \frac{z}{L} & \text{inside} \\ 0 & \text{outside} \end{cases} \tag{3}$$

where $\delta c_\bullet$ is a constant surface value, leads to the result

$$\frac{d\sigma}{d\Omega} = \left(\frac{4 L^2 k^2 \sin^2 \alpha}{1 + 4 L^2 k^2 \sin^2 \alpha} \right)^2 \left(2L \frac{\delta c_\bullet}{c} k^2 ab \frac{J_1(2k \cos \alpha A)}{2k \cos \alpha A} \right)^2, \tag{4}$$

where

$$A^2(\theta) = a^2 \sin^2 \theta + b^2 \cos^2 \theta, \tag{5}$$

with a and b the semiminor and semimajor axes of the ellipse. For low frequency and high frequency limits, the appropriate forms for the Bessel function are

$$J_1(x) \rightarrow \begin{cases} x/2, & |x| \ll 1 \\ \sqrt{\frac{2}{\pi x}} \cos(x - \frac{3\pi}{4}), & |x| \gg 1. \end{cases} \tag{6}$$

The resulting low and high frequency backscatter cross sections are

$$(kA\cos\alpha, kL\sin\alpha) \ll 1 : \frac{d\sigma}{d\Omega} \rightarrow 16L^4 a^2 b^2 k^8 (L\frac{\delta c_s}{c})^2 \sin^4\alpha$$

$$(kA\cos\alpha, kL\sin\alpha) \gg 1 : \frac{d\sigma}{d\Omega} \rightarrow \frac{ka^2 b^2}{\pi A^3}(L\frac{\delta c_s}{c})^2 \sec^3\alpha \cos^2(2kA\cos\alpha - \frac{3\pi}{4}) \tag{7}$$

$$\rightarrow \frac{ka^2 b^2}{2\pi A^3}(L\frac{\delta c_s}{c})^2 \sec^3\alpha$$

The second high frequency expression comes from $k-$ averaging over the rapidly varying phase angle $2kA\cos\alpha$. One notices from (7) that in the low frequency limit the amplitude of the backscattered field is proportional to the area of the elliptical patch, while in the high frequency limit, amplitude is proportional to the square root of one of the linear dimensions of the ellipse, and involves the horizontal propagation direction through the quantity $A(\theta)$.

For data comparison, we maximize the phase averaged high frequency cross section with respect to horizontal orientation θ. This gives $\theta_{max} = 90°$ and $A = a$, so that the incident wave's horizontal component hits the ellipse broadside. The resulting backscatter cross section is

$$\frac{d\sigma}{d\Omega}(\theta_{max}) \rightarrow \frac{kb^2 \cos\alpha}{2\pi a}\left(L\frac{\delta c_s}{c}\right)^2, \tag{8}$$

where a must exceed $\lambda/6\pi$ (with λ the acoustic wavelength) to justify the high frequency limit. The specific definition for the high frequency limit used here is that the argument of the Bessel function in (4) should exceed 3. We also assume that the grazing angle is large enough that $Lk\sin\alpha \geq 1$. For parameters L=1.5 m and acoustic frequency 3.5kHz, this gives $\alpha \geq 2.6°$.

B. Wind Speed Dependence

For wind speed between about 3 and 20 m/s, we use an empirical whitecap area coverage factor fit to Fig. 2.4 of Krause[6]:

$$F_{area} = 3.5 \times 10^{-3}(U_{10} - 3) + 4.2 \times 10^{-4}(U_{10} - 3)^2, \tag{9}$$

where the wind speed U_{10} is in m/s. F_{area} from (9) is shown in Figure 6. For purposes of later comparison with data, the backscatter cross section expressions of eqs. (2) - (8) are averaged over an area of the ocean surface containing several bubble patches. This is accomplished in the present approach by dividing $d\sigma/d\Omega$ by the area πab of the patch, then multiplying by the area coverage factor (9).

Backscatter data [2,9-11] give the backscatter cross section from one average square meter of ocean surface. The comparable quantity from the present model expressed in decibels relative to one square meter is the backscatter strength

$$BS = 10\log_{10}\left(\frac{F_{area}}{\pi ab}\frac{d\sigma}{d\Omega}(\theta_{max})\right) \qquad (10)$$

Sample results from the present model are given in Figure 7, which shows the wind speed dependence of (10) as a function of wind speed. The upper and lower curves are for $\delta c_s/c = .01$ and $.005$, respectively. Parameter values used are summarized in Table 1. These values result in reasonable agreement with recent data, although they fall short of the Chapman - Harris (1962) values represented in Figure 1.

TABLE 1. Parameters used in backscatter calculation.

c	1500m/s	L	1.5m
$\delta c_s/c$	.005, .01	a	.33m
α	15°	b	1.0m
θ	90°	f	3.5kHz

3.3 Bubble Resonances

The backscatter estimates developed here have assumed nonresonant scattering. It must be acknowledged that even a few resonant bubbles can have a large impact on acoustic scatter. It has been estimated[2] that the difference in backscatter between the curves marked "scattering theory" and "acoustic data" in Fig. 1 could be made up by hypothesizing a small number of resonant bubbles per unit area of water column. That number is $2 \times 10^{-4}m^{-2}$ at a frequency of 1kHz, and increases somewhat faster than the fourth power of frequency up to about 80kHz. Although this may seem to be a trivial number of resonant bubbles, the estimate is accompanied by a caveat that "... bubbles having radii larger than 500μ have not been observed in sea tests..." This means that for acoustic frequencies below about 7kHz, resonant scattering does not appear a likely explanation for observation.

4. Summary

Born approximation scatter from horizontally elliptical patches whose sound speed defect decreases exponentially with depth shows some agreement with near - surface

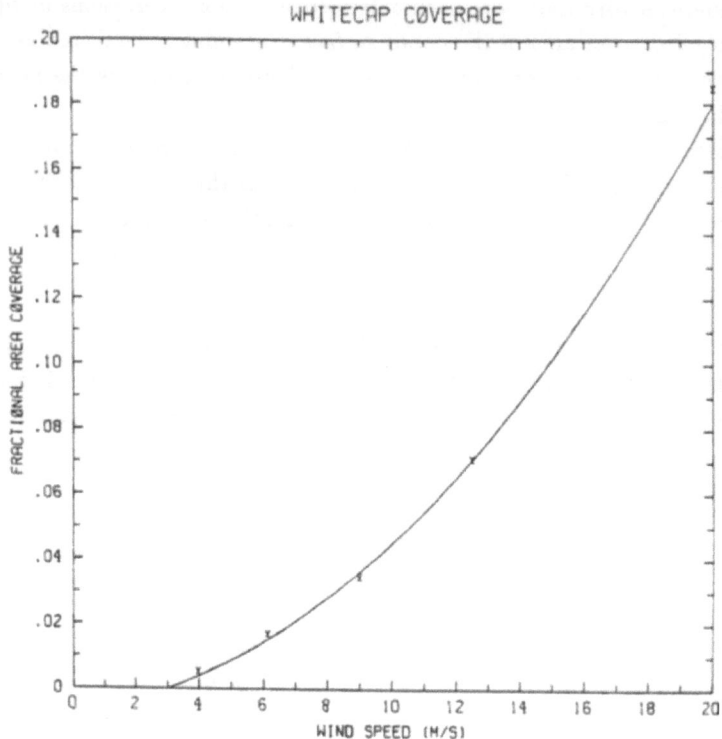

6 Whitecap coverage fraction as a function of wind speed from fig. 2.4 of ref. [6].

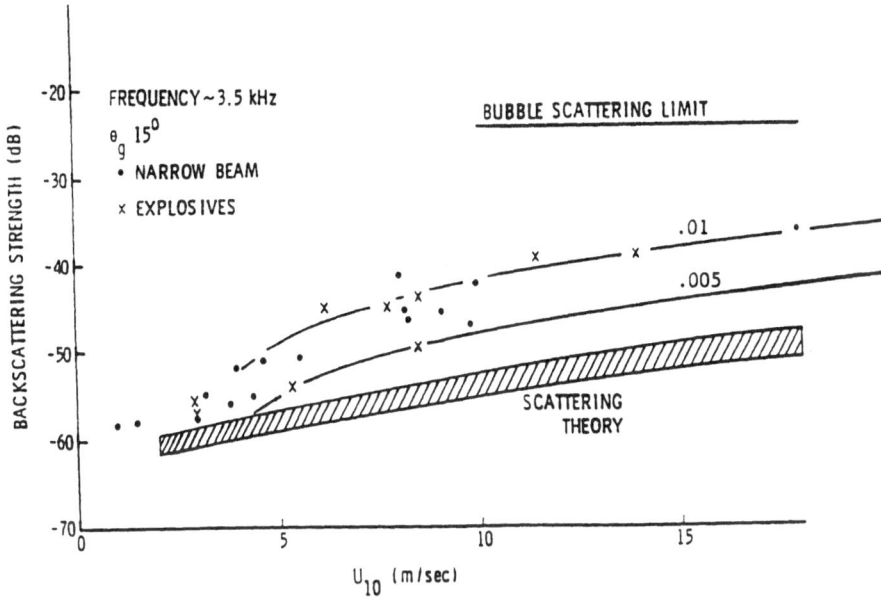

7 Present model results from eq. (11) superimposed on acoustic data set from Figure 1. Curve labels .01 and .005 refer to the maximum sound speed defect c_*/c assumed for the scattering patches (Figure courtesy S. McDaniel[2]).

acoustic backscatter data as a function of wind speed for parameters given in Table 1. It is premature, however, to conclude validity of the current model based on agreement for a limited parameter set. Backscatter predictions must be compared with data as a function of a number of parameters. Among these are grazing angle, acoustic frequency, elongation of the patches, and horizontal beam direction relative to patch orientation. This work is in progress and will be reported elsewhere.

5. Acknowledgment Work supported by ONR Code 11250A. The author wishes to thank Dr. Suzanne McDaniel for helpful discussion and data presentation in Figures 1 and 7.

6. References

1 B. F. Kur'ynov, "The Scattering of Sound Waves at a Rough Surface with Two Types of Irregularity,", *Sov. Phys. Acoust.*8, 252 (1963).

2 S. T. McDaniel, "High Frequency Scattering from the Sea Surface: Recent Progress," invited paper, *J. Acoust. Soc. Am.* 84, S121 (1988).

3 D. M. Farmer and S. Vagle, "Waveguide Propagation of Ambient Sound in the Ocean - Surface Bubble Layer," *J. Acoust. Soc. Am.* 86, 1897 - 1908 (1989).

4 L. Zedel and D. M. Farmer, "Bubble Clouds and Langmuir Circulations: Acoustical Observations During Ocean Storms," *EOS Trans. A. G. U.* 71, 73 (1990).

5 J. Smith and R. Pinkel, "Velocity Structure in the Mixed Layer During MILDEX," *J. Phys. Oceano.* 17, 425 - 439 (1987).

6 E. B. Krause, *Atmosphere - Ocean Interaction*, Oxford University Press (1972).

7 A. Sommerfeld, *Mechanics of Deformable Bodies*, Academic Press (1964).

8 B. E. McDonald, "Backscatter from Near- Surface Bubble Clouds," *Proc. Conf. on Natural Physical Sources of Underwater Sound*, Cambridge University 1990 (*in prep.*).

9 W. Bachmann, "A Theoretical Model for the Backscattering Strength of a Composite Roughness Sea," *J. Acoust. Soc. Am.*54, 712 (1973).

10 P. A. Crowther, "Acoustic Scattering from Near - Surface Bubble Layers," in *Cavitation and Inhomogeneties in Underwater Acoustics* ed. W. Lauterborn (Springer, New York, 1979).

11 B. Nutzel, H. Herwig, J. M. Monti, and P. D. Koenigs, "The Influence of Surface Roughness and Bubbles on Sea Surface Acoustic Back - Scattering," NUSC Tech. Rep. 7955, Naval Underwater Systems Center, New London CT (1987).

NONLINEAR EFFECTS IN WIND-WAVE GENERATION

Robert H. Mellen
Kildare Corp.
95 Trumbull St.
New London CT 06320

ABSTRACT. Wave dynamics play a critical part in the analysis of sea-surface scattering. Wave models are generally based on linear theory. However, both backscatter Doppler and wave-flume measurements show that wind-waves are non-dispersive in the gravity-capillary régime, indicating that nonlinear effects are involved. Experiments were carried out in a small laboratory flume to compare measurements of Doppler spectra and surface motion in the initial phases of wind-wave growth. Wave-gauges were used to measure waveheight vs. time and the frequency spectra were computed by FFT. In addition to autospectra, cross-spectra of signals from gauge-pairs were computed to investigate the statistical relationship between frequency and wavenumber. The results are modeled by a method similar to that used in the analysis of turbulent flow.

1. Introduction.

Wave dynamics play an critical part in the analysis of both sonar and radar scattering by the sea surface. Scattering strength calculations require the wavenumber spectrum of surface roughness [1,2]. Wavenumber spectra can be difficult to measure at the high wavenumbers involved in scattering and are generally inferred from frequency spectra. The assumption is that wavenumber and frequency are delta-function related by the dispersion equation and that linear theory provides a complete description of the wave dynamics. However, both wave-flume and Doppler measurements show that wind-waves tend to be non-dispersive at the higher wavenumbers, which indicates the importance of nonlinearity.

Evidence for the anomaly is quite extensive. Results of field measurements of acoustic backscatter Doppler were apparently first reported by the author [3]. Similar effects were observed in the acoustic experiments of Zel'dis, et al. [4,5] and Konrad, et al. [6] and in the radar experiments of Wright and Keller [7]. Wave-flume experiments were reported by Shemdin [8] and included a theoretical analysis involving the nonlinear effects of the drift layer. Flume experiments by Ramamonjiarisoa, et al. [9] at larger fetch showed dispersion only near the dominant wave frequency 2 Hz. Despite the evidence, there has been little or no effort to incorporate the findings in the wave models used in scattering theory.

Experimental evidence is presented in this paper that further confirms the anomaly and supports the hypothesis that nonlinear effects are important in wind-wave dynamics at the higher wavenumbers. The statistical relations between frequency and wavenumber are also investigated by correlation methods and the results are modeled by a method similar to that used in the analysis of turbulent flow [10].

J. Potter and A. Warn-Varnas (eds.), Ocean Variability & Acoustic Propagation, 187–198.
© 1991 *Kluwer Academic Publishers.*

2. Wave-Flume Measurements.

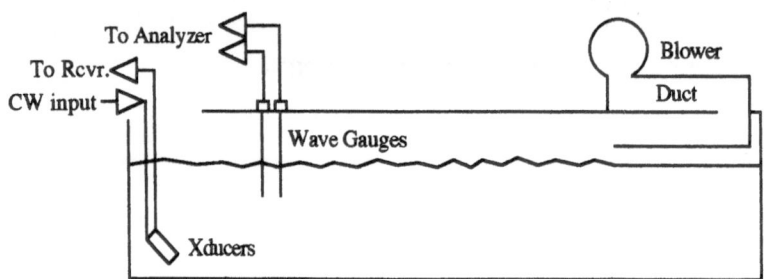

Figure 1. Wind-wave flume

The investigations of wind-wave dynamics were carried out in the small laboratory flume shown in Fig. 1 at approximately 1 m fetch with wind speeds up to 5 m/s. Conductance wave-gauges were used to measure waveheight vs. time. The gauge signals were digitized at 128/s and analyzed by Fast Fourier Transform (FFT). Pairs of gauges were used for the correlation measurements.

For comparison with acoustic backscatter measurements, separate identical transducers were used for transmit and receive in order to minimize the effects of acoustic cross-talk. The highly directional beams were oriented to illuminate a common surface area and the signals were also digitized and analyzed by FFT to compute the Doppler spectra.

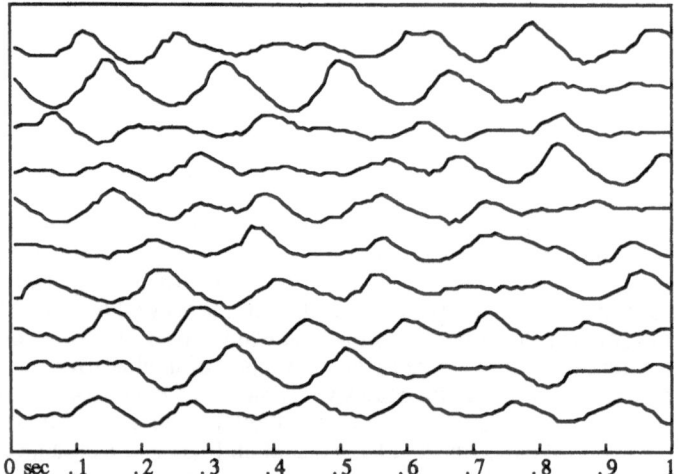

Figure 2. Wave amplitude vs. time for 5 m/s wind speed.

Figure 2 shows a portion of a typical recording of the signal from a single gauge after the waves have reached equilibrium. Time increases continuously from top to bottom. The rms amplitude is about 1 mm. The steep wavefronts in the quasi-sinusoidal regions are evidence of the finite-amplitude overtaking effect.

Frequency spectra were computed from the recordings by FFT. A Hanning window was used to suppress start/stop transients, which otherwise cause errors at higher frequencies. The autospectra were averaged for 100 samples or more.

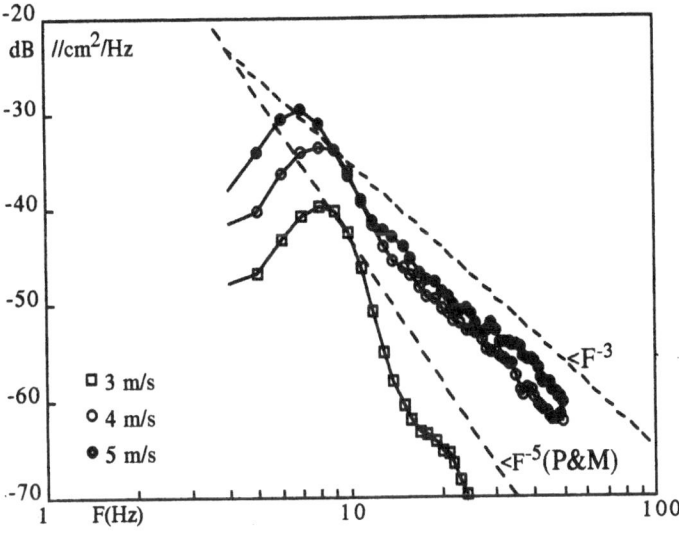

Figure 3. Autospectra of waves at 1 m fetch.

The autospectra for wind speeds 3, 4 and 5 m/s in Fig. 3 show that the spectrum tends to constant slope at higher frequencies as the wind speed is increased, which is similar to the spectrum of Pierson and Moskowitz (P&M) [11] except that the asymptotic dependence is F^{-3} rather than F^{-5}. Also, at 5 m/s wind speed, the frequency of the spectral peak at 1 m fetch is roughly 8 Hz compared to 0.3 Hz for the fully-developed sea conditions of P&M.

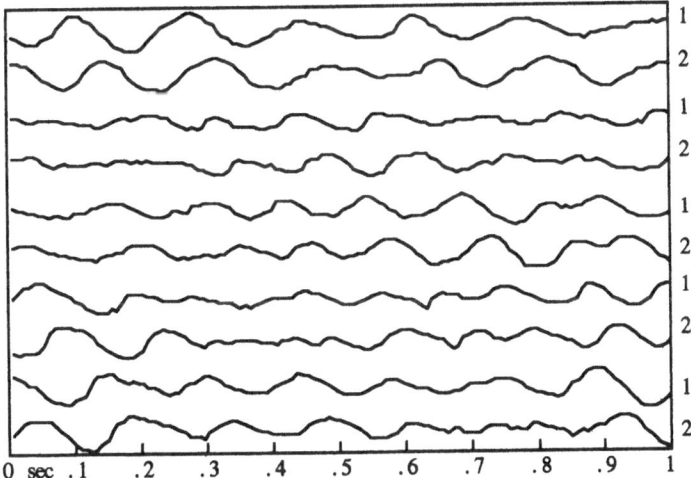

Figure 4. Comparison of gauge-signals with 1.5 cm. longitudinal separation.

Figure 4 compares the signals recorded from two gauges at wind speed 5 m/s. The time scale is the same as in Fig. 1 and the signal pairs have been displaced vertically for clarity. Signal 1 is from the upwind gauge and signal 2 shows the effects of decorrelation as well as time delay.

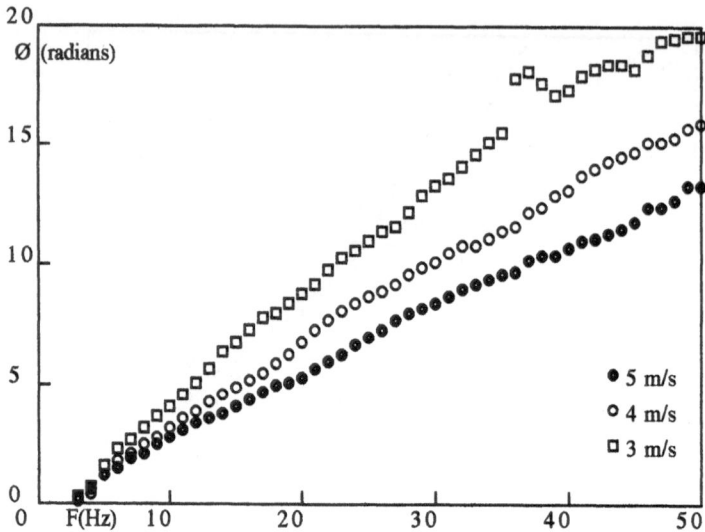

Figure 5. Phase vs. frequency of the longitudinal cross-spectra.

Figure 5 shows the computed phase-dependence Ø(F) for the 3 wind speeds. The gauge spacing 1.5 cm. was chosen as a compromise between decorrelation and cross-talk. Phase shifts are calculated from the formula:

$$\text{Ø}=\arctan(Si/Sr)+2n\pi, \quad n=0,1,2... \tag{1}$$

where Sr and Si are the real and imaginary components of the cross-spectra and n increases stepwise with frequency at each zero crossing. The phase speed C(F) is calculated from the relation Ø=Kd=2πFd/C, where d is the gauge spacing.

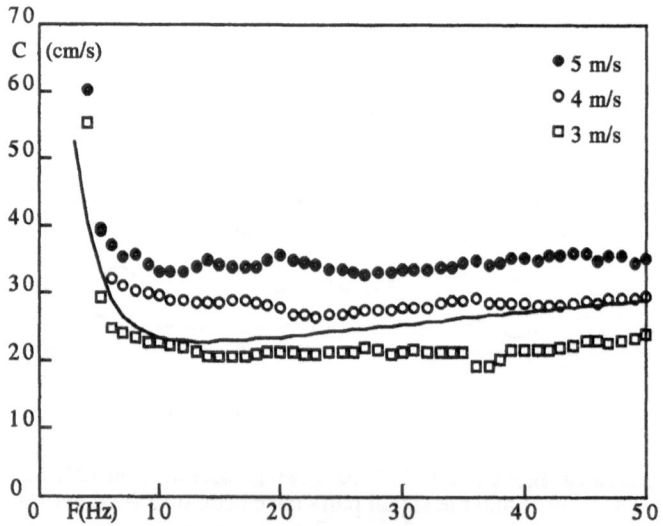

Figure 6. Longitudinal phase speeds vs. frequency.

Figure 6 shows longitudinal phase speed C(F) calculated from the phase data of Fig.5. The solid curve is the (linear) dispersion relation. The experimental data show wind-speed dependence but no evidence of dispersion except for frequencies well below the spectral peaks in Fig. 3. The dispersion equation is:

$$(C/C_o)^2 = K_o/K + K/K_o, \quad K_o = (\rho g/\sigma)^{1/2}, \quad C_o = (g/K_o)^{1/2} \tag{2}$$

where K is the surface wavenumber, g is the gravity constant and σ is the surface tension. The solution for $K(\omega)$, where $\omega = 2\pi F$, is:

$$K(\omega) = K_o[x + (x^2 + 1/27)^{1/2}]^{1/3} + [x - (x^2 + 1/27)^{1/2}]^{1/3}, \quad x = \omega^2/2K_o g \tag{3}$$

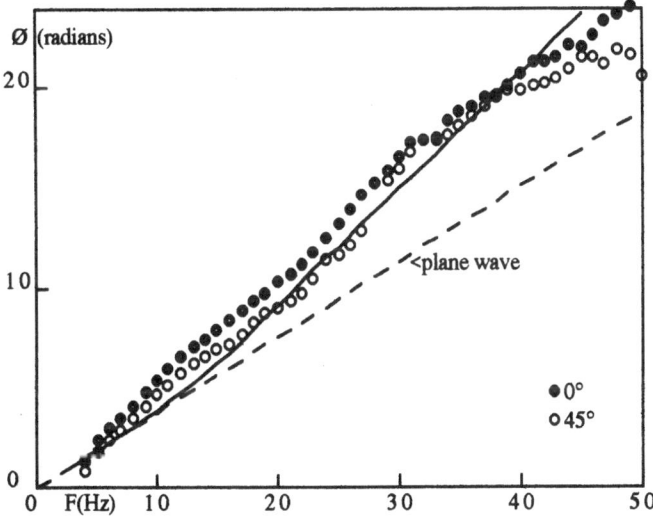

Figure 7. Phase vs. frequency in the longitudinal and diagonal directions.

Phase measurements in the diagonal (45°) and longitudinal (0°) directions for wind speed 5 m/s and 1.5 cm. gauge-spacing are compared in Fig. 7. In plane-wave theory, the slope at 45° is less than that of the 0° case by the factor cos(45°), as shown by the dashed line. However, the data are closer to the 0° values, which indicates angular spreading.

3. Theory.

Turbulent flow is modeled as random fluctuations convected by a steady current [10]. Cross-spectrum measurements show that the phases increase linearly with longitudinal spacing, indicating plane-wave propagation. Magnitudes of the normalized longitudinal and transverse cross-spectra decay exponentially in proportion to wavenumber times distance.

To model the wavenumber/frequency relationship, consider the longitudinal (x) and transverse (y) axes with respective decay coefficients α and β and take the iso-amplitude contours of the spatial Fourier window as elliptical. For propagation in the x direction, the space/frequency correlation function becomes:

$$\Psi(\omega,x,y)=\exp\{iK_\omega x-[(\alpha x)^2+(\beta y)^2]^{1/2}\}, \qquad K_\omega=\omega/C \tag{4}$$

The wavenumber/frequency spectrum is then given by $S(\mathbf{K},\omega)=S(\omega)\,T(\mathbf{K},\omega)$, where $\mathbf{K}$ is the wavenumber vector and $S(\omega)$ is the frequency spectrum. By Fourier transformation of Eq. 4 [12]:

$$T(\mathbf{K},\omega)=(\alpha\beta)^2/2\pi[(\alpha\beta)^2+(\alpha K_y)^2+\beta^2(K_x-K_\omega)^2]^{3/2} \tag{5}$$

where K_x and K_y are the axial components of $\mathbf{K}$ and 2π is the normalization factor.

The space/frequency correlation function for an angular distribution of plane waves can be taken as:

$$\Psi(\omega,r,\phi_o) = {}_{-\pi/2}\!\int^{\pi/2}d\phi\, D(\phi)\, \exp[-\alpha(\phi)r+iK_\omega(\phi)r\cos(\phi-\phi_o)] \tag{6}$$

where $r=(x^2+y^2)^{1/2}$, ϕ is the azimuth angle and $D(\phi)$ is the normalized directivity function. Fourier transformation of Eq. 6 gives:

$$T(\mathbf{K},\omega)=(\alpha/2\pi)\,{}_{-\pi/2}\!\int^{\pi/2}d\phi\, D(\phi)/[\alpha^2+(K_y-K_\omega\sin\phi)^2+(K_x-K_\omega\cos\phi)^2]^{3/2} \tag{7}$$

where $K_x=K\cos\phi_o$ and $K_y=K\sin\phi_o$.

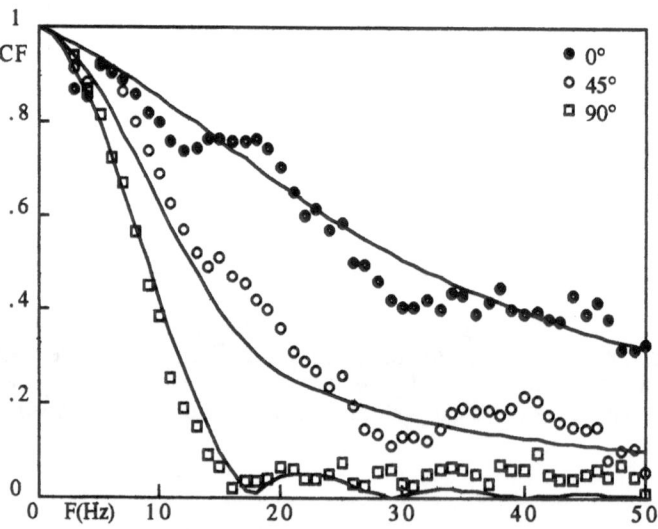

Figure 8. Magnitudes of normalized cross-spectra at angles $\phi_o=0°$, 45° and 90°.

The normalized magnitudes of the cross-spectra data for 5 m/s wind speed are compared to the spreading model in Fig. 8. The solid curves are numerical evaluations of Eq. 7 for $D(\phi)=(8/3)(\cos\phi)^4$, with C=33 cm/s and $\alpha=0.04K_\omega$ both independent of ϕ. The analysis of acoustic backscattering will be based on these approximations.

4. Acoustic Measurements.

Model predictions are compared with acoustic backscattering data by perturbation scattering theory [1,2]. The resonance conditions are $K_x=2k\cos\theta_o\cos\phi_o$ and $K_y=2k\cos\theta_o\sin\phi_o$, where k is the acoustic wavenumber and θ_o is the grazing angle. The measurements were made in the upwind direction ($\phi_o=0$) at the acoustic frequency f=550 kHz with 5 m/s windspeed and 1 m fetch. The Doppler spectrum vs. frequency-shift Δf is therefore:

$$SD(\Delta f)\approx4(k\sin\theta_o)^4\ S(\Delta f)\ T(2k\cos\theta_o,\ 2\pi\Delta f) \tag{8}$$

where $T(2k\cos\theta_o,\ 2\pi\Delta f)$ is obtained by numerical integration of Eq.7. Scattering strength is obtained by numerical integration of Eq. 8 over Δf.

The backscattered CW signals were detected by means of a radio receiver capable of both SSB and AM modes. Both sidebands were examined and the lower-sideband spectrum was found to be negligible. Since the upper sideband and AM modes gave the same results, the latter mode was used in order to eliminate frequency-drift problems. The received signals were recorded digitally at 1024/s and then analyzed by FFT (N=1024).

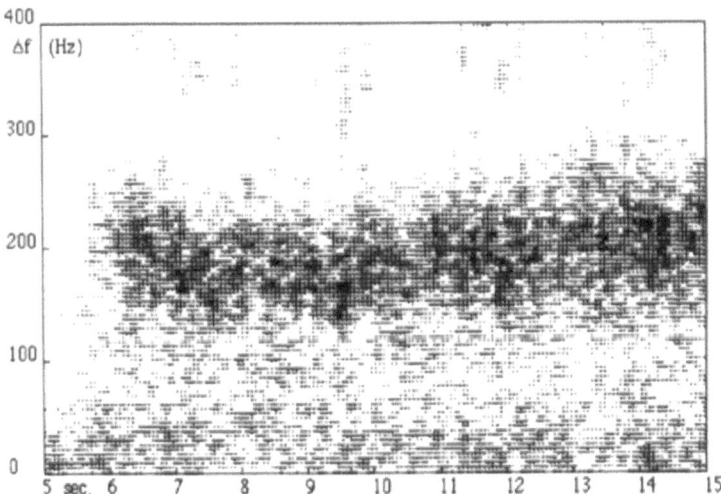

Figure 9. Sonagram of the Doppler spectrum.

The digital frequency/time spectrum (sonagram) in Fig. 9 shows the development of the Doppler spectrum with time. The frequency scale Δf is the shift relative to the CW carrier. By use of a Gaussian time-window, time resolution is reduced to 1/4 sec. and the window stepped in 1/16 sec. increments. The grey-scale consists of 3×3 pixels, each corresponding to an intensity factor of two. The overall dynamic range is then roughly 27 dB from all white to all black. The start-time of the sonagram is 5 secs. after turning on the blower and it is seen that the spectrum takes about 6 secs. to build up.

The sonagram of the wave spectrum in Fig. 10 is shown for comparison. The gauge signals were digitized at 128/s and analyzed by FFT (N=256) with a Hanning window, which gives roughly 1 sec. time resolution. The spectrum has been "whitened" to bring the high frequencies into the grey-scale range. Comparison with Fig. 9 indicates that the high frequencies of the Doppler spectrum take slightly longer to build up.

194

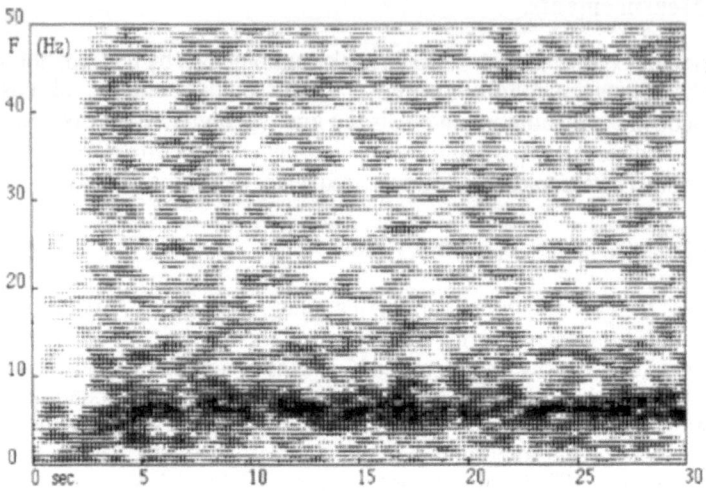

Figure 10. Sonagram of the wave spectrum.

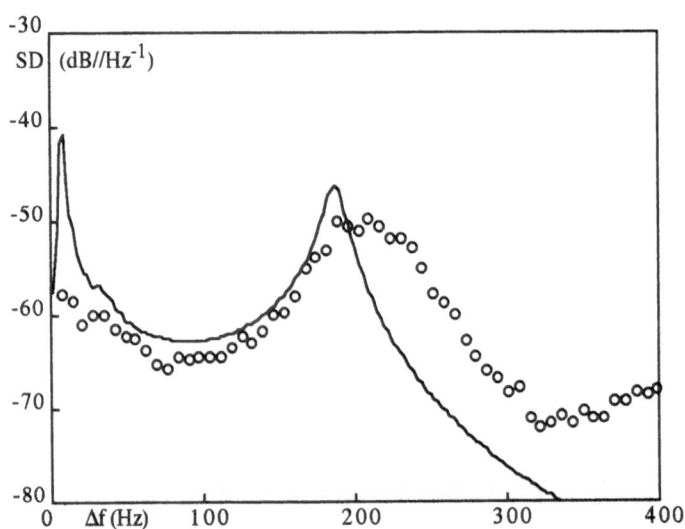

Figure 11. Comparison of measured and theoretical Doppler spectra.

The 60 sec. average Doppler spectrum for wind speed 5 m/s and $\theta_0=40°$ is compared with theory in Fig. 11. The system was calibrated by the pulse-echo method with a 1" dia. stainless-steel sphere at the same range as that of the experiments. At normal incidence, the effective areas would be approximately equal. The circles are measured levels corrected for grazing angle and signal strengths relative to the target strength of the sphere [13].

The solid curve is the spreading model of Eq. 7. The values used in numerical evaluation of Eq. 8 are k=23 cm-1, and C=33 cm/s. The wave spectrum S(F) used in Eq. 8 employs measured values from Fig. 3 up to 50 Hz and is extrapolated asymptotically as F^{-3} above 50 Hz.

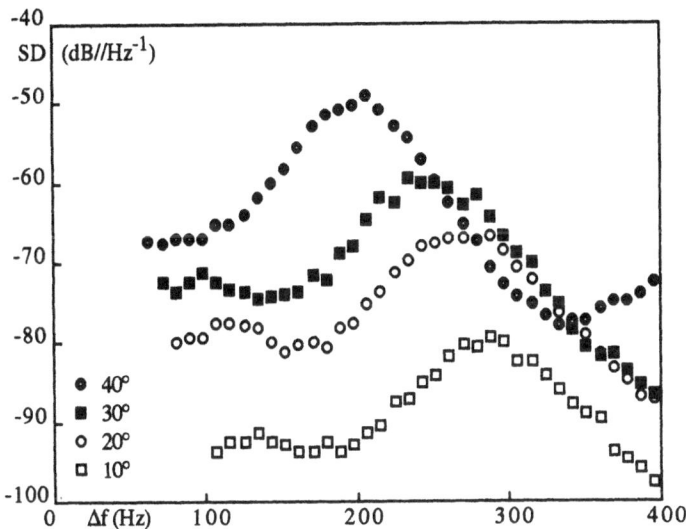

Figure 12. Grazing-angle dependence of the Doppler spectrum.

The Doppler spectra for selected grazing angles θ_o are compared in Fig. 12, showing the decrease in amplitude and the upward shift of the spectral peak with decreasing grazing-angle. Data points are omitted in the low-frequency range where extraneous reverberation is apparently dominant.

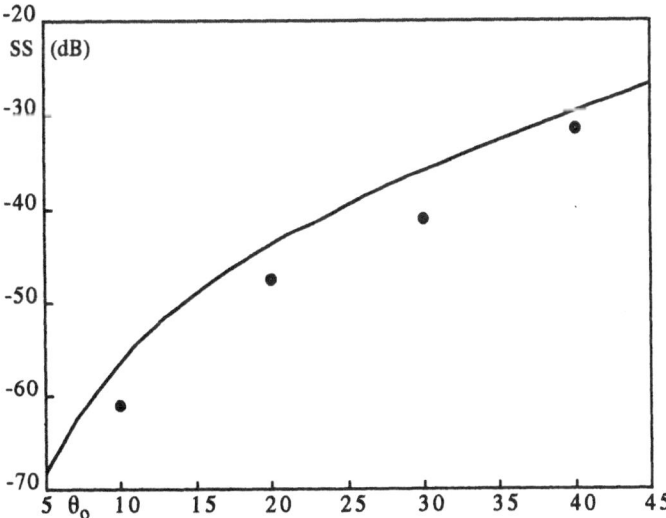

Figure 13. Scattering strength vs. grazing angle.

Calculations of scattering-strength vs. grazing angle θ_o are compared with experimental data in Fig. 13. The curve is the analytic result obtained by numerical integration of Eq. 8. The experimental points are calculated by numerical integration of the Doppler spectra of Fig. 12.

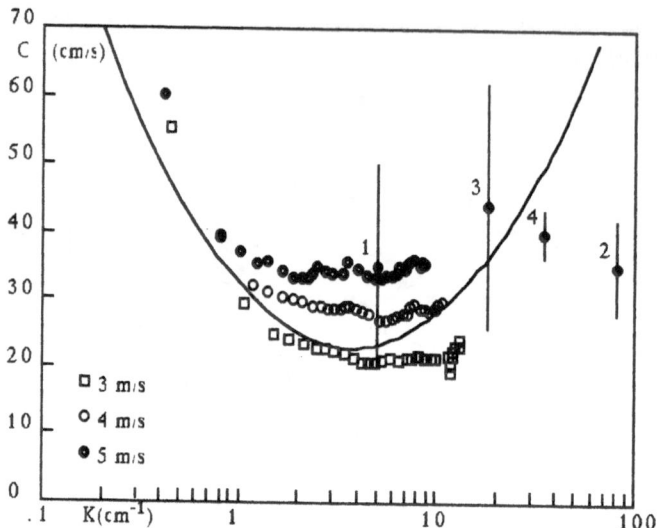

Figure 14. Comparison of measured phase velocities with experimental Doppler data:
Doppler data: location, ref., frequency, grazing angle, wind speed.
1: Thames River [3], 85 kHz, 45°, 10 m/s; 2: Thames River [3], 1400 kHz, 45°, 10 m/s;
3: Lake Seneca [6], 250 kHz, 30°, 7m/s; 4: present work, 550 kHz, 40°, 5m/s.

Results of laboratory and field measurements of phase velocity are compared in Fig. 14.
The wave data of Fig. 6 are plotted here vs. wavenumber $K=2\pi F/C$. The Doppler data are
for either upwind or downwind-looking conditions and the phase velocities are calculated
from the Doppler formula $C=c_o|\langle\Delta f\rangle|/2f\cos\theta_o$, where $\langle\Delta f\rangle$ is the mean frequency-shift of
the Doppler spectrum and c_o is sound speed. The acoustic wavenumbers are calculated by
the resonance formula. The numbered points are the mean values and the bars indicate the
rms spread arising from the combined effects of normal fluctuations, modulation by gravity
waves and finite beamwidths. The solid curve is linear dispersion theory of Eq. 2.

5. Discussion.

The autospectra in Fig. 3 show slight undulations above the spectral peak. Similar results
were reported by Leykin, et al. [14] which, according to the nonlinear wave-theory of Lake
and Yuen [15], they ascribe to harmonics of the dominant wave-frequency. The existence
of coherent harmonics is evident from the steep wavefronts of the quasi-sinusoidal regions
of the time series in Figs. 2 and 4. The smooth asymptotic trend at higher frequencies was
not observed because their measurements were limited to less than 10 Hz.

The phase data of Fig. 7 and the cross-spectra of Fig. 8 also show undulations, but the
significance is still not clear. Discrepancies between theory and data, particularly in phase,
suggest that the model may not be adequate.

One problem with the model is the poor agreement with the measured Doppler spectrum
at high frequencies, as seen in Fig. 11. The theoretical spectral peak is lower in frequency
and the bandwidth is less than the measurements, while total energies are nearly the same.
Taking Q as the ratio of mean frequency shift to effective bandwidth, it is seen that $Q\approx3$ for
the measurements. From Eqs. 7 and 8, $Q\approx2k\cos\theta_o/\alpha=25$ for the model.

Another problem is the high peak near 8 Hz in Fig. 11. Although the energy is small, it should be observed, if present, unless attenuated by the spatial-filter effect of the acoustic aperture and then obscured by extraneous reverberation.

While the low-frequency waves do not contribute much to scattering strength, they do cause frequency modulation. Sonagram analysis of the Lake Seneca data shows that the large rms spread in Fig. 14 is due to slow displacement of the scatterers by gravity waves, while the short-term spread is comparable to laboratory values. This and similar results by Zel'dis, et al [5] suggest that laboratory values apply to open-sea conditions if modulation by the orbital motion of gravity waves is taken into account. However, modulation effects apparently do not account for the bandwidths of the current experiments.

The grazing-angle dependence of the spectral peaks in Fig. 12 is in good agreement with the Doppler formula and indicates a longitudinal phase speed of roughly 40 cm/s compared to 35 cm/s for the model. The phase measurements in Fig. 7 also show some indication of increasing speed at the higher frequencies, which might explain this discrepancy.

The fair agreement between the measured and predicted scattering-strengths in Fig. 13 is evidence that perturbation theory is valid for the current experiments. The measurements of Zel'dis, et al. [4,5] also show good agreement at lower frequencies.

The uniformity of the average phase speeds of the Doppler data of Fig. 14 is further evidence that the dispersion relation does not hold over the entire wavenumber range.

6. Conclusion.

General consistency between laboratory and field measurements clearly demonstrates the importance of nonlinearity in wind wave dynamics. This conclusion is further supported by other investigations by the author [16] relating to the growth mechanism. By "seeding" with mechanically-generated waves, a number of nonlinear phenomena were revealed by sonagram analysis similar to that in Fig. 10. For 8 Hz excitation, strong quasi-coherent amplification was observed initially, followed by more erratic lock-on behavior. For 16 Hz excitation, the initial amplification was followed by a sudden jump to a broadband response at 8 Hz, indicating a weak type of period doubling. For 5 Hz excitation, amplification was weaker and frequency doubling dominated. These results indicate that synergetic feedback between wind and waves is an important part of the nonlinear growth mechanism.

Nonlinearity implies that the frequency/wavenumber relation must be statistical and not deterministic, as in linear theory. Zaslavskiy and Leykin [17] interpret wave measurements in the gravity régime as a "blurring" of the linear dispersion relation and conclude that the measured value $Q \sim F/\Delta F \sim 1$ indicates that short waves have dispersion characteristics similar to turbulence.

Nonlinear wavefront-steepening also means that wave dynamics can not be considered a random process. Laybo and Leykin [18] investigated the bi-spectra of gravity waves and found significant correlation between harmonics. Investigations involving the generation of ambient noise by fluctuations of the Bernoulli-pressure $\rho U^2/2$ (U=vertical particle-velocity) were also carried out by the author [19]. Here, the measured spectrum of U^2 was found to be asymptotically 3 dB higher than that obtained by convolution of the measured spectrum of U. Since convolution assumes that all phases are random, this result is also indicative of harmonic coherence.

It is clear that a nonlinear wave model is essential for the description of wave dynamics involved in scattering. An effort has been made here to incorporate some of the observed effects in a heuristic model. However, despite the rough qualitative agreement between predictions and experiment, the many obvious discrepancies indicate that the model is still far from satisfactory.

198

7. References.

[1] Brekhovskikh, L. and Lysanov, Y. (1982) Fundamentals of Ocean Acoustics, Springer-Verlag, Berlin, Ch. 9.
[2] Mellen, R. H. (1989) 'On underwater sound scattering by surface waves', IEEE J. Oceanic Eng. 14, 245-247.
[3] Mellen, R. H. (1964) 'Doppler shift of sonar backscatter from the sea surface', J. Acoust. Soc. Am., 36, 1395-1396.
[4] Zel'dis, V. I., Leikin, I. A., Rozenberg, A. D. and Ruskevich, V. G. (1973) 'Amplitude characteristics of sound signals scattered by water waves', Sov. Phys. Acoust., 19, 118-122.
[5] Zel'dis, V. I., Leikin, I. A., Rozenberg, A. D. and Ruskevich, V. G. (1974) 'Phase characteristics of sound signals scattered by water surface waves', Sov. Phys. Acoust., 20, 145-149.
[6] Konrad, W. L., Browning, D. G. and Mellen, R. H. (1981) 'Doppler spectra of sea-surface backscatter at high acoustic frequencies', NUSC TM 6735.
[7] Wright, J. M. and Keller, W. C. (1971) 'Doppler spectra in microwave scattering from wind waves', J. Phys. Fluids, 14, 466-474.
[8] Shemdin, O. H. (1972) 'Wind-generated current and phase speed of water waves', J. Phys. Oceanography, 2, 411-419.
[9] Ramamonjiarisoa, A., Baldy, S. and Choi, I. (1978) 'Laboratory studies on wind-wave generation, amplification and evolution', in A. Favre and K. Hasselmann (eds.), Turbulent Fluxes Through The Sea Surface, Plenum Press, New York, pp.403-420.
[10] Smol'yakov, A. V. and Tkachenko, V. M. (1983), The Measurement of Turbulent Fluctuations, Springer-Verlag, Berlin.
[11] Pierson, W. J. and Moskowitz, L. (1964) 'A proposed spectral form for fully developed wind seas based on the similarity theory of S. A. Kitaigordski', J. Geophys. Res., 69, 5181-5190.
[12] Gradshteyn, I. S. and Ryzhik, I. M. (1965) Tables of Integrals, Series and Products, Academic Press, New York, Eqs. 3.914, 6.699(12).
[13] Urick, R. J. (1983) Principles of Underwater Sound, McGraw-Hill, New York.
[14] Leykin, I. A., Pokazeyev, K. V. and Rozenberg, A. D. (1984) 'Growth and saturation of wind waves in a laboratory channel', Izv. Atmos. Oceanic Phys., 20, 313-318.
[15] Lake, B. M. and Yuen, H. C. (1978) 'A new model for nonlinear wind waves. Part 1: Physical model and experimental evidence', J. Fluid Mech., 88, 33-62.
[16] Mellen, R. H. (1990) 'Laboratory study of nonlinear phenomena in wind-wave generation', IEEE J. Oceanic Eng. 15, 130-132.
[17] Zaslavskiy, M. M. and Leykin, I. A. (1981) 'Interpreting measurements of wind wave dispersion characteristics', Izv. Atmos. Oceanic Phys., 17, 10-19.
[18] Laybo, A. B. and Leykin, I. A. (1981) 'Experimental investigation of nonlinear interactions in the wind wave spectrum', Dokl. Earth Sci., 258, 212-215.
[19] Mellen, R. H. (1990) 'Experimental investigation of the wave interaction mechanism for ambient noise', IEEE J. Oceanic Eng. (to be published).

MULTICHANNEL ACOUSTIC REFLECTION PROFILING OF OCEAN WATERMASS TEMPERATURE/SALINITY INTERFACES

J.D. PHILLIPS and D.F. DEAN
The University of Texas at Austin
Institute for Geophysics
Austin, Texas 78759
USA
(512) 471-6156

ABSTRACT. Multichannel seismic reflection data acquired off the Grand Banks have been analyzed to detect, enhance and map reflections from near-surface acoustic impedance interfaces in the ocean watermass. Temporal high pass (25 Hz) and spatial F-K filtering techniques were used to suppress the dominant, outgoing direct wave of the airgun primary pulse and its bubble wavetrain in order to recover weak reflections across a 48-channel, 2400 meter long hydrophone streamer. Comparison of the velocity-depth profiles obtained from normal moveout (NMO) stacking analyses of the common depth point (CDP) processed reflection data with hydrographic profiles, derived from expendable bathythermograph measurements made during the seismic survey, suggest that the observed reflection horizons emanate from 1) the base of a warm seasonal surface duct, 2) a thin layer of less-saline Labrador sea water overlying the Gulfstream 3) the top of the main thermocline, and 4) an interfaces just above the deep sound channel (SOFAR) axis.

1. Introduction

1.1. BACKGROUND

Hydrophone receiver arrays and pulse-type sound sources (Figure 1) have been used in marine seismic research and offshore oil exploration since about 1960 (Savit et al., 1962). Digital computer signal processing of towed, multichannel seismic (MCS) array data in the 10-100 Hertz (Hz) frequency band have been providing high resolution, deep penetration, common depth point (CDP) reflection images and precise velocity-depth profiles of the sub-seafloor geologic structure for more than 20 years (Mayne, 1962; Sheriff and Geldart, 1982).

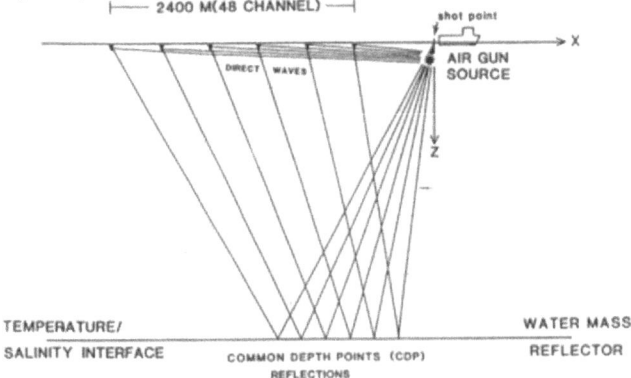

Figure 1. Typical 48-trace multichannel seismic profiling system, showing schematic raypaths of direct waves and reflections to representive hydrophone channels. See text for explanation.

J. Potter and A. Warn-Varnas (eds.), Ocean Variability & Acoustic Propagation, 199–214.
© 1991 *Kluwer Academic Publishers.*

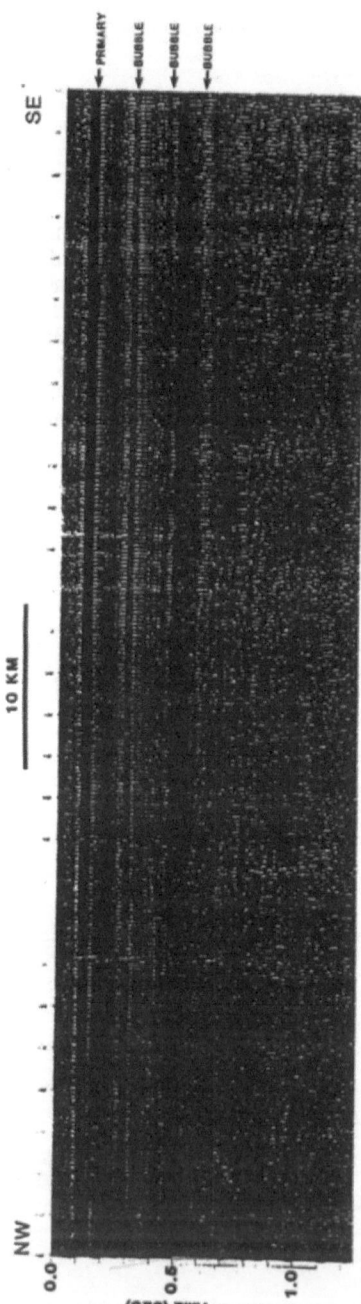

Figure 2. Initial 1.2 second record time of a 24-fold common depth point (CDP) stacked section from conventionally processed multichannel seismic profile NB21, showing strong direct wave of primary airgun pulse and its reverberant bubble wavetrain arrivals (horizontal banding).

More recently, high frequency (>100 KHz) acoustic back scattering and doppler sonar reflection methods have been used to detect suspended matter and internal waves, and to measure current motions in the shallow upper ocean (Orr and Hess, 1978; Pinkel, 1981; Rowe and Young, 1979; respectively). Also, towed multichannel hydrophone arrays are now utilized extensively in passive, low frequency naval surveillance applications (Dugdale, 1986). Most recently, French scientists have used multichannel seismic methods to detect what were believed to be internal waves (Gonella and Michon, 1989).

Despite the major contributions of marine reflection profiling for examining the solid earth beneath the seafloor, virtually no effort has been made to use the method to determine the temperature and salinity properties of the ocean water mass that would be responsible for acoustic impedance contrasts that might cause reflections. This situation has largely resulted from the belief by geologists and geophysicists that the acoustic waves commonly observed during the first few hundred milliseconds (ms) of travel time in deep ocean seismic profiling were direct waves associated with variations in the primary and bubble pulse radiation from the sound source (Figure 2). Such high energy, low frequency (6-30 Hz) radiation would overwhelm weak signals that may be reflected and refracted from shallow interfaces.

Ocean acousticians and physical oceanographers, on the other hand believed that the low frequency sound source spectrum used in MCS reflection studies inherently lacked the resolving power and sensitivity to detect the fine scale, velocity and/or density contrasts caused by temperature and salinity variations in the upper ocean water mass, except in such unusual circumstances as the near-bottom brine pools in the Red Sea and Gulf of Mexico (Hunt et al., 1967; Bouma et al., 1983). Also, spatial filtering by the long hydrophone group intervals commonly used in MCS streamer arrays (typically 50 meters) of low incidence angle, high frequency, sound waves reflected from very near-surface interfaces would limit observations to frequencies less than 30 Hz.

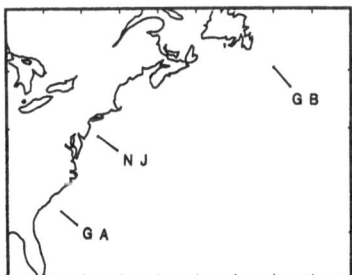
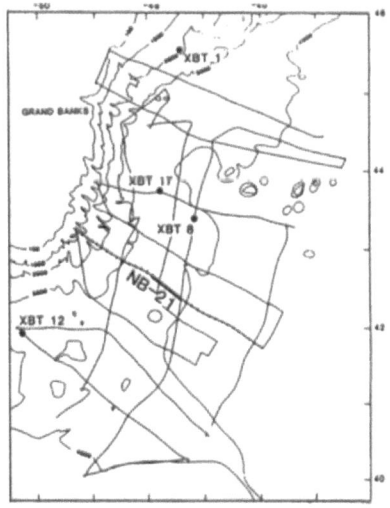

Figure 3. Index map showing general location of multichannel seismic profiles (MCS) where ocean watermass reflections have been observed (left) and detailed map of the Grand Banks grid survey for the NB-21 profile reported here (right). Expendable bathythermograph (XBT) stations are also shown.

1.2. CURRENT WORK

The past reluctance to use the multichannel reflection method for investigating watermass structure appears to have been unfortunate. Recent studies by Stark (1986) of the use of direct wave arrivals to monitor the effects of airgun source variations for constructing calibrated, seismic profiles demonstrated that weak coherent reflections from watermass interfaces are observed during the initial 500-1500 ms of the traveltime record. More recently, we have examined several deep water, airgun source, MCS profiles across the Gulfstream in the northwest Atlantic off Georgia, New Jersey and Grand Banks to detect potential reflections from shallow-ocean watermass interfaces (Figure 3, left). For this report we have re-processed a small portion of a Grand Banks MCS profile, NB-21 (Figures 2 and 3, right) to enhance and map continuous acoustic reflections which appear to emanate from temperature and/or salinity interfaces between 40 and 900 meters (m) depth.

Line NB-21 can be considered a typical deep ocean, marine seismic profile (Figure 2). The primary direct wave and its reverberatory bubbles clearly dominate the initial few hundred milliseconds as would be expected for a large volume airgun source (Kramer, et al, 1968). Note the uniform temporal spacing of the strong direct wave of the airguns' primary pulse and its first two bubble pulses across the entire profile; however, there is considerable temporal variability in the later bubble pulse arrivals after 500 ms. While this lateral variability could simply be due to source pressure/depth fluctuations as the ship traverses along the profile, it could result from reflection arrivals interfering with the direct waves.

We believe that the latter alternative better explains the data. Accordingly, the purpose of this report is to demonstrate that true watermass reflection information can be obtained from conventional, low frequency (<50 Hz) marine seismic data and to suggest that much finer scale information, useful for hydrographic and ocean acoustic research, could be acquired with existing higher frequency (>150 Hz) MCS systems specifically designed for watermass reflection profiling.

2. Method

2.1. STUDY AREA

Figure 3 (right) shows the location of the NB-21 seismic profile off the Grand Banks. Note that this is but one profile line in a 2500 nm grid geophysical survey conducted over the Newfoundland Basin during August, 1984 aboard R.V. Conrad. Expendable bathythermograph (XBT) observations were also made during the grid survey for purposes of calibrating the ship's multibeam swath bathymetry system. These XBT profiles with their computed, smoothed velocity-depth profiles are shown in Figure 4. Note the sharp temperature/velocity inversion at 50-150 m depth that is probably caused by the cold, less saline waters of the Labrador current which is over running and mixing with the main thermocline waters of the Gulf Stream (Fuglister, 1960). The warm surface duct is a result of summer seasonal heating effects (Schulkin, 1969).

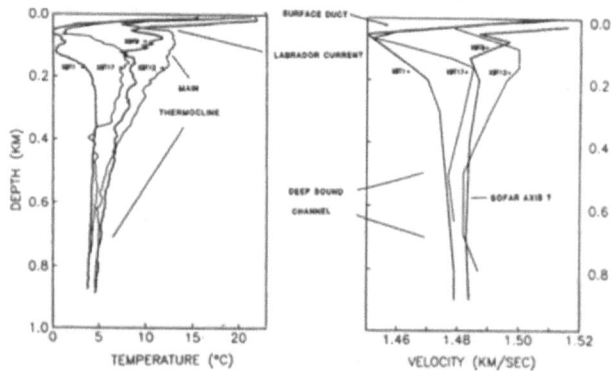

Figure 4. Depth profiles of water temperature and computed, smoothed sound velocity for XBT stations shown in Figure 3 (right).

2.2. DATA ACQUISITION

The MCS data used in this analysis were acquired with a 2400 meter long, 48-channel (trace), 50 meter group interval streamer and a pneumatic sound source array consisting of four (4) BOLT airguns (Model 1500C), 466 in^3 volume each, operating at 2000 psi. A Texas Instruments DFS-IV was used to record the field data at a 4 ms sample rate. The airgun's primary pulse peak energy was centered at 26 Hz with the bubble pulse energy peak at about 8 Hz. Both the streamer and airgun arrays were towed at a nominal depth of 13 meters at a speed of 9 km/hr (5 knots). The airgun array was fired at a 20 seconds repetition rate. This provided a 48-trace data record every 50 meters along the ship's track. These data were sorted into 25 meter, 24-fold (trace multiplicity) CDP bin gathers.

3. Results

3.1. OBSERVED DATA

The specific portion of NB-21 profile data used for our detailed analysis was selected after careful visual inspection of several near-trace (300 meter offset) record sections of the MCS grid survey to detect lateral variability in the bubble pulse wavetrain. A portion of the single channel, near-trace record from the NB-21 profile is shown in Figure 5. Generally, these near-trace profiles showed several discontinuous events between 200 and 600 ms record time which can not be easily dismissed as being due to airgun source depth/pressure fluctuations, surface ghost reflections or gun misfires. Also the possibility that the events were related to "overprinting" by previous shot ghosts and water bottom multiples is considered unlikely since the nominal 20 second shot repetition rate was randomized within a 1 second window.

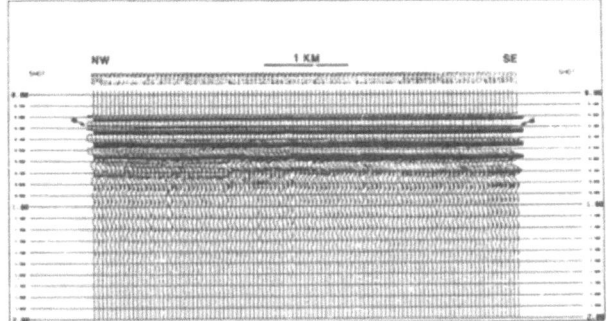

Figure 5. Near-trace display (300m offset,channel 48) of a portion of MCS line NB-21

The postulated reflection event in Figure 5 occurring at about 290 ms, is particularly distinctive. This event arrives between the primary pulse and its first bubble pulse, but not always at the same time. Other distinctive arrivals are seen at about 600 and 700 ms, and appear to disrupt the third and fourth bubbles of the direct wave, respectively. Note that all these events appear to have a much shorter period (25 ms), than the primary source pulse (~40 ms) and the bubble pulses (~125 ms), further suggesting that they are not part of the direct wave, airgun source signature.

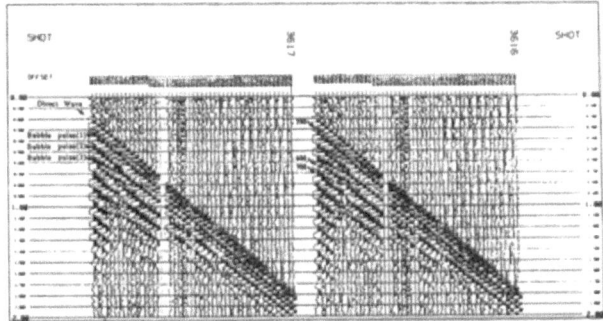

Figure 6. Raw shot gathers from representative shots along the Figure 5 profile showing the dominant direct wave, airgun source energy blast The traces are only low pass filtered (62 Hz).

3.1.1. *Reflection Wavetrain Identification* . The best test to determine if the postulated events are reflections or simply airgun source generated "noise", travelling as direct waves to the streamer is to examine common shot gathers of arrivals over the entire streamer length for evidence that the event arrival times show normal move out (delay) as a function of offset (Figure 6). The prime requirement for all reflections emanating from an interface beneath an end-fired, point source-linear receiver array is for the reflections to show a curvilinear, hyperbolic wavetrain arrival, time delay pattern. Whereas, source generated waves that travel directly down the streamer length show a simple, linear wavetrain arrival, time delay pattern. Note the straight linear time delay pattern of the high amplitude direct waves. However, the postulated 290, 600 and 700ms reflection events are weakly discernable as curvilinear wavetrains.

To help discriminate between the curvilinear reflection wavetrains and the straight direct waves the traces in each gather have been static time shifted to simulate the linear moveout delay with offset for direct wave events traveling at 1500 m/s down the streamer (Figure 7). This results in a more clear visual display by making the direct wave events horizontal (flat), while revealing the reflection wavetrains to curve upward as they become asymptotic with the direct waves at far offsets. Note that the direct wave's source pulse including its bubble pulses is very high amplitude over the entire streamer length.

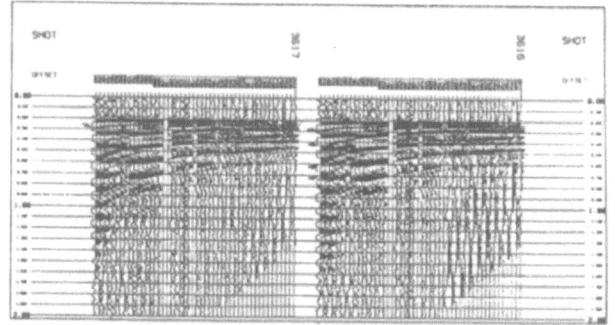

Figure 7. Same shot gathers of Figure 6 but after application of a 1500 m/sec reduction velocity that serves to align the direct wave arrivals horizontally across the 48-trace array.

The direct wave totally obscures observation of most curvilinear hyperbolic event arrivals occurring within the first 300 ms of record time in the near offset traces (300-1100 meters offset). However, note that a curvilinear hyperbolic event is clearly evident at 290 ms in the near traces between the primary pulse and the first bubble pulse. This event appears to be the same 290 ms event seen in the single channel, near-trace profile (Figure 5). In fact several more, weak curvilinear hyperbolic wavetrains can now be seen in Figure 7, beginning at about 400 ms. Unfortunately, on the far traces (1900-2650 meters offset) the low frequency, direct wave of the source and first two bubble pulse arrivals still partially mask the curvilinear wavetrains as they become nearly horizontal and parallel with the direct waves. Also note the apparent increase in high frequency events, particularly within the 200-400 ms record interval in the far traces beyond 1100 meters offset. These events may not be real and are probably noise due to spatial aliasing by the streamer's 50-meter long hydrophone groups on low incidence angle, high frequency sound waves reflected from shallow interfaces.

3.2. REFLECTION DATA ENHANCEMENT

3.2.1. *Temporal Filtering*. Careful inspection of the raw shot gather data in Figure 7 suggests that weak curvilinear, high frequency, reflection arrivals may be present but they are obscured by the high amplitude, ow frequency direct primary wave and its bubble wavetrain. Accordingly, we calculated the power spectrum of the data from several near trace records (300m offset) for the initial second of traveltime (Figure 8, left). The dominant energy is associated with the low frequency direct wave of the first bubble pulse at about 8 Hz and its associated harmonics at 16 and 24 Hz. The weak peak at 26 Hz may represent energy from both the direct wave of the airgun source pulse and reflected waves Only reflection energy is probably represented by the higher frequency spectral peaks at 35, 42 and 50 Hz, since direct wave events with frequencies greater than about 30 Hz are spatially rejected by the end-fire source-receiver geometry of the 50 meter long, individual hydrophone receiver groups.

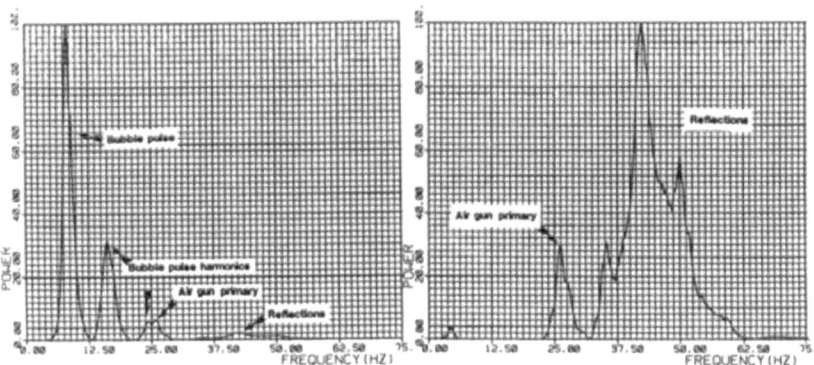

Figure 8. Power spectra of initial 1000 ms traveltime of raw (left) and high pass filtered (right) near-trace records in Figure 5.

Clearly, the low frequency direct wave arrivals can be suppressed using simple high-pass filtering that sharply attenuates the low frequency arrivals below 25 Hz but that does not itself introduce severe side lobe energy into the higher frequency reflection data spectrum. For this we designed a steep roll-off (72 db/octave), 25 Hz high pass filter. Figure 8 (right) shows the power spectrum for the near trace data after application of this filter. Note the enhancement of the 26-50 Hz band and almost total elimination of the low frequency energy below 24 Hz.

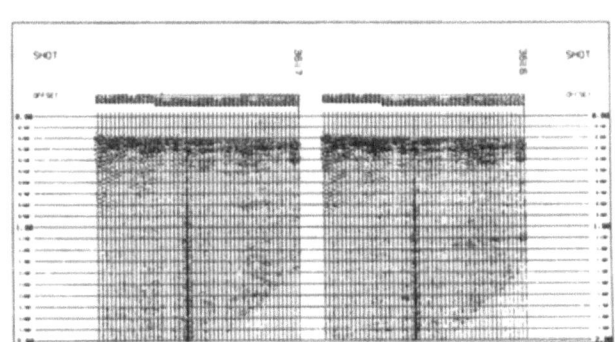

Figure 9. Shot gathers shown in Figure 7 but after application of a 25 hz high pass filter.

In Figure 9, the high pass filter has been applied to the shot gathers shown previously in Figure 7. Note that virtually no low frequency direct wave bubble pulse arrivals are seen beyond 500 m offset during the first 300 ms of record time. In fact, several weak, high frequency curvilinear hyperbolic wavetrains can now be followed continuously from the near offset traces between 500 and 900 ms record time to the far traces offset beyond 1800 m.

3.2.2. *Spatial Filtering*. Despite the effectiveness of temporal filtering, strong horizontal, high frequency direct wave arrivals from the airgun source primary pulse are still observed in Figure 9 during the initial 300 ms traveltime, especially in the near traces(<500m offset). To further attenuate this direct wave energy, we designed a near-zero dip (±1 ms/trace), spatial F-K filter (Figure10). This filter was applied after the static time shifting of the traces using the 1500 m/s linear moveout or reduction velocity as in Figure 9. Careful comparison of Figures 10 and 9 shows that the amplitude of the horizontal direct wave arrivals are sharply reduced.

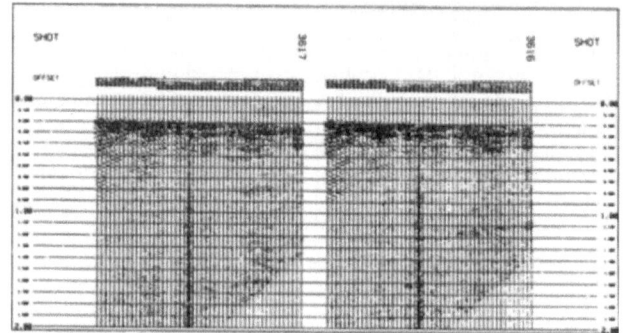

Figure 10. Shot gathers of Figure 9 but after applying an F-K velocity filter of ±1 ms/trace.

3.3. COMMON DEPTH POINT PROCESSING

Careful consideration of Figure 1 will reveal that redundant shot-receiver raypath geometries result as the ship proceeds along the shooting line since the reflection mid-point for each shot-receiver offset pair shifts to the next adjacent shotpoint . This means that the signal to noise (S/N) ratio of the reflection image profile can be markedly increased simply by adding together or "stacking" several data traces after appropriate manipulations to account for the shot-receiver geometry (Mayne, 1962). The number of traces summed determines the stack fold (multiplicity).

First, all shot-receiver offset pairs which have the same subsurface reflection point are sorted into common groups, or as more frequently termed, common depth point (CDP)gathers. The component traces are then algebraically summed after applying a traveltime correction to account for the time delay or normal moveout (NMO) of arrivals as a function of the source-receiver offset . This correction is a non-linear, hyperbolic function based on the velocity of the material traversed and causes traveltime "stretching" distortion. However, the effect is minor when the offset is small compared to the depth of the reflecting interface.

The S/N ratio improvement in CDP stacking is proportional to n, the number of traces summed and approaches the theoretical $n^{1/2}$ enhancement limit for random noise since all component shot-receiver raypaths within each gather are focused and thus traverse essentially the same small portion of subsurface materials. This focusing effect serves to reduce coherent noise contributions that might be introduced by lateral heterogeneity along.the profile line.

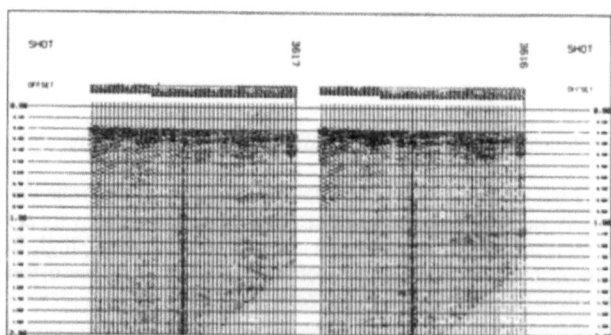

Figure 11. Common Depth Point gathers (CDP 4345/50). Note enhancement of the curvilinear reflection wavetrains after sorting trace included in the shot gathers shown in Figures 7-10.

.3.3.1. *Sorting*. Figure 11 shows several common depth point (CDP) gathers which derive their component traces from the shot gathers shown in Figures 9-10. They have been processed with the same temporal high pass filtering and spatial F-K filtering as the shot gathers. Note the improved continuity of the curvilinear hyperbolic reflection pattern of the CDP gathers compared to shot gathers shown in Figure 10.

3.3.2. *Normal Move Out*. Figure 12 shows the CDP 4345/50 gathers of Figure 11 but after applying normal move out (NMO) with a uniform 1.5 kilometers/second (km/s) over the entire travel time interval. Note the general horizontal alignment of the strong events at 200 ms, 400 ms, 700 ms and 1100 ms second travel time. Close alignment might be expected since variation of the true water velocity about 1.5 km/s are relatively small over this ocean depth interval. However, significant deviations from horizontal alignment appear in the near-traces within the 200-500 ms time interval. Here several events are seen to dip steeply downward with increasing offset. This probably results from a sharp velocity decrease over this travel time (depth) interval relative to the assumed 1.5 km/s velocity.

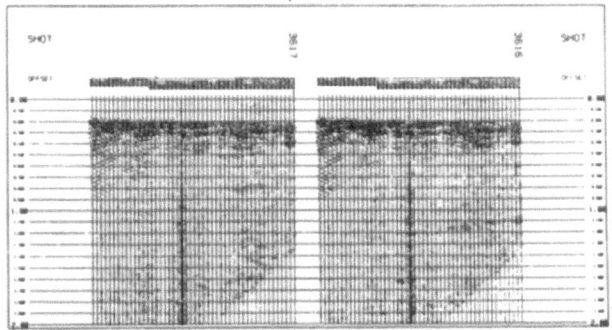

Figure 12. Common depthpoint gathers (CDP 4345/4350) after normal move out (NMO) correction with a uniform 1.5 km/s velocity.

The downward slope of the trace terminations between 450 and 2650 m offset results from the truncation (mute) of those portions of the far trace data trace records that have been severely distorted by the "stretching" of the NMO correction and spatial-aliasing. Offset muting is a standard CDP seismic data processing procedure to eliminate contaminated data that would otherwise degrade the S/N ratio of the final, summed trace for each gather.

3.3.3. *Velocity Analysis*. Very precise horizontal alignment of the reflection wavetrains shown in the CDP 4345/50 gathers of Figure 12 can be accomplished by carefully selecting a sequence of NMO velocities as a function of traveltime to make each reflection wavetrain horizontal (Figure 13).

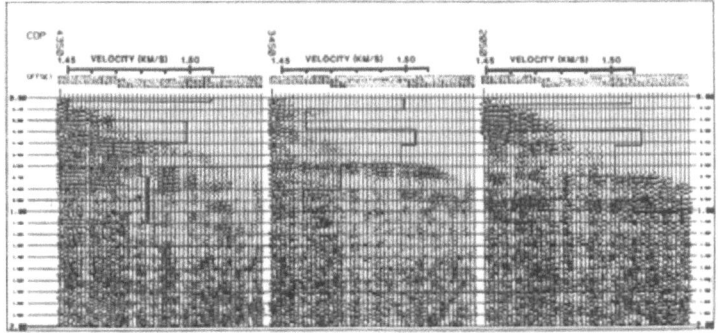

Figure 13. CDP gathers 4350, 3450 and 2050 after iterative NMO velocity analysis

208

This is an iterative procedure, commonly termed, velocity analysis, in CDP seismic data processing, and is routinely used to produce exact alignment of a gather's entire reflection wavetrain pattern before summing. This maximizes the S/N ratio of the trace sum. The results of such iterative NMO velocity adjustments for CDP 4350 as well as other selected gathers along the NB-21 profile are shown in Figure 13. Note the closer horizontal alignment of the CDP 4350 wavetrains as compared to Figure 12. .Interval velocity-depth profiles, comparable to sound velocity profiles,have been calculated from the NMO velocity-time functions (Dix, 1955). are superimposed (Figure 13).

The XBT-derived observations of sound velocity and depth made in the Grand Banks area appear to confirm that the curvilinear, wavetrain arrivals are true reflections from watermass interfaces. That is, the strong reflection events used to derive the average NMO velocity/traveltime functions provided distinctive interval velocity/depth profiles which correlate well with the nearby XBT-derived velocity profiles (Figure 14). Note the general similarity in the shape of the XBT and NMO-derived velocity-depth profile observations. Also, the correspondence of the sharp velocity inversions is most apparent. Both the NMO and XBT profiles show a sharp low velocity inversion zone extending form just below the surface to about 200m depth and a velocity maxima in the upper part of the main thermocline.

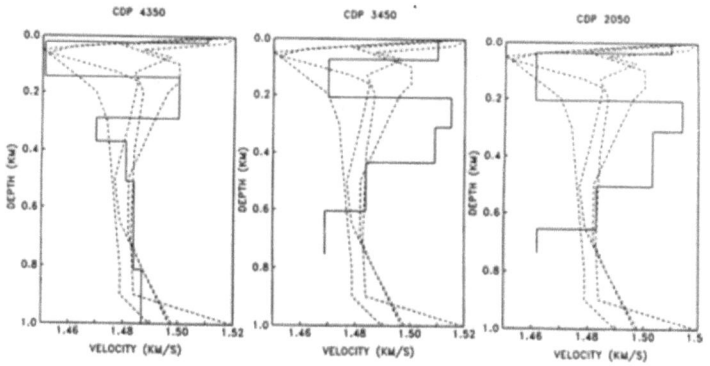

Figure 14. Comparison of velocity-depth profiles obtained from CDP gathers 4350, 3450, 2550 and 2050 (solid lines) with XBT-derived velocity -depth profiles shown in Figure 4 (dashed).

3.3.3. *Stacking.* A high resolution image of the configuration of the subsurface reflection interfaces along a ship's track is constructed by summing the muted, NMO-corrected traces in each CDP gather into a single "stacked" trace and arranging the stack traces sequentially to form a 2-dimensional horizontal distance vs traveltime profile section. The close, equal spacing of the stack traces usually provides adequate visual continuity of waveform phase/amplitude to allow stratagraphic correlation of geologic reflection interfaces over several kilometers. Careful inspection of the stacked CDP reflection image section for the NB-21 profile shown in Figure 15 (top) suggest similar continuity for ocean watermass reflection interfaces may also be possible.

The top of Figure 15 shows the stacked CDP section of NB-21 after it has been "acoustic" processed as described above using high pass and spatial filtering to suppress the direct wave. The bottom of Figure 15 shows the original "seismic" processed NB-21 profile of Figure 2, except its horizontal and vertical (time) scales are now the same as the acoustic processed NB-21 profile. Four NMO velocity analyses were performed at nominal 900m intervals along the profile (CDP 2050, 2550, 3450 and 4350) and were linearly interpolatated to moveout and stack each CDP gather in Figure 15 (top). Note the virtual elimination of the strong horizontal direct waves of the source and its bubbletrain, especially during the initial 0.5 seconds traveltime, and the much finer temporal resolution overall.

The interval velocity-time functions derived from the acoustic processing NMO velocity analysis of the selected CDP gathers are shown superimposed on the acoustic processed stack profile to demonstrate the correlation of the major reflection horizons with interval velocity depth (time)

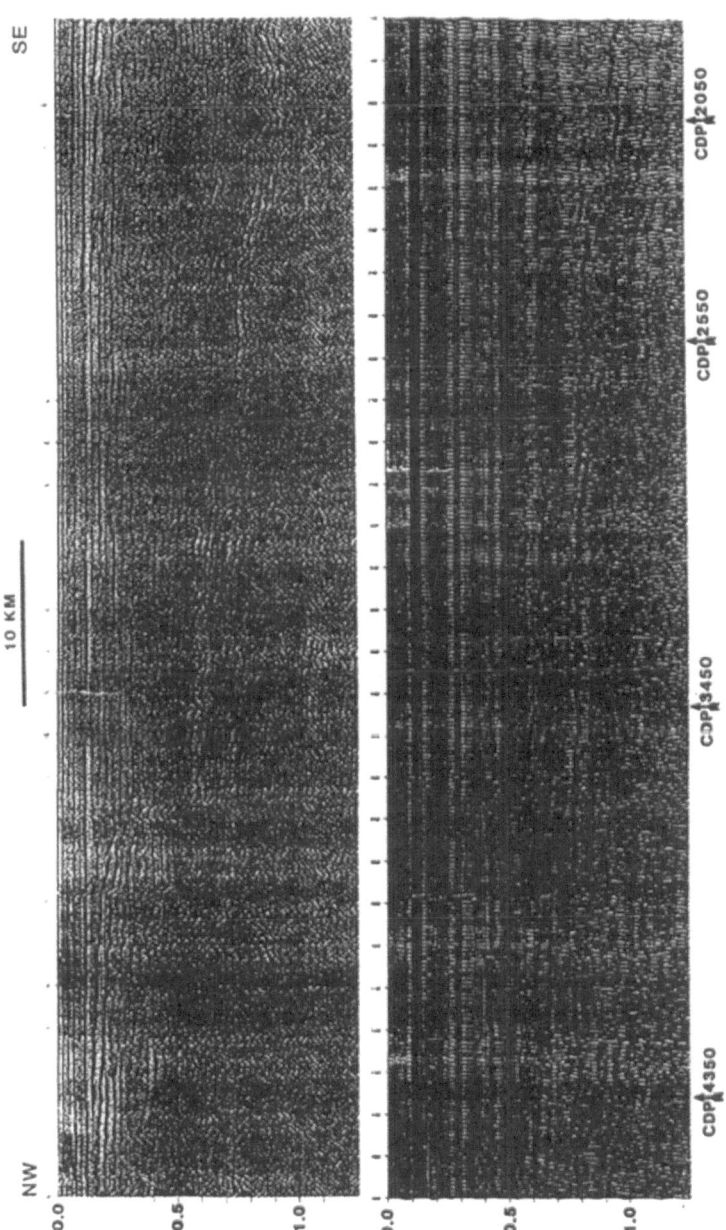

Figure 15. (Top) 24-fold common depth point (CDP) stacked section for MCS profile NB-21.after acoustic-processing to suppress the direct wave of the the airgun source pulse and to enhance shallow watermass reflections, 25 m CDP spacing.

(Bottom) Conventional, seismic-processed common depth point (CDP) stacked section for MCS profile NB-21.shown in Figure 2 except redisplayed with same scales as acoustic-processed section shown at top.

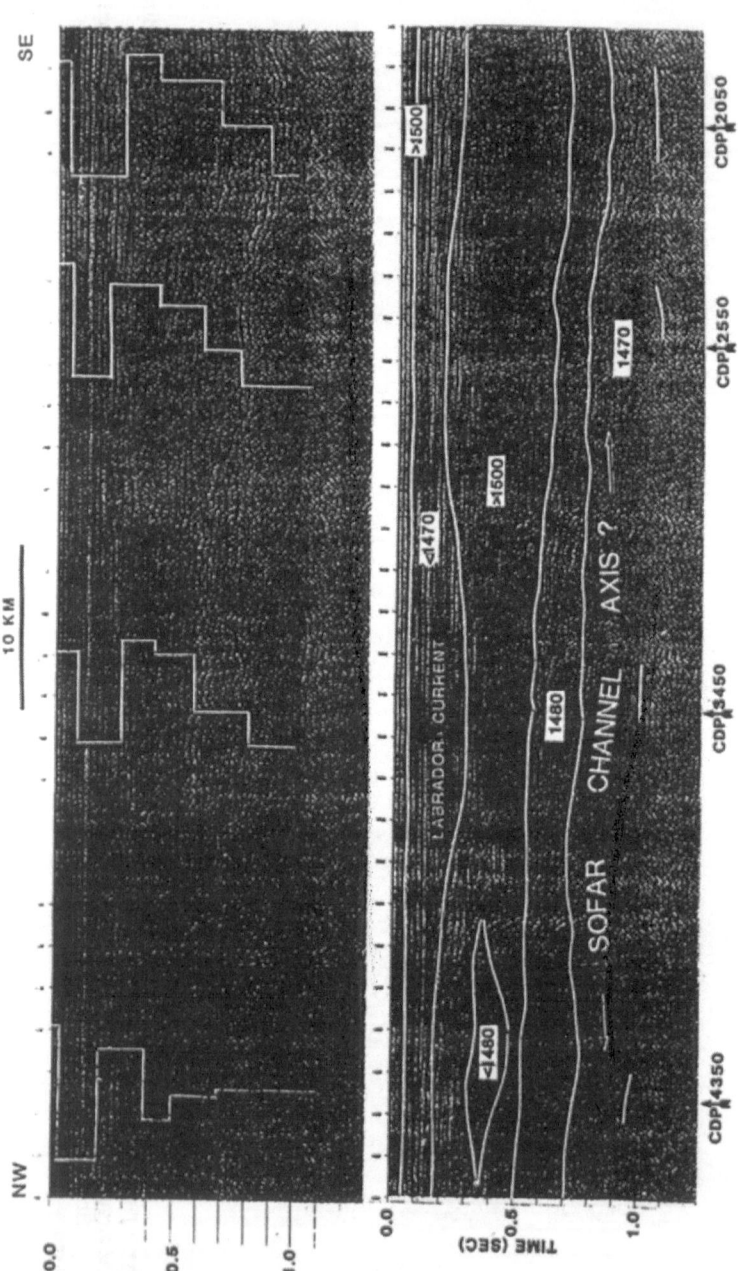

Figure 16. (Top) 24-fold common depth point (CDP) stacked, reflection image section of Profile NB-21 (Figure 15,top) with superimposed interval velocity-time profiles derived from velocity analyses of CDP gathers (Figures 13 and 14). Note correlation of interval velocity boundaries and strong continuous reflections especially for the dipping reflector horizons.

(Bottom) Interpretation of CDP stacked, reflection image section (Figure 16, top)) showing low velocity inversion layer (<1470 m/sec) associated with cold Labrador current overlying the warmer Gulf Stream waters (>1500 m/sec) of main thermocline. Note the lens of low velocity (<1480 m/sec) water near CDP 4350 which appears to be detached and may be mixing with the main thermocline waters. The base of the surface duct is nearly horizontal at about 80 ms TWT(~60 meters depth) but rises to about 60 ms near the center of the profile. The depth of the SOFAR channel axis is estimated from the XBT observations and appears to lie within the layer having 1470 m/sec velocity.

changes (Figure 16, top). Note that Figures 16 and 15 (top) show relatively low resolution CDP stack sections. A much higher resolution stacked section could have been constructed if many more NMO velocity analyses of very closely-spaced gathers (i.e., at 25 CDP intervals) were used to moveout and sum each CDP gather. In any case, continuous reflection horizons can be traced across the entire section over regional distances of 50 km. Particularly distinctive are continuous reflections that appear to emanate from 1) the base of a 1500m/s, surface layer, 2) the base of a low velocity (<1470 m/s) zone which forms a near surface inversion layer showing considerable depth relief across the entire profile , 3) the top and 4) base of a 1500-1480 m/s layer which appears to gradually rise toward the northwest end of the profile.

4. Discussion

Close examination of the acoustic-processed reflection image section in Figures 15 and 16 (top) reveals considerable detailed information about the reflection structure of the upper ocean water mass. Whereas, there is little evidence of any horizons showing depth relief that might be suspected to be reflections in the same seismic-processed NB-21 section (Figure 15,bottom). In fact, none of the strong reflection horizons seen in the initial 300 ms of the acoustic-processed section are evident in the seismic processed section. They are completely obscured by the direct wave's primary pulse and its first two bubble pulses. Only the much later, steeply dipping horizons at the southeast end (CDP 2050) of the acoustic section between 800-1000 ms are weakly depicted in the seismic processed section (Figure 15, bottom). The 2nd bubble pulse at the northwest end (CDP 4350) of the seismic profile seems to be superimposed on a near horizontal reflector at 400 ms record time. Clearly, acoustic-processing to suppress the source direct wave provides a superior reflection profile image that is useful for making quantitative observations about the velocity-depth structure of the upper ocean watermass.

Before we discuss the hydrographic implications of the reflection profile image shown in Figure 16, it must be pointed out that although the same fundamental tenets of Huygen's Principle and Snell's Law govern the reflection of sound waves (rays) from acoustic impedance interfaces in both sea water, a compressible fluid and rock, a compressible solid, the factors which control acoustic impedance, the velocity-density product, in each medium are quite different. In the oceans, temperature is the primary factor affecting the variability of sound velocity and hence acoustic impedance since the spatial variability of sea water density is very small. However, local salinity (composition) anomalies are known to produce sharp changes in density as well as minor velocity changes. In crustal rocks, on the other hand, temperature changes are relatively unimportant in affecting velocity or density at seismic depths of interest. Lithologic composition appears to be the major factor controlling both velocity and density changes.

4.1. SURFACE DUCT

While the cause of deeper reflectors probably result from combined temperature/salinity impedance contrasts, the almost continuous reflection horizon seen at about 50-60 ms (two-way traveltime, TWT) below the surface in Figure 16 (bottom) is most easily explained as a temperature-induced velocity contrast at the base of a warm isothermal surface or mixed layer. The mixed layer forms as a result of solar heating of the normally cold Labrador Sea water during late summer (Schulkin, 1969; Urick, 1982). This heating coupled with wind stirring, produces a warm, high velocity surface layer or surface duct, 40-50 meters thick.

4.2. NEAR-SURFACE LOW VELOCITY LAYER

The reflection at the base of the near-surface, 1470 m/s low-velocity inversion layer is particularly distinctive at the northwest end of the NB-21 profile (Figure 17 CDP 4400). Here its depth gradually decreases toward the southeast from about 160 meters (200 ms) to 200 meters (280 ms) at CDP 3700 over a distance of about 20 km. A similar steeply dipping reflection from the layer is seen at the southeast end of the profile (CDP 1900). In the center portion of the profile (CDP 3400) the base of this near surface low-velocity layer appears to reach its maximum depth of 225 m (300 ms). The top of the low-velocity layer is marked by a nearly continuous reflection at about 45 meters depth (60 ms TWT). This horizon is only absent at the southeast end of the profile (CDP 2000).

The good correlation of the XBT derived velocity-depth profiles with the NMO-derived profiles shown in Figure 14 suggest that cold, low salinity Labrador Sea surface water, which is known to overrun the much warmer, more saline Gulf Stream watermass during the late summer period of the Grand Banks MCS survey (Fuglister, 1960), is responsible for the near-surface, 1470 m/s low-velocity inversion layer.

4.3. MAIN THERMOCLINE

The reflection horizons associated with the 1500 m/s and1480 m/s layers within the main thermocline are even more continuous than the near surface, low-velocity layer reflections (Figure 16). In fact, the horizon at the base of the 1480 m/s layer can be traced continuously across the entire profile. The top of the 1500 m/s layer can also be traced across most of the profile; however, it appears to be interrupted at the northwest end of the profile (CDP 4200). Significantly it is here that there appears to be an overlying lens of low-velocity (<1470 m/s) water within the high velocity (>1500 m/s) layer at the top of the thermocline. Apparently this is a region of strong mixing of the near-surface low-velocity, Labrador Sea water with higher velocity Gulfstream deep water. Turbulent mixing may cause the watermass interfaces to become locally rough, thus making them poor reflectors.

4.4. DEEP SOUND CHANNEL

The strong reflector at the base of the deep 1470 m/s layer in Figure 16 may be associated with the top of the deep sound channel (Ewing and Worzel, 1948). Northrup and Colburn (1974) have shown the general depth of the deep sound channel (SOFAR) velocity minima to be about 600-900 meters depth off the Grand Banks. Our XBT profiles shown in Figure 14 support this observation. However, neither our stacked reflection profile.image nor our NMO-derived velocity profiles show reflection horizons deep enough to measure a true velocity minima for the 600-900 m depth interval (800-1000 ms, TWT). In fact, the SOFAR channel's depth (time) interval is probably within the deepest layer we detected, the 1470 m/s layer at the base of the thermocline. Although this layer is the minimum interval velocity layer we observed (1470 m/s) within the main thermocline, we cannot confirm a velocity increase at greater depths. The signal to noise ratio for the deeper reflector events is too low to obtain a reliable average NMO velocity to compute an interval velocity.

It appears that the base of the 1480 m/s layer may be a continuous interface just above the SOFAR channel axis here. The small 7.5 m (10 ms) changes in depth over distances of 250 meters (CDP 3000-3400) in the 1480 m/s layer's lower reflection boundary may be indicative of the surface relief along the top of the layer which includes the SOFAR channel. Such detailed information about local roughness of deep sound channel structure may be important for assessing the effectiveness of the SOFAR channel to act as a waveguide for long range sound propogation (Emery et al., 1979).

4.5. CONCLUSIONS

Conventional multichannel, common depth point seismic reflection methods using airgun sources (6-30 hz peak energy band) and long (50m) receiver group, hydrophone streamers can recover ocean watermass reflections in the 25-60 hz frequency band from velocity contrast impedance interfaces on the order of 10 m/s over 10 m depth intervals, down to depths of 1000 m. Both positive and negative velocity contrast produce reflections. Sharp density contrasts may also produce reflections

Regional mapping of fine scale changes in the velocity-depth structure of near surface waters as well as the upper surface of the deep sound channel has been done. High resolution reflection images showing velocity-depth relief as small as a few meters were observed over horizontal distances of a few tens of meters which implies that ocean temperature variability can be measured with similar horizontal resolution. Heretofore, only towed thermistor chains, rapidly deployed temperature/salinity sensing devices (CTD/STD) from drifting platforms (FLIP) and yo-yo tow systems, or closely spaced XBT stations could produce observations with comparable resolution; and only over a more limited depth range, generally less than 500 meters (Marmorino et al., 1985; ;Katz, 1973; Pinkel, 1975, 1981).

Although no reflections were recovered from below the deep sound channel where impedance contrasts are expected to be very small, the use of higher frequency multichannel CDP systems may allow detailed mapping of interface reflections across the deep SOFAR channel depth interval as well as from the shallower, near surface watermass structure. In fact, with higher frequency non-reverberant sources velocity and distance, resolution in the meter/second and meter range respectively, might be attained with multichannel reflection systems specifically designed for recovering shallow ocean water mass reflections. By adapting current high frequency (>150hz peak energy band), non-reverberatory sound sources such as tuned airguns and watergun arrays with short (<10m) receiver group, hydrophone streamers watermass reflection images from velocity contrast impedance interfaces on the order of 2 m/s over 2 m depth intervals.may be possible. With shot repetition and data record lengths on the order of 2 seconds, discrete velocity-depth profiles to more than 1000 meters depth could be measured directly at 5 meter CDP intervals along the ship's track!

5. Acknowledgements

We thank Tracy Stark and Arthur Maxwell for useful discussions about processing the data and for initially suggesting this study.. Also James Austin, a co-chief scientist aboard R/V CONRAD cruise 25-10 in the Newfoundland Basin, kindly provided the multichannel data used in this study. Milo Backus, Jan Garmany, Tom Shipley, Eli J. Katz and Worth Nowlin reviewed the manuscript.

6. References

Bouma, A. H., C. E. Stelting, and M. H. Feeley, "High Resolution Seismic Reflection Profiles", in Seismic Expression of Structural Styles, Edited by A. W. Bally, American Association of Petroleum Geologists, Tulsa, 1983.

Dugdale, Donald, "Navy Plays a Listening Game in Search for Soviet Subs", Defense Electronics, vol. 18, no. 3, pp. 68-74, March 1986.

Dix, C. H. "Seismic Velocities from Surface Measurements", Geophysics, vol. 20, p. 68-86, 1955.

Emery, W. J., T. J. Reid, J. A. DeSanto, R. N. Baer and J. P. Dugan, "Mesoscale Variations in the Deep Sound Channel and Effects on Low-frequency Propagation", J. Acoustical Soc. Amer., vol. 66, pp. 831-841, 1979.

Ewing, M., and J. L. Worzel, "Long Range Sound Transmission", Geological Society of America, Memoir 27, 1948.

Fuglister, F. C., "Atlantic Ocean Atlas, Temperature and Salinity Profiles from the IGY", Woods Hole Oceanographic Institution, Woods Hole, 1960.

Hunt, J. M., E. E. Hays, E. T. Degens, and D. A. Ross, "Red Sea, Detailed Survey of Hot-Brine Areas", Science, vol. 56, p. 514-16, 1967.

Katz, E. J., "Profile of an Isopycnal Surface in the Main Thermocline of the Sargasso Sea", Journal of Physical Oceangraphy, vol. 3, p. 448-457, 1973.

Kramer, F. S., R. A. Peterson, W. C. Walter, "Seismic Energy Sources 1968 Handbook", Society of Exploration Geophysicists, 38th Annual Meeting, Denver, United Geophysical Corp., 57 pages, 1968.

Marmorino, G. O., L. J. Rosenblum, J. P. Dugan, and C. Y. Shen, "Temperature Fine Structure Patches Near an Upper Ocean Density Front", Journal of Geophysical Research, vol. 96, pp. 11799-11810, 1985.

Mayne, W. H., "Common Reflection Point Horizontal Stacking Techniques, Geophysics, vol. 27, pp. 927-938, 1962.

Northrop, J., and J. G. Colborn, "SOFAR Channel Axial Sound Speed and Depth in the Atlantic Ocean", Journal of Geophysical Research, vol. 79, pp. 5633- 5642, 1974.

Orr, M., and F. Hess, "Remote Acoustic Monitoring of Natural Suspensate Distributions, Active Suspensate Resuspension, and Slope/Shelf Water Intrusions", Journal of Geophysical Research, vol. 83, p. 4062-4068, 1978.

Pinkel, R., "Upper Ocean Interval Wave Observations", Journal of Geophysical Research, vol. 80, no. 27, pp. 3892-3912, 1975.

Pinkel, R., "Observation of the Near Surface Internal Wavefield", Journal of Physical Oceanography, vol. 11, p. 1248-1257, 1981.

Sheriff, R. E., and L. P. Geldart, "Exploration Seismology: History, Theory and Data Acquisition", vol. 1, Cambridge Univ. Press, Cambridge, 253 p., 1982.

Schulkin, M., "Propagation of Sound in Imperfect Ocean Surface Ducts", Underwater Sound Laboratory, Report 1013, New Haven, 1969.

Stark, T. J., "Information Extraction from Deep Water Seismic Reflection Data: LASE Line 2", University of Texas Ph.D. Thesis, 1986.

Urick, Robert J., "Sound Propagation in the Sea", Peninsula Press, Los Altos, Chapters 6 and 7, 1982.

ACOUSTIC VARIABILITY DUE TO INTERNAL WAVES AND SURFACE WAVES IN SHALLOW WATER

DAVID RUBENSTEIN
MICHAEL H. BRILL
Science Applications Intl. Corp.
P.O. Box 1303
McLean, VA 22102
USA

ABSTRACT. During a recent shallow water experiment, acoustic propagation of a continuous 400 Hz signal was measured over a 18.5-km range, from a fixed source to a broadside array of hydrophones. Transmission loss varied significantly over a wide range of time scales during certain periods, and was relatively uniform during other periods. During intermittent intervals, acoustic features appeared to propagate across the array of hydrophones, at a speed of about 0.75-1.0 m/sec. This speed is similar to the propagation speed of internal soliton waves, as predicted by internal wave theory.

In this paper we examine two candidate mechanisms that may be responsible for these fluctuations. First, we test the hypothesis that internal wave packets are responsible for these intermittent periods of high acoustic variability. A Parabolic Equation (PE) model was set up with time- and range-dependent sound speed fields. Measured temperature time series were advected at a constant speed in an appropriate direction, to simulate a dispersionless train of internal waves. Second, we examine the hypothesis that scattering off the surface waves drains energy out of the specular component—which retains the carrier frequency. Our analysis included statistical scattering theory and a simulation of several realizations of the random surface using the PE model to represent the acoustic wave.

1. Experiment

1.1 EXPERIMENT OVERVIEW

The experiment took place during July and August, 1988, in 140 m water off the Washington coast. An acoustic source was deployed at 45 m depth. Acoustic data was collected primarily from a bottom-mounted array of hydrophones, located approximately 18.5 km south of the source. The array was in an approximate east-west alignment, broadside to the source. The bathymetry profile between the source and receiver was nearly flat. Some data were also collected at 7.4 km range, along the same radial track.

Several times over the course of the experiment, the acoustic source projected a 400 Hz CW tone (and occasionally a 115 Hz tone) continuously over several hours. During these times, received signals were acquired, digitized, demodulated, low-passed filtered, and subsampled to a 1.47 Hz sampling rate. Unfortunately, due to a nonuniform phase-drift in

J. Potter and A. Warn-Varnas (eds.), Ocean Variability & Acoustic Propagation, 215–228.

the source driver, phase information is contaminated in the frequency range of 0.4-1.6 x 10^{-2} Hz. Careful analysis shows that amplitude information is not affected. Therefore, all results discussed in this paper rely only on amplitude.

A 15-element thermistor chain was deployed over the hydrophone array, during a 10-day period. Thermistors were situated at uneven intervals between depths of 20 and 97 m, with closest spacing nearest the surface. Temperature was simultaneously sampled from the thermistors at 16-second intervals. Thermistors were calibrated to provide absolute accuracies to 0.1°C, and provided relative accuracies to 0.01°C. Comparisons with CTD profiles confirmed these thermistor accuracies. An Endeco waverider buoy collected surface wave height data with a 1 Hz sampling rate, continuously over 17 minute periods, at 3 hr intervals.

1.2 SURFACE WAVES

Figure 1 shows a time series of a wave height record, along with its envelope. The significant wave period is 7 s. This particular time interval (August 1 at 7:00 Z) was chosen for presentation because it represents some of the strongest waves of the experiment. For the two previous days, the wind had been coming from 300° at 10-12 m/s. The wave height has been put through a band-pass filter, which passes periods ranging from 4.7 to 12.8 s. The method of filtering is described by Longuet-Higgins (1984).

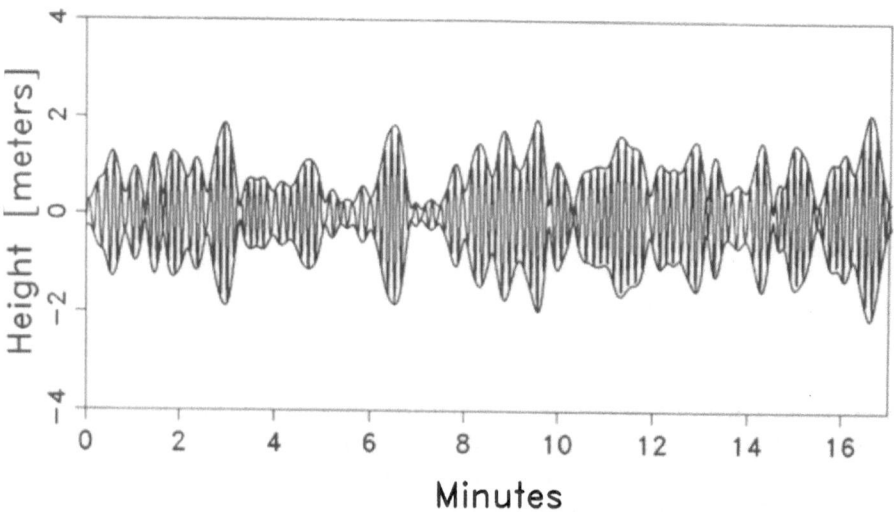

Fig. 1: Measured wave height record and envelope. A band-pass filter passes periods within the range from 4.7 to 12.8 s.

1.3 INTERNAL WAVES

Figure 2 shows profiles of temperature, sound speed, and Brunt-Vaisala frequency, from a representative CTD cast. The sound speed is not strongly affected by salinity. Figure 3 shows a time series of temperature at 20 m depth, over a 5 day interval. The temperature signal is made up of two components. First, a semidiurnal internal tide has a strong signature. Second, intermittent bursts of high-frequency waves occur, roughly in phase with the tidal component. These bursts correspond to a soliton-like wave packets riding on the shallow density interface. These waves have frequencies very near the buoyancy frequency, in the range 6-10 cph. These wave packets have been observed in Seasat SAR images off the coast of Washington. SAR images show that the packets are about 5 km in width, and 17 km apart, implying a propagation speed of 40 cm/sec.

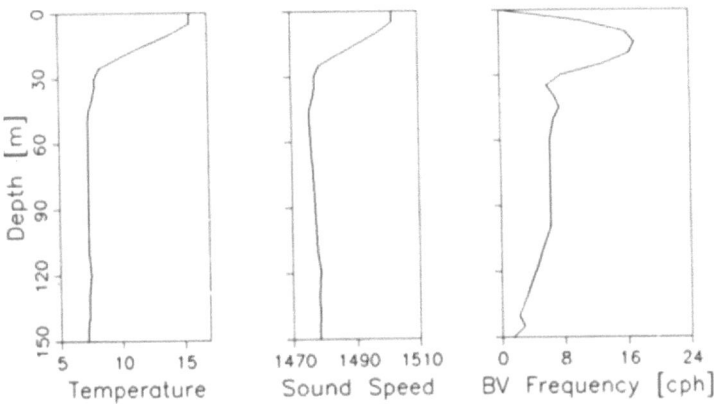

Fig. 2: Temperature (°C), sound speed (m/s), and Brunt-Väisälä frequency (cph), from a representative CTD cast.

An empirical orthogonal function (EOF) analysis was applied to the 15 thermistor channels. We found that the first mode overwhelmingly dominated. Approximately 90, 5, and 2% of the variance is contained in the first three modes, respectively.

1.4 ACOUSTIC VARIABILITY

Figure 4 shows two time series of intensity, separated in the cross-track direction by 13 m. The signal has been demodulated with respect to the carrier frequency $f_0 = 400$ Hz, low-pass filtered, and sampled at a rate of 1.47 Hz. The obvious 10 s variability is due to scattering by surface waves. Interestingly, this 10 s variability is uncorrelated over this separation. The similarity between the surface wave spectrum and the acoustic spectrum is obvious from Figure 5. We defer discussion of the mechanism until Section 3.

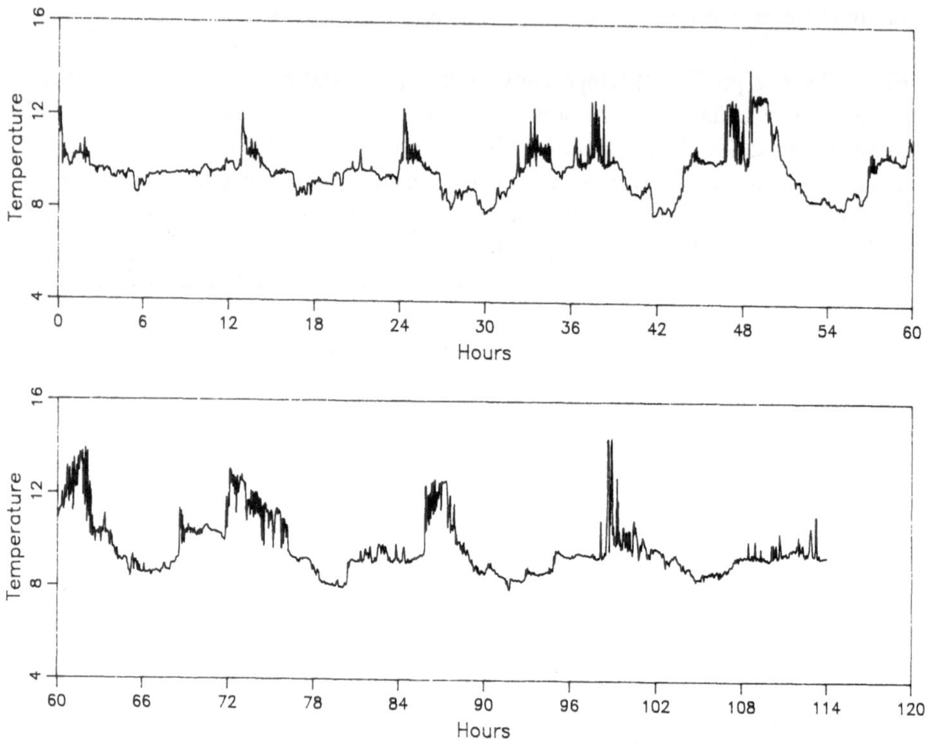

Fig. 3: Temperature (°C) at 20 m depth, over a 5 day interval.

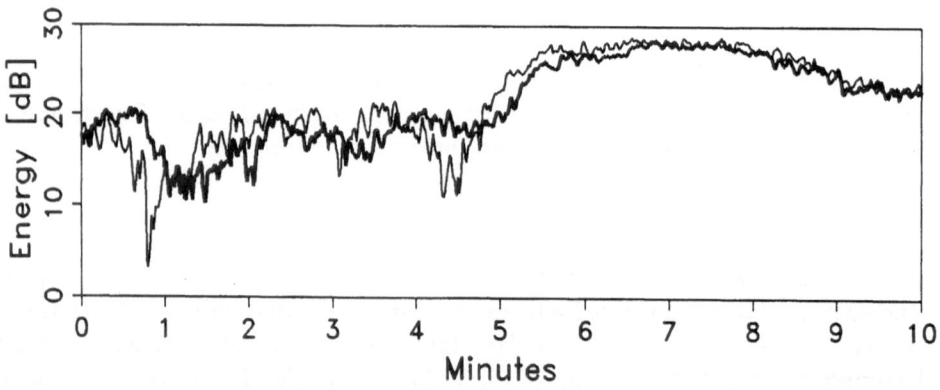

Fig. 4: Acoustic energy from two hydrophones, separated in the cross-track direction by 13 m.

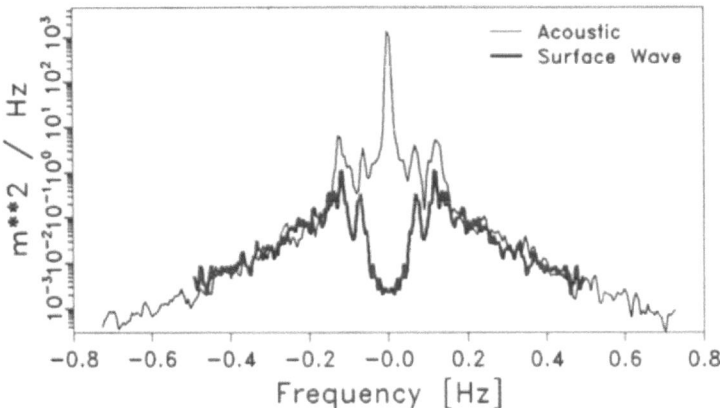

Fig. 5: Power spectra of acoustic signal (thin curve) and surface wave height (thick curve). The vertical scale is appropriate to the surface wave spectrum. The acoustic spectrum is plotted on an arbitrary vertical scale, such that its level in the skirts coincides with the surface wave spectrum level.

For now, we discuss the internal wave contribution to acoustic variability. To isolate the low-frequency contribution, the complex-valued, demodulated data has been low-pass filtered, with a 3 dB level bandwidth of 25 mHz. Figure 6 shows a set of time series of normalized intensity, $I/<I>$, from the hydrophone array. The separation between adjacent hydrophones is 7.5 m. The signals from adjacent hydrophones are successively offset by 0.5 in ordinate, to give a waterfall appearance. The uppermost trace in the figure belongs to the west end of the array, and the lowest to the east end. The total separation is 232.5 m.

The most interesting aspect of Fig. 6 is the propagation of acoustic features from west to east, at approximately 0.75-1.0 m/s. We speculate that this propagation is due to long-crested internal soliton-like wave packets propagating landward toward shallow water, up the continental shelf. While this propagation effect was not always observed, when it was observed, the propagation direction was generally landward, as shown schematically in Figure 7. The acoustic record in Fig. 6 corresponds to Hour 33 in the temperature record in Fig. 3, and coincides with the onset of an internal wave packet.

If one looks at Fig. 6 closely during the time range from 15-27 minutes, one can see a few hints of propagation in the opposite direction. This effect may indicate that a very weak internal wave is propagating seaward.

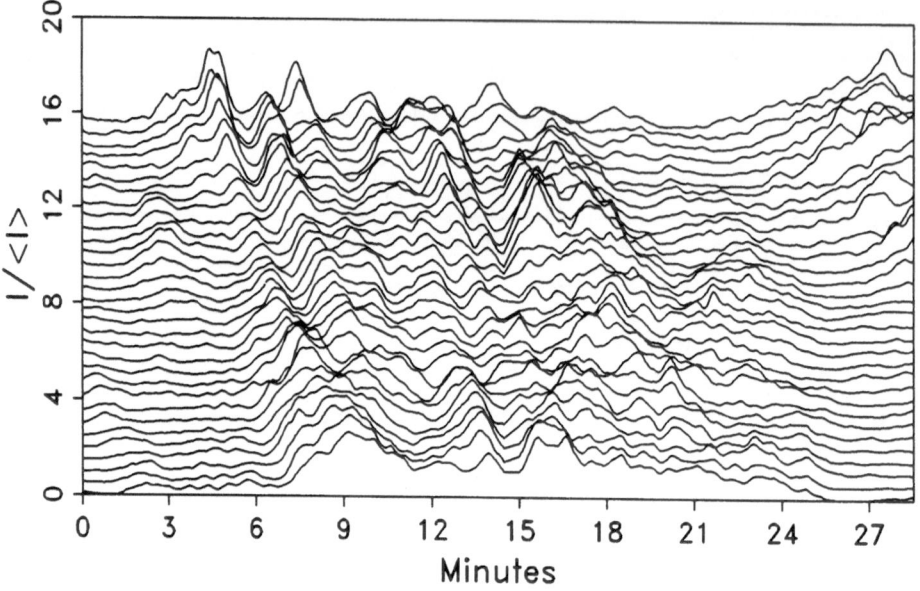

Fig. 6: Normalized acoustic intensity, from cross-track aligned hydrophone array at 18.5 km range. Adjacent hydrophones are separated by 7.5 m. Signals have been demodulated with respect to the central $f_0 = 400$ Hz carrier, and low-pass filtered. Traces are successively offset by 0.5, to give a waterfall effect.

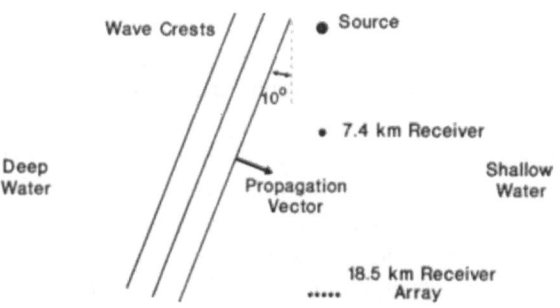

Fig. 7: Schematic diagram of source and receivers, and packet of wave crests propagating obliquely, from deep to shallow water.

2. Model of Internal Wave-Induced Acoustic Fluctuations

There is a considerable body of literature dealing with the subject of internal wave-induced acoustic fluctuations. For example, the reader is referred to Flatté (1983) for a review. Most of the work on this subject deals with deep ocean environments, where the Garrett-Munk type of spectrum is observed. In this paper, we are concerned with shallow water, where long-crested internal wave packets present us with a quasi-deterministic environment.

In order to model the acoustic fluctuations, we make a basic assumption which allows us to relate temperature measurements at a point to a range-dependent field. We assume that the temperature field is made up of a series of soliton-like internal waves. Because solitons are dispersionless, we can advect temperature measurements $T(z,t)$ with a phase velocity u, to yield a range-dependent field, $T(z,x-ut)$. Here, x is the range coordinate from the source to the receiver, and u is the projection of velocity vector u onto this coordinate. We use deviations of $T(z,x-ut)$ from the time-averaged profile $<T(z)>$ to provide an implicit measure of vertical displacement. We interpolate from the time-averaged sound speed profile $<c(z)>$ to obtain a range and time-dependent sound speed field $c(z,x-ut)$. The phase velocity u can be in an arbitrary direction. To make the model as realistic as possible, we choose a direction such that internal wave crests propagate upslope, from deep toward shallow water. Wave crests are aligned $10°$ from the source-receiver track, and they propagate obliquely, from the source to the receiver. We choose the magnitude of u to be 0.75 m/s, so that its projection is $u = |u|\csc(10°) = 4.32$ m/s.

We use the split-step PE model to simulate acoustic transmission. This model is chosen because it allows the environment to be rapidly range dependent. We set up model runs as follows. We use thermistor chain data to set up a snapshot of the range-dependent sound speed field $c(z,x-ut_1)$, at time t_1. We integrate the PE model to obtain the pressure field in the x-z plane, and from this we compute amplitude and transmission loss at a set of fixed points out to 18.5 km range, at the bottom, $z = 128$ m. Then we advance to the next time t_2, compute $c(z,x-ut_2)$, and continue. Time steps are separated by the thermistor chain sampling interval, 16 s. By chaining together the computed acoustic amplitudes, we simulate a time series.

Figure 8 shows normalized intensity computed by PE, as a function of time and cross-track location. Two receiver ranges are shown; 16.7 km (top panel) and 18.5 km (bottom panel). The eight traces in each panel are radials separated by $0.1°$ (approximately 33 m in the cross-track direction); altogether they subtend approximately the same cross-track distance $\Delta y = 231$ m as the hydrophone array. Each trace is successively offset by 0.5 to give a waterfall appearance. Three-dimensional effects tend to cause asymmetric propagation, which one can see in the variety of cross-track slopes of the various acoustic features. These slopes correspond to phase speeds which range from 0.66 to 1.65 m/s. This range of phase speeds is somewhat broader than the observed range (0.75-1.0 m/s).

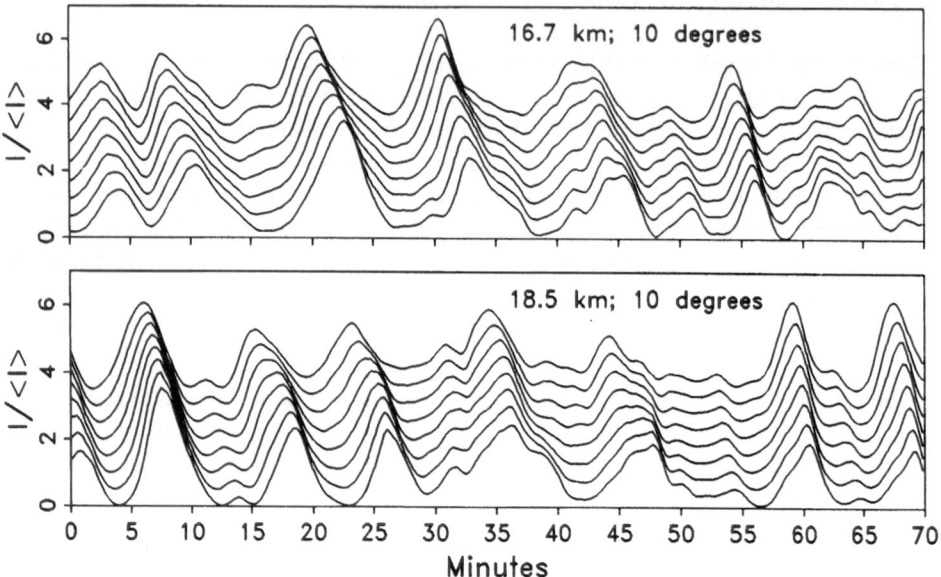

Fig. 8: Modeled normalized acoustic intensity. Receivers are at 16.7 (top) and 18.5 km (bottom) range, and span a 231 m aperture in the cross-track direction. Successive traces are offset by 0.5, to give a waterfall effect. The measured temperature field is advected, to simulate plane waves whose crests are aligned 10° with respect to source-receiver track.

We have analyzed several long sequences (up to 9 hr) of acoustic data, and several long (up to 14 hr) of model predictions of acoustic variability. Comparisons show that it is not possible to correlate individual features of acoustic variability with individual internal wave features. Broad patterns of acoustic variability on multi-hour time scales appear to be modulated with semi-diurnal periodicity. The disjoint nature of the acoustic data set prevents us from making strong assertions about the correlation between acoustic variability and the internal wave packets.

Figure 9 shows a comparison between observed and modeled spectra of normalized intensity. The model underestimates the variance at frequencies above 20 cph. One can see this effect directly from the time series, by comparing Figs. 6 and 8. The data contains more small-scale structure. A possible reason is that the frozen fluid hypothesis is not 100% valid. Aircraft photographs show that internal waves are not exactly planar, but have curvatures on the order of 20 km. The dispersion of internal waves, and the presence of temperature finestructure, introduces additional acoustic variance that the model does not incorporate.

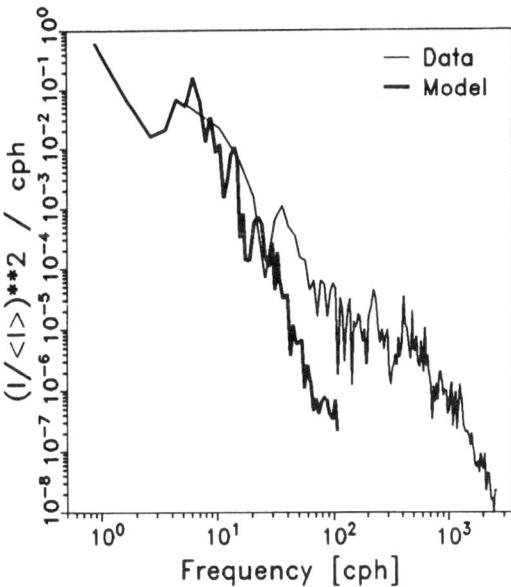

Fig. 9: Observed and modeled spectra of normalized intensity, computed from time series shown in Figs. 6 and 8, respectively.

The character of the acoustic variability is somewhat sensitive to the angle of internal wave propagation. Figure 10 shows PE predictions of normalized acoustic intensity, but with wave crests aligned 30° from the source-receiver track. The acoustic variability is significantly different from the 10° case shown in Fig. 8. Variability is weaker in Fig. 10. Fade-outs are weaker, and peaks are smoother. The reason is that internal wave crests intersect the source-receiver track at a wider angle, and therefore sound speed fluctuations are less coherent over the track.

3. Effects of Surface Wave Packets

Surface waves are organized in propagating packets. We hypothesize that inhomogeneous surface scattering due to these packets may produce bottom absorption that is inhomogeneous in space and time. This inhomogeneity would then show up as acoustic variability on short time scales. The duration of a wave packet is typically about three wave periods. Therefore acoustic variability might be expected on time scales on the order of 20-40 seconds.

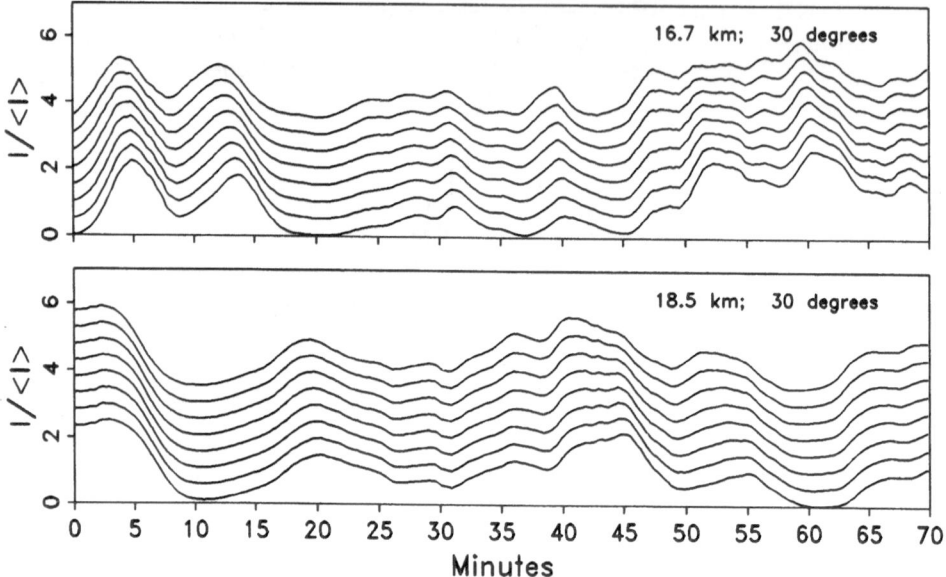

Fig. 10: Same as Fig. 8, but crests are aligned 30° with respect to source-receiver track.

To test this hypothesis, we performed two sets of runs with the PE model. The first set of runs contains only the effects of internal waves. The second set of runs also contains the loss effects of surface wave packets. As an approximation, the reflection coefficient R represents the fraction of specularly reflected energy that remains after a single reflection from a random rough surface is

$$R = \exp(-\rho^2) \quad ; \quad \rho = \frac{4\pi f_0}{c} \sigma \sin\theta \quad .$$

Here, ρ is the Rayleigh parameter for the surface interaction, σ is the rms height of the surface, θ is the incident grazing angle with respect to the mean surface, c is the speed of sound in the water, and f_0 is the carrier frequency of the incident sound wave.

A range-dependent reflection coefficient R was implemented in the PE model. The wave envelope $H(t)$ shown in Fig. 1 was used to specify a local rms height $\sigma = H(x-vt)/2^{1/2}$, where $v=5.6$ m/s is the group velocity for the dominant 7 s wave, and the projected velocity is $v=|v|\csc\varphi$. We set the angle between wave crests and the source-receiver track to be $\varphi=10°$. Here again, as with internal waves, we assume that surface waves are not dispersive. This approximation is surely not justified, but the simplification allows us to generate a simple wave envelope field which, although not quantitatively accurate, has a representative amplitude, and representative space and time scales.

Figure 11 shows results of a set of PE runs, with and without surface waves. Transmission loss is shown at two ranges, 7.4 and 18.5 km. At both ranges, the effect of surface waves on transmission loss is generally less than 0.5 dB. There is negligible change to the acoustic variability. For certain ranges (e.g., 5.5 km, not shown here), much more acoustic variability occurs in the simulations because of an anomalous abundance of surface-interacting paths. The effect is not systematic with range.

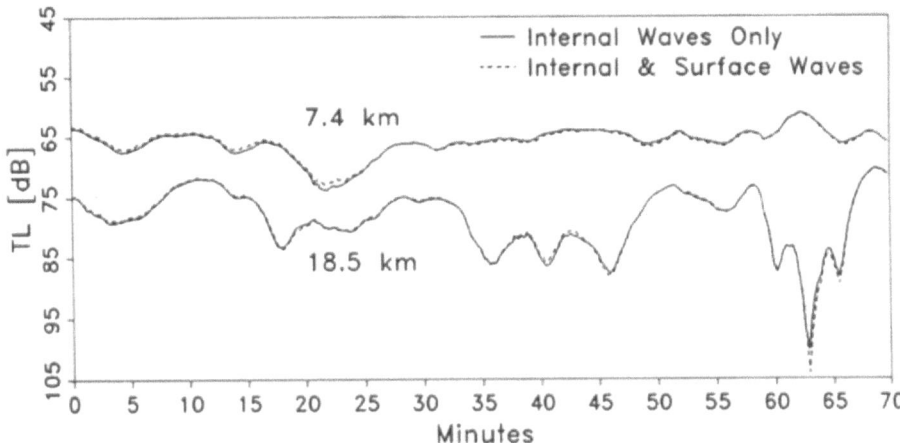

Fig. 11: PE model predictions for transmission loss, with receiver ranges at 7.4 and 18.5 km. In one set of cases, only internal wave effects were included (solid curves), while in another set, surface wave effects were also included (dashed curves).

Increasing range does not systematically increase the surface-wave contribution to acoustic variability for the following reason. As range increases, the number of paths with many surface bounces increases, but these paths interact more often with the bottom as well. Hence these many-bounce paths do not contribute much to the energy of the received signal. On the other hand, the paths with few bounces contribute energy that is influenced only a few times by the surface waves, and the number of these *effective* interactions tends to be range independent.

This explanation has an interesting corollary that relates to the parcelling of received energy between specular and nonspecular paths (of which, incidentally, only the latter are perturbed in frequency due to the moving-surface interaction). As severely as bottom bounces attenuate the specular energy, nonspecular attenuation will be even more severe because nonspecular scattering angles are generally steeper than specular. Because energy that is nonspecularly scattered tends to be absorbed by the bottom before the next surface bounce, nonspecular energy from only the most recent bounce is received.

The results of the simulation—and the explanation above—are borne out by frequency-spread acoustic data at the ranges 7.4 and 18.5 km. For an example of a plot of energy versus frequency, see Fig. 5. In this figure, a surface wave spectrum is plotted in registration with the carrier-demodulated acoustic spectrum that is scattered off the surface at the time the surface spectrum was recorded. Clearly the major peaks of the surface-wave spectrum also appear in the received acoustic spectrum for frequency offsets greater than 0.1 Hz. The surface-wave spectrum in fact modulates the transmitted acoustic wave at these frequencies, although the spectra are not exact replicas of each other. For frequency offsets below about 0.1 Hz, internal wave effects are responsible for broadening the transmitted spectrum.

Figure 12 shows that the frequency spreads such as appear in Fig. 5 are substantially independent of range, being the same for 7.4 and 18.5 km. Because the scattered acoustic frequency is correlated with scattering angle, this phenomenon can be understood in terms of the "angle-stripping" action of bottom absorption described above. Nonspecular energy from only the last bounce is received without such attenuation. As a result, the frequency spectrum recorded at the receiver from each surface-interacting path depends only on the distance from the last bounce, not on the total distance from source to receiver.

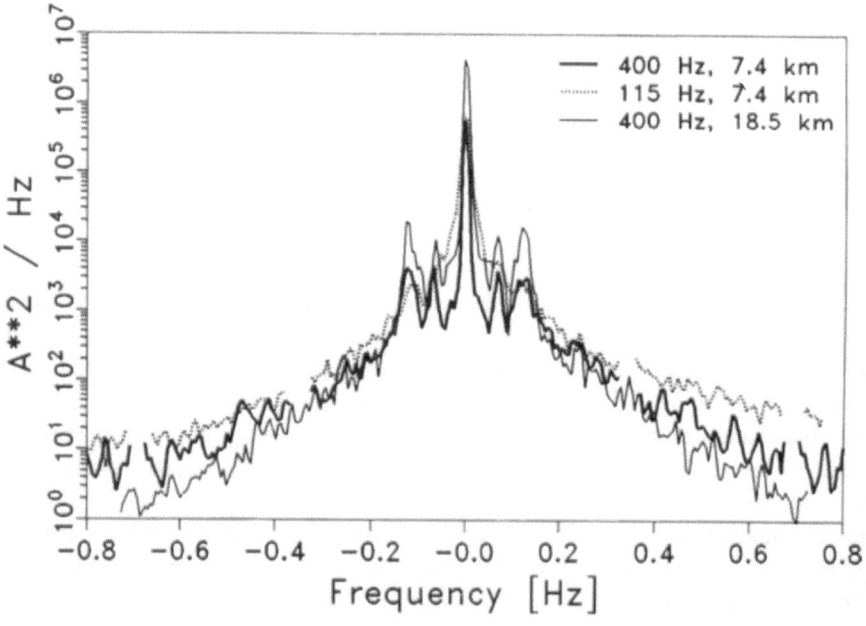

Fig. 12: Simultaneous acoustic spectra, referenced to carrier frequency, for carrier frequencies 115 and 400 Hz, and ranges 7.4 and 18.5 km.

Although the major peaks of the received frequency spectrum are at the same frequencies in Fig. 12, there is substantial variability in the heights of the peaks and in the detailed structure of the smaller peaks. Examination of data from intermediate ranges between 7.4 and 18.5 km shows that these variations are not systematic in range.

Another noteworthy property of the frequency-spread data is that it is substantially independent of the transmitted frequency f_0 between 115 and 400 Hz (see Fig. 12). This independence is also a property of first-order statistical scattering theory for low Rayleigh numbers and forward scatter (Parkins, 1967; Scharf and Swarts, 1974). According to that theory, for $\rho^2 << 1$ (as is most often encountered in our experiment), a "slightly rough sea" limiting formalism applies, whereby the acoustic Doppler-spread sidebands are copies of the surface-wave power spectrum, and independent of f_0. As was the case for the range independence, the f_0-independence seems to apply to the locations of the large peaks and to the general trends of the surface-scattered spectrum, but not to other details of the spectrum. However, examination of data for f_0 between 100 and 400 Hz shows that even the variable features of the spectrum display no systematic trends with changing f_0.

In summary, the acoustic frequency-spread data are substantially independent of range (number of bounces) and of transmitted frequency f_0. The range-independence can be attributed to absorption of nonspecular rays by the bottom until the last bounce, and the f_0-independence can be attributed to the properties of the statistical scattering theory at low Rayleigh parameters.

4. Discussion

Our numerical experiments showed that non-dispersive, plane wave propagating internal wave packets can explain a portion of the acoustic variability of the energy contained in a 400 Hz carrier signal. The model generates low-frequency (<10-15 cph) acoustic variability, which has characteristics qualitatively similar to those observed:

- At 18.5 km range, acoustic features propagate across the 232 m cross-track array with a phase speed which is similar to that predicted by internal wave theory.

- Acoustic features do not propagate at a single phase speed. Three-dimensional effects cause the observed speed to vary somewhat within a range from 0.75-1.0 m/s, and the modeled speed from 0.66-1.65 m/s.

- Acoustic features tend to be coherent over the cross-track array, although their shapes can become distorted.

The model generates significantly less acoustic variability than that observed, in the frequency range above the Brunt-Väisälä frequency, ~10 cph. Perhaps by including internal wave dispersion, and internal wave crest curvature, the modeled acoustic variability may be increased. This will be the subject of future research.

A simplified model of incoherent surface reflection loss was examined. This model propagated a realistic wave envelope of surface roughness across the source-receiver track. The model provides a negligible amount of additional high-frequency acoustic variability. The reason is that surface-interacting energy propagates rather steeply into the partly-absorbing bottom, where it is attenuated.

The surface loss model makes an approximation, that all of the non-specularly reflected, incoherent energy is scattered into steep angles, and therefore lost by bottom absorption. This approximation is not accurate. A portion of the reflected energy is incoherently scattered into shallow angles and nearby frequencies. In future work, we plan to look into this effect in more detail. A useful approach may be to use the PE Rough Surface model (Dozier, 1984), which incorporates the explicit profile of the sea surface.

5. Acknowledgements

This work was supported by the Office of Naval Research under contracts N00014-86-D-0701 and N00014-88-C-0225. We thank Lewis Dozier and Paul Stokes for their helpful discussions.

6. References

Beckmann, A. and A. Spizzichino (1963) *The Scattering of Electromagnetic Waves from Rough Surfaces.* New York: MacMillan. p. 93.

Dozier, L. B. (1984) PERUSE: A numerical treatment of rough surface scattering for the parabolic wave equation. *J. Acoust. Soc. Am.,* **75,** 1415-1432.

Flatté, S. M. (1983) Wave propagation through random media: Contributions from ocean acoustics. *Proc. IEEE,* **71,** 1267-1294.

Longuet-Higgins, M. S. (1984) Statistical properties of wave groups in a random sea state. *Phil. Trans. R. Soc. Lond. A,* **312,** 219-250.

Parkins, B. E. (1967) Scattering from the time-varying surface of the ocean. *J. Acoust. Soc. Am.* **42,** 1262-1267.

Scharf, L. L. and R. L. Swarts (1974) Acoustic Scattering from a stochastic sea surface. *J. Acoust. Soc. Am.* **55,** 247-253.

OBSERVATIONS OF OCEAN INHOMOGENEITIES

J. SELLSCHOPP
Forschungsanstalt der Bundeswehr
für Wasserschall- und Geophysik
Klausdorfer Weg 2-24
2300 Kiel 14
Germany

ABSTRACT.

Ocean stratification data are presented showing the spatial variability of sound velocity profiles. From an acoustic point of view the irregularities must be separated into features, which have to be treated as deterministic phenomena, and random structures in a statistically homogeneous stratification.

In a Baltic Sea example an ocean front appears as an oblique plane separating two adjacent homogeneous water masses with different temperature, salinity, and density. For sound waves it acts as a prism. North of the Faroes a double structured front was found at a mesoscale eddy. The isotherms and isovelocity lines do not coincide with the isopycnals. Therefore a proper prediction of the density field gives a bad estimate for acoustics in this case.

Aside from fronts, when the ocean stratification can be regarded as statistically homogeneous, internal waves are responsible for variations in the sound velocity profile. Profile variation is also caused by alternate intrusions of different water masses. Temperature differences are balanced in respect to density by salinity differences. In many cases thermohaline effects are more important than internal waves.

1. Introduction

The manuals for sonar operators contain mean sound velocity profiles for the area and season of interest, which shall serve for a proper estimation of sonar conditions. These profiles represent an average of available historical data mixing up significantly different weather conditions. The historical profile can diverge from an actual situation even in the sign of the sound velocity gradient at the source or receiver depth. Then its value for a first guess is questionable. Variability of the sound velocity field hence cannot be defined as the difference between an actual situation and a historical profile. Something more realistic has to be taken for

229

J. Potter and A. Warn-Varnas (eds.), Ocean Variability & Acoustic Propagation, 229–236.
© 1991 *Kluwer Academic Publishers.*

the representation of the constant field instead.

The definition of the undistorted sound velocity field gives rise to greater problems than visible at the first glance, because the ocean is not a homogeneous medium with some irregularities, but is stratified. Therefore a constant field means, that the depth dependence of sound velocity is the same in the area of interest, horizontal gradients of the constant field are identically zero. The vertical profile representing the constant field does not only depend on the position in the ocean and the season, but reflects also the latest weather history.

The constant field may be defined as an average of profiles belonging to the same situation. An ensemble average can be replaced by a temporal or horizontal mean. As a desired result of the averaging process random irregularities are smoothed out, in addition an unwanted consequence of averaging is, that sound velocity gradients are weakened, if a layer boundary alters its depth.

If it were internal wave motion only, that causes variability, then by chance the primary profile may be picked from a profile series or it could be calculated by inverse modelling to avoid the unfavourable effects of profile averaging. Mostly however internal waves are not the only source of sound speed variability.

Hence in many situations, even if the amount of measured data is excellent, it proves difficult or impossible to establish a confident definition of the undisturbed sound velocity field and divide it from the variable field. In some studies it may be more appropriate to take a representative profile rather than the mean for a characterisation of the constant part of the sound velocity field. Even the result of a one-dimensional ocean model adding the actual influence of the atmosphere to historical data can provide for an appropriate constant profile.

According to the length scales of sound transmission ocean variability can be treated as a random process or must be respected as deterministic.

2.   A simple front in the Baltic Sea

Evidently do sudden changes of the sound velocity at ocean fronts belong to those variations, which have to be subject of deterministic treatment in sound transmission calculations. The eddy front in the Bornholm basin of the Baltic Sea shown in figure 1 is an example of a well defined boundary between different water masses in the surface layer. This front was examined during several days not only by means of a thermistor chain, but also by CTD probes. It is also visible on a satellite image of surface radiation temperature.

At the frontal boundary salinity changes simultaneously with temperature. So the slope of the temperature front coincides with that of sound velocity and density as well. The warmer water simultaneously has higher salinity, so that the temperature difference is more than balanced in respect to density. The density anomaly of the wedge of warm water is 0.5 units greater than that of the overlying colder surface water. This means that temperature is an insufficient indicator for density structures in the Baltic Sea. There may be no temperature signal at all connected with a dynamically active front. A numerical model of the eddy front used as a prerunner to acoustical prognosis needs information about salinity in addition to possibly available infrared images of the surface.

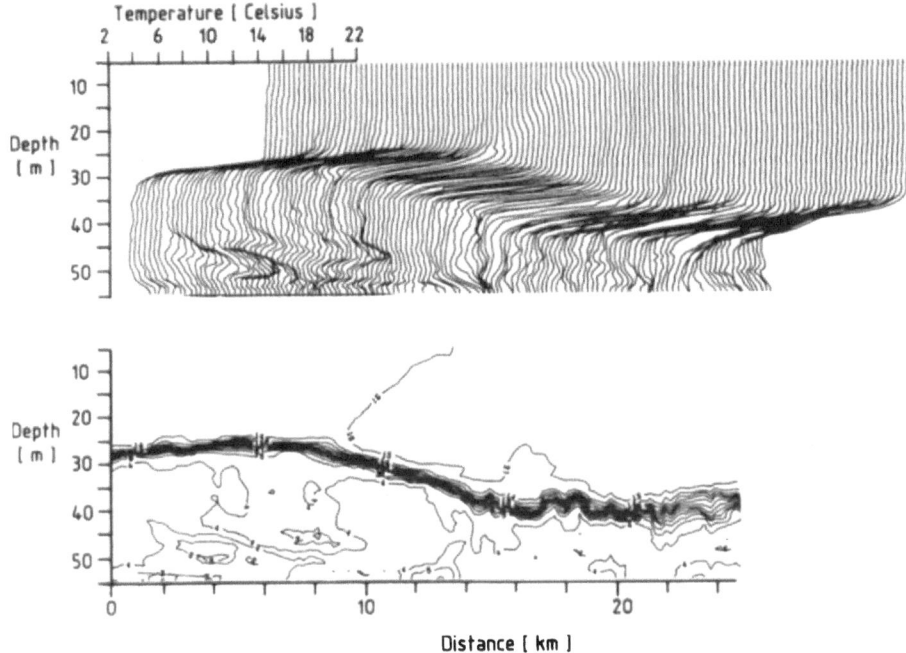

Figure 1.   Eddy front in the Bornholm basin with different homogeneous water masses in the top layer. Salinity is higher in the warmer water. Both depictions are plotted from the same set of thermistor chain data.

Sound radiated towards the front of figure 1 in the colder surface layer will be guided between the water mass boundary and the surface and will accumulate at the intersection point at the surface. From there only a small portion of sound energy can be expected to be sent into horizontal direction and stay in the surface layer. Similarly the front acts as a barrier for sound transmitted by a source located in the surface layer on the warmer side.

3.   An eddy front in the Norwegian Sea

North of the Faroe islands one of the warm core eddies was crossed, which are found at the meandering Iceland Faroe front. The northern edge of the eddy displayed in figure 2 has a doubled temperature front. In the depth range down to 150 m, which is covered by towed thermistor chain measurements, a cold dome of less than one kilometre horizontal extension precedes the permanent change to smaller temperature. The slope of the isotherms is between 30 and 40 m/km in both separated fronts. The shape of isovelocity lines, which have to act as an input to sound spreading calculations, is identical to that of the isotherms.

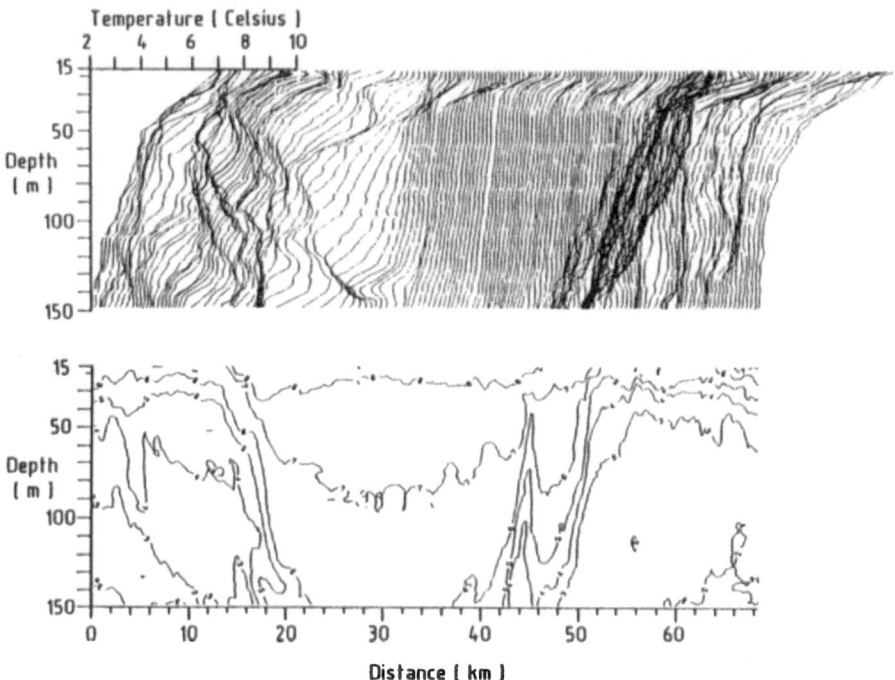

Figure 2.   Warm core eddy north of the Faroe Islands. The double front on the right is present in the sound velocity field as well. In contrast the density field does not contain a footprint of the cold intrusion, because temperature is balanced by salinity.

        Simultaneously with the thermistor chain an undulating fish carrying a CTD package was used to supply the salinity and density field. It turns out that temperature and salinity are balancing so much in the dome, that the dome vanishes in the representation of the density structure. Isopycnals have a moderate slope of 4.4 m/km, which is one order of magnitude smaller than the slope of thermoclines and isovelocity lines.
        A situation like this demonstrates, that investigations of ocean dynamics can have results very different from those necessary for acoustic studies. The sound velocity field contains details, which balanced in respect to density have no complement in the forcing of currents or wave motion. Discrepancies between temperature an density fronts were repeatedly observed northwest of the Faroes during a cruise in July 1989.

4.   Areas with exceptionally high variability

The highest variability of sound velocity profiles was detected in the
domains of surface- or undercurrents differing in salinity from the surrounding water. Examples are the Bornholm channel and the Arkona basin
in the Baltic or the Skagerrak south of Norway. Figure 3 with data from
the Norwegian coastal current at 62°N 3°E demonstrates, that even small
separations of two nautical miles between neighbouring profiles can be too
large for a correct interpolation of the stratification in between.

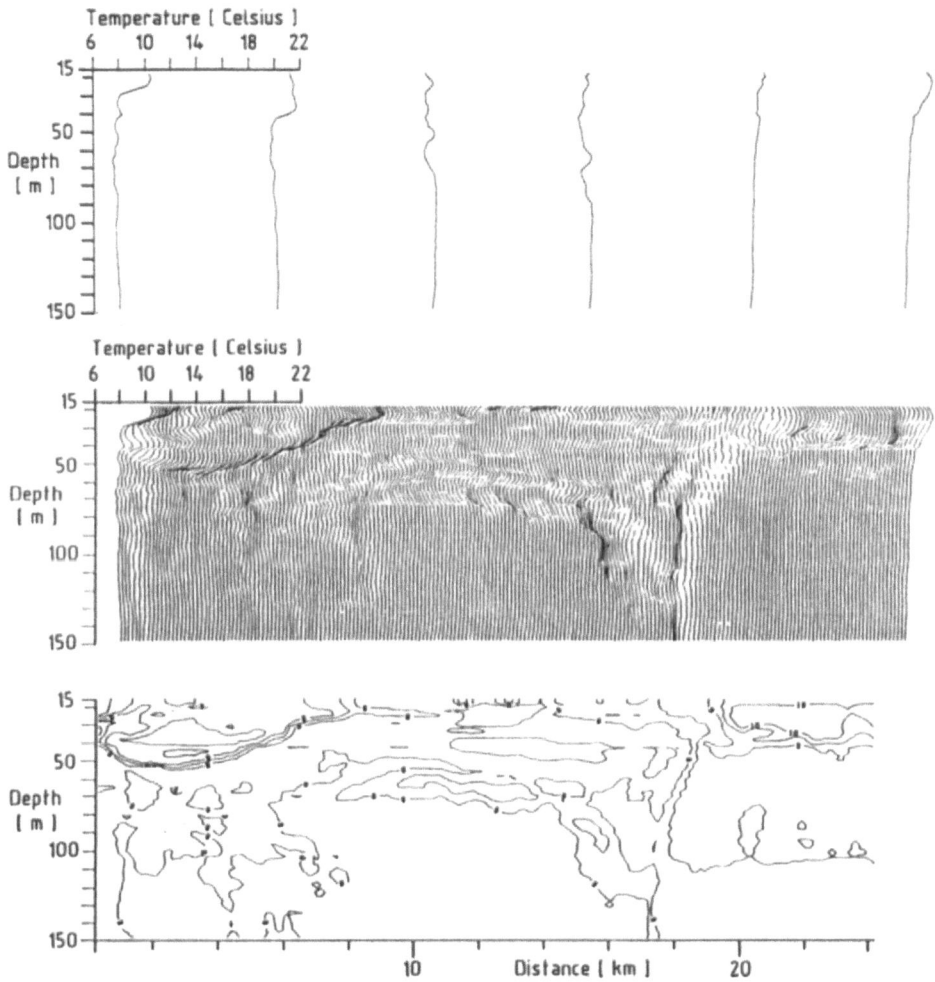

Figure 3.   Temperature section in the Norwegian coastal current. Valid
horizontal interpolation is impossible with 5 km profile separation. Much
higher resolution is necessary to realise the trace of local sound channels
and reflecting layers.

234

Inversions in the sound velocity profile can be an indication of advected intrusions of colder less saline water into equally dense warmer water. They may as well only reflect small inhomogeneities of the water compound. Two dimensional measurements with sufficiently narrow spacing are necessary to distinguish between sound ducts and random refraction cells. Profile averaging is a first step not only to data reduction, but also to separation of ongoing profile distortions from random variations. In order to smooth out random scatterers and preserve intruded layers the average should be taken in horizontal direction rather than within a single profile.

Even for highly variable ocean areas like the Norwegian coastal current statistical stratification parameters can hopefully be found, which describe the quasi random fluctuations of sound velocity and are suitable to be adopted into acoustical calculations. Today there is a deficiency in the knowledge of the parameters needed for ocean acoustics and in the way to establish an adapted parameter set from measurements in such irregular areas. Therefore it seems reasonable to take original stratification data as an input to acoustical models, the data being of high resolution in respect to both horizontal and vertical spacing.

In the everyday forecasting practice it is impossible to rely on fine meshed stratification data, they are not available. But accessible high resolution measurements should be used in sensitivity studies to divide less significant sound velocity structures from important and to define the scales of ocean inhomogeneities required for acoustic modelling.

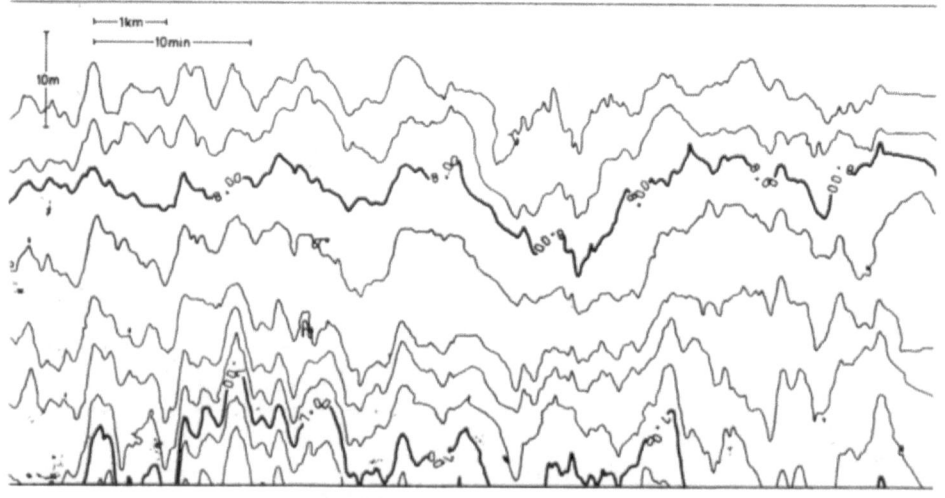

Figure 4.   Isotherms in the depth range from 5 to 55 m in the northern North Sea. The situation can be regarded as statistically homogeneous. The wave length of the oscillations depends on wind speed.

5.   Stationary random structures and internal waves

The isotherms in figure 4 are results of measurements in the northern North Sea during the warming period in early June. Profiles were very similar throughout the 100 by 100 km area of the experiment. There were changes due to solar heating and turbulent mixing within several weeks covered by thermistor chain measurements, and there was the permanent oscillation of the upper layer, which is seen in figure 4.

On account of their length scale these oscillations slip through the meshes of usual CTD casts. It is known from previous cruises, that the significant wave length of the oscillations decreases with increasing wind speed. During the measurement of the data presented in figure 4 wind speed was between 20 and 25 knots.

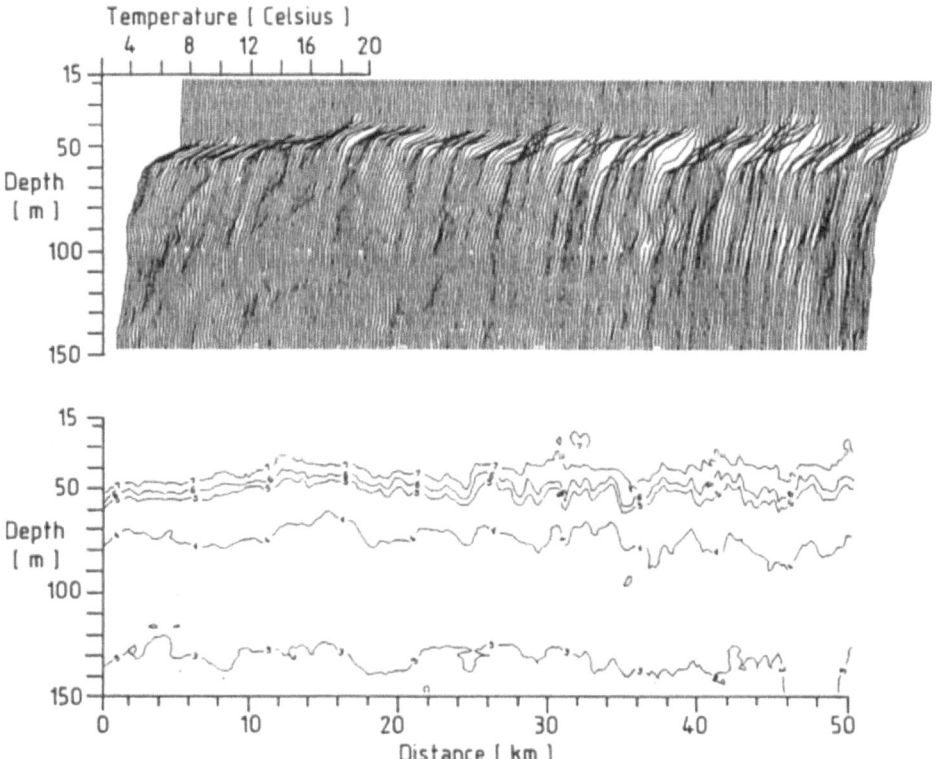

Figure 5.   Temperature section in the Norwegian Sea. The characteristic of thermocline oscillations is disrupted in the middle of the section.

Finally an example of internal wave motion visible as vertical deviations of the thermocline is shown in figure 5. It stems from the central Norwegian Sea. By inspection of temperature profiles there is no apparent reason, why the amplitude of short waves is much greater in the right part of the section. Data of the fish, which was towed at the same time, indicate a small change of salinity at this position. So the discontinuity of internal wave activity seems to be related to current shear.

6.   Conclusion

Two dimensional measurements of ocean stratification carried out in the Baltic, the North Sea, and the Norwegian Sea in the course of many years lead to the impression, that purely statistical distortions of a mean sound velocity profile with the statistics preserved along a sound track, are merely unusual. Internal waves and random microscale scatterers, which are evidently present in the investigated sections, very often if not in most cases appear intermingled with profile deformations related to currents, fronts, thermohaline intrusions, and similar events. The latter are neither statistically homogeneous, nor are they predictable in detail.

As long as no high resolution data of ocean stratification are available, assumptions have to be made in acoustic modelling about the sound velocity profile and its expected variation. The influence of unknown and unpredictable variations may be negligible in favourable cases or it may be substituted by a surplus to a better known kind of variability. The decision of how to treat existing sound speed variations requires simultaneous measurements of sound transmission and stratification.

THE PROBLEM OF CREATING A SYNTHETIC APERTURE IN A NON-ISOTROPIC OCEAN

Stergios STERGIOPOULOS
SACLANT Undersea Research Centre
Viale San Bartolomeo 400
19026 La Spezia, Italy

ABSTRACT   This study is concerned with the limitations imposed by the characteristics of the ocean on attempts to create a synthetic aperture by using towed-array measurements. The multi-element synthetic or physical aperture of a passive array operates in a non-isotropic noise field and in a medium of limited spatial coherence length. In particular, it is assumed that due to distant shipping the noise field is partially directive and imposed upon an isotropic noise background, and in the rough bounded transmitted medium many multipaths exist which are closely spaced in arrival time and arrival angle. Results from real-data applications of the theoretical development of this study indicate that the coherence properties of the ocean can be sufficient for effective long (80λ) towed-array applications and that the formation of a very long (250λ) synthetic aperture in an anisotropic medium can be successful. The algorithm that is used to create a synthetic aperture is the Extended Towed Array Measurements (ETAM) technique.

I. INTRODUCTION

The formation of a passive synthetic aperture in an anisotropic ocean is a complex problem that includes two main areas of investigation. The first one is related to the limitations imposed by the temporal and spatial coherence properties of the medium on the size of an effective synthetic aperture, and the second area of investigation is related to the development of an algorithm that will synthesize coherently the successive spatial information of a small moving line array into an extended aperture.

By the *coherence* of acoustic signals in the sea we mean the degree to which the noise pressures are the same at two points in the sea located a given distance and direction apart. Acoustic pressures received at these two points will have identical outputs if the received acoustic signals are perfectly coherent; if the two sensor outputs are totally dissimilar then the signals is said to be incoherent. Thus, the loss of spatial coherence results in an upper limit on the useful aperture of a receiving array of hydrophones. In other words, an acoustic signal that propagates through the ocean will interact with the transmitting medium microstructure and the rough boundaries. As a consequence of this interaction, a point source detected by a high-angular resolution receiver is perceived as a source of finite extent. This limitation in angular resolution is due either to the angular spread of the incident energy about a single arrival as a result of the scattering

237

J. Potter and A. Warn-Varnas (eds.), Ocean Variability & Acoustic Propagation, 237–250.

phenomena or to the multipaths and their variation over the aperture of the hydrophone array. A model for the spatial coherence length of an underwater medium has been suggested by Wille and Thiele. [1] A thorough discussion of the relative importance of these effects on the underwater acoustics measurement problem has been given by Carey and Moseley. [2] Investigation of these effects, however, is beyond the scope of this paper and we refer the reader to some excellent reviews. [3,4] Consequently, the knowledge of the angular uncertainty of the signals caused by the medium is considered essential in order to define quantitatively the influence of the medium on the array gain, which is also influenced significantly by a partially directive anisotropic noise background. Therefore, for a given non-isotropic underwater medium it is desirable to estimate the optimum towed array length and array gain for sonar and synthetic aperture applications. The problem of creating a passive synthetic aperture in an anisotropic ocean is also examined in this paper. Previous studies [5-7] have shown that this problem is centered on the estimation of a phase correction factor, which is used to compensate for the phase differences between successive in time towed array measurements in order to synthesize coherently their space information into a synthetic aperture. When the estimates of the above phase correction factor are correct, then the information inherent in the synthetic aperture is the same as that contained in an equivalent in size physical aperture. In this paper, signal processing techniques that can provide successful extension of the aperture of an actual towed array are very briefly discussed.

The aim of the present investigation is to determine numerically the dependence of the array gain on the coherence length of the medium and on the directivity pattern of a noise field. Experimental estimates of the array gain for a signal embodied in a partially directive noise field are compared with array gain predictions and an indication of the coherence length of the medium is derived. This is basically an approximate solution of the inverse problem for estimates of the coherence length of the ocean by using towed array measurements. The physical processes defining the coherence proporties of the medium are not examined in this study. Models, however, describing these properties have been considered here in order to estimate the array gain predictions for a non-isotropic underwater environment.

II.   THEORY

II.1 Spatial Coherence

Let us consider a line array in an anisotropic underwater medium. The array has N equally spaced hydrophones with δ being their spacing and receives an acoustic signal from a distant source with bearing θ. The angle θ is measured from the broadside of the horizontal line array. The signal is sampled at time increment Δt with $t_i = i\Delta t$, where $i = 1, 2, ..., M$, M being the number of data samples for each one of the hydrophone time series. It was discussed before that the coherence of the ocean acoustic pressure field in space and time imposes a limiting angular resolution on a line array of hydrophones. In this study, the term coherence means the statistical response of a line array to the acoustic field. This response is the result of the multipath and scattering phenomena discussed before. Models [1,2] have been suggested to relate the spatial coherence with the physical parameters of the underwater medium for measurement interpretation. In these models, the interaction of the acoustic signal with the transmitting medium is considered to result in superimposed wavefronts of different directions of propagation. The received signal by the nth hydrophone of a line array is expressed by

$$x_n(t_i) = \sum_{l=1}^{K} A_l exp[-j2\pi f_l(t_i - \frac{\delta(n-1)}{c}\theta_l)] \tag{1}$$

where $n = 1, 2, ..., N$, A is the amplitude, $X_n(f_o)$ is the Fourier transform of $x_n(t_i)$ at the frequency f_0, $l = 1, 2, ..., K$; K is the number of superimposed waves, and the crosscorrelation between two sensors is reduced to [2]

$$R_{nm}(f_o, \delta_{nm}) \simeq \tilde{X}^2(f_o)exp[-(\frac{\delta_{nm}}{L_o})^k], \qquad k = 1, \ 1.5, \ 2 \tag{2}$$

where L_o is the horizontal coherence length of the medium and $\tilde{X}^2(f_o)$ is the mean acoustic intensity of a hydrophone time sequence of the array. The exponential form of the above expression for $k = 1.5$ has been suggested by Beran and McCoy [8] and the one for $k = 2$, which is the Gaussian form, has been proposed by Wille and Thiele. [1] A review of these models is given elsewhere. [2] It is assumed in the present study that the acoustic field is Gaussian and this is translated here as a Gaussian distribution of the directions of propagation with mean directional standard deviation σ_θ, called *angular uncertainty of the medium*, for the superimposed waves received by the sensors of the line array. A more explicit expression for the Gaussian form of the Eq. (2) is[1]

$$R_{nm}(f_o, \delta_{nm}) \simeq \tilde{X}^2(f_o)exp[-(\frac{2\pi f_o \delta_{nm} \sigma_\theta}{c})^2/2] \tag{3}$$

and the crosscorrelation coefficients are given from $\rho_{nm}(f_o, \delta_{nm}) = R_{nm}(f_o, \delta_{nm})/\tilde{X}^2(f_o)$. At the distance $L_c = c/(2\pi f_o \sigma_\theta)$, called "coherence length", the correlation function in Eq. (3) will be 0.6. This critical length is determined from experimental coherence measurements plotted as a function of δ_{nm}. Then a connection between the medium angular uncertainty and the measured coherence length is derived as, $\sigma_\theta = 1/L_c$, and $L_c = 2\pi \delta_{1m} f_o/c$ where δ_{1m} is the critical distance between the first and the mth sensors that the coherence measurements get smaller than 0.6. The above parameter definition will be used in this study to interpret experimental measurements of the angular uncertainty of the medium, which influences the performance of a line array. The angular scattering function $\Phi(f_o, \theta)$ of the medium can be derived [9] from Eq. (3) and it is expressed by

$$\Phi(f_o, \theta) = \frac{1}{\sigma_\theta \sqrt{2\pi}} \exp[-\frac{\theta^2}{2\sigma_\theta^2}]. \tag{4}$$

II.2 Array Gain

The performance of a line array to an acoustic signal embodied in a partially directive noise field is characterized by the *array gain* which is defined by [10]

$$G = 10 \log \frac{\int [\Psi_S(\theta) B(\theta) \cos \theta] d\theta / \int [\Psi_S(\theta) \cos \theta] d\theta}{\int [\Psi_N(\theta) B(\theta) \cos \theta] d\theta / \int [\Psi_N(\theta) \cos \theta] d\theta} \tag{5}$$

where $\Psi_S(\theta)$, $\Psi_N(\theta)$ are the signal and noise power per unit angle respectively, and $B(\theta)$ is the power beam pattern of the line array. It is assumed here that the above directivity power patterns (i.e. $\Psi_S(\theta)$, $\Psi_N(\theta)$, $B(\theta)$) have rotational symmetry about the axis of the horizontal line array. The integrals $\int [\Psi_S(f_o, \theta) \cos \theta] d\theta$, $\int [\Psi_S(f_o, \theta) \cos \theta] d\theta$ are equal [9] to $\tilde{X}_S^2(f_o)$, $\tilde{X}_N^2(f_o)$ for the signal and the noise respectively. When the directivity pattern of the source is a delta beam, it has been shown [9] that the above Eq. (5) for the array gain is modified to the following form

$$G = 10 \log \frac{\sum_{n=1}^{N} \sum_{m=1}^{N} \sum_{g=-G/2}^{G/2} [\Phi(f_o, \theta_g) \exp(\frac{-j2\pi f_o \delta_{mn} \sin(g\Delta\theta)}{c}) \cos(g\Delta\theta)]}{\sum_{n=1}^{N} \sum_{m=1}^{N} \sum_{g=-G/2}^{G/2} [\Psi_N(f_o, \theta_g) \exp(\frac{-j2\pi f_o \delta_{mn} \sin(g\Delta\theta)}{c}) \cos(g\Delta\theta)]} -$$

$$- \; 10 \log \frac{\tilde{X}_S^2(f_o)}{\tilde{X}_N^2(f_o)}, \tag{6}$$

where the directivity pattern of the received signal $\Psi(f_o, \theta)$ has been replaced by the angular scattering function $\Phi(f_o, \theta)$ of the medium.

It is important to note here that in the expression (6) the array gain includes the spatial characteristics of the medium together with the power directivity pattern of the noise field. This expression will be used later to derive predictions for the spatial coherence length of a medium by assuming that the angular directivity pattern of the acoustic noise field is known.

II.3 The Problem of Creating a Synthetic Aperture

Equation (1) describes an acoustic signal received by the nth element of a towed array. The frequency f, however, of the above received signal includes the Doppler shift that is due to the combined movement of the array and the source radiating this signal. Let V_T denote the speed of the array along a straight-line course. For geometrical simplicity, it is considered here that the relative speed of the source with respect to the towed array is negligible. If f_o is the frequency of the stationary field, then the received frequency f is expressed by $f = f_o(1 - \frac{V_T \sin \theta}{C})$, and an approximate expression [6] for Eq. (1) is given by

$$x_n(\theta; t_i) = A \exp[-j2\pi f_0[t_i - \frac{(V_T t_i + \delta(n-1))\sin\theta}{C}] + \varphi]. \tag{7}$$

τ seconds later the array has moved by τV_T. By proper choise of the parameters V_T and τ we have $\tau V_T = q\delta$, where q represents the number of hydrophone positions that the array has moved, and the received signal $x_n(\theta; (t_i + \tau))$ is expressed by

$$x_n(\theta; (t_i + \tau)) = \exp(-j2\pi f_o\tau)A \exp[-j2\pi f_0[t_i - \frac{(V_T t_i + (q + (n-1))\delta)\sin\theta}{C}] + \varphi]. \tag{8}$$

If the term $\exp(+j2\pi f_o\tau)$ is used to correct the towed array measurements shown in Eq. (8), then the spacial information included in the successive set of measurements at $t = 0$ and $t = \tau$ is equivalent to that derived from an array with $q + N$ hydrophones.

This idealized expression (i.e. $2\pi f_o\tau$) for the phase-correction is used in the conventional *radar synthetic aperture technique*. These phase-correction estimates, however, for synthetic aperture applications require knowledge of the speed, V_T of the towed array and accurate estimates for the frequency of the received signal. An additional restriction is that the synthetic aperture processing techniques have to compensate also for the disturbed paths of the towed array, which influences the synthetic aperture process very badly. Recently, there have been only two passive synthetic aperture techniques [5-7] and an MLE estimator, published in the open literature, that deal successfully with the above restrictions. Their real data application results have indicated that the formation of a synthetic aperture in the sea is feasible when the spatial coherence length of the medium is at least equal to the synthetic aperture length.

It has been also shown [5] that in one of the above synthetic aperture schemes, (i.e. ETAM algorithm), the phase correction estimates will introduce an increase in the variance of the phase term, which includes the directional information of the source. This increase in the phase variance follows a random-walk iterative procedure, which must break down at some maximum value of a synthetic aperture length. More explicitly, if the aperture of a N-hydrophone towed array is extended to a synthetic aperture with qN hydrophones, the ETAM algorithm requires $2(q - 1)$ iterations to synthesize this qN element array. During the μth iteration the phase variance increases by $\sigma_{\psi\mu}^2 = \sigma_\psi^2(\frac{2\mu}{N})$, where σ_ψ^2 is the phase variance of one hydrophone phase term of the N-element physical array during the first iteration. At the end of the synthesizing procedure, which is the $2(q-1)$th iteration, the phase variance will be, $\sigma_{\psi^{2(q-1)}}^2 = \sigma_\psi^2(\frac{4(q-1)}{N})$. Let us consider for example the case of a medium with 128λ spatial coherence length and a towed array with a physical aperture equal to 16λ, where λ is the wavelength of the received acoustic signal. In this case, applications of the ETAM algorithm to extend the physical aperture of the array to 128λ require 14 iterations. Then the phase variance σ_ψ^2 at the 14th iteration will be $\sigma_{\psi^{14}}^2 \simeq \sigma_\psi^2$, which is not a dramatic increase. It is important to note here, that in the above example the increase during each iteration of the phase variance will be linearly distributed along the synthetic aperture. In other words, each 8λ subaperture of the 128λ synthetic array will have a phase variance increase by $0.06\sigma_\psi^2$. This example suggests that the breakdown of a synthetic aperture procedure will be caused by the limited spatial coherence length of the medium and not by the random-walk iterative procedure of the ETAM algorithm, since the value of 128λ is considered as an upper limit for the spatial coherence length of the medium.

HORIZONTAL DIRECTIONALITY

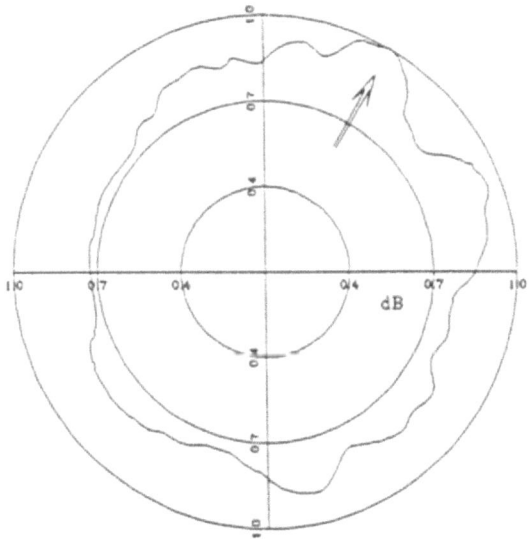

Figure 1: Horizontal directionality pattern of the acoustic noise field at 50 Hz due to distant shipping for a specific area in the Mediterranean sea. The arrow at $30°$ bearing shows the broadside direction of a 64-hydrophone towed array, which is considered for simulations.

The above discussion leads to a conclusion that a synthetic aperture scheme will induce an increase in the variance of the phase terms, which include the directional information of the source along the hydrophones of the synthetic array. This is equivalent to a decrease in the values of the coherence coefficients between hydrophones of the synthetic array and this decrease results in a reduction of the array gain. The determination of a line array's response to hydrophone time series with a given phase variance has been thoroughly discussed by Shifrin.[11] His results include an analytical evaluation of the angular power pattern of a line array as a function of the phase variance and the size of the correlation length, which in our case is the spatial coherence legth L_c of the medium. Thus, the results of the theoretical discussion in the previous sections can be used in synthetic aperture applications. In the present study, the ETAM algorithm has been selected for synthetic aperture applications (reported in Sec. IV) because the performance and the increase in the phase variance induced by this algorithm are known.

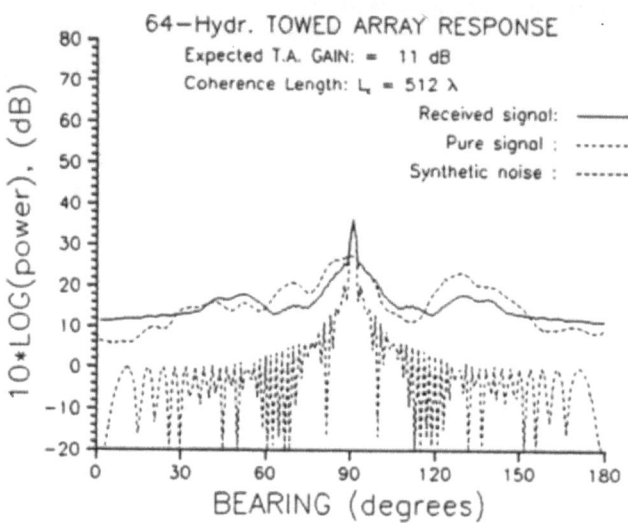

Figure 2: A 32λ array response for (1) a synthetic source signal propagating in a medium with 512λ spatial coherence length which is shown by the dotted line; (2) a real non-isotropic noise field the same as in Fig. 1 which is shown by the dashed line; and (3) the received signal including the source signal and the non-isotropic noise is shown by the solid line. The source signal and the anisotropic noise have been scaled according to the desired value of signal-to-noise ratio, which for this case is $\varsigma = 10$ dB in the frequency bin. The expected array gain for the above source signal is 11 dB.

III. SIMULATIONS

In an another study[9], a processing scheme has been developed for array gain estimates derived from the directivity azimuthal pattern of a received signal. In this scheme the noise directivity pattern is assumed to be known or that it can be derived from time averaged broad band bearing estimates of the ambient noise acoustic field. It is important to note here that in the above array gain processing scheme the value of signal-to-noise ratio is needed to be higher than 5 dB in the frequency bin, which (i.e. 5 dB) is the threshold value of this method.

In order to test with simulations the above theoretical discussion it has been chosen in this study to use real anisotropic noise, where the correlation coefficients for the noise directionality patern are known from experimental measurements.

Figure 1 gives the horizontal directionality of the acoustic noise field due to distant shipping for a specific area in the Mediterranean sea. The arrow at 30° bearing shows the broadside direction of a 64-hydrophone towed array, which is considered for the following set of simulations. This directionality pattern is for 50 Hz frequency and was derived using a technique developed by Wagstaff. [12] For the above noise angular pattern the correlation coefficients of the anisotropic noise field for the 64-hydrophone towed array have been derived using a processing scheme that has been discussed elsewhere [9]. In one set of simulations the correlation coefficients of a synthetic signal propagating in a medium with 512λ coherence length were generated by using Eq. (3). The correlation coefficients of this signal were added to those of the anisotropic noise following a scaling according to the value of signal-to-noise ratio, which was considered to be $\varsigma = 10$ dB in the frequency bin. This operation provided the correlation coefficients of a received signal including real anisotropic noise and synthetic source signal for a medium with known spatial coherence length.

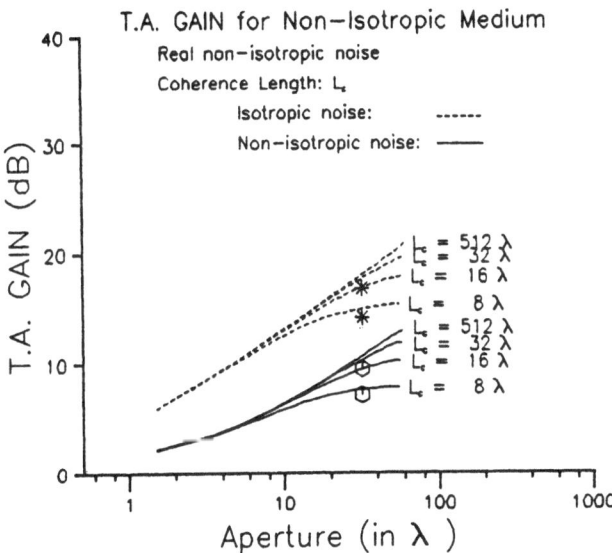

Figure 3: Shown by the solid lines are the array gain estimates for the anisotropic noise field of Fig. 1, for different values of spatial coherence length of the medium (i.e., $L_c = (512, 32, 16, 8)\lambda$). The dashed lines give the array gain predictions for isotropic noise field and for the same values of spatial coherence length as before. The data points indicated by a star symbol correspond to array gain estimates for isotropic noise field, for a source signal propagating in a medium with 8λ, 512λ coherence length, and for 32λ line array . Shown by the hexagon-shaped data points are array gain estimates for the real anisotropic noise field of Fig. 1, for a signal propagating in a 8λ, 512λ coherence length medium, and for 32λ line array

Figure 2 shows the response of a 64-hydrophone towed array with $\frac{1}{2}\lambda$ spacing, for the above received signal, the real anisotropic noise and the synthetic signal for a medium with 512λ coherence length. The expected array gain is 11 dB and it has been estimated from the correlation coefficients of the signal and the noise since they are known from the above simulation procedure. Another estimate also of the array gain has been derived by applying the array gain processing scheme [9] on the received signal and the ambient noise directionality patterns. This new estimate is 9.9 dB and it is smaller than its expected value since the signal-to-noise ratio is rather small.

For the above ambient noise directivity pattern, the array gain predictions are derived from Eq. (6) and are shown in Fig. 3 by the solid lines. These lines are for different values of spatial coherence length of the medium (i.e., $L_c = (512, 32, 16, 8)\lambda$). The dashed lines in Fig. 3 present the array gain predictions for white noise and for the same values of coherence length as before. The data points indicated by a star symbol correspond to array gain estimates using the above processing scheme for a received signal in a medium with 512λ and 8λ coherence length. However, the consideration in this section about the acoustic field due to distant shipping to be taken as anisotropic noise is arbitrary and for others this noise field can be treated as signal. Then for the above set of simulated received signals and for the case of a white noise field the new values of the array gains are estimated in the same way as before, assuming that the correlation coefficients of the noise are $\sum_{n=1}^{N} \sum_{m=1}^{N} \rho_{N_{nm}}(f_o, \delta_{nm}) = N$. These new estimates are indicated by the hexagon-shaped data point, which have been plotted in Fig. 3.

If the broadside direction of the towed array is considered to be at 270° bearing in the anisotropic noise field that is shown in Fig. 1, then the estimates of the array gain should improve because the noise field is not as directive as in the case of 30° bearing that was examined before. The above expectation is correct and it has been demonstrated by the improved array gain estimates and predictions of another set of simulations that included a broadside direction for the line array at 270° bearing. It is important to note here that a comparison in Fig. 3 between the array gain predictions and estimates provides an approximate estimate of the simulated medium coherence length.

IV. EXPERIMENTS

Two kinds of experimental setups were used to apply the theoretical development of the present study on real data. The first one included a 64-hydrophone line array towed by a research vessel. Ambient noise measurements obtained with this line array were used to estimate the array gain of a monochromatic signal radiated by a distant ship and to create long synthetic apertures. In the second setup, long towed array measurements from a 256-hydrophone geophysical-type streamer were used to test the validity of the above synthetic aperture results. These long towed array measurements have been provided by the U.S. Office of Naval Research and they originate from experiments carried out by Yen and Carey. [7] The objective in their study was to test a passive synthetic aperture technique that they had developed. Their reported synthetic aperture results are successful and any comparison in performance between the ETAM algorithm and their technique is beyond the scope of the present study.

IV.1   64-Hydrophone Towed Array

The experimental setup included a receiving line array with 64 hydrophones spaced at 1 m. This array was towed at 100-m depth by a vessel moving along a straight-line course at 5 kn. The water depth was 2000 m and aerial surveillance in the area indicated the presence of few ships beyond the 28 km range. The data-acquisition and control system included amplification of the hydrophone signals, bandpass filtering, digitization, and recording on a high-performance digital recorder for off-line processing of the time series. The acoustic field generated by the distant shipping was recorded continuously for a period of 6 min by the towed array while the

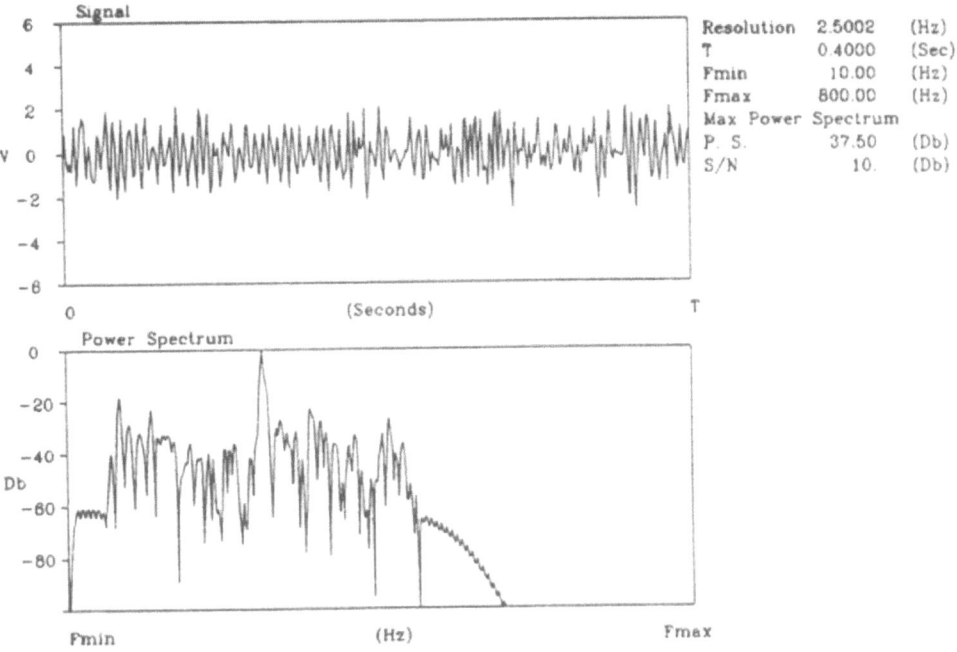

Figure 4: Shown in the upper part is the signal received by the hydrophones of the towed array. The digitized time series have been windowed and filtered to include, in this case, a bandpass frequency regime (250 - 650 Hz) centered at 400 Hz. In the lower part, the power spectrum of the above decimated time series is shown. The received monochromatic signal at 400 Hz is clearly shown in the above spectrum.

tow vessel was moving along a straight-line course.

Presented in Fig. 4 is the power spectrum of one of the hydrophones received broad band signals, which was processed with a numerical bandpass filter centered at 400 Hz. Clearly shown in the above spectrum is the presence in the sampled acoustic field of a monochromatic signal at 400 Hz and of unknown origin. At this frequency, the averaged signal-to-noise ratio along the sensors of the line array was $\varsigma = 21$ dB in the frequency bin. The monochromatic signal that is embodied in the noise field of the distant shipping has been used in this study to test the theoretical development and the array gain processing scheme, which have been discussed in Sec. II.

Shown by the solid line in Fig. 5 are the bearing estimates obtained from beamforming at 400 Hz the 512-hydrophone synthetic array derived from the 64-hydrophone physical array using the ETAM algorithm. For comparison, the dotted line in the same figure gives the bearing estimates at 400 Hz from the 64-hydrophone physical array. The averaged azimuthal pattern at 395 Hz of the broadband acoustic field due to distant shipping is given by the dashed line in Fig. 5. Apparently, the above results show a great consistency between the bearing estimates obtained from a long 512-hydrophone synthesized aperture and those derived from the 64-hydrophone physical array. The power levels of the bearing estimates indicate that the 512-hydrophone extended aperture was synthesized coherently. However, the question of whether the 512-

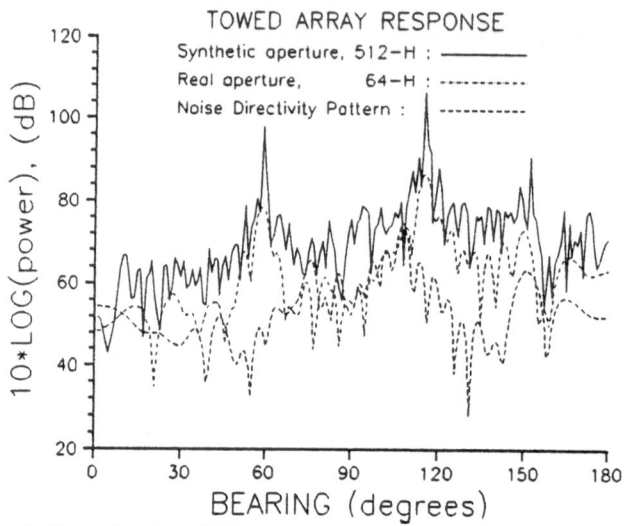

Figure 5: Shown by the solid line are the bearing estimates of the source obtained by beamforming the 512-hydrophone extended towed array derived from a 64-hydrophone towed-array time series with $T = 175$ *seconds* observation period by using the *ETAM* algorithm. For comparison, the bearing estimates from a 64-hydrophone fully populated array are given by the dotted line. The dashed line shows the average azimuthal power pattern of the noise field due to distant shipping at a frequency regime near 400 Hz that does not include the monochromatic 400 Hz line.

element aperture was synthesized in a medium with coherence length at least equal to that of the extended aperture could be addressed by applying our array processing scheme on these measurements.

Figure 6 shows the results of array gain estimates using the above processing scheme for the physical 64-hydrophone array, and for 128, 256 and 512-hydrophone extended apertures. Predictions of the array gain estimates for 273λ and 17λ medium coherence lengths, and for isotropic and anisotropic noise fields are also shown by the dashed and solid lines respectively. In order to derive the above array gain predictions and estimates, the power directivity pattern of the anisotropic noise field was considered to be the average azimuthal pattern of the noise field due to distant shipping, which is shown by the dashed line in Fig. 5. The meaning of the above values of array gain for this kind of anisotropic noise field is related with the array performance to discriminate against distant shipping noise in favor of the desired source signal. Clearly in Fig. 6 the array gain estimates agree with the predictions for 273λ medium coherence length. This indicates that the coherence properties of the underwater environment in our experimental area are sufficient for long towed array and synthetic aperture applications.

IV.2   256-Hydrophone Towed Array

A geophysical-type streamer, consisting of 256-hydrophones with 2.5-m spacing, was towed at a normal depth of 226 m by a vessel moving along a straight-line course at 2.8 kn and at approximately 200 km from a moored CW source. The water depth was 3200 m and the depth of the source was 300 m. Towed array data were obtained with a 256-channel digital data

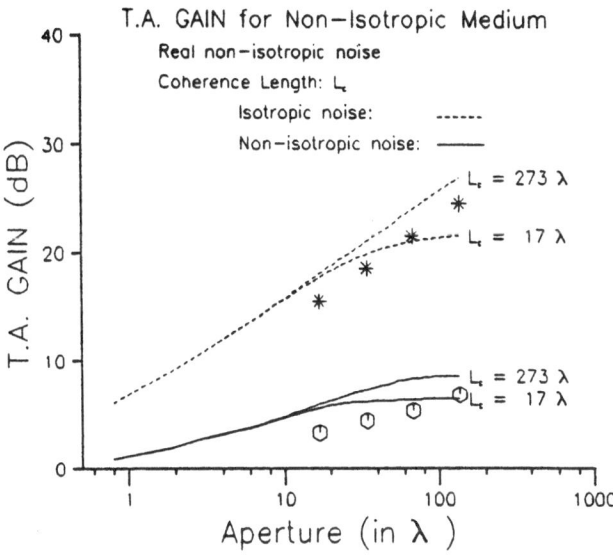

Figure 6: Shown by the solid lines are the array gain estimates for the anisotropic noise field of Fig. 5 due to distant shipping, for different values of spatial coherence length of the medium (i.e., $L_c = (273, 17)\lambda$). The dashed lines give the array gain predictions for isotropic noise field and for the same values of spatial coherence length as before. The data points indicated by a star symbol correspond to array gain estimates for isotropic noise field, for the 400 Hz monochromatic signal, for the 64-hydrophone real array, and for (128, 256, 512)-hydrophone extended apertures derived from the 64-hydrophone physical array using the ETAM algorithm. Shown by the hexagon-shaped data points are array gain estimates for the real anisotropic noise field of Fig. 9, for the 400 Hz signal and for the same apertures (i.e., (64, 128, 256, 512)-hydrophone) as before.

acquisition system. These data were Hann windowed and FFT processed in a 0.125 Hz band. Measurements were taken every 13 s and during this period it was expected that the line array had moved by approximately 7.5 hydrophone spacings (i.e., 18.7 m). A detailed description of this experimental setup is given elswhere. [7]

In the present study, the processing method for bearing estimates of the hydrophone array measurements was the conventional beamforming technique without shading, which is an optimum estimator. [5] Shown in Fig. 7 are the bearing estimates obtained from beamforming the 256-hydrophone towed-array data at 174.5 Hz. The presence in this figure of the CW signal at 43° bearing is very clear. The signal-to-noise ratio was approximately $\varsigma = 3$ dB in the frequency bin. The above results are considered here as a basis for comparison with those from the extended aperture measurements derived from applications of the ETAM algorithm on subapertures of the above long line array.

Presented by the solid line in Fig. 8 are the bearing estimates obtained from beamforming a 256-hydrophone synthesized array at 174.5 Hz, which is derived by the ETAM algorithm from a subarray of 32 hydrophones of the actual array. For comparison, the results obtained from beamforming the equivalent physical aperture of the 256-hydrophone array are given by the dashed line.

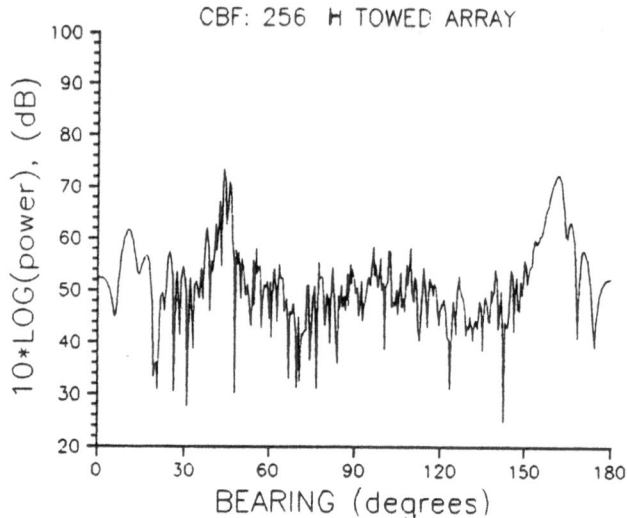

Figure 7: Bearing estimates of an active at 175 Hz source from real data received by a 256-hydrophone line array with 2.5 m spacing. The range between the projector and the receiver was approximately 200 km, the water depth was 3200 m, and the depth of the receiver and the projector was 220 and 300 m, respectively.

V. CONCLUSIONS

In this paper, it has been shown that array gain estimates, which are derived from the directivity power pattern of the received signal by assuming that the power pattern of the noise field is known, are in agreement with array gain predictions when the signal-to-noise ratio of the received signal is higher than 5 dB in the frequency bin. This value of 5 dB is an embirical threshold value of this method derived from simulations.

A comparison of experimental estimates of towed array gains with gain predictions derived according to Eq. (6) can provide approximate estimates of the spatial coherence length of the medium when the signal-to-noise ratio of the received signal exceeds the threshold value of 5 dB. When the agreement between the experimental array gain estimates and the predictions is not very good due to very poor signal-to-noise ratio, then the approximate estimates of the spatial coherence length of the medium can be derived from the array gain estimates directly. In other words, the experimental array gain estimates will linearly increase up to an aperture size that is equivalent to the spatial coherence length of the medium, and all the other gain estimates for longer apertures will asymptotically reach a constant array gain value.

The results presented in this paper (Fig. 8) have shown that the broadside bearing estimates from a 256-hydrophone synthetic aperture derived from a 32-hydrophone subarray are nearly identical with those from an equivalent 256-hydrophone fully populated physical array.

At this point, it is important to discuss the bearing results for the broadband source at the endfire of the towed array shown in Fig. 8. It has been demonstrated that for the case of broadband signals the performance of the ETAM algorithm is successful. [6] The endfire bearing estimates in Fig. 8 from the extended aperture measurements, however, suggest that the ETAM

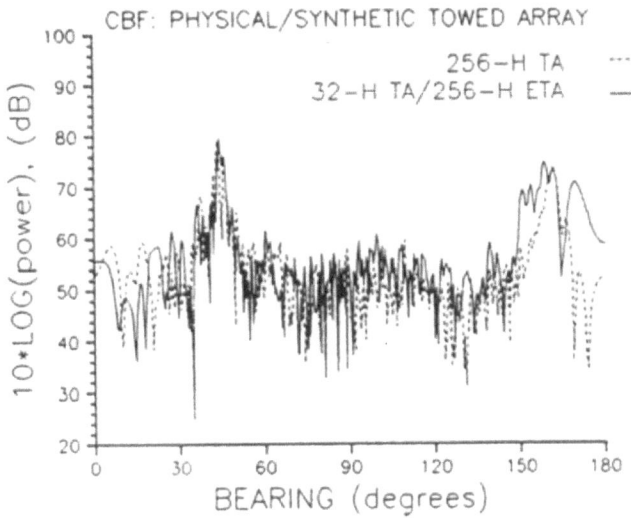

Figure 8: Shown by the solid line are the bearing estimates of the same source as in Fig. 12 from 256-hydrophone extended-towed-array measurements derived from a 32-hydrophone physical subarray using the ETAM algorithm. For comparison, the bearing estimates from the equivalent fully populated physical array are given by the dashed line, which has the same results as those in Fig. 7.

algorithm in this case does not provide sharply defined bearing results. The cause of this failure may be due to the array deformation at the endfire or to an artifact created by the algorithm itself. Future experiments, including ships as broadband sources and signals, have been planned to clarify this ambiguity.

In conclusion, real data applications have indicated that the coherence properties of the ocean can be sufficient for effective long towed array operations and the extension of the physical aperture of a conventional towed array by more than one order of magnitude can be successful. One question, however, that is needed to be addressed here is related to the effects of the ocean variability in long towed array applications. It was discussed before that the overall effect of the ocean variability in underwater acoustics is expressed quantitatively by the spatial coherence properties of the ocean acoustic pressure field. Therefore the conclusions of this study about the upper limit of the spatial coherence properties of the ocean in forming a long synthetic aperture indicate that the overall effects of the ocean variability are not very critical for long towed array applications of the order 100λ. However, the acoustic coherence properties of an underwater medium are influenced by the water depth and the underwater geographical morphology of an area. As a result, the above conclusions about the upper limit of an effective long aperture are related to deep oceans only. In order to generalize these limits, there is a requirement for more experimental work in shallow and deep sea areas of interest.

REFERENCES

1 P. Wille and R. Thiele, *Transverse Horizontal Coherence of Explosive Signals in Shallow Water*, J. Acoust. Soc. Am. **50** 348-353 (1971).

2 W.M. Carey and W.B. Moseley, *Space-time processing, environmental-acoustic effects*, Progress in Underwater Acoustics, (Plenum, 1987), p.743.

3 S.M. Flatte, R. Dashen, W.H. Munk, K.M. Watson, F. Zachariasen, *Sound transmission through a fluctuating ocean*, (Cambridge Univ. Press, New York, 1979).

4 S.M. Flatte, *Wave propagation through random media: Contributions from ocean acoustics*, Proc. IEEE, **71**(11) 1267 (1983).

5 S. Stergiopoulos and E.J. Sullivan, *Extended towed array processing by an overlap correlator*, J. Acoust. Soc. Am. **86** 158 (1989).

6 S. Stergiopoulos, *Optimum bearing resolution for a moving towed array and extension of its physical aperture*, J. Acoust. Soc. Am. **87** 2128 (1990).

7 N.C. Yen and W. Carey, *Application of synthetic-aperture processing to towed-array data* J. Acoust. Soc. Am. **86**(2) 754 (1989).

8 M.J. Beran and J.J. McCoy, *Propagation through an anisotropic random medium*, J. Math. Phys. **15**(11) 1901 (1974). Also in J. Acoust. Soc. Am. **56**(6) 1667 (1974).

9 S. Stergiopoulos, *Limitations on towed array gain imposed by a non-isotropic ocean*, submitted Saclantcen Memorandum, SACLANT Undersea Research Centre, La Spezia, Italy, (1990).

10 R.J. Urick, *Principles of Underwater Sound for Engineers*, (McGraw-Hill, New York, 1967), p.49.

11 Y.S. Shifrin, *Statistical Antenna Theory*, (The Golem Press, Boulder, Colo., 1971), p.40.

12 R.A. Wagstaff, *Iterative technique for ambient-noise horizontal directionality estimation from towed line-array data*, J. Acoust. Soc. Am. **63** 863 (1978).

PREDICTION OF COASTAL OCEAN THERMAL VARIABILITY

DONG-PING WANG
Marine Science Research Center
State University of New York
Stony Brook, New York 11794, USA

ABSTRACT. The thermal variability in coastal oceans is mainly induced by the upper–ocean frontal motion and the internal tidal motion. A time–dependent three–dimensional limited–area primitive–equation model was developed for prediction of the coastal thermal structures. The general circulation model includes an embedded one–dimensional turbulence closure submodel for simulation of the mixed layers. The model has been used to simulate the data from the Coastal Ocean Dynamics Experiment and the Gibraltar Strait Experiment. Excellent agreement between the predicted and the observed density fields was obtained in both studies. This suggests that the coastal thermal variability can be predicted for interface with range–dependent acoustic model and that the remotely sensed acoustic data may be used for validation of coastal circulation model.

1. Observations

In coastal ocean the large–scale temperature structure is mainly generated by the seasonal heating cycle. In winter, the coastal water usually is well–mixed due to surface cooling and wind stirring. The thermocline starts to develop in early spring when the ocean surface becomes a net gainer of the solar heating. The density contrast in water column continues to increase, and the thermocline reaches minimum depth in summer. The thermocline is rapidly eroded in early fall, and water column again becomes well–mixed. Theory for the thermocline evolution in upper ocean is fairly well–understood (Price, *et al.*, 1987). There are two basic approaches in parameterization of the turbulent mixing; these include the bulk mixed–layer model which scales vertical entrainment to the wind stress frictional velocity, and the turbulence–closure model which relates vertical mixing to the shear instabilities. These same two approaches have been extended to the shallow coastal ocean by including a bottom mixed layer associated with the tidal currents. For example, Bowers (1984) used a bulk mixed–layer model for the Celtic Sea and Chen *et al.* (1988) used a turbulence–closure model for the Long Island Sound. In the latter study, Chen *et al.* found that the sudden breakdown of density stratification in early fall was caused by the combination of surface cooling, wind stirring, and spring tide.

In addition to the large–scale thermocline structure, the coastal ocean has considerable spatial and

J. Potter and A. Warn-Varnas (eds.), Ocean Variability & Acoustic Propagation, 251–259.

252

temporal variability due to fronts. On the west coast, the warm stratified season coincides with the upwelling season. Consequently, the nearshore water actually is cooled during active upwelling. The advection of cold water counters the surface heating, leading to regular rises and falls of water temperature through the upwelling season. Figure 1 shows the temperature time series at five depths in the mid–shelf during CODE 2 off central California (start date = 13 April 1982). During active upwelling, *e.g.*, at days 133–143, water was cool and homogeneous, but, water was rapidly warmed and re–stratified following the wind relaxation. Since the advective effect tends to confine to the nearshore zone, a sharp upwelling front, *i.e.*, a narrow band of strong horizontal temperature gradient, was present most time at the mid–shelf. Theory for coastal upwelling is based on the familiar Ekman dynamics that the offshore transport in the surface layer is proportional to the alongshore wind stress. The Ekman theory which does not include the thermodynamics, however, cannot predict the thermal frontal structure.

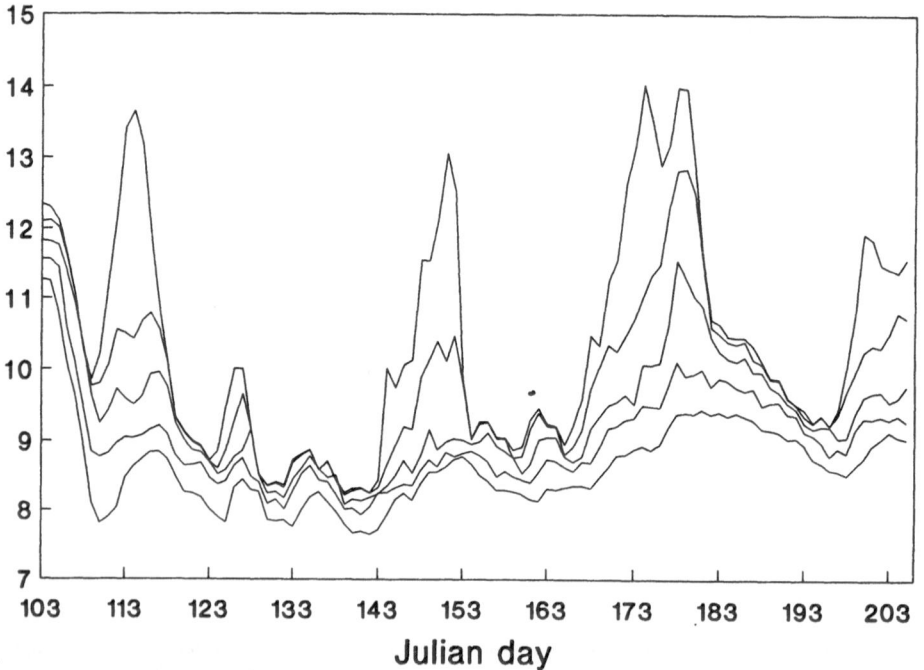

Figure 1. Time series of the observed temperature (in degree C) at mooring C3 in depths of 0, 10, 20, 35, and 53 m, of the CODE 2.

The coastal upwelling front is mainly caused by the differential advection. Fronts in coastal ocean may also be generated by the differential mixing. For example, tidal fronts on the northwest European shelves are generated by the spatial variation in tidal mixing (Simpson and Hunter, 1974). In deep water and in regions of modest tidal currents the thermocline is maintained by surface heating and wind stirring. On the other hand, in shallow tide–dominated water the thermocline is destroyed by tidal mixing. Thus, in regions of rapid change of tidal currents, tidal front is formed at the transition between stratified and well–mixed water. Theory for tidal front has been based on the energy balance criterion that a water column becomes homogeneous when the effective work done by tidal currents exceeds the stabilization by solar heating. The energy balance argument which is empirical, however, cannot predict the evolution of thermal fronts.

Since wind forcing has dominant effects on advection and mixing in coastal water, the temporal variability of coastal fronts usually has time scales similar to those of the synoptic wind. In the case of tidal fronts, there is also evidence of spring–neap variation associated with the modulation of tidal currents. In addition, large vertical displacements of the seasonal thermocline at tidal periods often occur over steep continental slopes. The internal tides which are generated near the shelf break, may radiate out in sharp surges. For example, Heathershaw, et al. (1987) showed strong internal tides at the Celtic Sea shelf edge that temperatures dropped abruptly when tidal currents reversed from off-shelf to on–shelf. Perhaps the most striking internal tidal surges occurred in the Gibraltar Strait (Lacombe and Richez, 1982); the density contrast in the strait is due to the salinity difference between the Atlantic Ocean and the Mediterranean Sea. Figure 2 shows the density interface at the sill over a spring–neap cycle, obtained from the Gibraltar Strait Experiment. Strong internal surges with amplitudes as large as 200 m were generated over the sill when tidal currents reversed from outflow (into the Atlantic Ocean) to inflow. The internal wave form is severely distorted because the barotropic tidal current speed is comparable to the internal wave phase speed (Hibiya, 1988).

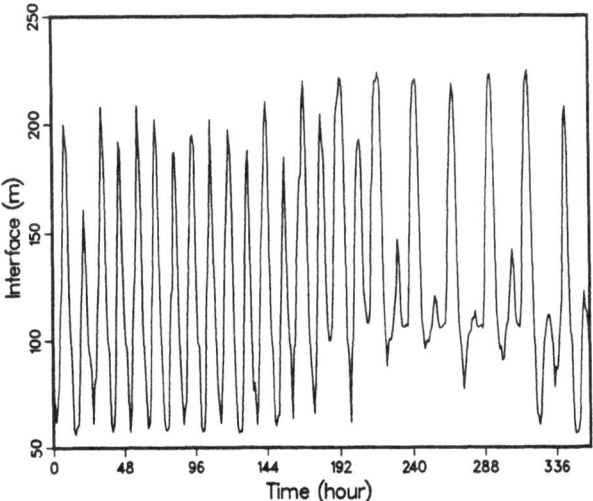

Figure 2. Time series of the observed interface depth at the sill. (start time = 25 October 1985)

2. Model

A three–dimensional time–dependent limited–area coastal general circulation model has recently been used to examine the feasibility of predicting the thermal variability associated with coastal fronts and internal tides. The model uses a flux–corrected transport scheme in temperature and salinity advective terms to reduce the artificial (numerical) diffusion commonly found in general circulation models. This numerical implementation is essential in frontal simulations, since an initially sharp front may otherwise be smeared out quickly. In the application to limited area, the model uses Orlanski radiation boundary condition. The radiation condition is effective in transmitting the interior disturbances through open boundaries. However, the Orlanski condition may not work properly if the wind forcing is present at the open boundary. In such condition, the numerical solution at the boundary is split into local and nonlocal parts with the radiation condition applied only to the nonlocal part (Chen, 1990). Since the spatial scale of wind forcing typically is larger than the model domain, the implementation of a forced radiation condition is essential.

The coastal general circulation model retains the free surface motion; in contrast, most general circulation models use a rigid–lid approximation which filters out the surface tide. To achieve computation efficiency, the model is solved through two–mode integration scheme in which the external mode (free surface and transport) is updated with short time step and the internal mode (vertical velocity shears, temperature, and salinity) is updated with long time step. Typically, the long time step is 10–100 times larger than the small time step. The vertical eddy coefficients can be specified from a Richardson–number dependent function, *e.g.*, the Munk–Anderson formula. Alternatively, the boundary mixing can be derived from an embedded one–dimensional mixed–layer submodel. In the latter case, the mixed–layer submodel is based on Chen *et al.* (1989) which uses the Mellor–Yamada level–2 turbulence closure formulation. Eddy coefficients computed by the mixed–layer model are passed into the general circulation model, whereas the vertical density profile used in the mixed–layer model is interpolated from the general circulation model result. The embedded–model approach adds extra 15% computation effort to the general circulation model, yet it is able to fully resolve the boundary layer structure. The surface boundary condition includes wind stress and heat flux (long– and short–wave radiations) and the bottom boundary condition includes a logarithmic tidal boundary layer. For internal tide simulation, the tidal volume flux also is specified at the open boundaries. The model code has been vectorized, and is currently running on a IBM 3090–600E at the NSF–sponsored Cornell National Supercomputer Facility (CNSF).

3. Applications

3.1. CODE SIMULATION

The CODE 2 experiment in summer 1982 is by far the most comprehensive coastal study program. Our model simulation started with a two–dimensional case neglecting the alongshore variations (Chen and Wang, 1990). The model domain is a vertical plane from the coast to 40 km offshore which includes the entire shelf/slope area. The model simulation covered the CODE 2 period of 103 days, using the observed heat flux and wind stress as the driving force. The initial temperature distribution was derived from the hydrographic data, and the initial flow field was set at rest. Figure 3 shows the predicted temperature time series at mid–shelf that correspond to the observed temperature time series of Figure 1. The model prediction reproduces all major episodes. Table 1 lists comparison of mean, variance, correlation, and regression coefficient between the observed and the modeled temperature time series. Agreement between observed and modeled mean temperatures is excellent. For the temperature fluctuations, the correlations are high, and the regression coefficients are close to unity when the temperature variance is large. Similar statistics also are found at other mooring locations

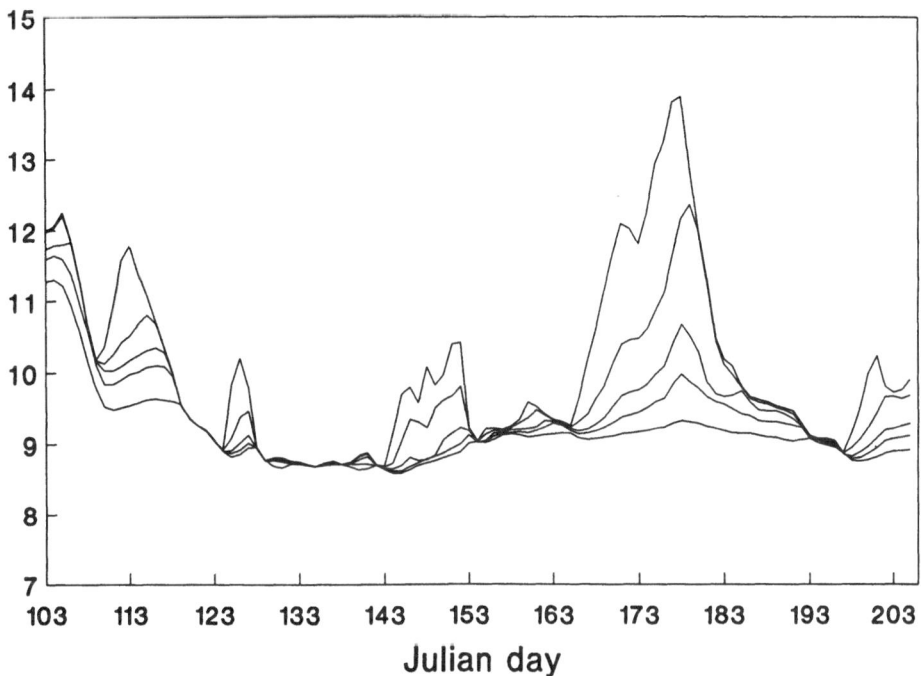

Figure 3. Time series of the modeled temperature (in degree C) at mid–shelf in depths of 2, 10, 20, 30, and 50 m.

covering the entire shelf region (Chen and Wang, 1990). This study has clearly demonstrated the model's quantitative skill. For brevity, the spatial structure of the predicted coastal thermal structure is not shown. In general, a sharp front is present in the upper layer during active upwelling and is temporally destroyed during relaxation.

The two–dimensional model, however, under–estimates the warm episodes. During the wind relaxation the intrusion of warm surface water from the south had significant contribution to the heat budget in the nearshore zone (Send et al., 1987). This process was investigated with the full three–

Table 1. A comparison of observed (o) and modeled (m) mean temperature (T), standard deviation (σ), correlation (γ), and regression coefficient (β) at Mooring C3.

Depth	T_o	T_m	σ_o	σ_m	γ	β
0 m	10.4	10.2	1.6	1.4	0.88	1.0
10 m	9.8	9.7	1.0	0.9	0.94	1.1
20 m	9.4	9.4	0.9	0.8	0.89	1.1
35 m	9.0	9.2	0.7	0.6	0.81	1.0
53 m	8.7	9.0	0.7	0.5	0.77	1.1
70 m	8.6	8.9	0.6	0.5	0.72	1.3
83 m	8.4	8.8	0.6	0.5	0.73	1.4

dimensional model using an idealized wind condition characteristic of the wind relaxation sequence (Chen, 1990). The predicted scenario indicated the expected warm water intrusion, in good agreement with the current meter and satellite observations. On the other hand, because the spatial coverage of the wind forcing is inadequate, a realistic three–dimensional model simulation cannot be undertaken. This points to the need of a better knowledge of the coastal wind field.

The general circulation model also has been used to simulate tidal fronts in the Celtic and Irish Seas (Wang, *et al.*, 1990). The two–dimensional (along the axis of St. Georges Channel) model was run for a period of 43 days using the observed atmospheric and tidal forcing. The model correctly predicted the location of tidal front and the temporal variation of the surface mixed layer. In particular, the model results suggested that the upper–ocean front is controlled by the atmospheric forcing and is not sensitive to the spring–neap tidal variation. However, since the available data is limited, the model's predictive skill cannot be fully evaluated. The UK North Sea Project should provide a good data base for assessing the model's utility in the tide–dominated coastal seas.

3.2. GIBRALTAR SIMULATION

The Gibraltar Strait Experiment in 1985–1986 is a major field study of the tidal strait. The three–dimensional model was used previously to study the mean gravitational circulation and the internal tidal motion for idealized forcing (Wang, 1989). Our recent effort has been towards the realistic simulation. The model's computation domain is a rectangular channel of 100 km wide and 200 km long, centered at the strait. The model's initial condition is a lock–exchange experiment setup; the salinity is 36.2 ppt in the Atlantic Ocean and 38.2 ppt in the Mediterranean Sea. The model's forcing condition is the prescribed tidal transports at open boundaries with the semidiurnal (M_2 and S_2) and diurnal (K_1 and O_1) tidal components obtained through the harmonic analysis of the observed volume transport data. The model was run with the harmonic tidal forcing for a spring–neap cycle, and the predicted tidal constants were compared with the observations.

The interface is defined at the salinity equal to 37.5 ppt. Table 2 shows comparison of semidiurnal tidal harmonic constants at the sill and the eastern exit, of the upper–layer transport, the lower–layer transport, the total transport, and the interface depth. The modeled total transport at the sill matches the data, indicating that the barotropic tidal forcing at the sill is well calibrated. However, the predicted transport at the eastern exit is about 15% less than the observed. Since the transport data for the eastern exit is based on only two current meter measurements and is likely to be overestimated, no effort is made to recalibrate the model. The modeled upper and lower layer transports both agree well with the data. The upper layer transport decreases from about 2.1 Sv at the sill to 0.2 Sv at the eastern exit, whereas the opposite trend occurs in the lower layer. Furthermore, the mean (averaged over the simulation period of 15 days) layer transport of 0.7 Sv and the mean salt transport of 1.5 in Sv unit (above the basic Atlantic water salinity of 36.2 ppt) predicted by the model are identical to the observed transports.

Table 2. A comparison of observed (o) and modeled (m) semidiurnal amplitude (H) and phase (Θ) for upper, lower, and total transports (in Sv) and interface depth (in meter) at sill and eastern exit.

		H_o	H_m	Θ_o	Θ_m
	upper	1.50	1.43	−116	−108
	lower	1.03	1.05	−99	−117
sill	total	2.51	2.47	−109	−112
	interface	40.5	45.1	−156	−155
	upper	0.32	0.21	−181	−174
	lower	2.97	2.49	−107	−109
exit	total	3.07	2.58	−112	−113
	interface	15.3	15.3	123	116

The change in the layer transport mainly reflects the shallowing of the interface depth towards the Mediterranean Sea. The modeled mean interface depths of 107 m at the sill and 64 m at the eastern exit compare well with the observations of 120 m and 74 m. The total transports are in phase between the sill and the eastern exit, that is, the barotropic tide is a standing wave in the strait. In contrast, the interface oscillations show a significant phase difference of about 90°, or 3 hr, between the sill and the eastern exit. Also, the interface oscillations are much larger at the sill than at the exit. These results are consistent with the idea that internal tides are generated at the sill and propagating eastward into the strait. The predicted amplitude and phase for interface oscillations at the sill and at the eastern exit agree well with the observations. While the harmonic constants provide an objective comparison, the internal tidal energetics in the strait can be better appreciated by viewing the time series. Figure 4 shows the modeled salinities at the sill during transition from spring to neap tide. The internal surges are clearly marked by the sudden drops of the interface during tidal flow reversals. Modulations of internal tides at the spring–neap cycle and at the diurnal inequality are quite obvious.

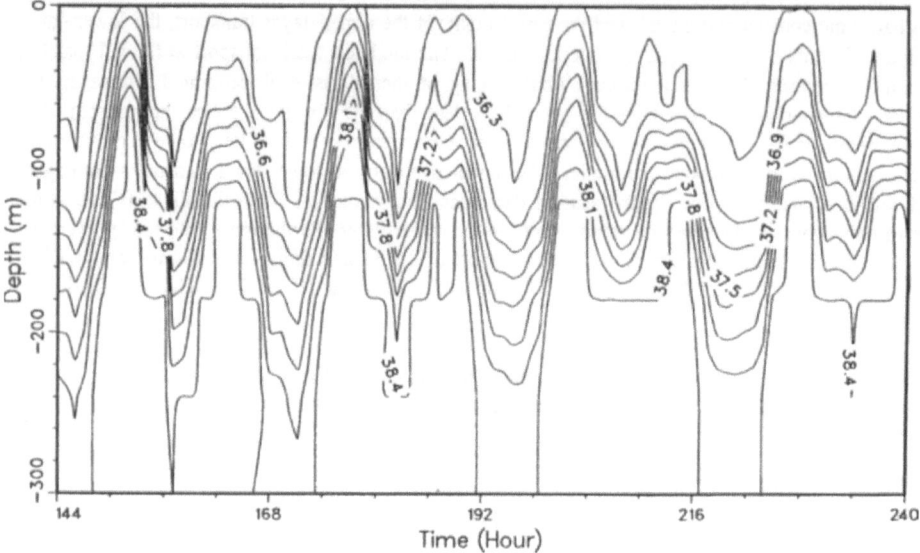

Figure 4. The modeled salinity contours at the sill.

4. Discussion

The circulation model has been advanced to the point that, given the external forcing, the model is capable of predicting the spatial and temporal variability of thermal structures in coastal ocean. We have demonstrated the model's ability in simulating the temporal variability through verification with the moored time series data. It is also implied in this study that the model is capable of resolving the spatial scales comparable to the mooring array. As the wind and tidal forcing in coastal region can be monitored, it appears now feasible to predict the temperature structures in coastal ocean.

Interface of coastal ocean model to a range-dependent acoustic model should be quite interesting. The thermal variability can have important effects on the acoustic propagation over continental shelf/slope. For example, the presence of surface duct depends on the mixed-layer thickness which varies through an upwelling cycle (Fig.1). Also, the acoustic path, and hence the transmission loss, depends on the location of the interface (thermocline) which fluctuates through a tidal cycle. Thus, the sound velocity variability in coastal ocean can be better understood with a more reliable estimate about the thermal variability. It also appears that the acoustic remote sensing may be used to validate the coastal ocean prediction. For example, a simulated temperature time series can be generated by the model for analysis of the sensitivity of acoustic propagation. Based on the simulated results an acoustic-based monitoring network may be implemented to monitor the interior temperature variations for validation of the model predictions.

5. Acknowledgement

The CODE data is obtained from Robert Beardsley of Woods Hole Oceanographic Institution, and the Gibraltar Strait Experiment data is from Clinton Winant of Scripps Institution of Oceanography. This research is partially supported by the National Science Foundation and the Cornell National Supercomputer Facility.

6. References

Bowers, D.G. (1984) A two-layer model of the seasonal thermocline and its application to the Celtic Sea. UCES Report U84-4. University College of North Wales.

Chen, D., S.A. Horrigan, and D-P. Wang (1988) Vertical nutrient mixing in late summer in Long Island Sound. J. Mar. Res., 46, 753-770.

Chen, D., and D-P. Wang (1990) Simulating the time-variable coastal upwelling during CODE 2. *to appear in* J. Mar. Res.

Chen, D. (1990) Dynamics of the time-variable coastal upwelling. Marine Sciences Research Center, State University of New York, Stony Brook, Ph.D. dissertation.

Heathershaw, A.D., A.L. New, and P.D. Edward (1987) Internal tides and sediment transport at the shelf break in the Celtic Sea. Cont. Shelf Res., 7, 485-517.

Hibiya, T. (1988) The generation of internal waves by tidal flow over Stellwagen Bank. J. Geophys. Res., 93, 533-542.

Lacombe, H., and C. Richez (1982) The regime of the Strait of Gibraltar. in "Hydrodynamics of semi-enclosed seas," J. Nihoul, editor, Elsevier Oceanography Series 34, 13-73.

Price, J.F., E.A. Terray, and R.A. Weller (1987) Upper ocean dynamics. Rev. Geophys., 25, 193-203.

Send, U., R.C. Beardsley, and C.D. Winant (1987) Relaxation from upwelling in the Coastal Ocean Dynamics Experiment. J. Geophys. Res., 92, 1683-1689.

Simpson, J.H., and J.R. Hunter (1974) Fronts in the Irish Sea. Nature, 250, 404-406.

Wang, D-P. (1989) Model of mean and tidal flows in the Strait of Gibraltar. Deep Sea Res., 36, 1535-1548.

Wang, D-P., D. Chen, and T.J. Sherwin (1989) Coupling between mixing and advection in shallow sea fronts. Cont. Shelf Res., 10, 123-136.

Flatté and Munk

The papers presented in session 2 gave rise to a number of questions each of which could have been fruitfully discussed for the hour allowed in the final open discussion:

1. Chaos in ray theory – when is it important?

2. Where in the ocean is density-compensated temperature/salinity structure important?

3. How can one best separate the 'deterministic mean' profile from ocean variability? Is this separation always meaningful?

4. What are the relative merits of impulse/time domain experiments versus CW experiments?

5. How do we bring the two points of view – acoustical oceanography and ocean acoustics – into a cooperative spirit?

6. Can 'reflection seismology' techniques give useful images of ocean interfaces between water masses?

The actual discussion was spent solely on the first question of chaos in ocean acoustics. The discussion was not always clear, mirroring the title of the subject.

Kevin Smith pointed out that the question seemed to be one of proper numerical technique with appropriate accuracy for the ray integration being done; people should be aware of the pitfalls that might arise in using ray theory. Questions came from the audience indicating some confusion as to whether chaos is 'fundamental' or only dependent on finite numerical accuracy. Alfred Osborne suggested that chaos theory is of limited utility in geophysics because we are not dealing with systems with few (e.g. three) degrees of freedom.

Stan Flatté pointed out two observable effects in long-range pulse transmission experiments that could be related to ray analysis involving chaos theory.

First, a particular arrival is observed to be 'split' into several arrivals spaced by a few tens of milliseconds at 3,000 km. The question arises as to what the vertical structure of these multiple arrivals might be. Flatté suggested the existence of miniwave-fronts in a sawtooth patterns, whose sharp points are caustics, induced by small-scale structure in the medium (e.g. internal waves).

Second, the data of the SLICE89 experiment show significant energy very late and at depths up to several hundred meters displaced from the sound-channel axis, out of the region illuminated by rays calculated from the smooth range-independent profile. (The last – slowest – energy to arrive at a particular range from an axial source should come along the axis, and hence arrive at the axis depth.) Hans Schneider suggested that small-scale medium fluctuations could explain this vertically displaced energy, by a 'scattering' process. Robert Laval pointed out that to obtain a late steep arrival calls for propagation very near the axis for most of the range with a scatter into a steeper path near the receiver.

J. Potter and A. Warn-Varnas (eds.), Ocean Variability & Acoustic Propagation, 261.
© 1991 *Kluwer Academic Publishers.*

Session 5

*Stochastic modelling
in oceanography and
acoustics*

TREATMENTS OF INCOHERENT SCATTERING FOR THE PARABOLIC EQUATION AND ASTRAL PROPAGATION MODELS

L. B. Dozier
J. S. Hanna
C. R. Pearson
Science Applications International Corporation
1710 Goodridge Drive
McLean, Virginia 22102
U. S. A.

ABSTRACT. Upward-refracting sound speed profiles, such as surface ducts and most Arctic profiles, induce repeated scattering from a rough ocean surface. Propagation models that treat only surface loss will rarely give the right answer, because the lost energy is discarded from the problem rather than being redistributed into other propagating, small angles. In this paper we present a coupled mode approach which can potentially be implemented numerically into many such models to carry out such a redistribution of energy; we have done this for two range-dependent propagation models, ASTRAL and the parabolic equation model (PE). The coupled mode approach requires as inputs two functions of angle, a reflection coefficient and a scattering kernel. Obtaining these inputs from a particular scattering model, we compare propagation model results with measured Arctic data.

1. Introduction

Upward-refracting sound speed profiles, such as surface ducts and most Arctic profiles, induce repeated scattering from a rough ocean surface, or from the underside of rough Arctic ice. For typical roughnesses and low to moderate acoustic frequencies, much of this scattering is incoherent, and from one small forward angle into another. Such incoherently scattered energy is thus a significant contributor to the acoustic field even out to long ranges. Unfortunately, theoretical difficulties associated with incoherent scattering have proved formidable, often causing researchers to discard the incoherently scattered field altogether. On the other hand, propagation models based on such theories, which of necessity treat only surface *loss*, will rarely give the correct answer for upward-refracting environments.

The approach taken here is to modify the numerical algorithms of two recognized (U. S. Navy Standard) range-dependent propagation models, ASTRAL (a hybrid ray-mode model) and the parabolic equation (PE) model. These modifications are strongly motivated by theories of mode coupling which redistribute modal energies continuously in range. The redistribution is done statistically, as described in Section 2, according to two input functions of angle, a reflection coefficient and a scattering kernel which describes the probability of scattering from one angle into another. The particular scattering kernel used here is described in Section 3.

Results of the modified numerical algorithms are compared with Arctic transmission loss data reported by Diachok (1976). A reasonable choice of two ice parameters, mean keel spacing and mean keel depth, leads to a good fit of ASTRAL results to the data at four widely spaced frequencies, as shown in Section 4. A similar discussion is given for the PE results in Section 5.

J. Potter and A. Warn-Varnas (eds.), Ocean Variability & Acoustic Propagation, 265–281.

2. Coupled Mode Model

A coupled mode model which accounts for the continuous exchange of energy with range among the modes is outlined below. It begins with the intensity sum of modes given by

$$I(r,z) = \frac{1}{r} \sum_m \phi_m^2(z_0) \phi_m^2(z) e^{-r\alpha_m} e^{-r\gamma_m} \,, \tag{1}$$

where z and z_0 are source and receiver depths and ϕ is the mode depth function. The $\exp(-r\alpha_m)$ term contains the volume and bottom scattering losses, and the $\exp(-r\gamma_m)$ term contains the surface scattering losses in the specular direction *only*. As it stands, Eq. (1) is not conservative with respect to the surface scattered energy; the energy scattered out of mode m into non-specular directions is lost from the problem. We now modify Eq. (1) to keep track of this scattered energy by reapportioning it among other modes corresponding to these non-specular directions. To begin, we replace the exponential decay factors by a more general energy function $E_m(r)$:

$$I(r,z) = \frac{1}{r} \sum_m \phi_m^2(z_0) \phi_m^2(z) E_m(r) \,. \tag{2}$$

$E_m(r)$ includes two fundamental mechanisms of surface scattering, specular reflection and incoherent scattering. After characterizing each mechanism in terms of decay constants below, we will be in position to derive a system of ordinary differential equations for the $E_m(r)$. It should be noted that the incoherent scattering will in general include a component in the specular direction.

Experimentally, specular reflection is typically described by a loss-vs-angle function $L(\theta)$ given in dB. $L(\theta)$ is then simply related to a specular reflection coefficient $R(\theta)$ by $L=-10\log(R)$. Now let θ_m be the ray-equivalent angle of mode m. $R(\theta_m)$ is a fraction between 0 and 1 that relates incident and reflected intensities for a ray incident on the surface at angle θ_m from the horizontal by

$$I_r(\theta_m) = I_i(\theta_m) R(\theta_m) \,. \tag{3}$$

From the modal point of view, we then interpret this loss as equivalent to the decay term $\exp(-r\alpha_m)$ accumulated over a total range of X_m, the ray cycle period for the ray equivalent of mode m. That is,

$$e^{-\gamma_m X_m} = R(\theta_m) \,, \tag{4}$$

or equivalently, the decay constant can be written as (Ln denotes the natural logarithm)

$$\gamma_m = -Ln[R(\theta_m)]/X_m \ .$$ (5)

A single incoherent scattering event at the surface can be described by a scattering kernel $S(\theta_n,\theta_m)$. Analogous to $R(\theta_m)$ above, $S(\theta_n,\theta_m)$ is a fraction between 0 and 1 that quantifies the portion $I_s(\theta_n,\theta_m)$ of incident energy scattered incoherently from mode m to mode n, or from the equivalent ray angle θ_m into angle θ_n:

$$I_s(\theta_n,\theta_m) = I_i(\theta_m)S(\theta_n,\theta_m).$$ (6)

We interpret the scattering of energy from mode m to mode n as having accumulated over the ray period X_m via a decay function analogous to the above, of the form $\exp(-r\beta_{mn})$. Then the energy gained by mode n over the distance X_m due to incoherent scattering from mode m is

$$I_s(\theta_n,\theta_m) = I_i(\theta_m)[1-e^{-\beta_{nm}X_m}] \ .$$ (7)

Using Eqs. (6) and (7), the decay constant β_{nm} is given by

$$\beta_{nm} = -Ln[1-S(\theta_n,\theta_m)]/X_m \ .$$ (8)

We now return to Eq. (2) and derive a system of ordinary differential equations for the energy functions $E_m(r)$. Consider an incremental range step Δr. The total energy in mode m at range $r+\Delta r$ is equal to the energy $E_m(r)$ which was there at range r, multiplied by the attenuation factors due to surface and bottom scattering, and incremented by the energy gained from incoherent scattering:

$$E_m(r+\Delta r) = E_m(r)e^{-(\alpha_m+\gamma_m)r} + \sum_n [1-e^{-\beta_{mn}\Delta r}]E_n(r) \ .$$ (9)

Subtracting $E_m(r)$ and dividing through by Δr gives

$$\frac{E_m(r+\Delta r)-E_m(r)}{\Delta r} = \frac{e^{-(\alpha_m+\gamma_m)\Delta r}-1}{\Delta r}E_m(r) + \sum_n \frac{[1-e^{-\beta_{mn}\Delta r}]}{\Delta r}E_n(r) \ .$$ (10)

Taking the limit as $\Delta r \rightarrow 0$ and using the power series expansion for the exponentials yields a system of ordinary differential equations for the $E_m(r)$:

$$\frac{dE_m}{dr} = -(\alpha_m + \gamma_m)E_m + \sum_n \beta_{mn}E_n \ , \tag{11}$$

or in matrix form,

$$\frac{dE}{dr} = BE \ , \tag{12}$$

where the coefficient matrix B is given by

$$B = \begin{vmatrix} \beta_{11}-(\alpha_1+\gamma_1) & \beta_{12} & \cdots & \beta_{1N} \\ \beta_{21} & \beta_{22}-(\alpha_2+\gamma_2) & \cdots & \beta_{2N} \\ \vdots & \vdots & \ddots & \vdots \\ \beta_{N1} & \beta_{N2} & \cdots & \beta_{NN}-(\alpha_N+\gamma_N) \end{vmatrix} \tag{13}$$

and is usually independent of range over large intervals in range, changing only where a different ice province or region of sea surface roughness is specified. Over any such range interval, the solution to Eq. (12) can be evaluated as a matrix exponential. For example, given an initial vector of intensities E(0), the solution over such a range interval can be written as

$$E(r) = e^{Br} E(0) \tag{14}$$

The coupled power equations (11) or (12) are reminiscent of master equations obtained earlier by Dozier and Tappert (1978) to describe statistically the interchange of energy among modes due to scattering by oceanic internal waves. Just as in the earlier case, Eqs. (11) and (12) obey a conservation of energy law which insures the stability of solutions of the form given in Eq. (14); there is no possibility that the norm of the intensity vector E(r) will grow exponentially with range. To show this, we first note that the sum of the coherent, specularly reflected energy remaining in mode m plus the energy incoherently scattered out of mode m cannot exceed the energy originally incident in mode m:

$$I_r(\theta_m) + \sum_n I_s(\theta_n, \theta_m) \le I_i(\theta_m) \ , \tag{15}$$

with equality holding if bottom and volume scatter are ignored ($\alpha_m=0$) and if the sum is over *all* angles which receive scattered energy, i.e., including backscatter if any. Using Eqs. (3) and (6), Eq. (15) yields

$$R(\theta_m) + \sum_n S(\theta_n, \theta_m) \le 1 \quad . \tag{16}$$

Next, observe that for any set of numbers S_n, $0 \le S_n < 1$, such that also

$$0 \le \sum_n S_n < 1 \quad , \tag{17}$$

it follows that

$$Ln[1 - \sum_n S_n] \le \sum_n Ln[1 - S_n] \quad . \tag{18}$$

This can be seen by forming a function of n variables S_n equal to the right side minus the left. This function and all its first partial derivatives vanish at the origin (all $S_n=0$) in n-dimensional space, and all its second partials with respect to each S_n are positive at the origin, so that this function is zero at the origin and is positive whenever at least one S_n is positive. Then from Eqs. (16) and (18) we obtain immediately that

$$Ln[R(\theta_m)] \le \sum_n Ln[1 - S(\theta_n, \theta_m)] \quad , \tag{19}$$

from which Eqs. (5) and (8) immediately yield that

$$-\gamma_m + \sum_n \beta_{nm} \le 0 \quad . \tag{20}$$

Then summing Eq. (11) over m, we obtain

$$\frac{d}{dr} \sum_m E_m = \sum_m [-(\alpha_m + \gamma_m) + \sum_n \beta_{nm}] E_m \le 0 \quad , \tag{21}$$

since all E_m are nonnegative intensities. Equivalently, one can note that Eq. (20) implies that the eigenvalues of the matrix B in Eq. (13) are nonpositive. Thus the norm of the intensity vector can only decay with range, and exponential stability is assured.

3. Arctic Scattering Kernel and Reflection Coefficient

To apply the coupled mode model of Section 2, scattering kernel and reflection coefficient inputs are needed. We have exercised the coupled mode model against Arctic data reported by Diachok (1976) using a scattering kernel and reflection coefficient developed by Rubenstein (1990), who made the following assumptions:

1) The under-ice surface is flat with cylindrical bosses of elliptical cross section;
2) Ice keels have a constant size and a half-width-to-depth ratio of 1.6 (Diachok (1976));
3) Ice keels are randomly oriented;
4) Ice keels are randomly spaced and uniformly distributed along a track;
5) Keel depth is a Rayleigh-distributed random variable; and
6) Pressure release boundary conditions apply.

The resulting relations that are used in this paper for keels of random depth are summarized as follows (here Rubenstein uses the formalism of Twersky (1957) and Burke and Twersky (1966)):

$$<Z> = \frac{n}{k \, \sin u} \, <f(u,u)> \quad , \tag{22}$$

$$<\sigma(\varphi,u)> = \frac{2n}{\pi k} \frac{<|f(\varphi,u)|^2>}{|1-<Z>|^2} \quad , \tag{23}$$

$$<R> = \left|\frac{1+<Z>}{1-<Z>}\right|^2 . \tag{24}$$

Here n is the number of keels per unit length, φ is the scattered angle, u is the incident angle, and k is the acoustic wavenumber. Also, the quantity $<R>$ is the reflection coefficient, and $<\sigma(\varphi,u)>$ is the incoherent scattering cross section per unit length. The "$<..>$" means expected value in terms of the Rayleigh probability density function for keel depth.

Numerical evaluation of the scattering function $f(\varphi,u)$ is computationally intensive, so to use the above relations in practice, Rubenstein (1986) introduced a simple functional fit involving a table look-up for three parameters. The functional form was motivated by exact calculations of $\sigma(\varphi,u)$ which showed that plots of the scattered energy contained primary and secondary lobes, the latter becoming more evident at higher frequencies, and increasingly complex at higher grazing angles. However, the secondary lobes can be neglected since the great majority of this energy is scattered into angles greater than the critical sound channel angle and is thus absorbed by the bottom. Therefore, only the primary lobe needed to be included, which allowed a simple parameterization easily integrated to yield a similar parameterization for the expected value $<|f(\varphi,u)|^2>$ for incident and scattered angles between 2.5° and 22.5°:

$$<|f(\varphi,u)|^2> \approx B \cos[\beta(\varphi-\varphi_0)] \quad , \tag{25}$$

where B is defined as

$$B \sim \frac{\beta}{2} \int_{\varphi_1}^{\varphi_2} <|f(\varphi,u)|^2> d\varphi \quad , \tag{26}$$

where

$$\beta \sim \cos^{-1}(1/2) / \Delta_{1/2}\varphi = 60^{\circ}/\Delta_{1/2}\varphi \quad , \tag{27}$$

where $\Delta_{1/2}$ is the half-width at half-power of the primary lobe defined for a field of random depth keels. The limiting angles φ_1 and φ_2 of the lobe are defined by

$$\varphi_1 = \varphi_0 - \frac{90^{\circ}}{\beta} \quad and \quad \varphi_2 = \varphi_0 + \frac{90^{\circ}}{\beta} \quad , \tag{28}$$

where φ_0 is the central angle of the lobe. The parameters β, B, and φ_0 were computed as functions of incident angle u and a parameter h which relates the elliptical keel size to an acoustic wavelength. Finally, the exact scattering cross section $<\varphi,u)>$ was computed and fit using the functional form in Eq. (25).

To obtain the scattering kernel function $S(\theta_n,\theta_m)$, we note that, as in Rubenstein (1990), conservation of energy is expressed in terms of expected values of the Rayleigh distribution as follows:

$$<R> + csc(u) \int_0^{\pi} <\sigma(\varphi,u)> d\varphi = 1 \quad , \tag{29}$$

where u is the incident angle and φ is the scattered angle. Comparing Eqs. (16) (with equality holding) and (29), we obtain

$$S(\theta_n,\theta_m) = (csc\,\theta_m) \int_{\Delta\varphi_n} <\sigma(\varphi,\theta_m)> d\varphi \quad , \tag{30}$$

where $\Delta\varphi_n$ is the continuous angle interval associated with the discrete mode n. The evaluation of the integral in Eq. (30) is now done explicitly by using the cosine approximation of the primary lobe given in Eq. (25), yielding the scattering kernel

$$S(\theta_n,\theta_m) = (csc\,\theta_m) \frac{2nB[\sin\beta(\theta_n^u-\varphi_0)-\sin\beta(\theta_n^l-\varphi_0)]}{\pi k\beta \,|1-<Z>|^2} \quad , \tag{31}$$

where the angles θ_n^u and θ_n^l, which are the upper and lower angle boundaries of the bin $\Delta\varphi_n$, are given by

$$\theta_n^u = \theta_n + \Delta\varphi_n/2 \quad and \quad \theta_n^l = \theta_n - \Delta\varphi_n/2 \; . \tag{32}$$

4. ASTRAL Model Implementation and Results

The ASTRAL (ASEPS Transmission-Loss) model was originally developed to meet the need for an accurate, high-speed fully automated model capable of predicting range-smoothed (over 30-40 nautical miles) propagation loss in a range-dependent environment. Along with the parabolic equation (PE) model used in Section 5, ASTRAL has been designated a U. S. Navy Standard range-dependent propagation loss model.

The usual ASTRAL algorithm computes the acoustic intensity precisely as in Eq. (1), as a sum over "smodes" ϕ_n which are actually hybrid ray-modes, each representing a bin, or interval, in ray angle space. The mode amplitudes decay exponentially in range according to assumed volume and boundary scattering losses. Boundary losses along any given path actually occur at discrete interactions with the surface or bottom. The exponential boundary decay factor, accumulated over one ray cycle length, yields the specular reflection loss.

The modified ASTRAL algorithm computes the intensity as in Eq. (2), with the mode amplitudes $E_m(r)$ being determined from Eq. (14). The initial intensities $E_m(0)$ for each mode are extracted from the usual ASTRAL model just before modal attenuation begins, i.e., at a range of 2 nm, the end of ASTRAL's near-field bathymetry region. The matrix exponential e^{Br} in Eq. (14) is computed efficiently via a single diagonalization of the coefficient matrix B for each range interval over which the input scattering kernel is constant. Typically, there is only one kernel for an entire problem, and the extra computation does not significantly impact the very fast run times for which the ASTRAL model is famous.

Using the EISPACK software package (Smith *et al.* (1976) and Garbow *et al.* (1977)), the matrix B is diagonalized in the form

$$B = PDP^{-1} \; , \tag{33}$$

where D is the diagonal matrix of eigenvalues and the columns of P are the eigenvectors of B. Since B is not symmetric (even if the scattering kernel S is; cf. Eq. (8)), the inverse of P is not simply the transpose of P, but rather is computed using the LINPACK software package (Dongarra *et al.* (1979)). The matrix solution E(r) in Eq. (14) is then expressed as

$$E(r) = P e^{Dr} P^{-1} E(0) \tag{34}$$

and the matrix e^{Dr} is easily evaluated.

The modified ASTRAL model was exercised in a comparison with Arctic data reported by Diachok (1976). The propagation geometry and a single-profile, flat-bottom environment were set to match the Diachok cases at acoustic frequencies of 40, 50, 200, and 400 Hz. A Snell's-law calculation was done to determine the maximum number of modes to process in order to avoid bottom effects. A reflection coefficient input was determined from Eq. (24), and a scattering kernel from Eq. (31), by the specification of two parameters, mean keel spacing and mean keel depth. By searching parameter space near Diachok's nominal values of 100 m spacing and 4 m mean depth, a good fit to transmission loss data at all four frequencies was obtained for 80 m spacing and 5.8 m depth, as illustrated by the solid curves in Figures 1 to 4. Since only two parameters can be varied, but fits at four frequencies are obtained, the coupled mode algorithm appears well validated. On the other hand, the dashed curves in Figures 1 to 4 were obtained for the case of no incoherent scattering by evaluating Eq. (34) with all $\beta_{nm}=0$, so that the matrix B is at once diagonal, i.e., B=D. Not only do the dashed curves not fit the data for all four frequencies, but no such fit could be found for any reasonable mean keel depth and spacing.

The difference between the coherent-only (dashed) and incoherent-plus-coherent (solid) curves is small at the lower frequencies (40 and 50 Hz) in Figures 1 and 2 because the scattering kernel at low frequencies puts almost all the incoherently scattered energy into fairly high angles that are then absorbed by the bottom. On the other hand, for incident angles in a range including 10-15 degrees, scattering loss rises rapidly with frequency up to about 200 Hz, and an increasing amount of energy is redistributed to other low angles (Rubenstein (1990)). This accounts for the much wider separation between the coherent and incoherent curves at 200 Hz. As shown in Figure 3, the redistribution of the incoherent energy into other small, forward-propagating angles is necessary in order to obtain a good fit to the Diachok data, illustrated by the plotted squares. Beyond about 200 Hz, incoherent scattering loss for 10-15 degrees incident angle actually decays somewhat with frequency (Rubenstein (1990)), accounting for the fact that the 400 Hz coherent and incoherent curves, shown in Figure 4, are actually closer together than the 200 Hz curves. As at 200 Hz, the incoherent curve is a better fit to the reported data.

5. Parabolic Equation Model Implementation and Results

The split-step Fourier numerical algorithm used in the PE model was recently modified to include loss by Moore-Head et al. (1989). After completion of the usual split-step Fourier algorithm, involving Fourier (or sine) transforms over the entire depth (water plus bottom) of the problem, at each range step a transform is taken locally near the surface, over a layer of depth 2d involving only a small fraction of the total depth mesh points in the problem. The results of the transform give a good approximation of the directional pattern of the energy grazing the surface at that range. In this Fourier (angle) space, a loss-versus-angle curve (assumed known) is applied, normalized for the length of the range step relative to an effective horizontal ray cycle distance R_s within the near-surface layer. In fact, $R_s=2d_{eff}/\tan(\theta)$ where $d_{eff}=(1+2/\pi)d$ is the "effective" depth of the surface layer. (There is an error in Moore-Head et al. where it is stated that $d_{eff}=1.5d$; the depth integral of the function w(z) is incorrect.) The surface layer actually consists of two parts, the upper half of depth d in which the full force of the loss is applied, and the lower half of depth d in which a smoothing function w(z), which goes from 1 at depth d to 0 at depth 2d, is first applied to the transformed PE complex pressure in order to avoid an abrupt transition in the pressure field at the bottom of the layer.

We have enlarged upon this algorithm to include incoherent scattering. The loss-versus-angle value L is related to the reflection coefficient R in Eq. (3) simply by $L(\theta)=-10\log(R(\theta))$. The scattering kernel $S(\theta_n,\theta_m)$ is an additional input that must be provided. The coefficients γ_m and β_{nm} are then computed from Eqs. (5) and (8) exactly as for ASTRAL except that now the ray cycle period X_m is taken to be the effective horizontal range R_s that the ray spends within the

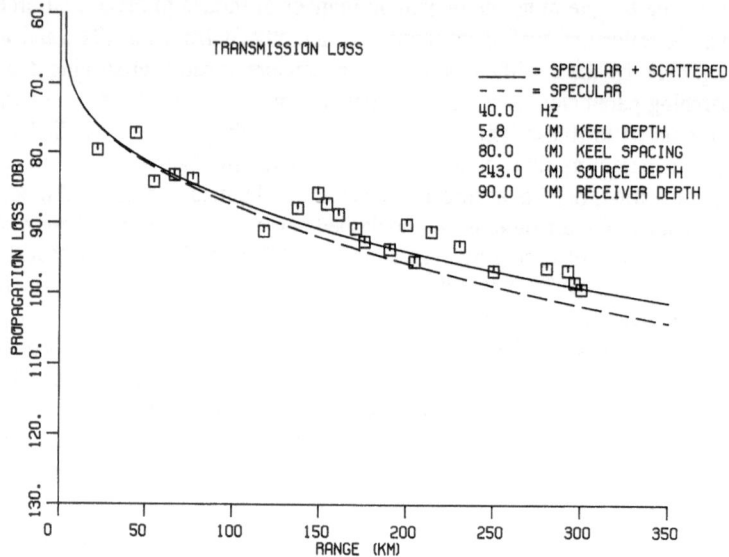

Fig. 1: Comparison of Diachok Arctic data at f=40 Hz (squares) with results from the loss-only ASTRAL algorithm (dashed curve) and the loss-plus-incoherent scatter ASTRAL algorithm (solid curve) for mean keel depth of 5.8m and mean keel spacing of 80m.

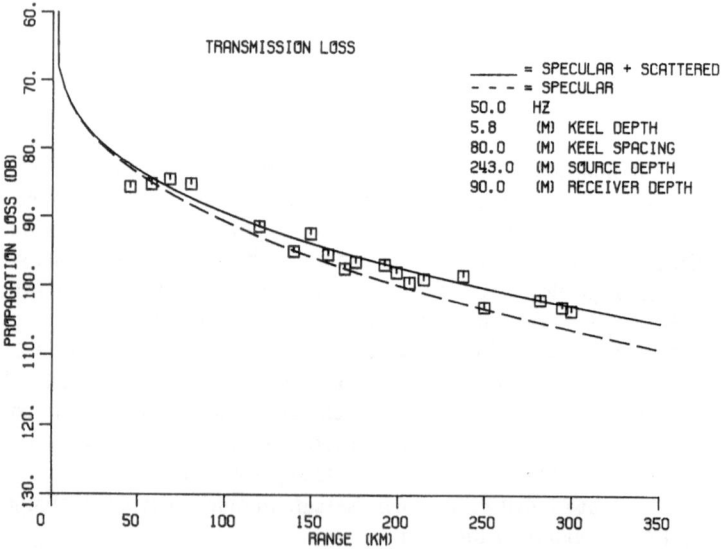

Fig. 2: Comparison of Diachok Arctic data at f=50 Hz (squares) with results from the loss-only ASTRAL algorithm (dashed curve) and the loss-plus-incoherent scatter ASTRAL algorithm (solid curve) for mean keel depth of 5.8m and mean keel spacing of 80m.

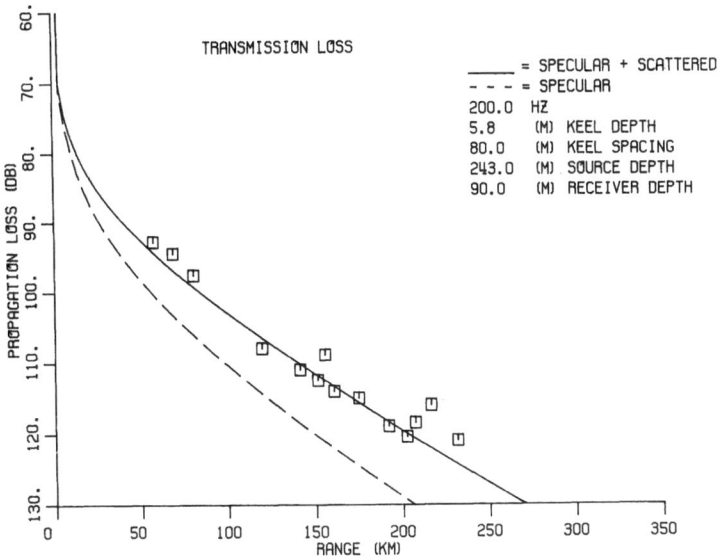

Fig. 3: Comparison of Diachok Arctic data at f=200 Hz (squares) with results from the loss-only ASTRAL algorithm (dashed curve) and the loss-plus-incoherent scatter ASTRAL algorithm (solid curve) for mean keel depth of 5.8m and mean keel spacing of 80m.

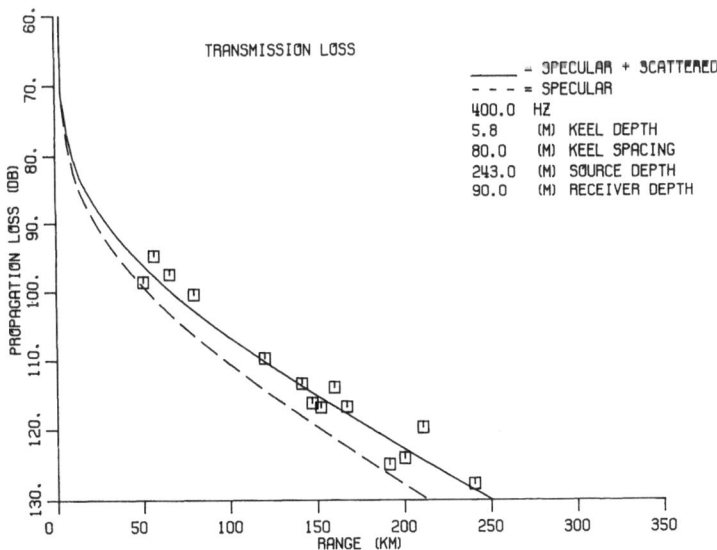

Fig. 4: Comparison of Diachok Arctic data at f=400 Hz (squares) with results from the loss-only ASTRAL algorithm (dashed curve) and the loss-plus-incoherent scatter ASTRAL algorithm (solid curve) for mean keel depth of 5.8m and mean keel spacing of 80m.

surface layer, as defined above. The coefficient matrix B is now defined, and Eq. (14) is applied in the form

$$E(r+\Delta r,k) = e^{B\Delta r} E_1(r+\Delta r,k) \quad , \tag{35}$$

where $E_1(r+\Delta r,k) = |DFT[w(z)\Psi_1(r+\Delta r,z)]|^2$ is the *intensity* vector in angle space, where Ψ_1 is the result of the standard split-step algorithm as in Eq. (2) of Moore-Head *et al.* (1989). Here DFT denotes the discrete Fourier transform over the near-surface layer. The complex vector in angle space is then recovered from the magnitude of each component of $E(r+\Delta r,k)$ combined with the original phase of the corresponding component of $DFT[w(z)\Psi(r+\Delta r,z)]$. The inverse DFT is then applied, thus completing our modification of Eq. (2) of Moore-Head *et al.* (1989).

Figures 5 and 6 illustrate the application of this algorithm in an isovelocity ocean to a narrow-beam source emitting energy concentrated primarily between 13 and 16 degrees above the horizontal and directed upward toward the surface from a depth of 1800 ft. Figure 5 shows the resulting pressure contours when only the original loss algorithm of Moore-Head *et al.* (1989) is applied, with a loss of 3 dB (i.e., one-half the energy) per bounce. As expected, the beam reflects specularly off the surface. In Figure 6, however, we included a scattering kernel designed to redistribute the lost energy from the 13-16 degree aperture into an aperture between 5 and 8 degrees. The result is a pronounced splitting of the beam, with now a substantial amount of energy also scattering off the surface into the 5-8 degree aperture. At a range of about 2.5 miles, the most intense remaining energy level contour, corresponding to transmission loss of 70 to 73 dB, dies out in both beams, indicating an approximate initial equipartition of energy, as desired. However, the scattering kernel used here is symmetric, meaning that there should be reverse scattering from the 5-8 degree aperture back into the 13-16 degree aperture. Before the shallower 5-8 degree energy entirely escapes from the near-surface layer, some of the energy it has just gained apparently scatters back into the 13-16 degree aperture, with the result that the 13-16 degree aperture maintains more energy at longer ranges beyond 2.5 nm.

To exercise the model for more realistic ocean cases, we returned to the Diachok Arctic data discussed earlier in Section 4. Using the same reflection coefficient and scattering kernel (i.e., same mean keel depth and spacing), we ran the same cases using the modified PE model. The results, range-averaged over an 8-nm sliding window, are given in Figures 7-10. Just as for ASTRAL, we see that there is negligible difference in the PE results for the coherent and incoherent cases at 40 and 50 Hz, significantly more difference at 200 Hz, but again less difference at 400 Hz. The PE loss-only curves agree quite well with their ASTRAL counterparts. At 40 and 50 Hz, the incoherent scatter is negligible, and its inclusion makes little difference. However, at 200 Hz and 400 Hz, its inclusion provides a needed correction in the direction of better agreement with the data. The correction is not as much as for the incoherent ASTRAL curves, which are an even better fit to the data for this choice of mean keel depth and spacing. For other choices of ice inputs, or for other measured data sets, the PE curves might well fit better. We did verify that the PE results at 200 Hz and 400 Hz were insensitive to varying the surface layer thickness between 16 and 32 PE mesh points in depth.

6. Summary

We have presented a general coupled mode approach for adding a treatment of incoherent rough surface scattering to existing propagation models. Such treatment is needed for accuracy in

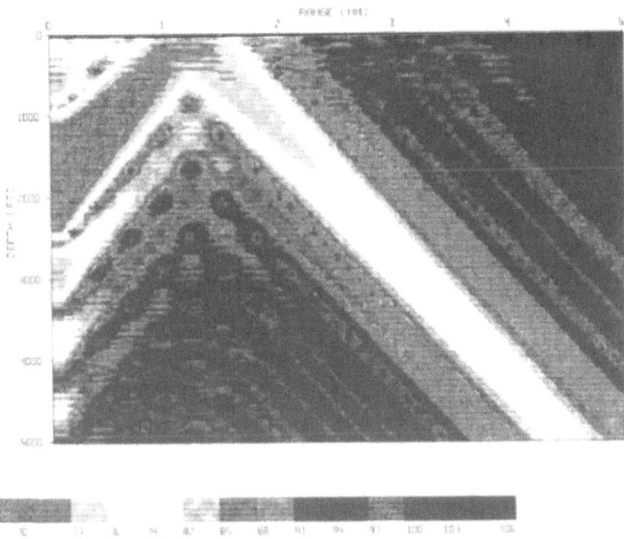

Fig. 5: Intensity contour plot from unmodified (loss-only) PE model for a narrow-beam (13°-16° aperture) source directed upward toward the surface from a depth of 1800 ft. The input loss curve is 3 dB over the aperture. Only specular reflection is observed.

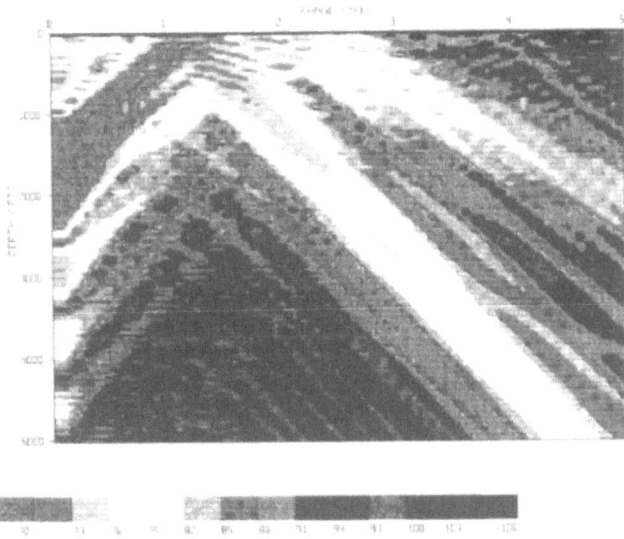

Fig. 6: Intensity contour plot from modified (loss plus incoherent scatter) PE model for a narrow-beam (13°-16° aperture) source directed upward toward the surface from a depth of 1800 ft. The input loss curve is 3 dB over that aperture, and over a 5°-8° aperture. A symmetric scattering kernel exchanges the energy lost from each aperture with the other aperture. The initial beam splits into two beams, one for each aperture.

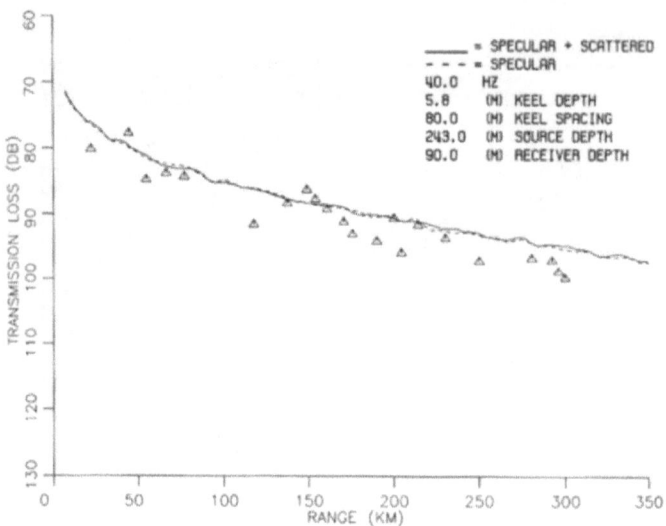

Fig. 7: Comparison of Diachok Arctic data at f=40 Hz (squares) with results from the loss-only PE algorithm (dashed curve) and the loss-plus-incoherent scatter PE algorithm (solid curve) for mean keel depth of 5.8m and mean keel spacing of 80m.

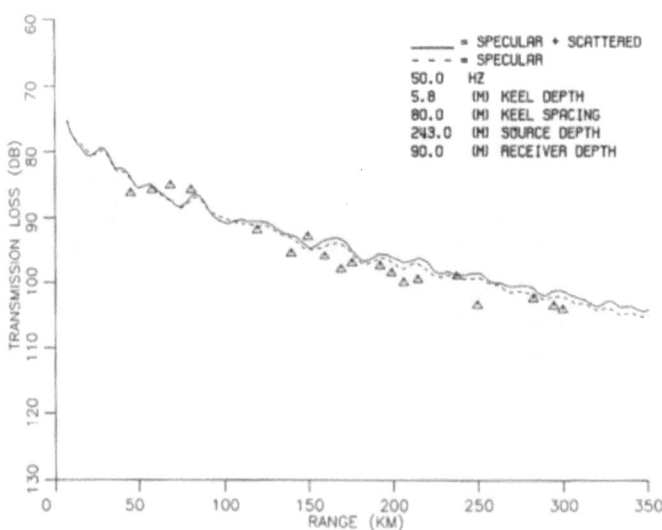

Fig. 8: Comparison of Diachok Arctic data at f=50 Hz (squares) with results from the loss-only PE algorithm (dashed curve) and the loss-plus-incoherent scatter PE algorithm (solid curve) for mean keel depth of 5.8m and mean keel spacing of 80m.

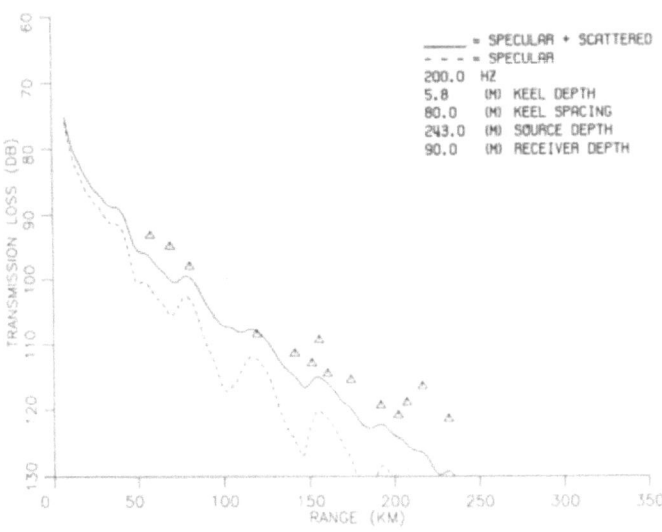

Fig. 9: Comparison of Diachok Arctic data at f=200 Hz (squares) with results from the loss-only PE algorithm (dashed curve) and the loss-plus-incoherent scatter PE algorithm (solid curve) for mean keel depth of 5.8m and mean keel spacing of 80m.

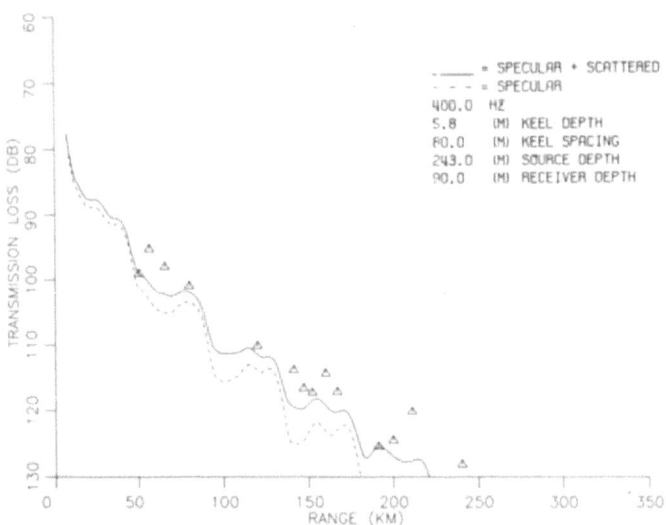

Fig. 10: Comparison of Diachok Arctic data at f=400 Hz (squares) with results from the loss-only PE algorithm (dashed curve) and the loss-plus-incoherent scatter PE algorithm (solid curve) for mean keel depth of 5.8m and mean keel spacing of 80m.

upward-refracting environments which induce repeated scattering from rough ocean or ice surfaces. We have implemented our treatment into the two U. S. Navy Standard range-dependent propagation models, and we have exercised these models against measured Arctic data. Results for a single set of only two input parameters, mean keel depth and mean keel spacing, validate our treatment by showing good agreement with data across a rather wide range of frequencies.

Our treatment of incoherent scattering can also be used to investigate the sensitivity of acoustic transmission loss to ocean variability, as described by the input scattering kernel. In this case, the kernel is modeled from mean keel depth and spacing; in a non-ice-covered environment, it might be modeled from a sea surface spectrum. Multiple runs of a model, each with different kernel inputs, can be used to bound the effects of ocean variability. For example, while using the ASTRAL implementation to search the input parameter space for agreement with data, we observed significant variations in transmission loss between our final input values, 5.8 m keel depth and 80 m keel spacing, and Diachok's nominal values of 4 m keel depth and 100 m keel spacing. Diachok's values yield significantly less loss than the measured data, e.g., at 40 Hz and a range of 200 km, the solid curve in Figure 1 shows a loss of about 94 dB, as compared with 88 dB (not shown) for Diachok's values. At higher frequencies the differences are even greater, e.g., at 200 Hz, 120 dB in Figure 3 versus 107 dB (not shown) for Diachok's values, and at 400 Hz, 123 dB in Figure 4 as compared with 113 dB (not shown) for Diachok's values. Thus, environmental variability of under-ice scattering effects is quite significant.

Since several assumptions were used in formulating the scattering kernel, we cannot claim that the above results are accurate in an absolute sense. It would be an overstatement, for example, to assert that our results show conclusively that the mean keel depth for the Diachok experiment was closer to 5.8 m than 4 m. On the other hand, because the kernel includes most, if not all, of the relevant physical mechanisms, the degree of variability should be well indicated by the above. Indeed, models implementing our treatment can be used to further refine the scattering kernel formulation, when run in conjunction with more detailed under-ice measurements, taken simultaneously with measurements of acoustic transmission loss.

7. Acknowledgments. This work was supported by the ASW Environmental Acoustic Support Program of the U. S. Office of Naval Research under contract N00014-86-D0137. We thank David Rubenstein, Ami Gilad, and Pete Council for their helpful discussions.

8. References

Burke, J. E. and Twersky, V. (1966) "Scattering and reflection by elliptically striated surfaces", *J. Acoust. Soc. Am.* **40**, 883-895.

Diachok, O. A. (1976) "Effects of sea-ice ridges on sound propagation in the Arctic Ocean", *J. Acoust. Soc. Am.* **59**, 1110-1120.

J. J. Dongarra, C. B. Moler, J. R. Bunch, and G. W. Stewart (1979) *LINPACK User's Guide*, SIAM, Philadelphia.

Dozier, L. B. and Tappert, F. D. (1978) "Statistics of normal mode amplitudes in a random ocean. I. Theory", *J. Acoust. Soc. Am.* **63**, 353-365.

Dozier, L. B. and Tappert, F. D. (1978) "Statistics of normal mode amplitudes in a random ocean. II. Computations", *J. Acoust. Soc. Am.* **64,** 533-547.

B. S. Garbow, J. M. Boyle, J. J. Dongarra, and C. B. Moler (1977) *Lecture Notes in Computer Science* No. 51, *Matrix Eigensystem Routines -- EISPACK Guide Extension*, Springer-Verlag, Berlin.

M. E. Moore-Head, W. Jobst, and E. S. Holmes (1989) "Parabolic-equation modeling with angle-dependent surface loss", *J. Acoust. Soc. Am.* **86,** 247-251.

Rubenstein, D. R. (1990) "Modeling the acoustic scattering by under-ice ridge keels", submitted to *J. Acoust. Soc. Am.*

Rubenstein, D. R. (1986) "Acoustic scattering kernels from Arctic Sea Ice, Part 2: Rayleigh Distribution of Keel Sizes", unpublished manuscript.

B. T. Smith, J. M. Boyle, J. J. Dongarra, B. S. Garbow, Y. Ikebe, B. C. Klema, and C. B. Moler (1976) *Lecture Notes in Computer Science* No. 6, *Matrix Eigensystem Routines -- EISPACK Guide*, 2nd ed., Springer-Verlag, Berlin.

Twersky, V. (1957) "On scattering and reflection of sound by rough surfaces", *J. Acoust. Soc. Am.* **29,** 209-225.

AVERAGE SOUND INTENSITIES IN RANDOMLY VARYING SOUND-SPEED STRUCTURES

H. G. SCHNEIDER
Forschungsanstalt der Bundeswehr für Wasserschall- und Geophysik
Klausdorfer Weg 2-24, D-2300 Kiel
Fed.Rep.Germany

ABSTRACT. The acoustic field due to a deterministic ocean environment may significantly change, if it is influenced by an additional random variability of the the sound-speed structure. If only the sound intensity has been measured, as in most propagation loss experiments, then the influence of the random variability may be detected in basicly two ways: either by compairing the measured loss with that due to the deterministic ocean or by measuring the scattered energy directly as e.g. in illuminated shadow zones. The first approach requires to define the deterministic environment which may not be a trivial task, while the second approach could discern the presence of substantial variability without relation to the determistic ocean. This will be discussed utilizing two acoustic experiments with considerably different sound speed variability. The two approaches will briefly be related to the so called 'honest' and 'dishonest' solutions of the stochastic wave equation, and the propagation loss data will be compared to results from stochastic ray tracing.

1. Introduction

The acoustic field due to a deterministic ocean environment may significantly change, if it is influenced by an additional random variability of the the sound-speed structure. This applies to the fluctuations in amplitude and phase as well as to the average sound intensity.

These fluctuations are clear indications of the stochastic sound speed component, and in case of weak scattering it may be possible to even deduce the nature of the variability and to estimate the average sound speed conditions which are mostly considered as the unperturbed or deterministic oceanic conditions.

However, in many propagation loss experiments only the average sound intensity has been measured, and the influence of the stochastic environment often becomes only apparent, if the measured acoustic data differs from the modeling results utilizing a deterministic description of the environment. This necessitates to define an appropriate deterministic sound speed layering which may not be a trivial task. This is especially true for shallow water, where the sound speed variability may have a profound influence over the entire waterdepth. Additionally, the effect of the sound speed variability on the propagation loss is often coupled closely to losses introduced by the bottom, so that both influences are difficult to separate.

J. Potter and A. Warn-Varnas (eds.), Ocean Variability & Acoustic Propagation, 283–292.
© 1991 *Kluwer Academic Publishers.*

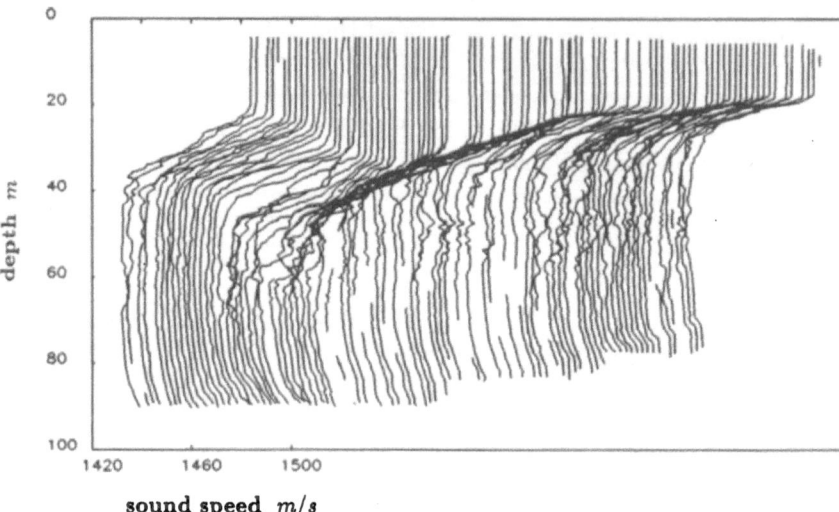

Figure 1   Spatial series of sound speed profiles displayed with constant offset. The scale relates to the leftmost profile. Total range is 38 nm (70 km).

The favourable conditions for such experiments in shallow water are therefore a ducted propagation entirely within the water column, i.e. propagation paths without bottom contact. The interpretation of such acoustic data is more conclusive, if the scattered sound field can be measured directly as e.g. in illuminated shadow zones. This will be discussed utilizing two propagation loss experiments and briefly be related to the so called 'honest' and 'dishonest' solutions of the stochastic wave equation.

2. Experiments

The two sets of propagation loss data presented in the following were measured with explosives and cannot be used for signal anlysis. They differ in their geographic location within the Baltic Sea and exhibit considerably different sound speed variability.

2.1. Experiment 1

Fig. 1 displays a spatial series of sound speed profiles over a range of 38 nm (70 km) which has been measured by a towed up and down diving body. Unfortunately the profiles end about 20 m above the bottom and therefore do not show the increase in sound speed at larger depths which gives rise to a midwater sound channel (see profile in Fig. 2). Two items are noteworthy, first the small scale variability from profile to profile and second the increase in the mean thermocline depth at about 1/4 of the range. The horizontal resolution is not sufficient to draw any quatitativ statistical conclusions.

Propagation loss data for the source at midwater depth are shown by the symbols in Fig. 2 for two receiver depths. Beyond a range of 12 km the two loss curves run parallel with a

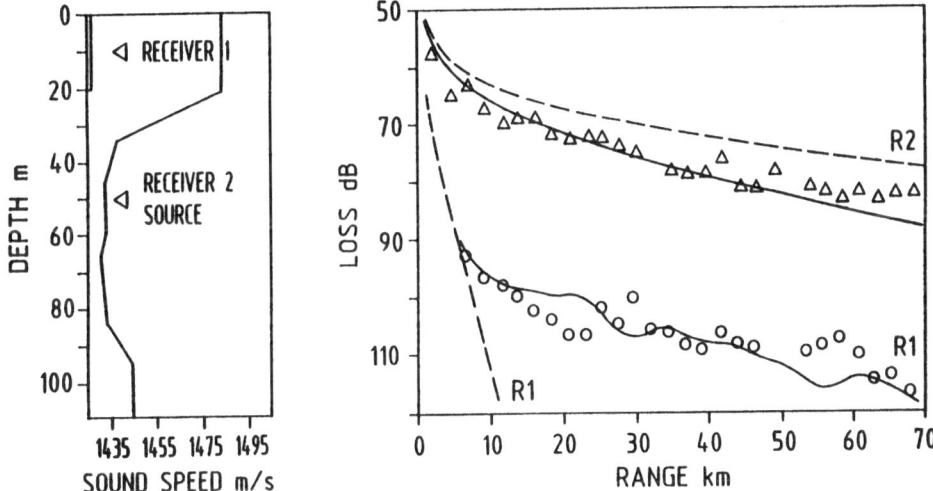

Figure 2   Left: constant gradient sound speed profile for modeling
Right: Propagation loss for source in the sound channel and two receiver depths R1, R2
    symbols : measured data
    dashed line : computed loss for deterministic sound speed layering
    continuous line : computed loss with stochastic sound speed variability

considerable offset. From this measurement alone no conclusion can be drawn on the importance of the sound speed variability on the propagation. This can only be decided on either
by additional measurements or by compairing the experimental results with corresponding
computations from a suitable deterministic environment which may be range dependent.

The latter approach requires to define an appropriate deterministic sound speed profile.
Rather than averaging stochasticly range dependent sound speed values at constant depth
which may result in unrealistic sound speed gradients, a good guess from the practical point
of view is to chose the sound speed profile at the receiver. This is a member of the stochastic
ensemble of profiles and governs the local sound field at the receiver. In the present case in
which we neglect also the determistic range dependence of the sound speed structure this
choice is straight forward and the results from this deterministic computation based on one
profile is given by the dashed lines in Fig. 2.

For propagation inside the sound channel the additional loss at channel axis relative to this
modeling result provides no clue about the mechanism causing this additional loss, and
being a small correction to a large energy it may easily be overlooked.
More significant is the loss decrease or energy increase relative to the deterministic modeling at the shallow receiver. The modeling defines this depth as high loss region, and the
occurance of energy in a high loss region suggests quite naturally to assume a scattering or
diffraction mechanism. For the frequency of 4 kHz ($\lambda = 37$ cm) it can be excluded that
diffraction is the dominating effect.

Beyond an initial range of about 12 km the two measured loss curves in Fig. 2 run parallel,

286

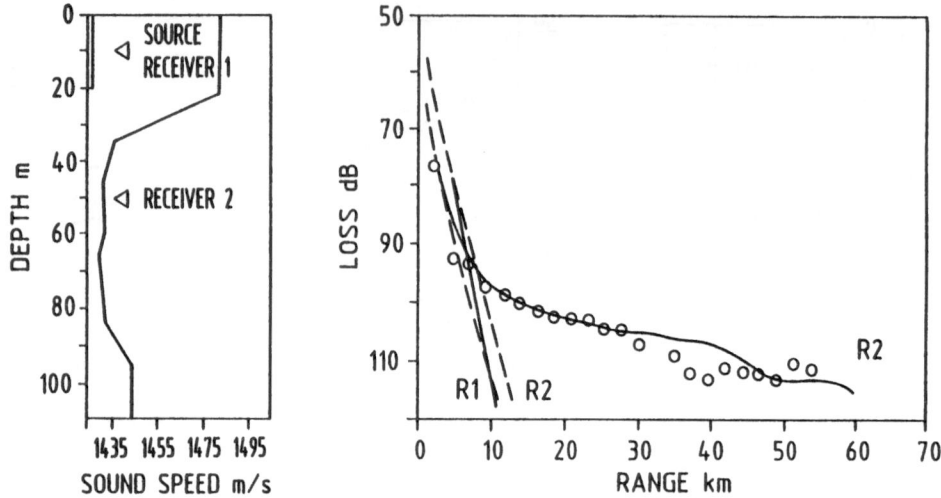

Figure 3   Left: constant gradient sound speed profile for modeling
Right: Propagation loss for source at shallow depth and two receiver depths R1, R2
        symbols : measured data
        dashed line : computed loss for deterministic sound speed layering
        continuous line : computed loss with stochastic sound speed variability

thereby indicating that stationary propagation conditions have been reached, which requires that a constant fraction of the energy in the channel is deviated out of the channel per unit range interval. The separation of these loss curves is governed by the scattering rate and the loss due to the soft bottom. This means in terms of rays, that the angular energy density distribution over ray angles has become stationary up to sufficient large angles necessary to penetrate the thermocline.

The second approach to discern the variability from propagation loss features utilizes an additional measurement with reciprocal geometry in which the source is at shallow depth. These propagation loss data are displayed in Fig. 3 by symbols. Because of the soft bottom and the large sound speed excess the losses are exspected to be high, and could not be measured sucessfully at the shallow receiver. Instead energy occured at channel axis, and this loss curve is within measuring accuracy identical with the loss at shallow depth in in Fig. 2, although this is an independent experiment.

To study the effect of the sound speed variability this receiver/source combination seems to be the more conclusive one, because no deterministic modeling is required to realize the stochastic effect. With the experimental evidence that propagation at shallow depth suffers from high losses, the only explanation for the energy propagating at midwater depth is that path or mode conversion to ducted propagation has occured. Since the deterministic changes of the sound speed structure are not sufficient for this conversion this is due to the stochastic variability.

For this environment it is certainly equally appropriate to measure the scattered energy

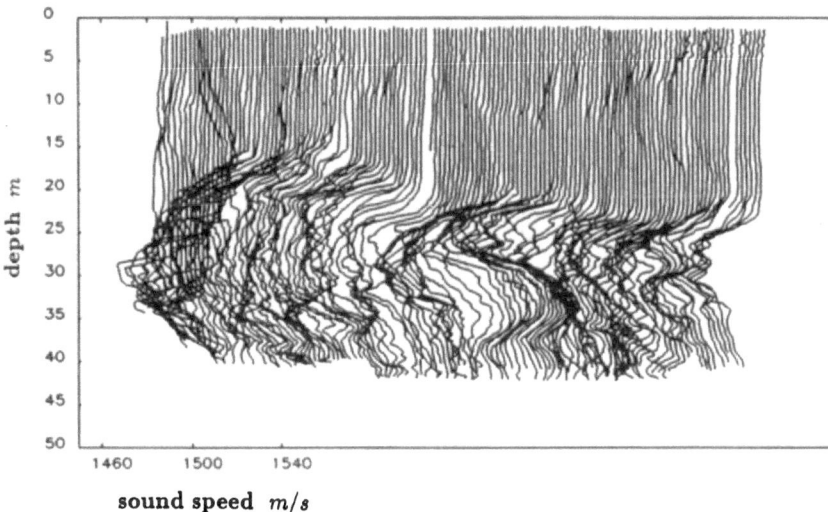

sound speed *m/s*

Figure 4   Spatial series of sound speed profiles displayed with constant offset. The scale relates to the leftmost profile. Total range is 40 nm (74 km).

directly or to deduce this quantity by comparing the loss in the channel with a deterministic loss model. The latter method is applicable only because the so called unperturbed or deterministic sound speed conditions are well defined and the absorption losses within the medium are known. Indeed the difference in slope off the measured loss curve relative to the spreading and absorption losses are then sufficient to suggest a scattering loss [1,2] which amounts here to about 0.07 dB/km.

However, nature is not always so kind and the stochastic as well as range dependent deterministic sound speed variations cannot always be nicely separated, as illustrated by the following experiment.

2.2. Experiment 2

The spatial series of sound speed profiles, depicted in Fig. 4, covers a range of 40 nm (74 km), the propagation range starts with the 7'th profile. The sound channel in the lower part of the water column is of varying sound speed excess as well as of varying extension in depth.

For this set of profiles it is problematic to define appropriate deterministic sound speed conditions which could serve as reference or unpertubed conditions. Those horizontal scales in this environment which are small relativly to the propagation distance should be regarded as stochastic and larger scale variability should be treated as deterministic. However, for detailed analysis of the sound speed variability the oceanographic data is not sufficiently sampled and averaging at constant depth would destroy the sound channel. As before, the profile measured at the receiver position is chosen and the computed results should be judged cautiously.

288

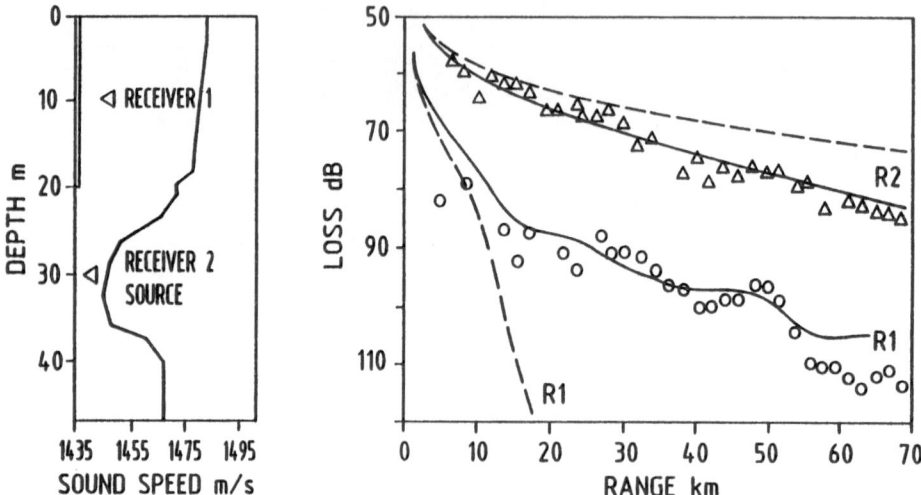

Figure 5   Left: constant gradient sound speed profile for modeling
Right: Propagation loss for source in the sound channel and two receiver depths R1, R2
         symbols : measured data
         dashed line : computed loss for deterministic sound speed layering
         continuous line : computed loss with stochastic sound speed variability

Fig. 5 corresponds to Fig. 2 with source at channel axis and the symbols denoting the measured data. Again the loss for ducted propagation is increased relativly to the deterministic
computation and the loss at shallow depth has markedly been decreased.

Fig. 6 corresponds to Fig. 3 with source at shallow depth, but here the loss at the shallow
receiver could be measured additionally. For ranges larger than 15 km the loss at both
receiver depths displays roughly the same slope which is also identical with that of the
data in Fig. 5. Since the separation of the loss curves for the shallow and deep receiver are
identical as well for both source depths, the same stationary propagation conditions have
been reached for both source depths.

In this example, with a higher sound speed variability and a harder bottom, the consecutive
scattering from one propagation domain into the other could be followed over two steps.
First the energy from the shallow source is scattered into the sound channel, propagates
therein and is subsequently scattered out of the sound channel to the shallow receiver,
where it was measured at ranges beyond 15 km.

The qualitativ similarity of the propagation loss data of both acoustic experiments suggests
to consider this mess of highly irregular profiles in Fig. 3 as one stochastic ensemble. This
implies, that effects due to the deterministic component are of minor importance which is
of course dependent on the acoustic frequency and might be considered realistic for this
frequency of 4 kHz ($\lambda = 37$ cm).

In spite of the considerable differences in the sound speed structures of both experiments

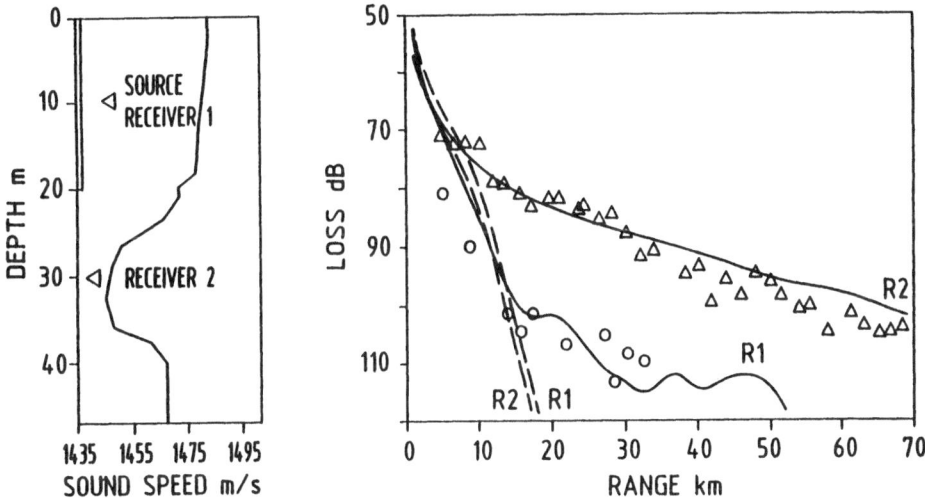

Figure 6   Left: constant gradient sound speed profile for modeling
Right: Propagation loss for source at shallow depth and two receiver depths R1, R2
        symbols : measured data
        dashed line : computed loss for deterministic sound speed layering
        continuous line : computed loss with stochastic sound speed variability

the type of stochastic acoustic field is the same, so that ducted propagation may be assumed even for the second experiment. The scattering loss in the sound channel, that is the loss increase relativ to the deterministic or unperturbed environment, is about 0.07 dB/km for the first and 0.14 dB/km for the second experiment. The latter is however a very rough estimate since the deterministic sound speed channel is not well defined.

Now how can this be related to the oceanography in order to compute the stochastic sound field ?

3. Theoretical Considerations

With respect to computing measurable components of the acoustic field two approaches may be distinguished to solve the stochastic wave equation. In terms of Keller [3] these are the "dishonest" and the "honest" solution.

3.1. Dishonest Approach

Here all those situations can be considered, where the deterministic ray path is experimentally retrievable and may be verified. The stochastic acoustic field is most often discussed in relation to the deterministic one, or the deterministic part is subtracted out and difference measurements are considered. The deterministic path is the optimal path and any stochastic quantity will on the average be less optimal and thus the application is for problems as increased loss and signal decorrelation.

Possibly this is a sufficient approach for weak scattering (as internal waves). Following Ishimaru [4, chapter 20-3] for an index of refraction $n(\vec{r}) = < n > [1 + \mu(\vec{r})]$ with $< n >$ denoting the ensemble average and $< \mu(\vec{r}) > = 0$ the stochastic wave equation in the parabolic approximation reads

$$2ik\frac{\partial}{\partial x}U(\vec{r}) + \nabla_t^2 U(\vec{r}) + k^2\mu(\vec{r})U(\vec{r}) = 0 \tag{1}$$

with $U(\vec{r})$ being a solution to the the deterministic wave equation and the index t is for the transverse coordinates. Averaging leads to

$$2ik\frac{\partial}{\partial x} < U(\vec{r}) > + \nabla_t^2 < U(\vec{r}) > + k^2 < \mu(\vec{r})U(\vec{r}) > = 0 \tag{2}$$

The dishonest approach is made by introducing a functional

$$< \mu(\vec{r})U(\vec{r}) > = g[\mu(\vec{r})] < U(\vec{r}) > \tag{3}$$

and solving for $< U(\vec{r}) >$

$$2ik\frac{\partial}{\partial x} < U(\vec{r}) > + \nabla_t^2 < U(\vec{r}) > + k^2 g[\mu(\vec{r}] < U(\vec{r}) > = 0 \,. \tag{4}$$

To be exact, this requires μ to be Gaussian with a delta correlation in propagation direction. The solution has the form

$$< U(x,\vec{\rho}) > = U_0(x,\vec{\rho})\, exp(-\alpha_0 x). \tag{5}$$

This clearly requires an average solution to exist to which the stochastic environment adds a small change. This approach can be utilized for the additional scattering loss α_0 relativ to the sound field U_0 of the unperturbed or deterministic environment.

A similar approximation is made in the path integral formulation for the average intensity $< \psi >$ in the saturated regime of multi micro paths. Following Flattè [5] the average intensity may be computed by integration over all unpertubed paths:

$$< \psi > = N \int d(paths)\, exp[\, iq_0\, S_0(path) - \frac{1}{2}iq_0^2 < (\int_{path} \mu\, dx)^2 >] \tag{6}$$

Thus only ray paths are considered which already exist in the non variable environment, and any new paths created by the variability, as those which illuminate a deterministic shadow zone, are outside the scope of this approach.

3.2. Honest Approach or Numerical Modeling

The scattered energy can only be calculated within the honest solution of the wave equation, where the stochastic equation or an approximation thereof is solved first and averaging is performed later. This is done in the transport equation or may be achieved within numerical models based on coupled modes, the parabolic equation or stochastic ray tracing.

Whatever method could be applied, there is no satisfactory description for the stochastic variability of the environments considered here [6]. To input the measured profiles and matching the loss data does not help to forecast the propagation loss for different environments of this type. However, since the propagation becomes stationary a simple approach should suffice in this stationary limit. The computational results displayed in Fig. 2, 3, 5, 6 were achieved by a stochastic ray tracing scheme utilizing an approach analogous to ray diffusion.

According to Chernov [7], in a homogenious stochastic medium the probability density for the angular deviation $\Delta\phi$ of a ray from its initial angle of propagation is Gaussian with

$$\overline{\Delta\phi} = 0 , \quad \overline{\Delta\phi^2} = \sigma^2 = 2D_0 S , \quad D_0 = \sqrt{\pi}\,\frac{\overline{\mu_0^2}}{a} , \tag{7}$$

S denotes the path length and D_0 is the diffusion constant $\overline{\mu_0^2}$ is the variance of the index of refraction and a the horizontal coherence length.

However, these environments are by no means homogenious and a theory of similar simplicity for layers of finite extent containing gradients has not yet been developed and would have to include the statistics of the horizontal gradients [8]. So an empirical approach was used for the depth dependent sound speed profile $c(z)$ by assuming a depth dependence of the diffusion constant D

$$D(z) = D_0(1 + \frac{1}{g}|\frac{\partial c}{\partial z}|)^2 \tag{8}$$

The gradient g is scale variable with numerical value 1. In layers with small gradients this reduces to the diffusion in homogenious media, otherwise the gradient term is dominating and the diffusion constant is increased. Layers with large gradients have only a small vertical extension, so that ray paths in those layers are short, and the the scattering angle is small.

Since no quantitativ information on μ or the horizontal scale lengths were available the value for D_0 had to be fitted. With $D_0 = 10^{-7}\ m^{-1}$ the computed loss agrees exellent with the measured data as may be infered from Figs. 2, 3, 5, 6.

It is astonishing that the same diffusion constant could be used for both experiments which indicates that the gradient term in D is the important one. It is noteworthy that the sucessfull modeling of the stationary propagation conditions, including the consecutive scattering over two propagation states, implies to maintain a stationary distribution of energy versus propagation angle, and because small angle scattering is assumed, the angular energy density has to extend continuously from small grazing angles in the channel to sufficient large ones for propagation outside of the channel.

4. Conclusion

Two experiments with substantial different sound speed variability leading to very similar stationary acoustic propagation were discussed. Only in one experiment an average or unpertubed profile can meaningfully be defined, while in the other one this is not the case, and hence any relation to the propagation in an average environment is ambigous. To discern the influence of the sound speed variability on the propagation loss, the measurement with the source depth outside the sound channel proves to be more conclusive, since no relation to average propagation conditions are required. The theoretical treatment of this configuration necessitates to solve the stochastic wave equation before averaging is carried out, or the application of numerical models. The propagation loss data could sucessfully be modeled by an empirical stochastic ray tracing approach. This is not very satisfying, but the empirical approach is the only way to forcast this type of propagation loss at present. Only appropriate oceanographic measurents and the mathematical description of the oceanic phenomena could lead to better justifyed approaches.

5. References

[1] Mellen, R. H., Browning, D. G. and Ross, J. M. (1974) 'Attenuation in randomly Inhomogenius Sound Channels', J.Acoust.Soc.Am. **56**, pp.80-82

[2] Mellen, R. H. and Schneider, H. G. (1977) 'Diffusion Loss in Refractive Sound Channels: Bottom-Loss Effects', J.Acoust.Soc.Am. **62** , pp.1038-1041

[3] as cited by DeSanto, J.A. (1979) 'Theoretical Methods in Ocean Acoustics' in DeSanto, J. A. (Ed.), Ocean Acoustics, Springer-Verlag, Berlin

[4] Ishimaru, A. (1978) Wave Propagation and Scattering in Random Media, Academic Press, New York

[5] Flattè, S. M. (Ed.),(1979), Sound transmission trough a fluctuating ocean, Cambridge University Press, Cambridge

[6] Thiele, R. (1990) 'Modeling of Sound Propagation in a Randomly Varying Ocean by Stochastic Mode Coupling', this volume

[7] L.A.Chernov, L. A. (1960) Wave Propagation in a Random Medium, McGraw-Hill, New York

[8] J.Sellschopp, J. (1990) 'Stochastic Raytracing in Thermoclines', this volume

STOCHASTIC RAY TRACING IN THERMOCLINES

J. SELLSCHOPP
Forschungsanstalt der Bundeswehr
für Wasserschall- und Geophysik
Klausdorfer Weg 2-24
2300 Kiel 14
Germany

ABSTRACT. Modifications of the sound field due to random sound velocity fluctuations are often successfully considered by a ray diffusion approach. It is shown, that the constraints of ray diffusion theory are violated in many ocean stratification situations. A different theory is necessary primarily for the domain of high gradient thermoclines.

The direction change of a ray in a thermocline depends on the angle of incidence and the sound velocity in both layers, but not on the actual depth or width of the thermocline. For simplicity the thermocline may be treated as a discontinuity of sound velocity. Random inclinations of the thermocline cause modifications of ray directions after refraction. For a presumed ray the probability distribution of directions is given by the probability of slopes met by the ray, the effective slope distribution. The latter differs from the ordinary slope distribution of the thermocline by an enhanced probability of slopes facing the incident ray and the occurrence of shadowed thermocline portions. The effective slope distribution contains the incident ray direction as a parameter. Few weak assumptions are necessary to derive the effective slope distribution, shadowing function included, from the ordinary distribution of thermocline slopes.

Mean effective slopes are different from zero. Therefore in contrast to diffusion theory mean ray directions after refraction differ from those calculated for a smoothly stratified ocean. The action of thermocline undulations is to redistribute sound energy across layer boundaries. Channel leakage as well as sound trapping in a channel is satisfactorily explained by the effective slope approach.

The theory is well suited for Monte Carlo type numerical models. Some example computations are carried out for a realistic shallow water environment. Transmission loss data measured in a Baltic Sea experiment showing sound trapped in a channel were compared with theoretical results. They prove the validity of the theory.

J. Potter and A. Warn-Varnas (eds.), Ocean Variability & Acoustic Propagation, 293–312.

1. Introduction

Existing measurements of ocean stratification were used to distil parameters of ocean variability, which are suited as an input to acoustic models. The first three paragraphs deal with the unsuccessful attempt to quantify the diffusion constant for random modifications of ray directions.

When in a vertical plane with x and z being the horizontal and vertical coordinates and β the angle with the x-axis an acoustical ray starts at x_0, z_0, β_0, its trace may be calculated by simultaneous integration of the ray equations

$$\frac{dx}{dt} = c \cos \beta$$

$$\frac{dz}{dt} = c \sin \beta \tag{1}$$

$$\frac{d\beta}{dt} = \frac{\partial c}{\partial x} \sin \beta - \frac{\partial c}{\partial z} \cos \beta$$

Sound velocity c is a function of x and z. An ocean area is regarded as statistically homogeneous, if the sound velocity field can be split up into a constant profile C(z) and a random distortion c'(x,z) with zero mean when averaged in horizontal direction. Note that vertical integration of c' need not give a zero result. The simplest case where it does not, is a two layered ocean with an undulating interface. Here vertical integration leads to values different by up to sound speed difference multiplied by the height of the interface wave.

Neglecting the random sound velocity variations simplifies ray tracing. Since the horizontal derivative of the mean velocity field is identically zero, the rays obey Snell's law: $\cos \beta / C(z) =$ const. The ray direction at a selected depth is given by initial direction and initial depth. The mean profile C(z) may be modelled by linear segments and hence the rays themselves by segments of circles.

If random sound speed variability is introduced into ray tracing, it seems adequate to think of rays, which statistically deviate from those calculated be means of constant sound velocity profiles. The variance of that. deviation will increase with time or distance passed. This method to deal with sound speed inhomogeneities is the ray diffusion approach. In the following two sections it will be shown that ray diffusion is not always an appropriate tool. Then, in section 4, an alternative theory is presented , which is suited to solve the problem at the location of the greatest sound velocity fluctuations, the thermocline.

2. Ray diffusion

Sound velocity is split up into mean profile and random field as before: $c(x,z) = C(z) + c'(x,z)$. We denote the solution of the constant profile case with X(t), Z(t), B(t) and the corrections caused by sound speed variation with x'(t), z'(t), β'(t). The equations for the ray corrections are

$$\frac{dx'}{dt} = c' \cos B - \beta' C \sin B$$

$$\frac{dz'}{dt} = c' \sin B - \beta' C \cos B \tag{2}$$

$$\frac{d\beta'}{dt} = \frac{\partial c'}{\partial x} \sin B - \frac{\partial c'}{\partial z} \cos B$$

The first two of these equations are correct to the first order of β', while the third equation is zero order in β'. The speed of the location corrections x' and z' is slow compared with sound velocity, but the change of the angular correction can be much faster than the direction change of the undisturbed ray. A valid solution for the modified ray can only be obtained, if the right hand side of the third of the equations above changes its sign rapidly enough. The correlation length of sound speed fluctuations must be sufficiently small.

Ray angles B to deal with in a horizontally spreading sound field are small. Moreover vertical gradients of random sound velocity structures in the ocean use to be much greater than horizontal gradients. By these reasons the equation for β' is dominated by its second term. The vertical gradient of the random sound velocity field may replace the gradient transverse to the ray direction.

A formal lowest order solution for the random deviation from the ray direction in the constant profile case is given by

$$\beta_1' - - \int_0^{t_1} \frac{\partial c'}{\partial z} \cos B \, dt = - \int_{x_0}^{x_1} \frac{1}{C} \frac{\partial c'}{\partial z} \, dx \tag{3}$$

The integration has to be carried out along the ray path, where not only $\partial c'/\partial z$ varies, but also depending on z the mean sound velocity C and even the statistics of $\partial c'/\partial z$ may change. As long as the actual ray path does not deviate too much from the ray path through an undisturbed ocean, the integration may proceed along the curve X(t), Z(t). Only then is it possible to calculate the mean angular deviation $\langle \beta_1' \rangle$ by a horizontal average with interchanged order of integration. The horizontal average of the integrand is zero by definition of C and c'. So with the stated restrictions the expectation value of β_1' becomes $\langle \beta_1' \rangle = 0$.

The same considerations hold for higher moments of β_1'. Horizontal averaging can only be interchanged with integration along the actual ray, if the path of the mean ray can be taken instead. Then the variance of β_1' will be

$$\langle \beta_1'^2 \rangle = \int_{x_0}^{x_1} \int_{x_0}^{x_1} R_z(u-v) \, du \, dv \tag{4}$$

$R_z(u-v)$ is the correlation function of the expression $1/C \cdot \partial c'/\partial z$ taken along the path from x_0 to x_1. The index z indicates that it depends on depth. In addition it depends on the direction B of the path.

If $R_z(0)$ is small and decreases to zero in a distance L_k much smaller than the integration path, then the integration will yield the ray diffusion solution

$$\langle \beta_1'^2 \rangle \approx L_k \cdot R_z(0) \cdot (x_1 - x_0) \tag{5}$$

The next task will be to extract realistic correlation functions from measured data of ocean stratification.

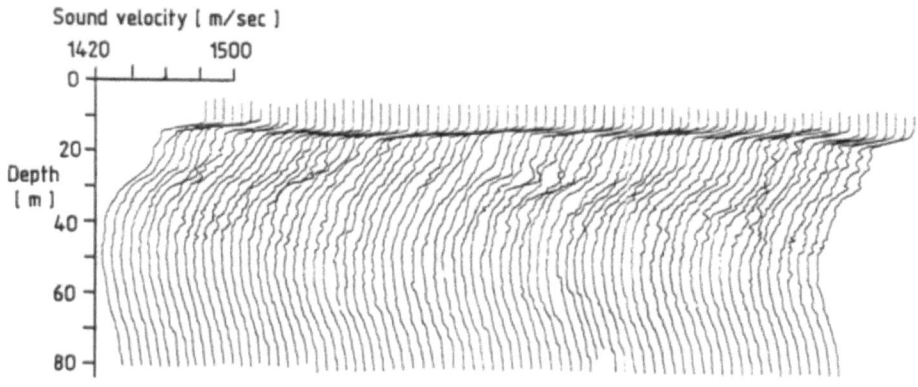

Figure 1.   Sound velocity profiles in a Baltic Sea section of 30 nautical miles range obtained by an undulating fish in August 1982.

3.   Failure of perturbation theory

Figure 1 shows sound velocity profiles in the Baltic Sea measured by a towed fish on a 53 km long course. Simultaneously the temperature field was measured with 0.8 m vertical and 20 m horizontal resolution. Temperature was converted to sound velocity by means of the salinity stratification known from towed fish data.

Correlation functions $R_z(x)$ as defined in the previous section were computed from this data set consisting of 60 by 2650 data points. As the examples in figure 2 show, the deviations from the mean sound velocity gradient are correlated over some kilometres of horizontal distance, which is as large as typical skip distances of acoustical rays in that area. Horizontal autocorrelation functions in different depths may have different shapes as seen in figure 2. They also have different amplitudes. While in the top mixed layer $R_z(0)$ is less than 10^{-3} km^{-2}, it is almost 5 km^{-2} in the vicinity of the thermocline.

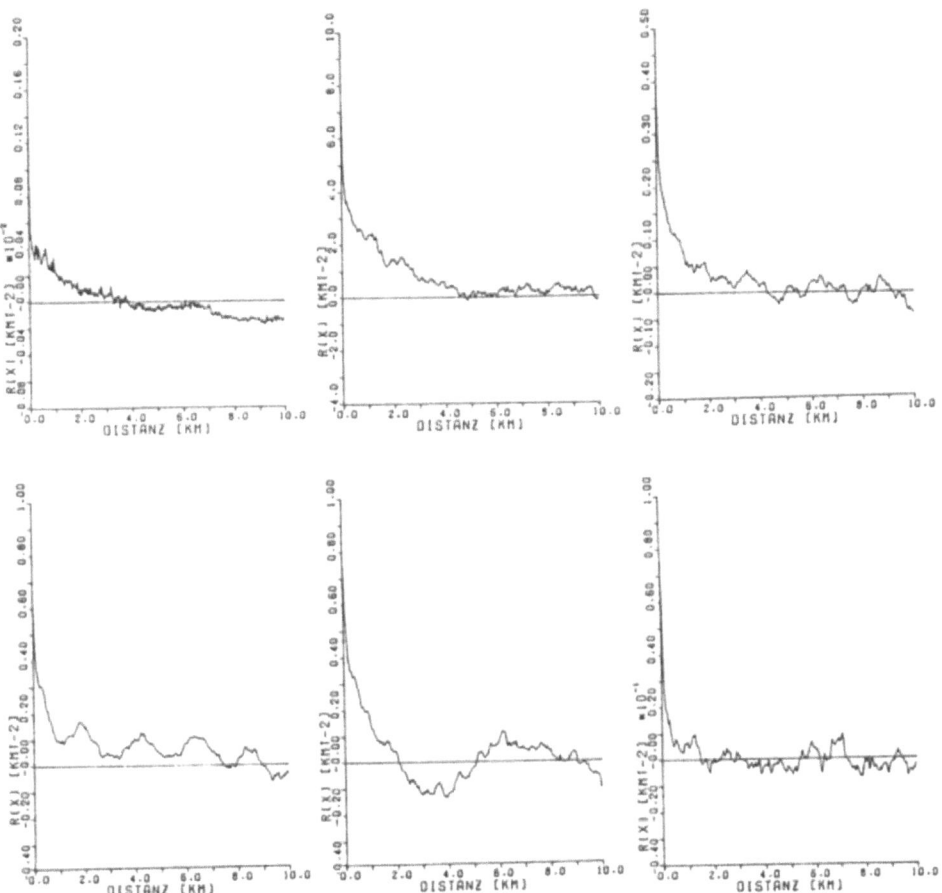

Figure 2.   Autocorrelation functions of the quantity $1/C \cdot \partial c'/\partial z$ in different depths. $C(z)$ is the mean sound velocity profile, $c'(x,z)$ the deviation from the mean profile.
Top left: z = 10 m.   Top middle: z = 13 m.   Top right: z = 24 m.
Bottom left: z = 26 m.   Bottom middle: z = 29 m. Bottom right: z = 46 m.

Also vertically the fluctuations are highly correlated over distances of several metres. The crosscorrelation of the random part of the sound velocity gradient taken from some separate depths can be negative. Especially a steep gradient of the sound velocity profile below the mean thermocline is coupled with a small gradient above and vice versa. So R_z changes its sign when the thermocline depth is crossed.

It turns out that the required integration along a mean ray in formula (4) cannot be done for four reasons.

– The correlation function R_z along a ray depends on depth and ray direction. It has no simple and universal structure. Therefore it cannot be estimated from an easily obtainable data set.

– An actual ray which deviates only by few metres is influenced by a sound velocity statistics different from that of the mean ray. In 30 m depth after a path length of 100 m the uncertainty of the ray direction is already 0.07 and hence the mean vertical deviation about 3 m. In the neighbourhood of the thermocline the situation is even worse.

– Usually the integration path in (4) is selected such that its length is much greater than the correlation length, but small enough that the result of the double integration is small compared to 1. These conditions cannot be simultaneously fulfilled with the correlation functions presented in figure 2. The only depth interval where it might be possible is the top mixed layer.

– If in a certain depth the random component of the sound speed gradient is still correlated after the ray returned from its upper or lower turning point, this means that long range features of the stratification are contained in the random sound velocity field which must not be treated statistically. Even in the surface duct the correlation function does not decrease to zero within a skip distance. If sound velocity data are high pass filtered to avoid large correlation lengths, the structure of the correlation function will depend on the applied filter.

Nevertheless ray diffusion theory with matched diffusion constants proved as a powerful tool even in shallow water situation with high variability of sound velocity. The reason is that the physical mechanism must be similar even if the restrictions of the theory are violated. Only at depth intervals with large sound velocity gradients a theory different from ray diffusion seems to be more adequate.

4.   Neglect of the vertical thermocline extension

Contemplations will start with just the extreme case, where ray diffusion theory is least acceptable, namely with narrow interfaces between layers of different sound velocity. In the calculation of sound rays the result is scarcely influenced by the magnitude of the sound velocity gradient in a narrow interface. Only the difference of the sound velocity in the adjacent layers is important. Figure 3 demonstrates the similarity of ray traces in an environment typical for a summer situation in the Baltic Sea. Skip distances and ray curvatures in the layer interfaces are influenced by different gradients. But the specifics, the penetration into another layer, the grazing angle at the surface, the possible extinction at the ocean bottom, they are all conserved if the sound velocity gradient is changed.

This situation permits us to reduce an already narrow thermocline to a zero width boundary and perform ray tracing in adjacent homogeneous or gently modifying layers. When a ray meets the boundary, it is refracted or totally reflected according to Snell's law. Partial reflection will not be considered bearing in mind, that the sound velocity jump is not real, but was obtained by an abstraction.

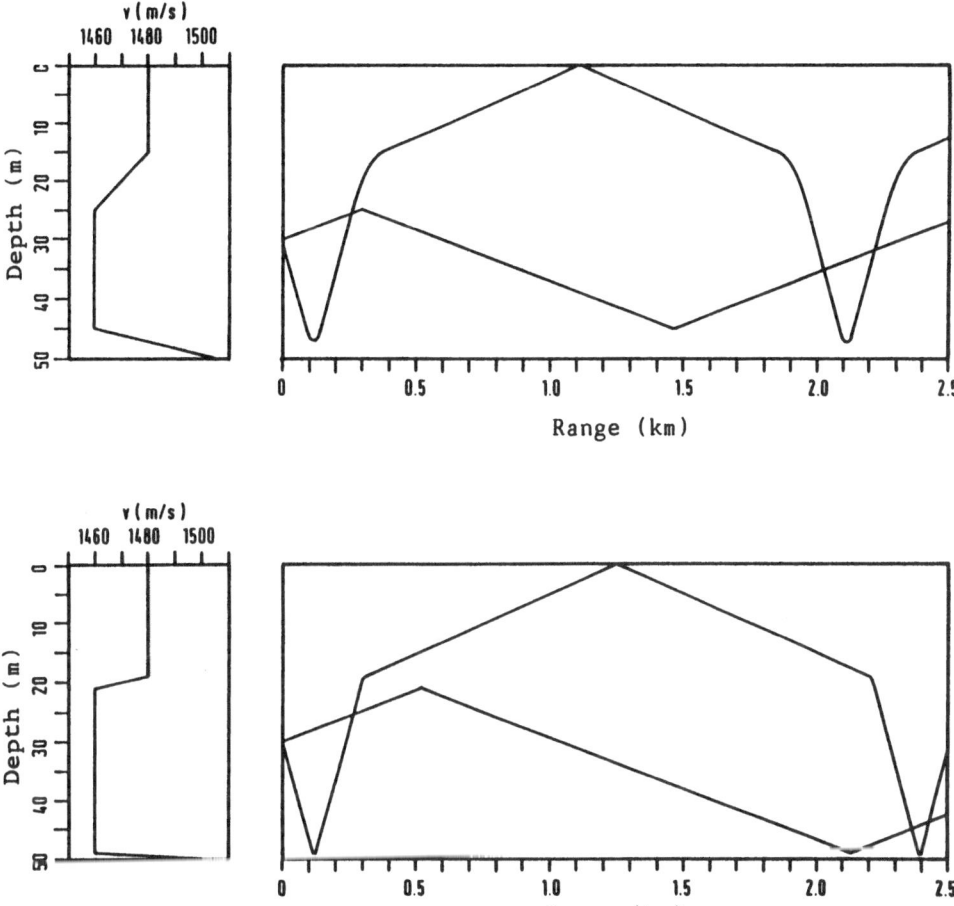

Figure 3. Rays in a typical Baltic Sea environment. The sound velocity gradient in the thermocline is of no greater importance than small deviations of the thermocline depth.

In the ocean the depth of a thermocline varies with time and location. As mentioned above the vertical position itself has minor effect on sound spreading. The importance of the undulations emerges from the boundary slopes connected with depth variations. If a ray with grazing angle β_1 meets the thermocline at a spot with slope angle α, it will leave the thermocline with angle β_2, which is given by

$$c_1 \cos(\beta_2 - \alpha) = c_2 \cos(\beta_1 - \alpha) \tag{6}$$

c_1 is the sound velocity in the medium of the incident ray and c_2, where the ray leaves. In equation (6) they are presumed to be such, that β_2 is real. Otherwise the ray is reflected to the direction $2\alpha - \beta_1$.

For every ray with direction β_1 the resultant ray direction β_2 is determined by the slope angle α of the thermocline facet met. A random distribution of α implies a resultant distribution of β_2. The width of that distribution depends on the incidence angle β_1. For grazing incidence the standard deviations of β_2 and α are the same, for vertical rays the standard deviation of β_2 is smaller by the factor $(1-c_2/c_1)$. The hardly influenced steep rays are of no interest here, because only rays with small grazing angles are involved in long distance sound transmission. It will soon be seen, that in contrast to usual ray diffusion theory the emerging probability distributions are far from being symmetrical with zero mean. If $c_1 < c_2$ the probability distribution of β_2 may even split into two constituents, when depending on α the ray is either refracted into the adjacent layer or totally reflected.

In the following considerations it is convenient to use the distribution of thermocline slopes s instead of that of the slope angles.

$$s = \tan \alpha \qquad (7)$$

If the statistics of slopes at the spot is known, where an incident ray hits the thermocline, then the statistics of the new ray direction in both the refraction or reflection case is the result of a transformation.

$$p(\beta_2;\beta_1) \ d\beta_2 \ = \ p(s,\beta_1) \ ds \qquad (8)$$

$p(\beta_2;\beta_1)$ is the probability density of ray directions β_2 provided the incident ray has the direction β_1, while $p(s,\beta_1)$ is the probability density of facet slopes met by rays which have an angle β_1 relative to the x-axis. The relation between β_2 and s, which is necessary to carry out the transformation is under consideration of equ. (7) given by equ. (6) for refracted rays and otherwise by the equation of reflection $\beta_2 = 2\alpha - \beta_1$.

It must be emphasised that $p(s,\beta_1)$ is not identical to the probability distribution of thermocline slopes, which is expected to be an even function with zero mean. In contrast $p(s,\beta_1)$ must be zero for slopes, which cannot be reached by an almost horizontal ray and therefore lie in a shadow. Even if the boundary is illuminated, the probability to meet a facet which faces the incident ray is greater than to hit a facet with reversed slope. Not the proper slope distribution of the thermocline does influence an incident ray, but the portion of it apparent for that ray. $p(s,\beta_1)$ may be named the effective slope distribution.

5. Evaluation of the effective slope distribution

The arguments in this paragraph will closely follow those of Lynch and Wagner (1970) with two exceptions. While they immediately intended to find an expression for the angular distribution of scattered energy multiple scattering included, here as an intermediate stage a formula for the effective slope distribution is desired. When this can be established, the next step to the computation of the distribution of refracted and reflected rays implies no principal hurdles. Secondly it will be shown, that the shadowing function can be easily obtained by a simple argument. No numerical tests are necessary for verification.

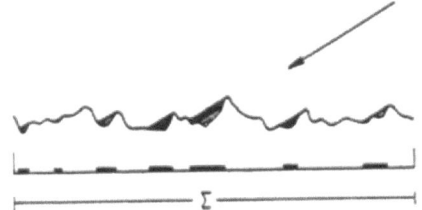

Figure 4. When rays with a distinct small grazing angle are sent to a rough boundary, it will partly lie in a shadow. The boundary portion under consideration is Σ, the sum of illuminated parts Ω.

At first a portion Σ of the thermocline is chosen. It shall be sufficiently long to represent the slope statistics. As indicated in figure 4 a bundle of parallel rays divides Σ into shaded and unshaded sections. The total of the unshaded parts shall be named Ω. Σ is split up into small portions of length δx. Coupled with any δx there is a facet with size $\delta x / \cos \alpha$, where α is the slope angle of that facet. If the facet is illuminated, then the portion of rays out of the bundle, which hit the facet, is equal to the projection of the facet to a plane vertical to the incident rays divided by the bundle aperture. The contribution of δx to the effective slope distribution is easily obtained by geometry sketched in figure 5.

$$p(s;m) \ \frac{\delta x}{\Sigma} = q(s) \ \frac{\sin(\beta - \alpha) \ \dfrac{\delta x}{\cos \alpha}}{\Sigma \ \sin \beta} \qquad (9)$$

$$= \left(1 - \frac{s}{m} \right) q(s) \ \frac{\delta x}{\Sigma} \qquad \text{if } \delta x \text{ in } \Omega$$

$$s = \tan \alpha, \quad m = \tan \beta$$

$q(s)$ is the probability density, that for δx the facet slope is s. In other words $q(s)$ is the proper slope distribution of the boundary. Now we define a function $\varepsilon(x;m)$.

$$\varepsilon(x;m) = \left\{ \begin{array}{ll} 1 & \text{if } x \text{ in } \Omega \\ \\ 0 & \text{otherwise} \end{array} \right. \qquad (10)$$

If in (9) the distribution $q(s)$ is replaced by the combined probability distribution $q[s, \varepsilon = 1]$ then (9) holds for any δx in Σ. The thermocline undulations are presumed to be a stationary random process. So there are no preferred points and $q[s, \varepsilon = 1]$ is independent from x. Now integration of (9) is trivial and yields

$$p(s;m) = (1 - \frac{s}{m}) \, q[s, \; \varepsilon(m) = 1] \tag{11}$$

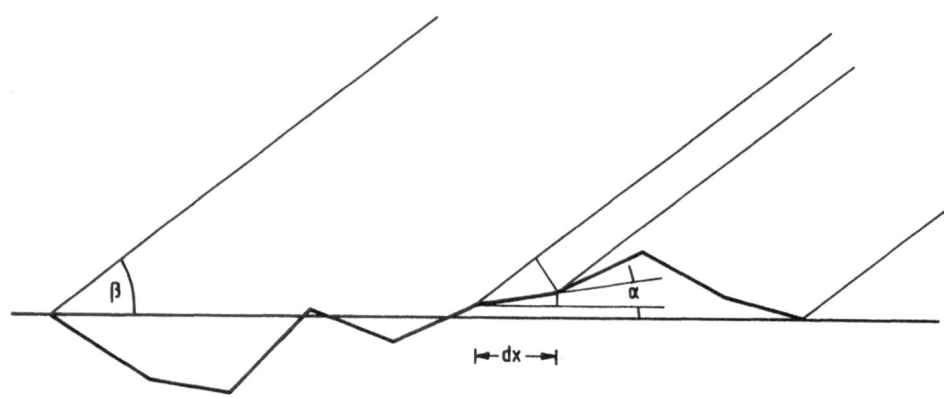

Figure 5. The percentage of rays with direction β, which illuminate the boundary facet associated with the horizontal segment δx depends on the slope angle α.

The combined probability distribution q[s, ε = 1] can be written as a product of the proper slope distribution q(s) and a conditional probability that a point is illuminated by a ray provided the slope at the point is s.

$$p(s;m) = (1 - \frac{s}{m}) \, q(s) \, P[\varepsilon(m) = 1 \mid s] \tag{12}$$

The term $P[\varepsilon(m) = 1 \mid s]$ is usually called the shadowing function. It must vanish for slopes s greater than the incident ray inclination m, when s and m have the same sign. These facets cannot be illuminated. All other facets may be shadowed or illuminated. For very small grazing angles most of the rough boundary lies in a shadow. If it is accepted, that the probability of slopes at a point is independent from the location on the undulating boundary, then the shadowing function will be the same for all facets, which can be illuminated. It will be 1 for vertical incidence and decrease with decreasing grazing angles.

$$P[\varepsilon(m) = 1 \mid s] = \left\{ \begin{array}{ll} S(m) & \text{for } s/m < 1 \\[2mm] 0 & \text{for } s/m > 1 \end{array} \right. \tag{13}$$

Now the leftover part of the shadowing function can be obtained by normalising the effective slope distribution p(s;m) or, as it was defined, the probability density for a ray with inclination m to encounter the slope s at the spot, where it hits the rough boundary. The probability to find any slope between $-\infty$ and ∞ is 1. Integration of (12) for m > 0 using (13) results in

$$\int_{-\infty}^{\infty} p(s;m) \; ds \;\; = \;\; S(m) \int_{-\infty}^{m} (1-s/m) \; q(s) \; ds \;\; = 1 \tag{14}$$

For m < 0 the integration is carried out from m to ∞ instead. So the shadowing function S(m) turns out to be the inverse of a weighed integral over the distribution of boundary slopes q(s). If the slope distribution q(s) is known, then the effective slope distribution p(s;m) is also maintained by equations (12) to (14).

6.   Realistic slope distributions

The slope distribution of thermoclines is expected to be a normal distribution.

$$q(s) \;\; = \;\; \frac{1}{\sqrt{(2\pi)} \; \sigma} \;\; e^{-\frac{s^2}{2 \, \sigma^2}} \tag{15}$$

Then the integration in (14) can be carried out to get the shadowing function and the effective slope distribution.

$$S(m) \;\; = \;\; \frac{2}{\frac{1}{\sqrt{\pi} \; V} \; e^{-V^2} + 1 + erf(V)} \tag{16}$$

$$p(s;m) \;\; = \;\; \frac{(1-s/m) \;\; e^{-\frac{s^2}{2 \, \sigma^2}}}{\frac{\sigma^2}{|m|} \; e^{-V^2} + \sqrt{(\pi/2)} \; \sigma \; [1 + erf(V)]} \qquad \text{for } s/m < 1$$

$$V \;\; = \;\; \left| \; \frac{m}{\sqrt{2} \; \sigma} \; \right|$$

Thermistor chain data already mentioned in section 3 were used to view the distribution of thermocline slopes and either verify or abandon the theoretical formula (16). A portion of temperature data is shown in an isoline depiction in figure 6. The 15°C isotherm was taken representative for the thermocline. It was calculated as a polygon by linear interpolation between vertically separated data points. The slope distribution of that isotherm is given in figure 7. It is well approximated by a Gauss-distribution having the same standard deviation σ = 0.023.

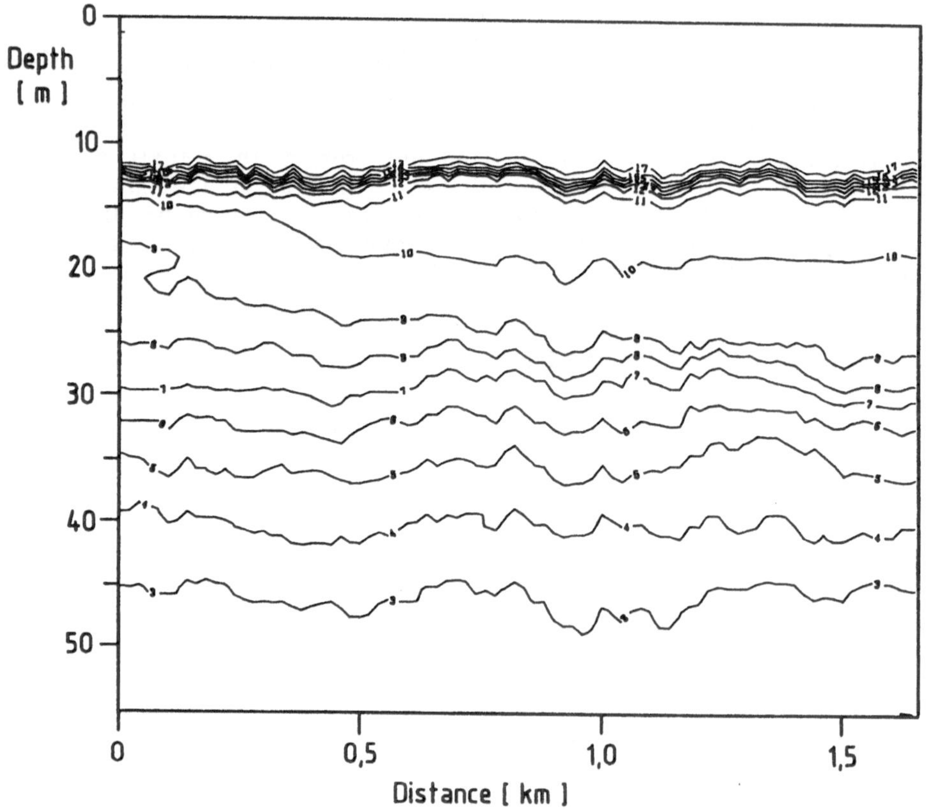

Figure 6.   Isotherms measured by a towed thermistor chain representing a small portion of the Baltic Sea section in figure 1.

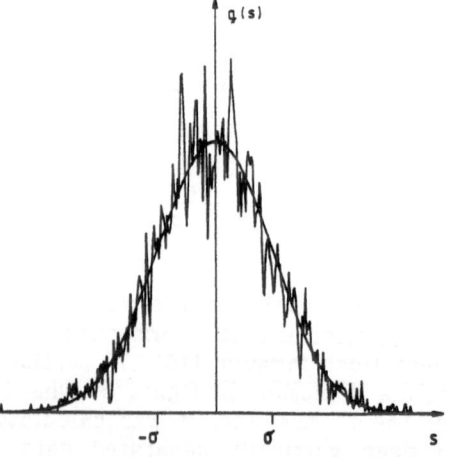

Figure 7.   Distribution of thermocline slopes obtained from towed thermistor chain measurements on the course shown in figure 1 and partly in figure 6.
The smooth curve represents a normal distribution with standard deviation σ = 0.023.

The effective slope distribution was evaluated experimentally by a numerical procedure. A multitude of equally spaced parallel rays was constructed and their first intersection point with the thermocline polygon determined. The thermocline slopes at the respective intersection points were aligned into classes and the events in each class counted. Distributions achieved by this method compared with the theoretical result are shown in figure 8. In spite of the fact, that with decreasing grazing angles the number of concerned facets drastically decreases, the agreement between theoretical and experimental curves is satisfying even for very small ray inclinations m = 0.1 σ.

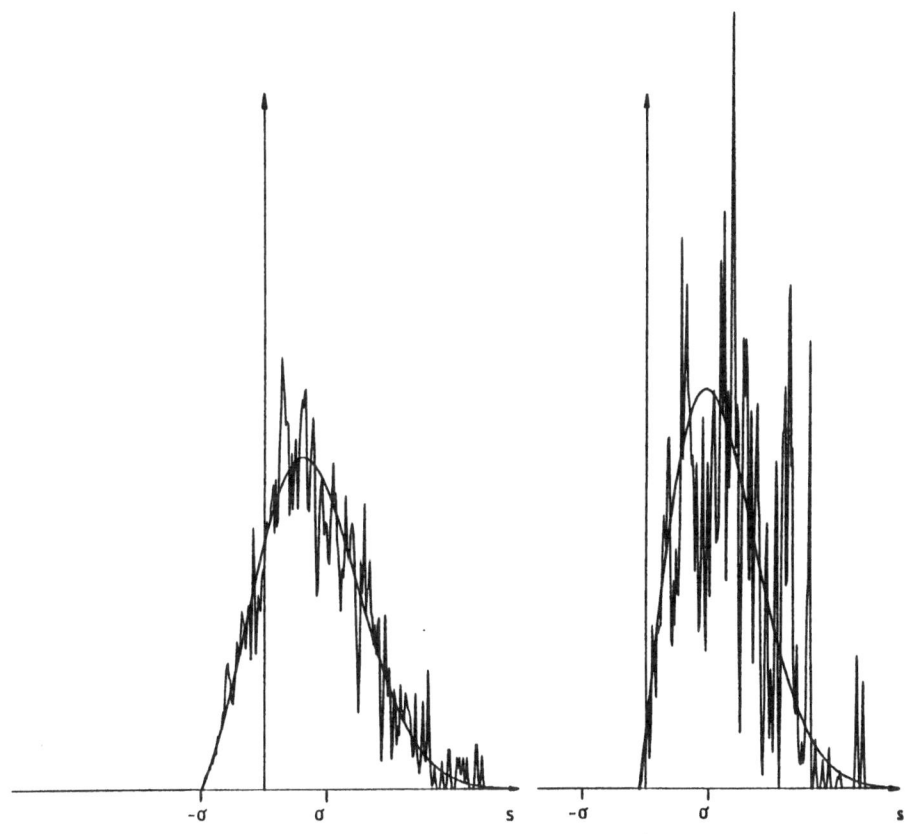

Figure 8. Effective slope distributions of the thermocline for rays with small grazing angles compared with the theoretical result.
Left: m = −σ.   Right: m = −0.1 σ.

The moments of the effective slope distribution are obtained by integration of $s^n \cdot p(s;m)$ from minus infinity to m. The first two moments are

$$\langle s(m) \rangle = - \frac{\sigma^2}{m} \frac{1 + erf(V)}{\frac{1}{\sqrt{\pi} \, V} e^{-V^2} + 1 + erf(V)} \tag{17 a}$$

$$\langle s^2(m) \rangle = \sigma^2 \frac{2 \, e^{-V^2} + \sqrt{\pi} \, V \, [1 + erf(V)]}{e^{-V^2} + \sqrt{\pi} \, V \, [1 + erf(V)]} \tag{17 b}$$

The mean effective slope $\langle s(m) \rangle$ is zero only for vertical incidence. Its value increases with decreasing grazing angles having the opposite sign and reaches $\sqrt{(\pi/2)} \cdot \sigma$ for horizontal rays. The variance of the effective slope distribution $\langle s^2(m) \rangle - \langle s(m) \rangle^2$ has its maximum σ^2 for vertical rays. When the ray inclination m decreases the variance slightly recedes to be still $(2 - \pi/2) \cdot \sigma^2$ at m = 0. The curves for the mean effective slope and its standard deviation are given in figure 9. It shall be stressed, that for the present slope statistics there is no specular reflection in the limiting case of vanishing grazing angles, as it would be with sinusoidal boundary waves. In the contrary boundary roughness has its greatest influence on rays with small grazing angles.

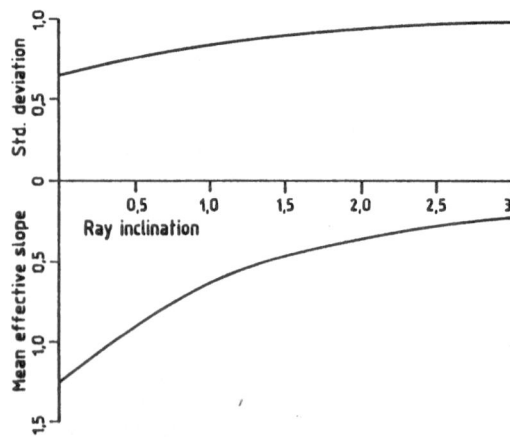

Figure 9. Mean effective slope and standard deviation of the effective slope scaled in units of the standard deviation σ of the slope distribution of a random rough surface.
The abscissa is given in units of $V = m/(\sqrt{2} \, \sigma)$ with m the ray inclination.

7. Results from mean effective slopes

The effective slope concept has the advantage, that it can be applied to reflecting interfaces like the ocean surface or a sound channel boundary as well as to refracting layers. First insights to the slope action are already obtained if the expectation value of the effective slope (17 a) is used instead of the full distribution.

As an example a shallow water environment shall be considered with a smooth reflecting bottom at depth h, a totally mixed water column, and a rough surface, σ being the standard deviation of the surface slope. A ray with direction β_n is expected to be reflected at a surface spot, where according to (17 a) the slope is $\langle s_n \rangle = -\sigma^2/m_n = -\sigma^2/\tan \beta_n$. The second term in (17 a) can be neglected if $|m_n| > 2\sigma$. The slope angle $\langle a_n \rangle$ necessary for the calculation of the reflection angle is small and therefore equal to the surface slope $\langle s_n \rangle$. When the ray returns from the bottom to interfere with the surface again its direction is

$$\beta_{n+1} = \beta_n + 2\ \sigma^2/\tan \beta_n \qquad (18)$$

The distance between two points of surface contact, the skip distance, is $\delta x_n = 2h/\tan \beta_n$. The remarkable consequence of equation (18) is a linear relation between the expected change of ray directions and horizontal distances passed.

$$\delta\beta = \sigma^2/h\ \delta x \qquad (19)$$

This formula may be used to estimate the distance after that initially small grazing angles become so great, that controlled by surface waves bottom absorption will be dominant, or that caused by undulations of its boundaries a sound channel looses energy.

In a second example the situation is regarded similar to that in figure 3. The stratification of the shallow water environment consists of two layers. While there is specular reflection at the smooth surface and bottom, now the thermocline is supposed to be rough. An almost horizontal ray below the thermocline remains in the bottom sound channel until after several contacts with the thermocline its grazing angle increases so much, that it can proceed to the upper layer satisfying the refraction law. The distance where this occurs can be calculated by equation (19). The ray inclination m is much smaller in the top layer than below. Therefore corresponding to (17 a) the ray has to expect a steeper thermocline slope when approaching from above than from below. That is the reason why the trace in figure 10 is not symmetrical. When the ray is back in the bottom layer again, it is less steep than is was before it left. An almost horizontal ray that left the surface layer will be trapped in the bottom layer for a while. So an equilibrium ray angle is perceived to exist in this environment. But it needs only little surface roughness to destroy this balance.

Figure 10. For the incident ray angle β_1 the mean effective thermocline slope is smaller than for the grazing angle after a specular surface reflection. Therefore $\beta_2 < \beta_1$.

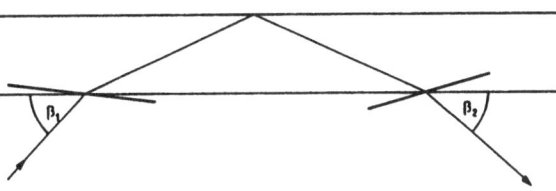

Deterministic ray calculations like in this section, even if correct expectation values are used for the slopes of layer boundaries, can produce unrealistic shadow zones and cut off distances. If the sound energy shall be calculated, which has to be expected after many random ray direction changes, then the full probability distribution of effective slopes has to be taken into account.

8.   Application of distribution functions

Computations not to be detailed here were carried out for an undirected source and receiver located on opposite sides of the thermocline. All possible rays crossing the thermocline once through facets of any slope were considered, when the expected intensity at the receiver was calculated. The application of the effective slope distribution led to the satisfying result, that transmission loss is conserved if the positions of source and receiver are interchanged. Compared with the smooth thermocline case there is significant intensity gain only if the device in the higher sound velocity layer is located near the rough thermocline.

In shallow water transmission over distances much larger than water depth many refraction and scattering events in the thermocline and at the surface require multiple integrations of distribution functions. This can be avoided by the Monte Carlo method. Usual ray tracing is done until a layer boundary is met. Then a sample is chosen from random numbers, whose probability density function is identical to the effective slope distribution for that ray. With that realisation of the random slope the refracted or reflected ray is calculated and propagated. This is repeated until a given detector window is met or missed by the ray. Many random rays have to be constructed until the percentage of rays falling through the detector window becomes stable and yields a measure for the expected transmission loss.

Procedures are available for digital computers, which produce random numbers equally distributed in the interval from 0 to 1. They can be used for the generation of effective slope samples, if the effective slope distribution is transformed to the probability density function $p(Z)$. $p(Z) = 1$ for $0 < Z < 1$, otherwise 0. The relation between Z and s is obtained by integration of the identity $p(Z(s;m))\ dZ = p(s;m)\ ds$.

$$Z(s;m) = \int_{-\infty}^{s} p(s';m)\ ds'$$

$$= \frac{e^{-W^2} + \sqrt{\pi}\ V\ [1 + \mathrm{erf}(W)]}{e^{-V^2} + \sqrt{\pi}\ V\ [1 + \mathrm{erf}(V)]} \tag{20}$$

$$V = \left| \frac{m}{\sqrt{2}\ \sigma} \right| \qquad\qquad W = \frac{s}{m}\ V \qquad \text{for } -\infty < \frac{s}{m} < 1$$

For a given random number Z between 0 and 1 the root $s(Z;m)$ of equation (20) will quickly be obtained by a numerical procedure. Newton's rule may be applied starting at the only inflection point $W_0 = \frac{1}{2}[V-\sqrt{(V^2+2)}]$.

When ray traces were investigated in the section before, that were influenced only by the most probable boundary slopes, every ray could immediately leave the boundary after refraction or reflection. Since now the whole distribution is applied, even slopes very close to the inclination of the incident ray occur. Reflections at these slopes do not sufficiently alter the ray direction to route them back to the layer, from where it came. The ray will hit the boundary once more after a very short distance. If it is accepted that the statistics of slopes met after a short distance is independent from the slope just left, then the numerical treatment of these cases in the Monte–Carlo scheme is straightforward and without problems.

A ray leaving a boundary in a small grazing angle may collide with a hump of the same boundary before it comes clear. The probability, that this happens, is equal to the probability, that a ray travelling to the opposite direction does not illuminate the starting point. That means, the shadowing function S(m) from equation (16) has to be applied to the new ray direction, too. In Monte Carlo ray tracing a random number between 0 and 1 has to be sampled. If it is greater than S(m), m being the ray inclination after interference with the boundary, then another boundary contact has to be regarded.

In order to get an impression, how much sound transmission is influenced by thermocline undulations, transmission loss calculations were carried out for a shallow water environment. In the examples in figure 11 water depth is 30 m with a well reflecting sand bottom. Sound velocity is 1480 m/s above and 1460 m/s below the 15 m deep thermocline. The source is located in 5 m depth. Transmission loss curves are presented in full lines for the smooth thermocline case. Application of random thermocline slopes with standard deviation 0.007 resulted in the dashed curves.

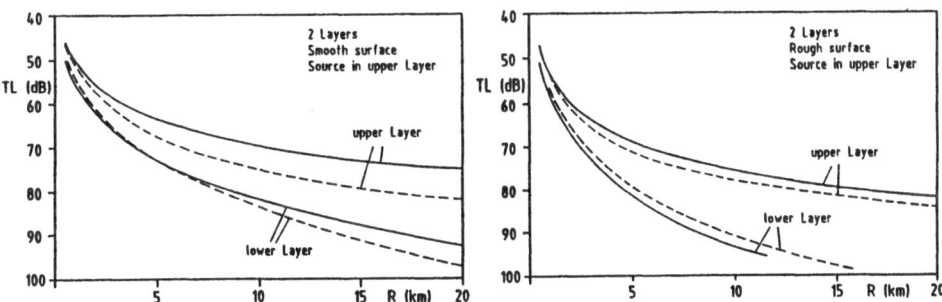

Figure 11. Transmission loss in a two layer shallow water area. The dashed lines take thermocline roughness into account, the solid lines disregard it.

On the left side a smooth surface is assumed. The effect of the rough thermocline is to increase transmission loss, because it continuously enlarges grazing angles of many rays leading to higher bottom absorption.

310

Sound intensity is diminished in both layers. Only for ranges smaller than 5 km intensity is slightly enhanced in the lower layer by those rays, which are trapped below the thermocline as described in section 7.

Mostly the roughness of the surface will even be greater than that of the thermocline. For the curves on the right side a standard deviation 0.07 was taken for the surface slope, ten times greater than for the thermocline. Transmission loss is obviously greater than with a smooth surface. It is interesting to see, that this time a rough thermocline leads to higher intensity in the lower layer than with a smooth thermocline. Some rays are prevented to return to the upper layer and therefore are not scattered at the surface.

The magnitude of the rough thermocline effect in the model environment greatly depends on the reflectivity of the bottom. In another computation a narrow bottom layer with sound velocity 1480 m/s as in the upper layer was inserted producing a sound channel in between. With a smooth surface the consequence of thermocline variations is a redistribution of sound energy. After 5 km distance from the source the intensity is higher in the sound channel than in the top layer. The total energy in the water column is nearly the same as in the smooth thermocline case.

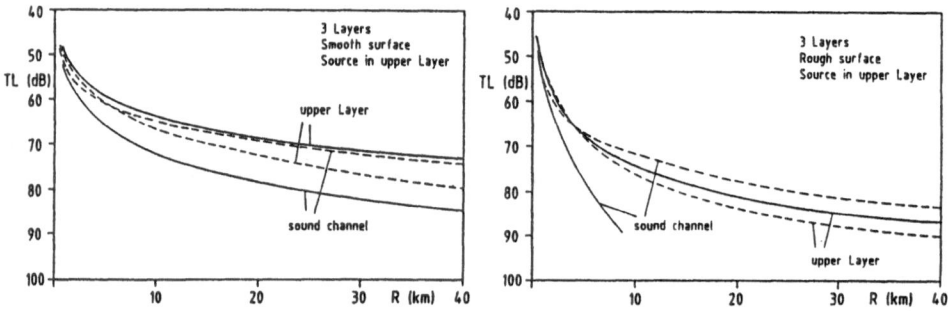

Figure 12. Transmission loss in a three layer shallow water area. The dashed lines show the effect of sound trapping by thermocline roughness.

On the right side of figure 12, where surface roughness was added, the action of an undulating thermocline is even more pronounced. With the assumption of a smooth thermocline all rays coming from a source located in the upper layer cross the sound channel in a relatively steep direction, therefore the intensity in the channel is much smaller than in the upper layer. If however the thermocline oscillations are taken into account, trapping into the sound channel leads to totally different results. Now after 5 km range there is more energy in the channel than in the top layer, even than there was in the smooth thermocline case.

Experimental results from sea trials like that shown in figure 13 clearly demonstrate the importance of trapping in the sound channel. A hindcast of the transmission loss was done by stochastic ray tracing. From this experiment in the Baltic Sea no hydrographic section data from towed instruments, but only single sound velocity profiles at the position of the receiving ship are available. The sound velocity profile shown in figure 13 was used and some reasonable values for the variance of the thermocline slope tested. A slope variance $\sigma^2 = 10^{-5}$ led to the most satisfying agreement, but fortunately computations are quite robust against changes of σ^2.

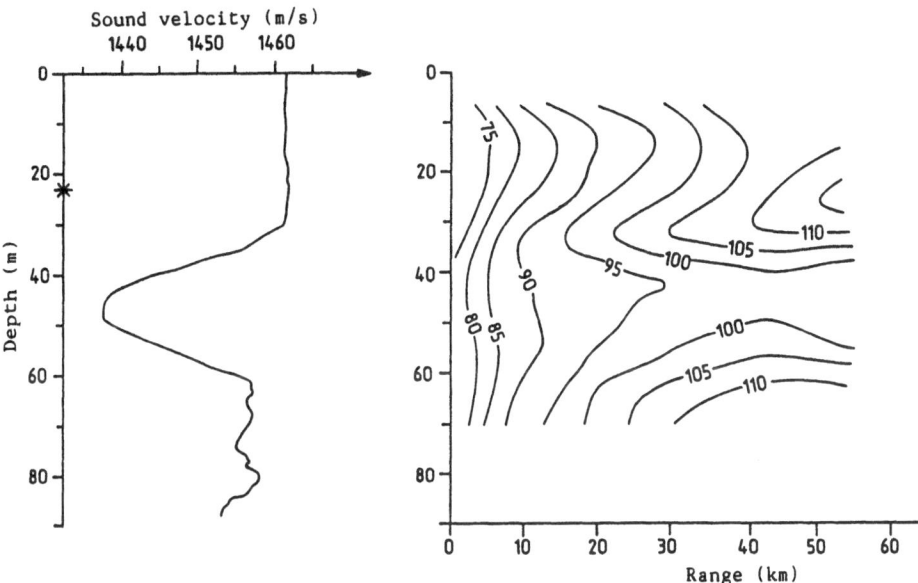

Figure 13. Experimental results of transmission loss in the Baltic Sea given in decibels. The source is located in the surface layer. After few kilometres the sound intensity is higher in the channel than above and below.

9.   Summary

Sound speed variability in the ocean is usually described as the difference between the actual velocity field and a mean profile. It was shown, that at least in some shallow water areas this difference and its correlation lengths are not small enough to treat them as small random inhomogeneities. Moreover in an attempt to establish a ray diffusion approach the structure of the necessary correlation functions obstructs the calculation of simple and universal diffusion constants.

Ray diffusion theory is least appropriate for the depth range, where the sound velocity gradient is particularly high, the thermocline. In ray acoustics a thermocline of small vertical extension may be replaced by a discontinuity of sound velocity. Random modifications of ray directions at the discontinuity surface are produced by the slope statistics of the thermocline. They can be treated by means of "effective slope distributions", that are the probability density functions, that rays with a given incidence angle hit the surface at a spot with a certain slope.

The form of the effective slope distribution is derived by geometrical considerations. It is in general not symmetric, the mean effective slope has the sign opposite to the incident ray inclination. The shadowing function being one constituent of the effective slope distribution is automatically obtained by the argument, that integrating the distribution over all plausible slopes must yield 1.

Making use of the effective slope distributions simplifies sound spreading calculations, because no distinction is necessary in advance, whether reflection or refraction occurs at the surface. Only after a sample or mean effective slope is known this decision will be made. A numerical procedure to fetch sample slopes for stochastic ray tracing is shown to be very handy.

Even if thermocline slopes are small and can only slightly modify ray directions, the modifications sum up to deviations from undisturbed sound transmission, which must not be neglected in some environments like shallow water ducts or narrow sound channels. This is demonstrated by a transmission loss hindcast compared with field measurements.

References

Chernov, Lev A. (1960) Wave Propagation in a Random Medium, McGraw-Hill, New York.

Fechner, G. (1986) 'Ergebnisse vom Einsatz des Meßsystems Schleppfisch während der Forschungsfahrt Ostsee 82', Internal FWG Report 1986-4.

Lynch, P.J. and Wagner, R.J. (1970) 'Rough-Surface Scattering: Shadowing, Multiple Scatter, and Energy Conservation', Jour. Math. Physics 11, 3032-3042.

Schneider, H.G. (1977) 'Excess sound propagation loss in a stochastic environment', J. Acoust. Soc. Am. 62, 871-877

Sellschopp, J. (1985) 'OSTSEE 82, Zusammenstellung der Schlepp-kabelmessungen', Internal FWG Report 1985-15.

Sellschopp, J. (1987) 'Zum Einfluß der Sprungschichtneigung auf die Schallausbreitung im Meer', thesis, Hamburg.

Modeling of sound propagation in a randomly varying ocean by stochastic mode coupling

R. THIELE
Forschungsanstalt der Bundeswehr für Wasserschall- und Geophysik
Klausdorfer Weg 2–24
2300 Kiel 1
W.-Germany

ABSTRACT. The impact of ocean variabilities on sound transmission loss is successfully handled by stochastic ray theory in general. However, in cases of comparable wavelengths of the acoustic signals and the oceanographic variations, ray theory is not sufficient. For this a stochastic approach to coupled mode theory is applied, especially for situations with small scale variations and a range independent mean sound speed profile. The concept of the stochastic coupled mode approach, the problems about the collection of significant oceanographic input data, and the possibilities to achieve a reasonable statistical description of the received sound field are discussed.

1. Introduction

Fluctuations of sonar signals are well known and a severe problem for modern signal detection. In the past mainly fluctuations by reflections at the rough sea surface and the seafloor were studied. However, even in situations of direct path transmission variations may arise. This problem has been treated by many authors. The main flow of theoretical research in this field starts with a certain mathematical treatment, the separation of the mean sound speed profile and the random variability of it:

$$c(x,y,z) = c_0(z)\,[1 + \mu(x,y\mid z)] \tag{1}$$

with

$$\langle \mu(x,y\mid z)\rangle = 0$$

and

$$\langle \mu^2(x,y\mid z)\rangle = \sigma_c^2(z)$$

The mean sound field and the path of energy transport are calculated by the mean profile $c_0(z)$, and the fluctuations by the variability $\sigma_c(z)$ along the path.

The most significant phenomenon produced by sound speed variations we have observed is the change of mean intensity by forward scattering into the surface duct (mixed layer) or into the sound duct prevalent in the Baltic Sea. It is not handled adequately by the distinction between first order—i.e. mean sound intensity—and second order— i. e. fluctuations. This type of separation of input mean and variability for calculating output mean and variability is correct for linear systems. The relation of sound speed distribution to sound intensity distribution is extremely non-linear. For a non-linear system we should expect no reasonable second order output if we already know that the first order statistics are wrong.

Certainly the mean sound intensity is only changed by fluctuations if the mean sound speed is depth dependent. However, for a mean sound speed profile without depth dependence we should expect no severe sound speed variations anyway; at least no internal waves.

J. Potter and A. Warn-Varnas (eds.), Ocean Variability & Acoustic Propagation, 313–321.
© 1991 *Kluwer Academic Publishers.*

For these reasons we propose first to develope models which handle the non-linear nature of sound propagation adequately and then to compare fluctuation models with them.

Each range dependent propagation model can be used for this reason. Of the three well known range dependent modeling approaches, as there are ray tracing [1, 2], parabolic equation [3] and coupled mode theory [4], we prefer the latter, because for it the least simplifications and restrictions for long range propagation with strong fluctuations have to be applied. For the case of horizontally stationary sound speed profiles with a mean profile and constant water depth McDaniel [5, 6, 7] has provided a coupled mode approach where the statistical characterisation of the environment can be applied directly. She used this model for boundary roughness problems, where we have some doubt whether the needed simplifications will hold. Compared to rough surface scattering the scattering by sound speed variations is locally weak. It is a narrow angle forward scattering process well matched to McDaniel's numerical technique.

It turns out that in principle the model works fine. It exhibits the mentioned phenomena and it is numerically easy to handle. A simple co-program to the well known SNAP normal mode model [8] was written which increases the calculation time not significantly. Only the strategy will be shown. It appears to be rather simple to exploit further results on fluctuating sound field and to adjust some oversimplifications applied until now.

It turns out that the restrictions of the study are not given by oversimplifications but by the lack of oceanographic input data. Shortcomings of these data are discussed.

2. Stochastically coupled normal modes.

The mathematics about normal mode coupling is provided by McDaniel [5, 6, 7]. She applied it to rough boundary scattering. In [4] the author has given a formulation for horizontal sound speed variations. There we posed the problem of zero correlation length of the horizontal gradients of variations. This may be avoided by application of an other formulation provided by McDaniel [6], giving a coupling term of

$$Q_{mn} = \frac{1}{4} \int_{-\infty}^{\infty} \frac{\sqrt{k_n k_m}}{k_n^2} \langle C_{mn}(x + \xi) C_{mn}(x) \rangle e^{i\xi(k_n - k_m)} d\xi. \tag{2}$$

When applying the relation from Milder [9]

$$C_{mn} = 2 \int_0^H u_m \frac{\partial u_n}{\partial x} dz = \frac{2}{k_n^2 - k_m^2} \int_0^H u_m \frac{\partial (k^2)}{\partial x} u_n dz \tag{3}$$

we get a Fourier integral on the sound speed gradients by placing Eq.(3) into Eq.(2) with

$$\int_{-\infty}^{\infty} \left\langle \frac{\partial c(x + \xi, z)}{\partial x} \frac{\partial c(x, z)}{\partial x} \right\rangle e^{i\xi k} d\xi = k^2 S_c(k, z). \tag{4}$$

We have to emphasize that we assumed at this place in [4] that the vertical correlation length L_z is small compared to the change of $u_n(z \mid x, y)$ with z. This may be wrong. But whenever the true vertical correlation is known, it can be treated adequately with only some additional numerical effort. Applying the horizontal wave number spectrum of the sound speed and assuming L_z being small we get the very simple formulation for the coupling coefficients

$$
Q_{mn} \approx \left[\frac{k}{(k_n - k_m)c_0}\right]^2 L_z \int_0^H u_m^2 u_n^2 \int_{-\infty}^{\infty} \left\langle \frac{\partial c(x + \xi, z)}{\partial x} \frac{\partial c(x, z)}{\partial x} \right\rangle e^{i\xi(k_n - k_m)} d\xi dz
$$

$$
\approx \left(\frac{k}{c_0}\right)^2 L_z \int_0^H u_m^2 u_n^2 S_c(k_n - k_m) dz. \tag{5}
$$

These coupling coefficients Q_{mn} describe the change of energy of mode m by coupling to the mode n in an incoherent way by

$$
\frac{\partial P_n}{\partial x} = \sum_{m=1}^N Q_{mn} P_m. \tag{6}
$$

where P_n is the total energy of mode n. This system of linear first order differential equations forms a symmetric eigenvalue problem easy to be solved by library routines. The eigenvalues are attenuation coefficients and the eigenvectors are sets of mode energy values. The smallest eigenvalue represents the asymptotic far field attenuation and the attributed eigenvector a stationary mode distribution. In case of no coupling, only the diagonal of the coupling matrix is non-zero and the eigenvalues are the attenuation coefficients of the individual modes; the eigenvectors have each only one non-zero element, representing the individual mode.

The off-diagonal elements are the coupling coefficients. The diagonal

$$
Q_{nn} = -\sum_{m \neq n} Q_{mn} - 2\alpha_n \tag{7}
$$

represents the loss of the mode by attenuation α_n and by coupling to other modes. When zeroing all off-diagonal elements in Eq. (6) we get the coherent part of the sound energy. By the comparison with the complete solution the coherence of the received signal can be evaluated as shown in [4].

316

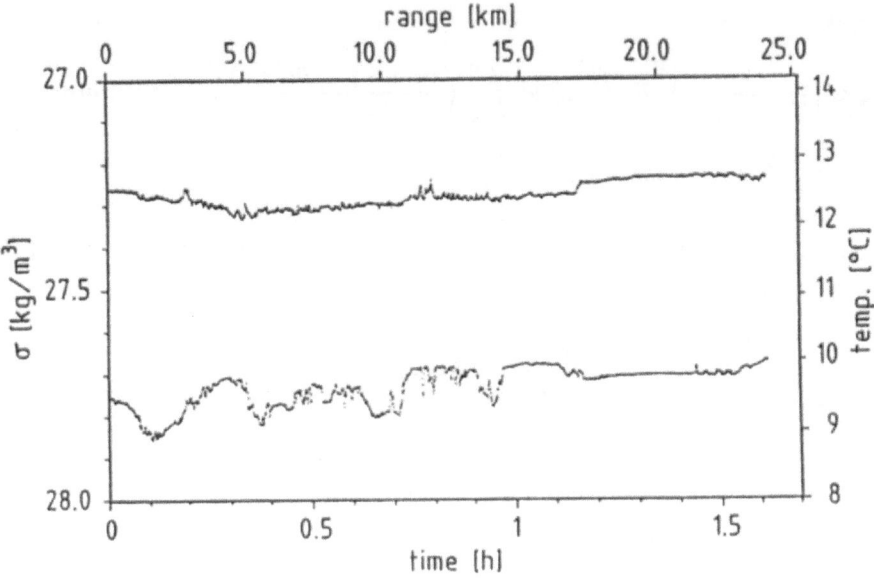

Figure 1. Density anomaly (upper curve) and temperature (lower curve) at constant depth. The scale factors are adjusted for identical depicted variations at constant salinity.

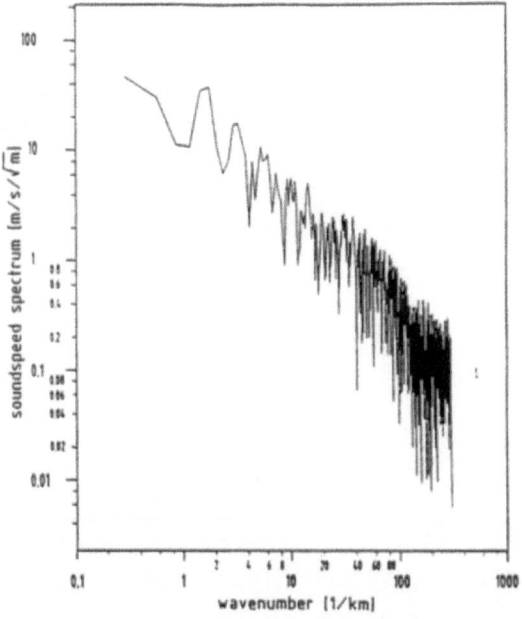

Figure 2. Amplitude density spectrum of sound speed versus horizontal wave number $k = 2\pi/\lambda$.

3. The horizontal wave number spectrum.

Wave number spectra are not often measured, at least not in connection with adequate sound propagation fluctuation experiments. Nevertheless, such acoustic experiments are rather often accompanied by a lot of sound speed measurements. It is tempting to apply some internal wave theory for extending these measurements to wave number spectra of the sound speed. For this approach we must at first check, whether the density provides an adequate description of the sound speed, because the internal waves are produced by density gradients.

Fig. 1 shows an example of temperature and density, taken by a "batfish", a depth controlled towed CTD, in this example kept in a constant depth. The scale is chosen so that—with a constant salinity—the curves of density and temperature should be similar, only separated by a possible offset. It is clearly visible that the temperature variations are an order of magnitude stronger than the density variations. And even the weak density variations there appear to be more influenced by inclination due to current shear than by internal waves. Obviously the temperature variations are not produced by internal waves but by thermohaline effects, meaning that there were incompletely mixed water masses of cold fresh water in warm, more salty water in equilibrium.

The attributed horizontal wave number spectrum of the sound speed, fig. 2, exhibits a k^{-2}-law. This is in agreement with spatial internal wave spectra, but that is no surprise. It means nothing else a type of fractal behaviour with the structure independent of scale. Indeed, this is again a difficulty, as in the given figure we observe no lower wave number limit of the spectral decay. Possibly there is no limit, indicating an instationary behaviour of the variations. This is the impression of the horizontal temperature behaviour as Sellschopp [10] has observed with many towed-thermistor-chain measurements. But with instationary changes of sound speed we will have problems to establish a mean sound speed profile.

The sound speed spectrum is very closely related to the temperature spectrum. Therefore we may apply temperature measurements for sound speed determination. However, the density spectrum, fig. 3, differs strongly. The minor slope is a result of noise in the measurements, as the data where calculated from conductivity and temperature which neutralize another, so that the relative noise is increased for the small signals.

We do not know whether an evaluation exists, how often sound speed variations are produced by internal waves and how often they are of the demonstrated nature. The author has only a limited experience in oceanography, but certainly the high amount of two-dimensional sound speed profiles collected in the Baltic, the North-Sea and the Norwegian Sea are dominated by non-internal wave effects, as we observed also in the Mediterranean [11].

Even Chernov [12] reports a paper of Beranek from 1949, stating that the sound speed variations are typically an order of magnitude higher than to be expected from density variations.

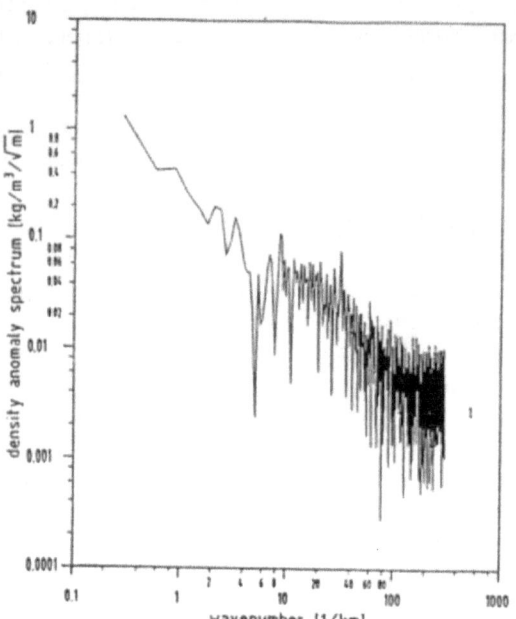

Figure 3. Amplitude density spectrum of water density versus horizontal wave number.

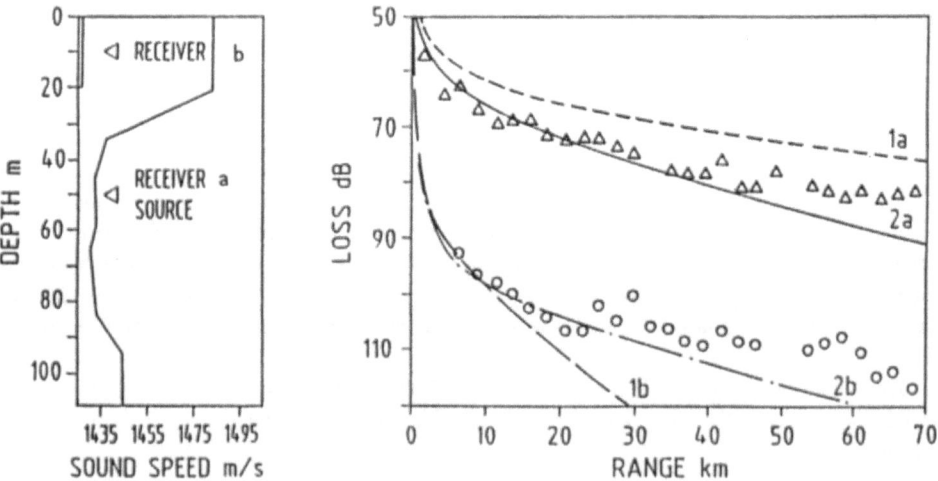

Figure 4. Comparison of the transmission loss of measured data (symbols) model calculations (lines). Triangles and curves a) are for a receiver depth 50 m, circles and curves b) are for receiver depth 10 m.

Curves 1 are from range independent deterministic mode calculations by the model SNAP, curves 2 are random mode coupling calculations with $S_c = 1k^{-2}m^{-2}$ and $L_z = 0.1m$

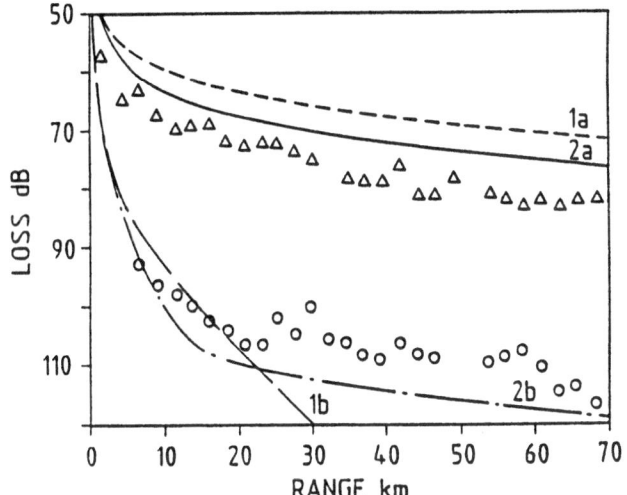

Figure 5. As fig. 4., but with $S_c = 0.01k^2\text{m}^{-2}$

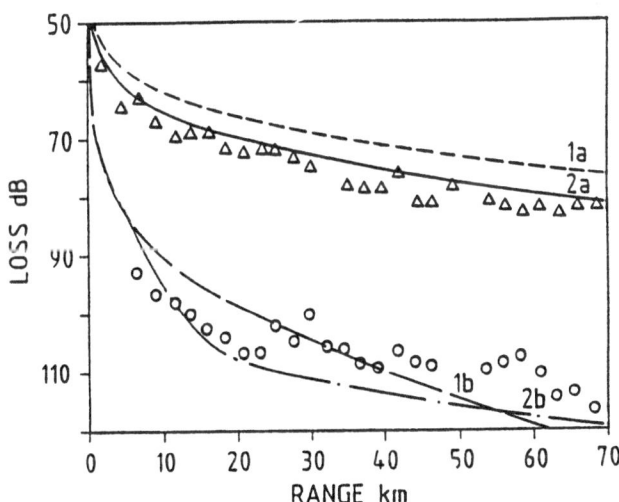

Figure 6. As fig. 4., but with $S_c = 0.01k^2\text{m}^{-2}$ and a bottom sound speed at the seafloor interface of 1485 m/s instead of 1482 m/s

4. Example of calculation.

We have chosen an example from the Baltic Sea already shown by Schneider [13] with the sound source in the duct and each a receiver inside and outside the duct (fig. 4). It is not difficult to achieve a rather good agreement. Even the lower attenuation of the data at greater distances are not surprising, because it is known that the bottom becomes harder after some distance in this experiment.

The main difficulty is that we have no horizontal wave number spectrum of the data available. The result is tuned with a spectrum of the $S_c = 1k^{-2}m^{-2}$. In fig. 3 we measured $S_c \approx 0.01k^{-2}m^{-2}$. This means that the effective sound speed variation is ten times so high. We have further assumed a vertical correlation length of 0.1 m. Possibly this is to small, but when the true vertical correlation is significantly larger, its effect should not increase linearly as it appears in Eq. (5), but less. This is a further knob to twiddle. Certainly the Baltic Sea is well known for strong sound speed variations. But we definitely do not know if the variations will be sufficient to produce the observed intensities. In fig. 5 we have reduced S_c by a factor of 10 compared to fig. 4. A better agreement is achievable by reducing the bottom attenuation, what we got by an increase of the sound speed in the bottom by 3 m/s in fig. 6.

5. Conclusion

With modern computers it is not difficult to calculate the influence of forward scattering sound transmission by variations of the sound speed profiles. All range dependent model approaches may be applied. The coupled mode theory is presented as an example. The results may produce excellent agreement; however a prove can not be given.

The problem is the lack of adequate oceanographic information. Theories on internal wave spectra are no general escape, so long as no prove is available that the sound speed variations have close relation to the density variations. An example is presented where this relation is completely missing. We have no indication that this example should not be typical.

Certainly the presented acoustic experiment is rather special. In most cases the influence of the sound speed variability on the mean intensity will be weak.

But just at weak effects which can be observed only by highly sophisticated acoustic experiments we should be reluctant to take some agreement as a validation of models when the oceanographic input is not definitely known.

Not the models form the critical path of ocean-acoustic fluctuation studies but the lack of well defined ocean-acoustic experiments.

6. References

[1] Schneider, H. G. (1977) 'Excess propagation loss in a stochastic environment', J. Acoust. Soc. Am. **62**, 871–877

[2] Sellschopp, J. (1990) 'Stochastic ray tracing in thermoclines', in this volume

[3] Schneider, H. G. and Sellschopp, J. (1981) 'Transmission loss computations from sound speed data of high horizontal resolution with strong variability', FWG-Report 1981-2, Kiel, Forschungsanstalt der Bundeswehr für Wasserschall- und Geophysik

[4] Thiele, R. (1987) 'The application of coupled-mode theory to propagation in shallow water with randomly varying sound speed', SACLANTCEN SM-199

[5] McDaniel, S. T. (1976) 'Coupled power equation for cylindrically spreading waves', J. Acoust. Soc. Am. **60**, 1285–1289

[6] McDaniel, S. T. (1977) 'Mode conversion in shallow-water sound propagation', J. Acoust. Soc. Am. **62**, 320–325

[7] McDaniel, S. T. (1978) 'Calculation of mode conversion rates', J. Acoust. Soc. Am. **63**, 1372–1374

[8] Jensen, B. F. and Ferla, M. C. (1979) 'SNAP: The SACLANTCEN normal-mode acoustic propagation model', SACLANTCEN SM-121

[9] Milder, D. M. (1969) 'Ray and wave theory invariants for SOFAR channel propagation', J. Acoust. Soc. Am. **46**, 1259-1263

[10] Sellschopp, J. (1990) 'Observations of ocean inhomogeneities', in this volume

[11] Uscinski, B. J., Potter, J. R. and Akal T. (1989) 'Broadband acoustic transmission fluctuations during NAPOLI 85, an experiment in the Tyrrhenian Sea: Preliminary results and an arrival-time analysis', J. Acoust. Soc. Am. **88**, 706–715

[12] Chernov, L. A. (1960) Wave propagation in a random medium, McGraw-Hill Book Company, New York e.a., p. 50

[13] Schneider, H. G. (1990) 'Average sound intensities in randomly varying sound speed structures', in this volume

Summary of Session 5

McCoy and Schneider

All papers in this session considered the influence of a deterministic background profile and stochastic environmental fluctuations on the acoustic intensity. The presentations unanimously emphasized the importance of the scattered field for intensity computations and stressed that the complete redistribution of scattered energy over all normal modes or equivalently all rays supported by the medium must be included into the calculation.

In the discussion the authors of both mode coupling approaches declared that the respective coupling matrix could be worked out for different types of ocean variability, if the appropriate statistics could be provided.

Dozier referenced a paper of Dozier and Tappert (1978) concerning the treatment of internal waves in an acoustic normal mode approach.

Thiele explained that his approximation concerning the vertical correlation of the sound speed variability had been made only because appropriate data had not been available, otherwise the approximation was not necessary.

The discussion included remarks to the effect, that the importance of a strong coupling of an inhomogeneous background profile and a fluctuating environmental component for determining the averaged intensity is not expected to be as important for determining the statistics of the signal fluctuations. it was also noted that radiative transport theory is available, which can be used to estimate the average intensity in the presence of both an inhomogeneous background and fluctuations in the environment.

For the highly variable sound speed stratification in the Baltic, it was suggested that EOF's (Empirical Orthogonal Functions) might be useful in separating the deterministic and stochastic components. The fundamental idea appears to define the stochastic component of the environment in terms of the 'statistical' homogeneity of that part of the environment which remains after that described by a few EOF's is removed. Together with the statistics of the EOF's this could then be used to improve the global statistical description of the environment, e.g for the ray diffusion approach.

J. Potter and A. Warn-Varnas (eds.), Ocean Variability & Acoustic Propagation, 323.
© 1991 *Kluwer Academic Publishers.*

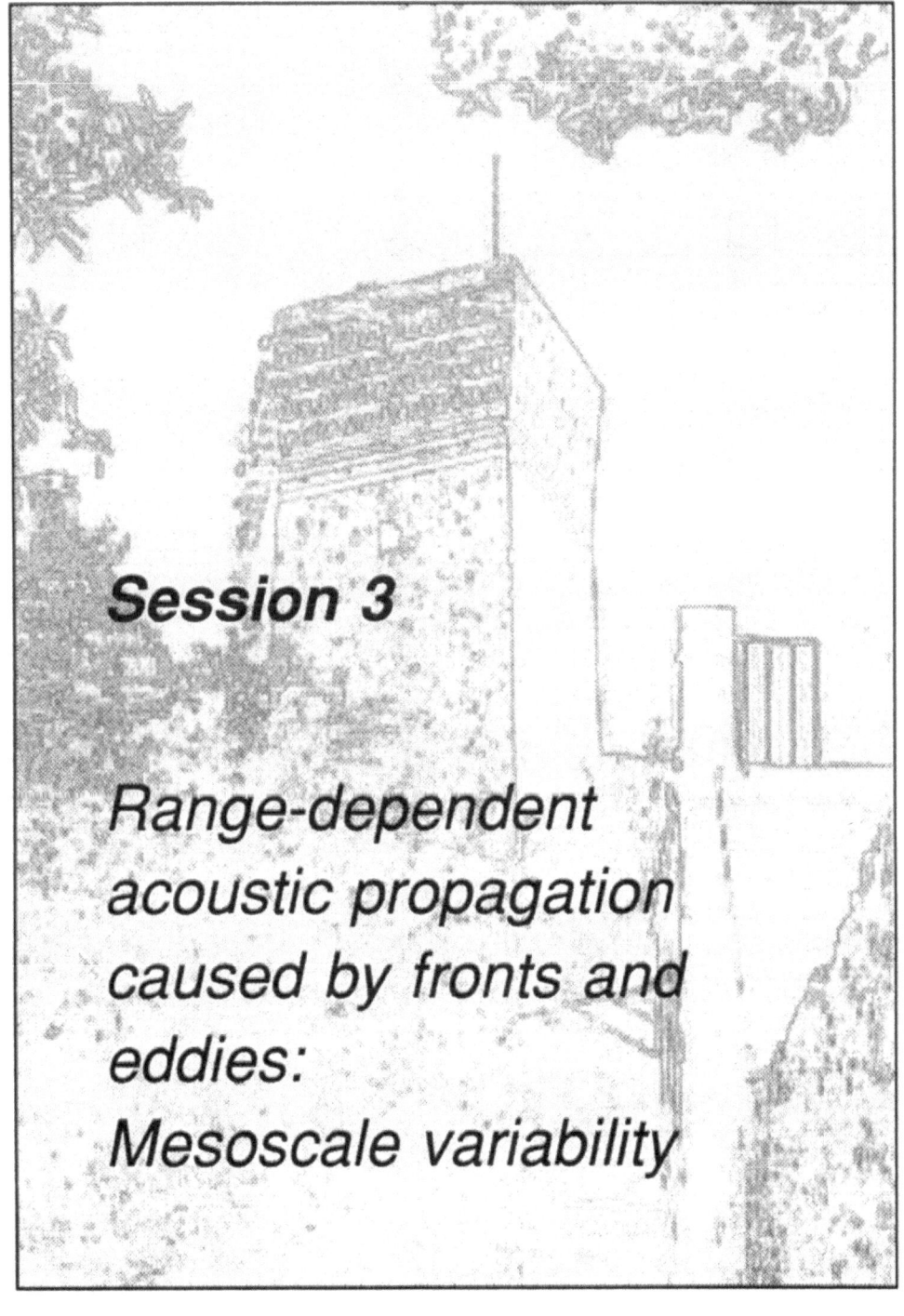

Session 3

*Range-dependent
acoustic propagation
caused by fronts and
eddies:
Mesoscale variability*

Radar Altimetry and Acoustic Prediction

by

Christophe BOISSIER* and Hubert BOUXIN**

ABSTRACT

An evaluation of the PARQUES acoustic propagation model to take into account the oceanic mesoscale variability as observed during the ATHENA experiment and an assessment of the GEOSAT altimeter for observing this variability and predict acoustic propagation are presented. The ATHENA experiment took place in summer 1988 in the north East Atlantic around 53°N - 25°W. The main objectives for this in-situ experiment conducted by the french Service Hydrographique et Océanographique de la Marine (SHOM) were:
+ investigate the oceanic mesoscale variability in this region of the Atlantic Ocean,
+ validate the GEOSAT altimetry for quantitatively observe this variability,
+ validate a quasi-geostrophic model in this area and evaluate assimilation techniques,
+ study the impact of the variability on sound propagation through simulations.
Impact of the horizontal gradient of acoustic properties of the ocean onto acoustic propagation is assessed using the PARQUES acoustic model which is a wide angle parabolic approximation based on a quadratic approximation of the square root operator. To assess the improvement due to the GEOSAT altimeter, comparisons are made with the true field of propagation loss (using the in-situ data), the field using only the first in-situ data and the "altimetric" field using the restitution from GEOSAT data. In most cases the improvement is relevant.

---

* SHOM (Service Hydrographique et Océanographique de la Marine)- 3 Ave Octave Gréard- 75200 PARIS Naval.
** GERDSM (Groupe d'Etude et de Recherches en Détection Sous-Marine)- Le Brusc-83140 SIX FOURS LES PLAGES.

J. Potter and A. Warn-Varnas (eds.), Ocean Variability & Acoustic Propagation, 327–342.

INTRODUCTION

The purpose of this paper is the evaluation of the PARQUES acoustic propagation model to take into account the oceanic mesoscale variability as observed during the ATHENA experiment and the assessment of the GEOSAT altimeter for observing this variability and predict acoustic propagation. The ATHENA experiment has been conducted by the "Service Hydrographique et Océanographique de la Marine", (the french Navy hydrographic Service), in order to investigate the mesoscale variability and to develop tools for its observation, analysis and prediction. GEOSAT data were provided in real time by JHU/APL. Model testing was performed in cooperation with the Groupe de Recherches et de Géodésie Spatiale (GRGS, Toulouse) and the Ocean Modeling Group, Harvard University, which took part in the experiment. Acoustic studies based on Athena experiment are performed by GERDSM, Toulon.

In this topic, data collected during the experiment are used in PARQUES model (wide angle parabolic approximation) to measure the effects of the oceanic variability on the acoustic propagation.

Some computations are presented in 2D and 3D environmental situations, with trues acoustic parameters (sound velocity profile measured in situ) and with the same parameters deduced from the altimeter measurements. Deep fields are estimated from the surface measurements using statistical extrapolations based on vertical EOF's.

I- The Athena experiment

The Athena experiment took place in the North-East Atlantic during the summer 1988 (fig. 1); it was a local experiment just east of the Mid-Atlantic Ridge, based on CTD, XBT, current-meters, mooring and drifting buoys.

The main objectives for Athena where:

+ investigate the oceanic mesoscale variability in this region of the Atlantic Ocean,

+ validate the GEOSAT altimetry for quantitatively observe this variability,

+ validate a quasi-geostrophic model in this area and evaluate assimilation techniques,

+ study the impact of the variability on sound propagation through simulations.

At 53°N, GEOSAT ground tracks have a 100 Km horizontal spacing; the along track resolution of the altimeter is 7 Km. The Athena experiment was designed to have Geosat passes coincident with in-situ data over a 200 Km diamond shaped domain (fig. 2).

II- Oceanic mesoscale variability observed in-situ

Global statistics have been performed over the whole space-time domain (200Km*200Km, 30 days, 0-2000m) for temperature, salinity and acoustic celerity.

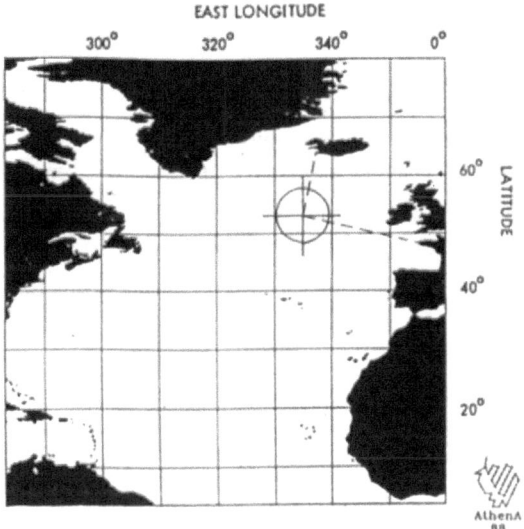

Fig 1

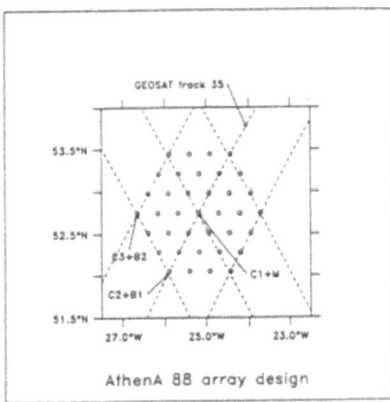

Fig 2

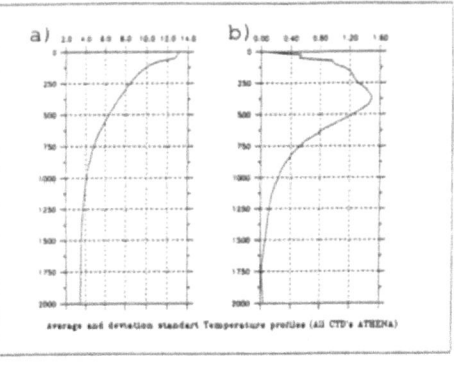

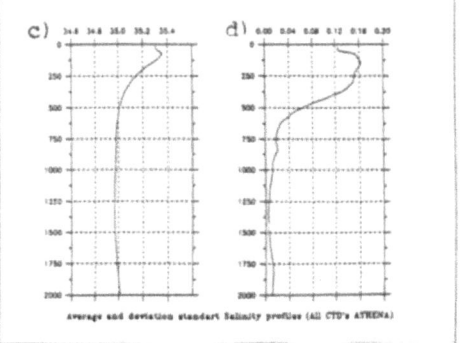

Fig 3

The mean temperature profile (fig. 3-a) shows the main thermocline spread out between 200 and 800 meters. It corresponds to a maximum in temperature standard deviation of 1.5 °C rms at 400 meters (fig 3-b).

The halocline is found to be shallower (fig 3-c) between 200 and 400 meters showing a maximum in salinity standard deviation of 0.16 ppm at 200 meter-depth (fig 3-d).

Synoptic views of temperature and salinity across an West-East vertical section evidence the structure of the variability (fig 4). Horizontal changes occur over typical distances of 100 Kilometers; the 7 °C isotherme rises from 600 a meter-depth up to 200 meters.

The salinity structure is coherent with the temperature structure except for the 0-200 meters layer where the variability appears to be more complex.

Combination of temperature and salinity effects yields a mean celerity profile which shows up a deep sound chanel at 750 meters (fig 5-a). The maximum horizontal variability is found between 350 and 400 meters and reaches 6 m/s rms (fig 5-b).

A synoptic view of the same vertical West-East section as for temperature and salinity shows maximum horizontal changes of 18 m/s at 400 meters (fig 6-a).

Over the same distance of 100 km, the deep sound chanel depth varies from 900 meters to 350 meters and its minimum value from 1485 m/s to 1478 m/s. The deep celerity field, below 1500 meters is not affected by this variability (fig 6-b).

The mesoscale variabilty in temperature, salinity and celerity, because of its influence on the density field, has a sea surface topography signature.

Drifting buoys trajectories tracked during the Athena experiment allows to verify the geotrophic hypothesis, and the simple geotrophic relation between the deep density gradients and the sea surface topography is valid:

$$\overrightarrow{\nabla_H}\,\eta = \frac{1}{\rho_0} \int \overrightarrow{\nabla_H}\,\rho(z)\,dz$$

where η is the surface elevation, and ρ the density at depth z.

Dynamic height computation from the in-situ vertical profil and objective analysis of sea surface topography anomalies allows the mapping of the mesoscale variability in terms of dynamic height.

Mesoscale structures are found to be present everywhere over the experiment domain; typical horizontal scales are 100 kilometers from crest to through and typical amplitude is from + 10 cm for the anti-cyclonic structure to -6 cm for the cyclonic structure.

A fairly good temporal stability is observed at 17-day interval (fig. 7).

III- GEOSAT altimeter measurements

The US-Navy Geosat altimeter flew over the domain in coincidence with the in-situ experiment for two cycles at a 17-day interval. Because of on-board recording problems due to spacecraft attitude

Fig 4

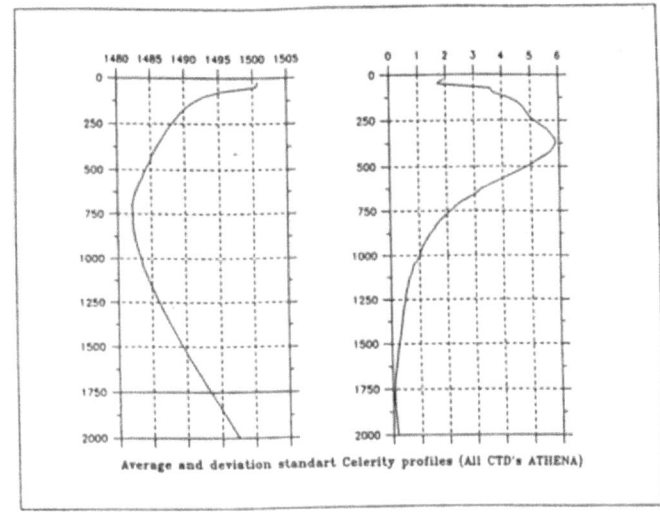

Average and deviation standart Celerity profiles (All CTD's ATHENA)

Fig 5

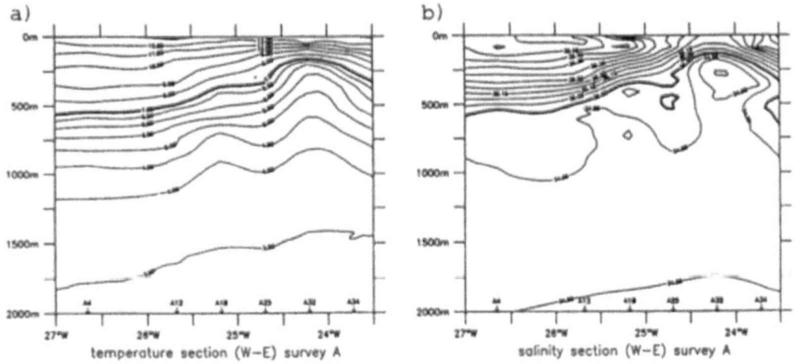

a)

temperature section (W-E) survey A

b)

salinity section (W-E) survey A

Fig 6

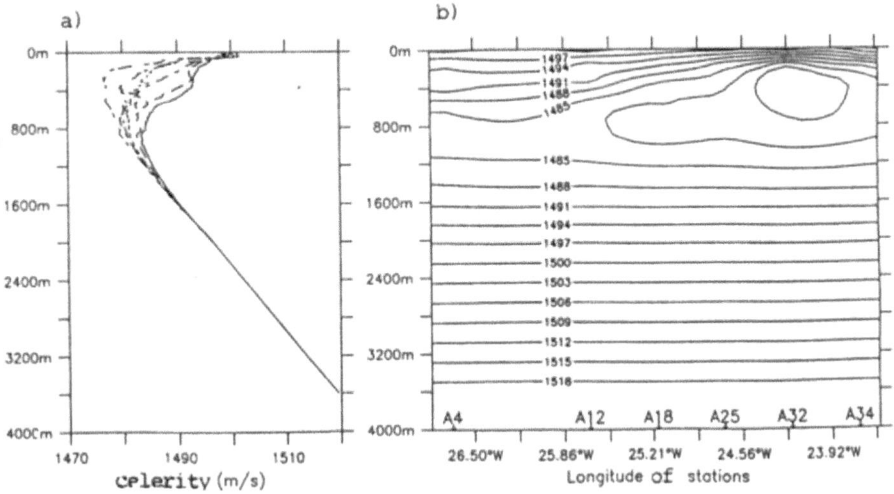

a)

celerity (m/s)

b)

Longitude of stations

control, radar altimeter data has been available only over one ground track crossing the domain center from North-East to South-West.

Sea surface topography anomalies are computed using the classical repetitive track technique.

A mean profile has been computed using two years of Geosat data, forming our best estimate of the geoid. Orbital errors are accounted for by adjusting a tilt and bias over 1500-kilometer long arcs on the residuals between each individual pass and the mean profile.

Figure 8-a shows these residual profiles after ajustement for cycles 25 through 39. Cycles 37 and 38 correspond to the two legs of the Athena in-situ experiment. Typical sea surface variation over the domain are found to be of the order of 30-40 cm over distances of 100 kilometers.

Comparison with in-situ data (fig. 8-b&c) shows a fairly good agreement between horizontal variations in dynamic height as computed from in-situ data (no motion level at 1000 meters) and horizontal variations as observed by the Geosat altimeter.

The front separating the anti-cyclonic structure to the west from the cyclonic eddy to the east is well observed by the altimeter as well as its dispacement and its strenghtening from one cycle (July 29) to the next (August 15).

IV- Bidimensional analysis and vertical extension

Space-time objective analysis is performed using all altimeter data within a 20 days temporal influence radius and 100 kilometers spatial influence radius. A priori statistics, such as space and time correlation functions based on 2 years of GEOSAT data [1], are used for optimal mapping of the sea level anomaly at a given date [2]. Such maps confirm the fact that the mesoscale eddies and fronts are everywhere present along the north Atlantic drift (fig. 9).

The oceanic variability observed during the experiment was found to be very coherent over the vertical. Empirical Orthogonal Functions (EOF) analysis confirms this high coherence. In celerity for instance the first mode accounts for 92% of the variability. Moreover this first empirical mode in celerity accounts for 96% of the variability of the dynamic height as computed from the in-situ data. These results where obtained by computing multivariate modes, the first variable being the dynamic height at the surface relative to 1000 m, all other variables being the series of the celerity at various depths [3].

When this first mode is used to extend over the vertical the surface information provided by the dynamic height, the standard deviation of the error on the estimation of the celerity field (1.8 m/s) is much lower than the actual variability of the celerity (6 m/s).

The estimated celerity field using the first empirical mode to extend the surface information shows the same general structure as the actual celerity field (fig. 10-a&b). The acoustic part of the work was designed to evaluate the relevance of the description on an acoustic propagation point of view.

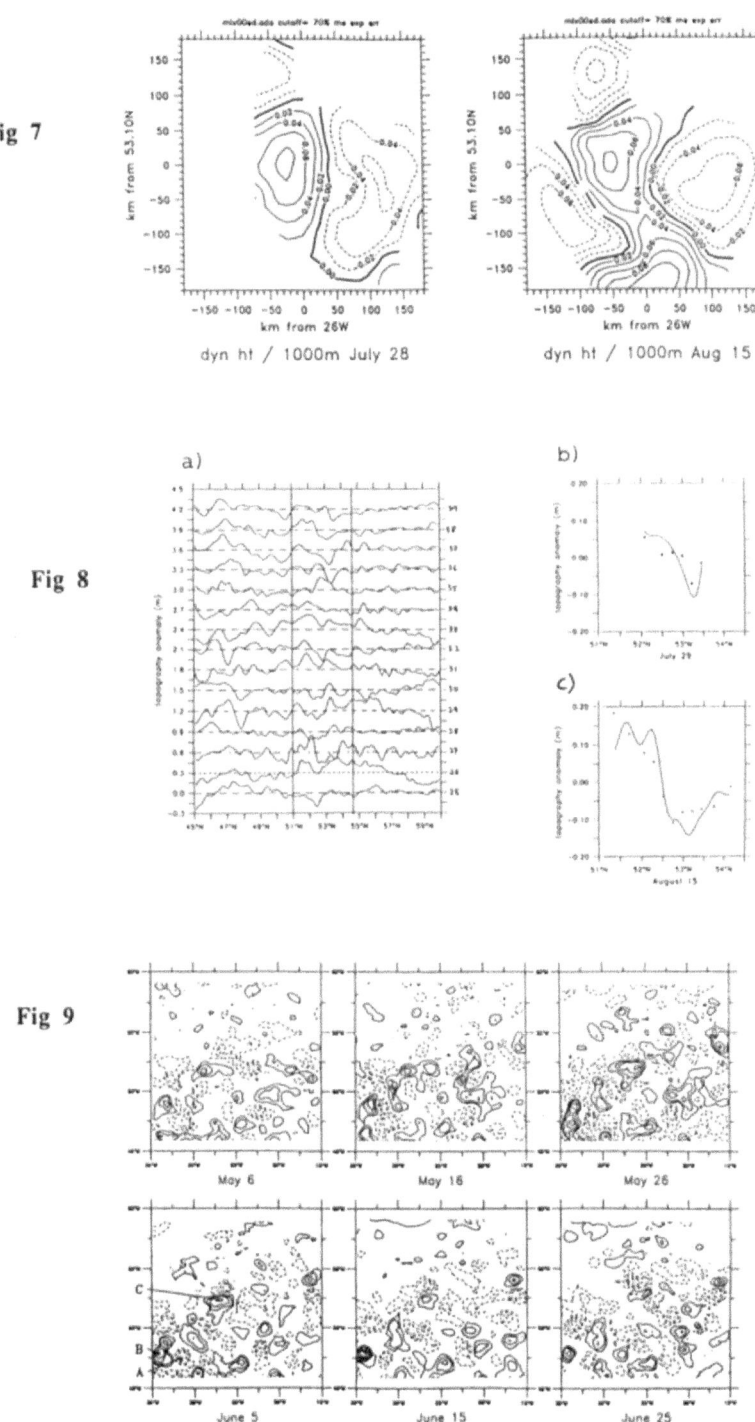

Fig 7

dyn ht / 1000m July 28   dyn ht / 1000m Aug 15

Fig 8

a)

b)

c)

Fig 9

May 6   May 16   May 26

June 5   June 15   June 25

334

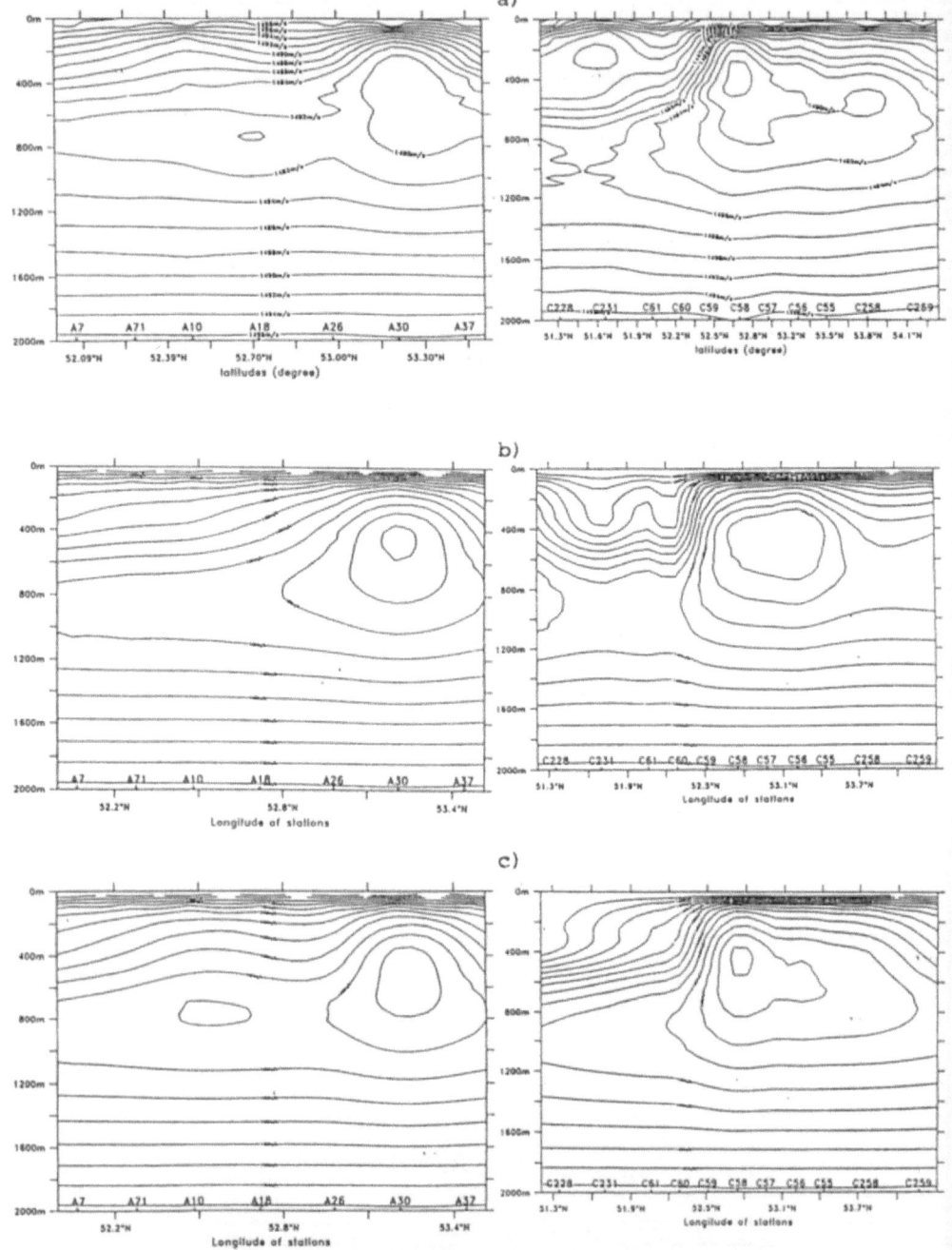

Fig 10: Actual celerity field (a), extrapolated with altimetry and EOF (b), extrapolated with dynamic Height and EOF (c) - section SW NE - array A (left) and array C (right).

V- The PARQUES acoustic propagation model

The study of the impact of the variability on sound propagation was conducted at GERDSM through simulations using the PARQUES (2D and 3D) acoustic model.

The standard parabolic equation (PE) introduced in the underwater acoustics community by Tappert [4] has been widely used. One of the advantages of the PE model is its ability to handle range-dependent environments.

The Helmholtz equation is given by

$$\Delta p + k_0^2 n^2 p = 0$$

where p is the acoustic pressure, k0 is the reference wavenumber and n the refraction index.
With

$$p(r,z) = \frac{1}{\sqrt{r}} \psi(r,z)$$

$$\Psi(r,z) = u(r,z)\, e^{ik_0 r}$$

$$p(r,z) = \frac{e^{ik_0 r}}{\sqrt{r}}\, u(r,z)$$

The parabolic equation can be written as:

$$(\mathbb{P} - i k_0 \mathbb{Q})\, \Psi(r,z) = 0$$

where,

$$\mathbb{P} = \frac{\partial}{\partial r}$$

$$\mathbb{Q} = \sqrt{1 + (n^2 - 1) + \frac{1}{k_0^2} \frac{\partial^2}{\partial z^2}}$$

The PARQUES acoustic propagation model (developed by B. Grandvuillemin at GERDSM [5]) is a wide angle parabolic approximation based on a quadratic approximation of the square root operator $\mathbb{Q}$.

VI- Impact of the mesoscale structure onto acoustic propagation - 3D computations

The in-situ measurements made during the Athena experiment yield a fairly good description of the mesoscale oceanic variability on the whole domain. Those measurements have been used as input data of the 3D propagation model running on a CRAY computer. These computations are compared with computations done with only the sound speed profile measured at the position of the emittor, supposing the ocean to be homogeneous on the horizontal. The difference is an estimate of the impact of the mesoscale activity onto the acoustic propagation. The magnitude of the difference is

336

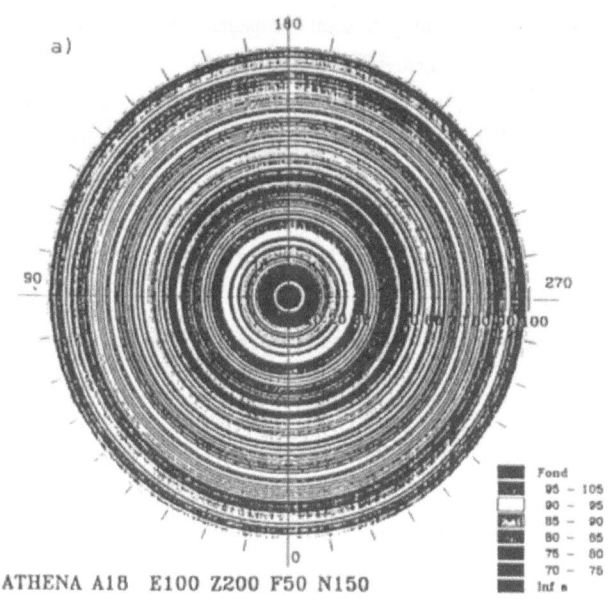

ATHENA A18   E100 Z200 F50 N150

	Fond
	95 – 105
	90 – 95
	85 – 90
	80 – 85
	75 – 80
	70 – 75
	Inf a

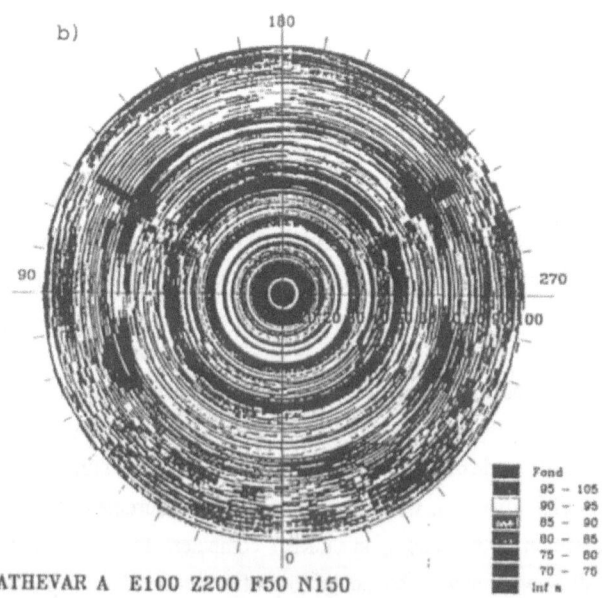

ATHEVAR A   E100 Z200 F50 N150

	Fond
	95 – 105
	90 – 95
	85 – 90
	80 – 85
	75 – 80
	70 – 75
	Inf a

Fig. 11: Computation with only one sound speed profile (a) and with all profiles (b). {Emitor depth: 100m, Receptor: 200m, f=50 Hz}.

found to be as high as 18 dBs in some places. Nevertheless, this result has to be considered with care: a small displacement of the convergence zone can sometimes yield such a difference even if the operational impact is not very important. However, the convergence zones generaly tends to vanish with the mesoscale activity (fig. 11).

VII- Impact of the mesoscale structure and improvement due to the GEOSAT altimeter- 2D computations

1) Simulations

Acoustic propagation simulations using the 2D PARQUES model running on a CONVEX computer have been performed along a South-West -> North-East section and along a North-East -> South-West section of the first and second legs of the experiment in 4 situations:

a) The "true situation":

The sound speed profiles used are those measured during the experiment. Concerning the South-west -> North-east section for the second leg, the mesoscale structure causes a variation of the depth of the deep sound channel from 900m to 400m. The dividing line between cold and warm water is approximately 120 km away from the emittor.

b) The "present situation":

An horizontally homogeneous celerity field characterized by the sound speed profile measured at the emittor position is considered.

c) The "altimetric situation"

The estimated celerity field using the first empirical mode to extend the surface information provided by GEOSAT is taken into account.

d) The "dynamic situation"

The estimated celerity field using the first empirical mode to extend the theoretical dynamic height computed from the in-situ data is taken into account.

The isovelocity fields concerning the a), c) and d) situations are presented on figure 10.

Propagation loss fields have been computed for a 100m and 300m deep emittor and for a 50 Hz and 200 Hz frequency. Propagation loss as a function of range have been computed for a receptor depth of 50, 100, 200, 300 and 400m (so we have 80 cases).

2) Propagation loss fields

Analyzing the fields of propagation loss leads to the following conclusions:

For the first leg (A-leg), the impact of the mesoscale structures does not appear to be very important.

On the other hand, the impact for the second leg (C-leg) is much more important;

338

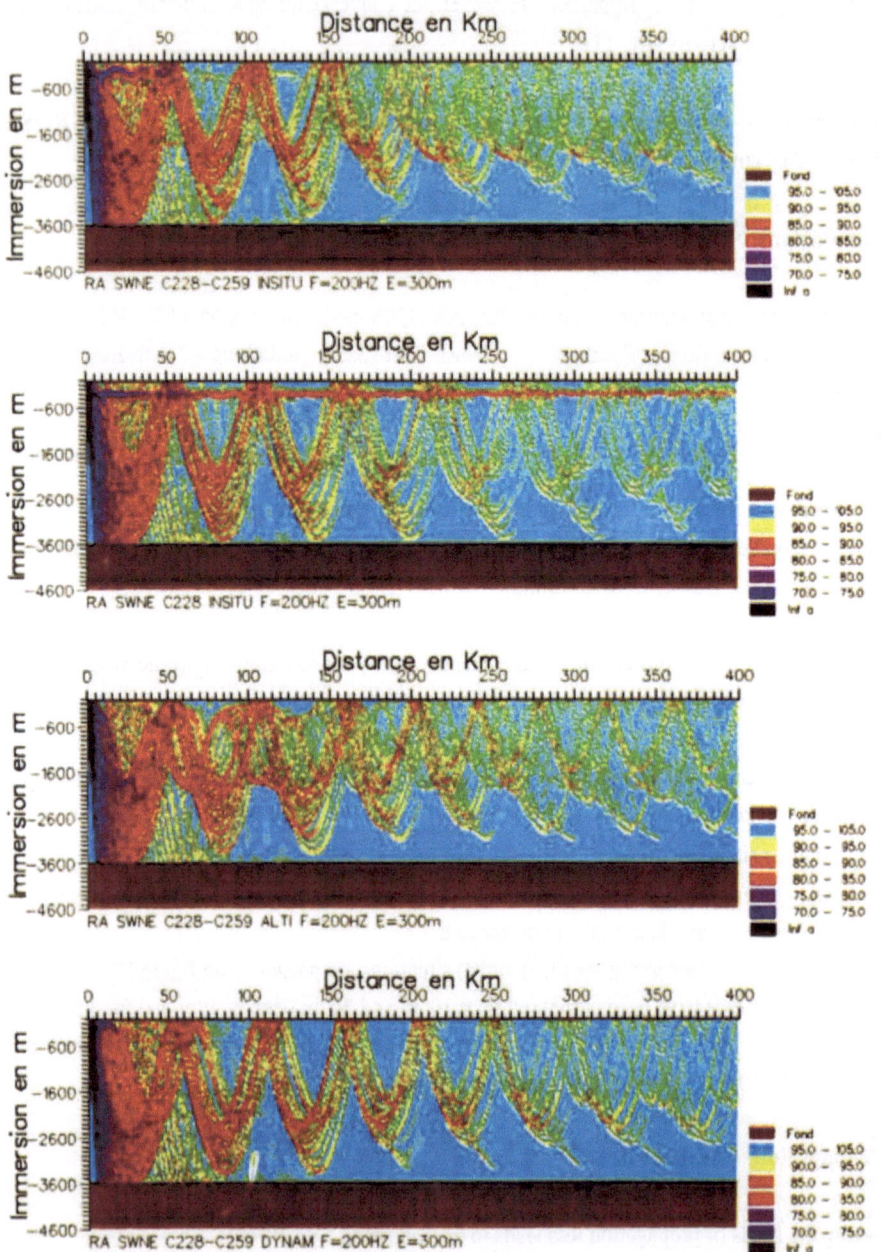

Fig 12: The "present situation" (b) would have predict a deep sound speed channel that, in fact, disappears on the "true situation" (a) 50 km from the emittor. The "altimetric" (c) and the "dynamic situation" (d) would have predict the right structure of the propagation field. Emittor depth =300 m, f =200 Hz, SWNE - Array C.

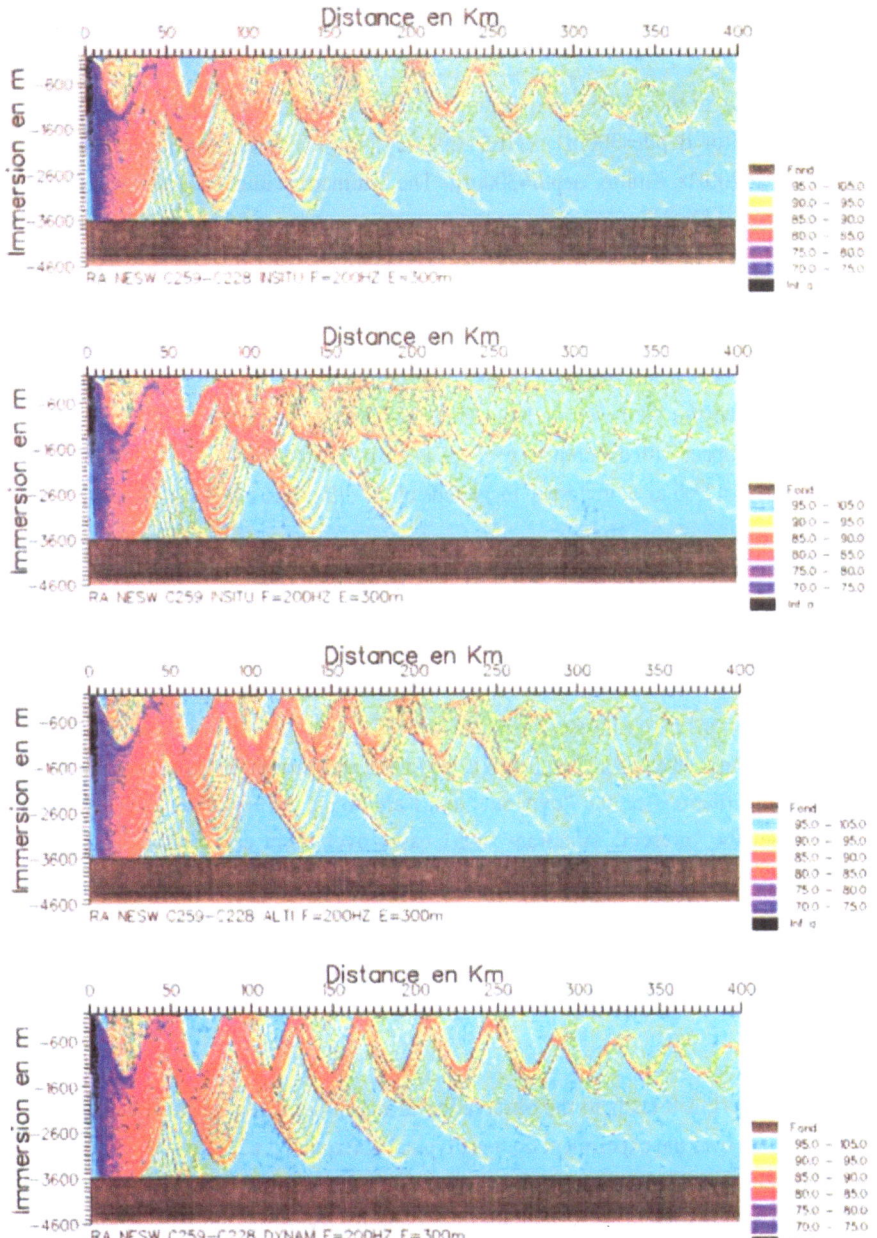

Fig 13: The plunging down of the sound speed channel is observed on this case: {NESW, C-leg, f=200Hz, Emittor depth=300m} (a). The "altimetric" (c) and the "dynamic situation" (d) roughly describe this plunging down; the "present situation" (b) does not.

- In the case {C-leg, SWNE, f=50 or 200Hz, emittor depth=300m}, the "present situation" would have predict a deep sound speed channel that, in fact, disappears on the "true situation" 50 km from the emittor. In this particular cases, the "altimetric" and the "dynamic situation" would have predict the right structure of the propagation field, consequently the improvement due to altimetry is relevant (fig. 12).

- The plunging down of the sound speed channel is observed on the following cases: {NESW, C-leg, f=50Hz, Emittor depth=300m}, {NESW, C-leg, f=200Hz, Emittor depth=100m}, and {NESW, C-leg, f=200Hz, Emittor depth=300m}. The "altimetric situation" roughly describe this plunging down (fig. 13).

- The impact of the mesoscale structures seems to be more significant for the 200 Hz frequency than for 50 Hz, and for an emittor 100-m and at 300-m depth.

The representation based on propagation field allows us to assess qualitatively the impact of the oceanic variability, when the structure of the field is modified. The improvement due to the altimeter seems to be -for the cases we have analysed- significant. On the other hand, to assess quantitatively the impact with differences of fields is difficult because a small displacement of the convergence zone can lead to an important magnitude of the difference that is not very significant.

3) Propagation loss as a function of range

The outputs of the PARQUES propagation model have been filtered with a Barlett filter over 2 Km (fig. 14). The differences to the "true situation" have been plotted for the "present", "altimetric" and the "dynamic" situations for each cases described at § VII-1 (fig. 15). The mean square difference has been computed in each cases and is chosen as criteria of comparison for the different situations: The "altimetric" situations or the "dynamic" ones are found to better describe the acoustic propagation than the "present" situations in 51 cases over 80,in comparison with the "true situation". At this point of the work, the "altimetric" situations and the "present" ones are approximately equivalent. The method used to restitute the estimated celerity field could be modified to better take into account the acoustic propagation properties.

VIII- Conclusion

The ATHENA experiment leads to the observation of mesoscale structures that are found to be present everywhere over the experiment domain; typical horizontal scales are 100 kilometers from crest to through and typical amplitudes are 20 cm.

GEOSAT altimeter data have been validate by comparison with in-situ data. A fairly good agreement between horizontal variations in dynamic height as computed from in-situ data and horizontal variations as observed by the Geosat altimeter is found.

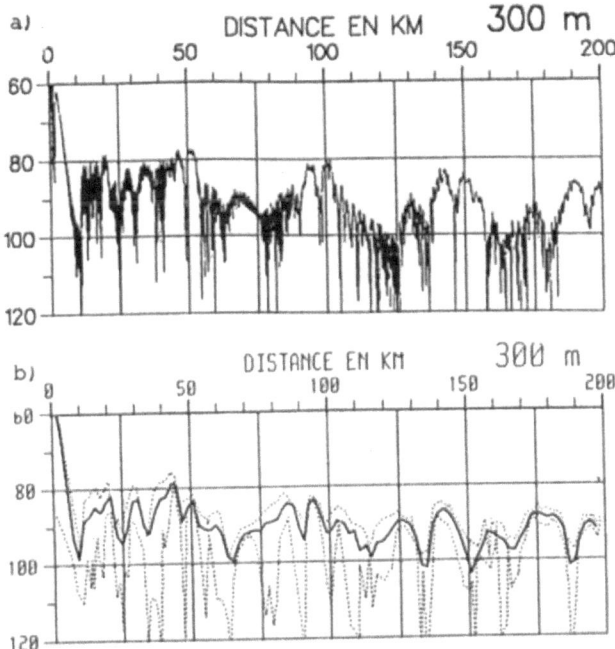

Fig 14: The outputs (a) of the PARQUES propagation model have been filtered with a Barlett filter over 2 Km (b).

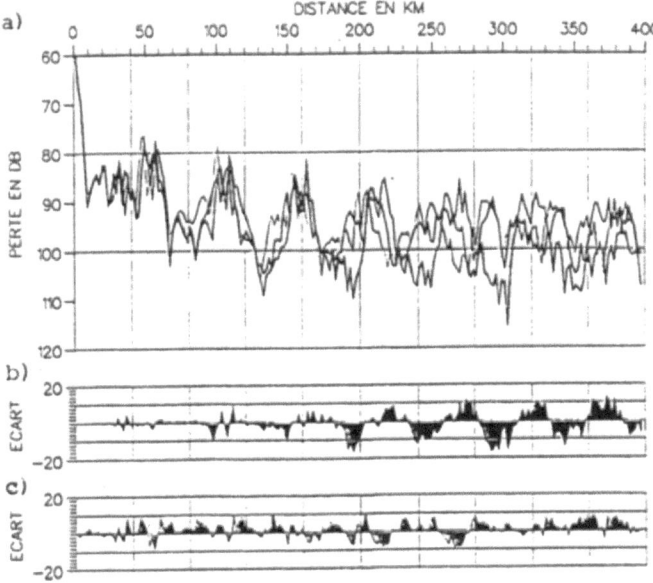

Fig 15: Propagation loss for the "true", "present" and "altimetric" situations on the SWNE array C (a) and the differences to the "true" situation for the "present" (b) and the "altimetric" (c). {Emitor depth: 100m, Receptor: 400m, f=50 Hz}.

The estimated celerity field using the first empirical mode to extend the surface information shows the same general structure as the actual celerity field.

The measurement of the impact of the mesoscale activity onto the 3D acoustic propagation may have a magnitude that is found to be as high as 18 dBs.

The representation with 2D propagation field allows us to assess qualitatively the impact of the oceanic variability, when the structure of the field is modified. The improvement due to the altimeter seems to be -on the cases we analysed- significant.

The "altimetric" situations or the "dynamic" ones are found to describe better the acoustic propagation than the "present" situations in 51 cases over 80, in comparison with the "true situation".

Nevertheless a small displacement of the convergence zone can sometimes yield such a difference even if the operational impact is not very important. However, the convergence zones generaly tends to vanish with the mesoscale activity.

At this point of the work, the "altimetric" situations and the "present" ones are approximately equivalent. The method used to restitute the estimated celerity field could be modified to better take into account the acoustic propagation properties; in particular one could combine both in-situ data and altimeter data to improve the efficiency of the sattellite information.

References

[1] Le Traon, P.-Y., M.-C. Rouquet, and C. Boissier, 1990: Space scales of mesoscale variability in the North Atlantic as deduced from GEOSAT data, accepted for publication in J. Geophys. Res.

[2] De Mey, P., A. F. Robinson, 1987: Assimilation of altimeter eddy fields in a limited area quasi-geostophic model, J. Phys. Oceanogr., 2280-2293.

[3] Le Squere, B. et C. Boissier, 1990: Estimation de la structure verticale de l'ocean à partir de l'altimetrie et d'une connaissance statistique de la variabilité in-situ, Note technique de l'EPSHOM, Brest.

[4] Tappert, The parabolic approximation method. Lecture notes in physics 70. Wave propagation and underwater acoustics. Springer Verlag.

[5] Grandvuillemin, Application de l'approximation parabolique à l'acoustique sous-marine. Thèse de doctorat de troisième cycle (1985).

A RANGE-DEPENDENT ANALYSIS OF ACOUSTIC TRANSMISSION ACROSS A COLD FILAMENT IN THE CALIFORNIA CURRENT

LCDR Lawrence M. Jendro USN
Office of Naval Research, European Office
London, England

Robert H. Bourke
Naval Postgraduate School
Monterey, CA

Steven R. Ramp
Naval Postgraduate School
Monterey, CA

ABSTRACT. CTD data were taken in an area where satellite imagery had detected a cold water filament to frequently recur in the California Current System in order to determine the temporal and spatial variability of the hydrographic and velocity fields. Sound speed profiles were constructed from this data and predicted sonar ranges (PSRs) were computed for passive sonar using a range-dependent parabolic equation model. Analysis of model results applied to tactical scenarios showed the acoustic advantage between two adversaries to change as their positions relative to the front and to each other were changed. An investigation of the acoustic mechanisms involved in the variations of PSRs showed that small variations in surface temperature were enough to cause significant changes in PSRs. Changes in temperature of sufficient magnitude to effect PSRs were found also in SST imagery of cold filaments in other eastern boundary currents around the world.

I. Introduction

Recent investigations, aided by satellite imagery, indicate that eastern boundary regions are far more dynamically active than previously thought, containing frequent mesoscale eddies and cold filaments (Huyer and Kosro, 1987; Flament et al., 1985). These filaments originate in the cold, upwelled, nutrient rich waters along the coast and, crossing the shelf, may extend hundreds of kilometers offshore (Brink, 1983). They present significant gradients of temperature and salinity at their boundaries and may encompass regions of high biological productivity (Lutjeharms, 1987). The filaments are visible from satellites using infrared (IR) imagery from the Advanced Very High Resolution Radiometer (AVHRR), which produces sea surface temperature images; or near-surface ocean color imagery from the Coastal Zone Color Scanner (CZCS) which produces chlorophyll images (Nyjkjaer, 1986). The filaments have been studied in the offshore regions of California, Peru, Portugal, Northwestern Africa, and the west coast of South Africa.

These filaments have been the subject of many investigations, but their forcing mechanisms are not yet well understood. An extensive study of cold filaments off Pt. Arena was conducted in the California Current System under the auspices of the Coastal Transition Zone (CTZ) program. The preliminary efforts of this group provided a new concept of the structure and kinematics of the flow in the coastal transition zone during the upwelling season. A continuously meandering upwelling jet is now envisioned. A segment of this jet, a "generic" offshore filament, can be identified as a strong baroclinic jet about 40 km wide, flowing offshore at a peak speed of at least 0.5 m/s, embedded in a field of cyclonic and anticyclonic eddies at a scale of order 100 km (the CTZ Group, 1989). The sampling plan for 1988 was designed to provide more detailed quasi-synoptic spatial and temporal coverage of roughly one meander of the coastal current than had previously been available.

Data acquired during the CTZ88 observation program were used in this investigation. These data were taken from northwest to southeast transect which crossed a cold filament extending offshore from Pt. Arena, Ca. An acoustic propagation model based on the parabolic equation (PE) was run on this data to determine what effect, if any, the cold thermal anomaly might have on the propagation of sound within its boundaries or in the surrounding waters.

The results indicate that the intrusion of the cool waters of the filaments into the warmer oceanic waters presents a moderate acoustic front. Substantial variation was found between the predicted sonar ranges

343

J. Potter and A. Warn-Varnas (eds.), Ocean Variability & Acoustic Propagation, 343–358.
© 1991 *Kluwer Academic Publishers.*

calculated when a 50, 200, or 800 Hz source was moved from station to station. An analysis of the mechanisms resulting in these variations, coupled with descriptions of filaments in other areas (such as Fiuza et al., 1989; Lutjeharms et al, 1987; and Brink, 1983), indicates that filaments of similar dimensions and with similar sea surface temperature gradients could possibly provide similar acoustic frontal features and, hence, be of tactical significance.

II. Data and Methodology

The observations for this study were taken between 6 and 12 July 1988, spanning a time period when historical satellite imagery of this area of the California Current off Point Arena subsequently showed the occurrence of a cold filament. Figure 1 shows the satellite sea surface temperature image from 9 July 1988 at 2317 Z, with the CTZ88 station numbers superimposed. Data used in this study were taken at the stations numbered 33 through 39, and offshore transect in deep ($\sim$3000 m) water that makes an approximately normal transect across the filament.

A. *Data Acquisition*

The conductivity, temperature, and depth (CTD) data were taken from the surface to 500 m using a Neil Brown Mark IIIB CTD. These data were used to determine the fields of temperature, salinity, and density and to construct their vertical profiles. Sound speed was calculated at 5 m increments using the Chen-Millero algorithm (Unesco, 1983).

B. *Sound Speed Profiles*

Sound speed profiles (SSPs) to 500 m were constructed from the CTZ88 CTD data. Each SSP had to be extended to a depth consistent with the bottom depth in the region, about 3000 m, to allow the deep sound transmission paths to be computed. This was accomplished by computing an average deep profile produced from 11 deep (3000 m) casts acquired in the area of this study during the OPTOMA 11 cruise (Reinecker et al., 1984). This deep SSP was then appended to the bottom of each of the SSPs calculated from the CTZ88 data.

C. *The Parabolic Equation Acoustic Model*

The parabolic equation model is a range-dependent model which determines the propagation of sound through a changing acoustic environment (Mellberg et al., 1987). It allows varying environmental conditions to be entered at positions along the path of sound propagation. The sound radiates along one line of direction away from the position of the source. The attenuation of acoustic energy is computed at closely spaced range increments ($\sim$0.1 km) at distances measured from the source. The distance between stations (SSPs) for the CTZ88 grid was 25 km. The PE model must consider two complex boundaries, the sea surface and the sea bottom. The sea surface was assumed to be smooth. The bottom was assumed to be fully absorbing. A constant depth of 3000 m was specified. The model allows variations of source depth, receiver depth, and source frequency.

A total of 252 model runs were performed with the input parameters varied. The model was run with the source depth at 5 m to simulate noise emanating from a surface ship and at 100 m to simulate a submarine-radiated noise source. Frequencies of 50, 200, and 800 Hz were used. The selection of receiver depths at 5, 100, and 200 m supported the ability of both platforms to listen at various depths. The model was run with sound projecting down the line of stations from north to south and vice-versa to determine any asymmetry in the propagation of sound across the filament. Model runs from each source depth, each receiver depth and each frequency were done at each station.

D. *Predicted Sonar Ranges*

By computing predicted sonar ranges (PSRs), the output of the PE model was organized in a way which allowed direct comparison between model runs and produced operationally meaningful results. A figure of merit (FOM), the maximum amount of transmission loss which can occur and still allow detection, of 80 dB was chosen. The range to this level of transmission loss, defined as the predicted sonar range (PSR), was determined graphically from the model output.

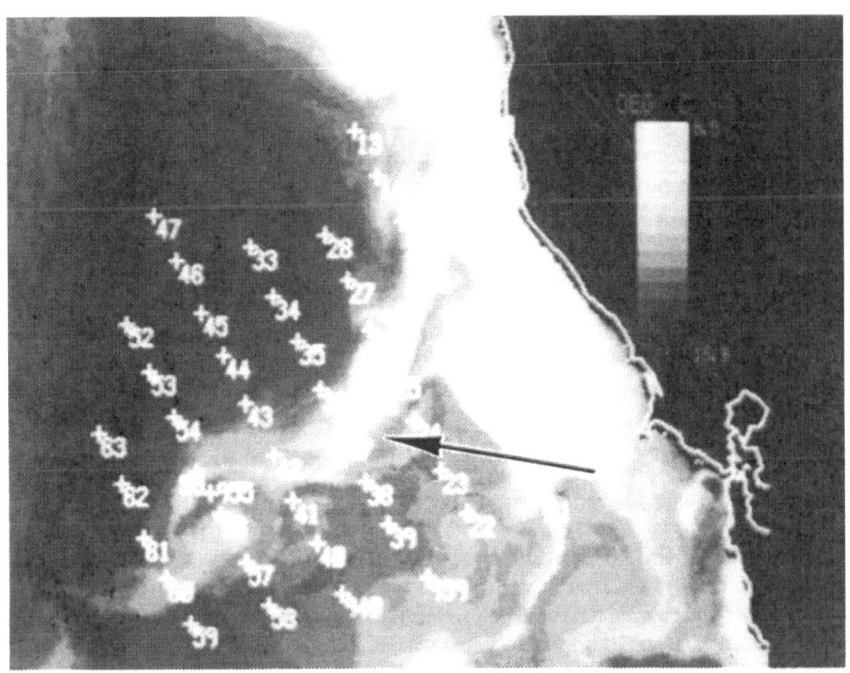

Figure 1.
An AVHRR Image of the cold filament investigated: This image was taken on 9 july 1988. The OTZ88 station numbers are superimposed. Stations 33, 34, 35, 36, 37, 38, 39 correspond to stations 133, 134, 135, 136, 137, 138, 139 in this report.

Table 1.

MIXED LAYER DEPTH AND LOW FREQUENCY CUTOFF AT EACH STATION

STATION ID	133	134	135	136	137	138	139
MLD (m)	43	71	71	31	41	26	41
F_{oo}(Hz)	709	335	335	1464	764	1508	761

III. Results

A. *Vertical Cross Sections*

Vertical cross sections of temperature, salinity, density anomaly and geostrophic velocity were constructed for the transect, with station 133, the northernmost station on the left and station 139, the southernmost station on the right (looking onshore).

The temperature cross section (Figure 2) shows that the core of the cold filament was located at station 137 as indicated by the surface temperature minimum of < 11 degrees C. The effect of the cold filament on temperature was limited to 75 km horizontally (station 135-138) and to the upper 250 m. The gradient in temperature from station 137 was slightly stronger toward the south (3 degrees C/25 km) than toward the north (2 degrees C/25 km). Also apparent is the elevation in the depth of the mixed layer when progressing from station 133 towards station 137, and its subsidence from station 137 toward station 139. An important feature appeared at station 135 at a depth of 75 m. Interleaving of the cold water from the filament with the warmer oceanic water occurred here at the margin of the filament. This situation was conducive to enhanced focusing of sound in the surface duct near this station.

B. *Sound Velocity Profiles*

Figure 3 illustrates sound speed profiles derived from the CTD data at stations 133 through 139. Significant variation appeared in the parameters determined from the various sound velocity profiles, the mixed layer depths (MLD)s, and the low frequency cutoff (F_{co}) values appear in Table 1. The existence of a significant mixed layer depth at some stations indicates that surface ducting is possible for source frequencies greater than the cutoff value. For the frequencies used in this study (50, 200 and 800 Hz), it is expected that only the 800 Hz signal will exhibit significant surface ducting. Substantial changes in critical depth were also observed.

All of the sound speed profiles were overlayed (Figure 3) to demonstrate the spatial variability of the SSPs. The SSPs fell into three groups, identified as A, B, and C. Group A consisted only of the SSP at station 137 which stood alone with the lowest surface sound speed. This occurred near the core of the cold water filament. Here the surface sound speed was approximately 18 m/sec slower than the highest surface sound speeds indicated on the Group C SSPs. At station 137 convergence zone transmission was expected at all source depths.

The stations in Group B included stations 136, 138, and 139. Stations 136 and 138 are in the filament margins where isotherms slope sharply upwards toward the surface resulting in a shallow mixed layer depth. The mixed layer depth at station 139 to the south was also shallower than those found to the north. The shallow mixed layer depth caused this group of stations to exhibit weak surface duct direct path transmission while the relatively warm temperatures provided CZ transmission for a 5 m source.

The SSPs to the north of the filament, identified as Group C, were stations 133, 134, and 135. These SSPs were similar in their relatively deep mixed layer and relatively high near-surface sound speeds. The deep mixed layer served to extend direct path ranges at 800 Hz while the high surface temperatures acted to drive down the critical depth and preclude CZ transmission for a 5 m source.

To summarize, the sound speed profiles for stations 133 - 139 clearly were affected by the presence of the cold filament, as indicated by variations in surface temperature, mixed layer depth, and critical depth. Between stations 136 and 137 critical depth changed by 650 m in 25 km, indicating that the cold filament in this study did present an acoustic front.

C. *Resulting Predicted Sonar Ranges*

In order to make the PSR data easy to interpret, the table data which best characterized the effects of the acoustic front were graphically displayed as plan-views which show the PSRs in the context of the geographical separation of the stations (Figures 4 and 5). The stations are labeled across the top and bottom of each figure. The solid circle below the station number indicates that the source originates at that station. Extending from each dot are arrows pointing in both horizontal directions. These arrows

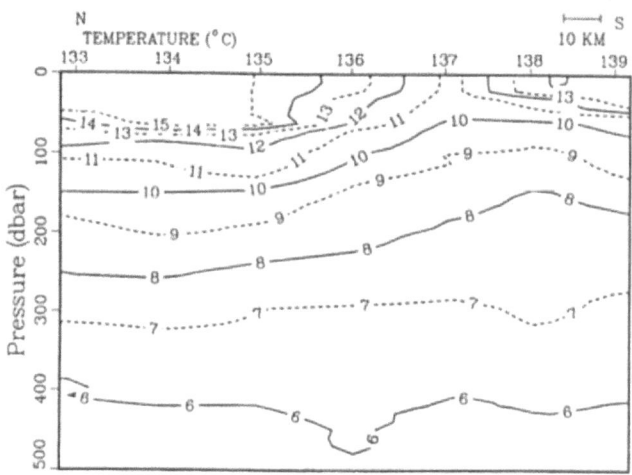

Figure 2.   The temperature cross section, from the northern station, 133, to the southern station, 139.

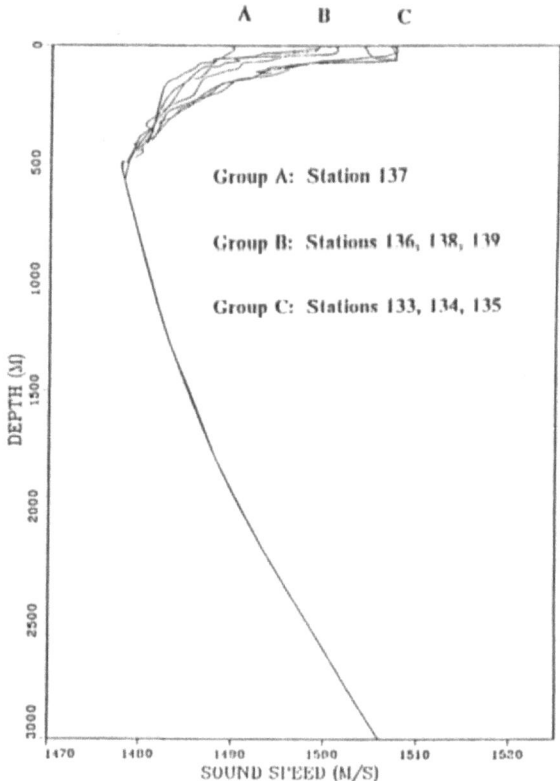

Figure 3.   An overlay plot of the SSPs taken at all the stations. The deep sound channel axis is at 575 m.

Predicted Sonar Ranges

Geographic Representation

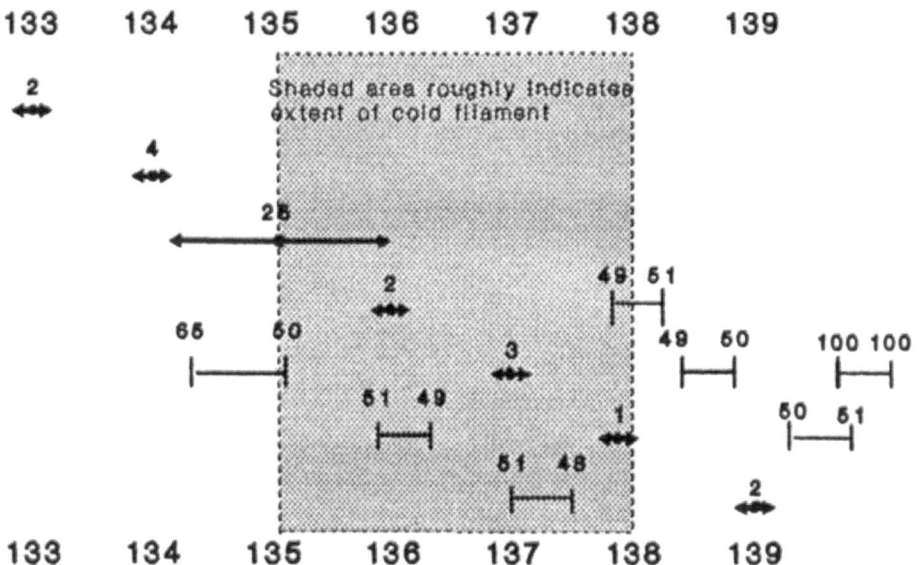

Figure 4.
   A geographic representation of the predioted sonar ranges: Arrows
indicate the direct path range projected north or south from the
source. The source position is centered between the arrowheads.
Bracketed values (I--I) indicate the range to the inner and outer
edge of the convergence zone (CZ) when present. Direct path and
convergence zone ranges (km) are indicated above each graphic
symbol.

Source: 5m, Receiver: 5m, 800 Hz

Predicted Sonar Ranges

Geographic Representation

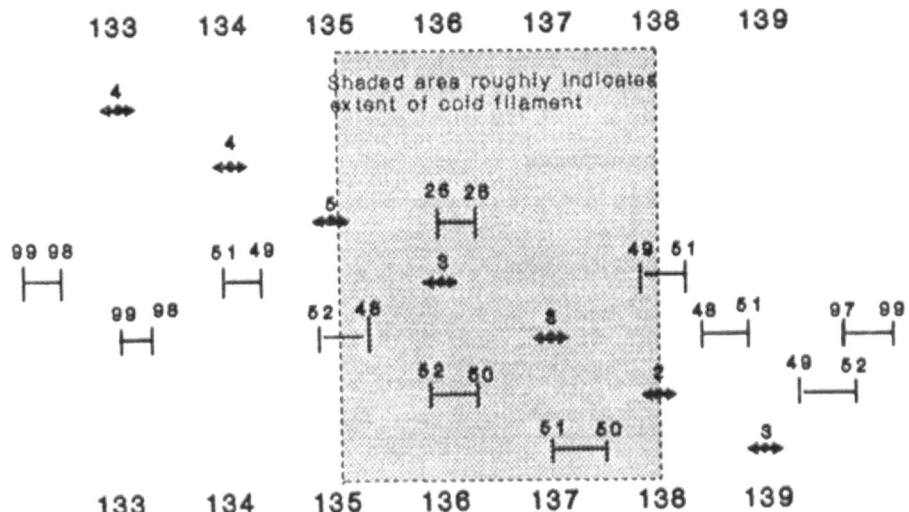

Figure 5.
   A geographic representation of the predicted sonar ranges: Arrows
   indicate the direct path range projected north or south from the
   source. The source position is centered between the arrowheads.
   Bracketed values (I--I) indicate the range of the inner and outer
   edge of the convergence zone (when present). Direct path and
   convergence zone (CZ) ranges (km) are indicated above each graphic
   symbol.

Source: 5m, Receiver: 100m, 800 Hz

represent the direct path ranges available at each station with the actual distance in kilometers indicated above each dot. The distance to the inner and outer edges of the convergence zones are indicated by vertical bars (with actual distances shown above) connected with horizontal dashed lines, i.e., the source can be heard (to the 80 dB FOM) within the bars.

1. Variation of PSRs with frequency

a. The longest direct path ranges from a shallow source occurred at the highest frequency, 800 Hz. This was because the surface mixed layer was sometimes very effective at trapping the 800 Hz signal.

b. When the source and receiver were at the same depth, convergence zones occurred most frequently at the lowest (50 Hz) frequency. This was an expression of lower spreading losses at lower frequencies (50 Hz).

2. Variation of PSRs with receiver depth

The longest ranges for shallow sources of all frequencies occurred at the deepest receiver depth. This was probably due to image interference at the surface.

3. Variation of PSRs with mixed layer depth

Longer direct path ranges occurred for shallow sources located at stations in the filament and to the north (Figures 4 and 5). This was due to the warm, salty intrusion at station 135 and the deeper mixed layer at the other stations which trapped more energy in the surface duct.

4. Variation of PSRs with surface temperature

Second convergence zone ranges were available for sources near the surface in the cold waters of the filament but not in the surrounding warmer waters (Figures 4 and 5). This was the result of the raising of the critical depth by a decrease in surface temperature to the point where fewer bottom losses occurred.

5. Variation in PSRs due to shallow SSP structure

The longest direct path surface duct ranges occurred at station 135 (Figure 4). The shape of the transmission loss curve from the source to 26 km (Figure 6) indicated that the surface ducting occurred due to the unique shape of the SSP at station 135. The 26 km range may be erroneous. This feature was not resolved by the 25 km station spacing.

6. Variations in PSRs showing acoustic asymmetry across the front

The CZ transmission toward the north from station 137 resulted in a convergence zone annulus 16 km wide (Figure 4). This was significantly wider than the width of the corresponding band at the station to the south, station 138, which had a convergence zone 3 km wide. The transmission loss curve for station 137 toward the north (Figure 7) showed evidence of surface ducting. The shape of the curve and its similarity to the direct path conditions at station 135 (Figure 6) suggested that the sound transmitted from station 137 reached the surface at station 135 via a convergence zone path where it was then trapped in the mixed layer and traveled via the surface duct.

The mechanism of expansion of the depth span of the convergence zone ray path was indicated in another asymmetric situation. When projecting southward across the front from station 133 (Figure 8), no convergence zone reception occurred for a shallow receiver at a distance of 50 km (near station 135) but a CZ did occur at 99 km.

In summary, the effect of the cold filament on PSRs is to cause substantial variation in sonar range capabilities, particularly at shallow depths, depending on the positions of both the source and the receiver relative to the cold water filament. These variations, while explained by several different mechanisms, are all due to variations in the SSPs in the upper 100 m brought about by the incursion of the cold water filament into the warmer oceanic waters.

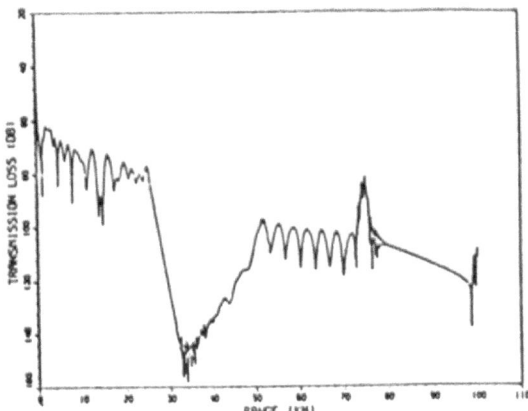

Figure 6.   The PE Acoustic Model run from station 135 southward to
station 139: The source depth was 5 m, the receiver depth
was 5 m, and the source frequency was 800 11z. The sharp
increase in transmission loss at 26 km indicated the abrupt
change in acoustic conditions at the transition between the
SSP at station 135 (with its surface duct) and the SSP at
station 136.

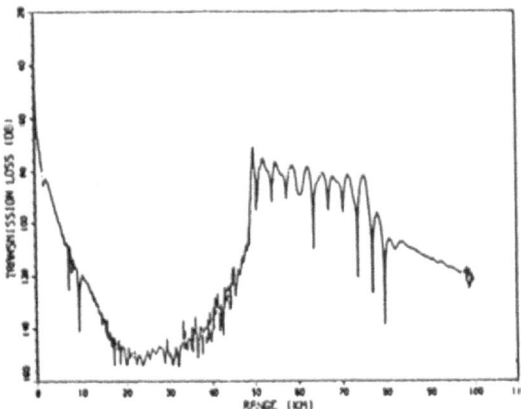

Figure 7.   The PE Acoustic Model run from station 137 to station 133:
The source depth was 5 m, the receiver depth was 5 m, and
the source frequency was 800 11z. The extended first
convergence zone was probably due to the trapping at the
surface of the first convergence signal by the surface
duct at station 135.

352

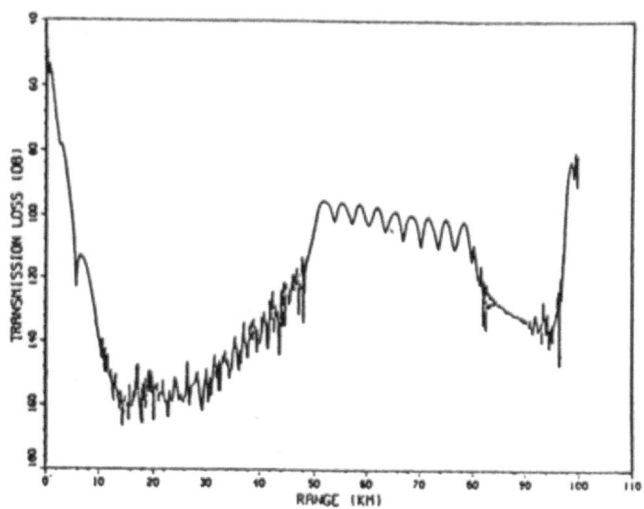

Figure 8.   The PE Acoustic Model run from station 133 to station 139:
            The source depth was 100 m, the receiver depth was 5 m, and
            the source frequency was 800 11z. The lack of a convergence
            zone at 50 km (while one is present at 100 km) was probably
            due a range dependent contraction of the depth span of the
            convergence rays in the vicinity of station 135 due to
            increased surface temperatures. The subsequent expansion
            of the depth span of the convergence rays in the vicinity
            of station 137 due to decreased surface temperatures
            promoted the convergence zone at 100 km.

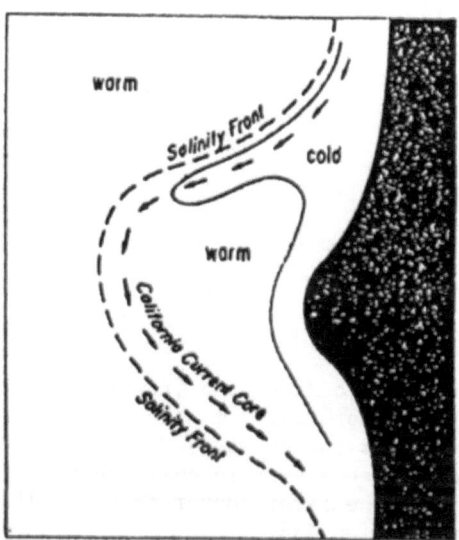

Figure 9.   A schematic diagram of an offshore cold filament. (From the
            CTZ Group, 1988)

IV. Discussion

A. *Oceanography*

The California current system is the most intensely studied eastern boundary region in the world. The basic state consists of an equatorward mean flow driven by wind stress over the eastern North Pacific. Coastal upwelling, driven by equatorward wind stress and the resulting offshore Ekman transport of surface waters (Huyer et al, 1987), is a seasonal feature, occurring between the months of May and October. This upwelling creates an upwelling front and a geostrophically balanced equatorward current jet flows along the front (Brink, 1983). Recent observations (CTZ group, 1988) suggests that this equatorward jet begins to meander at some point, which allows the cold water to extend far offshore within the meander (Figure 9) in the form of a cold filament. As it travels farther offshore the water in the jet is warmed by insolation and mixing with warmer surrounding waters and gradually loses its identity. Additionally, there is some evidence of subsidence of the cold salty water along the upwelling front as the filament moves offshore (The CTZ Group, 1988). The satellite observation of cold filaments of similar scale in other eastern boundary regions suggests that equatorward wind stress, coastal upwelling, and inshore coastal geometry and bottom topography are all essential ingredients necessary for their formation. More detailed dynamical comparisons must await better in-situ data sets from other parts of the world.

B. *Acoustics*

1. The Nature of the Front

The overwhelming influence of temperature on the speed of sound makes the acoustic front strongly dependent on horizontal variations in temperature. The temperature cross-section (Figure 2) indicates a surface temperature gradient of 1.6 degrees C/25 km between the cold filament in this study and the warmer water to the north. This probably represents an underestimate since the 25 km station spacing was not adequate to resolve the maximum gradients. Other studies using a continuous-sampling thermosalinograph (Snow, 1988) have demonstrated maximum gradients of 1.6 degrees C/km. The vertical extent of the thermal front in Figure 2 was relatively shallow, less than 300 m.

2. The Tactical Significance of Variations in PSRs

To emphasize the importance of this variation, three tactical scenarios were constructed and the PSRs used to determine the acoustic advantage of various relative positions across this cold filament. For all scenarios a surface ship and a submarine were considered, each with an FOM of 80 dB and each listening for an 800 Hz signal. The surface ship as a source was constrained to the surface but could optionally lower a hydrophone to listen at 100 m. The submarine was required to operate and listen at 100 m.

In the first scenario (Figure 10), the surface ship was positioned at station 134, listening at 100 m, with the submarine at station 136. In these positions, the surface ship's direct path signal could be detected at 4 km and there were no convergence zone paths. The submarine's signal, however, did produce CZ ranges which allowed initial detection to occur in the second CZ at 99 km with another convergence zone at 52 km. The direct path signal could be detected by the surface ship at 4 km. The acoustic advantage in this scenario went to the surface ship.

Next we placed the surface ship at station 137, again listening at 100 m, and moved the submarine to station 135 (Figure 11). The surface ship's signal could be heard via direct path to 4 km, with convergence zones at 51 and 99 km. The submarine's signal was now detectable at 4 km direct path, with convergence zones at 52 and 100 km. Both had essentially equal acoustic capabilities. Neither vessel had an acoustic advantage.

The final scenario placed both the surface ship and the submarine near station 135 (Figure 12). The direct path signal of the surface ship extended to 26 km. The submarine can be detected via direct path to 4 km. There are no convergence zone ranges listed because the distance between the adversaries was

Tactical Scenario #1

Geographic Representation

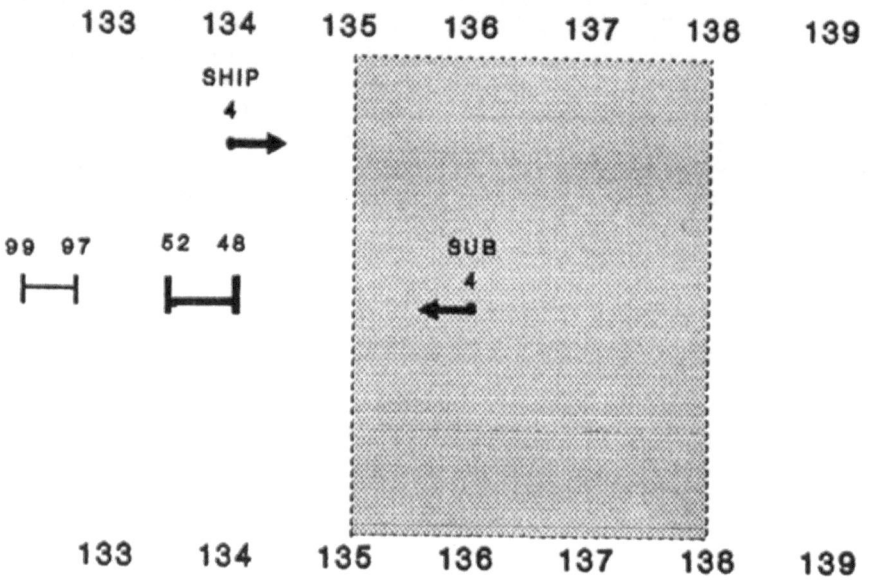

Figure 10.
Tactical Scenario 1: A surface ship at station 134 at 5 m with a hydrophone at 100m and a submarine at station 136 at 100 m. Arrows indicate the direct path range projected from the source towards the adversary. Source position is identified by a solid circle. Bracketed values (I--I) indicate the distance (km) to the inner and outer edge of a convergence zone (when present).

Advantage: Surface Ship

Tactical Scenario #2

Geographic Representation

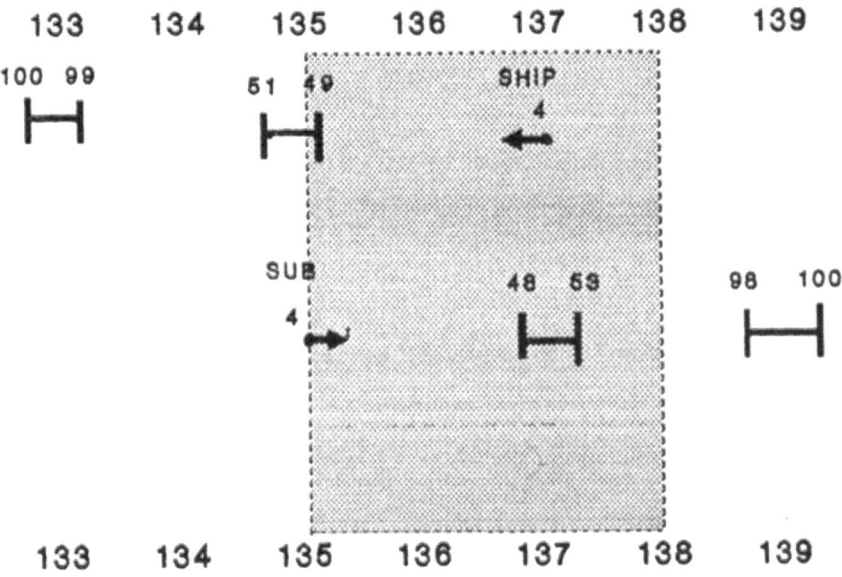

Figure 11.
Tactical Scenario #2: A surface ship at station 137 at 5m with a hydrophone at 100m and a submarine at station 135 at 100 m. Arrows indicate the direct path range projected from the source towards the adversary. Source position is identified by a solid circle. Bracketed values (I--I) indicate the distance (km) to the inner and outer edge of a convergence zone (when present).

No Acoustic Advantage

Tactical Scenario #3

Geographic Representation

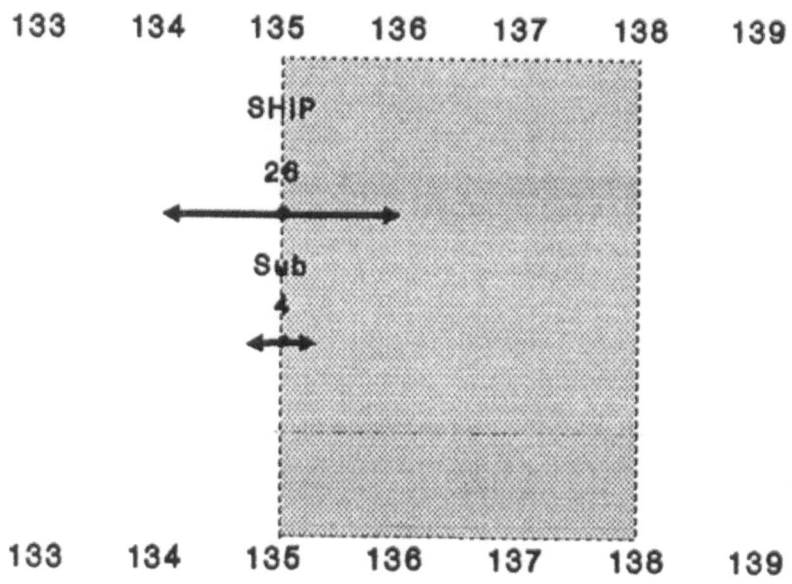

Figure 12.
    Tactical Scenario # 3: A surface ship at station 135 at 5m with a
    hydrophone at 100m and a submarine at station 135 at 100m. Arrows
    indicate the direct path range projected from the source towards the
    adversary. Source position is identified by a solid circle. Bracketed
    values (I--I) indicate the distance (km) to the inner and outer edge
    of a convergence zone (when present).

Advantage: Submarine

less than the range to the first convergence zone. In this case the submarine had the acoustic advantage (the degree of the advantage is not accurately described by the surface ship's direct path range of 26 km for reasons stated earlier).

The significance of these scenarios is that this cold water filament produced an acoustic front sufficient to cause a change in acoustic advantage based on relative position with respect to the front. For this cold water filament a surface ship steaming offshore would be far less detectable if it stayed to the north of the cold zone, rather than transversing within it.

3. Applicability to Other Eastern Boundary Regions

Substantial variation in predicted PSRs occurred across this filament in the California Current System. This filament has a greater surface temperature change across it (5 degrees C) than some filaments observed (based on satellite SST) in other parts of the world. Perhaps the cold filaments in other regions do not represent enough of a temperature gradient to produce a substantial variation in sonar ranges. Actually, not much of a temperature difference is needed. The sea surface temperature differences between station 134 and 135 was less that 1 degree C/25 km, yet there was a substantial increase in predicted sonar range at station 135 because of the structure of the shallow SSP which occurred at the margin of the filament. A change in sea surface temperature of 1 degree C/25 km at the margin is commonly observed for other filaments measured by AVHRR imagery, but their vertical structure and degree of interleaving is largely unknown. The shoaling of the critical depth which permitted convergence zone transmission from near-surface sources was a function only of temperature at the surface. A temperature gradient of 1.1 degrees C/25 km was enough to create the difference seen in having convergence zone transmission at station 136 and not having it at station 134. The cold filaments in other regions discussed above had temperature gradients which exceeded 1.1 degrees C/25 km, and could have provided similar PSR variations.

V. Conclusions

Cold filaments recur off prominent coastal features in strategically important coastal areas in oceanic eastern boundary regions around the world. They are detectable by satellite using either infrared or color imagery from either the AVHRR or CZCS instruments.

The cold filament investigated in this study presented an acoustic front in the upper 300 m of the water column which was sufficient to change the acoustic advantage between two adversaries operating in that depth band depending on their positions relative to each other and to the acoustic front.

None of the acoustic mechanisms involved required strong horizontal temperature gradients. Temperature differences of close to 1 degree C/25 km were sufficient to determine whether or not CZ transmission occurred. All of the filaments reviewed in this investigation had such gradients. This implies that similar acoustic fronts, which are sufficient to effect a tactical advantage, could be presented by cold filaments wherever they are detected by satellite imagery.

List of References

Brink, K. H., 1983: The Near Surface Dynamics of Coastal Upwelling. Progress in Oceanography, 12, 223-257.

CTZ Group, 1988: The Coastal Transition Zone Program. EOS, Transactions of the American Geophysical Union, July 5,1988, 698-707.

Fiuza, A.F.G. and F.M.Souza, 1989: Preliminary Results of a CTD Survey in the Coastal Transition Zone off Portugal During 1 - 9 September 1988. Coastal Transition Zone Newsletter, 4, 1, 2-10.

Flament,P., L. Armi and L. Washburn. 1985: The Evolving Structure of an Upwelling Filament. Journal of Geophysical Research, 90, 11, 765-778.

Huyer, A. and P. M. Kosro, 1987: Mesoscale Surveys over the Shelf and Slope in the Upwelling Region Near Pt. Arena, California. Progress in Oceanography, 12, C2, 1655-1681.

Lutjeharms, J. R. E. and P. L. Stockton, 1987: Kinematics of the Upwelling Front off Southern Africa. The South Africa Journal of Marine Science, 5, 35-49.

Mellberg, L.E., and O. M. Johannessen. D. N. Conners, G. Botseas, and D. Browning. 1987: Modeled Acoustic Propagation Through and Ice Edge Eddy in the East Greenland Sea Marginal Ice Zone. Journal of Geophysical Research, 92, C7, 6857-6868.

Nyjkjaer, L. L. Van Camp and N. Hojerslev, 1986: Remote Sensing of the Northwest African Upwelling Area. II, University of Copenhagen Press, Copenhagen, N. Denmark.

Snow, R.L., 1988: Sea Surface Temperature and Salinity Structure in Cold Upwelling Filaments Near Point Arena as Observed Using Continuous Underway Sampling Systems. M.Sc. Thesis, Naval Postgraduate School, Monterey California.

ACOUSTIC EFFECTS OF THE ICELAND-FAEROE FRONT

FINN B. JENSEN, GIANCARLO DREINI and MARK PRIOR [1]
SACLANT Undersea Research Centre
19026 La Spezia
Italy

ABSTRACT. The effect of a front on long-range sound propagation in the ocean is investigated theoretically. A numerical model of the parabolic equation type is used for simulating propagation through a real front observed on the Iceland-Faeroe Ridge. This front separating warm Atlantic water from cold Arctic water has horizontal sound-speed changes of 30 m/s over a range of 50 km. Oceanographic data (CTD, XBT, thermistor chain) collected in the frontal area over the last decade are used as input to the acoustic model. Three different transects of the front are considered, each corresponding to a propagation path of 150 km length. The acoustic effects of the front are in all cases found to be significant (> 10 dB), but with strong dependence on environmental parameters as well as on source/receiver depth and frequency.

1. Introduction

Fronts and eddies are mesoscale oceanographic features which separate or enclose water masses of different origin, and therefore of different temperature and salinity. They are the oceanic equivalent of weather systems in the earth's atmosphere, and are observed in most ocean areas [1]. They are, however, much more persistent than their atmospheric counterparts, with time scales of the order of months or even years. Some of the stronger fronts are easily identified on infrared satellite images of the sea surface, with temperature changes of as much as 5–10 °C over horizontal distances of 5–50 km. However, even the strongest fronts have horizontal sound-speed gradients that are one order of magnitude smaller than the sound-speed gradient in depth due to pressure effects alone (∼16 m/s per km).

While front and eddy structures have been studied extensively in the oceanographic community, their effect on long-range sound propagation in the ocean is much less explored. However, some interesting modelling results were reported by Lawrence [2] in 1983 concerning propagation across the edge of a warm-core eddy in the Tasman Sea. He observed a drastic change in propagation conditions as ducted sound near the surface in the eddy was converted into convergence zone (CZ) propagation outside the eddy. The same effect was observed by Akulichev [3] in experi-

---

[1]Currently at Admiralty Research Establishment, Portland, UK

J. Potter and A. Warn-Varnas (eds.), Ocean Variability & Acoustic Propagation, 359–374.

mental data collected recently in the Northwest Pacific across the Kuroshio current. The simulation studies by Heathershaw. et al. [4,5] addressed propagation through simplified frontal structures (no topographical effects), again noting the possibility of propagation changing from surface ducting to CZ-type propagation across a front. Finally, Mellberg et al. [6] modelled the time variability of convergence-zone propagation through the Gulf Stream and its eddies, observing that within a few days the changing ocean structure can cause significant changes in both the location and the sound levels of convergence zones. Thus, the first CZ was seen to move as much as 10 km in range with level changes of up to 5 dB. Again, bottom effects were ignored in the study.

The present investigation is an attempt to include the full environmental complexity (including topography) in the modelling of propagation through the Iceland-Faeroe front. With the front being located in less than 1000 m of water, we expect propagation to be heavily influenced by the sea floor and, hence, we should observe propagation characteristics quite different from those seen in previous simulation studies on deep-water fronts.

2. The Ocean Environment

The geographical area of interest is the southern part of the Norwegian Sea just north of the Iceland-Faeroe Ridge (Fig. 1). Here we find a permanent frontal feature separating warm Atlantic water to the south from colder Arctic water to the north. The oceanographic conditions in this mixing region are extremely complicated and highly variable in both space and time. However, the front has been observed at regular intervals over the past 20 years, always being located 100–200 km north of the Iceland-Faeroe Ridge [7].

In selecting oceanographic data sets for the acoustic studies, three issues were considered important: 1) to have complete data available (CTD rather than XBT) and with good range and depth coverage for computing the two-dimensional sound-speed structure needed as input to the acoustic models; 2) to have data available along tracks with different bathymetry in order to study bottom effects on propagation; 3) to have a high-resolution data set available (thermistor chain) to investigate the issue of the minimum horizontal profile sampling required.

Three data sets satisfying one or more of the above criteria were selected (Fig.1). Collectively these data represent some typical environmental conditions found in the Iceland-Faeroe frontal region. Transect ARE-86 represents thermistor chain data interspersed with XBT casts collected in June 1986 by the Admiralty Research Establishment [8]. The bottom slope is gentle with depths varying between 500 and 1000 m over the 150 km track. Transect FA-80 represents CTD casts collected by the US Naval Oceanographic Office in October 1980 [9]. Here the bottom slope is steeper with depths varying between 500 and 2500 m. Finally, transect GIN-89 represents CTD data collected by SACLANTCEN in June 1989 [10]. This track starts north of the Faeroe Islands in a water depth of just 200 m but with the bottom sloping steeply down to 3000 m.

Additional environmental information was obtained from Podeszwa [11], who provides representative (monthly) mean profiles for both the Atlantic water to the south of the front and the Arctic water to the north of the front. Since the historical pro-

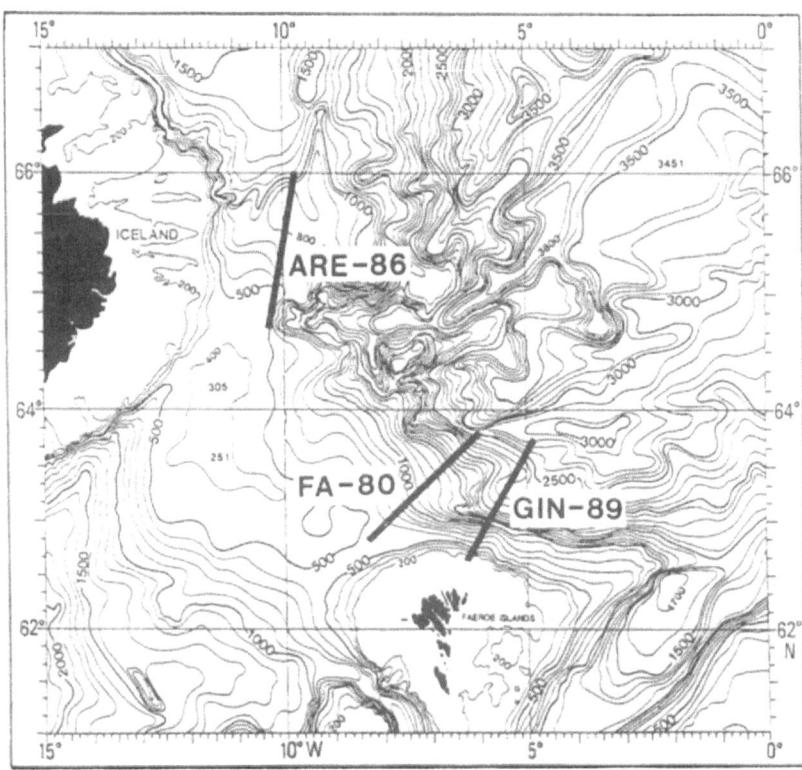

Figure 1. Three experimental tracks in the Iceland-Faeroe frontal region. Oceanographic information was collected in October 1980 along track FA-80, in June 1986 along track ARE-86, and in June 1989 along track GIN-89.

files were found to bracket the actually measured profiles in the mixing region, they were used to represent extreme situations in the acoustic studies, i.e. propagation entirely within warm Atlantic water or entirely within cold Arctic water.

3. Acoustic Modelling Results

The acoustic simulations were performed with a well-tested numerical propagation model of the parabolic equation type [12,13]. This model treats propagation in a 2-D geometry with arbitrary variations of sound speed with depth and range. Moreover, the model treats bottom effects realistically allowing for both sound speed and density contrasts at the seafloor. The assumption of a 2-D propagation geometry could be a problem considering the environmental complexity of the test area. We feel, however, that a meaningful qualitative assessment of environmental factors influencing propagation predictions in this frontal area can be carried out with a 2-D acoustic model. An assesment of 3-D effects not only requires much more environmental information (plus ground-truth acoustic data), but also requires considerably more computational effort for the simulations.

Environmental inputs (sound-speed profiles, bathymetry) were interpolated linearly in range. The seabed was assumed to be homogeneous with the following geoacoustic parameters: $c=1600$ m/s, $\rho=1.7$ g/cm^3 and $\beta=0.5$ dB/λ. Both sea surface and sea floor were assumed to be smooth. Propagation loss going from shallow into deep water was calculated for several source/receiver combinations and for source frequencies between 25 and 400 Hz. To facilitate the comparison of predicted acoustic mean levels for different environmental situations, the rapidly-varying multipath interference structure was removed from propagation-loss curves by performing a spatial averaging over a 3 km range window.

In addition to propagation-loss calculations done by the parabolic-equation code, a series of ray plots [14] were generated to illustrate qualitative features of propagation through the front.

3.1. TRACK FA-80

A series of six sound-speed profiles were obtained along the 150 km track, where bottom depths vary between 500 and 2500 m (Fig. 2a). The front separating warm Atlantic water from cold Arctic water is located at a range of approximately 70 km, its position being determined by a maximum in the horizontal sound-speed gradient at the surface. The composite plot of sound-speed profiles in Fig. 2b shows that we have sound speed changes of nearly 30 m/s between profiles 1 and 2 at a depth of 500 m. Shown here are also representative mean profiles for the month of October for both Atlantic and Arctic waters [11]. An alternative representation of the sound-speed variation through the front is given in Fig. 2c. Note the change in thermocline depth from around 500 m on the ridge to less than 100 m at longer ranges.

A qualitative depiction of the propagation conditions along track FA-80 for a source depth of 100 m is given by the ray diagrams in Fig. 3. The upper graph is based on the full frontal information while the other graphs are generated using either the Atlantic mean profile (Fig. 3b) or the Arctic mean profile (Fig. 3c) over the entire track. It is evident that the presence of the front affects both surface-duct

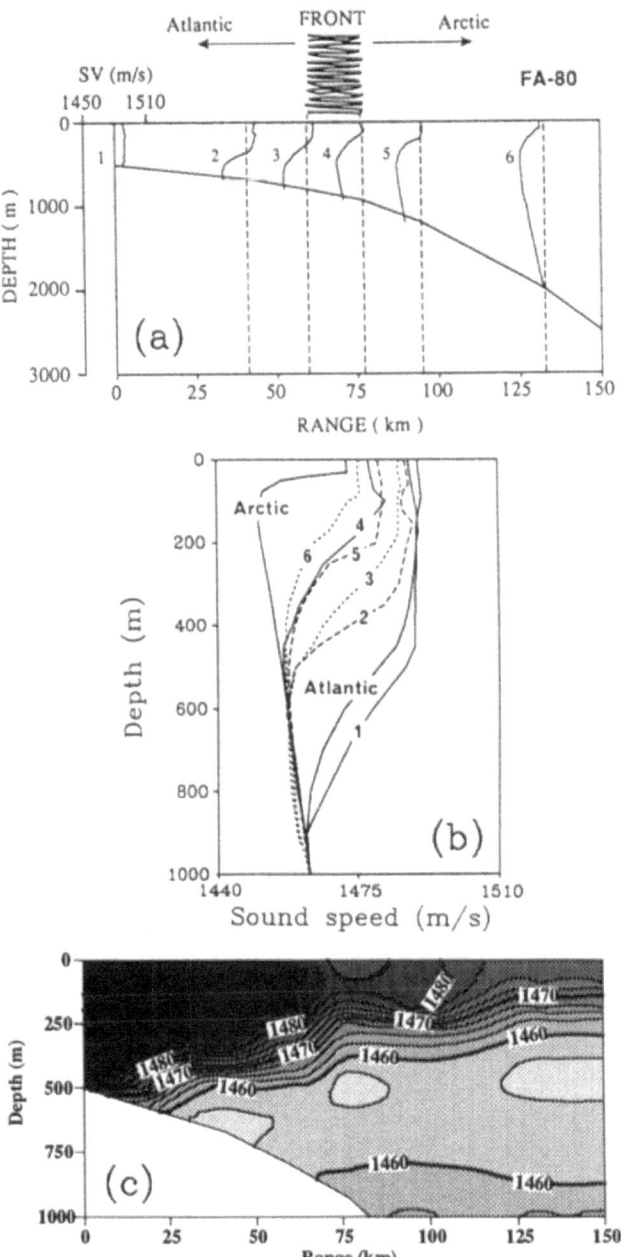

Figure 2. Measured sound-speed profiles along track FA-80. (a) Profile positions and bathymetry; (b) Comparison of measured profiles with representative mean profiles for both Atlantic and Arctic water for the month of October; (c) Contoured sound-speed field in the upper 1000 m of the water column.

364

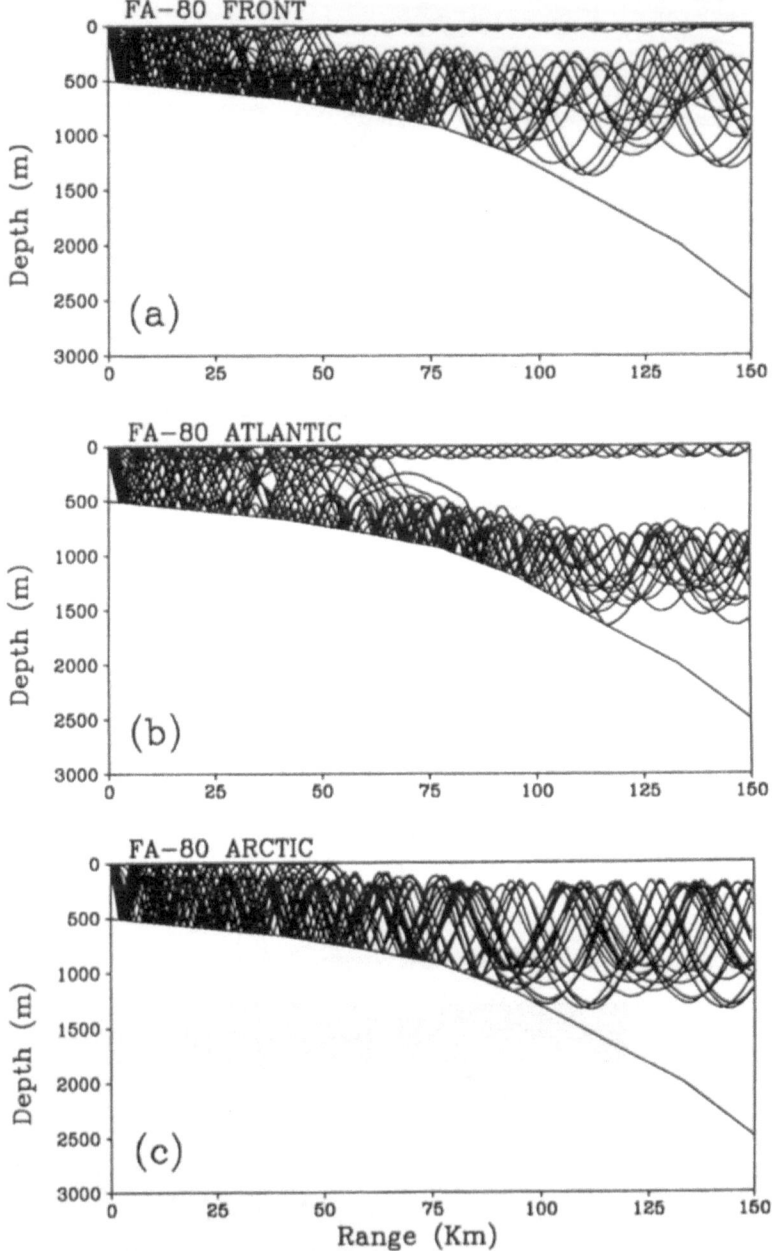

Figure 3. Ray diagrams showing qualitative differences in propagation conditions along track FA-80 for different oceanographic conditions. (a) Full frontal information (6 profiles); (b) Single Atlantic profile; (c) Single Arctic profile. The source depth is in all cases 100 m.

and deep sound-channel propagation. In particular, the front causes sound to be channeled at a much shallower depth than is the case for the Atlantic profile. We also note that there are no convergence-zone paths in this bottom-limited environment.

Examples of computed propagation loss at 400 Hz for two different receiver depths are given in Fig. 4. It is clear that the full frontal information is required for accurate loss predictions at both receiver depths. Note that propagation conditions are quite stable over the first half of the propagation track, independent of the profile chosen. Beyond 75 km we have large level differences, particularly for the shallow receiver (15 m). We note that the dependence of the results on profile and receiver depth is quite complicated, with the Atlantic profile giving too low a loss for the shallow receiver and too high a loss for the deep receiver.

In summary, if we ignore the presence of the front and just use a single profile (historical or measured) in a propagation prediction, then level errors of 10–20 dB can occur at 400 Hz. The effect of the front at lower frequencies is less important, but it is still significant at 100 Hz, as shown in Fig. 5.

The two important issues of bottom effects on propagation and horizontal sampling requirements for accurate acoustic predictions in frontal regions can both be addressed to some extent with the present data set. First we remove bottom effects by moving the seafloor down to 3000 m depth over the entire track. As seen in Fig. 6 this completely changes the propagation situation. Thus the ray diagram (Fig. 6a) now shows a distinct convergence-zone pattern which was absent in the bottom-limited case (Fig. 3a). Moreover, we note that propagation-loss differences for the three environmental situations considered (Fig. 6b) are much smaller than observed in the bottom-limited case (Fig. 4a). This result clearly shows that bottom effects are important in this situation and that meaningful acoustic predictions cannot be made without taking into account the detailed bathymetry and reflectivity of the seafloor.

The second issue, dealing with the required horizontal profile sampling, is addressed in Fig. 7. We again consider propagation at 400 Hz for a source at 100 m depth and a receiver at 15 m. Two propagation-loss curves are shown: The full line was obtained by linear interpolation between the six measured profiles across the front to generate a profile update every 500 m along the propagation track. This is considered our reference solution. The dashed line was obtained by linear interpolation between just the two end-profiles, which leads to propagation-loss errors of nearly 10 dB at longer ranges. Hence, it appears that a horizontal profile sampling every 25 km is required for accurate propagation predictions at 400 Hz. More evidence in support of this conclusion will be presented in connection with analysis of the thermistor chain data (ARE-86).

3.2. TRACK GIN-89

Seven CTD profiles were taken in June 1989 along this 150 km track, which is positioned to the east of the FA-80 track analyzed above. The bottom depth varies between 200 and 3000 m (Fig. 8a). The composite plot of sound-speed profiles in Fig. 8b shows that we have sound speed changes of nearly 25 m/s between profiles 3 and 5 at depths between 200 and 400 m. Also note that the mean profiles for the month of June for Atlantic and Arctic waters bracket the measured profiles across the front. The sound-speed contours in Fig. 8c again show a characteristic change

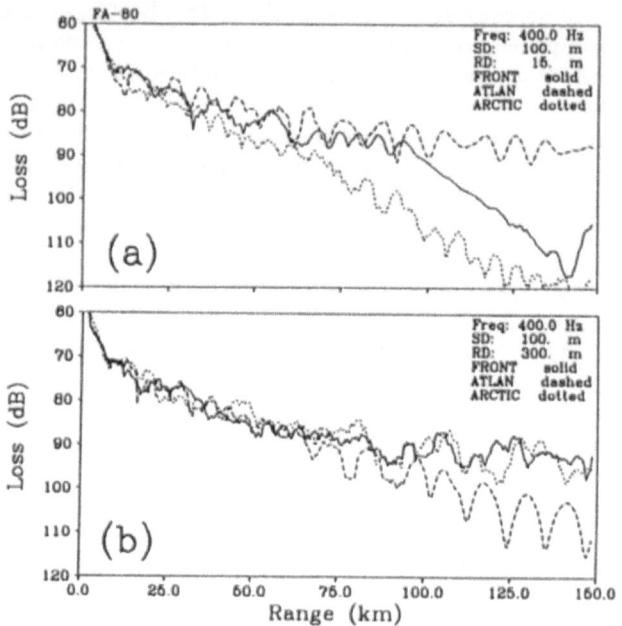

Figure 4. Predicted propagation losses at 400 Hz for a source at 100 m depth and receivers at (a) 15 m and (b) 300 m. The three curves in each figure refer to: full frontal information (solid line), single Atlantic profile (dashed line), and single Arctic profile (dotted line).

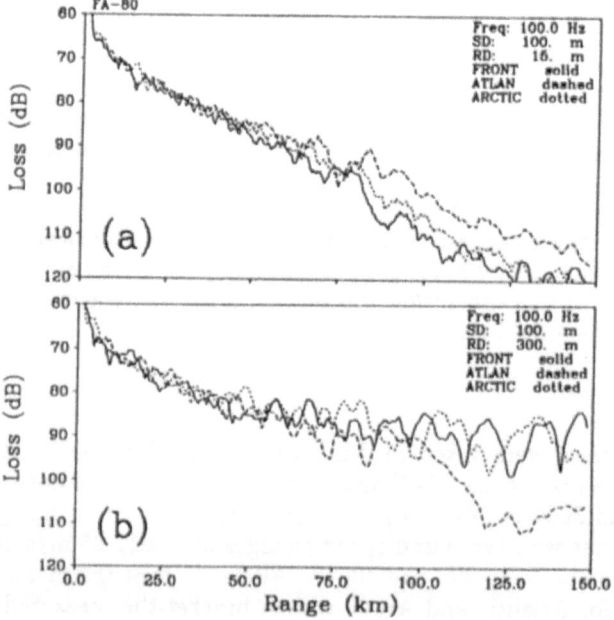

Figure 5. Predicted propagation losses at 100 Hz for similar situations as shown in Fig. 4.

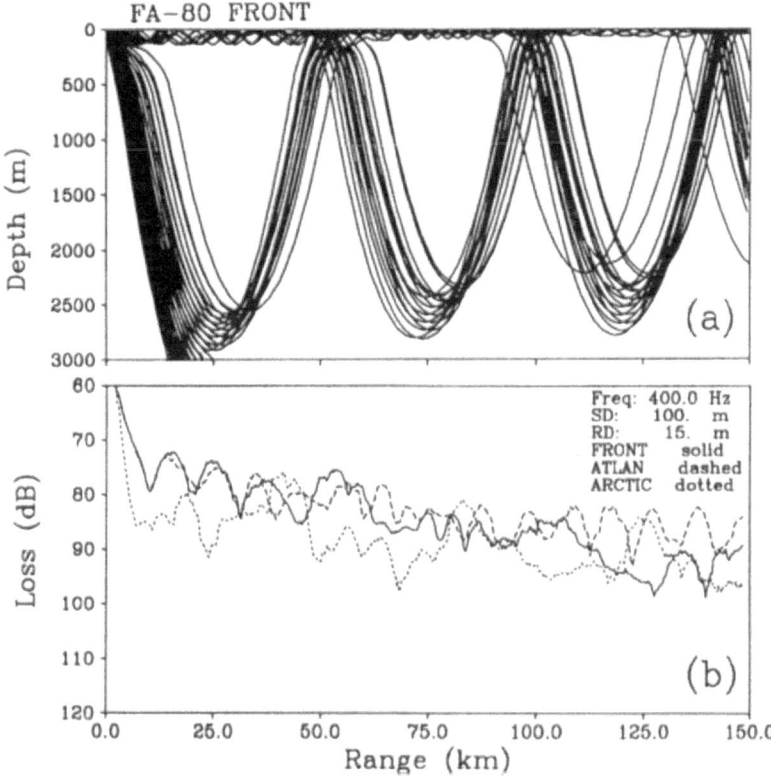

Figure 6. Predicted propagation losses along track FA-80 when bottom effects are neglected. (a) Ray diagram showing both convergence-zone and surface-duct propagation from a source at 100 m depth; (b) Propagation losses for the three oceanographic conditions described in Fig. 4.

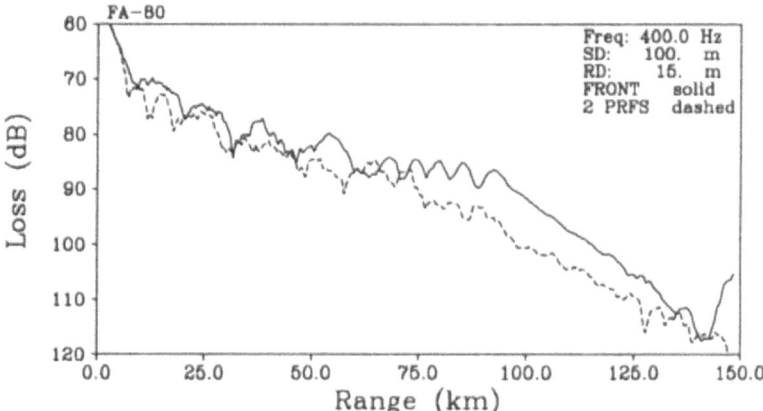

Figure 7. Sensitivity of propagation predictions to lateral profile sampling. Both curves involve profile updates every 0.5 km in range obtained by linear interpolation between either the 6 measured profiles (solid line) or the two end-profiles (dashed line).

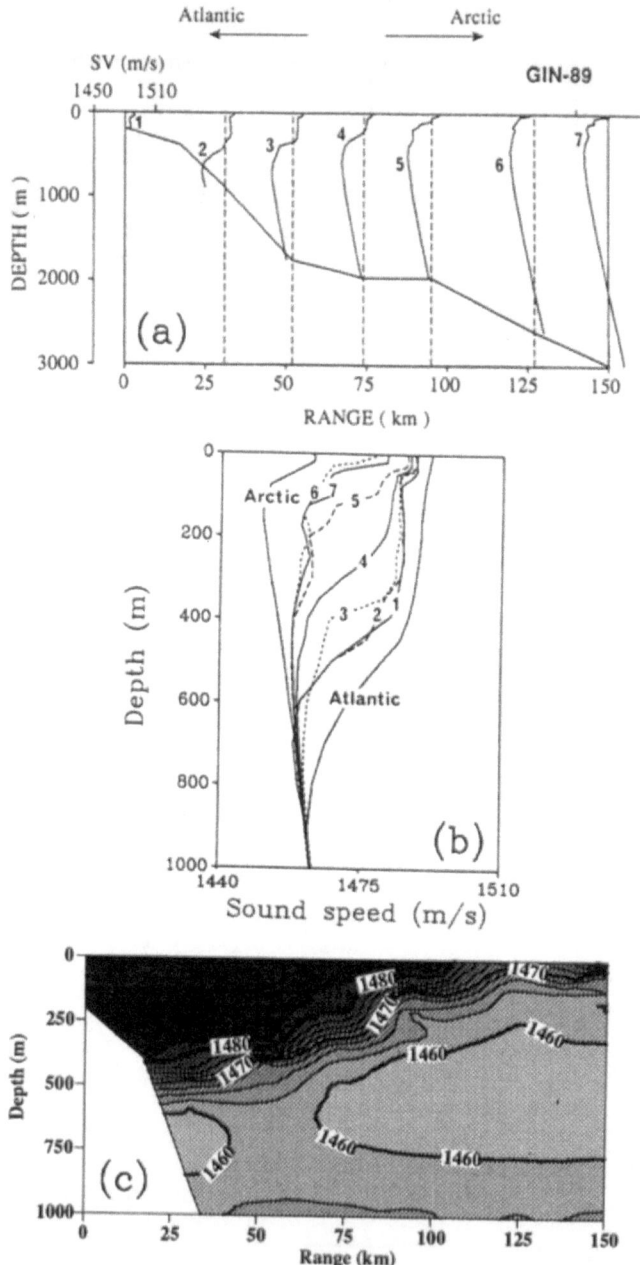

Figure 8. Measured sound-speed profiles along track GIN-89. (a) Profile positions and bathymetry; (b) Comparison of measured profiles with representative mean profiles for both Atlantic and Arctic water for the month of June; (c) Contoured sound-speed field in the upper 1000 m of the water column.

in thermocline depth with range, starting deep (≈ 500 m) on the ridge and moving towards the surface at longer ranges.

The three ray pictures in Fig. 9 illustrate the change in propagation conditions with changing environmental conditions. The upper graph is based on the full frontal information while the other graphs are generated using either the Atlantic mean profile (Fig. 9b) or the Arctic mean profile (Fig. 9c) over the entire track. It is evident that significant changes in the acoustic pattern are caused by the front in connection with the particular bathymetry encountered along the track.

Examples of computed propagation loss at 400 Hz for three different receiver depths are given in Fig. 10. We again note that the full frontal information is required for accurate loss predictions. As before the dependence of the results on profile and receiver depth is seen to be complicated, with the Atlantic profile giving too low a loss for the 100 m receiver and too high a loss for the 300 m receiver. However, this data set confirms that if we ignore the front and use just a single profile (historical or measured) in a propagation prediction, then level errors of 10–20 dB can occur at 400 Hz. We again find that the effect of the front at lower frequencies is less important.

3.3. TRACK ARE-86

This data set was obtained with a towed thermistor chain covering the upper 300 m of the water column. In addition several XBT casts were done in order to provide information on the temperature structure in the lower part of the water column. A set of computed sound-speed profiles with a 1 km spacing is displayed in Fig. 11a. A composite plot of sub-sampled profiles every 25 km given in Fig. 11b again shows cross-frontal sound speed changes of approximately 25 m/s at depths between 100 and 300 m. The sound-speed contours in Fig. 11c show an even more rapid change in thermocline depth with range than observed in the other data sets.

The issue of minimum required profile sampling in range can be thoroughly addressed with this high-resolution data set. In fact, we have computed propagation loss for both a 1 km profile spacing and a 25 km profile spacing, with linear interpolation in range in both cases to provide the acoustic model with profile updates every 500 m. Propagation loss results at 400 Hz for two different receiver depths are given in Fig. 12. Here the heavy line is obtained with the dense 1-km profile sampling. We note that there is little change in computed propagation loss whether using the dense profile set (1 km sampling) or the sparse set (25 km sampling), indicating that a 25 km sampling is sufficient at 400 Hz. This conclusion was confirmed by loss calculations at several other source/receiver depths.

We again note that a 7-profile frontal structure (25 km spacing) is required for accurate loss predictions at both receiver depths. However, the potential prediction errors are here somewhat smaller than found for the previous data sets. This effect can be shown to be due to a different bathymetry along this track. Thus if we substitute the current bathymetry with that of track FA-80, we again obtain large prediction errors of 10–20 dB (Fig. 13). This, in turn, means that the observed strong site dependence of propagation loss across the Iceland-Faeroe front is due, not so much to the variability in water column properties, but primarily to changes in bottom topography along the different tracks.

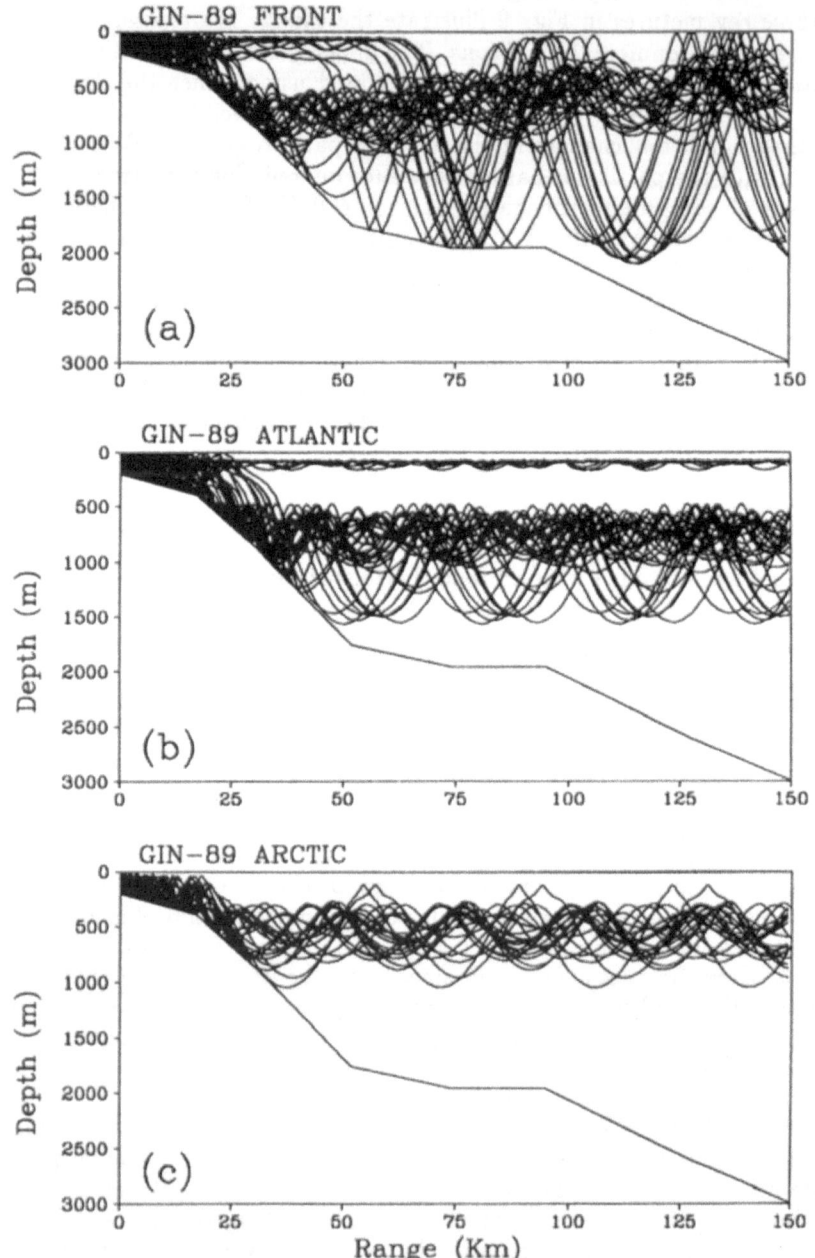

Figure 9. Ray diagrams showing qualitative differences in propagation conditions along track GIN-89 for different oceanographic conditions. (a) Full frontal information (7 profiles); (b) Single Atlantic profile; (c) Single Arctic profile. The source depth is in all cases 100 m.

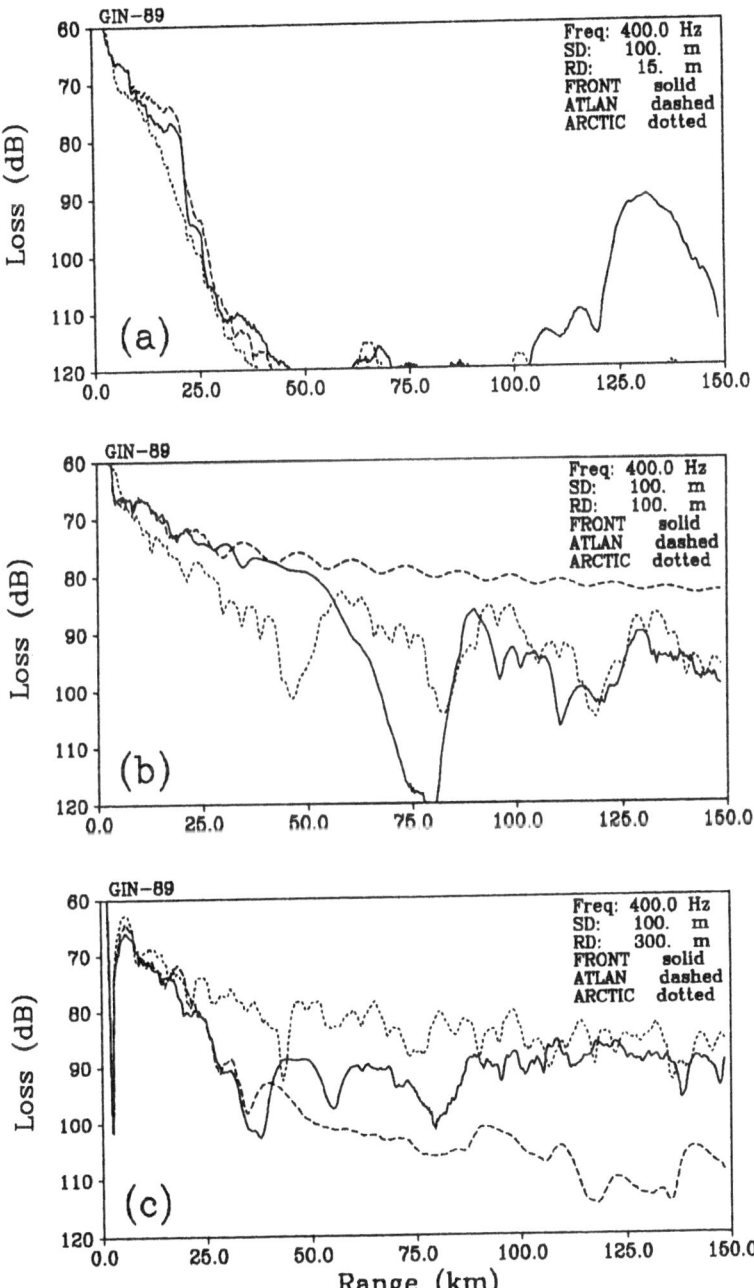

Figure 10. Predicted propagation losses at 400 Hz for a source at 100 m depth and receivers at (a) 15 m, (b) 100 m, and (c) 300 m. The three curves in each figure refer to: full frontal information (solid line), single Atlantic profile (dashed line), and single Arctic profile (dotted line).

372

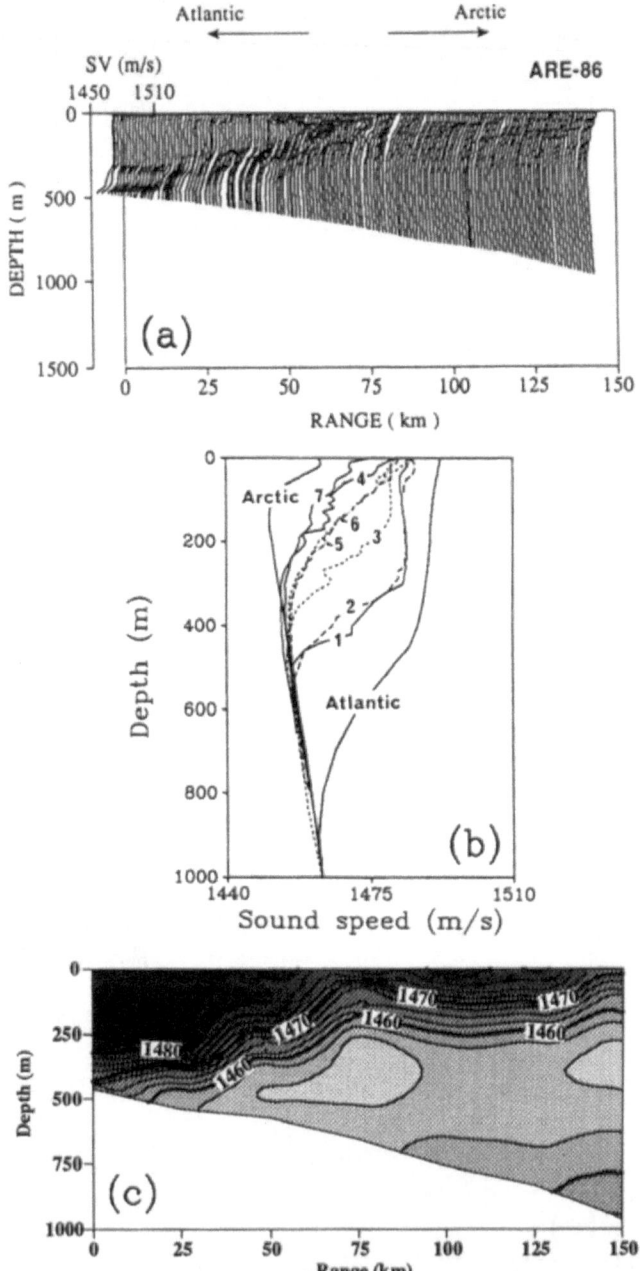

Figure 11. Measured sound-speed profiles (thermistor chain) along track ARE-86. (a) Display of profile structure sampled every 1 km; (b) Comparison of measured profiles every 25 km with representative mean profiles for both Atlantic and Arctic water for the month of June; (c) Contoured sound-speed field in the upper 1000 m of the water column.

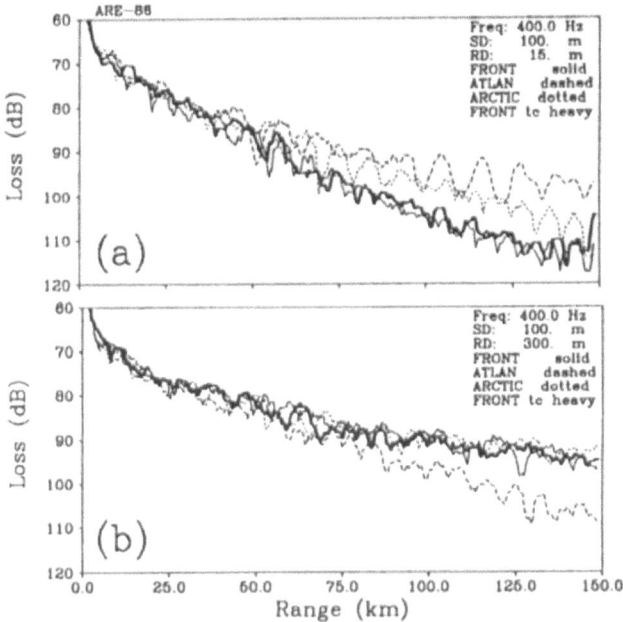

Figure 12. Predicted propagation losses at 400 Hz for a source at 100 m depth and receivers at (a) 15 m and (b) 300 m. The four curves in each figure refer to: full frontal information sampled every 1 km (heavy line), frontal information sampled every 25 km (solid line), single Atlantic profile (dashed line), and single Arctic profile (dotted line).

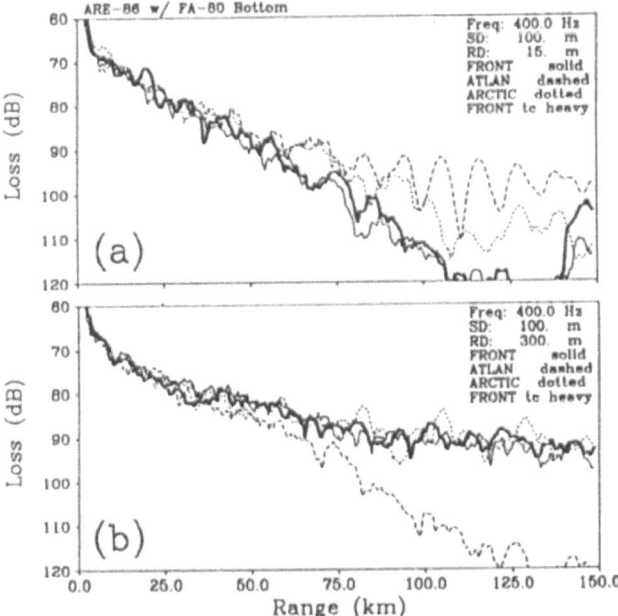

Figure 13. Same information as displayed in Fig. 12, but with bathymetry from track FA-80.

4. Summary and Conclusions

It is clear from this study that the acoustic effect of the Iceland-Faeroe front can indeed be significant (10–20 dB at 400 Hz), but with the effect being strongly dependent upon 1) the bottom topography in the frontal region, 2) the lateral sound speed variation across the front, 3) the depth of source and receiver, and 4) the source frequency. Moreover, we find that the required profile sampling in range for accurate propagation predictions is around 25 km for frequencies up to 400 Hz.

References

[1] Johannessen, O.M. (1975) "A review of oceanic fronts," in W. Bachmann and R.B. Williams (eds.), Oceanic Acoustic Modelling, Rep. CP-17, pt. 5, SACLANT Undersea Research Centre, La Spezia, Italy, pp. 28-1 to 28-33.

[2] Lawrence, M.W. (1983) "Modeling of acoustic propagation across warm-core eddies," J. Acoust. Soc. Amer. **73**, 474–485.

[3] Akulichev, V.A. (1989) "The study of large-scale ocean water inhomogeneities by acoustic methods," in Proceedings of 13th International Congress of Acoustics, Vol. 5, Sava Centar, Belgrade, Yugoslavia, pp. 117–128.

[4] Heathershaw, A.D., Maskell, S.J., Cooper, W. and Hillman, R.C. (1988) "Studies of sound propagation through a front using an eddy resolving ocean model," Rep. TM-88144, Admiralty Research Establishment, Portland, U.K.

[5] Heathershaw, A.D. and Hillman, R.C. (1989) "Sound propagation through ocean eddies: Some experiments using simple sound speed parameterizations and range dependent acoustic models," Rep. TM-89101, Admiralty Research Establishment, Portland, U.K.

[6] Mellberg, L.E., Robinson, A.R. and Botseas, G. (1990) "Modeled time variability of acoustic propagation through a Gulf Stream meander and eddies," J. Acoust. Soc. Amer. **87**, 1044–1054.

[7] Hopkins, T.S. (1988) "The GIN Sea: Review of physical oceanography and literature from 1972," Rep. SR-124, SACLANT Undersea Research Centre, La Spezia, Italy.

[8] Scott, J.C. (1990), personal communication.

[9] Teague, W.J. (1981) "CTD profiles in the Northeast Atlantic and Norwegian Sea," U.S. Naval Oceanographic Office, Stennis Space Center, MS.

[10] Perkins, H. (1990), personal communication.

[11] Podeszwa, E.M. (1979) "Sound speed profiles for the Norwegian Sea," Rep. TD-6035, Naval Underwater Systems Center, New London, CT.

[12] Jensen, F.B. (1988) "Wave theory modeling: A convenient approach to CW and pulse propagation modeling in low-frequency acoustics," IEEE J. Oceanic Engn. **13**, 186–197.

[13] Jensen, F.B. and Martinelli, M.G. (1985) "The SACLANTCEN parabolic equation model (PAREQ)," SACLANT Undersea Research Centre, La Spezia, Italy.

[14] Cornyn, J.J. (1973) "GRASS: A digital-computer ray-tracing and transmission loss prediction system," Rep. 7621, Naval Research Laboratory, Washington, DC.

DEEP HYDROGRAPHIC FLUCTUATIONS IN THE NORTH-EAST ATLANTIC MEDITERRANEAN OUTFLOW : INFLUENCE ON ACOUSTIC PROPAGATION

J.M. DARRAS - R. LAVAL
Société AERO - Paris
F.R. MARTIN-LAUZER
EPSHOM - Brest

ABSTRACT

Some general characteristics of the mesoscale variability of the deep sound velocity profile in North-East Atlantic, due to the intrusion of Mediterranean Water through the Straight of Gibraltar are described, together with its influence on sound propagation.

In the Canarian basin the effect takes the form of isolated lenses (MEDDIES) of about 100 km diameter. In an area off the Portugal Coast and up to 60° N the Mediterranean Water extends as a deep layer, progressively extinguishing as going North and West.

In all cases, the presence of Mediterranean Water introduces a local maximum at about 1 000 m depth on the sound velocity profile. Recent measurement of the SHOM indicates that the value of sound velocity at the deep local maximum does present strong spatial oscillations.

From the acoustic point of view, the presence of a deep maximum divides the principal sound propagation channel into two new subchannels. It is shown that the principal characteristics of sound propagation in such a variable medium can be explained by the exchange of acoustic energy between the various channels resulting from the variations of the value of sound velocity at the maximum. Knowledge of the variations of this sound velocity value on a grid scale of about 25 km would be adequate for good deterministic predictions of sound propagation.

Some general rules are formulated to describe the effect of interchannel energy transfers on sound propagation as a function of the source and receiver depth and their horizontal position with respect to the "fronts".

Several examples of sound field computations are given.

Introduction

A realistic prediction of low frequency sound propagation waves at long range has to take into account the mesoscale variations of the sound velocity profile. This is particularly the case in North-East Atlantic where the intrusion of Mediterranean Water creates a situation of high variability of the deep sound velocity profile.

A number of problems have to be resolved in order to develop coherent methods of acoustic predictions in this type of environment.

1.- Define the type of oceanographic description which is "necessary and sufficient" to carry out reasonably good predictions of the sound propagation characteristics.

2.- Dispose of simple and fast range-varying propagation models which can be easily used to execute a sufficient number of sound field computations corresponding to different source location and different axes of observation.

3.- Develop a good understanding of the general qualitative properties of sound propagation in a variable medium, in order to be able to extract practical conclusions from a

375

J. Potter and A. Warn-Varnas (eds.), Ocean Variability & Acoustic Propagation, 375–390.

combination of "a priori knowledge" and a reasonable number of carefully selected sound field computations.

4.- Be able to combine a deterministic and a probabilistic description of the main propagation characteristics when a precise deterministic description of the oceanic variations is not available.

We have tried to give a beginning of reply to these various questions in the course of a study concerning the oceanic and acoustic mesoscale variability in North East Atlantic executed for the "Etablissement Principal des Services Hydrographiques et Océlanographiques de la Marine" (EPSHOM).

1. Oceanic description

The Mediterranean Water issued from the Gibraltar straight is extending over a very large area in the North East Atlantic.

Being warmer and more saline than the original Atlantic water, it founds a density equilibrium around a 1 000 m depth, thus creating a deep maximum on the sound velocity profile.

For a number of causes which are not yet completely understood (presence of currents with turbulent edges, bottom relief, pulsations of the outflow jet...) the diffusion of the Mediterranean water in the Atlantic is far from being a continuous process with a regular decrease when going away from the straight. The mixing of the two types of waters, on the opposite, generates a situation of very high spatial and temporal variability of the sound velocity distribution.

We have considered two different areas in the North East Atlantic where the Mediterranean Water extension takes completely different forms.

The South West Area, corresponding to the Canarian Basin (fig. 1 a) is characterised by the presence of compact and isolated "lenses"which are anticyclonic eddies of Mediterranean Water (also called MEDDIES) surrounded by pure Atlantic water. The characteristics of these lenses have been described by several authors [réf. 1 to 10].

A specimen observed by ARMI and ZENK is shown on fig. 1.d. The MEDDIES are typically centered between 1 000 and 1 100 m depth, with a vertical extension around 900 m. Horizontally they have a circular shape with a diameter of 100 km. The sound velocity at their center is 12,5 m/s higher than the surrounding waters at the same depth. The total lifetime of a MEDDY seems to be at least two years. During the first year, the horizontal and vertical sizes are progressively reduced, but the sound velocity at the center is not affected. During the second year the sound velocity at the center as well as the size are decreasing but they are keeping a coherent shape.

The density of MEDDIES in the Canarian Basin is about one per 300 000 km^2. The probability to find one between a source and a receiver at 250 km distance from each other is about 0,08.

The second area is located at the North and West of the Gibraltar straight. Here, the Mediterranean water extends as a progressively evanescent and strongly oscillating layer, still centered around a 1 000 m depth. The spatial oscillations of this layer have been revealed by the hydrographic measurements conducted by EPSHOM [réf. 11]. The oscillations are essentially affecting the value of the deep maximum velocity, the mean value and the amplitude modulation of this maximum decreasing progressively toward North and West.

The effect has been observed up to a 60° latitude point (North of Scotland). Two sound velocity sections illustrating the above properties are presented in fig. 1.a and 1.b. It can be seen that the depth of the maximum sound velocity is regularly decreasing when going North : 1 300 m at the exit of the Cadix Gulf less than 1 000 m beyond 50° latitude North. The sound velocity at the maximum (the mesoscale fluctuations being filtered) is also

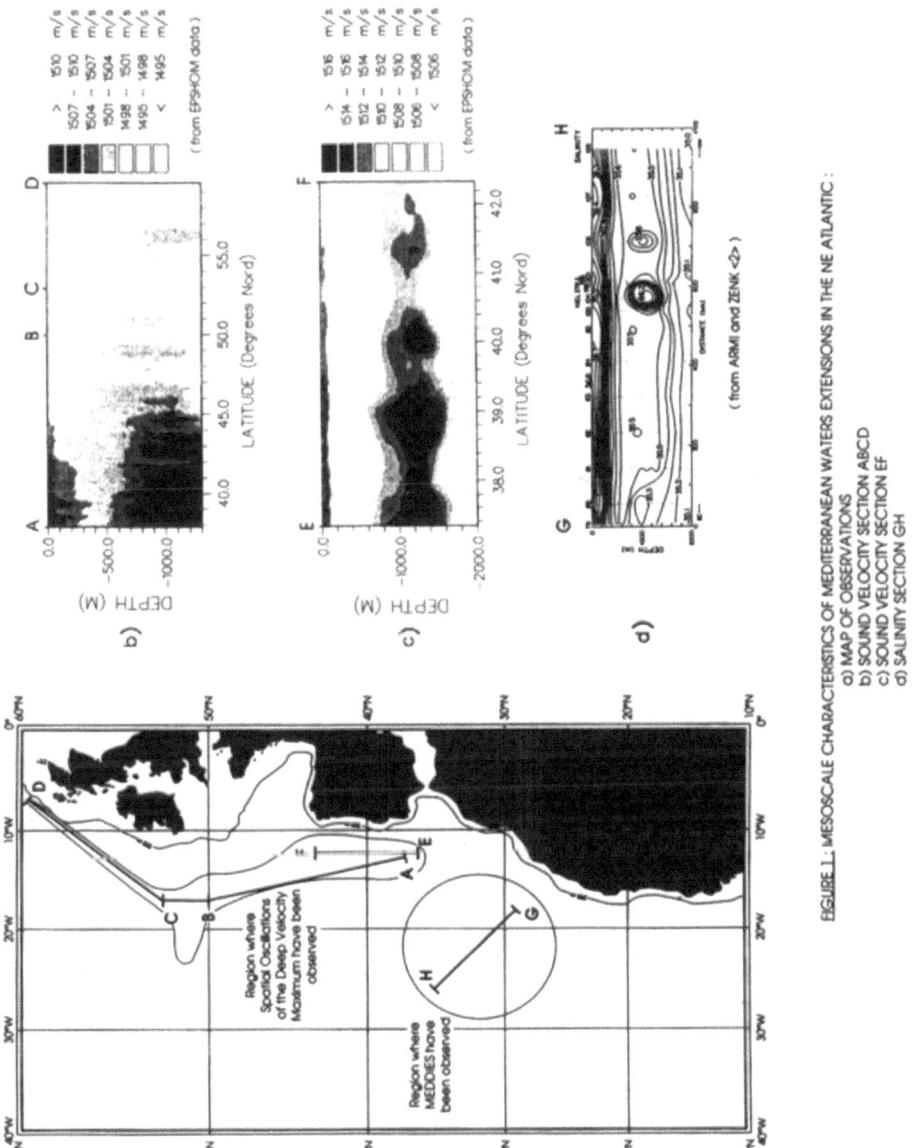

b) _{DEPTH (M)}

(from EPSHOM data)

	1510	m/s
>	1510	m/s
1507 –	1510	m/s
1504 –	1507	m/s
1501 –	1504	m/s
1498 –	1501	m/s
1495 –	1498	m/s
<	1495	m/s

c)

(from EPSHOM data)

	1516	m/s
>	1516	m/s
1514 –	1516	m/s
1512 –	1514	m/s
1510 –	1512	m/s
1508 –	1510	m/s
1506 –	1508	m/s
<	1506	m/s

d)

(from ARMI and ZENK <2>)

FIGURE 1 : MESOSCALE CHARACTERISTICS OF MEDITERRANEAN WATERS EXTENSIONS IN THE NE ATLANTIC :

a) MAP OF OBSERVATIONS
b) SOUND VELOCITY SECTION ABCD
c) SOUND VELOCITY SECTION EF
d) SALINITY SECTION GH

decreasing. From the hydrological surveys which are shown, the average slopes going North are about :
- 0,4 m/km for the depth of the maximum and - 0,01 m/s/km for the sound velocity between 38 ° N and 42° N. The peak to peak amplitude of the mesoscale sound velocity fluctuations of the maximum is about 6 m/s, with a spatial period between 100 and 200 km.

At a higher latitude, or more West the amplitude of the fluctuations is probably lower. The spatial period in the East-West direction has only be measured at 50° N where its average value is about 400 km.

The geographical position of maximum and minimum values of the deep maximum sound velocity seems to be stable over a period of one year.

2. Influence of the Mediterranean outflow on acoustic propagation

In the Canarian Basin, in absence of Mediterranean Water Lenses, the sound velocity profile has no deep maximum, then offering a single sound propagation channel extending over the whole water depth. Let's call it the "Principal channel" (PC). In such a channel, the sound rays issued from a shallow source are focalized on a succession of Convergence Zones (CZ) near the surface at ranges regularly spaced of 60 km. (fig. 2.a)

The presence of Mediterranean Water introduces a deep maximum which divides the principal channel into two sub-channels : An Upper Channel (UC), above the maximum, and a Lower Channel (LC) below, the principal channel being also excited as long as the sound velocity at the bottom is higher than the one at the deep maximum.

If the sound velocity profile is invariant with range, the subchannels extend over constant depth intervals, between the depth of maximum velocity and the high or low "image depths" where the sound velocity is equal to the sound velocity of the deep maximum (svm). (If the sound velocity at the surface is lower than the s.v.m. the U.C. begins at the surface).

As long as the source depth is above the upper image depth of the deep maximum, in other terms as long as the sound velocity at the source (svs) is higher than the sound velocity at the maximum (svm), only the principal channel (PC) is excited (fig. 2.b). The only effect of Mediterranean Water is to defocus the convergence zones, which are broader as compared to the pure Atlantic water case.

If the source is in the UC (which means that svs < svm) all the acoustic energy which is transmitted by the source within the vertical angular domain comprised within $\theta = 0$ and

$$\theta = \cos^{-1}\left(\frac{svs}{svm}\right)$$ is trapped inside the UC. The energy transmitted within the angular do-

main $\theta > \cos^{-1}\left(\frac{svs}{svm}\right)$ remaining in the PC (fig. 2c)

The excitation of the UC first increases with the source depth and reaches its maximum value when the source is in the axis of the channel (fig. 2d). It would then decrease to become zero when the source is at the depth of maximum velocity. If the source depth is near the edge of the UC channel a succession of convergence and shadow zones is observed, the spatial periodicity for this channel being about 40 km. If it is at the channel axis or in its vicinity the channel insonification is practically continuous as a function of range, with a maximum at (or near) the channel axis and a progressive decrease when going from the axis to the edges of the channel (fig. 2.d).

If the source is below the s.v.m. depth, it is the LC which will be excited. Only one of the two subchannels can be excited by a single source.

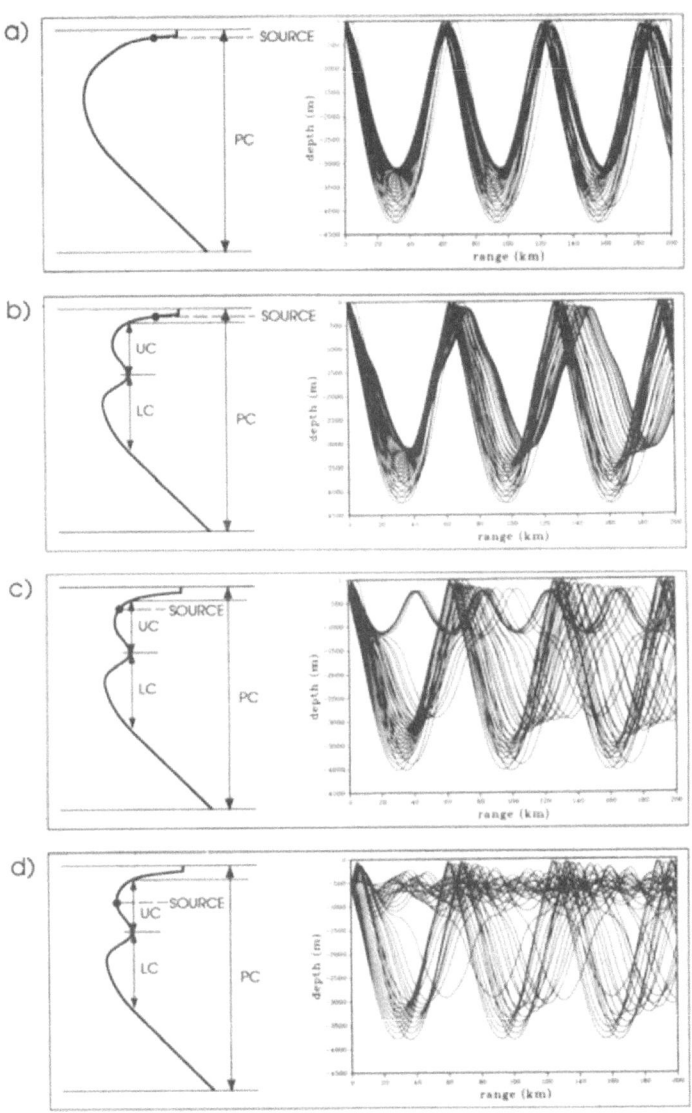

FIGURE 2 : RAYS TRACING IN RANGE INVARIANT MEDIA

a) WITHOUT MW AND FOR A 150 m SOURCE DEPTH
b) WITH MW AND FOR A 150 m SOURCE DEPTH
c) WITH MW AND FOR A 250 m SOURCE DEPTH
d) WITH MW AND FOR A 350 m SOURCE DEPTH

For what we have seen in part 1, however, the characteristics of the Mediterranean Water intrusions are rarely stable over ranges longer than 100 km, (except, perhaps, for some particular East-West sections). As long as propagation ranges of a few hundred kilometers are being considered, it is necessary to take into account the mesoscale variability of the medium.

A number of ray tracing and propagation loss vertical sections has been executed for an ensemble of typical combinations of the various parameters, such as source depth, horizontal distance with respect to the position of a MEDDY or of a single front, horizontal source position inside an oscillating s.v.m. area etc...

A special range varying ray tracing program including the computation of propagation losses averaged over a finite depth interval (thus avoiding to find infinite values on caustics) has been specially developed for this application.

The principal conclusions which can be deduced from this analysis can be formulated as an ensemble of practical guide rules :

The most important effect of the mesoscale variations due to the intrusing of Mediterranean Waters is to provoke transfers of energy between the three channels which have been defined (UC, LC, PC).

The mechanism of these transfers is entirely governed by the horizontal variations of the value of the deep maximum sound velocity, the detailed shape of the deep sound velocity profile or the small variations of the depth of this maximum having very little influence on the resulting sound field.

As long as acoustic applications are concerned, it is sufficient to describe the deep mesoscale variations in the form of a map giving the s.v.m. value in function of the geographical position.

When considering the variations of sound velocity profile along a particular vertical section, it can be schematically represented as a succession of deep sub-surface "fronts" which are defined as the range intervals where the s.v.m. increases or decreases in an appreciable way. The fronts are usually separated by other range intervals which can be considered as being more or less range invariant.

The crossing of a MEDDY, for instance, can be treated as the crossing of two opposite fronts, with invariant Atlantic waters outside and an invariant Mediterranean water layer inside.

The area of oscillating values of the s.v.m. can be represented by an irregular alternance of positive and negative fronts separated by small quasi-range invariant areas.

Interchannel energy transfers occur at the crossing of the front.

When crossing a front in the increasing s.v.m. direction, part of the energy which was originally contained in the principal channel (PC) is transferred in the upper channel (UC), in the lower channel (LC) or in both of them. If the transition zone is straight (not larger than 20 km) the commutation of the transferred energy toward the UC or the LC depends on the horizontal distance between the source and the front. If the front is centered on a distance where the bundle of rays issued from the source in the PC is in a low position, part of the rays will not be able to cross the deep maximum and will remain in the LC. This happens if the distance between the source and the front is around 30 km ± 15 km (half the CZ distance) or an odd multiple of 30 km. If it is centered on a distance where the bundle is in high position the rays leaving the PC will remain in the UC. This happens if the front starts just after the source, or if it does coincide with the position of a convergence zone, i.e. when the distance is around 60 km, or a multiple of this distance. Between these distances intervals a part of the rays leaving the PC will be transferred into the UC and another part into the LC.

When crossing a front in the decreasing s.v.m. direction, part of the energy (or eventually the whole) which was contained in the UC and/or in the LC is transferred into the PC. This energy reappears in the PC as a new bundle of rays which has the same 60 km spatial periodicity as the one directly issued from the source. Depending upon the entering position of this new bundle, it may either tend to create new CZ between the preceeding ones, or reinforce the existing ones.

Crossing several fronts in succession, or crossing one or two fronts with a large transition range (bigger than 30 km) will produce a complex mixture of different types of interchannel energy transfers. A detailed prediction of the resulting sound field can only be done if the variations of the s.v.m. is effectively known on the complete range, starting from the source location.

The critical range scale generating the exchanges of energy between channels being half of the PC convergence zone period, the mesoscale variability should be described with slightly higher precision in order to obtain correct deterministic predictions from a sound propagation model. Giving a deep sound velocity profile every 20 to 25 km would represent a correct sampling rate for the input data delivered to the acoustic models.

Examples of sound propagation losses vertical sections are presented in fig. 3 and 4 as an illustration of the properties which have just been enumerated.

Fig. 3 shows the propagation across a MEDDY for a source depth of 550 m and for three horizontal distances between the source and the MEDDY center. On fig. 3.a where the source is at 100 km from the center and at 60 km from the center of the first front, energy is transferred from PC to UC at the input front, and back to the PC at the output. On fig. 3.b where the source is 20 km closer, the LC instead of the UC receives the transferred energy. On fig. 3.d, the source is located at the center of the lens, thus exciting directly the UC, the corresponding energy then going to the PC outside the lens limits. This position of the source gives a particularly regular coverage as the spatial oscillations of the ray bundle transferred from UC to PC are in phase opposition with the ones directly excited in the PC.

Fig. 4 shows examples of sound propagation through the idealized situation of a regular oscillation of the s.v.m., with a 140 km spatial periodicity. Three different source locations are presented corresponding to high, low and middle (on the increasing side) positions of the s.v.m. value. The multiplicity of transfers creates a complicated structure at long range. The coverage near the surface is better when the source position coincides with a high or a middle (increasing) value of the s.v.m. than when it is at a low value. In this last case the energy which is leaving the PC is first transferred into the LC and it does reappear in the PC at 120 km, provoking at this range a reinforcement of the second convergence zone.

The quantity of acoustic energy which is transferred from one channel to another at the crossing of a front depends on the ratios between the sound velocity at the source (s.v.s.), and the two sound velocities of the deep maximum, on the high side (s.v.m. $_{max}$) and on the low side (s.v.m. $_{min}$) of the front.

- If s.v.s. > s.v.m.$_{max}$ there will not be any transfer, all the energy being contained in the PC on both sides of the front.

- If s.v.m.$_{min}$ < s.v.s. < s.v.m.$_{max}$, the UC and/or the LC will only be excited on the high side of the front. All the energy on the low side will be contained in the PC.

- If s.v.s. < s.v.m. $_{min}$ the part of energy which is issued from the source at low grazing angles will remain in the UC on both sides of the fronts. This situation never exists if one side of the front has no deep maximum (case of the MEDDIES).

If the source depth is comprised between 0 and 600 m, it does remain inside the seasonal or the permanent thermoclines, which are practically not affected by the mesoscale variations of Mediterranean Water. If D_1 et D_2 are the image depth of svm$_{max}$ and svm$_{min}$ the depth of the source (above D_1, between D_1 and D_2, or below D_2) determines the position of svs with respect to svm$_{max}$ and svm$_{min}$. D_1 and D_2, are in fact extremely variable as a function of both the season and the geographical position.

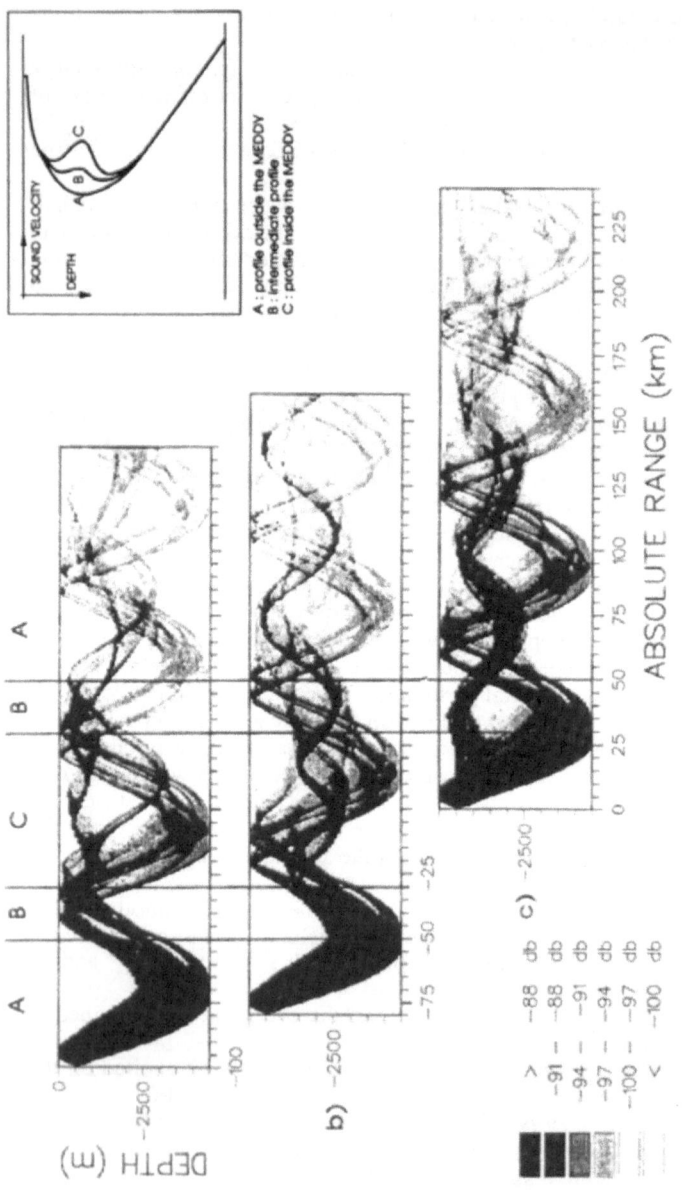

frequency = 200 hz; absorbent sediment

FIGURE 3 : SOUND PROPAGATION THROUGH A MEDDY FOR A 550 m SOURCE DEPTH:

a) SOURCE AT 100 km OF THE MEDDY CENTER
b) SOURCE AT 80 km OF THE MEDDY CENTER
c) SOURCE AT THE MEDDY CENTER

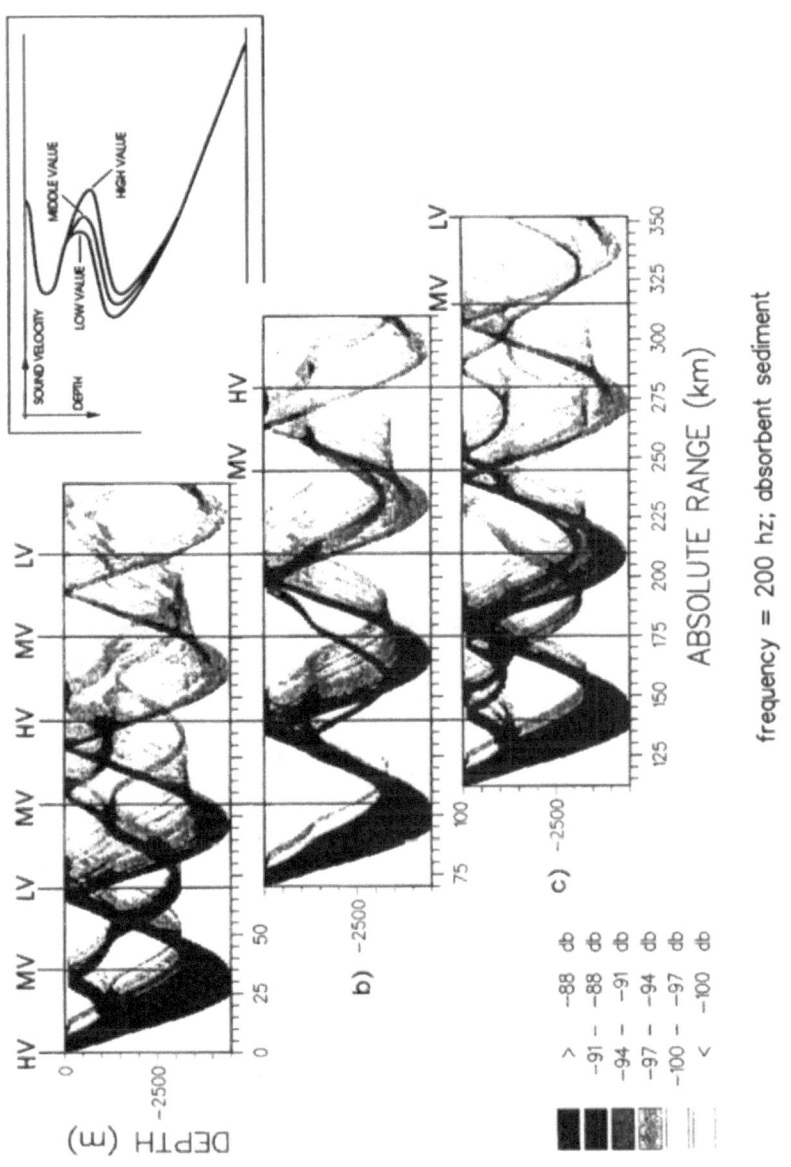

DEPTH (m)

ABSOLUTE RANGE (km)

frequency = 200 hz; absorbent sediment

> - -88 db
-91 - -88 db
-94 - -91 db
-97 - -94 db
-100 - -97 db
< -100 db

FIGURE 4 : SOUND PROPAGATION THROUGH SPATIAL PERIODIC OSCILLATIONS OF
THE SOUND VELOCITY DEEP MAXIMUM FOR A 100 m SOURCE DEPTH

a) SOURCE AT A HIGH VALUE OF THE DEEP MAXIMUM
b) SOURCE AT A LOW VALUE OF THE DEEP MAXIMUM
c) SOURCE AT A MIDDLE VALUE OF THE DEEP MAXIMUM

384

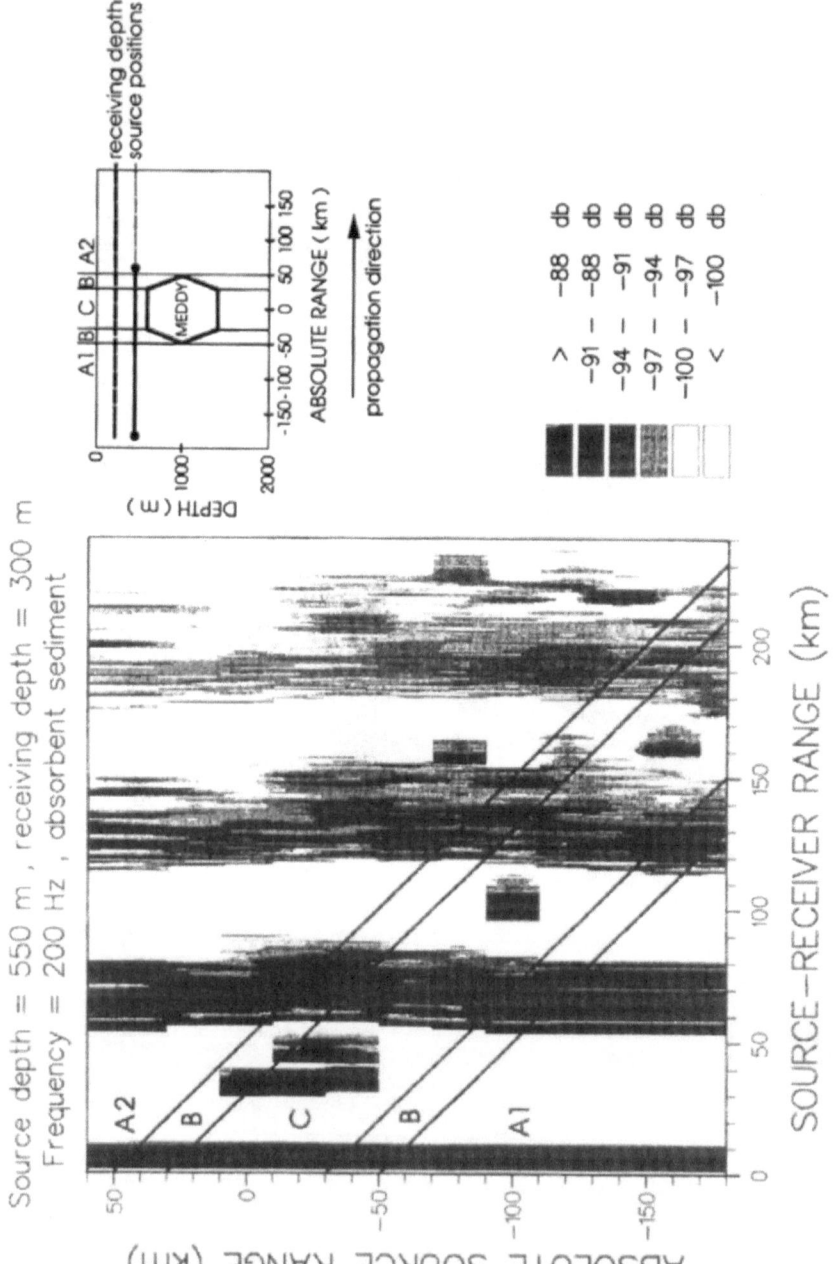

Source depth = 550 m , receiving depth = 300 m
Frequency = 200 Hz , absorbent sediment

FIGURE 5: SOUND PROPAGATION THROUGH A MEDDY, FOR A FIXED
RECEIVING DEPTH, WHEN THE DISTANCE BETWEEN
THE SOURCE AND THE MEDDY CENTER VARIES

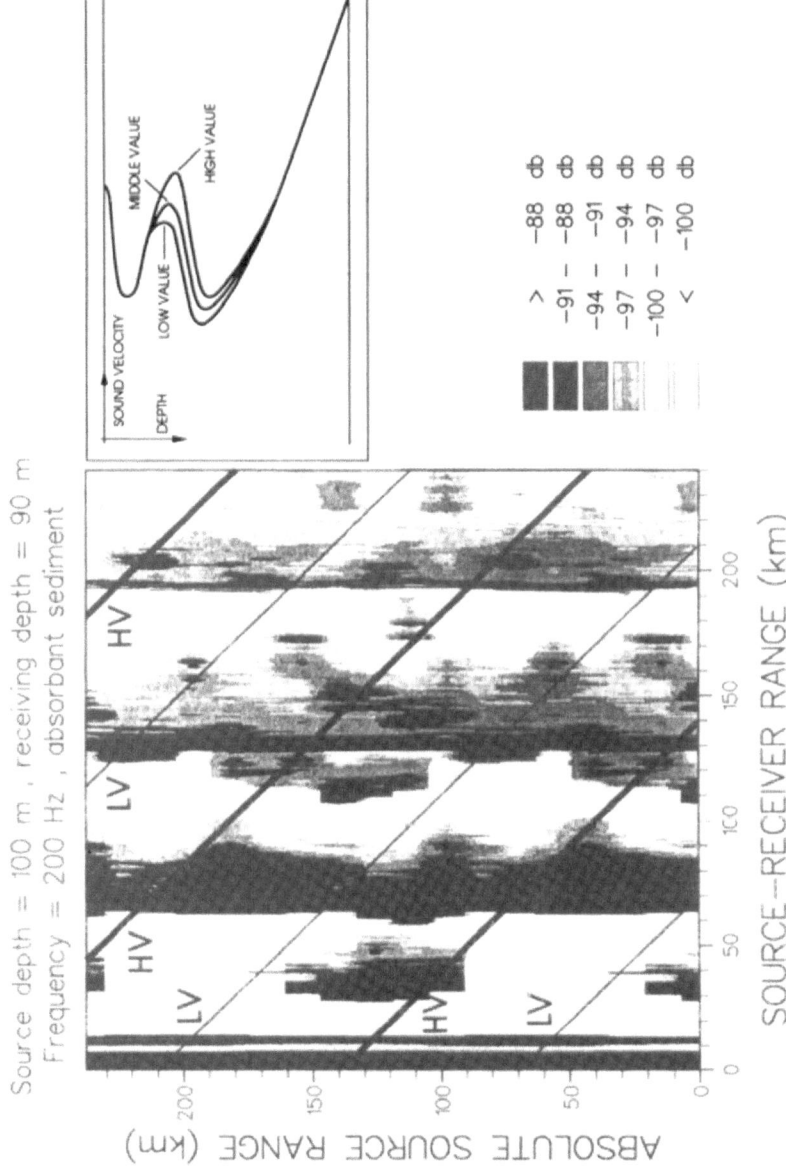

Source depth = 100 m , receiving depth = 90 m
Frequency = 200 Hz , absorbant sediment

> -88 db
-91 — -88 db
-94 — -91 db
-97 — -94 db
-100 — -97 db
< -100 db

FIGURE 6 : SOUND PROPAGATION THROUGH 140 km SPATIAL PERIODIC
OSCILLATIONS OF THE SOUND VELOCITY MAXIMUM. FOR A FIXED
RECEIVING DEPTH, WHEN THE ABSOLUTE SOURCE RANGE VARIES

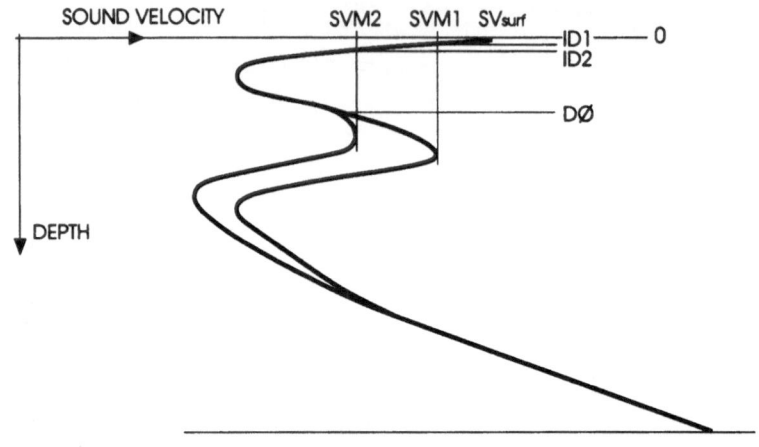

<u>FIGURE 7</u> : EXTREME SOUND VELOCITY PROFILES AND CHARACTERISTIC DEPTHS INTERVALS

		SOURCE DEPTH		
		O - ID1	ID1 - ID2	ID2 - DØ
RECEIVING DEPTH	O - ID1	CZF	CZF	CZF
	ID1 - ID2	CZF	CZF + ICET	CZF + ICET
	ID2 - DØ	CZF	CZF + ICET	Range Continuity

CZF : Convergence Zone Fluctuations

ICET : Inter Channels Energy Transfers

<u>TABLE 1</u> : PROPAGATION EFFECTS ASSOCIATED WITH DEPTHS (SEE FIG. 7)

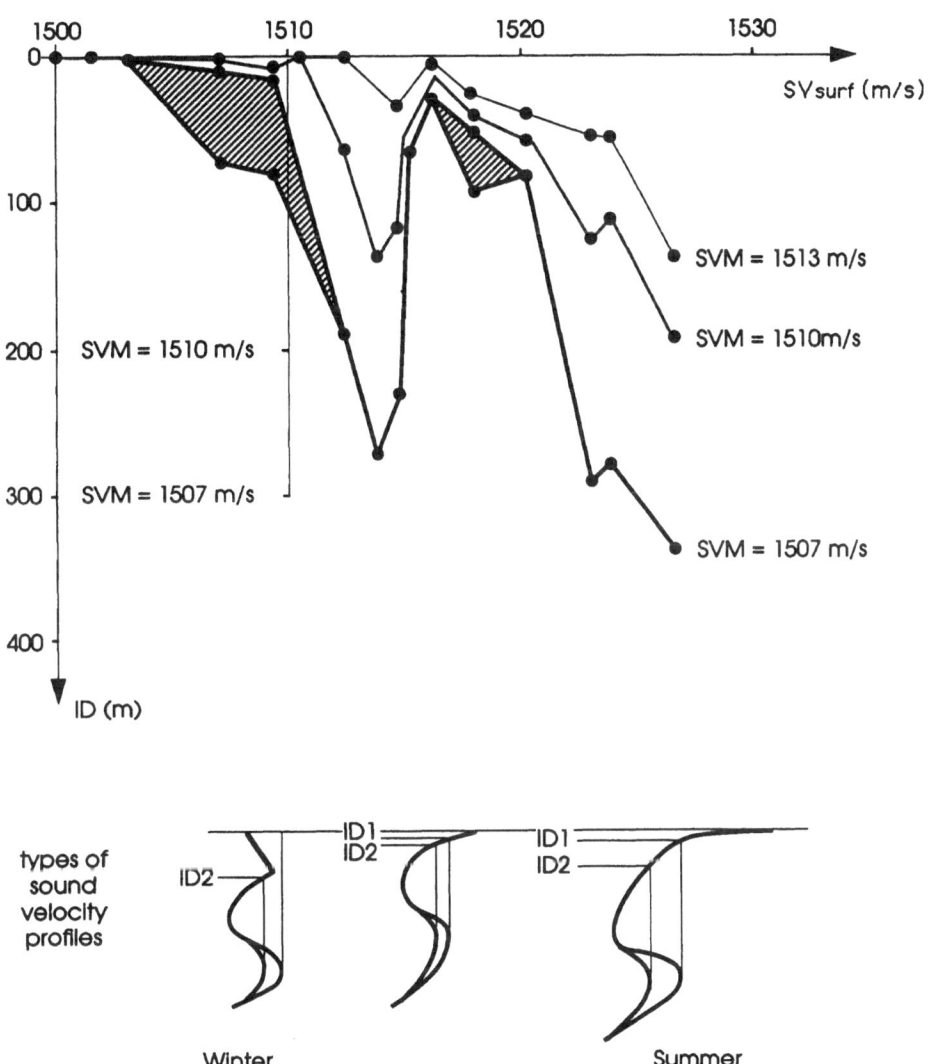

FIGURE 8 : IMAGE DEPTHS OF SOUND VELOCITY DEEP MAXIMUM

Using the data contained in [12] for the area corresponding to fig. 1c extended to 50° N the approximate image depths corresponding to three values of svm have been traced on fig.8 as a function of the surface temperature (which is itself depending on the latitude and season).

Although if they are only indicative (the data base being reduced and partly extrapolated) the curves clearly show that the depth intervals between values of D_1 and D_2 are extremely variable :

The interval between the surface and D_1 does not exist all the time (except in late summer).

The interval between D_1 and D_2 can be very large or extremely reduced. In this last case (corresponding to the presence of a strong gradient below the surface) small variations affecting the source depth, the receiver depth, or the thermocline depth around this critical interval could have more influence on the propagation losses than the mesoscale variations of svm.

The reasoning concerning the influence of a source depth can be extended to the receiver depth by applying the reciprocity principle.

Table 1 indicates the main propagation effects resulting from the deep mesoscale variability as a function of the depth intervals occupied by the source and by the receiver.

If one at least of the two depths is between the surface and D_1 there will not be any effect of interchannel energy transfers. The deep mesoscale variations will just introduce some modulation on the ranges, widths and peak values of the main convergence zones (noted czu).

If the two depths are between D_1 and D2, or if one is between D_2 and D2 and the other below D_2, InterChannel Energy transfers (ICET) will perturbate the propagation law.

If the two depths are below D_2, ICET effects will still be present, but a quasi regular range coverage will be insured by the UC (range continuity).

To a lesser extent, the mechanism of energy transfer is also depending on the horizontal distance between the source and the front. This is due to the fact that the "ray parameter" k,

defined by the ratio $k = \dfrac{c}{\cos \theta}$ (where c is the sound velocity and θ the grazing angle)is no more a constant along a ray travelling in a range dependent medium. In first approximation, the variations of k are equal to the integral of the horizontal gradients encountered along the ray. For a ray crossing a narrow front at the depth of the maximum, the variation of k is practically equal to | s.v.m. $_{max}$ - s.v.m. $_{min}$ |, increasing when going from the low to the high side, decreasing in the opposite sense. Inversely, a ray which is in a high or a low position at the crossing of a narrow front does not practically encounter any horizontal gradient, and the value of its k is unchanged.

Some effects of these variations of k can be seen on fig. 3.c and 3.c : for both cases, a bundle of rays which is transferred from the UC to the PC at the crossing of a decreasing front has an upper turning point which is deeper than the source depth, as if it was issued from a deeper source than the real one.

Under very particular circumstances, the variations of k along a ray can produce strange effects, in contradiction with the rules which have been stated about the influence of source (or receiver) depth on ICET : A ray suffering a reduction of its k value when crossing at intermediate depth a first (decreasing) front and then crossing a second (increasing) front at high or low position (this time without any variation of k) may be trapped in the UC or in the LC, even if the source depth is above D_1.

3. Practical conclusions

A reasonably precise deterministic prediction of the sound propagation losses between two points in the NEA (corresponding to the source and the receiver positions) whose depth is comprised between the surface and a few hundred meters can be done provided the variations of the sound velocity at the deep maximum can be described on an horizontal grid scale of about 25 km spacing.

The prediction also requires the knowledge of the sound velocity profile below the surface down to the maximum possible depth of the source and receiver.

A complete computation of the propagation losses for any possible position of the two points would have to be done in a 6 dimension space, which is completely unrealistic, both from a computtation time point of view and from the capability of a useful interpretation of the huge data base which would be produced for solving the practical problems under concern.

The general rules which have been given concerning the mechanisms of interchannel energy transfers can be used to give a qualitative description of the acoustic variability resulting from the deep mesoscale variations of s.v.m., thus reserving the propagation models to compute a limited number of carefully selected sound field sections.

When the s.v.m. variations are not known with precision, it can be attempted to give a statistical description of the propagation properties. The form of this description will depend on the practical problem to be considered.

As an example one may give the probability for the propagation losses to remain below a certain threshold (corresponding to the level which is necessary for the detection of a given target by a given sonar) as a function of range. The results can be modulated or not as a function of the source (or receiver) position and/or the direction of observation, depending of the partial knowledge which can be available.

From the above point of view, the MEDDIES represent a particular case. If their position is unknown, their existence can be practically ignored as the probability to find one between source and receiver is rather low.

If the position of a MEDDY is known with precision, it can be shown that a receiver positioned above its center, at a depth between D_1 and D_2 will benefit from rather favourable listening conditions. This is illustrated on fig. 5 which gives a two dimension display of the propagation losses between a source and a receiver at constant depths. The vertical axis represents the distance between the source and the center of the MEDDY, the horizontal axis the source-receiver distance. A given position of the receiver corresponds to a line at 45° on the picture. It can be seen that the line corresponding to the receiver being situated at the center of the MEDDY crosses all the peaks of energy of the upper channel, which are just imbricated between the peaks resulting from the convergence zones of the principal channel.

A similar presentation has been done for the propagation across oscillating values of s.v.m. (fig. 6). It does appear that placing the receiver at the vertical of an area of high value of the s.v.m. may give a more continuous coverage than placing it above a low value (at least until the first CZ).

The last asumption, however, would have to be carefully examined for different scales of oscillations, and within the context of the real problem to be solved (threshold, range and depth of interest).

If the position of the maxima and minima of the s.v.m. is not known, the variations of the values observed along each vertical line gives an idea of the statistical distribution of the propagation losses for the corresponding source-receiver distance.

BIBLIOGRAPHY

[1] L. Armi, D. Hebert, H. Oakey, J. Price, Ph. L. Richardson, Th. Rossby,
R. Ruddick - The history and decay of a Mediterranean salt lens
Nature, vol. 333, pp. 649-651, June 1988.

[2] L. Armi, W. Zenk - Large lenses of highly saline Mediterranean Water
JPO, Vol. 14, pp. 1560-1576, 1984.

[3] A. Beckmann, R.H. Kase - Numerical simulation of the movement of a
Mediterranean Water lens
Geoph. Res. Letters, Vol. 16, N° 1, pp. 65-68, January 1989.

[4] R.H. Kase, A. Beckmann, H.H. Hinrichsen - Observational evidence of salt lens
formation in the Iberian basin.
JGR, Vol. 94, N° C4, pp. 4905-4912, April 1989.

[5] R.H. Kase, J.F. Price, Ph. L. Richardson, W. Zenk - A quasi-synoptic survey of
the thermocline circulation and water mass distribution within the Canary Basin.
JGR; Vol. 91, N° C8, pp. 9739-9748, August 1986.

[6] R.H. Kase, W. Zenk - Reconstructed Mediterranean salt lens trajectories
JPO, Vol. 17, pp. 158-163, 1987.

[7] J. Kielmann, R.H. Kase - Numerical modeling of meander and eddy formation in
the Azores current frontal zone;
JPO, Vol. 17, pp. 529-541, 1987.

[8] J. Marshall - Submarine salt lenses
Nature, Vol. 333, pp. 594-595, June 1988.

[9] W.H. Munk - Horizontal deflection of acoustic paths by mesoscale eddies
JPO, Vol. 10, pp. 596-604, April 1980.

[10] N.S. Oakey - Epsonde : an instrument to measure turbulence in the deep ocean
IEE Journal of oceanic engineering, Vol. 13, N° 3, pp. 124-128, July 1988.

[11] EPSHOM -
Radiales :   D'ENTRECASTEAUX (mai 1987
                  CDT Birot (décembre 1987)
                  LAPEROUSE (février 1988)
                  SUROIT (mars, avril 1988)
                  CHARCOT (juin 1988)
                  CHARCOT (mai 1988)
                  D'ENTRECASTEAUX (juillet 1988)
                  D'ENTRECASTEAUX (août 1988)

[12] SHOM- Album de champs sonores N° 716, Atlantique Nord-Est, tome A, 1978.

ASPECTS OF OCEANOGRAPHIC VARIABILITY OBSERVED FROM THERMISTOR CHAINS ON FREE-DRIFTING BUOYS.

P. J. MINNETT and T. S. HOPKINS
Applied Oceanography Group
SACLANT Undersea Research Centre
Viale San Bartolomeo 400
I-19026 San Bartolomeo
La Spezia
Italy

ABSTRACT. Two drifting Meteorological-Thermistor Chain Buoys were deployed in the Færoe-Shetland Channel area in June 1987. These buoys consisted of a small surface spar buoy supporting a surface meteorological package and a 300-m long thermistor chain with 15 sensors. The temperature measurements are analysed to characterise the thermal and sound speed variability in this region of complex oceanographic structure. Ancillary data from CTD surveys and a nearby current meter mooring are used to aid the interpretation of these measurements.

1. Introduction

During 1987 the SACLANT Centre conducted the Atlantic Inflow Experiment designed to provide synoptic data on the forcing, water-mass structure and variability of the Norwegian Atlantic Current (NwAtC) in the region from the Færoe-Shetland Channel to the Lofoten Basin. During the spring an extensive hydrographic cruise was conducted aboard the HNLMS TYDEMAN (Hopkins et al., 1990a) followed by a mooring-deployment cruise on the R/V MARIA PAOLINA (Hopkins et al., 1990b). Three months later another hydrographic sampling was made on board the R/V BELGICA (Hopkins et al., 1990c) and the moorings were recovered from the GNS FEHMARN. As an observational component of the Experiment, two drifting Meteorological-Thermistor Chain Buoys (Met Buoys) were deployed to obtain meteorological data and quasi-Lagrangian data on the temporal structure of the upper (300 m) layer of entering Atlantic Waters. In this work we summarize the results of the data from these Met Buoys as they relate to the variability in the region.

2. Oceanographic Setting

The Norwegian Atlantic Current (NwAtC) is a poleward eastern boundary current of the Arctic Ocean that transports northwards the inflow from the Atlantic Ocean through the Norwegian Sea to the Polar Sea. The transported water mass, the Norwegian Atlantic Water (NwAtW), is the parent water mass for the entire Arctic Ocean and its importance

J. Potter and A. Warn-Varnas (eds.), Ocean Variability & Acoustic Propagation, 391–406.

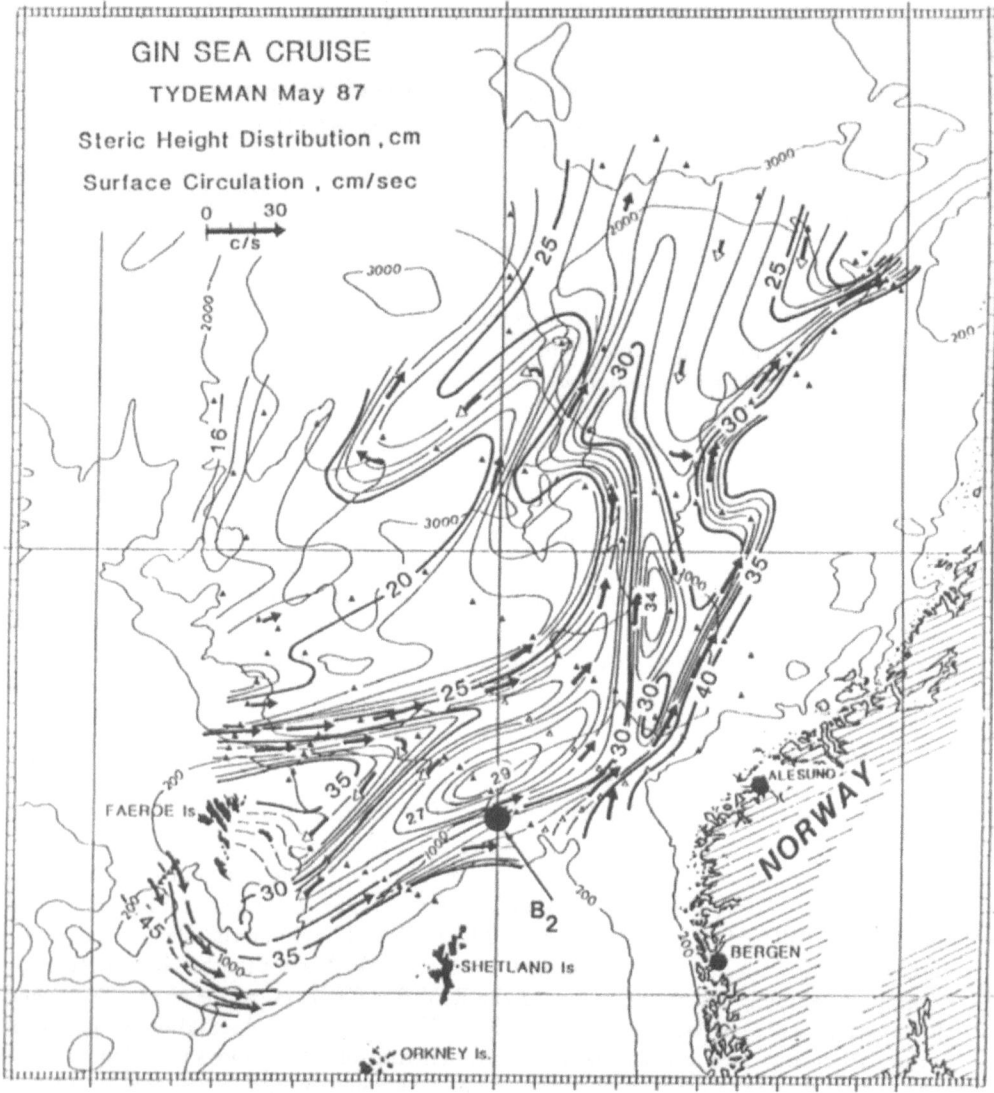

Figure 1. The distributions of steric height anomalies and the associated surface geostrophic circulation derived from CTD profiles in the area of the Atlantic Inflow into the Norwegian Sea. The conditions for May 1987.

to the oceanography and meteorology of 'northern European Seas' has long been recognized (cf. Helland-Hansen and Nansen, 1909; Hopkins, 1988). The amount of inflowing Atlantic Water (5-8 Sv) creates an extensive water-mass interface with the resident Arctic waters. Where this interface intersects the surface it is referred to as the Arctic Front. Underneath the NwAtC it is about 400-600 m deep. Helland-Hansen (1934) described the large (100-300 m) vertical displacements of the interface delimiting the NwAtW from the underlying Norwegian Sea waters. Helland-Hansen's supposition that these were the result of elongated geostrophic vortexes was substantiated in a study reported by Dickson (1972), in which he showed the interfacial variations (2-5 days) to be primarily stationary deformations associated with north- or south-going branches of the NwAtC. Energy at the 2-5-day band in both the currents and winds was also noted by Meincke and Kvinge (1978); and Mysak and Schott (1977) suggested that an observed ∼3 day energy peak could be attributed to baroclinic instabilities in the flow field.

It was in this background of spatial and temporal variability that the Atlantic Inflow Experiment was conducted. The surface circulation from the spring hydrographic cruise is shown in Figure 1 using the method of steric height anomaly (Hopkins et al., 1990d). The steric heights are computed relative to the deepest point within the region of sampling and the velocities by the geostrophic approximation. The entrance of the inflow both through the Færoe-Shetland Channel and over the northern Færoe Slope is clearly indicated. The latter tends to contribute to the western branch of the NwAtC that approximately follows the 2000-m isobath and the former to the eastern branch that follows the Norwegian Slope (<1000m). Hopkins and Mouchet (1990) have suggested that both of these inflows vary in magnitude alternatively out of phase with each other, and that both vary in position such that the entire zone of confluence is subjected to strong variability as these two branches of the NwAtC wax and wane in strength and oscillate laterally in position. As seen in Figure 1, a portion of the Færoe-Slope inflow turns south and recirculates within the Færoe-Shetland Channel, as demonstrated previously by Dooley and Meincke (1981). Figure 2 gives an example of the cross-sectional structure in the water masses and in the velocity at the position of one of the current meter moorings (B2). Of particular importance to the following discussion are the large amplitude variations in the isotherms and the horizontal shear structure of the flow field.

The experimental objective of the Met Buoy deployment was to locate one in each of the two inflow locations in an attempt to obtain a quasi-Lagrangian record of the advected water mass as it enters the Norwegian Sea and proceeds northward with the NwAtC.

3. Drifting Buoys

The Met Buoys or, more properly, 'Air-Sea Interaction Drifting' Buoys are manufactured by the Polar Research Laboratory, and consist of a small surface spar buoy supporting a surface meteorological package and a 300-m long thermistor chain of 15 sensors. The data from all of the sensors are block averaged over 8 minutes and broadcast to the Data Collection and Location System on the NOAA series of polar orbiting weather satellites. The position of the buoys and the time of data reception are added to the data, which are distributed by *Service ARGOS*. Figure 3 provides a schematic representation of the Met Buoy configuration and the data return during the experiment.

394

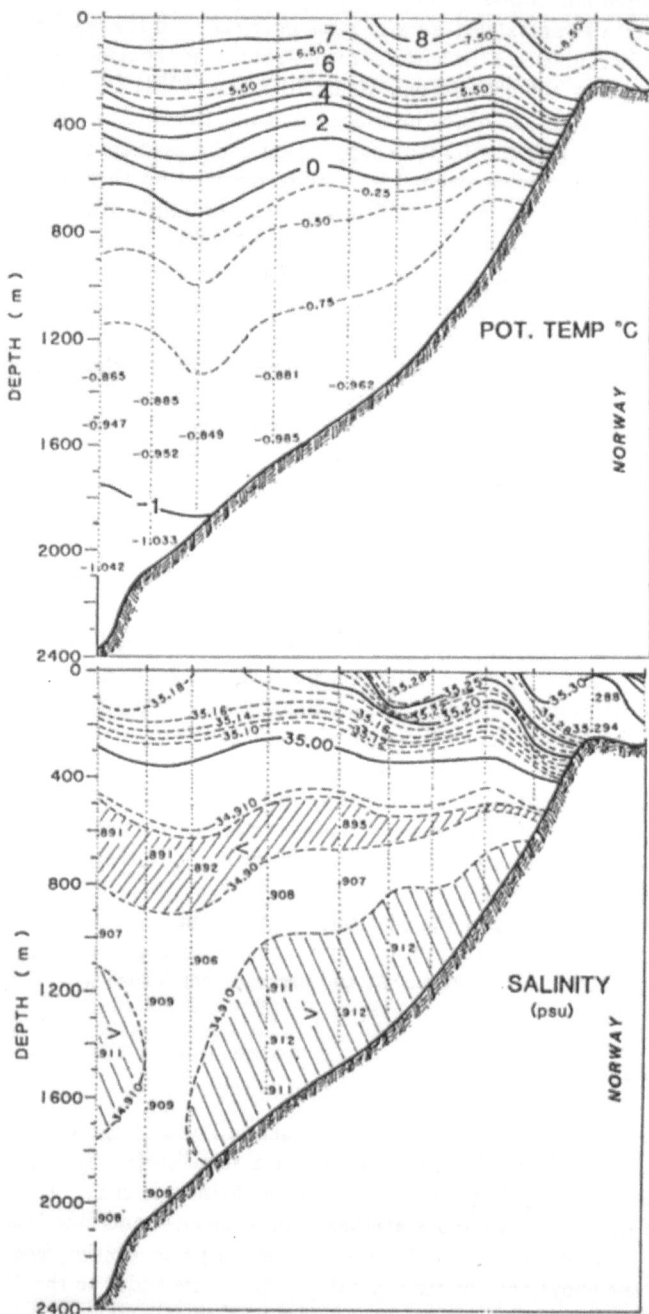

Figure 2. Sections of potential temperature in °C (a), salinity in psu (b), and geostrophic currents in cm s^{-1} (c), derived from CTD data across the Atlantic Inflow at the northern end of the Færoe Shetland Channel close to the location of mooring B2 indicated in Figure 1.

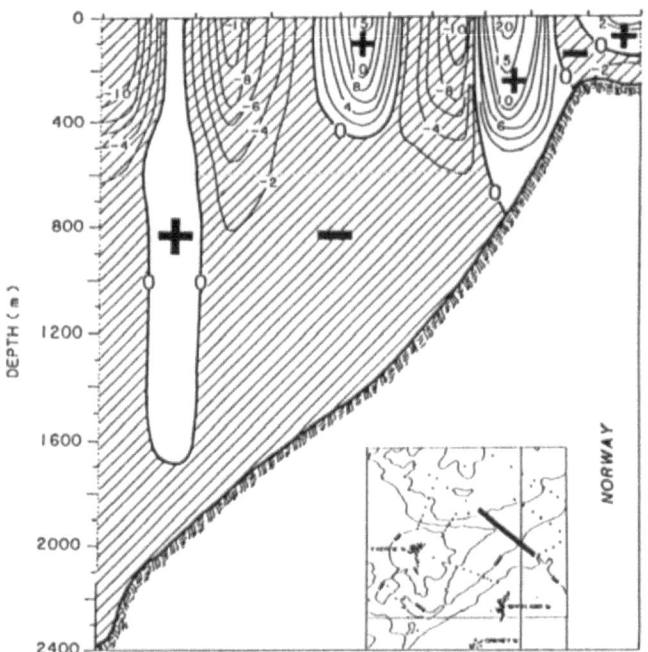

Figure 2c.

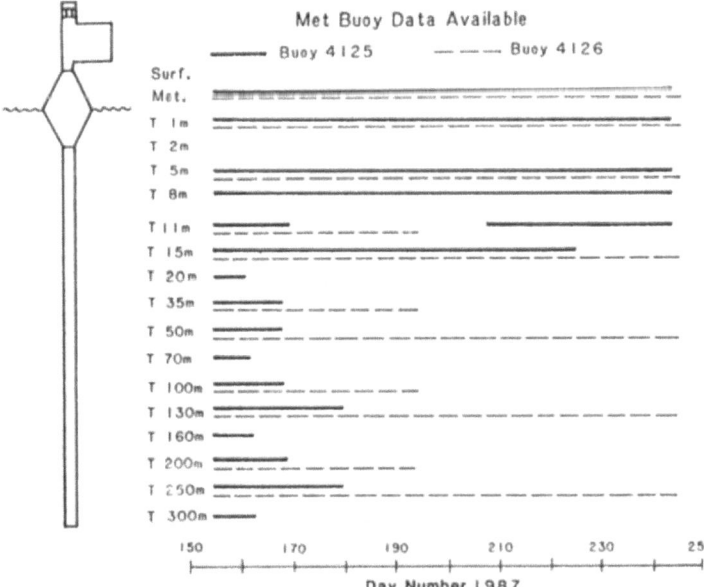

Figure 3. Schematic representation of the Met Buoys and the data return from the thermistors and surface meteorological package from each Met Buoy.

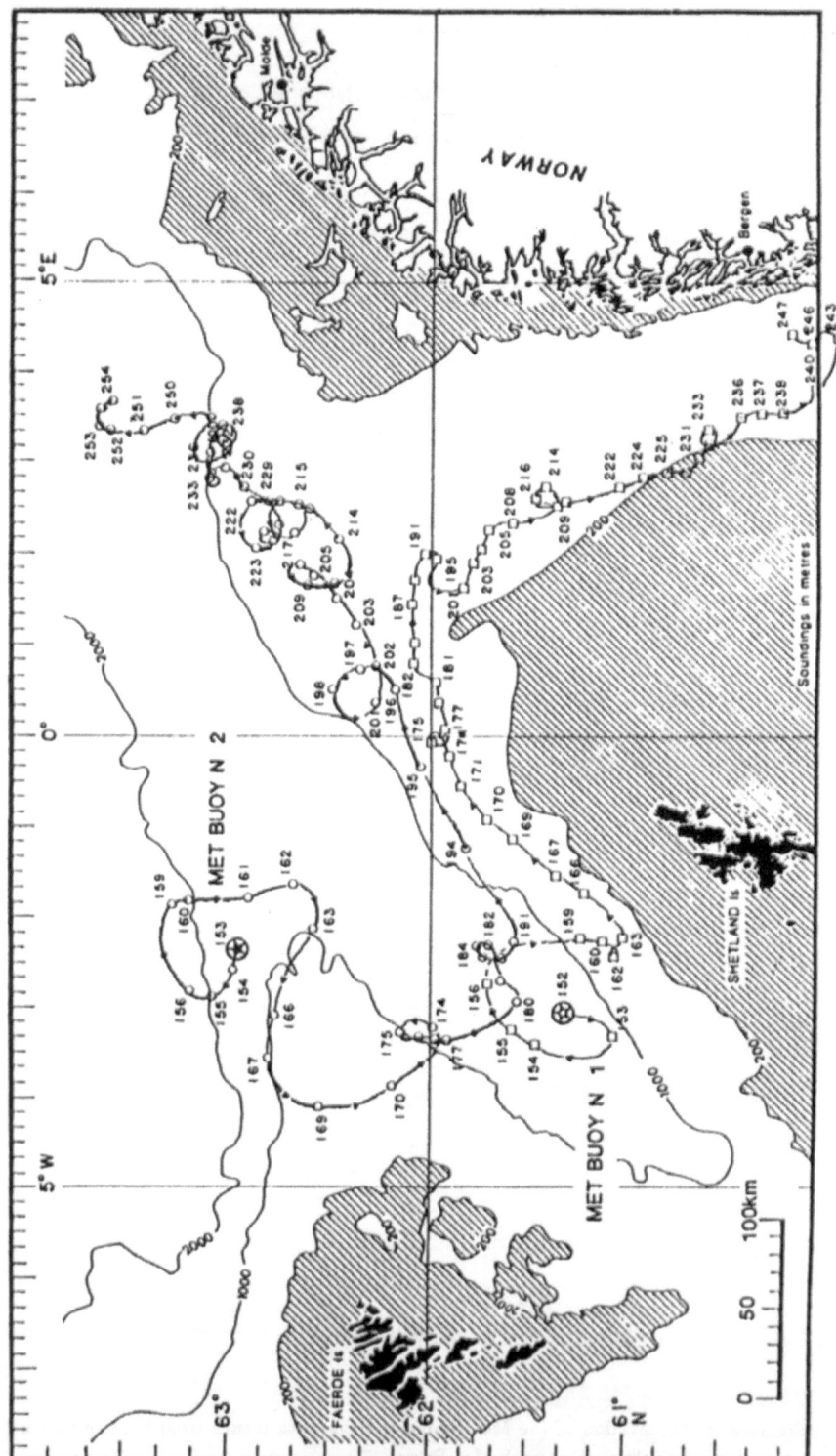

Figure 4a. Trajectories of the Met Buoys. The positions at the start of each day are marked.

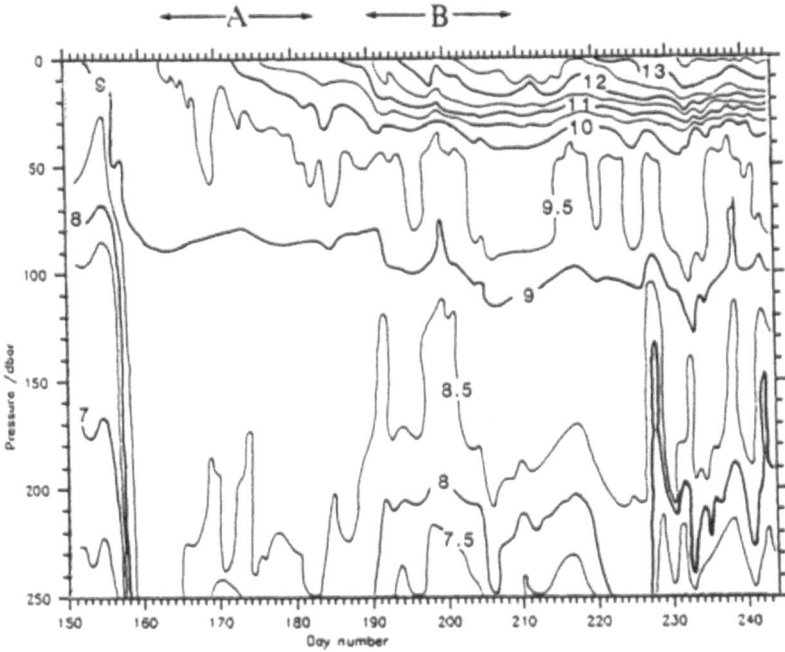

Figure 4b. Contours of sea temperature derived from the thermistor chain data of Met Buoy 1. The contour interval is 0.5K. The two distinct oceanographic regimes are marked as 'A' for the Atlantic Inflow and 'B' for the entrance to the Norwegian Trough.

Met Buoy 1 (serial number 4126) was launched on 1 June 1987 (day 152) at 61°19'N, 3°07'W and Met Buoy 2 (4125) on 2 June 1987 (day 153) at 62°58'N, 2°33'W. Their subsequent trajectories are shown in Figure 4. Six of the 15 thermistors on Met Buoy 1 failed immediately on deployment, and a further four after about 40 days. Thus there are about 91 days of data from 5 thermistors, plus sea-surface temperature, from Met Buoy 1. One sensor from Met Buoy 2 failed immediately, and all of those below 15 m failed in the following few weeks; only two thermistors survived as long as the surface sensors, for the full 90-day deployment.

4. Results

The initial movement of both the buoys (Figure 4a) was counter to that expected, due, it is assumed, to very strong anomalous westward winds prior to their deployment. This was followed by a southeastward motion for both buoys until day 162 whence Met Buoy 1 was swept into the core of the inflow and Met Buoy 2 was caught in the recirculation pattern of the Færoe-Shetland Channel. At approximately day 191, Met Buoy 1 entered the Norwegian Trough and Met Buoy 2 entered into the core of the inflow, by now located further west. Note that about 1 Sv of the eastern-most portion of the NwAtC makes the

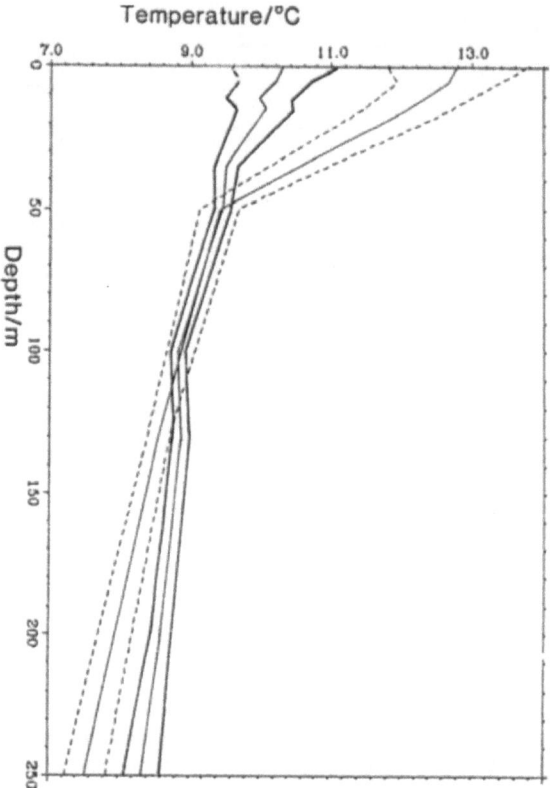

Figure 5. Profiles of mean temperature bracketed by the envelopes of ± 1 standard deviation, derived from measurements from Met Buoy 1 for the two periods: days 163–183 in the Atlantic Inflow and days 190–210 for the entrance to the Norwegian Trough, indicated as 'A' and 'B' in Figure 4b.

deviation into the North Sea in the form of a recirculation pattern over the Norwegian Trough (cf. Hopkins, 1988), as indicated by Met Buoy 1 after day 240.

Only the thermistor data of Met Buoy 1 will be used in the following discussions because of its longer time series. The Lagrangian section (Figure 4b) shows the transition during days 157 and 159 from colder to warmer waters as it was advected across the 1000-m isobath and into the main core of the NwAtC. Prior to day 157 the water column was much colder and more stratified throughout the 250 m measurement window. During the transition the 8.5°C isotherm plunges from about 20 m depth to below 250 m and the 9°C from close to the surface to about 90 m. After day 192 the more stratified waters of the Norwegian Trough are evident in the upper (<35 m) thermistors.

We now look at two 20-day intervals: days 163–183 while the buoy was in the NwAtC and days 190–210 while it was entering the North Sea. The profiles of the mean temperature during these two intervals reveal several characteristic properties (Figure 5). The variability at the surface was a maximum and secondarily so at the bottom. The minimum

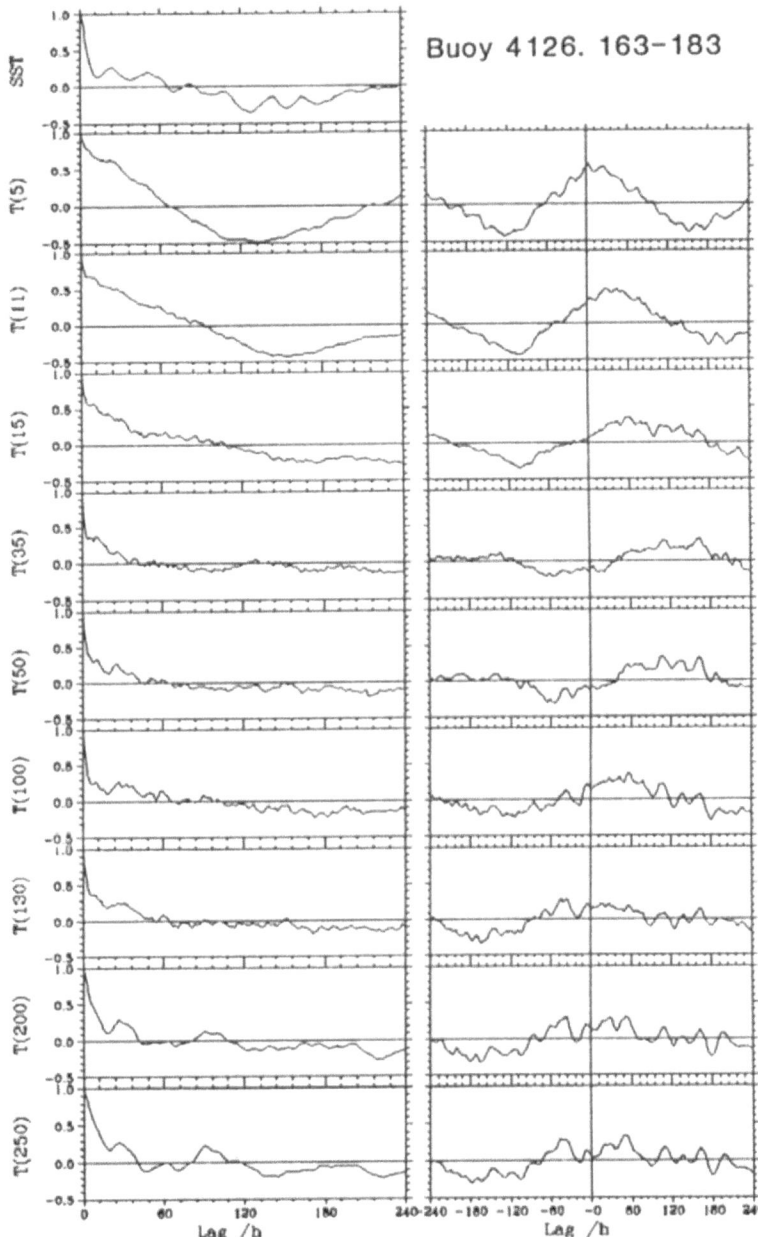

Figure 6. Time-lagged autocorrelation functions (left hand panels) for the thermistors on Met Buoy 1, and the time-lagged cross-correlation functions with respect to SST (right hand panels). These are calculated for the time interval from day 163 to 183, i.e. when the buoy was in the Norwegian Atlantic Current. Each time-series has 480 points.

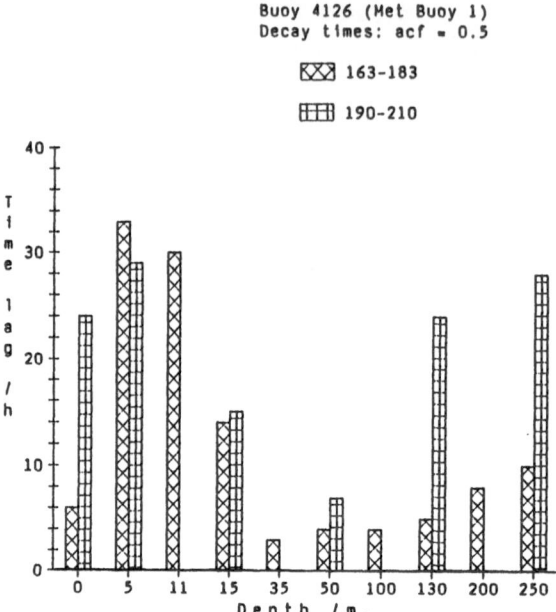

Figure 7. Decay times, in hours, of the autocorrelation functions (i.e. the time lag at which the acf = 0.5) of temperature fluctuations for each level of the thermistor chain on Met Buoy 1 for the two periods: days 163–183 in the Atlantic Inflow and days 190–210 for the entrance to the Norwegian Trough.

variability portions, from 35-100 m, also have the same mean values for the two intervals. This represents the depth range in which the buoy motion was most closely approximating Lagrangian. The large standard deviation at the surface is interpreted to be caused by both a seasonal warming trend and spatial surface variations in the water moving faster than the buoy. At depth the variations are considered to be caused by the buoy moving through the water which has spatial variations associated with the water-mass interface under the NwAtW.

Autocorrelation functions (acfs) calculated for these two intervals (shown in Figure 6 for the period of days 163–183) indicated shorter decay times (acf=0.5) on the order of 6 h or less for the Lagrangian middle section compared to the decay times of $\sim$30 h for the surface and deeper portions (Figure 7). At the very surface the decay times decreased somewhat, presumably due to a greater variability at higher frequencies. For example, the effect of diurnal heating is evident. Cross-correlations made with the surface temperature thermistor (during days 163–183) suggested a propagation of heat downwards corresponding to a vertical rate of warming of $\sim$8 m d^{-1} (Figure 6). For a diffusive length scale of 10 m this would give an eddy diffusion coefficient of $\sim$1 cm^{2}s^{-1}.

We return now to the question of the depth of zero drag, i.e. the portion of the thermistor chain that most closely followed the water movement and therefore rendered the best Lagrangian estimates of the temperature variability. To make this estimate we use the interval of days 163–183, from which the mean speed of the buoy was 11.6 cm s^{-1}. During this period the Acoustic Doppler Current Profiler located on the Shetland Slope at 800 m

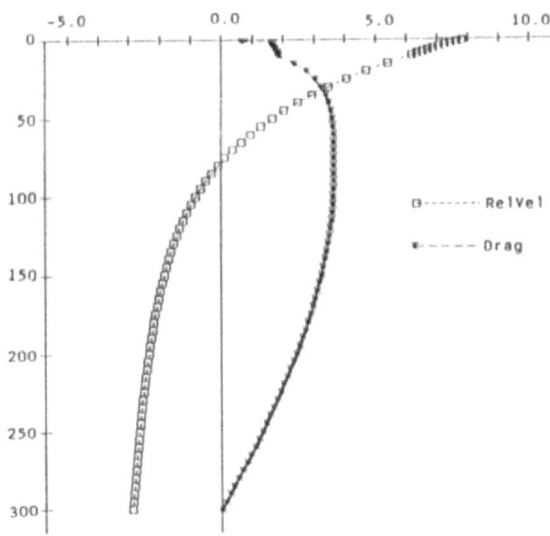

Figure 8. An example of a horizontal current profile, cm s^{-1}, that returns a zero depth-integrated drag on Met Buoy 1 for days 163–183, when the mean speed of the buoy was 11.6 cm s^{-1}. The depth of zero drag was 77 m.

recorded a mean vertical shear in the upper 100-300 m depth of 3.7 10^{-5} s^{-1}. We have assumed that the current speed profile has an exponentially decreasing surface component, and have determined this component by adding it to the observed mean speed and shear and iteratively estimating its magnitude until a depth-integrated drag of zero is obtained (Figure 8). The depth of zero drag for this instance was 77 m. Reasonable variations in the mean speed and shear give a range about this value of 50–130 m. Consequently we consider the sensors at 50, 100, and 130 m as providing the best Lagrangian estimates. This is supported by the above minimum in the autocorrelation functions, and more directly by the lack of visual variability in the sensors within this depth range (Figure 9).

Below the surface layer, in which atmospheric exchanges of heat and water induce changes in temperature (T) and salinity (S), the T-S relationship is fairly well behaved such that for short ranges of T and S it can be considered as linear. We have used this characteristic to construct linear T-S regressions valid for each of the thermistor ranges from the CTD data taken in the same water masses that the Met Buoys passed through. Thus, we have been able to convert the thermistor data to salinity and sound speed data. The resulting sound speed time series for the longer-lived sensors are shown in Figure 9.

Both Met Buoys passed close to the current meter mooring (B2) indicated in Figures 1b and 4a. Mooring B2 had an Aanderaa current meter at 130 m with temperature and conductivity sensors. We can make two useful comparisons. The first is to compare the time series of the sound speed at B2 with that of Met Buoy 1 (Figure 10). Despite the slightly warmer waters recorded by the Met Buoy, the difference in the variability of the two time series is obvious with the standard deviation of the Eulerian-B2 series (±2.4 m s^{-1}) being six times greater than that of the Lagrangian-Met Buoy series. The latter (±0.4 m s^{-1}) also contains a contribution from the effects of higher frequency Eulerian variability, such

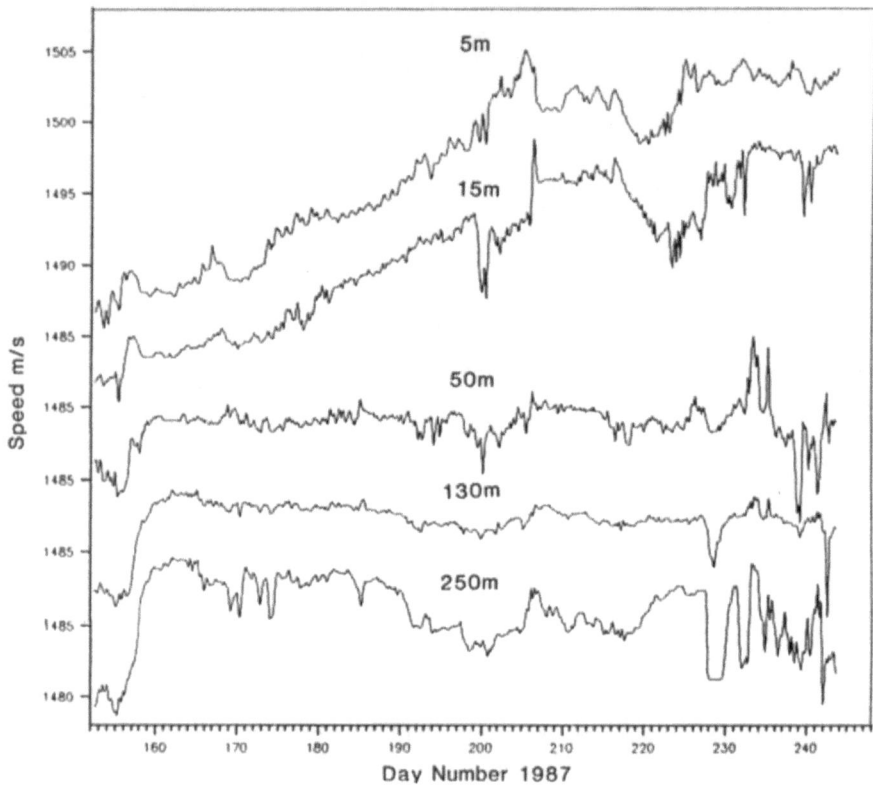

Figure 9. Sound speed time series calculated from the temperature measurements from Met Buoy 1 and a statisitical T-S relationship derived from nearby CTD stations. Adjacent traces are offset by 5 m s^{-1} and the axis labels refer to the values at 5-m depth. Note that the lowest level thermistor was dragged along the sea-floor from day 227 to day 230, and subsequent measurements from this sensor are of doubtful quality.

as might be generated by internal waves, the motion of which is too rapid for the Met Buoy to follow. The magnitude of the sound speed variations caused by tidal oscillations in the B2 series is <1 m s^{-1}. The major source of the Eulerian variability is in the mesoscale, i.e. motions of days to weeks or roughly 10 to 200 km. To demonstrate that this variability is caused primarily by horizontal and not by the vertical motions (i.e. such as might be forced by displacements of the interface below the Atlantic Inflow and deeper than our measurement window) we make the following simple calculations: to create a temperature change of 1 K d^{-1} using typical horizontal ($5 \ 10^{-7}$ K cm^{-1}) and vertical ($5 \ 10^{-5}$ K cm^{-1}) gradients as, for example, were found in the zone of B2 (Figure 2), would require a horizontal speed of 10 km d^{-1} normal to the temperature gradients, which tend to be alligned along the bathymetry, whereas it would require a vertical speed of 100 m d^{-1}. The standard deviation of the diabathic velocity component during the interval of days 163–183 was,

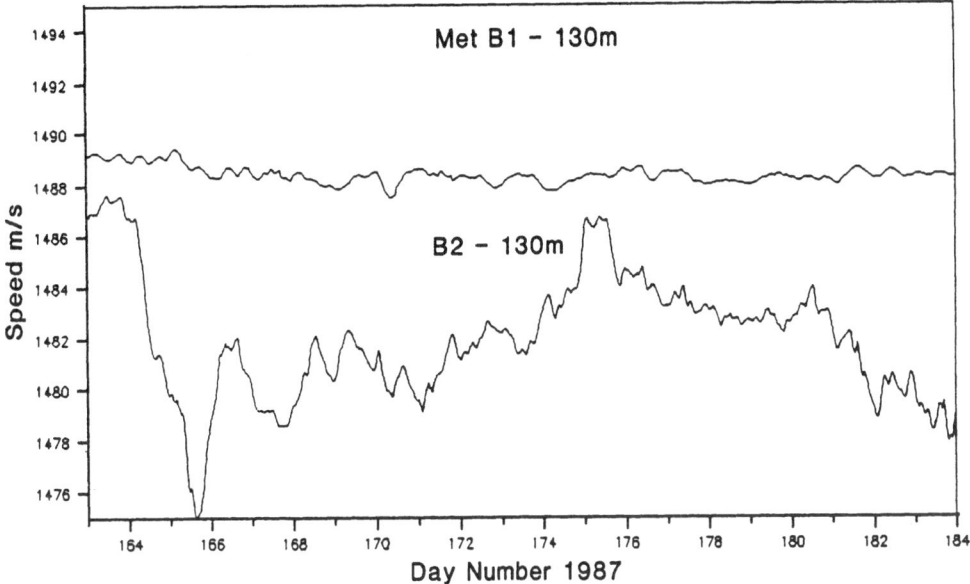

Figure 10. The sound speed time series from Met Buoy 1 for the five longest-lived thermistors. The calculation of sound speed from temperature was done using T-S relationships derived from accompanying CTD data.

for example, 11 km d^{-1}. The vertical speeds were not measured, but a vertical speed of 100 m d^{-1} (0.12 cm s^{-1}) over a period of a day or more seems quite unlikely.

As the second comparison, we consider the movement of Met Buoy 2 from days 196-202 when it made a loop of ~25-km diameter near the B2 mooring. The buoy trajectory along with the progressive vector diagrams from the three current meters of B2 and for the the same interval are shown in Figure 11. The Met Buoy trajectory gives the impression that it was caught in a circular eddy type of motion; however, the current meter data do not necessarily support this. Throughout the interval the flow at B2 was towards the southwestward at the 130-m depth suggesting that B2 was registering a countercurrent relative to the mean northeasterly flow experienced by the Met Buoy ~25 km to the east (also Figure 2c). If the velocity structure involved a countercurrent rather than an eddy (at ~130 m), we would ask how then did the buoy make the loop? There is some indication that the deeper motion, as registered by the current meters at 500 and 700 m, moved very suddenly offshore and then returned during days 197-200. We suggest that it is possible that the differential drag on the buoy was sufficient to cause it to move first westwards into the countercurrent and then back eastwards into the main northeasterly flow. This episode demonstrates that such drifting buoy trajectories should not be interpreted as providing even a visual image of the flow variability in regions of strong shear.

404

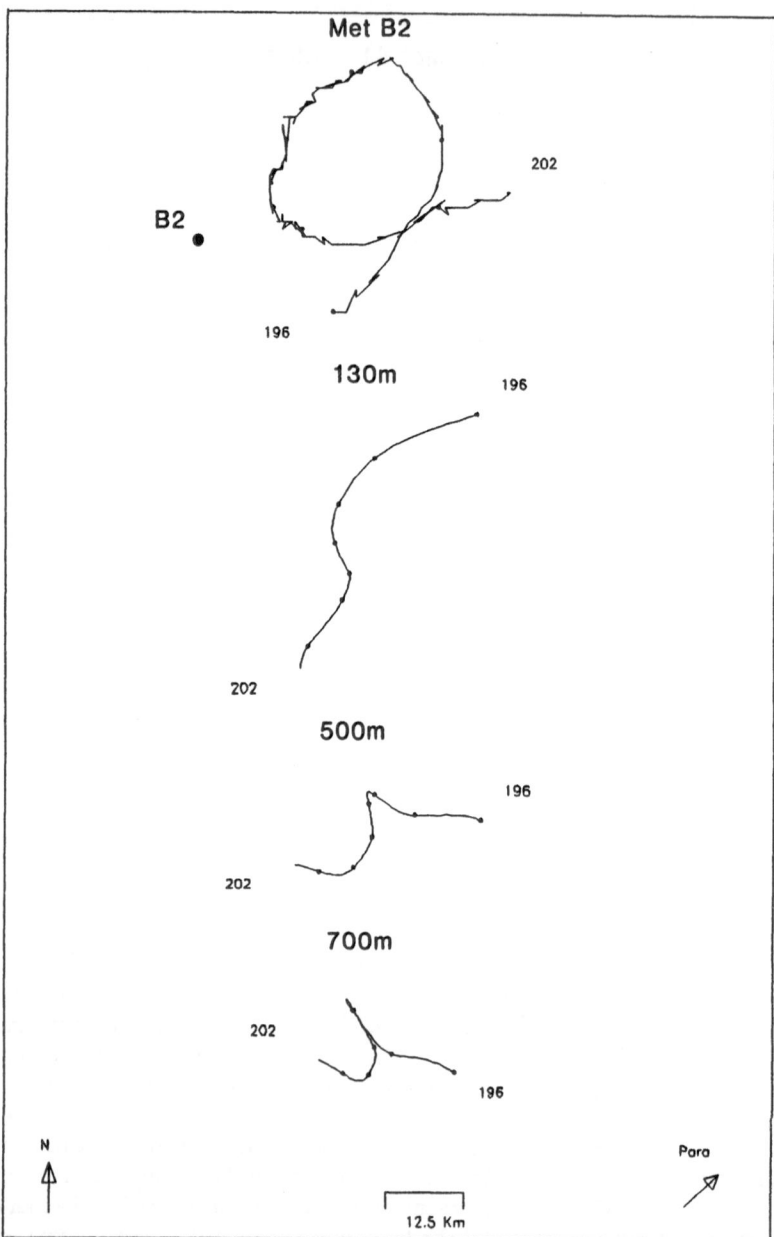

Figure 11. The trajectory of Met Buoy 2 (uppermost) and the progressive vector diagrams from the three current meters of the mooring B2 at the depths of 130, 500 and 700 m. The position of the B2 mooring relative to Met Buoy trajectory is also indicated, and the arrow marked 'Para' indicates the parabath, i.e. the tangent to the local isobath at the mooring site.

5. Conclusions

Satellite tracked free-drifting buoys supporting thermistor chains and using a satellite data relay are a novel and valuable source of oceanographic data over extended time intervals. Their obvious drawback is, of course, that after launch there is no control over their trajectories. It is tempting to analyse the data as Lagrangian time series, but, at least in the cases examined here, the Lagrangian signal is masked by the advective effects of differential shear acting on horizontal temperature gradients for all but an 80-m section of the thermistor chain. The Lagrangian data can be identified by much reduced variability and shorter correlation time scales, i.e. they can be identified without recourse to external data. In this case ancillary data were available and confirmed the depth of the thermistors for which the time series were least contaminated by advective effects. For such 'Lagrangian' time series the correlation times were ~ 6 h and the level of sound speed variability was smaller than that recorded at a nearby fixed mooring by a factor of ~ 6. Thus the oceanographic variability and sound speed variability in this area, probably in common with many other areas, is dominated by mesoscale structure which, apart from the surface temperature expressions discernible from satellite-borne imaging infrared radiometers (cloud cover permitting), is practically impossible to measure synoptically using conventional oceanographic tools. From the point of view of determining oceanographic variability for acoustic purposes a possible way forward is to use the acoustic measurements themselves in conjunction with numerical oceanographic mesoscale models.

References

Dickson, R.R. (1972). 'Variability and continuity within the Atlantic Current of the Norwegian Sea'. *Rapport et Procès-Verbaux des Réunions, Conseil International pour l'exploration de la Mer*, **162**, 167-183.

Dooley H. and J. Meincke (1981). 'Circulation and water masses in the Færoese Channels during OVERFLOW'73'. *Deutsche Hydrographische Zeitschrift*, **34**, 41-55.

Helland-Hansen, B. (1934). 'The Sognefjord section. Oceanographic observations in the northernmost part of the North Sea and the southern part of the Norwegian Sea'. James Johnstone Memorial Volume. Liverpool University Press, Liverpool, U.K., 257-274.

Helland-Hansen, B. and F. Nansen (1909). 'The Norwegian Sea, its physical oceanography. Based on the Norwegian researches 1900-1904'. Report on Norwegian fishery and marine-investigations, Bergen, **2**, 1-359.

Hopkins, T.S. (1988). 'The GIN SEA. Review of Physical Oceanography and Literature from 1972'. SACLANTCEN Report, SR-124, SACLANT Undersea Research Centre, La Spezia, Italy, 195 pp. Also in *Earth Sci. Rev.*, **27**. (In the press.)

Hopkins, T.S. and A. Mouchet (1990). 'The summer flow structure of the Norwegian Atlantic Current'. *Deep-Sea Research* (Submitted.)

Hopkins, T.S., G. Baldasserini, M. Ribera and P. Zanasca (1990a). 'The Atlantic Inflow Experiment. GIN Sea cruise 87-T : Part I TYDEMAN Hydrography'. SACLANTCEN Memorandum, SACLANT Undersea Research Centre, La Spezia, Italy. (In preparation.)

Hopkins, T.S., P. Giannecchini, L. Gualdesi, P.J. Minnett, A. Mouchet and P. Zanasca (1990b). 'The Atlantic Inflow Experiment. Part II: Circulation. A data report'. SACLANTCEN Memorandum, SACLANT Undersea Research Centre, La Spezia, Italy. (In the press.)

Hopkins, T.S., G. Baldasserini, A. Goffart, P. Povero and P. Zanasca (1990c). 'The Atlantic Inflow Experiment. GIN Sea cruise 87-B : Part III BELGICA Hydrography'. SACLANTCEN Memorandum, SACLANT Undersea Research Centre, La Spezia, Italy. (In the press.)

Hopkins, T.S., G. Baldasserini, G. Tognarini and P. Zanasca (1990d). 'The circulation and water mass structure in the southern GIN Sea: A summary of the Atlantic Inflow Experiment 1986-1988'. SACLANTCEN Report, SACLANT Undersea Research Centre, La Spezia, Italy. (In preparation.)

Meincke, J. and T. Kvinge (1978). 'On the atmospheric forcing of overflow events'. ICES C.M. C: 9. (Unpublished Document.)

Mysak, L. and F. Schott (1977). 'Evidence for baroclinic instability of the Norwegian Current'. *J. Geophys. Res.*, **82**, 2087-2095.

THEORETICAL DETERMINATION OF THE FRACTAL DIMENSION OF FLUID PARCEL TRAJECTORIES IN LARGE AND MESO-SCALE FLOWS

A. R. OSBORNE and R. CAPONIO
Istituto di Fisica Generale dell'Università
and Istituto di Cosmo-Geofisica del C.N.R.
Corso Fiume 4
10133 Torino
Italy

ABSTRACT. Recent work suggests that Lagrangian parcel trajectories in large and meso-scale flows can have a fractal dimension. Drifting buoy trajectories may be viewed as fractal curves in the plane of the ocean's surface and have typical fractal dimensions of about $D = 1.30 \pm 0.06$ in a range of scales normally attributed to geostrophic turbulence. Herein we address some of the theoretical issues and show that a simple Hamiltonian system with Eulerian velocity spectrum $k^{-\delta}$ ($\delta \sim 0.5$) can describe many of the measured properties of the drifter motions. Generally speaking the dimension D is related to δ, to the dispersion relation of the flow and to nonlinear dynamics. It is tempting to use the dimension D to describe some of the variability inherent in some oceanic motions; in particular we show that fluid parcels do not obey the classical diffusion law (Dt, D the diffusion coefficient) but instead diffuse anomalously ($Dt^{2/D}$). The mixing and transport of passive tracers are generally enhanced in flows of this type. We emphasize that the focus herein is on *dynamic fractality*, i.e. fractality arises as a consequence of the Hamiltonian equations of motion and hence may be used as a tool for probing dynamical behavior. This contrasts to more traditional studies in which fractality is often studied for its own sake or for artistic stimulation, independent of physical content.

INTRODUCTION

The study of variability in the oceans has recently been linked to the fractal dimension of drifter trajectories in the Kuroshio extension [1-6]. The drifter motions were found to have a fractal dimension of about 1.3, e.g. the trajectories themselves follow fractal curves in the xy-plane of the ocean's surface. Generally speaking fluid flows in which particle trajectories have a fractal dimension ($D > 1$) are much more irregular than those for which the dimension $D = 1$. Fractal particle motions are simply more difficult to predict because of the irregularities (which typically occur at subgrid scales in many models) and because chaotic effects may also be present. Flows of this type are often described by the simple equation [7-9]

$$\dot{x} = U + u(x,t) \tag{1a}$$

where $x(t) = [x(t), y(t)]$ is the position of a selected water parcel in the xy-plane of the ocean's surface, U is a constant zonal flow and $u(x,t)$ is the Eulerian velocity field. The particle motion is fundamentally nonlinear because x appears on both the left and right hand sides of equation (1a). Since the motion is planer the velocity field is expressible in terms of a (normalized) stream function, $u = u_{rms}(-\psi_y, \psi_x)$ and (1a) becomes

$$\dot{x} = U + u_{rms}\hat{k}\times\nabla\psi(x,t) = U + u_{rms}(-\psi_y, \psi_x) \tag{1b}$$

where u_{rms} is the rms flow velocity. The over-dot denotes time derivative, $\hat{k}$ is the unit vector perpendicular to this plane and $\nabla = (\partial_x,\partial_y,\partial_z)$. Equations (1b) are Hamilton's equations of motion in two dimensions; phase space and real space are seen to coincide [7].

Strictly speaking the velocity field u or stream function ψ are solutions to complex partial differential equations such as Navier-Stokes or the potential vorticity equation. One solves the primitive equations of motion and uses (1a) to get particle trajectories; this has been done by Babiano, et al [10] for the case of two dimensional turbulence. Here, for the sake of analytical tractability and clarity, we take, with Davis [9] and Pettini, et al [11] a different approach. We assume the stream function has the stochastic Fourier series:

J. Potter and A. Warn-Varnas (eds.), Ocean Variability & Acoustic Propagation, 407–416.
© 1991 *Kluwer Academic Publishers.*

$$\psi(x,t) = \sum_{m=-M}^{M} \sum_{n=-N}^{N} C_{mn} \cos(k_{mn} \cdot x - \omega_{mn} t + \phi_{mn}) \tag{2}$$

where $C_{mn} = [2P(k_{xm}, k_{yn}) \Delta k_m \Delta k_n]^{1/2}$, $k_{mn} = (k_{xm}, k_{yn})$ and $\omega_{mn} = \omega_{mn}(k_{mn})$ is the dispersion relation. The power spectrum of ψ is assumed to have the form of a power law, $P(k_x, k_y) = B(k) k^{-\gamma}$ ($k = |k|$) and the phases ϕ_{mn} are uniformly distributed random numbers on $(0, 2\pi)$. $B(k)$ is a constant between (k_o, k_N) (the "scaling" range), where $k_o = 2\pi/L$ corresponds to the largest spatial scale L (here ~ 640 km, the size of the domain in which the particles are numerically integrated) and k_N to the smallest (here ~ 10 km). We set $B(k)=0$ below k_o, while at high wavenumber we employ a cosine taper to place a smooth shoulder on $B(k)$ near k_N. This has the effect of rendering ψ differentiable in the fractal range $3 \leq \gamma \leq 4$ (see below). Equations (2) might be viewed as a solution to the simple linear wave equation. We demonstrate that (2) can describe a fractal surface which represents a class of solutions not previously considered in hydrodynamic applications.

We use the dispersion law:

$$\omega = U \cdot k + uk \tag{3}$$

where $U \cdot k$ describes "frozen-in" turbulence and u is a constant velocity. Stream function spectra $k^{-\gamma}$ correspond to Eulerian velocity spectra $k^{-\delta}$, $\delta = \gamma - 3$. The range ($3 \leq \gamma \leq 4$, $0 \leq \delta \leq 1$) is typical of many oceanic observations [1, 3, 6, 9].

The advantage of the choice (2) for the stream function is that the fractal and chaotic properties of a "realistic" flow (i.e. one for which the Eulerian velocity spectrum has a power law) can be computed using both analytical and numerical methods to show that fluid particle trajectory solutions of equation (1): 1) while differentiable and chaotic at small scale, can be *fractal space curves* at larger scale and 2) undergo *anomalous transport*.

There are several fundamental questions which we address with regard to these results:

(1) One normally thinks of solutions to partial differential equations as being reasonably smooth, well-behaved differentiable functions. How can it theoretically be possible for the *dynamical* motions of particle trajectories to *physically* be fractal curves? Is it unreasonable to think that such behavior might be possible in the context of fluid dynamical motions in general?

(2) What are the *physical implications* of fractal trajectories on fluid flows in general?

(3) What new, *unique physical information* can the fractal dimension give us with regard to large scale flows?

(4) Can the fractal dimension be of use in solving the *inverse problem*, i.e. determination of the general circulation from drifter trajectories?

(5) What relation does the fractal dimension have to fluid properties such as *anisotropic and anomalous diffusion* in fluids?

(6) What are the implications on *fractal front propagation* and on *fractal frontogenesis*?

These and other issues are addressed herein and elsewhere [4, 12].

DETERMINATION OF THE FRACTAL DIMENSION FROM THE EQUATIONS OF MOTION

The first step is to carry out a Galilean transformation on the equations of motion, $X=(x-Ut)/L$, so that (1) becomes:

$$dX/d\tau = \hat{k} \times \nabla \psi(X, \tau) \tag{4}$$

Each Fourier component of (3) then may be written

$$\psi_{mn} = \psi_{mn}(Lk_{mn} \cdot X - \mu L k_{mn} \tau + \phi_{mn}). \tag{5}$$

$\tau = u_{rms} t / L$ and $\mu = u/u_{rms}$ is the ratio of dispersion to nonlinearity in the flow.

We now address the case $\mu = 0$ for which the stream function does not vary in time; physically $\psi(x,0)$ has the appearance of a rugged mountain range (Figures 1, 2). Equation (4) becomes

$$dX/d\tau = \hat{k} \times \nabla \psi(X, 0) \tag{6}$$

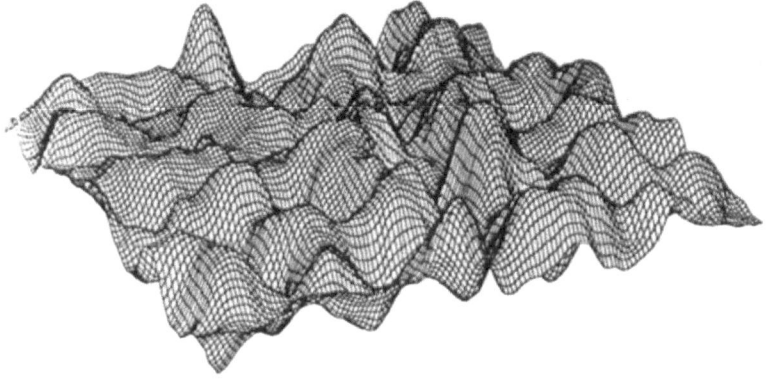

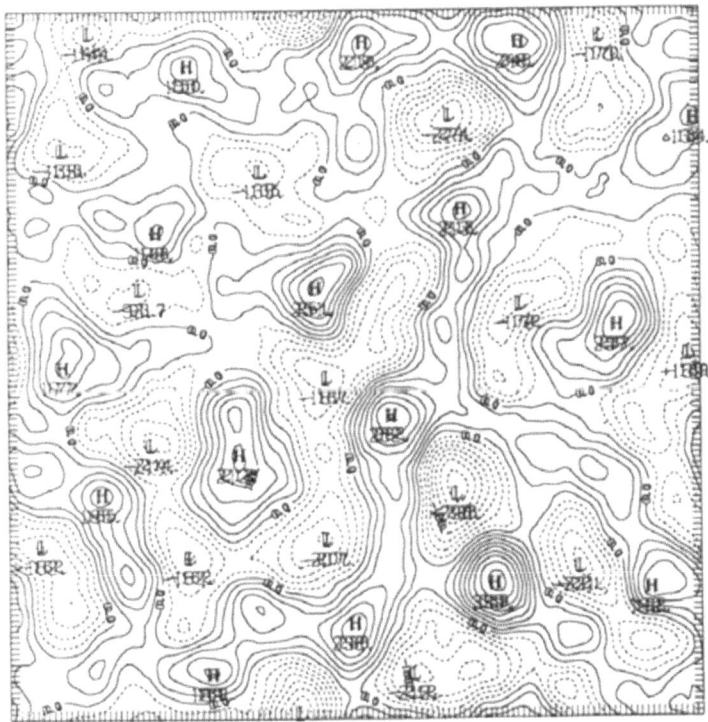

Figure 1. (a) Stream function surface for $\gamma = 6$, $\delta = 3$ as computed by equation (2) for uniformly distributed random phases. (b) Corresponding stream function contours. The stream function is nonfractal with dimension 2.0.

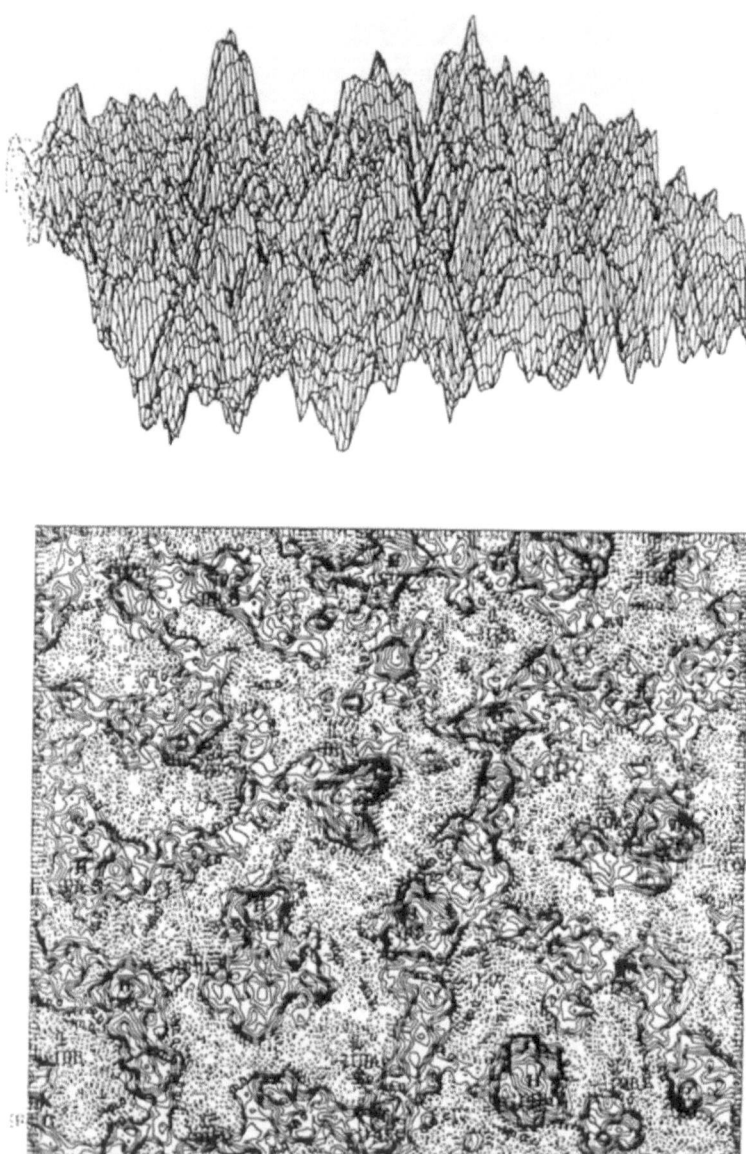

Figure 2. (a) Stream function surface for $\gamma = 3.5$, $\delta = 0.5$ as computed by equation (2) for uniformly distributed random phases. (b) Corresponding stream function contours. The stream function is a fractal surface with dimension 2.33.

which is quite nonlinear, but nevertheless mathematically integrable. This is because (6) generates trajectories which follow contours of $\psi(X,0)$. Selection of the initial conditions $X(t = 0) = X_0$ defines the particular contour and the motion is consequently perfectly periodic, albeit complicated, for all time.

In order to demonstrate that particle trajectories can be fractal curves in this case we first show that the Hamiltonian (3) (or stream function) $\psi(x,t)$ can be viewed geometrically as a fractal surface that is a function of x and y at some arbitrary fixed time t_0 (Figure 2). $\psi(x,t)$ is fractal only over a *finite range of physical scales* (i.e. it is a "natural" fractal [13]); the fractality extends from a small dissipation range up to the size of the physical system. The Hamiltonian is differentiable because we smooth it below small physical scales of interest (using a cosine taper at high frequency) and therefore Hamilton's equations remain valid, even in the presence of fractality. While the Hamiltonian is well behaved at small scales, it is extremely irregular and "non-smooth" over larger scales and generally has the appearance of a rugged, temporally varying, mountain range (compare Figure 1 (classical 2D turbulence, k^{-3}) to Figure 2 (fractal stream function, $k^{-0.5}$)). Chaotic fluid particle motions in the presence of a natural fractal Hamiltonian or stream function is the main subject of this paper. In particular we show that the Hamiltonian is a natural fractal surface of dimension $D'=\gamma/(\gamma-2) = (\delta+3)/(\delta+1)$, $2\leq D'\leq3$ in the scaling range (k_0,k_N) provided that $3\leq\gamma\leq4$ ($0\leq\delta\leq1$); it then can be shown that trajectories are fractal curves over a similar scaling range.

The proof of this assertion is motivated by the fact that for a time independent fractal stream function with dimension D', a trajectory is the *geometrical* intersection of a *horizontal plane* with ψ; it then naturally follows that the trajectory is a natural fractal curve with dimension $D=D'-1$. First consider the intersection of a *vertical plane* (assumed parallel to the x axis) with the surface $\psi(x,t_0)$ (t_0 fixed). The intersection has the (natural fractal) series [4, 12, 13]

$$\zeta(x,t_0) = \sum_n c_n \cos(k_n x - \omega_n t_0 + \phi_n) \tag{7}$$

where $c_n = [2P(k_n)\,\Delta k_n]^{1/2}$, which has the power spectrum

$$P(k) = \beta \left[\frac{1}{2} \cdot \frac{\gamma-1}{2}\right] k^{-\gamma+1} \tag{8}$$

where β is a beta function. The fractal dimension of (7) is easily computed to be [4, 12, 13]:

$$D_u = 2/(\gamma-2) = 2/(\delta+1), \qquad 1\leq D\leq2, \qquad 3\leq\gamma\leq4$$

$$D_u = 1, \qquad\qquad\qquad \gamma>4 \tag{9}$$

independent of t_0. The subscript "u" refers to the "upper limit" as the nonlinearity parameter $\mu^{-1} \to \infty$ ($\mu \to 0$). The dimension of the surface (2) is then $D'=D_u+1= \gamma/(\gamma-2)$, $2\leq D'\leq3$, for $3\leq\gamma\leq4$ and $D'=2$, for $\gamma>4$. We now note that the geometrical intersection of a *horizontal* plane with the Hamiltonian (2) is also given by (9); this is due to the relation $D'=D_u+1$, which is generally true for the intersection of a plane of *arbitrary* orientation with a fractal surface [13]. Hence for $\mu=0$ the motion $X(\tau)$ follows a contour and consequently eq. (9) is also the fractal dimension of the trajectory of a fluid parcel in the (time independent) stream function field (2). The fractal dimension of a trajectory in this limit ($\mu = 0$) is seen to depend solely on the exponent in the stream function or Eulerian velocity power spectra (e.g. γ or δ).

Surprisingly, for large μ, i.e. in the strict stochastic or linear limit of the particle motion, Eq. (9) is also found to hold [4, 12]; for intermediate μ more complex fractal dynamics can occur as seen below in the numerical simulations. The linear limit of the equations of motion in dimensionless variables is given by [12]

$$\dot{X} = U = \lambda \hat{k}\times\nabla_A\Psi(A,T; k_0, \gamma) \tag{10}$$

where the subscript on the gradient implies $\nabla_A = (\partial_A, \partial_B, \partial_C)$; A is the initial particle position $X_0 = A$. The solution of (1) (in dimensionless variables) is given by

$$X(T) = \sum_{m=-M}^{M} \sum_{n=-N}^{N} X_{mn} \cos[K_{mn}\mathbf{n}_{mn}\cdot A - K_{mn}T) + \phi_{mn}] \tag{11}$$

For the isotropic motion considered here the X and Y motions are statistically independent [12] and we may write

$$X(t) = \sum_{j=1}^{N} c_{xn} \cos(\omega_n t - \phi_{xn})$$

(12)

$$Y(t) = \sum_{j=1}^{N} c_{yn} \cos(\omega_n t - \phi_{yn})$$

where the Fourier coefficients are given by $c_{xn} = \sqrt{2P_X(\omega_n)\Delta\omega}$ and $c_{yn} = \sqrt{2P_Y(\omega_n)\Delta\omega}$, and the power spectra take the form of a power law

$$P_X(\Omega) = P_Y(\Omega) = B(\Omega) \, \Omega^{-\alpha}$$

(13)

where $B(\Omega)$ is constant in the range $(\Omega(K_o), \Omega(K_N))$. Note that the power spectra for the X and Y motions are the same; this results from the assumption that the motion is isotropic. The spectral exponent $\alpha(\gamma)$ is given by the relation

$$\alpha = \gamma - 1$$

(14)

which is strictly true *only in the linear limit*. In this limit the power law form of the trajectory power spectra (13) is a consequence of the power law form of the Eulerian velocity spectra. The conversion from wave number to frequency spectra is made using the dispersion relation.

It is possible to compute the fractal dimension of the motion give by equations (11, 12) [12]. In the linear limit ($\mu^{-1} = 0$) the fractal dimension depends 1) on δ and 2) on the assumed form of the dispersion relation, $\omega = uk$. This contrasts to the dimension computed above in the dispersionless limit, $\mu = 0$, where the dependence was solely on δ. Given the form of the power spectra (13) the dimension is found by [12]

$$D_l = \frac{2}{\alpha - 1} = \frac{2}{\delta + 1}$$

(15)

This is the "lower or linear dimension" D_l ($\mu \rightarrow \infty$) and it is identical to equation (9) above for the "upper or nonlinear limit" D_u ($\mu \rightarrow 0$). Consequently we have found the same dimension in the two limits $\mu \rightarrow 0$ and $\mu \rightarrow \infty$. Should the dispersion law differ from the simple form $\omega = uk$, then the lower dimension (15) would be altered. This possibility is explored elsewhere [12].

CHAOTIC DYNAMICS

Hamiltonian systems in which the Hamiltonian is itself a function of time, $\psi(x,t)$, are known to be candidates for chaotic motions. Systems of this type have a sensitive dependence on the initial conditions. Thus two particles, initially separated by a very small distance, will eventually diverge from each other at an exponential rate. Thus even particles so close together that they are separated by no more than the computer roundoff error (typically $\sim 10^{-16}$ in double precision) will, nevertheless, eventually separate exponentially in time. This is known as the "butterfly effect" and has been studied at length by many authors. Predictability in such systems over long time scales is not possible; hence the presence of chaos sets strict upper limits on how long chaotic dynamical motion can be accurately predicted. This time corresponds roughly to λ^{-1}, where λ is the maximum Lyapunov exponent of the flow. Chaotic motions are recognized as important in fluid dynamical motions (see [8] for a review) and it is therefore worthwhile treating them in detail in the present case.

We give a brief summary of the μ-dependent chaotic dynamics of the Hamiltonian system which we study here (1-3) [12]: (a) When $\mu=0$ the motion is highly nonlinear, but is nevertheless exactly integrable in a geometrical sense. For the initial conditions $X(0)=X_o$, the particle motion $X(\tau)$ lies on the contour given by $\psi(X_o)$=const. $X(\tau)$ can also be a fractal curve in the xy-plane if $0 \leq \delta \leq 1$. (b) For $\mu \ll 1$ and finite, the flow is chaotic and of *osculating KAM* (Kolmogorov-Arnol'd-Moser) *type*, i.e. the Hamiltonian consists of $\psi(X)$ (as a solution of $dX/d\tau = k \times \nabla\psi(X)$) plus a time dependent perturbation: $\psi(X,\tau)=\psi(X)+\mu\psi_p(X,\tau)$. The KAM "surfaces" are fractal curves around which stochastic layers form; renormalization of the perturbation theory is required after finite time to insure that the perturbation remains small. (c) For $\mu \gtrsim 1$, the flow is chaotic and lies in a *trapping regime*; *vortices* and *vortex hopping* are seen to occur. (d) For $\mu \gtrsim$

1, the flow is *fully stochastic* and may be described as a *nongaussian random walk* in the plane. These regimes are illustrated by numerical simulations below.

NUMERICAL SIMULATIONS

We now present examples of particle motions found by numerical integration of equations (2, 4, 5) [12]. We select 128^2 wavenumber components in (2) (therefore M=N=64), and the lower cutoff is chosen to be either $k_0=k_1$ or $4k_1$, where $k_1=2\pi/L$, for L=640 km. The upper cutoff is given by $k_N=64k_1$. Figure 3 gives an example in which the stream function does not change in time, $\mu=0$, $\gamma=7/2$ ($k_0=k_1$) for a time series of 8192 points. The stream function is fractal with dimension D'=7/3; the contours are thus fractal, D=4/3, and quite irregular in appearance (Fig. 3a). A trajectory follows precisely a fractal contour (Fig. 3(b)) and is also fractal with dimension 4/3.

In Figure 4(a) we give an example with $\mu=0.0014$, $\gamma=7/2$ ($k_0=4k_1$) in order to study chaotic aspects of the motion. For finite μ the particle is not constrained to follow steady stream lines but instead follows *chaotic* stream lines. In Figure 4(b) we give the case $\mu = 0.014$, $\gamma=7/2$, which is in the trapping regime where vortices and vortex hopping occur. The motion is nearly fully stochastic for $\mu=1.4$, $\gamma=7/2$ (Fig. 4(c)). Note the quite irregular appearance of the flow; the absence of vortices is an important feature of this motion which behaves as a kind of nongaussian random walk [12].

Note that by increasing μ the range over which fractality occurs is enhanced. This is seen in Figs. 4(a)-(c), where increasingly finer structure becomes visible at small scales for larger μ. The rich dynamical exploration of the scales of motion emphasizes profound differences between the *fractal Hamiltonian dynamics* of equation (4) and simple *fractal geometry* [13].

ANOMALOUS DIFFUSION

In the fully linear (stochastic) limit $\alpha = \alpha(\gamma) = \gamma - 1$, where equation (12) allows the temporal dependence of absolute diffusion to be computed analytically:

$$A(t) = <|x(t) - x(0)|^2> = D\ t^{2/D}$$

The dimension $D = 2/(\alpha-1)$, $2\leq\alpha\leq3$, $1\leq D\leq2$ and D is a "diffusion coefficient." For $\gamma=6$ we find D = 1 and the fractal contribution to absolute diffusion is purely convective $\sim t^2$. For the case $\gamma=7/3$, D = 4/3 and the time dependence is $t^{3/2}$, consistent with data [6]. Classical diffusion occurs when $A(t) \sim t$, which is seen to correspond to D = 2, e.g. Brownian motion.

DISCUSSION

We have shown that a simple dynamical model, previously considered in the context of large and mesoscale motions [5, 9, 12], predicts certain properties observed in floating drifter observations made in the Kuroshio Extension. The property of most interest here is that of *dynamic fractality* in which fluid particle motions are seen to be fractal curves in the xy-plane of the motion. This result, which at first seems surprising, arises because of the nonlinear structure in Hamilton's equations of motion and because the stream function (e.g. the sea surface elevation at large and meso-scales) is a fractal surface over a range of scales normally attributed to geostrophic turbulence. This class of stream functions has evidently not been considered previously. The fractal nature of the stream function occurs because the power spectrum has a power-law behavior $(k^{-\gamma})$, where $\gamma \sim 3.5$. Note that this value of γ differs from that normally assumed for geostrophic turbulence ($\gamma = 6$, $\delta = 3$); the reason for this experimentally determined discrepancy remains an open active area of research.

Fractality also depends on the form of the dispersion relation of the flow. This dynamical dependence offers the possibility of experimentally probing large and meso-scale motions for dispersive effects. Another prospect arises from the fact that fractality also varies according to the nonlinear dynamics of the flow; thus deviations of the dimension from the values in the linear and nonlinear limits should occur. In particular the dimension inside vortices should be $D \sim 1$. Other dynamical variations of D in the "clustering regime" [12] should also be observable.

Another active area of research relates to the inverse problem associated with the determination of the stream function or Eulerian velocity field from drifter measurements. The spirit of this quest can be seen best at the linear order of approximation as given by equation (11). The phases in (11) are the same as those

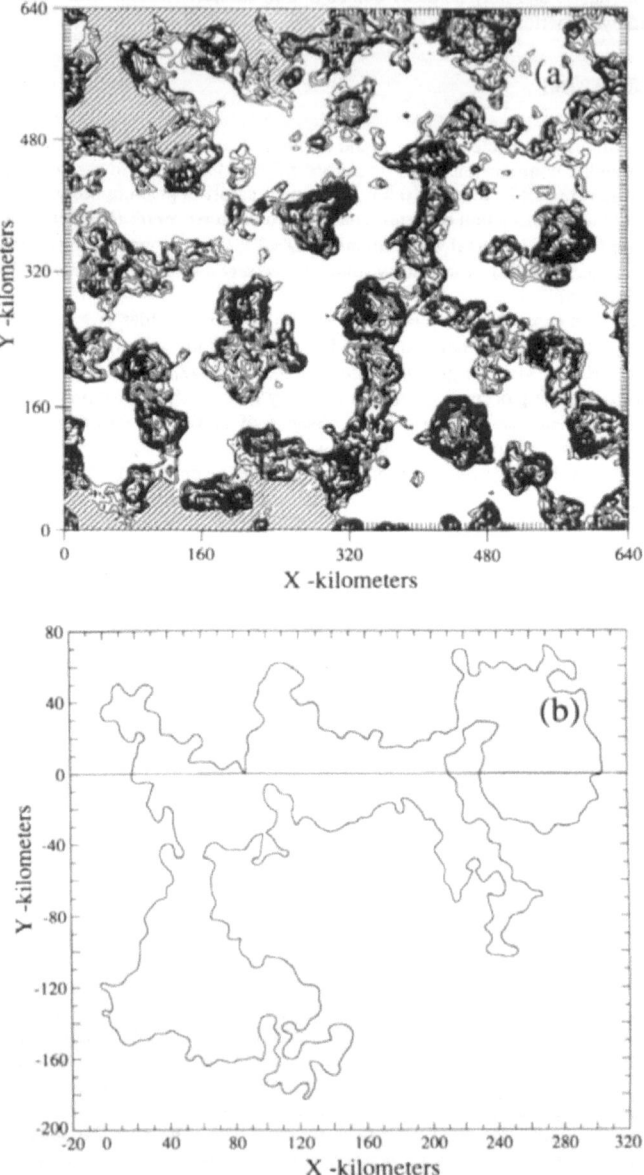

Figure 3. Stream function contours are fractal curves (D=4/3) for Eulerian velocity spectrum, $k^{-0.5}$ (μ=0.0) (a). A fractal (D=4/3) particle trajectory follows a stream line (b). Hashed region in (a) is bounded by the trajectory.

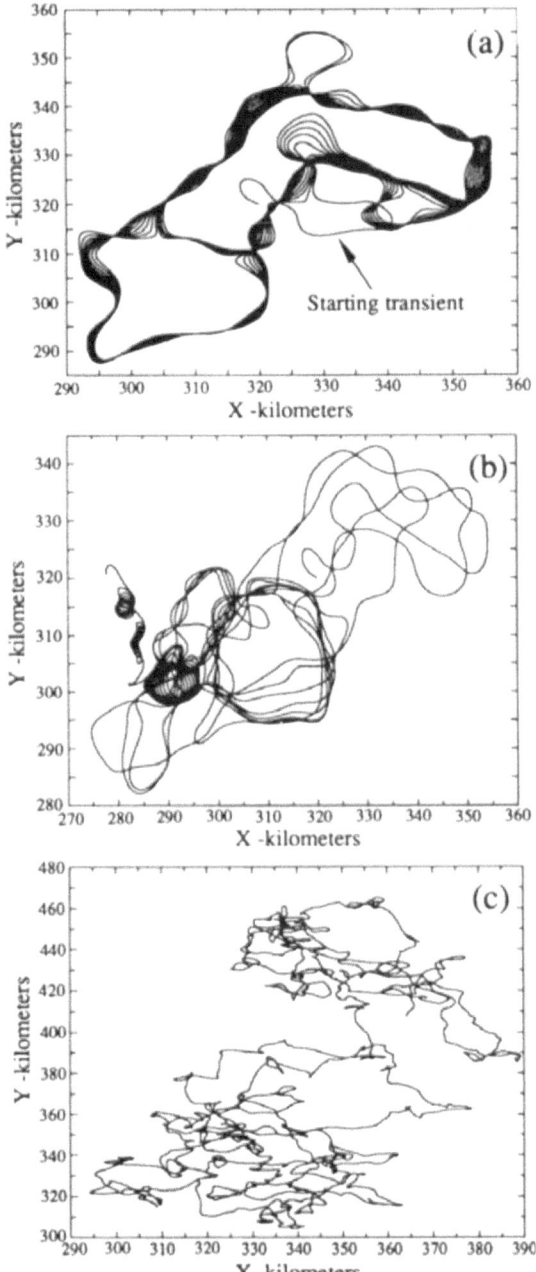

Figure 4. (a) Particle trajectory for nearly integrable Hamiltonian ($\gamma=7/2$, $k_o=4k_1$), $\mu=0.0014$, (b) Particle trajectory in trapping regime, $\mu=0.014$. (c) Particle trajectory in fully stochastic limit, $\mu=1.4$.

in the stream function and the directional spectrum of the trajectory is related to the stream function spectrum by a simple transfer function. Thus the inverse problem is easily accomplished in the linear approximation. The real challenge for future research is to determine the appropriate nonlinear corrections for this inverse procedure.

Two other interesting areas of research are that relating to the anisotropic and anomalous diffusion of passive tracers in the flow, mentioned briefly above, and to fractal front propagation in flows of the type considered herein. Both of these are pursued in detail elsewhere [12].

The problem of acoustic propagation in the ocean, and its association with the fractal dynamics of fluid parcels at large and meso-scales as described herein, is quite unexplored at the present time and it is therefore difficult to reach definite conclusions about what the possible influence of fractality on sound waves might be. Nevertheless we are of the opinion that a rather general framework can be developed for investigating these problems and this will be reported on in the future. Essentially, in its most general form, one separates the Navier-Stokes equation and the continuity equation into contributions at leading order (e.g. the linear wave equation) plus higher order nonlinear effects. The velocity field $u(x,t)$, density field $\rho(x,t)$ and the pressure field $p(x,t)$ are then described in terms of this dynamics, which at leading order are connected by linear transfer functions in the wave number domain. Using a power law for the velocity field $(k^{-\delta})$ then specifies all the fields in the Fourier sense. Propagation of sound waves may be viewed from this perspective as being influenced by power law spectra (from, say, currents at large and meso-scales, $\sim k^{-0.5}$, or from internal waves, $\sim f^{-2}$) and the techniques applied herein may then be used to search for fractal effects; multifractality would arise by considering higher order nonlinearities in the fundamental fields. Such a theoretical framework would then allow for an investigation of the influence of fractality on sound velocity and other ascoustic properties, and for specifically addressing fractal effects on various models used for predicting sound propagation in the ocean.

We thank Prof. C. Castagnoli and Prof. L. Bergamasco for their valuable aid, support and encouragement. Much of this work is taken from reference [12].

REFERENCES

[1] Osborne, A. R., Kirwan, Jr., A. D., Provenzale, A. and Bergamasco, L. (1986) 'A search for chaotic behavior in large and mesoscale motions in the Pacific Ocean,' *Physica* D 23, 75-83.

[2] Osborne, A. R. and Provenzale, A. (1989) 'Finite correlation dimension for stochastic systems with power-law spectra,' *Physica* D 35, 357-381.

[3] Osborne, A. R., Kirwan, Jr., A. D., Provenzale, A. and Bergamasco, L. (1989) 'Fractal drifter trajectories in the Kuroshio extension,' *Tellus* 41A, 416-435.

[4] Osborne, A. R., and Caponio, R. (1990) 'Fractal Trajectories and Anomalous Diffusion for Chaotic Particle Motions in 2-D Turbulence,' *Phy. Rev Lett.* 64, 1733-1736.

[5] Osborne, A. R., and Caponio, R. (1990) 'The Transition From Chaos to Stochasticity in 2-D Turbulence,' to appear in *Nonlinear and Turbulent Processes in Physics*, edited by A. G. Sitenko, V. E. Zakharov and V. M. Chernousenko, World Scientific, Singapore.

[6] Provenzale, A., Osborne, A. R., Kirwan, Jr., A. D. and Bergamasco, L. (1990) 'The study of fluid parcel trajectories in large scale ocean flows,' in A. R. Osborne (ed.), *Nonlinear Topics in Ocean Physics*, Elsevier, Amsterdam.

[7] Aref, H. (1984) 'Stirring by chaotic advection,' *J. Fluid Mech.* 143, 1-21.

[8] Ottino, J. M. (1990) *The kinematics of mixing: stretching, chaos, and transport.* Cambridge University Press, Cambridge.

[9] Davis, R. E. (1982) 'On relating Eulerian and Lagrangian velocity statistics: single particles in homogeneous flows,' *J. Fluid Mech.* 114, 1-26; (1983) 'Oceanic property transport, Lagrangian particle statistics, and their prediction,' *J. Geophys. Res.* 90, 4756-4772; (1985) 'Drifter observations of coastal surface currents during CODE: The statistical and dynamical views,' *J. Mar. Res.* 41, 163-194.

[10] Babiano, A., Basdevant, C., Legras, B. and Sadourny, R. (1987) *J. Fluid Mech.* 183, 379.

[11] Pettini, M., Vulpiani, A., Misguich, J. H., De Leener, M., Orban, J. & Balescu, R. (1988) *Phys. Rev. A* 38, 344.

[12] Osborne, A. R. and Caponio, R. (1990) 'Chaotic Lagrangian particle trajectories in two dimensional turbulence,' in preparation.

[13] Mandelbrot, B. B. (1982) *The Fractal Geometry of Nature.* Freeman.

How do eddies modify the stratification of the thermocline ?

J T Allen, R T Pollard and A L New

Abstract

Eddies play an important role in the mixing processes and transports of heat, mass and potential vorticity in the world's oceans. This role needs to be quantified in terms of eddy parameters and associated characteristic features. The eddies examined here, from all the major oceans, are diverse in character, with both cyclonic and anticyclonic examples, and having both simple and complex (double core) structures. Horizontal scales range from six kilometres to around two hundred and the depths at which the eddies are observed range from near the surface to close to the abyssal plain. With volume transports around the core of up to 40 Sverdrups and estimated lifetimes of 1 to 3 years, some of the eddies examined are clearly of significance from both the military and the commercial points of view. Observations clearly show anticyclonic (cyclonic) eddies to have decreased (enhanced) stratification at their core consistent with the theoretical structure of potential vorticity anomalies. Both anticyclonic and cyclonic eddies may display cold or warm surface temperature signatures.

Introduction

A literature survey has been carried out to extract a wide variety of well documented eddies from recent papers. Some two hundred papers have been examined and twenty eddies picked for closer study and comparison in this paper (Table 1). They include examples from all the major oceans and demonstrate the great variations in structure that may be observed. The variation in eddy scales is large, with diameters ranging from 6 km to 200 km. Vertically their characteristic distortion of pycnoclines can be felt over entire ocean depths or be limited to surface layers.

The objective of this work is to explore to what extent eddies can be identified and categorised in terms of typical properties, including their surface signatures. Such categorisation could help their assimilation into numerical models, and improve the predictive capability of acoustic models.

The most important property of an eddy is its potential vorticity (Rossby, 1940; Ertel, 1942), which involves its stratification and rate of spin. Potential vorticity is akin to angular momentum (Pollard et al 1990), and is conserved in an eddy except where the eddy intersects the surface or the ocean floor. Hoskins et al (1985) discuss potential vorticity at length and describe how it is partitioned between stratification and rate of spin. An eddy can be described as a potential vorticity anomaly (relative to its surroundings), knowledge of which is sufficient to specify completely the surrounding stratification (Fig 1). We shall show, as in Fig 1, that every anticyclonic eddy contains a region of decreased stratification in its core, while the core of every cyclonic eddy has increased stratification.

Biologically, eddies are important particularly with regard to the transport of nutrient rich waters into the surface layer and the movement of homogeneous lenses isolated great distances from their parent water masses. From a military point of view an eddy provides a change in acoustic environment (Petitpas and Browning 1986) in which sound channels and the surface duct will be very different from those in the surrounding and probably well forecast ocean area. A detailed model study of acoustic propagation through a cyclonic ice edge eddy (Mellberg et al, 1987) found that the sound speed structure within the eddy significantly altered the way propagation loss and acoustic modes were dependant upon receiver depth, frequency and source parameters. In a real time A.S.W. situation, the question "hunter or hunted ?" depends to a large extent on who has the best knowledge of the acoustic environment and can therefore avoid detection more effectively. However, this knowledge must rely on few (if any) in-situ observations. Therefore, the necessary sub-surface predictions require that historical study of eddy structures, associated characteristic features and related satellite interpretation will enable the assimilation of poorly identified eddies into numerical ocean circulation and acoustic models.

J. Potter and A. Warn-Varnas (eds.), Ocean Variability & Acoustic Propagation, 417–431.
© 1991 Kluwer Academic Publishers.

TABLE 1

AUTHOR	Cyclonic (C) Anticyclonic (A) D = double	Propagation Poleward (P) Equatorward (E)	Diameter (km)	Depth of Influence (m)	Depth of core (m)	Maximum tangential velocity (cms⁻¹)	Position	Surface Features
Feliks et al (1987)	A	E	100	1000	200	20	Mediterranean	None
Eliot et al (1986)	A	E	50	2000	1400	28	30°N 70°W	None
Newton et al (1974)	A	-	30	250	150	20	Arctic	Ice-covered
Le Groupe Tourbillon (1983)	A	Westwards	120	1500	500	30	47°N 15°W	None
Andrews et al (1976)	A	P	250	> 500	0	60 - 170	East Australian Current	Warm
Houghton et al (1986)	A	E	20	70	35	50	New England Cont. Shelf	Warm
Gordon et al (1986)	D A	E	70	5000	2700 / 3500	8	S W Atlantic	None
Brundage et al (1986)	D A	E	170	> 900	350 / 700	30	Sargasso Sea	Weak cold
Cresswell et al (1986)	D A	E	200	> 700	150 / 350	100	S E Australia	Weak cold
Bogdanov et al (1986)	D A	-	180	~ 1500	200 / 700	74	N W Pacific	Warm salty
Gründlingh (1987)	C	P	190	> 1000	250	50 - 100	S W Indian Ocean	Warm
Olson (1980)	C	-	100	> 2000	200	140	Gulf Stream	Warm
Schauer (1987)	C	P	70	2500	1800	10	N E Atlantic	Warm salty
Johannessen et al (1987)	C	Eastwards	20	400	125	20	Fram Strait	Fresh, cold, light
Richardson et al (1979)	C	P	150	5000	150	70	Gulf Stream	Cold, Fresh
Butnov et al (1986)	C	P - E	200	1000	250	50 - 80	Gulf Stream	
Lindstrom et al (1986)	C	E	60	1000	300	-	Gulf Stream	Fresh
Kupferman et al (1986)	C	P	70	> 900	200 - 500	40	Gulf Stream	Very cold
Johannessen et al (1987)	C	Eastwards	30	500	100	40	Fram Strait	Cold, dense
Manuito et al (1985)	C	-	6	80	10	50	Ligurian Sea	Cold

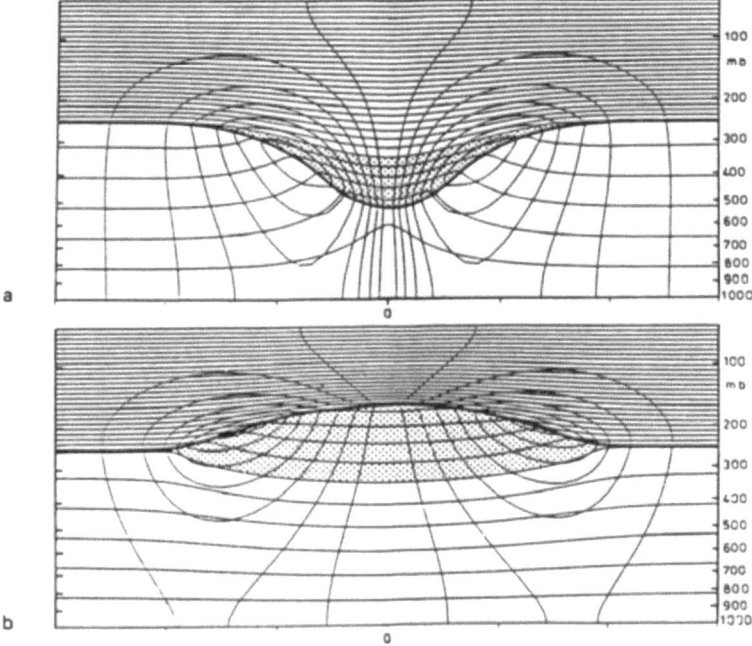

Fig 1 Idealized isentrope sections through (a) a cyclonic vortex
and (b) an anticyclonic vortex in the tropopause. Also shown are
lines of equal tangential velocity. [*Fig 15 Hoskins et al*]

I Anticyclonic Eddies

The classic anticyclonic eddy takes the form of a biconvex lens-like feature
in an isopycnal cross section. These homogeneous lenses of water may travel
considerable distances from their point of generation. Good examples are formed from
Mediterranean water outflowing over the Strait of Gibraltar sill; these appropriately
named "Meddies" carry high salinity water far into the North East Atlantic (Harvey
(1987)) and also six thousand kilometres westwards from their source to be observed
off the Bahamas (McDowell and Rossby (1978)). The propagation of isolated lenses
westwards is considered by Nof (1981,1985) and the analysis extended further by
Killworth (1983,1986). In the earlier case of a Meddy propagating northwards, as it
does so it will gain positive (cyclonic) planetary vorticity. Ignoring the physically
unlikely case of an eddy existing with negative absolute vorticity then to conserve
potential vorticity either its relative vorticity must become more negative (gain more
anticyclonic spin), or the isopycnals must spread farther apart and reduce the
stratification of the eddy core. Hence, if it were not for external frictional and mixing
processes acting on these eddies, they might be expected to become more intense
features as they moved northwards eventually reaching a point of dynamic instability
and break up.

Of the eddies closely examined for this paper, six appear to fit the classical
anticyclonic eddy structure. Feliks et al (1987) discussed a sub-thermocline anticyclonic
lens of high salinity water (Fig.2) in the eastern Levantine Basin formed off the Turkish
coast in a similar region as Mediterranean Intermediate Water and then having
propagated south and east to the point of observation. The eddy had a horizontal scale
of around 100 km in diameter, however, it appeared to be bounded above by the
seasonal thermocline and below the horizontal gradients died out at just under 1000m.

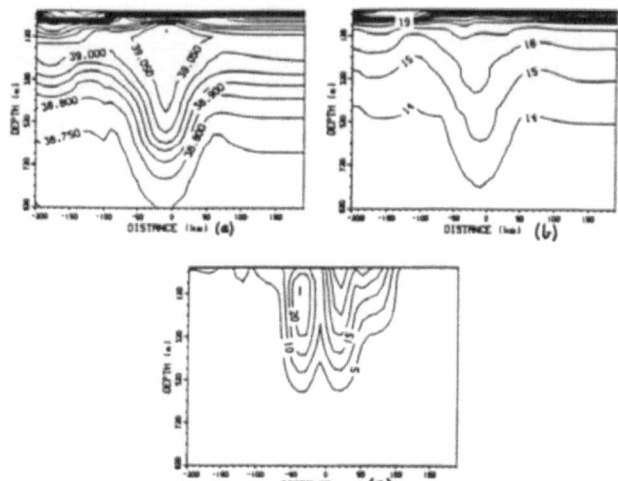

Fig.2 Vertical distribution of (a) temperature, (b) salinity and (c) geostrophic velocities ref. 920m [*Figs 4 and 5 Feliks et al 1987*].

The depth of the core of the eddy was at 200m and from geostrophic calculations relative to 920m it had a maximum tangential velocity of 20 cms[-1]. This was therefore a wide, squat and fairly slowly rotating feature compared to some of the other examples covered later in this paper. The eddy does not appear to significantly affect the thermocline although it is not clear whether the surface feature to the left of the eddy in Fig 2 has some connection. Therefore little or no visible or infra red surface signature could be expected, however some doming of the surface over the eddy might be detected by satellite altimetry.

The lens observed by Elliot et al (1986) at around 30°N, 70°W in the north west Atlantic was only some 50 km in diameter (Fig 3) but its distortion of isopycnals is noticeable from 900m down to 3000m with a core depth at around 1400m. Detailed current data showed the eddy to be propagating at up to 20 cms[-1] to the south west with a tangential velocity of 28 cms[-1] maximum at 1500 db. The eddy distorts the permanent thermocline at 1000 - 1100 metres but no surface signature would be expected apart from the possibility of detecting characteristic doming of the sea surface. According to water properties the core of the eddy is thought to have originated in the Labrador Sea. To conserve potential vorticity the eddy will have had to either gain positive (cyclonic) relative vorticity i.e.slow down or its vertical extent must decrease; this implies the eddy must become a less pronounced feature as it moves southwards.

In contrast, Newton et al (1974) investigated an eddy in the Beaufort Sea, from an ice camp during part of the 1972 Arctic Ice Dynamics Joint Experiment (AIDJEX) Pilot Study, which took the form of an anticyclonic lens - like feature only 25 - 30 km in diameter (Fig 4). The characteristic distortion of isopycnals extends from about 60 metres to a little under 300 metres and is remarkably symmetrical although this could be a false impression given by the sparse nature of the oceanographic stations. The core depth is around 150m and geostrophic calculations relative to a level of no motion below the ice at 20m give a tangential velocity of 20 cms[-1] whilst observations of surrounding ambient drift suggest a propagation speed of 2 cms[-1]. Covered by ice pack there could obviously be no surface signature and other physical balances in ice structure would obscure any small barotropic doming of the surface layers that might otherwise be expected. T - S correlations of these small scale eddies in the Beaufort Sea imply that they are not locally generated and the paper makes no attempt at suggesting an origin, however it seems likely that possible processes may include shelf edge spin up or the generation of negative (anticyclonic) relative vorticity in water masses moving north into the Arctic Ocean through the Bering Strait. The increase in planetary vorticity (cyclonic) would be expected to be balanced by a decrease in relative vorticity if the stratification remains constant and potential vorticity is to be conserved.

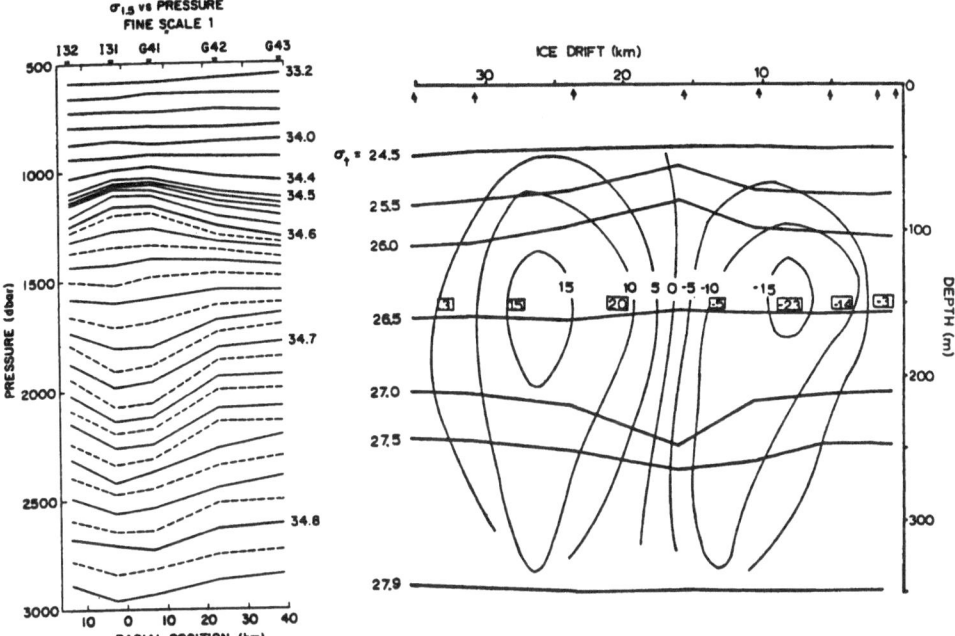

Fig.3 Fine scale survey section of potential density relative to 1500 db

($\sigma_{1.5}$) running east to west.
[*Fig 5 Elliot et al 1986*]

Fig.4 Section through under ice eddy
[*Newton et al (1974)*] isotachs are
computed geostrophic values relative
to 20 metres in cms^{-1}.

Le Groupe Tourbillon (1983) and later Arhan et al (1985) described an anticyclonic eddy observed during the Tourbillon Experiment 1979 at 47°N, 15°W in the North East Atlantic (Fig 5). Observations were made over a fifty day period during which time the eddy propagated westwards at 1.8 cms^{-1}; the core consisted of north east Atlantic central water and a tongue of Mediterranean water had been entrained around the circumference just above the permanent thermocline. The Tourbillon eddy was some 100 - 140 km in diameter and had a clear signature from just below the surface to 1500m. Once again the only signature detectable by satellite would have been surface doming as the eddy had little consistent correlation with surface temperature gradients during the survey, however, dynamic height maps indicate expected surface doming of the order of 10 - 20 cm.

Andrews and Scully - Power (1976) surveyed a warm core winter eddy generated off the East Australian Current. This anticyclonic feature broke the surface at or near its core leaving a distinct warm surface signature about 250 kilometres in diameter (Fig 6). Generally eddies formed in this region propagate south (polewards) at around 5 - 10 kilometres per day, however in this case the eddy is a surface breaking feature and air - sea interaction becomes important when considering spin down and mixing ; at such a boundary potential vorticity within the eddy will not necessarily be conserved. A surface breaking anticyclonic eddy such as this also produces a pronounced variation in mixed layer depth (Fig 7).

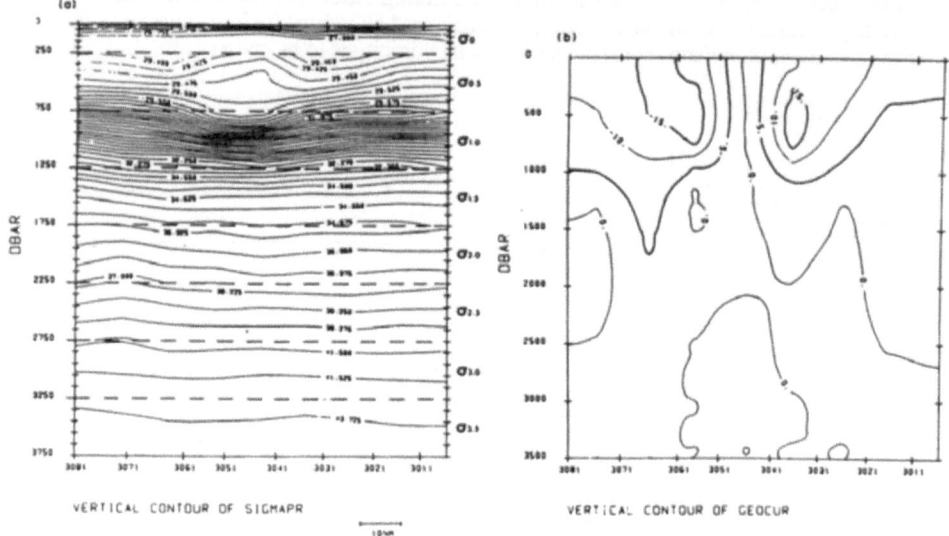

Fig 5   Sections of salinity, potential temperature, density
and geostrophic currents rel.3500db. [*Fig 16 Le
Groupe Tourbillon*].

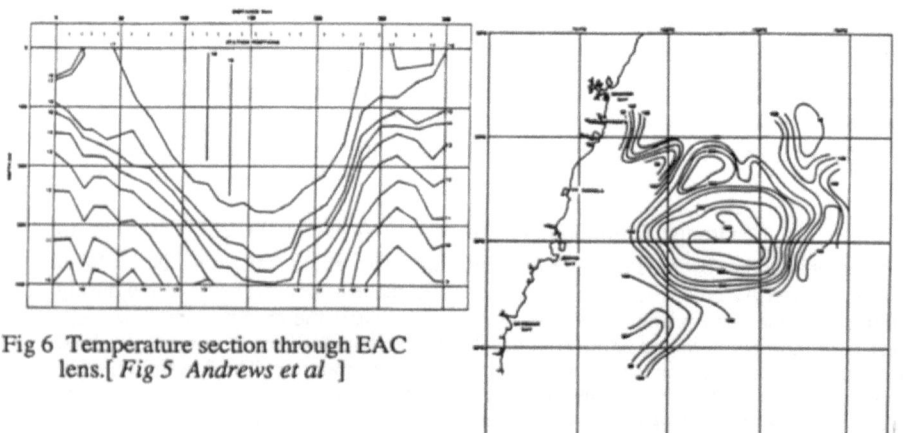

Fig 6 Temperature section through EAC
lens.[*Fig 5 Andrews et al*]

Fig 7   EAC eddy contours of mixed
layer depth (m). [*Fig 4 Andrews et al*]

      The last example of a simple classical anticyclonic eddy to be considered
here was formed off the continental shelf break south of New England (Houghton et al,
1986). This eddy was small compared to previous examples, 15 - 20 km in diameter
with a core depth of 35m. Its disturbance of the water column was apparent from the
surface to around 70 metres (Fig 8). During most of the survey the eddy remained
connected to a tongue of the shelf water from which it had been spun. It seems likely
from energetic considerations that the eddy was extruded by the passing of a warm core

Gulf Stream ring and developed its anticyclonic spin (max.0.5ms⁻¹) as a result of the compression of shelf water into the surface layers of the open ocean.

Four other anticyclonic eddies were examined and all demonstrate the other commonly occurring form of anticyclonic structure, namely a double biconvex lens. A deep water eddy in the southwest Atlantic, Gordon et al (1986), had a second lens sitting above and slightly off - centre (Fig 9). Modelled dynamics of conjugate vortex pairs demonstrate this eccentricity (Nof, 1985).The main eddy is centred at 3500 metres where it has a diameter of approximately 70 km and the whole double feature distorts isopycnals from above 2000m to the floor of the abyssal plain. Water properties suggest formation on the Falkland Plateau; it may have been generated by spin off from the Cape Horn or Falkland Currents and then carried down on the $\sigma_4 = 46$ isopycnal as it moved north.

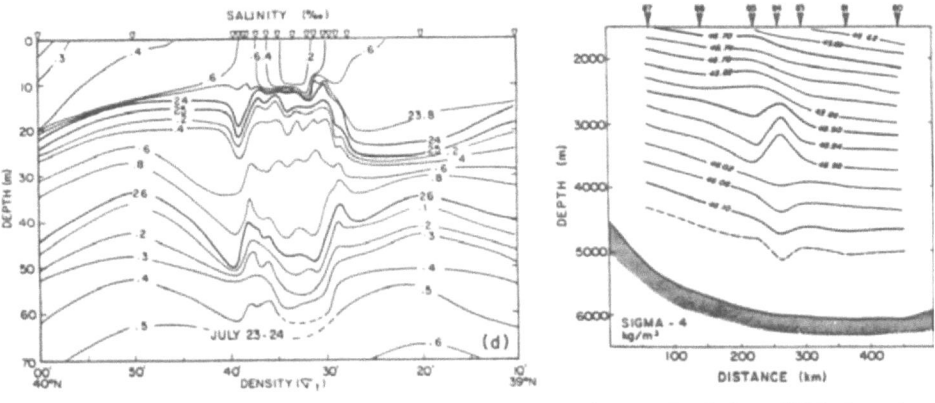

Fig 8   Density section through New
England shelf water eddy.
[*Fig 2   Houghton et al*].

Fig 9   Density (rel. to 4000m) section
across abyssal eddy.
[*Fig 2   Gordon et al*]

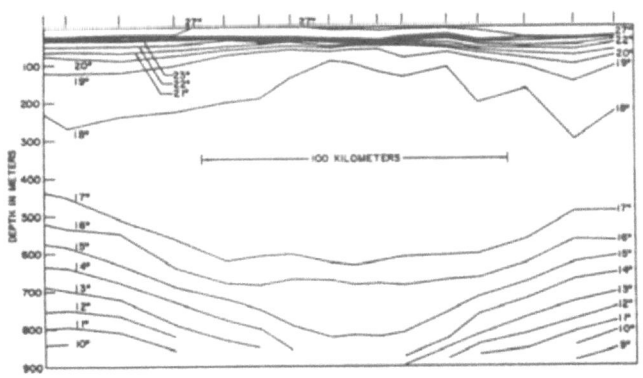

Fig 10   Temperature section across Sargasso Sea lens.
[*Fig 2   Brundage et al*]

A 170 km diameter eddy in the Sargasso Sea, Brundage et al (1986), centred at around 350 metres was found to have a second lens of water below it at 700m, (Fig 10). From expendable current profiler traces the eddy was found to have a maximum tangential velocity of 30 cms⁻¹ and was propagating at 4.4 cms⁻¹ to the south west. A similar although more pronounced double structure was found in eddy

424

"Mario", Cresswell and Legeckis (1986), off south eastern Australia.(Fig 11). Formed by the coalescence of two single eddies, "Maria" and "Leo", Mario also propagated equatorwards but had a higher tangential speed of 50 to 100 cms⁻¹. Both Mario and the Sargasso Sea eddy were expected to have lifetimes of at least many months according to their rate of decay during observation.

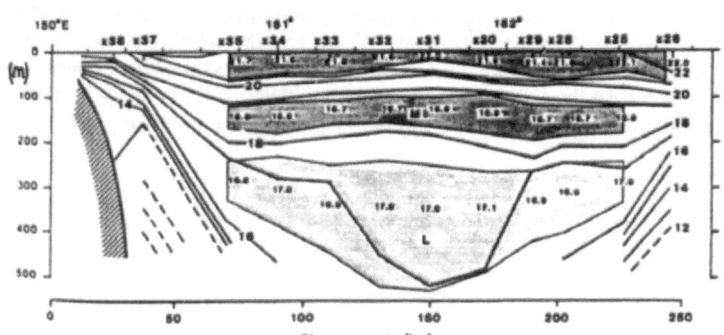

Fig 11 Temperature section across "Mario" showing clearly the components from "Maria" and "Leo". [*Fig 10 Cresswell et al*]

The fourth example of these complex anticyclonic eddies was observed by Bogdanov et al (1986) in the northwestern Pacific. Also around 180 km in diameter and having a high, up to 74 cms⁻¹, tangential velocity this eddy was centred around 200 - 250m with a warm saline surface breaking lens above and a smaller, less significant, warm saline lens below at 700m, (Fig 12). The main eddy is a homogeneous lens of warm Kuroshio water which appears to depress the thermocline over 300m. Note the

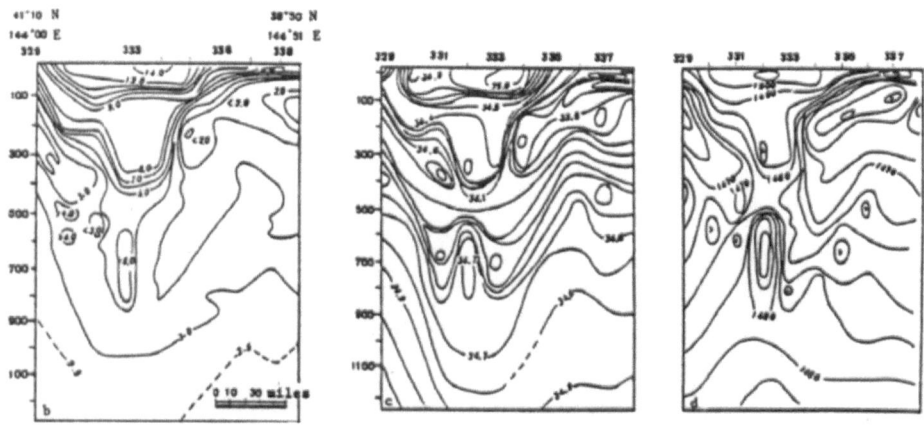

Fig 12 Temperature, Salinity and Sound Speed sections across anticyclonic eddy in northwestern Pacific. [*Fig 1 Bogdanov et al*]

large variations in sound speed profile across the eddy. Satellite images show the warm surface lens connected by a 'bridge' to the warm waters south adjoining the Kuroshio. From geostrophic calculations the volume transport around the core was approximately 34 Sverdrups and therefore perhaps indicates formation from a meander of its entire parent current, or 'root ' cut, see Kawai and Saitoh (1986).

II Cyclonic Eddies.

In the same way that the classical anticyclonic eddy takes the form of a biconvex lens in an isopycnal cross section, so it was expected for the classical cyclonic eddy to appear as a biconcave lens in a similar cross section. This was indeed found to be the case in the examples studied for this paper with the qualification that, unlike the anticyclonic lenses, these cyclonic 'lenses' are more frequently paired with characteristic features above, below or both.

Gründlingh (1987) studied a 190 km diameter cyclonic eddy spun from the Mozambique Ridge current. The core of eddy "John" was centred at around 250 db (Fig 13) below which lay a pronounced dome of upwelled cold water and above there was a surface lens of warm water. This resulted in a compressed thermocline and hence, although the maximum potential vorticity anomaly along an isopycnal occurs at the level of the eddy core, there is also a positive potential vorticity anomaly in the thermocline directly above the core. Eddy "John" was moving polewards (south) and from available potential energy and kinetic energy calculations Gründlingh predicted an expected lifetime of 1 - 3 years unless entrained into the Agulhas system. From geostrophic calculations relative to a level of no motion of 2000 db, the maximum around core volume transport was given as 40 - 50 Sverdrups.

Two similar diameter eddies, in the north west Atlantic, also demonstrated particularly large around core volume transports (30 - 50 Sverdrups or more) indicating formation from meanders of the entire Gulf Stream. The first of these cold core Gulf Stream rings examined here, named ring "Allen" (Richardson et al, 1979) was followed from September 1976 to April 1977, during which time it propagated in a north westerly direction at about 3 cms[-1]. The eddy core was centred between 120 and 200 metres (Fig 14) above which a relatively cold and fresh surface layer existed although this horizontal temperature gradient was small and Richardson et al suggest it would not have been detectable from satellite IR observations. There is evidence of upwelling of deep cold water at all levels below the eddy.

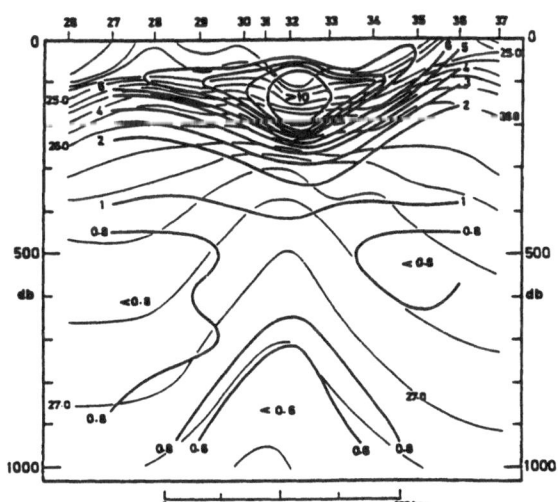

Fig 13 Eddy "John", faint lines are isopycnals, heavy lines
are lines of equal potential vorticity. [*Fig 8 Gründlingh*]

The second example of cold core Gulf Stream ring looked at here was observed in almost exactly the same place as eddy "Allen" and was surveyed three times in 1984 (Bubnov et al, 1986). This eddy showed an almost identical thermo-haline structure to that of eddy "Allen" and propagated poleward initially but later

426

during the second and third surveys turned south.During the first survey the eddy was one of two side by side joined vortices which later broke up, another similarity with eddy "Allen" which also spawned a vortex, eddy "Arthur", during its period under observation.

　　　　Remaining in the same geographical area but reducing the horizontal scale to a diameter of around 100 km, Olson (1980) discussed a cyclonic eddy generated from the Gulf Stream with which it merged and was then respun more than once during the five months of observation by the Cyclonic Ring Experiment. Ring "Bob", as it was christened, was centred around a 200 metre core depth (Fig 15) with a marked upwelling of cold water below observed as deep as 4000 db in the water column. With a maximum geostrophic tangential velocity of 140 cms^{-1} (relative to 2500 db) this was a very energetic ring as seems often to be the case with those that have interacted more than once with their parent current.

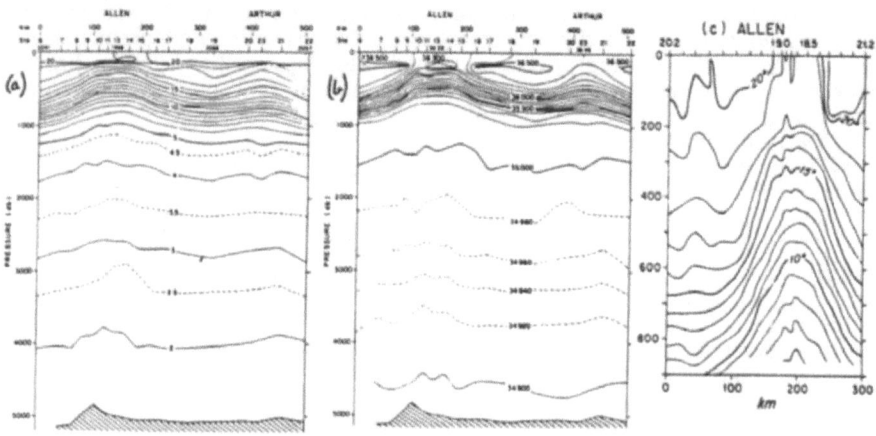

Fig 14　Sections of potential temperature (a), salinity (b) and temperature from XBT's (c) through eddy "Allen". December 7 - 13 1976. [*Figs 10 and 13　Richardson et al*]

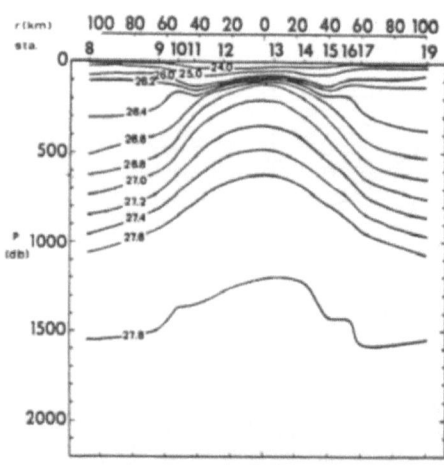

Fig 15　Potential density section across ring "Bob".[*Fig 1　Olson*]

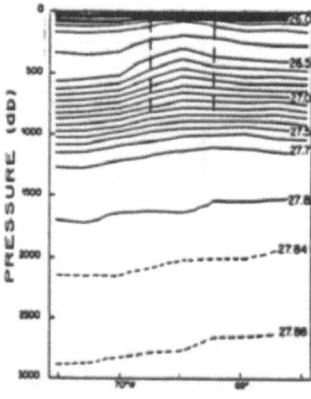

Fig 16　Section of potential density across POLYMODE LDE eddy.
[*Fig 3　Lindstrom et al*]

During much of the experiment ring "Bob" had a warm low density surface lens sitting on top which was identified as higher temperature surface Gulf Stream water that had spilled into the depression created over the ring as it spun off.

A year later, in 1978, a smaller, around 60 km diameter, cyclonic eddy was discovered and observed during the POLYMODE Local Dynamics Experiment (Lindstrom et al, 1986). In this example, no eddy signature was detected below a depth of around 1000m (Fig 16). However above the eddy rested a surface pool of low salinity water. This surface layer was considered too large to have been created by meteorological precipitation and it was suggested that its water properties identified it as shelf water from north of the Gulf Stream. Current meter data from six depth intervals to 839m clearly showed the passage of the eddy as it propagated south westerly at about 18 cms[-1].

Moving to the northeast Atlantic, two 70 km eddies show very different characteristics. Kupferman et al, 1986, described a cyclonic eddy with a core depth between 200 and 500 metres that was still attached to the Gulf Stream by a 300 km long tongue. Like other Gulf Stream cold core eddies, this displayed upwelling of cold water from great depths in the water column and in similarity with ring "Allen" in the western basin had an intense cold surface signature (Fig 17) with the 11°C isotherm raised from 400 metres to break the surface 15 km radius from the eddy axis. The eddy was propagating north east at approximately 3 km per day and from its T-S signature it was suggested that the water making up the eddy column had originated in the Newfoundland Basin.

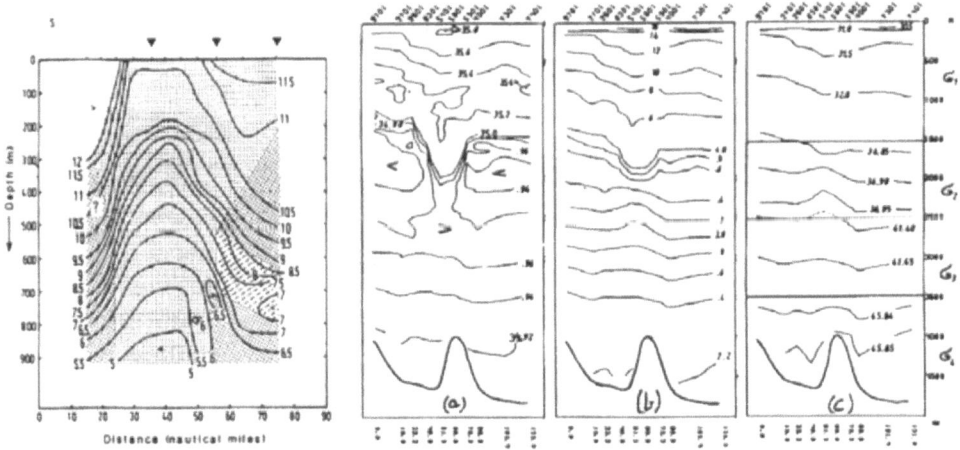

Fig 17 Temperature section through north
-east Atlantic cyclonic Gulf Stream eddy.
[*Fig 2 Kupferman et al*]

Fig 18 Sections of salinity (a), potential
temperature (b) and potential density
(c) across the deep cyclonic 'Meddy'.
[*Fig 3 Schauer*]

On the same horizontal scale but centred around a 1600 - 2000 metre core depth a cyclonic 'Meddy' (Schauer, 1987) was observed propagating westwards in the north east Atlantic at a speed of between 1 and 5 cms[-1]. The term 'Meddy' is usually attributed to anticyclonic Mediterranean lenses at a depth of around 1100 metres, however some Mediterranean water sinks deeper and is thought to propagate north and be a contributor to the high salinity of the North Atlantic Deep Water. In this example the eddy is characterised by a lack of Labrador Sea water normally found at 2000 metres. The eddy may have produced some upwelled signal in the deeper water although this is not too clear from the cross sections (Fig 18), however there appears to be a warm, saline, low density lens sitting above the eddy core in the top 100 - 200 metres. Relative to 4000 db geostrophic calculations, backed up by current meter data show only a moderate tangential velocity of around 10 cms[-1].

The Fram Strait presents another very active generation region for eddy structures and adds the further complication of ice edge interaction. Johannessen et al (1987), described in detail fourteen eddy structures in the Fram Strait, all but two of which were cyclonic. The first of these, 'E1', was 30 km in diameter and originated at a large meander in the ice edge, ice being entrained around a dense, warm and saline Atlantic water surface lens (Fig 19) as the eddy propagated southwards along the ice margin. A chimney of upwelled deep water existed below the eddy core but the resulting disturbance to the isopycnals died out below 1000 metres. E1 had a close neighbour, E3, indicating possible formation as an eddy pair. Satellite images showed the spiralling of lines of ice towards the eddy centre suggesting that the eddy's velocity structure had an ageostrophic component.

Another Fram Strait cyclonic eddy, 'E13', was a little smaller in diameter than E1 at just 20 km and propagated eastwards. E13 had an associated low temperature, low salinity, low density surface lens (Fig 20) but no significant upwelling below the eddy core was observed. E13 was first discovered at the ice edge and during its journey eastwards transported trapped ice and polar water into warmer regions. Johannessen et al considered the structure of this eddy to be typical of Atlantic water features and therefore proposed that it had been advected northwards into this region by the West Spitzbergen Current. After interaction with the ice edge E13 was observed to decay in about ten days. If E13 had moved northwards previously then, for potential vorticity to have been conserved, increasing planetary vorticity (cyclonic) must have caused either a decrease in relative vorticity i.e. the eddy span more slowly, or a reduced stratification of the eddy core; both of which would contribute to eddy decay.

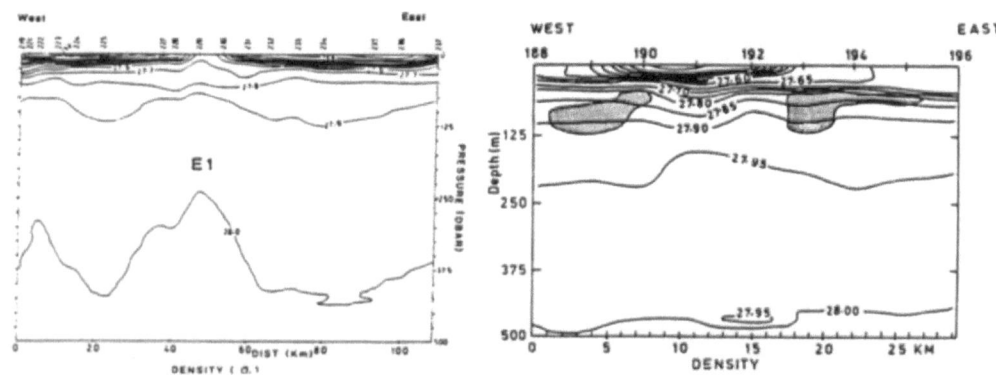

Fig 19   Density (σ_t) section through. eddy E1. [*Fig 4c Johannessen et al*]

Fig 20   Density (σ_t) section through E13. [*Fig10e Johannessen et al*]

The final eddy examined for comparison in this paper is also the smallest with a diameter of only 6 km; observed in the Ligurian Sea some 100 km north of the island of Corsica (Marullo et al, 1985) it had a core depth of just ten metres and no temperature signature below 80 metres. This region is well known for eddy generation where the wind induced cyclonic circulation of the Ligurian Sea combines with Atlantic water passing along the west coast of Corsica meeting Tyrrhenian Sea water to induce many small meanders through current shear. This feature is clearly confined to the surface layer however, with regard to its core depth and diameter scale, there was still a considerable upwelling of cold deeper water below the core (Fig 21). The thermal structure of the eddy shows a similarity with previously described eddies "Allen"

(Richardson et al) and the north east Atlantic Gulf Stream eddy (Kupferman et al), but scaled down by a factor of ten.

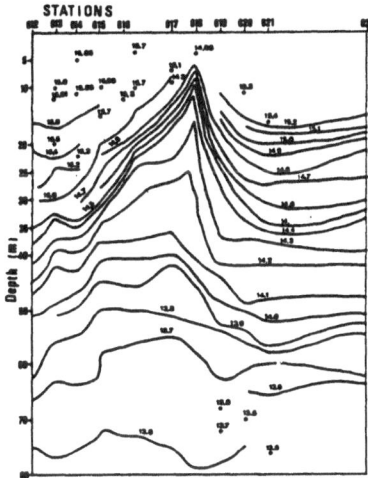

Fig 21   Thermal structure of the small Ligurian Sea cyclonic eddy. [*Fig 5 · Marullo et al*]

Discussion

Twenty oceanic eddies have been examined to see if their structures match those described by Hoskins et al (1985) for potential vorticity anomalies. In all cases, as predicted, anticyclonic (cyclonic) eddies have decreased (enhanced) stratification at their core. In all but one case, the core of the eddy, defined as the depth at which the isopycnal stratification anomaly is maximum, is subsurface. In a potential vorticity anomaly, potential vorticity must always be partitioned between stratification and relative vorticity (rate of spin), so there will also be a velocity maximum subsurface, around the eddy core.

In most cases, the core of the eddy lies within the top 500m (well within the range of XBTs), within or just below the seasonal thermocline. The reason for this is that potential vorticity anomalies can only be created where an isopycnal outcrops into the surface layer or intersects an ocean boundary. Like angular momentum, potential vorticity must be conserved, unless a torque is applied, which can only be done by air-sea interaction or bottom friction. Isopycnals slope upwards in general towards the poles, and where they pass through the thermocline and outcrop into the mixed layer, their stratification (i.e. potential vorticity) changes from a large value in the thermocline to small in the surface layer.

When water from the surface layer subducts by quasigeostrophic flow along a sloping isopycnal (Woods, 1985), it carries with it weak stratification, hence an anticyclonic anomaly. In Fig. 5, for example, the T/S relation of water in the core of an eddy west of the Bay of Biscay identifies it as winter mode water formed by deep winter mixing a few degrees further north. Cyclonic anomalies can only form by this mechanism if strongly stratified water from the seasonal thermocline advects gradually deeper, to weaker stratification. More commonly, cyclonic eddies form by breaking off a major current. It is noticeable in Table 1 that cyclonic eddies have often been described near the Gulf Stream. A possible explanation for their subsurface cores is winter formation followed by capping by seasonal stratification. Anticyclonic eddies have been found and described from a greater range of geographical locations and tend to have shallower depths of influence (relative to their diameter), in keeping with their more frequent generation by surface forcing than instability of a major current.

In a few cases, deep eddy cores are found. In Fig. 3, the eddy core is of Labrador SeaWater origin, so the eddy has propagated a long distance from a region where winter convection is very deep Fig. 18 is an example of a Meddy, deep because saline Mediterranean Water falls a long way in theNorth Atlantic to reach its density level. The eddy in Fig. 9 is the only example of an eddy generated at a deep boundary, the Falklands Plateau, but there are several examples of eddies that may have originated at a continental shelf (Arctic eddies, Figs 4, 19, 20; Fig. 8).

The occurance of several double-core anti-cyclonic eddies is noteworthy. Is it common for two eddies to coalesce (e.g. Fig. 11) as described by Cresswell and Legekis (1986)?

Finally, it should be noted that determination of an eddy structure from its surface signature is not straightforward. The generally subsurface core means that the surface temperature anomaly can be warm or cold for both anti-cyclonic and cyclonic eddies. This is because either warm or cold water may become trapped in the surface layer above the eddy core. Anticyclonic eddies often have no surface temperature anomaly, but could be identified remotely by altimetry.

Acknowledgements

This work has been carried out with the support of Procurement Executive Ministry of Defence.

References

Andrews, J.C. and Scully-Power, P. (1976) 'The structure of an East Australian current anticyclonic eddy', Journal of Physical Oceanography 6, 756-765.

Arhan, M. and Colin de Verdiere, A. (1985) 'Dynamics of eddy motions in the eastern North Atlantic', Journal of Physical Oceanography 15, 153-170.

Bogdanov, K.T., Il'ichev, V.I., Lobanov, V.B., and Medzhitov, R.D. (1985) 'A cyclonic eddy in the Northwestern Pacific',.Transactions (Doklady) of the USSR Academy of Sciences: Earth Sciences Sections 281, 202-204.

Brundage, W.L., and Dugan, J.P. (1986) 'Observations of an anticyclonic eddy of 18°C water in the Sargasso Sea', Journal of Physical Oceanography 16, 717-727.

Bubnov, V.A., Zubin, A.B., Moskalenko, L.V., and Osadchiy, A.S. (1985) 'Evolution of a cyclonic Gulf Stream eddy', Transactions (Doklady) of the USSR Academy of Sciences: Earth Sciences Sections 282, 223-225.

Cresswell, G.R. and Legeckis, R. (1986) 'Eddies off south eastern Australia', Deep Sea Research 33, 1527-1562.

Elliot, B.A. and Sanford, T.B. (1986) 'The subthermocline lens D1. Part 1: Description of water properties and velocity profiles. Part 2: Kinematics and dynamics.', Journal of Physical Oceanography 16, 532-548 and 549-561.

Ertel, Hans. (1942) 'On hydrodynamical vorticity equations', Meteorologische Zeitschrift 59.

Feliks, Y. and Itzikowitz, S. (1987) 'Movement and geographical distribution of anticyclonic eddies in the eastern Levantine Basin', Deep Sea Research 34, 1499-1508.

Gordon, A.L. and Greengrove, C.L. (1986) 'Abyssal eddy in the Southwest Atlantic', Deep Sea Research 33, 839-847.

Gründlingh, M.L. (1987) 'Anatomy of a cyclonic eddy of the Mozambique Ridge current', Deep Sea Research 34, 237-251.

Harvey, J. (1982) 'T-S relationships and water masses in the eastern North Atlantic', Deep Sea Research 29, 1021-1033.

Hoskins, B.J., McIntyre, M.E., and Robertson, A.W. (1985) 'On the use and significance of isentropic potential vorticity maps', Quarterly Journal of the Royal Meteorological Society 111, 877-946.

Houghton, R.W., Olson, D.B., and Celone, P.J. (1986) 'Observation of an anticyclonic eddy near the continental shelf break south of New England', Journal of Physical Oceanography 16, 60-71.

Johannessen, J.A., Johannessen, O.M., Svendsen, E., Shuchman, R., Manley, T., Campbell, W.J., Josberger, E.G., Sandven, S., Gascard, J.C., Olaussen, T., Davidson, K., and Van Leer, J. (1987) 'Mesoscale eddies in the Fram Strait marginal ice zone during the 1983 and 1984 marginal ice zone experiments', Journal of Geophysical Research 92, 6754-6772.

Kawai, H. and Saitoh, S. (1986) 'Secondary fronts, warm tongues and warm streamers of the Kuroshio Extension system', Deep Sea Research 33, 1487-1507.

Killworth, P.D. (1983) 'On the motion of isolated lenses on a beta plane', Journal of Physical Oceanography 13, 368-376.

Killworth, P.D. (1986) 'On the propagation of isolated multilayer and continuously stratified eddies', Journal of Physical Oceanography 16, 709-716.

Kupferman, S.L., Becker, G.A., Simmons, W.F., Schauer, U., Marietta, M.G., and Nies, H. (1986) 'An intense cold core eddy in the North-East Atlantic', Nature 319, 474-477.

Le Groupe Tourbillon. (1983) 'The Tourbillon experiment : a study of a mesoscale eddy in the eastern North Atlantic', Deep Sea Research 30, 475-511.

Lindstrom, E.J., Ebbesmeyer, C.C., and Brechner owens, W. (1986) 'Structure and origin of a small cyclonic eddy observed during the POLYMODE local dynamics experiment', Journal of Physical Oceanography 16, 562-570.

Marullo, S., Salusti, E., and Viola, A. (1985) 'Observations of a small-scale baroclinic eddy in the Ligurian Sea', Deep Sea Research 32, 215-222.

M^{C}Dowell, S.E. and Rossby, H.T. (1978) 'Mediterranean water: an intense mesoscale eddy off the Bahamas', Science 202, 1085-1087.

Mellberg, L.E., Johanessen, O.M., Connors, D.N., Botseas, G., and Browning, D. (1987) 'Modeled acoustic propagation through an ice edge eddy in the East Greenland sea marginal ice zone', Journal of Geophysical Research 92, 6857-6868.

Newton, J.L., Aagaard, K., and Coachman, L.K. (1974) 'Baroclinic eddies in the Arctic Ocean', Deep Sea Research 21, 707-719.

Nof, D. (1981) 'On the beta induced movement of isolated baroclinic eddies', Journal of Physical Oceanography 11, 1662-1672.

Nof, D. (1985) 'Joint vortices, eastward propagating eddies and migratory Taylor Columns', Journal of Physical Oceanography 16, 1114-1137.

Olson, D.B. (1980) 'The physical oceanography of two rings observed by the cyclonic ring experiment. Part 2 : Dynamics', Journal of Physical Oceanography 10, 514-528.

Petitpas, L.S. and Browning, D.G. (1986) 'Acoustic description of South Atlantic ocean warm core eddy', Presentation overheads.

Pollard, R.T. and Regier, L. (1990) 'Large potential vorticity variations at small scales in the upper ocean', submitted to Nature.

Richardson, P.L., Maillard, C., and Stanford, T.B. (1979) 'The physical structure and life history of cyclonic Gulf Stream ring Allen', Journal of Geophysical Research 84, 7727-7741.

Rossby, C.G. and collaborators. (1939) 'Relation between variations in the intensity of the zonal circulation of the atmosphere and the displacements of the semi-permanent centres of action', Journal of Marine Research 2, 38-55.

Schauer, U. (1987) 'A deep cyclonic meddy in the west european basin', ICES.

Woods, J.D. (1985) 'The physics of thermocline ventilation', in J.C.J.Nihoul (ed.), Coupled Ocean Atmosphere Models, Elsevier, pp.543-590.

THREE-DIMENSIONAL OCEANOGRAPHY AND ACOUSTICS

W. A. Kuperman and J. S. Perkins
Naval Research Laboratory
Washington, DC 20375
USA

ABSTRACT. Oceanography governs the structure of the acoustic index of refraction via sound speed and density which in turn controls underwater sound propagation. The strong connection between acoustics and oceanography becomes increasingly apparent when models of both are coupled in three and four dimensions. Acoustic fields, appropriately displayed reveal much of the oceanography while upper and lower limits to the necessary oceanographic resolution for meaningful acoustic prediction are also suggested by the coupling of these models. Furthermore, three-dimensional (3D) oceanographic complexity provides more opportunity than hindrance for extending the application of acoustics to an assortment of acoustics and oceanography problems.

1. Introduction

Ocean or "underwater" acoustics has traditionally been associated with the science and technology of sonar systems. Furthermore, sonar systems, with some exception, have traditionally been associated with the location of objects in the sea. Over the years, the science and technology of sound in the sea has developed to the state where sophisticated sonar systems exist and a growing body of knowledge has emerged with respect to the ocean environmental influence on sound. At this juncture, acoustics, oceanography and signal processing is merging into a new medium-dominated wave physics discipline. Consequently, new directions for future research involve exploiting oceanography for acoustics and vice versa. In the context of extracting signal from noise, the "signal" is no longer restricted to its traditional sonar type definition, but the signal may also be the ocean/geophysical parameters of the medium.

In this and a companion paper [1] we will address the transition to this generalized definition of signal. This will be done be series of simulation examples based on acoustic mode theory. However, the point of this paper is not the utility of mode theory, but rather the kind of insight one can obtain from a global and unified approach

433

J. Potter and A. Warn-Varnas (eds.), Ocean Variability & Acoustic Propagation, 433–448.
© 1991 *Kluwer Academic Publishers.*

to oceanography and acoustics. First we will review an application of mode theory to perform wide area ocean acoustic computations in complex three-dimensional (3D) environments. The model output suggests an image of the oceanography. This same theory is extended to model ambient noise in the same type of environment. We combine this capability with recently developed signal processing methods to simulate the search for a radiating object in this complex environment. In performing the latter process, we observe that: 1)The complex ocean environment can be utilized to be part of the effective aperture of the signal processing procedure and 2) The ambient noise "illuminates" (insonifies) the environment in a way that suggests that sophisticated processing schemes might be developed to image the environment using noise. Finally, in the companion paper [1] we examine the possibility of searching for an unknown object in an unknown ocean. Here we are led to a solution where the ocean is included in the focusing process and issues of the uniqueness of the focused ocean environment vis a vis localization of the unknown object emerge distinct from the parallel issues encountered in tomography.

2. Theory

Two-dimensional range-dependent wave-theory propagation modeling techniques were introduced into underwater acoustics with the split-step parabolic equation (PE) method [2]. The boundary value wave equation is approximated by the intial value PE and then rapidly solved by a multiple application of FFT's. Subsequent methods have altered the solution algorithm but the fundamental *marching* nature of the solution technique remains common to the PE. A straightforward extension to three-dimensional (3D) ocean acoustic modeling over large ocean areas takes us immediately to and beyond the limit of the present generation of computers.

The computational issues and difficulties central to the 3D ocean acoustic modeling problem have already been highlighted in previous papers [3,4,5,6,7]. Many of these issues arise from the marching nature of the algorithms employed; different source/receiver configurations relative to the environment require recomputation of the total acoustic wave equation solution. Three dimensional variation of the environment is associated with two features: the ocean bottom and oceanography. About these we make the following observations: topographical features are stationary and even the ocean medium as represented by the sound speed profile is typically static below certain depths depending on the particular oceanographic conditions. Traditional marching algorithms are in a sense redundant in that they fail to take advantage of the invariant parts of the problem. This redundancy is particularly unattractive when the environment remains constant and only the source/receiver positions change.

Adiabatic and coupled-mode theory is amenable to precalculations which can subsequently be used in a nonredundant manner to perform rapid three-dimensional acoustic field computations in a complex ocean environment. Algorithms have been developed [8,9] to take advantage of both horizontal and vertical precalculated quantities. We grid the ocean environment in terms of its local acoustic eigenvalues and normal modes. The acoustic field is then calculated by either an adiabatic coupled mode method; in the latter case, much of a coupled mode matrix computation can also be precomputed. All of the above precomputed quantities are independent of source/receiver configuration. This allows the acoustic field for a different source/receiver configuration to be recalculated with minimum effort as opposed to the total recalculation necessary in any marching algorithm. Speed-up over conventional marching algorithms is then accomplished by using this "spreadsheet" type approach of manipulating wave equation precalculations. We will confine ourselves to adiabatic theory since we are emphasizing the unity of combined oceanography, acoustics and signal processing. The details of the coupled mode extensions are given in [9]

2.1 ADIABATIC MODE THEORY

Our starting point is the Helmholtz equation in three dimensions. We have

$$\rho\nabla \cdot [\frac{\nabla p}{\rho}] + \frac{\omega^2}{c^2(x,y,z)}p = -\delta(x,y,z) \tag{1}$$

Here, ω is the circular frequency of the source, $c(x,y,z)$ is the ocean sound speed, ρ is the density and $p(x,y,z)$ is the acoustic pressure.

The normal mode solution when $c(x,y,z) = c(z)$, is

$$p(r,z) = \frac{i}{4}\sum_{n=1}^{N} u_n(z_s)u_n(z)H_0^{(1)}(k_n r), \tag{2}$$

or, using the asymptotic approximation to the Hankel function,

$$p(r,z) = \frac{i}{4}\sqrt{\frac{2}{\pi}}e^{-i\pi/4}\sum_{n=1}^{N} u_n(z_s)u_n(z)\frac{e^{ik_n r}}{\sqrt{k_n r}}, \tag{3}$$

where r denotes the range from the source and $u_n(z)$ simultaneously satisfies

$$\rho(z)\frac{d}{dz}[\frac{1}{\rho(z)}\frac{d}{dz}u_n(z)] + (\frac{\omega^2}{c^2(z)} - k_n^2)u_n(z) = 0,$$

$$u_n(0) + Z^T(k_n^2)\frac{du_n}{dz}(0) = 0,$$

$$u_n(H) + Z^B(k_n^2)\frac{du_n}{dz}(H) = 0,$$

$$\int_0^H [u_n(z)]^2/\rho(z)dz - \frac{1}{2k_n}\frac{d(\frac{1}{Z^T})}{dk}u_n(0) + \frac{1}{2k_n}\frac{d(\frac{1}{Z^B})}{dk}u_n(H) = 1. \tag{4}$$

The middle two equations represent a general impedance boundary condition with the eigenvalue k_n^2 appearing in the coefficients. In the simplest ocean acoustic problem the ocean surface is a pressure release surface and therefore Z^T is identically zero (Dirichlet boundary condition). In addition, a perfectly rigid ocean bottom model implies $Z^B \to \infty$ (Neumann boundary condition). The derivation of the normalization condition may be found in [10] and the algorithm used to solve Eq. 4 is taken from [11,12].

The generalization of this result to *range-dependent* problems is straight forward. The adiabatic mode theory result is,

$$p(r,z) = \frac{i}{4}\sqrt{\frac{2}{\pi}}e^{-i\pi/4}\sum_{m=1}^{M(r)} u_m(z_s;0)u_m(z;r)\frac{e^{i\int_0^r k_m(s)ds}}{\sqrt{k_m(r)r}}. \tag{5}$$

The sum is over the minimum number of propagating modes, $M(r)$, that exist between the source at range $r = 0$ and the receiver at range r. The modal sum involves the *local modes* at the source, $u_m(z_s;0)$, and at the receiver, $u_m(z;r)$. In practice, the calculation of modes can be somewhat computationally expensive; the environment is subdivided at points $r = r_j, j = 1,\ldots, J_p$ where J_p denotes the number of profiles. (In a physical problem such breaks might be placed at each range where a new CTD measurement has been made.) Then one solves for a set of modes at each r_j and uses linear interpolation to construct modes at ranges which lie between those r_j where the modes have been calculated.

The $N \times 2D$ generalization for a three-dimensionally varying environment is to solve the 3D problem on N azimuthal slices as if each slice of the sound speed profile and bathymetry profile were derived from a cylindrically symmetric problem. This approach has also been applied in parabolic equation modeling [5]. The adiabatic normal mode result, for example, is

$$p(r,z,\theta) = \frac{i}{4}\sqrt{\frac{2}{\pi}}e^{-i\pi/4}\sum_{m=1}^{M(r,\theta)} u_m(z_s;0,0)u_m(z;r,\theta)\frac{e^{i\int_0^r k_m(s,\theta)ds}}{\sqrt{k_m(r,\theta)r}}, \tag{6}$$

where the quantities in Eq. 6 also satisfy Eq. 4 parameterized with the azimuthal variable, θ.

Finally, we make one important observation concerning adiabatic mode theory. We note from Eq. 6 that the field depends on line integrals involving local eigenvalues, the latter quantities containing the local environment information. Therefore, if there is an environmental feature, the unique location of the environmental feature will not be evident a long distance from the feature; only the cumulative phase changes of the modes will propagate. From this we conclude that when adiabatic mode theory is valid, there is no unique ocean environment but rather many environments which will yield the correct long range acoustic result. There is no inconsistency here because the

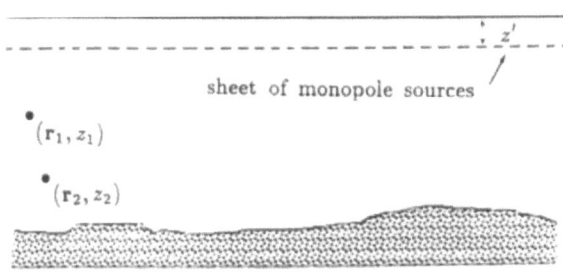

Figure 1: A vertical slice of the environment for the noise model.

adiabatic is only valid in the regime where there is interaction and hence averaging of an environmental feature over more than one acoustic cycle distance. Furthermore, there is no inconsistency with single or multiple slice tomography. This observation has profound consequences which will be discussed in the companion paper [1].

2.2 NOISE

The theory of the spatial correlation of surface generated noise in a range-independent ocean was developed in a wave-theoretical form in [13]. We begin our treatment of the range-dependent case in a similar manner to that work: we assume that the surface sources are monopoles of strength $s(\mathbf{r}', t)$ distributed over an infinite plane parallel to the surface and located at depth z' below the surface as shown in Fig. 1. The function $s(\mathbf{r}', t)$ is taken to be a random function. Equation 5 of Ref. [13] still holds, that is,

$$\phi(\mathbf{r}, z) = \int d^2\mathbf{r}' S(\mathbf{r}') G(\mathbf{r}, \mathbf{r}'; z, z'), \tag{7}$$

where $\phi(\mathbf{r}, z)$ and $s(\mathbf{r}', t)$ are the Fourier transforms (frequency domain) of the velocity potential and the monopole source strength, respectively; $G(\mathbf{r}, \mathbf{r}'; z, z')$ is the Green's function and the integral is taken over the source plane.

The details of the theory for the range-dependent ocean are in [14]. Here we summarize the derivation and final result. We assume the sound speed profile, bathymetry, bottom properties and surface source strength can vary with range. To calculate the integral in Eq. 7 we divide the area of the plane into two regions: a large circular area A_0 of radius R centered on the origin, and the remaining area outside, divided into N range independent subregions $A_\nu, \nu \neq 0$. Hence, rewriting Eq. 7, we have

$$\phi(\mathbf{r}, z) = \sum_{\nu=0}^{N} \int_{A_\nu} d^2\mathbf{r}' S(\mathbf{r}') G(\mathbf{r}, \mathbf{r}'; z, z'). \tag{8}$$

We form the cross-spectral density between two points, $C(r_1, z_1; r_2, z_2)$, by multiplying the field at one point (r_1, z_1) by the complex conjugate of the field at a second point (r_2, z_2) and taking the ensemble average. We get

$$
\begin{aligned}
C(r_1, z_1; r_2, z_2) &\equiv\ <\phi(r_1, z_1)\phi^*(r_2, z_2)> \\
&=\ \sum_{\nu, \mu} \int_{A_\nu} \int_{A_\mu} d^2r' d^2r'' [\ <S(r')S(r'')> \\
&\quad \times\ G(r_1, r'; z_1, z') G^*(r_2, r''; z_2, z')],
\end{aligned}
\tag{9}
$$

where the brackets indicate the ensemble average. Following the derivation in [13], we obtain

$$
C(r_1, z_1; r_2, z_2) = \frac{4\pi}{k^2} \sum_{\nu=0}^{N} \int_{A_\nu} d^2r' q_\nu^2 G(r_1, r'; z_1, z') G^*(r_2, r'; z_2, z'),
\tag{10}
$$

where q_ν^2 expresses the surface noise source strength in subregion A_ν. For the range-dependent case we use the adiabatic mode theory result of Eq. 6 which when fully evaluated results in a complicated expression for the cross-spectral density of the noise [14]. We present numerical results using this formulation later in the paper.

2.3 MATCHED FIELD PROCESSING

Matched-field processing (MFP) [15,17,16,18] is a generalization of plane-wave beamforming in which the plane-wave weighting vectors, i.e, replicas, are replaced by solutions to the wave equation for the particular environment of interest. Both linear, or Bartlett, and nonlinear, for example Maximum Likelihood Method (MLM), processing schemes have been previously studied. Matched-field processing in a three-dimensional (range and azimuthal dependent) environment provides the possibility of localizing a source in bearing, as well as in range and depth, with a purely vertical array since the environment itself breaks the azimuthal symmetry of the vertical array. This added dimension in aperture is a result of including the transverse environmental complexity in the matched-field process. For the purpose of these simulation we will search for the source location in range and horizontal position which is an extension of earlier range-depth matched-field processing [18]. For simplicity, we take the depth as known and suppress the z-coordinate of the source in the notation below. We also use the notation that a bold face upper case letter is a matrix while a bold face lower case letter is a vector.

The matched-field processing is done by computing ambiguity functions in the frequency domain for both the Bartlett processor,

$$
S_{Bart}(x, y) = w^+(x, y) K(x_T, y_T) w(x, y),
\tag{11}
$$

and the MLM processor,

$$S_{MLM}(x,y) = [\mathbf{w}^+(x,y)\,\mathbf{K}^{-1}(x_T,y_T)\,\mathbf{w}(x,y)]^{-1}, \tag{12}$$

In Eqs. 11 and 12, the normalized replica vectors (see [18]), $\mathbf{w}(x,y)$, for each set of search location parameters (x,y) is obtained from the adiabatic mode solution, Eq. 6. $\mathbf{K}(x_T,y_T)$ is the covariance matrix of the data on the array resulting from a source at its true location (x_T,y_T) and contains both signal and noise across the $ij-th$ pair of hydrophones,

$$K_{ij}(x_T,y_T) = \sigma_s^2\,\phi(x_T,y_T,z_i)\phi^*(x_T,y_T,z_j) + K_{N_{ij}}, \tag{13}$$

where σ_s^2 is the chosen source level, and the cross-spectral density of the noise, $K_{N_{ij}} \equiv C(\mathbf{r}_i,z_i;\mathbf{r}_j,z_j)$, is given by evaluating Eq. 10.

3. Simulation Results

In this section we present simulations of signal and noise fields and subsequent array processing for a three-dimensional complex ocean environment. We demonstrate that a numerical implementation of the theory outlined in the last section produces results which appear as acoustical "illumination" (insonification) of the ocean environment. Furthermore, we show that the environment can be effectively utilized to enhance the array processing localization procedure. We use the Gulf Stream at the edge of the North American continental shelf as the environment for the simulations. The steps of the simulations are:

1. setting up the environment;

2. computing the signal field;

3. computing the noise field;

4. simulating matched field processing with and without signal (only noise).

Item 2 indicates that a signal insonifying the ocean illuminates its features. Furthermore, the last part of item 4 demonstrates that ocean ambient noise alone, can do the same.

3.1 THE OCEAN ACOUSTIC ENVIRONMENT

For the purpose of this simulation we choose a complex 3D environment as shown in Fig. 2. The bathymetric contours were generated from the SYNBAPS [20] Data base. The approximate position of the Gulf Stream, as it was in March of 1983 [21] is shaded as are two cold-core eddies and a small seamount. Note the variable bathymetry with the continental rise to northwest. More detail on the specifics of the environment is given in Ref. [9,19].

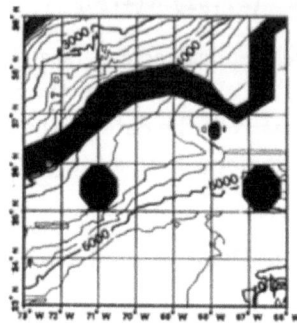

Figure 2: The Gulf Stream environment used for the simulations. The Gulf Stream, two cold-core eddies, and a small seamount are shaded over the bathymetric contours.

3.2 THE SIGNAL FIELD

For this simulation we use a 50 Hz source located at position $x = 333$ km, $y = 315$ km, and depth $z = 100$ m. The field was calculated using 65 radials spaced 5.625 degrees apart. Magnitude-squared pressure values (obtained from Eq. 6) were smoothed with a 5-km Gaussian-weighted window along each radial, converted to transmission loss (TL) and then interpolated onto a rectangular 1 km $\times$ 1 km grid over the entire 650 km $\times$ 650 km region. The resulting 3D acoustic field at a receiver plane of 217 m is shown in Fig. 3 provides a type of "acoustic image" of the environment. The region to the south-east is reasonably flat and has no oceanographic features; this part has a cylindrical symmetry as one would expect. The eddies break this symmetry, and the effect of the Gulf Stream is to produce generally greater loss on the far side. This does not mean that the Gulf Stream is acting like an acoustic "wall", the energy is being redistributed to other depths or being absorbed by the bottom.

We note (with details in Ref.. [9] that since all the modes are precomputed and gridded according to the environment, changing the source position requires minimum computation as compared to a marching type solution algorithm. In Ref.. [9] we also derive a mode coupling result using the same gridding and precomputation methods and compare the result to parabolic equation (PE) solution. In general, the agreement between adiabatic and PE is fairly good over the whole region but the agreement between couple mode and PE is much better. Furthermore, the precomputed coupled mode method required only about 50% more running time than the adiabatic solution.

3.3 THE NOISE FIELD

Since we are ultimately interested in the spatial structure of the noise field as it

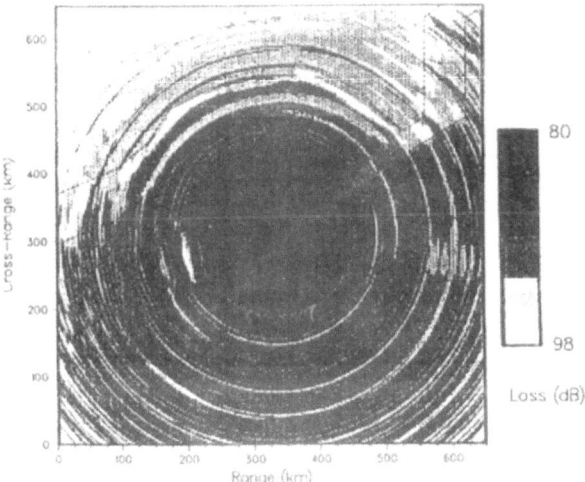

Figure 3: The acoustic field for a 50 Hz adiabatic mode calculation

appears at the output of a beamformer. Such matched field outputs we defer to the next subsection. Here we demonstrate that the theory as summarized by Eq. 10 with details given in [19] give physically understandable results. For this we examine the vertical structure of the noise field in the same Gulf Stream environment. The contributing areas of the surface noise, (the A_ν's) in Eq. 10 are 4 km squares. Storms are simulated by increasing the strength (the q_ν^2's) according to the size and strength of the storm. We consider two storms as shown in Fig. 4, one in deep water and one in shallow water.

In this simulation we use a twenty-one element vertical array, with 75 m element spacing, placed in the center of the region covering a depth extent of 100 m to 1500 m. The frequency of the simulation is 10 Hz. Plane wave beamforming reveals the vertical structure of the noise field. The results are shown in Fig 5. The deep storm shows a deep horizontal notch at the array whereas the storm in the shallow water region has significant arrival structure in the horizontal. Surface noise excites the higher order modes and hence it should tend to have a horizontal notch. However, if the noise originates in a shallow region and has to traverse a slope, steeper angles are converted to more horizontal angles [22]. Hence, in this rather complex case, we see that the noise theory yields results that are physically realistic. Below, we include the 3D complexity of the noise field into an array processing simulation.

3.3 3-D MATCHED FIELD PROCESSING

3.3.1 *Source Localization.* We use the same environment and array as above to

Range – Cross–Range Ambiguity Function

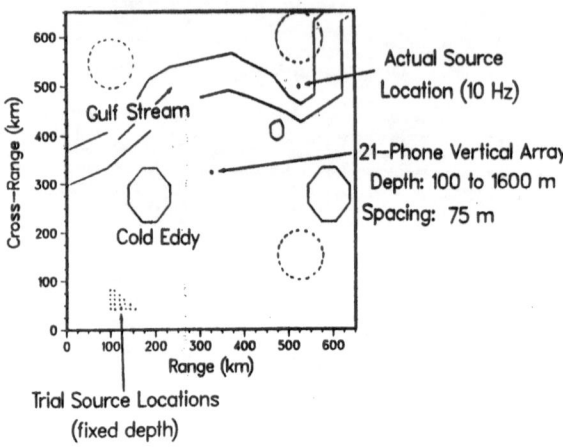

Trial Source Locations
(fixed depth)

Figure 4: The geometry and the environment for the noise and the matched field processing simulations.

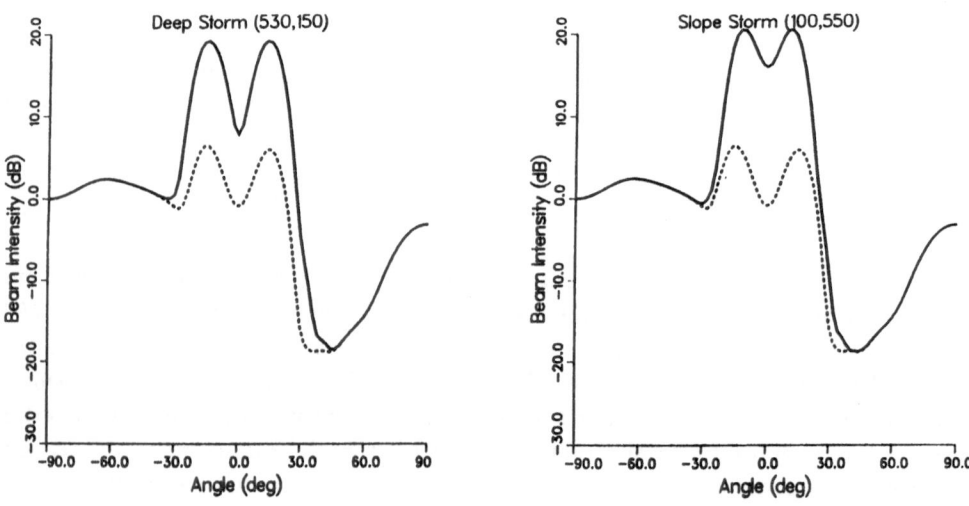

Figure 5: The vertical directivity of the noise field. The dashed line is for no storm. Negative angles are in the upward direction.

localize sources with matched field processing through simulation by implementing Eqs. 11-13. The inputs to these equations are Eqs. 6 and 10. The average signal-to-noise ratio at each hydrophone is approximately 0 dB. We use two source positions both at depths of 100 m. The first source is located southeast of the array toward a relatively homogeneous environment at position $x = 527.5$ km and $y = 152.5$ km. The second source is northeast of the array at position $x = 527.5$ km and $y = 502.5$ km (in the "kink" of the Gulf Stream).

The ambiguity functions we will discuss below were generated at a resolution of 5 km (the same resolution as our environment). Figure 6 shows the Bartlett and MLM results for the first source located to the southeast (position A, Fig. 2). For the Bartlett result there is a maximum at the correct source position, but there are circular ambiguities indicating the nonuniqueness of the environment in which this source is located. In a totally range-independent environment, a source at any azimuth, but the same range and depth as the true source, would propagate to the array exactly like the true source, and the ambiguities would be full rings. The "ring-ambiguities" in Fig.6 are true ambiguities, illustrating that when there is only mild range-dependance there may well be some locations which are not unique to the array. The higher resolution MLM processor reduces the ambiguities somewhat. This first location shows an ambiguous localization result.

The second location is in a rather unique environmental location–a kink in the Gulf Stream (position B, Fig. 2). Figure 7 shows the Bartlett and MLM results for this case. Here the Bartlett shows an absolute maximum at the true source location with some weaker though significant ambiguities which are more analogous to sidelobes. The MLM processor suppresses these sidelobe results for the unambiguous localization of the source. Hence, effective horizontal aperture is obtained with a vertical array and the symmetry breaking of the environment.

3.3.2 *Imaging the Environment with Noise.* Consider exactly the same process as above but without a source. Matched field processing on the noise alone also produces an ambiguity function. An MLM example is shown in Fig. 8. Here we distinctly see the Gulf Stream north wall and indication of the eddy structure. Of course, the result is circular in that the environment had to be known in order to generate the correct replicas for the processing. However, if, for example the location of the Gulf Stream was previously known, an iteration process could be used to determine its new location. Multi-parameter search methods (see [1]) could possibly carry this procedure even further, in effect, to a technique akin to "passive tomography."

4. Summary and Conclusion

By performing consistent simulations of acoustic propagation, noise, and array processing in a complex 3-D ocean environment we have graphically demonstrated

444

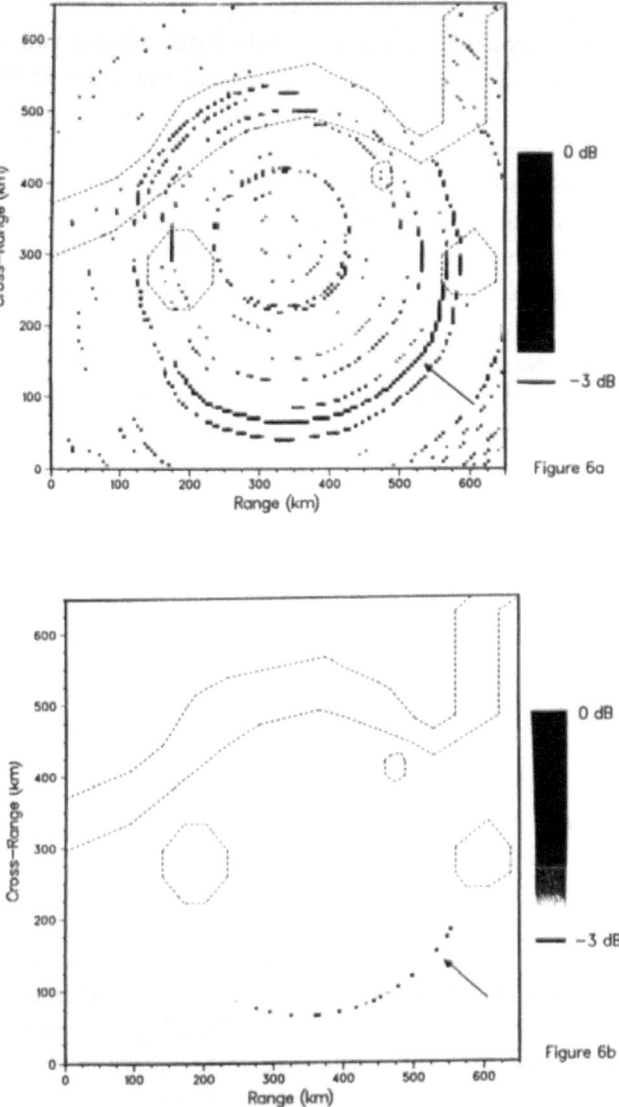

Figure 6: Bartlett (a) and MLM (b) ambiguity functions for the first source location x = 527 km, y= 152.5 km. Each function is shaded so that the dynamic range covers the top 3 dB.

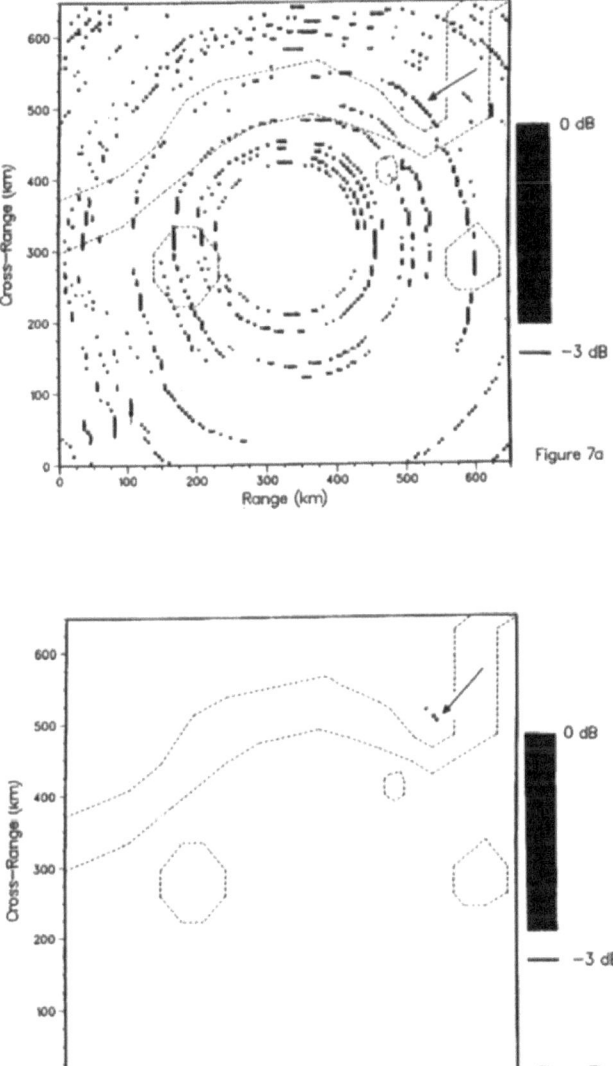

Figure 7: Bartlett (a) and MLM (b) ambiguity functions for the first source location x = 527.5 km, y= 502.5 km. Each function is shaded so that the dynamic range covers the top 3 dB.

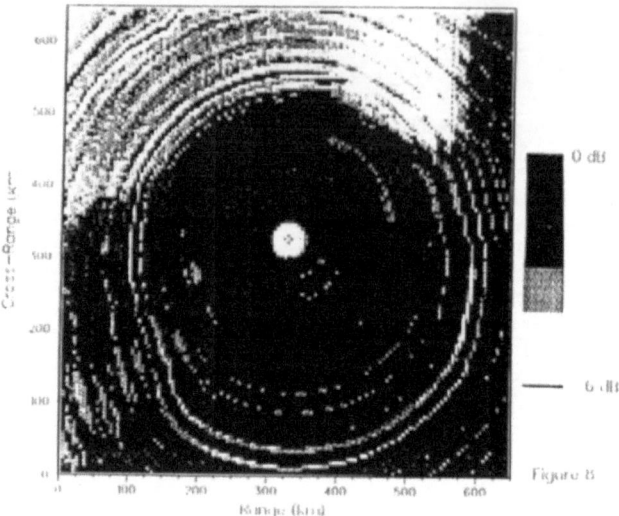

Figure 8: The MLM matched field result (6 dB dynamic range) for the noise alone showing a quasi-image of the oceanographic environment.

the connectivity between oceanography and wave propagation in a complex media. These simulations reveal that, in principle, a complex ocean environment, if known, can be used to enhance the signal processing. Furthermore, some of these processing methods seem to display some sort of capability to image the environment, even with just ambient noise. The latter suggests that new multi-parameter estimation methods may prove very productive in extending the procedures discussed in this paper to more general passive tomographic methods. Finally, we point to the companion paper [1] where a method for simultaneously searching for oceanographic parameters and acoustic sources provides further insights on the interdependence of acoustics and oceanography.

References

[1] M. D. Collins and W. A. Kuperman, "Environmental focusing and source localizaton in the deep ocean," this proceedings.

[2] F. D. Tappert, "The Parabolic Approximation Method," *Wave Propagation and Underwater Acoustics*, (eds.) J. B. Keller and J. S. Papadakis, Springer-Verlag, Berlin, (1977).

[3] Allan D. Pierce, "Extension of the method of normal modes to sound propagation in an almost-stratified medium", *J. Acoust. Soc. Am.* **37**:19–27 (1965).

[4] Henry Weinberg and Robert Burridge, "Horizontal ray theory for ocean acoustics", *J. Acoust. Soc. Am.* **55**:63–79 (1974).

[5] John S. Perkins and Ralph N. Baer,"An approximation to the three-dimensional parabolic-equation method for acoustic propagation", *J. Acoust. Soc. Am.* **72**:515–522 (1982).

[6] William L. Siegmann, Gregory A. Kriegsmann and Ding Lee, "A wide-angle three-dimensional parabolic wave equation," *J. Acoust. Soc. Am.* **78**:659–664 (1985).

[7] Michael D. Collins and Stanley A. Chin-Bing, "A three-dimensional parabolic equation model that includes the effects fo rough boundaries," *J. Acoust. Soc. Am.* **87**:1104–1109 (1990).

[8] W. A. Kuperman, M. B. Porter, J. S. Perkins and A. A. Piacsek, "Rapid three-dimensional ocean acoustic modeling of complex environments," In R. Vichnevetsky, P. Borne, and J. Vignes, editors, Proceedings, 12th IMACS World Congress on Scientific Computation, 231-233, Gerfidn, Villeneuve d'Ascq, France, 1988.

[9] W. A. Kuperman, M. B. Porter, J. S. Perkins and R. B. Evans, "Rapid computation of acoustic fields in three-dimensional ocean environments," submitted to *J. Acoust. Soc. Am.* Jan 1990.

[10] H. P. Bucker, "Sound propagation in a channel with lossy boundaries", *J. Acoust. Soc. Am.* **48**:1187–1194 (1970).

[11] Michael B. Porter and Edward L. Reiss, "A numerical method for bottom interacting ocean acoustic normal modes", *J. Acoust. Soc. Am.* **77**:1760–1767 (1985).

[12] M. B. Porter, *The KRAKEN Normal Mode Program*, Saclant Undersea Research Center Technical Report, 1990.

448

[13] W. A. Kuperman and F. Ingenito, "Spatial correlation of surface generated noise in a stratified ocean," *J. Acoust. Soc. Am.* **67**, 1988-1996 (1980).

[14] John S. Perkins, W.A. Kuperman, F. Ingenito and John Glattetre, "Modeling Ambient Noise in Three-Dimensional Ocean Environments," *J. Acoust. Soc. Am.* (in preparation).

[15] H. P. Bucker, "Use of calculated sound fields and matched field detection to locate sound sources in shallow water," *J. Acoust. Soc. Am.* **59**, 368-373 (1976).

[16] R. G. Fizell, "Application of high-resolution processing to range and depth estimation using ambiguity function methods," *J. Acoust. Soc. Am.* **82**, 606-613 (1987).

[17] M. J. Hinich, "Maximum Likelihood estimation of the position of a radiating source in a waveguide," *J. Acoust. Soc. Am.* **66**, 480-483 (1979).

[18] A. B. Baggeroer, W. A. Kuperman and H. Schmidt, "Matched field processing: source localization in correlated noise as an optimum parameter estimation problem," *J. Acoust. Soc. Am.* **83**, 571-587 (1988).

[19] J. S. Perkins and W. A. Kuperman, "Environmental signal processing: Three-dimensional matched-filed processing with a vertical array," accepted, *J. Acoust. Soc. Am.*, 1990.

[20] Roger J. Vanwyckhouse, *Synthetic Bathymetric Profiling System (SYNBAPS)*. Technical Report 233, Naval Oceanographic Office, 1973.

[21] J. W. Clark and S. Auer, "East Coast Ocean Features," *Oceanographic Monthly Summary*, III(3):19, March 1983.

[22] R. A. Wagstaff, "Low-frequency ambient noise in the deep sound channel-The missing component," *J. Acoust. Soc. Am.* **69**, p. 1009 (1981).

FRONTAL BOUNDARIES AND EDDIES ON THE ICELAND-FAEROES RIDGE

JOHN C.SCOTT and NICHOLA M.LANE
Ocean Science Division
Admiralty Research Establishment
Portland
Dorset DT5 2JS
UK

ABSTRACT. This paper presents detailed thermal structure data from
the Iceland-Faeroes frontal zone, from a series of towed thermistor
chain surveys in the period 1985-1988. The data allow the
deduction of the spatial scales of variability down to a few
metres. The results also show, in outline, the way the structure
of the frontal zone changes, both spatially within the zone and
temporally between seasons. Two aspects of the region are
described in some detail: the main frontal boundary of the zone,
and the cold eddies which are an important feature of the
intermediate zone.

1. Introduction

The Iceland-Faeroes Ridge is one of a system of shallow ridges
running between Greenland and the European Continental Shelf,
separating the warm, high salinity waters of the Atlantic Ocean
from the colder, less saline Nordic Seas. It is the longest ridge
in this system, but one of the shallowest, rising to about 300m and
reaching about 490m at its deepest part. Because of its relatively
shallow disposition it allows only a relatively insignificant
overflow of deep cold water into the Atlantic; only intermediate
water overflows the Ridge Southwards, and the quantity involved is
small compared with the deep water contributions of the Denmark
Strait and the Faeroe Bank Channel.

Several factors combine to make the Iceland-Faeroes frontal zone
particularly interesting: its considerable length, and the absence
of strong input currents associated with shelf features. These
factors make the zone differ significantly from those which exist
between Iceland and Greenland and between the Faeroes and the
Shetlands. It is common to all of these frontal zones, however,

449

J. Potter and A. Warn-Varnas (eds.), Ocean Variability & Acoustic Propagation, 449–461.
© *Her Majesty's Stationary Office, London 1991.*

that they are strongly constrained in position by their associated ridges.

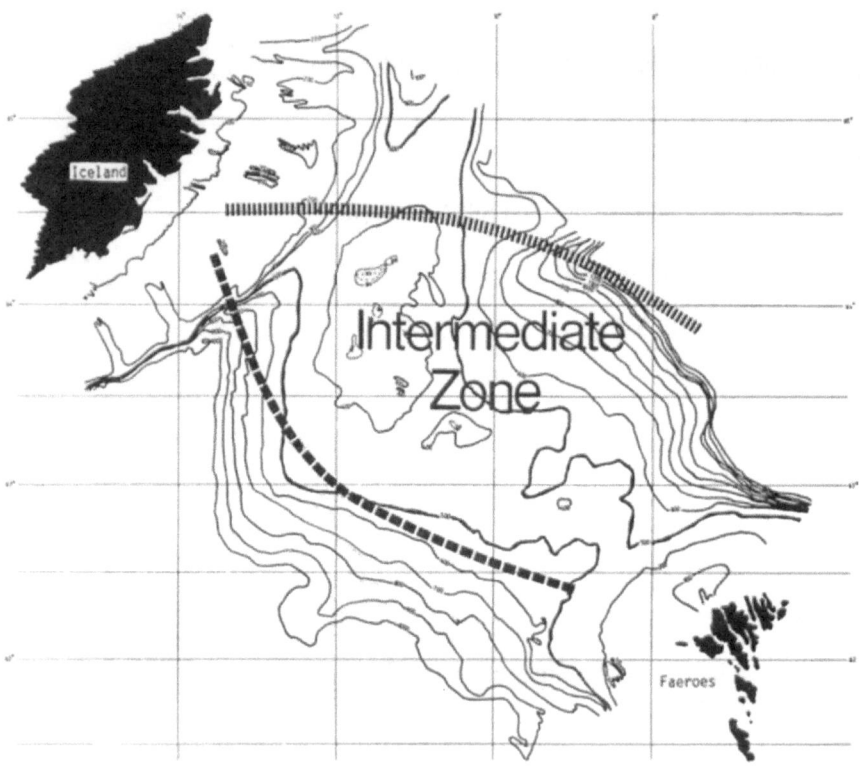

Figure 1. The general disposition of the Iceland-Faeroes frontal zone

The main Northern boundary of the frontal zone is 'attached' to the North-Eastern flank of the Ridge, and slopes Northwards, rising to meet the surface with gradients of a few percent. The boundary is steepest and most distinct at the Western end of the zone, where the zone is relatively narrow and the boundary is effectively pinned to the Icelandic Shelf.

Further Eastwards the zone broadens and becomes much more complex. The major boundary can normally be traced Eastwards as a Northern limit of the zone, but a significant 'Intermediate Zone' region can be observed, stretching Southwards from the major boundary to the region beyond the Ridge on its South-Western flank (figure 1). The Intermediate Zone is often found to have a weak Southern boundary.

Frontal activity can also be seen to the North of the major Northern boundary. This appears to have two principal forms: major Northwards excursions - 'warm intrusions' - of warm water over the boundary; and complex patterns of warm overflows limited to the near-surface layers.

It is the major Northern boundary, and the adjacent Intermediate Zone, which form the subject of this paper. Detailed structure measurements are presented which show the principal changes in the main boundary from West to East. We also present measurements showing the structure of the cold eddies which regularly occur in the Intermediate Zone, and which have their origin in cold jets which have been observed to penetrate the major boundary.

Section 2 of this paper briefly outlines the experimental technique used to obtain the data presented here; sections 3 & 4 report on the boundary structures and on the eddy structures. Section 5 considers the state of present knowledge of the region, and possible strategies for improving this.

2. The Measurements

The data presented here were obtained using the 400m ARE digital thermistor chain (Lane & Scott, 1986). This is a versatile and robust measuring instrument, variable in length and inter-sensor spacing, which samples 100 temperature sensors. Up to 25 pressure sensors are incorporated, allowing deduction of the chain shape as it is towed through the water. Towing is normally done at 4 knots (2m/s), which gives a sampled vertical aperture of about 300m, and sampling at 0.9s gives a horizontal resolution better than 2m. Equal spacing of the temperature sensors along the chain gives a vertical resolution of about 1.5m near the surface, increasing to nearly 4m at depth.

Other sensors can be physically attached to the chain, such as Neil Brown 'smart' CTD units, which give time series of salinity data at approximately fixed depths, in support of separately determined CTD station data. In practice it is found that the temperature-salinity (T,S) relationship in this region is sufficiently regular to allow the deduction of useful estimates of water column density from temperature data alone. To an even greater extent, the T,S relationship allows confident estimates of sound speed variability. The CTD units located on the chain also provide an independent continuous check on the calibrations of the chain's own temperature and pressure sensors.

The results shown here were obtained in four extensive measurement surveys, covering the three-year period 1985-1988, which examined the two main seasonal states of the water mass structures, typified by late Winter (April-May) and late Summer (July-September). The transitional early Summer (June) condition was also surveyed in this period. Some of the results from the 1986 survey were reported in (Scott et al, 1988).

Some observations about measurement sampling in this region are appropriate, since earlier surveys have typically under-sampled the variability, both temporally and spatially. Horizontally, the temperature may change by about 6C in less than 500m in certain locations, and although this variability represents an extreme rather than a typical state, it is likely that even spatial sampling at 1km significantly undersamples some features of interest. The problem is exacerbated by fact that the small-scale variability is rarely small compared with the large-scale trends, and undersampling can thus give a misleading picture. Thermistor chains are the only instruments presently capable of fully resolving the Iceland-Faeroes region.

Other sampling problems arise from the time variability of the observed structures. It has only proved possible to assess time variability with any confidence since recent improvements in spatial sampling allowed a separation of the two influences. An unknown advection of a spatially varying structure cannot be distinguished from a time variation. A second sampling factor present in this region is the possible presence of small detached cold eddies near the major boundary. If the sampling misses the warm water region situated between an eddy and the boundary, then the system may be misinterpreted as a locally distorted boundary. Even if the warm water region is partially sampled, the situation may still not be clear.

The thermistor chain data are shown here in the form of isotherm plots, a representation which gives a graphic picture of the overall scales of the features contained, and also an impression of the spatial variability at much smaller scales.

3. The Structure of the Major Boundary of the Frontal Zone

The temporal and spatial variations of the Iceland-Faeroes frontal zone will be illustrated here using three sections across different parts of the region. The sections shown were obtained in two different seasons, and where this factor affects the structure this will be pointed out. The seasonal variation will be illustrated further in section 4.

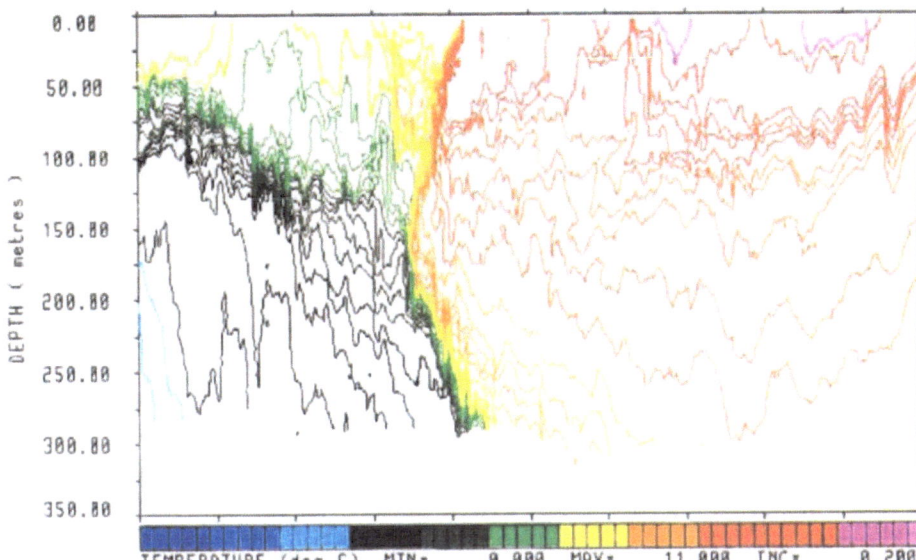

Figure 2. The thermal structure of major boundary of the Iceland-Faeroes frontal zone, at the Western end, close to 12°W. 0.2C contours are shown. The length of track shown here is about 55km.

The frontal zone is seen to contain an extensive Intermediate Zone, South of the major boundary and covering the top of the Ridge. Here, the thermohaline structure is determined by the two water mass inputs and the time history of their mixing. The results are consistent with mixing increasing from West to East, the main direction of the Iceland Current which provides the lower, colder component of the frontal system. The Intermediate Zone is approximately bounded to the North by the major boundary, and to the South by a much weaker boundary, or series of weak boundaries leading eventually to the Atlantic water.

The Intermediate Zone is characterized by a 200-300m thick upper layer of essentially Atlantic water which has been considerably cooled. This cooling appears to be the result of two processes: by contact (possibly involving entrainment) with the underlying layer of cold water, as this flows Southwards over the Ridge; and also through the mixing effect of cold eddies in the Zone. The upper layer here is considerably more stratified than that of its Atlantic source, as a result of the cooling from below, and even the surface layers may be noticeably cooler, although this effect is often masked to some extent in Summer by solar heating.

We are concerned in this section with the major boundary which forms the Northern limit of the Intermediate Zone. This boundary is usually the strongest boundary in the Iceland-Faeroes frontal zone. Figures 2, 3 & 4 show a series of sections through the boundary, progressing from West to East. All three sections have been drawn to the same horizontal scale for comparison.

Figure 2 shows a section taken in late Summer - September 1985 - close to 12°W, adjacent to the Iceland Shelf. The sharp nature of the boundary is immediately apparent, as is the complex 'overturned' structure. The sharpness is typical here, although the overturning is not typical. Well-developed summer thermoclines can be seen on both sides of the boundary, consistent with the season.

Observations in the same region at other times of year show that the abruptness of the boundary here is common to all seasons. In Winter conditions the extreme horizontal temperature gradients seen here persist all the way to the surface, changing by (typically) 5C in 500m.

At this extreme Western end of the frontal zone the width of the zone is quite small, the two impinging water masses undergoing very little modification during their approach. Further East, the situation changes quite quickly. Between about 11°W and 9°W the frontal zone bulges Northwards in the upper 250m, to give a feature known as the 'warm intrusion'. This bulge complicates the picture, giving boundaries further North than what we are calling here the major boundary. The warm intrusion feature is regularly observed, but it is not yet clear whether it appears and decays episodically, or whether it is more permanent.

Figure 3 shows a section near $10^{O}30'$which passes South into the Intermediate Zone from the warm intrusion. These data are from June 1986, and show an early stage in the transition to Summer conditions. The difference in the structures of the cold side of the boundary between figure 2 and 3 should be noted; the strongly stratified upper 250m of relatively warm water seen in figure 3 is a principal characteristic of the warm intrusion. The major boundary is much more diffuse than we observe further West, but it is clear from its location that it is essentially the same boundary, the Northern limit of the Intermediate Zone. The early Summer timing of the data shown here is evident in the weak and patchy appearance of the Summer mixed layer.

The boundary here is less steep, and is also strongly layered, with evidence of intense mixing and overturning. There are several

455

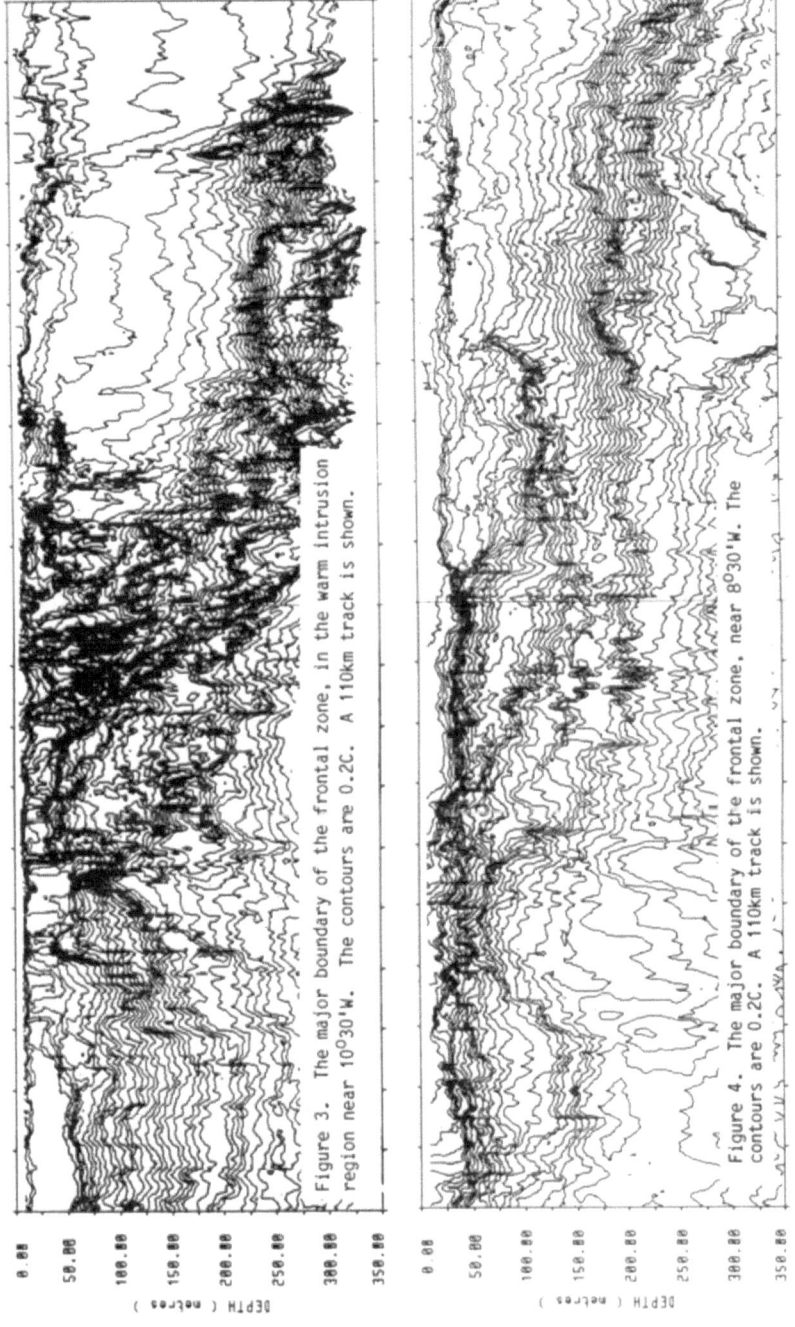

Figure 3. The major boundary of the frontal zone, in the warm intrusion region near 10°30'W. The contours are 0.2C. A 110km track is shown.

Figure 4. The major boundary of the frontal zone, near 8°30'W. The contours are 0.2C. A 110km track is shown.

DEPTH (metres)

0.00
50.00
100.00
150.00
200.00
250.00
300.00
350.00

examples of apparent upwelling of the cold water layer through the warmer upper layer. These small-scale features are typical of Summer conditions near the major boundary.

Moving further East again, the boundary becomes even more diffuse, and even less steep. Figure 4 shows a section, again taken in June 1986, at about $8^{\circ}30'W$. The Summer mixed layer is more evident in this section, particularly at the Southern end, where increasing winds were beginning to thicken the upper boundary layer.

Another typical Summer variation of the boundary structure, seen at least at the Western end, involves an outward bulge of cold water at the level of the Summer thermocline. This is associated with a steepening, even over-folding, of the boundary. An example of this Summer structure appears in the boundaries of the Summer cold eddy shown in figure 5.

4. Cold Eddies

It was indicated above that the Intermediate Zone was essentially composed of water of Atlantic origin, about 200-300m thick, overlying a cold overflowing layer. It is limited to the South by unmodified Atlantic water, along the South-Western flank of the Ridge, apparently at a position where the sinking cold water overflow can no longer influence the structure of the upper layer. Weak frontal boundaries can sometimes be found between the slightly different water masses in the upper layer.

The underlying cold water layer varies in thickness, depending on position on the Ridge and depending on the state of the overflow. Very little is known of the overflow, except that it appears to be normally restricted to about four relatively deep parts of the Ridge. The small number of current meter stations occupied along the Ridge do not allow the time variability to be separated from the spatial variability. Also, it is known that the currents, both warm and cold, have strong tidal signatures, and the interface between the layers probably rises and falls with tidal periodicity.

These factors would make the situation complex enough, but the existence of eddies in the Intermediate Zone makes the currents one of the least understood aspects of the frontal zone. Hansen and Meincke (1979) were the first to draw attention to these eddies.

The cold Iceland-Faeroes eddies have a distinctive structure, which depends on the season. In all cases they involve the raising of the cold overflow layer through the overlying warm layer - usually to the surface, or in the Summer, to the upper mixed layer. It appears that the eddies have only been detected where the cold

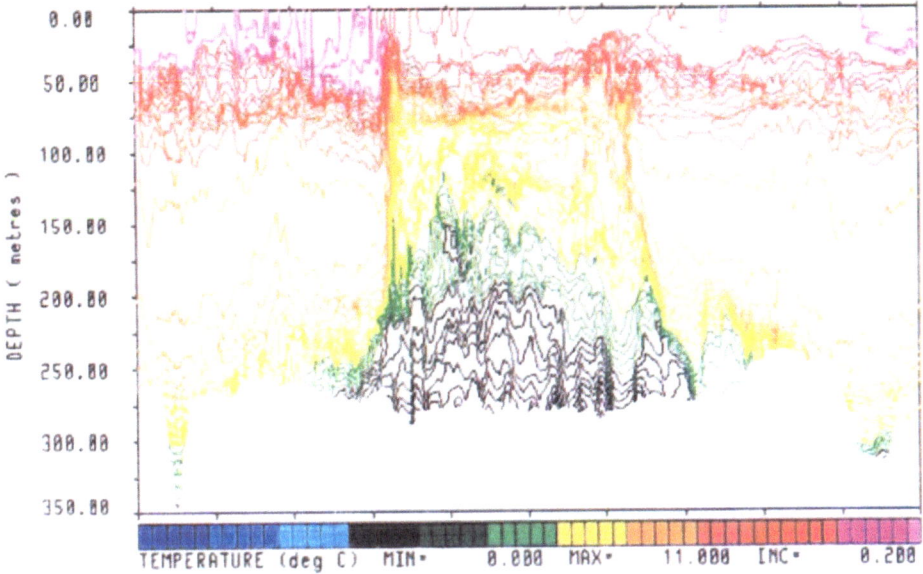

Figure 5. A cold eddy in the Iceland-Faeroes frontal zone, in Summer conditions. 0.2C contours are shown, the top contour being 7.0C. The length of track shown here is about 37km.

overflow layer is present, i.e. within the Intermediate Zone. The eddies have been observed, from sequences of satellite infra-red images and from acoustic Doppler current profiler (ADCP) data, to have a generally cyclonic (anti-clockwise) circulation, with currents greater than 50cm/s.

Cold eddies appear to be more common in the Intermediate Zone than are warm eddies. Although our series of surveys has positively identified more than twenty cold eddies, we have found only one warm eddy.

Figure 5 shows a track made through a cold eddy in July 1987. The well-developed Summer mixed layer is clearly evident both inside and outside the feature, although the temperature signature is very much smaller in this layer than it is below. The eddy boundary is very steep, vertical at some depths. This steepness appears to be a characteristic of the Summer condition, although it is possible that only relatively young eddies are like this. The bulge in the boundary at the level of the Summer thermocline is also typical,

458

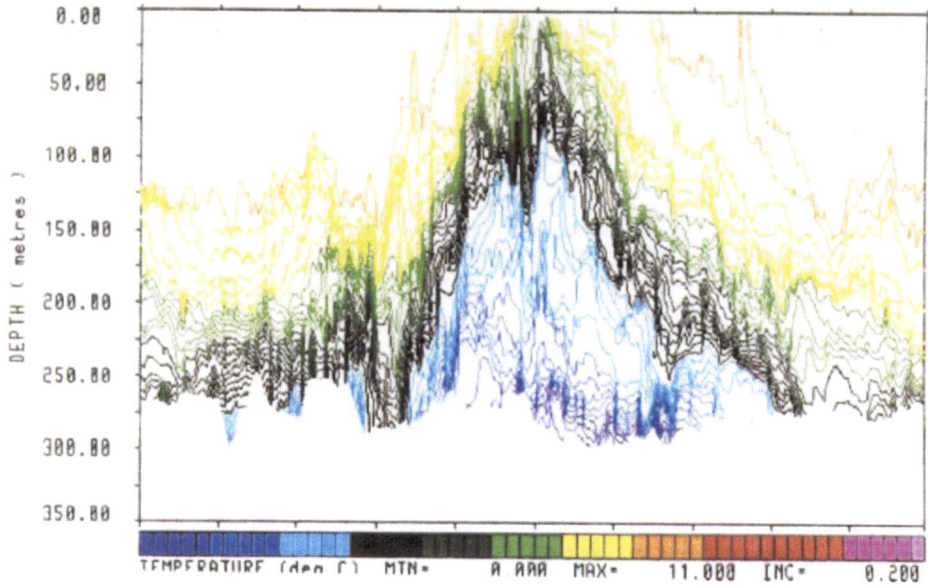

Figure 6. A cold eddy in the Iceland-Faeroes frontal zone, in late Winter conditions. 0.2C contours are shown. The mixed layer outside the eddy is at about 10.2C. The length of track shown here is about 37km.

and as was indicated above, these two characteristics can also be seen at the major boundary, at least at the Western end.

Figure 6 shows a similar eddy in April 1988, in the late Winter condition. The boundary slopes are much more gentle in Winter than they are in summer. The surface temperature signatures of eddies are more clearly related to their internal temperature variation than they are in Summer.

The accumulation of evidence over four years has indicated that cold eddies such as these are very common in this region, and re-evaluation of earlier published results supports this conclusion. It is possible that between five and ten eddies may be present in the Intermediate Zone at any time.

There appear to be two principal sources of the cold eddies. One source lies in the occasional occurrence of very broad jets of cold water passing Southwards through the major boundary. The evidence for these comes mainly from scanty satellite infra-red images, although there have been supporting indications from the behaviour of drifters. These broad jets can be several tens of km wide, and they appear to completely disrupt the normal structure of the frontal zone, passing over the top of the Ridge before they break up. Although it seems likely that their occurrence is related to atmospheric forcing in some way, the available evidence is very thin, and we have little indication of how frequent they might be. Drifter data suggest very large currents within the jets, approaching 1 m/s.

The other source is now relatively well understood, as a result of a fortunate coincidence of a two-day sequence of satellite infra-red images with a GEOSAT altimeter track and with a series of thermistor chain tracks (Scott & McDowall, 1990). These data, obtained in May 1988 during the survey series reported here, observed the formation of a cold eddy at the corner of the warm intrusion, at 64°30'N, 11°W. A narrow jet of cold water was observed to pulse through the major boundary at this position, and the water so passing through was found to form a cyclonic eddy.

The same survey detected at least one other eddy in the in-water data, and the synoptic images suggested the presence of several others. All of them could have had the same origin as the one whose genesis was directly observed.

Several other cold jets were observed within the warm intrusion, and there were indications that these, too, were forming eddies. However, it is not clear whether these eddies would pass through the major boundary which normally separates the warm intrusion from the Intermediate Zone.

It is possible that the broader, more energetic jets, apparently less frequent than those reported by Scott & McDowall, are exceptional examples of the same phenomenon.

5. Spatial and Temporal Development of the Frontal Zone

The results presented here show the detailed structures of the major boundary of the frontal zone, and also of the cold eddies which are commonly found just South of this boundary, in the Intermediate Zone. The nature of the spatial variability of the major boundary structure is indicated, as is the seasonal variation. The data presented have been extracted from a large data set which in its totality suggests that they are typical of the region.

Note that the seasonal variability deduced here is that of the structures, and not of the positions of features. It is now clear that attempts to deduce seasonal effects on boundary positions (e.g. Smart, 1984) have so far been made problematical, if not actually impossible, by the normal inability to separate spatial and temporal variation.

The detailed structure results allow us to deduce the effect of under-sampling the region spatially, which is important since it is likely that long-term monitoring of the region will have to rely on relatively sparse sampling techniques such as XBTs. We now have a clearer spatial context into which individual profiles and sections may be fitted.

The clearer spatial view of the frontal zone makes it possible to re-evaluate earlier data and improve our interpretation of it. This, in turn, gives us more confidence in using these earlier data to form a picture of the temporal variability, particularly as concerns the major boundary.

There are still many questions to be answered, both on the Intermediate Zone eddies and on the behaviour of the major boundary. We know very little about the generation, movement, and eventual decay of the eddies. Our knowledge of the instabilitites of the major boundary are also largely unknown, particularly as concerns the production of warm and cold intrusions. Synoptic monitoring of the region could clearly produce many of the answers, and it is regrettable that satellite infra-red is denied us by the near-continuous cloud cover.

Apart from remote sensing, however, we have found that only the most detailed in-water measurements are able to track the eddies. A survey pattern designed to cross backwards and forwards through an eddy (once discovered) in several directions can keep track of the feature, but this needs to be done over many tidal cycles. This puts a severe constraint on the ship in the event of rough weather. If it becomes necessary to leave the eddy for any length of time it can be difficult to re-locate it, and this introduces uncertainty about whether the newly located eddy is the same one as before.

The Iceland-Faeroes frontal zone is clearly a complex and interesting area, and although we have made considerable progress in understanding the modes its variability might take, it is clear that we still have insufficient data for a complete understanding. It will require a thorough combination of all possible resources: detailed structure measurement, current meter moorings, remote sensing, and ocean modelling, to deal properly with the problems outstanding.

Acknowledgements

The results presented here could not have been obtained without the concerted efforts of many people, particularly the sea-going teams,

but also those who have supported from the shore. We would
particularly like to thank Claire Burt, Norman Geddes, Geoff Kirby,
and Anne McDowall, whose contributions have been particularly
significant.

References

Hansen, B. and Meincke, J. (1979) 'Eddies and meanders in the
Iceland-Faeroes Ridge area', Deep-Sea Research 26, 1067-1082

Lane, N.M. and Scott, J.C. (1986) 'The ARE digital thermistor
chain', Proceedings, Marine Data Symposium, New Orleans, pp.127-
132, ISBN 0-933957-03-3

Scott, J.C., Geddes, N.R., Lane, N.M. and McDowall, A.L., (1988)
'Thermal structure and remote sensing measurements in a major
frontal region', Advances in Underwater Technology, Ocean Science
and Offshore Engineering, 16, 59-68

Smart, J.H. (1984) 'Spatial variability of major frontal systems in
the North Atlantic-Norwegian Sea area: 1980-81', Journal of
Physical Oceanography, 14, 185-192

UPPER OCEAN VARIABILITY ASSOCIATED WITH FRONTS

R. A. WELLER
R. M. SAMELSON
Woods Hole Oceanographic Institution
Woods Hole, Massachusetts 02543
U. S. A.

ABSTRACT. In January to June 1986 the cooperative Frontal Air-Sea Interaction Experiment (FASINEX) was conducted in the subtropical convergence zone in the North Atlantic southwest of Bermuda. Velocity, temperature, and conductivity variability associated with ocean fronts characteristics of the region was observed. The fronts maintained a southwest to northeast orientation and propagated to the northwest through the experiment site. The temporal and spatial variability of the upper ocean associated with the fronts and also with energetic sub-mesoscale features found in association with the fronts is described.

1. Introduction

Strong horizontal gradients in water properties and sound speed are found in the upper ocean at fronts (Figure 1). One region of the oceans where fronts are characteristically encountered is the Subtropical Convergence Zone (STCZ), roughly between 25°N and 30°N. To the north, westerly winds force southward transport of cool water; to the south, the Trade Winds force northward transport of warm water. Ship and aircraft surveys in the Atlantic southwest of Bermuda (Voorhis and Hersey, 1964; Voorhis, 1969; Katz, 1969) found fronts with large horizontal extent (over 1700 km), surface temperature gradients in excess of 2°C km^{-1}, and essentially geostrophic along-front velocities in excess of 60 cm s^{-1}. Repeated aircraft surveys suggested that these fronts had a lifetime of several months, though their configuration was observed to evolve with time (Voorhis, 1969).

More recently, interest in characterizing spatial variability in the upper ocean and in understanding its relation to air-sea interaction led to a new study of fronts in the western North Atlantic STCZ. During the Frontal Air-Sea Interaction Experiment (FASINEX) (Weller, 1990) data were collected from a variety of platforms on a variety of space and time scales. Satellite infrared imagery was collected from 1982 through the period of the field work in 1986. A closely-spaced array of surface and subsurface moorings was deployed straddling a front in January 1986 and recovered in June 1986. Ship and aircraft studies were conducted in February to March 1986. In this paper, we use data from FASINEX to focus on the variability associated with ocean fronts. That variability dominates the upper ocean in the STCZ. In contrast, data from a site 750 km to the north of FASINEX (see Figure 2 for a map showing the location of both sites) shows that local air-sea interaction processes control upper ocean variability.

Large space and time scales are discussed first, using data from the moorings and satellites to show the extent to which fronts dominate upper ocean variability in the STCZ. More information about the fronts at the FASINEX site, including descriptions of the velocity,

J. Potter and A. Warn-Varnas (eds.), Ocean Variability & Acoustic Propagation, 463–478.
© 1991 *Kluwer Academic Publishers.*

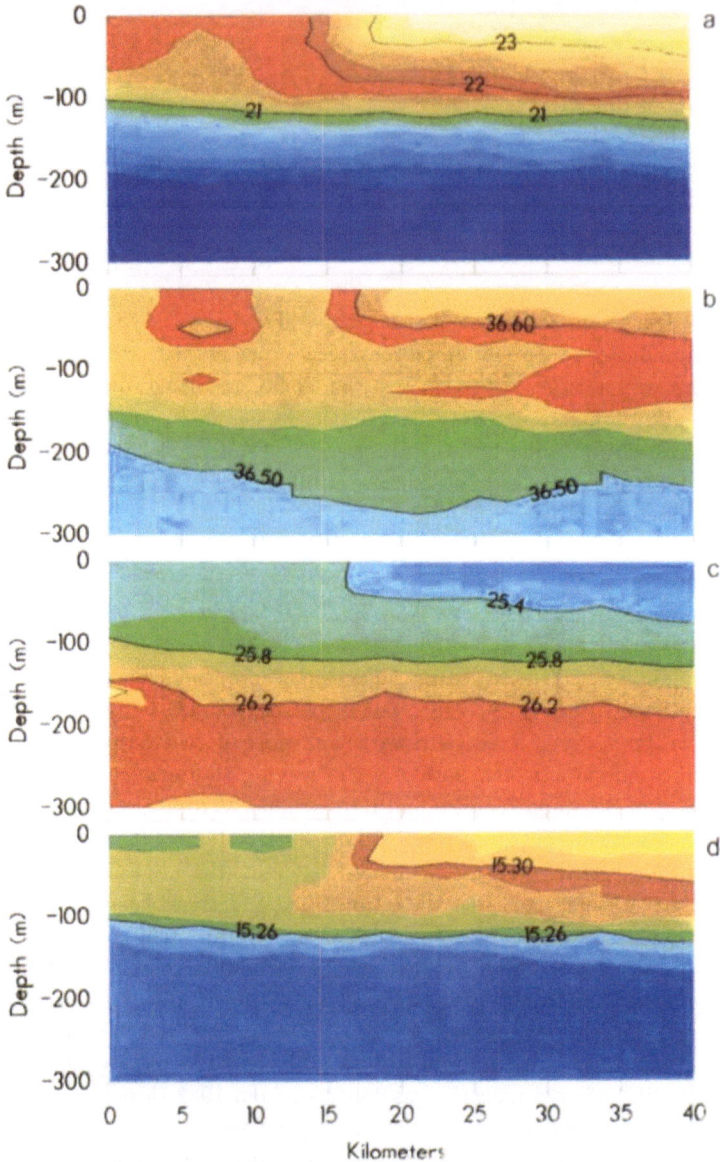

Figure 1: Cross-frontal sections of temperature (°C) (a), salinity (psu) (b), density (σ_θ, kg m^{-3}) (c), and sound velocity (divided by 100, m s^{-1}) (d) based on CTD data obtained during FASINEX. Sound velocity was computed using the algorithms of Chen and Millero (1977). Select contour intervals are labelled.

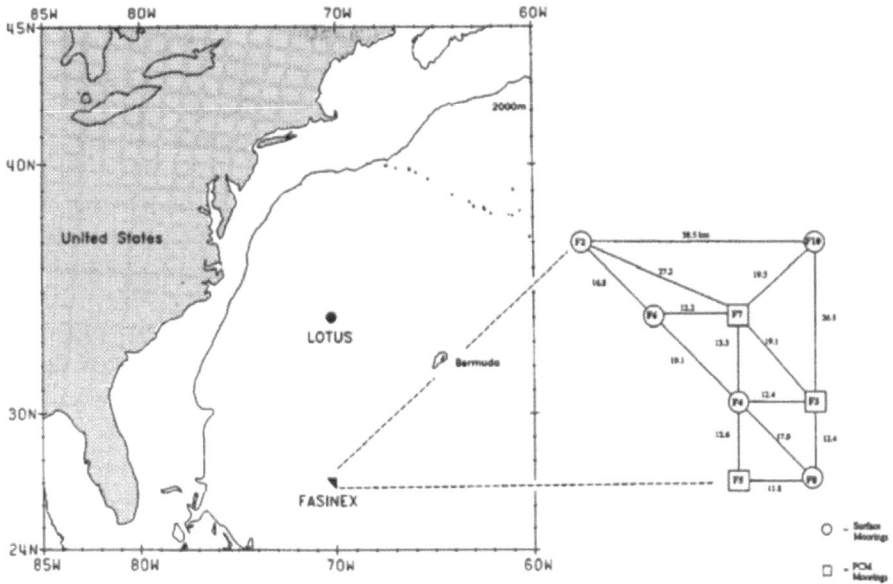

Figure 2: Map of the eastern North Atlantic with the locations of FASINEX and LOTUS indicated. One surface mooring was deployed during LOTUS. The inset shows the array of five surface moorings (o) and three subsurface Profiling Current Meter (PCM) moorings (□) deployed in FASINEX.

temperature, conductivity, and sound speed variability, are presented next. Unanticipated and strong variability on scales small compared to the width of the fronts was observed during the experiment, and a separate section is used to present examples of these structures. The dynamics of the observed variability is discussed, and a final section presents our conclusions.

2. Seasonal and Large Scale Variability

In many locations a significant fraction of the variability observed in the vertical structure of the upper ocean can be related to local atmospheric forcing. An annual cycle in upper ocean temperature structure is found that reflects seasonal change in heating and wind stress. For example, outside the STCZ, at 34°N, 70°W (site of the Long-Term Upper-Ocean Study or LOTUS (Briscoe and Weller, 1984)), a clear seasonal signal in the stratification of the upper ocean (Figure 3a) has been observed. Surface temperatures fell ~1°C from mid-January to early March, and in February and March the mixed layer was

466

cool and thick (over 200 m) following a winter of heat loss and wind mixing. The mixed layer temperature remained constant (~18.6°C) until mid-April. Following April, the net heat gain by the ocean in combination with reduced wind mixing led to restratification and a surface temperature rise of ~5.5°C by mid-June. Mixed layer depths decreased to several 10s of meters with additional transient shoaling associated with diurnal heating. Within the northern part of the STCZ at 70°W, the heat flux and wind stress histories are similar to those at the LOTUS site (Isemer and Hasse, 1987); yet, in contrast, a succession of ocean fronts, not local atmospheric forcing, provided the dominant source of upper ocean variability observed at the FASINEX moored array. Sea surface temperatures at the FASINEX site increased 3°C (from ~23.5°C to ~26.5°C) from mid-January to mid-June, but did not show a minimum in February to March. A deep mixed layer did not form in late winter in response to heat loss and wind mixing. Heating and restratification were not observed until mid-May. Instead, a sequence of step-like changes in upper ocean temperature were observed (Figure 3b).

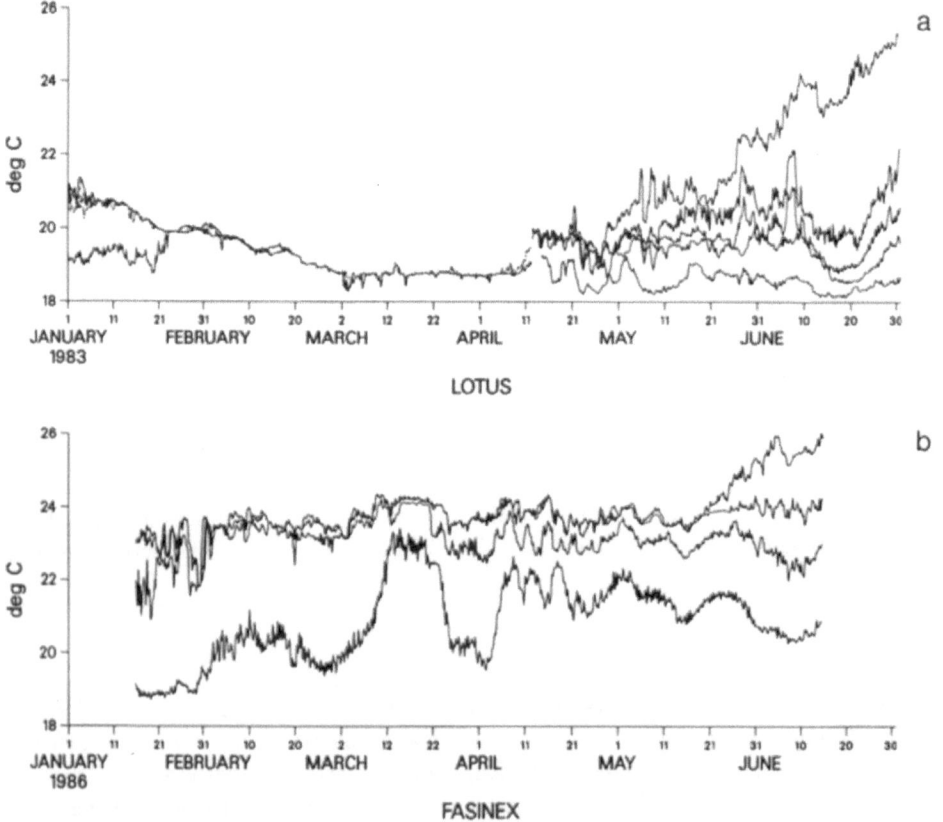

Figure 3: (a) Temperature versus time at 5, 50, 75, 100, 150, and 200 m depths at LOTUS for January to June 1983. (b) Temperature versus time at 10, 40, 80, 120, and 160 m depths at FASINEX for January to June 1986.

On the large scales of the mean surface wind field (>1000 km)) Ekman transport acts to bring water masses together within the STCZ. Because LOTUS was outside this region and FASINEX was inside, it is apparent that frontal variability is geographically localized. Halliwell *et al.* (1990) examined the large scale distribution of fronts using AVHRR (Advanced Very High Resolution Radiometer) imagery from before and during FASINEX. In the vicinity of 70°W, fronts were found from ~22°N to ~32°N, often with more than one essentially zonal front present in each image. Monthly composites of manually digitized frontal locations (Figure 4) showed that, in a given year, the fronts tended to be most often observed at certain preferred locations and also to have a characteristic orientation that persisted for months.

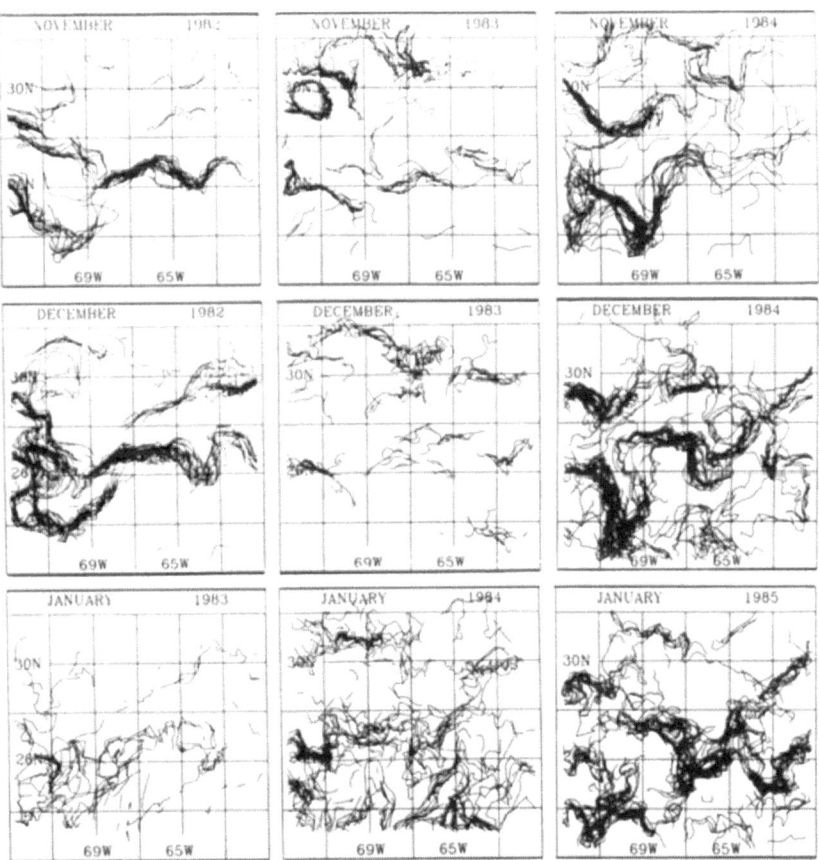

Figure 4: Monthly composites of frontal locations in the STCZ southwest of Bermuda. The locations were digitized from satellite infrared imagery, and these summaries of frontal location and frequency for November, December, and January of 1982 to 1985 were provided by P. Cornillon.

3. Fronts

The observation that the fronts in the FASINEX area took on a persistent, preferred orientation was used in the observational strategy. AVHRR images from December 1985 and early January 1986 showed a strong front aligned southwest–northeast near 27°N, 70°W. The array of moorings set in January 1986 was biased along the northwest–southeast direction, tuned to resolve cross-frontal structure under the assumption that fronts would continue to have the same orientation observed in early January.

The five-month record from the array showed the occurrence of a series of jet-like flows either toward the northeast or toward the southwest. Figure 5 shows temperature data from the upper 160 m from the five surface moorings beginning February 28 (Julian day 59) and ending March 14 (Julian day 73). The temperature gradient associated with a front was observed first at the southeastern end of the array, and progression of this feature through the array occurred at $\sim$10 km day^{-1} toward the northwesternmost mooring. This was characteristic for the length of the deployment. Eriksen *et al.* (1990) used maximum likelihood frequency–horizontal wavenumber spectral techniques to calculate propagation velocities for temperature and salinity features in the 40 to 120 m depth range and found them to be between 9 and 29 km day^{-1} to the northwest.

When encountered from a ship, the surface expression of the fronts was striking. Sargassum and other flotsam were drawn into a 1 to 10 m wide band extending to the horizon, and local cross-frontal surface temperature gradients of 1°C km^{-1} were observed. The total increase of $\sim$2°C associated with the fronts was observed across $\sim$10 km. The warm (southern) side of the fronts was saltier; and, above the seasonal thermocline, the front was a water mass boundary. Below the thermocline, neither salt nor temperature gradients were observed. The most saline water was typically found at depths of 100 to 160 m below the surface outcrop of the front; TS characteristics suggest that this water came from the surface toward the west (upstream) (Pollard, 1986).

Time series from one of the moorings from February 25 (Julian day 56) to April 5 (Julian day 95) are shown in Figure 6 in illustration of the variability associated with a front. The algorithms provided by Chen and Millero (1977) have been used to compute sound speed. During this time period first a warm front (cool to warm transition), then a cold front, and finally a warm front were observed. The largest cross-frontal contrasts in scalar properties were found at 120–160 m. Temperatures changed $\sim$2°C, and sound speed changed $\sim$6 m s^{-1}. At this depth, however, the changes occurred over a greater time (distance) than at the surface. Horizontal gradients near the surface were estimated to be 0.25°C km^{-1} for temperature and 0.6 m s^{-1} km^{-1} for sound speed; at 120–160 m these gradients were 0.05°C km^{-1} and 0.14 m s^{-1} km^{-1}, respectively. Conversion of time to distance in this case was done using the assumption that the fronts were frozen features carried to the northwest at the rate of 10 km day^{-1}. This assumption is valid to first order; spatial gradients computed from the time series agreed with spatial gradients computed from cross-sections of the fronts made by the ships. Evolution of the fronts as they progressed across the array was, however, evident. Figure 5, for example, shows some change in a warm front as it moved across the array.

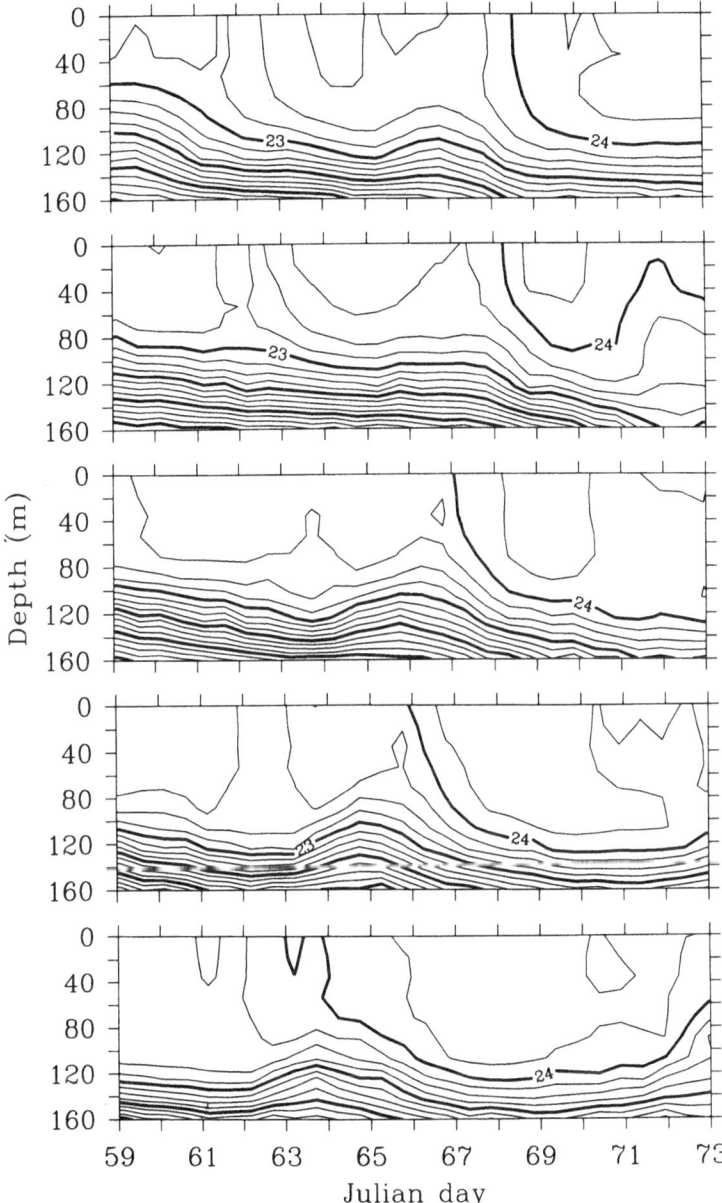

Figure 5: Contour plots of temperature (°C) versus time in the upper 160 m from the five surface moorings of the FASINEX array. During this period a warm front moved across the array. The plots are ordered with the southeastern-most mooring, F8, at the bottom. F4, F6, and F2 were each successively ~18 km further to the northwest and are shown in that order above the plot for F2. F10 was ~40 km north of F8; the plot for F10 is at the top.

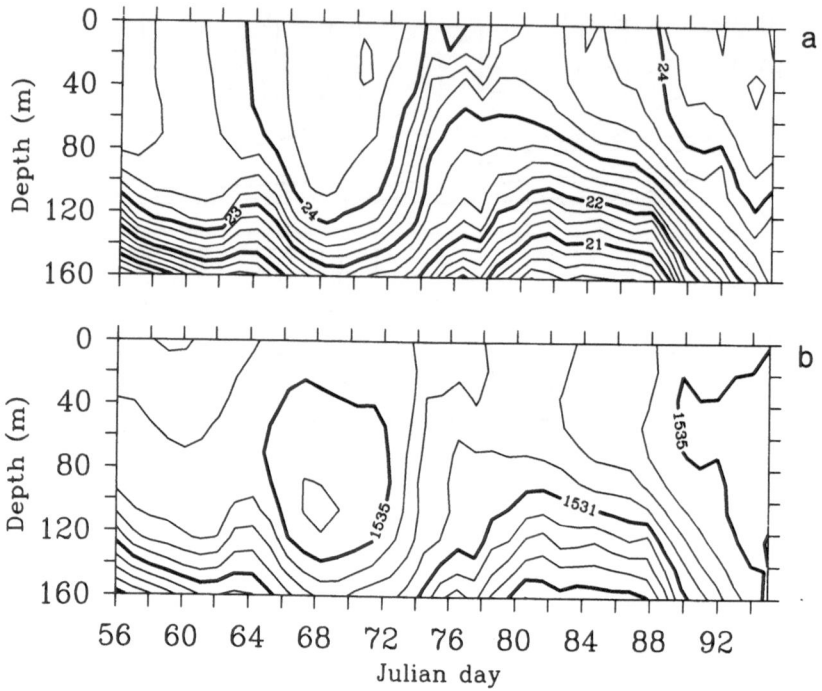

Figure 6: (a) Temperature (°C) and (b) sound speed (m s⁻¹) versus depth and Julian day at mooring F8.

4. Submesoscale Variability

In general, the fronts, with their associated density gradients and jet-like along-front velocity fields, were well-resolved by the combination of moored, shipboard, and remote-sensing techniques. However, unanticipated energetic variability was also observed at smaller scales. These "submesoscale" features were less well resolved by the sampling scheme. Moored instruments in the upper 120 m recorded several events in which surface and near-surface temperatures decreased 1.5 to 2.0°C within several hours, remained low for 24 to 36 hours, and then returned rapidly to near previous values. Freely-drifting instruments recorded spatially isolated intense localized downwelling of 20–30 m hr⁻¹, providing independent evidence for energetic submesoscale variability.

Temperature and sound speed versus depth and time are shown in Figure 7 for two events observed by the moored array. The temperature anomaly is mostly confined to the upper 80–100 m in each case, but in that range is nearly independent of depth. The features are roughly symmetric. Maximum sound speed anomalies exceed 5 m s⁻¹. Maximum flow velocities associated with these features exceeded 50 cm s⁻¹. All of the observed features had cold "cores," with cyclonic circulation. The first of the two events shown in Figure 7 occurred at mooring F2 (the northwesternmost mooring) on January 29–30. The surface

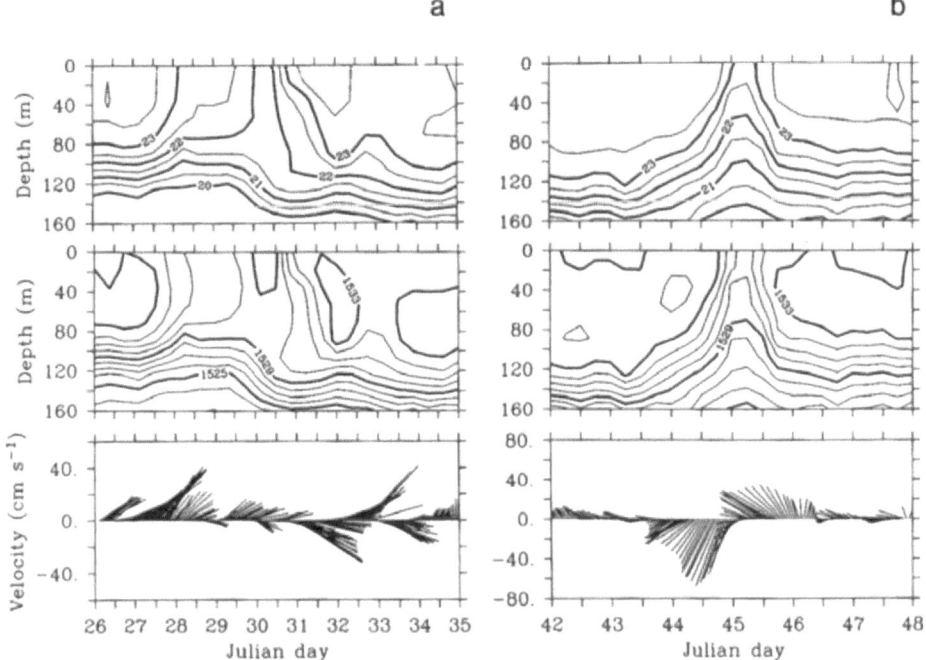

Figure 7: Temperature (°C) and sound speed (m s⁻¹) versus depth and time, and horizontal vector velocity (north is up) versus time for (a) January 27–February 1, F2, (b) February 12–15, F8.

expression of this feature is visible in AVHRR imagery (Böhm and Cornillon, 1987) as the southern tip of a V-shaped cyclonic frontal meander. The meander propagated northeastward as it passed the moored array. The meander was roughly 20 km wide, with measured flow velocities reaching 60 cm s⁻¹ and estimated relative vorticity roughly equal to f. The cold "core" of this feature was contiguous at the sea surface with the cold side of the front.

The second of the two events shown in Figure 7 occurred at mooring F8 (the southeasternmost mooring) on February 13–14 (Julian day 44–45). A similar event occurred nearly simultaneously at F5, 12 km to the west. This feature was obscured by clouds and could not be identified in satellite imagery, but it was evidently isolated at the sea surface from the cold side of the front, which was 100 km to the northwest. (The front was visible in AVHRR images on February 12 and 15, and no similar events occurred at moorings north of F5 and F8.) A "frozen-field" estimate of horizontal structure (see discussion below) indicates a width of 15–25 km for this feature. Careful examination of the salinity record at F5 indicates that the fluid in the core of this feature originated on the north side of the front. Salinities on the 25.5 σ_t-surface in the cold core of the feature were between 36.58 and 36.62, comparable to those on the north side, while those on the south side were consis-

472

tently between 36.65 and 36.75. This feature may have been formed by the "pinching-off" of a meander (see Figure 11 and discussion below).

Temperatures at fixed depths are shown in Figure 8 for a third event, which occurred on April 12–13 (Julian day 102–103) at mooring F10 (the northeasternmost mooring). Strong diurnal warming, comparable to that observed simultaneously at other moorings, occurs during the cold event, resulting in a 2.5°C/40 m vertical temperature gradient. The diurnal warming completely obscures the (already-weak) surface expression of this feature during daylight hours. AVHRR imagery from April 16 shows a complex pattern of surface isotherms and does not allow unambiguous identification, although the feature appears to be isolated from the front, which was 100–150 km north of the array.

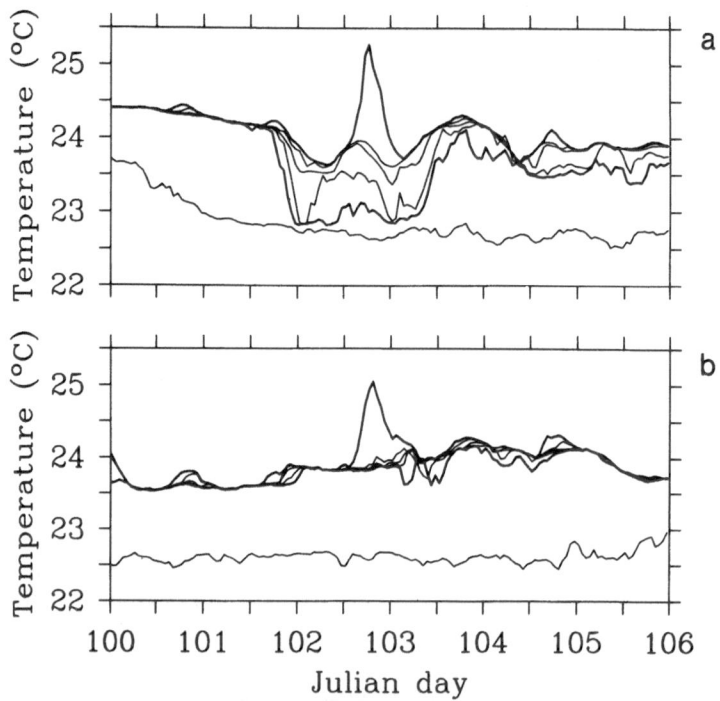

Figure 8: Temperature at 0, 10, 20, 30, 40, 120 m versus time for April 10–15 (thick lines — 0, 40 m). (a) F10, (b) F6.

In addition to the moored array, freely-drifting neutrally-buoyant Vertical Current Meters (VCMs) were deployed during FASINEX. They were released at depths of 90 to 150 m, close together as pairs or groups of three. A set of angled fins on their exterior cause them to rotate if there is vertical flow past the floats, and their records of depth and rotation are combined to give time series of vertical displacement. Two floats deployed on the cold side of the surface outcrop, on subsurface isopycnals sloping up toward the front, showed slow upward displacement, consistent with cold flow toward the outcrop (Figure 9a).

Two floats placed directly below the surface expression of the front, in water marked by high oxygen and thought to be carried down and under, had downward displacements (Figure 9b). One of these moved downward at 20 m day^{-1}. The other, less than 10 km away, showed downward displacements of 20–30 m hr^{-1} for several hours, with a total downward displacement of 160 m in 36 hours. No direct measure indicated strong upwelling. The strong downwelling, and its horizontal variability, provide another example of variability on subfrontal scales.

a   b

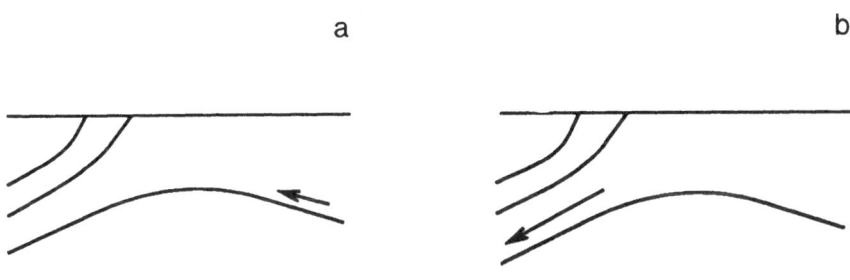

Figure 9: Schematic illustration of upward and downward motion along frontal isotherms from VCM's (a) cold side deployment, (b) warm side deployment.

5. Dynamics

Large scale Ekman convergence brings water masses together in the STCZ in the western North Atlantic. Within the STCZ other processes sharpen those gradients. The resulting frontal density gradients and jet-like along-front velocity fields, which were well-resolved by the moored array, were geostrophically balanced. Work is underway to identify the processes that produce fronts locally and control their orientation and location within the STCZ. Halliwell and Cornillon (1990) suggest a link between the location of the bands of fronts and large scale (700 to 800 km wavelength, 200 to 300 day period), westward propagating temperature anomalies.

Some information on the dynamics of the submesoscale features can be obtained from the moored observations. Two estimates of geostrophic shear indicate that their primary cross-stream momentum balance is geostrophic (Figure 10). The first was calculated from density differences at fixed levels at adjacent moorings (F6–F2, F8–F3 and F5–F4), while the second was calculated from a "frozen-field" estimate of horizontal structure. For the frozen-field estimate, mean velocities for current meters above 100 m over periods of 5–7 days surrounding each event were calculated, and the result used to convert each time series to a spatial "snapshot". Density and velocity records were averaged above and below 100 m for each mooring record, yielding estimates of horizontal structure at two fixed depths for each mooring. Figure 10 shows scatter plots of estimated geostrophic shear from each method versus measured vertical shear. For clarity, both geostrophic and measured shears have been normalized by the magnitude of the maximum geostrophic shear for each event. There is substantial scatter, but the majority of points fall in the first and third quadrants. The correlation (particularly for large shear) of geostrophic and measured shear is clear evidence that the primary cross-stream momentum balance in these features is geostrophic. Some of the scatter in the comparison of measured and adjacent-mooring geostrophic shears is likely due to insufficient horizontal resolution, since the distance between moorings is comparable to the scales of the features themselves. Strong and variable near-inertial motions were observed throughout FASINEX, and these also likely contributed to the scatter in the comparisons. The estimated residual (measured minus geostrophic) vertical shears from the frozen-field estimate are partially near-inertial in character. The observed shear $(10^{-2} - 10^{-3} \text{ s}^{-1})$ is not strong enough to satisfy a theoretical criterion (Emanuel, 1979) for linear inertial instability of steady geostrophic flow, consistent with the evidence for primary geostrophic balance.

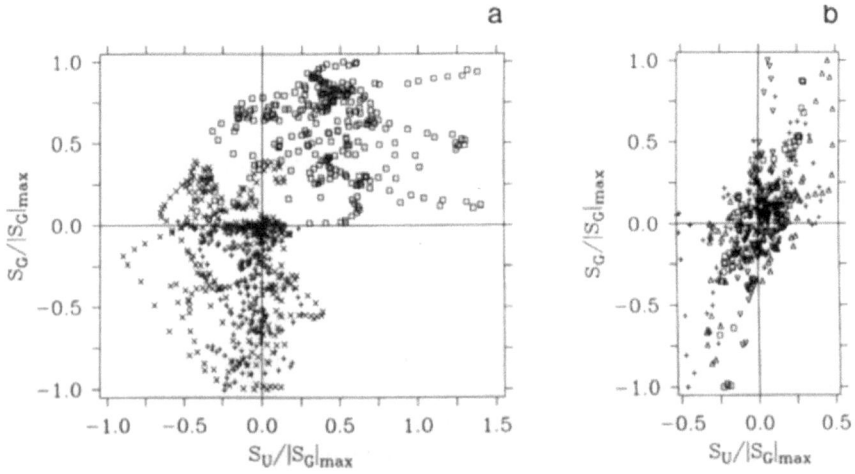

Figure 10: Estimated geostrophic vertical shear versus measured vertical shear. (a) Adjacent mooring estimate. (b) "Frozen field" estimate.

The cold core features likely originated as frontal waves or meanders. The first feature was identified as such in AVHRR imagery, while the cold core of the second was traced by a water mass argument to the cold side of the front. Figure 11 shows schematically a scenario for the deformation of a surface front by a developing baroclinic disturbance. The first two panels show a growing disturbance, with warm fluid moving northward and cold fluid southward. The third panel indicates divergence of warm fluid and convergence of cold fluid at the surface, with associated warm upwelling and cold downwelling. Such a pattern is characteristic of upper level flow in baroclinically unstable mid-latitude disturbances in the atmosphere (e.g., Palmen and Newton, 1969, Chapter 10). This picture is qualitatively consistent with several aspects of the observations. The warm divergence and cold convergence near the surface naturally leads to an asymmetry of the warm and cold features, with the cold surface features having shorter horizontal scales. This asymmetry is evident in both the moored (Figures 7 and 8) and remote observations. Figure 12 shows a particularly striking AVHRR image of thin cold features intruding into broader warm features. The cold features shown had evolved from broader features over several days. The "pinching-off" of a meander (Figure 11, panel 4) is a natural mechanism, akin to Gulf Stream ring formation, for the formation of isolated cold core features south of the front whose core fluid originated north of the front. The picture in Figure 11 is schematic, and the evidence for it fragmentary. It is not clear that purely baroclinic disturbances will have rapid enough growth rates, for instance. The two-dimensional numerical ocean frontogenesis model of Bleck *et al.* (1988) does show warm upwelling and cold downwelling, with frontal gradients growing significantly in a few days in the presence of a deformation field.

An alternative hypothesis for the generation of the cold-core features is that growing disturbances may induce cold upwelling by conservation of potential vorticity as relative vorticity changes rapidly locally. Observational and theoretical evidence for such processes in cold-dome Gulf Stream meanders has been found (Lee and Atkinson, 1983; Luther and Bane, 1985). This hypothesis appears less consistent with the observations, since the core fluid of the cold features evidently originated at the surface on the cold side of the front.

cold

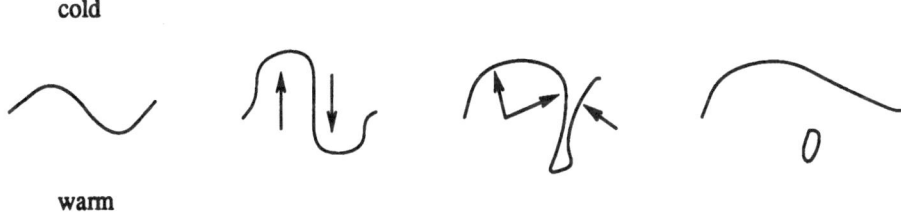

warm

Figure 11: Schematic scenario for deformation of a surface front by a baroclinic disturbance.

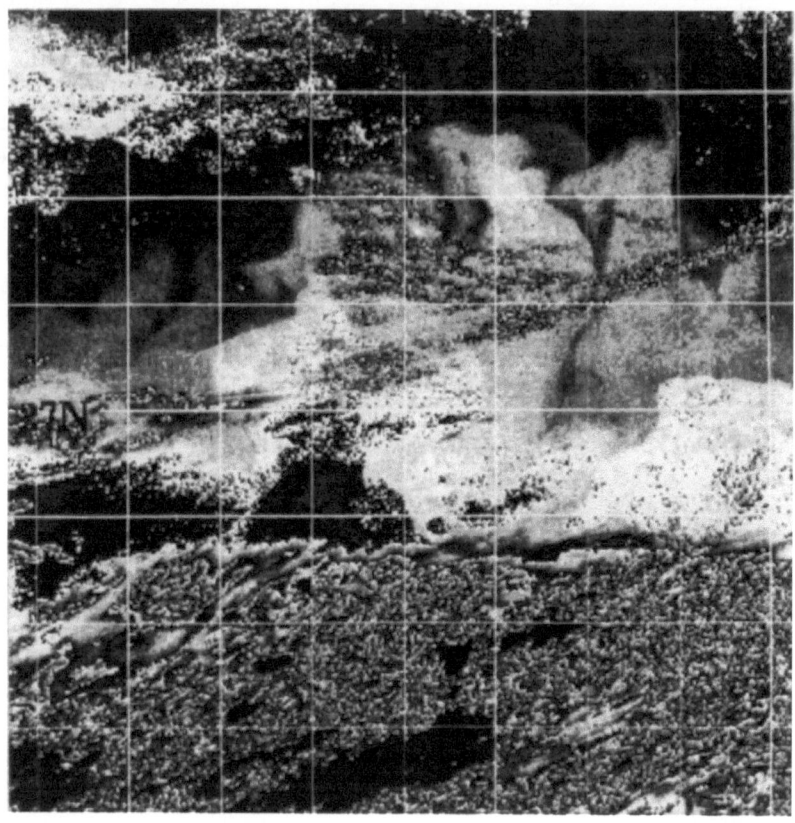

Figure 12: AVHRR image from April 26, centered at 27°N, 70°W, with a 1° grid. Lighter shades are warmer and clouds are stippled.

6. Conclusions

Over the five months of FASINEX and over the width of the STCZ upper ocean variability within the STCZ showed little relation to local forcing. Fronts, not the response to local heating and wind mixing, were the dominant source of the observed variability. The fronts were geostrophically balanced. Unanticipated energetic submesoscale variability was also observed. These features were also evidently geostrophic. They likely formed from frontal waves or meanders.

Associated with the fronts were sound speed variations in the upper 200 m of up to 6 m s^{-1} in 10 km. These rapid changes in sound speed can deflect sound rays significantly, particularly at high frequencies. Subsurface sound speed maxima occurred near some fronts (Figures 6 and 7), indicating the presence of a surface sound-duct in addition to the deeper sound channel. Shadow zones into which sound rays do not propagate can occur where sound speed increases rapidly. Sonar imaging of frontal features could potentially provide a rich source of oceanographic information and assist in the investigation of the structure and dynamics of ocean fronts, which remain imperfectly understood.

Acknowledgements:

Data from the Profiling Current Meter moorings were made available by Charles Eriksen (University of Washington) and were, because the PCMs provide conductivity as well as temperature and velocity as a function of depth, essential to our discussion of variability in sound speed. Peter Cornillon (University of Rhode Island) provided the AVHRR images that greatly assisted in studying the fronts seen in FASINEX and in further study of the submesoscale variability. Nancy Pennington assisted in this study. Support for this work came from the Office of Naval Research, Contract N00014-90-J-1470.

References

Bleck, R., Onken, R., and Woods, J. D. (1988) 'A two-dimensional model of mesoscale frontogenesis in the ocean', Quart. J. Roy. Meteor. Soc., 114, 347–371.

Böhm, E., and Cornillon, P. (1987) 'An atlas of AVHRR/2 images from the FASINEX experiment', Univ. Rhode Island Grad. Sch. of Oceanogr. Tech. Rep. 87-3, 75 pp.

Briscoe, M. G. and Weller, R. A. (1984) 'Preliminary results from the Long-Term Upper-Ocean Study (LOTUS)', Dyn. Atmos. Oceans., 8, 243–265.

Chen, C. T., and Millero, F. J. (1977) 'Speed of sound in sea-water at high pressures', J. Acoust. Soc. of Am., 62, 1129–1135.

Emanuel, K. (1979) 'Inertial instability and mesoscale convective systems, Part I: Linear theory of inertial instability in rotating viscous fluids', J. Atmos. Sci, 36, 2425–2449.

Eriksen, C. C., Weller, R. A., Rudnick, D. L., Pollard, R. T., and Regier, L. A. (1990) 'Ocean frontal variability in FASINEX', J. Geophys. Res., submitted.

Halliwell, G. R., Jr., and Cornillon, P. (1990) 'Large-scale SST anomalies associated with subtropical fronts in the Western North Atlantic during FASINEX', J. Mar. Res., submitted.

Halliwell, G. R., Jr., Cornillon, P., Brink, K. H., Pollard, R. T., Evans, D. L., Regier, L. A., Toole, J. M., and Schmitt, R. W. (1990) 'Descriptive oceanography during FASINEX: Medium to large-scale variability', J. Geophys. Res., submitted.

Isemer, H.-J., and Hasse, L. (1987) The Bunker Climate Atlas of the North Atlantic Ocean. Volume 2: Air-Sea Interactions, Springer-Verlag, New York, 252 pages.

Katz, E. J. (1969) 'Further study of a front in the Sargasso Sea, Tellus, XXI(2), 259–269.

Lee, T. N., Atkinson, L. P. (1983) 'Low-frequency current and temperature variability from Gulf Stream frontal eddies and atmospheric forcing along the southeast U.S. outer continental shelf', J. Geophys. Res., 88, 4541-4567.

Luther, M., and Bane, J. (1985) Mixed instabilities in the Gulf Stream over the continental slope', J. Phys. Oceanogr., 15, 3–23.

Palmen, E., and Newton, C. W. (1969) Atmospheric Circulation Systems, Academic Press, New York, 603 pp.

Pollard, R. (1986) 'Frontal surveys with a towed profiling conductivity/temperature/depth measurement package (SeaSoar)', Nature, 232, 433–435.

Voorhis, A. D. (1969) 'The horizontal extent and persistence of thermal fronts in the Sargasso Sea', Deep-Sea Res., 16, 331–337.

Voorhis, A. D., and Hersey, J. B. (1964) 'Oceanic thermal fronts in the Sargasso Sea', J. Geophys. Res., 69, 3809–3814.

Weller, R. A. (1990) 'FASINEX, A study of air-sea interaction in a region of strong oceanic gradients', J. Geophys. Res., submitted.

Summary of Session 3

Potter and Scott

The original intention of this session was to discuss range-dependent propagation caused by fronts and eddies, that is to say meso-scale features in the ocean. In practice, the presentations were united by a common oceanographic scale size, but involved very different types of work. A total of 10 papers were presented, varying between purely theoretical work in fractal dynamics, ocean-acoustic experimental and modelling results, and some remarkable pure oceanographic observations totally unsullied by acoustic observations or calculations.

As a theme then, the session was incoherent. As a cross-section of problems, both oceanographic and acoustic, but all of similar scale, it was an exciting mixture which brought out some promising cross-discipline benefits.

Some of the most interesting questions which arose are listed below. The topics either revolve around some new application or concept, or refer to a need to re-assess the role of some idea or technique.

1. To what extent is it useful to treat ocean-acoustic problems deterministically and stochastically?

2. How can we define what constitutes a 'necessary and sufficient' description of ocean features for acoustic purposes?

3. How can we make use of the qualitative behavioral advantages of fractal theory to help in quantitative improvement in ocean-acoustic understanding?

The discussion freely flowed between these areas and sometimes over the edge. Some of the most interesting, or most agreed upon, results follow:

With regard to deterministic versus stochastic.

Briscoe asked 'What is the definition of deterministic versus stochastic ocean variability?' The definition could be phrased in terms of time-invariance or of low wavenumber. Perhaps the most rational definition is to classify all resolved features as deterministic and the rest stochastic. Whatever is unresolved must be treated stochastically. In addition, features which are sufficiently sampled may also be treated stochastically by choice, whether they are mechanistically so or not.

Genuine chaotic behavior.

Laval maintained that acoustic and oceanographic understanding should be kept as deterministic as possible. He pointed out that the solution of an acoustic field is an 8-dimensional problem, if we include time and frequency-dependence. It is realised, however, that stochastic scattering causes significant distortion of any deterministic field. This will almost always be the case because we are always pushing up against the Rayleigh limits of frequency and range. It is precisely in this limit that stochastic scattering causes large fluctuations in the acoustic field and biases the mean.

J. Potter and A. Warn-Varnas (eds.), Ocean Variability & Acoustic Propagation, 479–481.
© 1991 *Kluwer Academic Publishers.*

Perhaps the most important consensus of opinion, although reached reluctantly, is that stochastic variability cannot be treated as separable from the deterministic field. It has been known for a long time that some conditions, such as two-channel profiles with a source in one channel, exhibit large deviations in the mean field from predictions which do not include stochastic scattering, due to energy leakage across the channel margin. It now appears that the mean fields predicted without stochastic volume scattering will almost always be significantly in error. This is a result of the extreme skewness of the intensity fluctuation probability distribution which has been recently evaluated for a large range of frequency and range parameters by Ewart (JASA, 1989).

With regard to a 'necessary and sufficient' oceanic description.

For the purpose of acoustic modelling, Munk pointed out that the method of interpolation was crucial. Munk described some early tomography interpretation in which it was discovered that the choice of interpolation scheme was more important than the propagation model. It is certainly a problem that any 'dumb' interpolation method which seeks to make a smooth transition from one observation to the next without reference to physical mechanism is bound to fail. The only exception is if the data is sampled sufficiently frequently to resolve all features of interest. This has never been known to occur.

The most satisfactory basis appears to be to use Empirical Orthogonal Functions (EOF's) to describe the features, so as to preserve the characteristics or 'features' during interpolation or extrapolation. This avoids the pitfall that time or space-averaged oceanic properties normally bear little resemblance to any single realisation of the ensemble. If the ocean can be described in terms of dynamically understood features, such as appear in Robinson's models, and EOF's, then we may be close to a sufficient description.

With regard to fractal theory.

Osborne was able to present some very interesting results from simulations of float trajectories with a fractal dimension of 1.3. Smith concurred that, in some work he did with Brown, they also found a fractal dimension of about 1.3 to reproduce characteristic behaviour of the ocean. Treating variability as having random phase is clearly fundamentally wrong, since such models never generate the correct 'appearance' of any real field. The reason is that there are always some special phase relationships. The fractal work appears to offer the potential of an alternative, and much improved, method of modelling variability in a way which is consistent with behaviour.

Conclusions

We have seen that there are some very difficult obstacles, but equally there are some promising prospects for advance. The most optimistic (and refreshing) note may have come from Kuperman, who turned the old problem of 'coping' with ocean variability in acoustics into one of actively profiting from the symmetry-breaking and other possibilities that it offers. In the future, ocean and acoustic developments must advance together; oceanography needs acoustics for imaging the ocean in order to identify the physical mechanisms, acoustics needs oceanography to model the medium, since it can never be sufficiently sampled. The task of truly bringing acoustics and oceanography together will be no mean feat. The relevant scientific communities have largely failed to do so for some 50 years. The hope that this can be changed comes about because each field now realises that it needs the other, and oceanography now has the tools (both mathematical and physical) to cope with the scales required. Of the wide range of specialisations present at the workshop, none was less than vital but very few present would claim full understanding of all of them. This session achieved much in bringing several aspects into collision: keeping them in contact will be a challenge in itself.

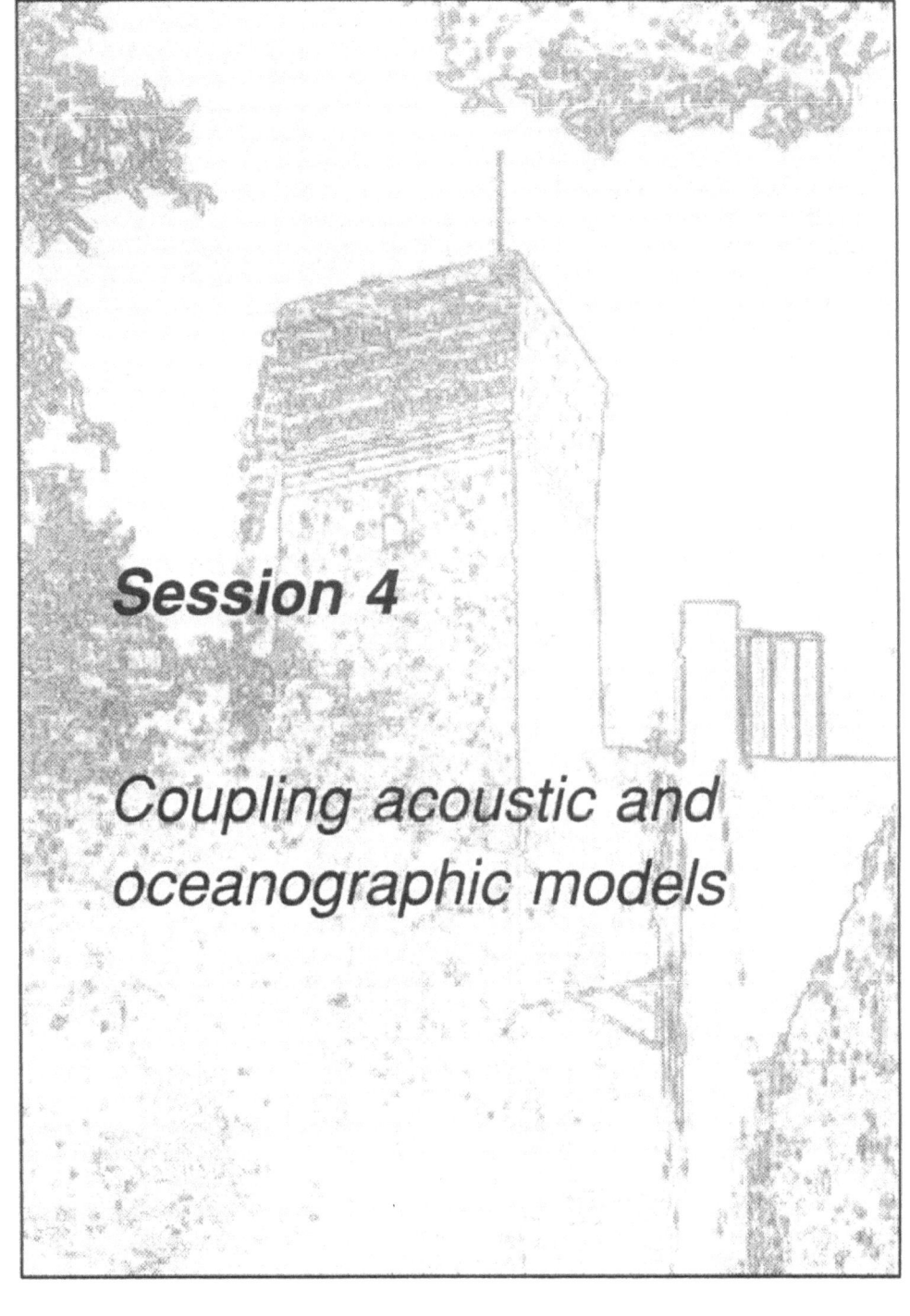

Session 4

Coupling acoustic and oceanographic models

A MIXED-LAYER MODEL FOR PREDICTING THE ACOUSTIC STRUCTURE OF SHALLOW SEAS

A. J. ELLIOTT and Z. LI
Unit for Coastal and Estuarine Studies
University College of North Wales
Marine Science Laboratories
Menai Bridge, Gwynedd LL59 5EY
United Kingdom

ABSTRACT. Tidal stirring can play an important role in determining the thermal structure of shallow shelf seas. A two-layer thermal model, which includes the effects of wind stirring and tidal mixing, has been used to reproduce the seasonal advance and retreat of frontal zones and to estimate the impact and persistence of anomalous seasons (e.g. a severe winter) on the thermal structure of the NW European shelf seas. The model simulations were tested against a database compiled from 45,000 XBT observations, and an agreement of better than 1 °C was obtained over most of the region. The poor agreement in some areas was due to the assumption of a two-layered structure and to the neglect of advection in the thermal model. The model and database results have been combined to form a package called MERMAID (Marine EnviRonmental Model And Integrated Database) for the simulation of the climatic temperature structure of the shelf seas. A forecasting scheme has been developed that uses the climatic model to provide the initial temperature distribution for a depth-resolving closure model. Two versions of the closure model have been developed: the first uses surface fluxes supplied by the British Meteorological Office atmospheric model, the second version has been developed for workstation application and is driven by surface flux data entered by an operator. Both closure models include realistic tidal stirring and can be used to forecast over periods of 6-48 hours. The workstation model has been interfaced with the GRASS acoustic model so that operational forecasts of acoustic structure can be made for the frontal zones of the NW European shelf seas.

1. Introduction

The coastal waters of NW Europe are characterized by their relatively shallow depths and strong tides; Flather (1976) has calculated that approximately 10% of all shallow water tidal dissipation occurs in these shelf seas. One consequence of this strong tidal stirring is the formation of surface thermal fronts in summer at the boundaries between the stratified water offshore and the cooler vertically-mixed water in the shallow regions of strong tidal currents. Two of these frontal regions are of particular importance to naval operations. They are the Islay front which develops north of Ireland to the west of the Clyde Sea region, and the Celtic Sea

485

J. Potter and A. Warn-Varnas (eds.), Ocean Variability & Acoustic Propagation, 485–500.
© 1991 *Kluwer Academic Publishers*.

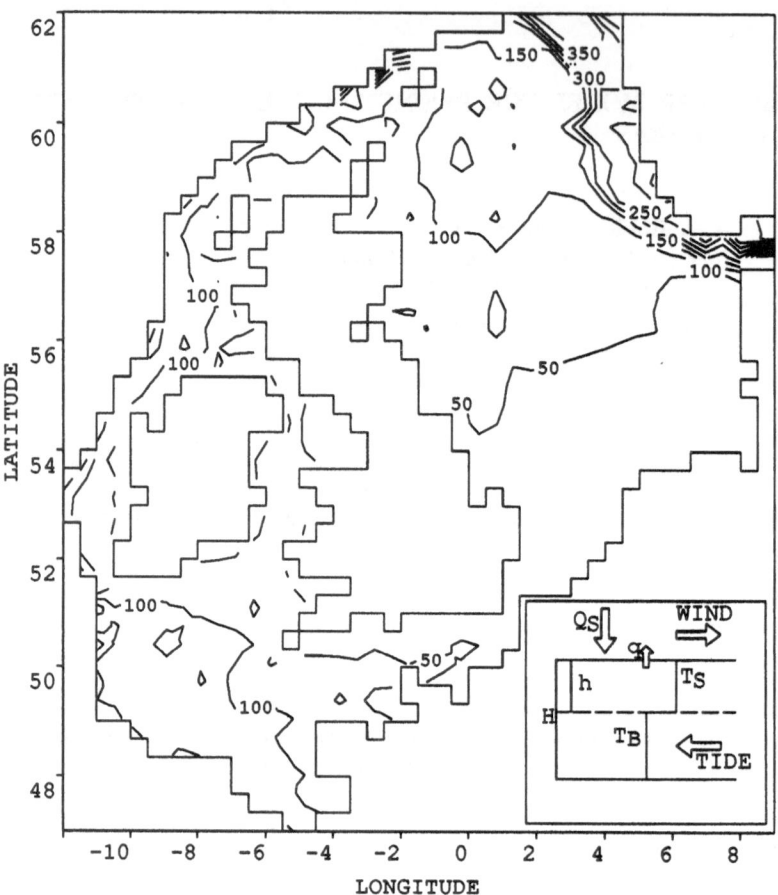

Figure 1.   Region covered by the model, inset is a schematic representation of the mixing mechanisms.

front which develops at the southern entrance to the Irish Sea on a line extending from the SE corner of Ireland towards the entrance to the Bristol Channel (Pingree and Griffiths, 1978). These two fronts are important because ships that enter the Clyde do so either by crossing the shelf region via the NW Approaches which means passing through the Islay front, or by entering the Irish Sea through the SW Approaches which involves passage through the Celtic Sea front. During the summer months, therefore, these frontal zones assume a strategic importance, and predictions of the thermal and acoustic characteristics of these regions are required. Simpson and Hunter (1974) showed that, if the surface heating rate is constant, frontal locations will depend on the ratio of H/U^3 where H is water depth and U is the tidal current amplitude. Simpson, Allen and Morris (1978) extended the tidal mixing approach to include the effects of wind mixing;

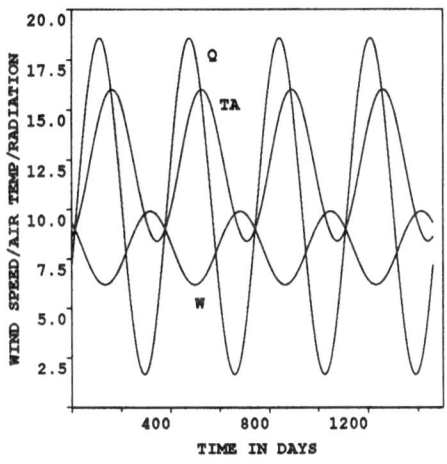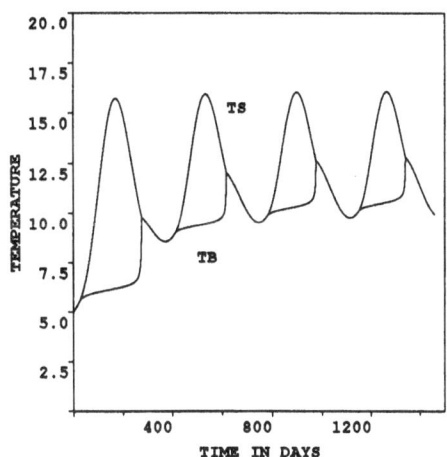

Figure 2.   Meteorological forcing and temperature predictions for a grid point in the Celtic Sea (units: radiation ($10 \times W/m^2$), air temperature (°C), wind speed (m/s), water temperatures (°C)).

while Simpson and Bowers (1981) estimated the fraction of the work done by tidal and wind mixing processes that is converted to potential energy of the water column, and (Simpson and Bowers, 1984) showed that the temperature of the sea surface can control the heat exchange between the atmosphere and the water. In a study of the Celtic Sea, Bowers (1984) developed a two-layer model that was computationally efficient but which produced relatively large errors in the simulated temperatures. Elliott and Clarke (1990a) improved his model by using an alternative set of heat flux parameterizations, and studied the seasonal stratification of the entire NW European shelf seas.

2.   The Climatic Model

A two-layer model can be used to predict the climatic thermal structure of the shelf seas, the model being applied to the region shown in Figure 1 (details of the computational method are given in Elliott and Clarke (1990a)). The grid spacing is 20 nautical miles (approximately 35 km), giving a grid size of 1/3° latitude by 1/2° longitude, and the grid covers the NW European shelf seas out to the shelf break where water depths approach 200 m. This grid spacing was chosen so that major topographic features (e.g. the English Channel) would be resolved and is adequate for defining the major frontal zones. Water depths are typically less than 50 m in the Irish Sea and English Channel; the deepest water (≈300 m) lies off the coast of Norway, while depths in the North Sea decrease from about 150 m in the north to less than 50 m in the southern part.

Two sets of input data, meteorological and hydrographic, are required to

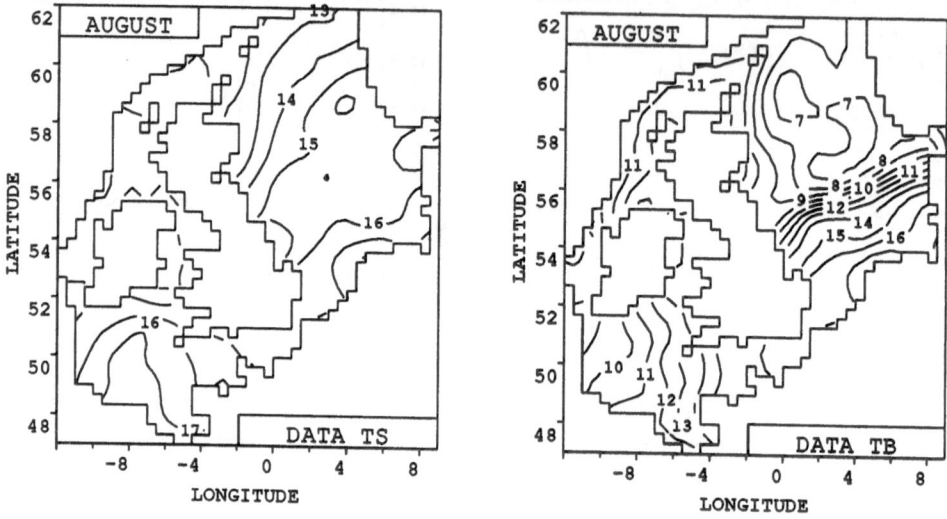

Figure 3.   Surface and bottom temperatures (°C) in August from the databank.

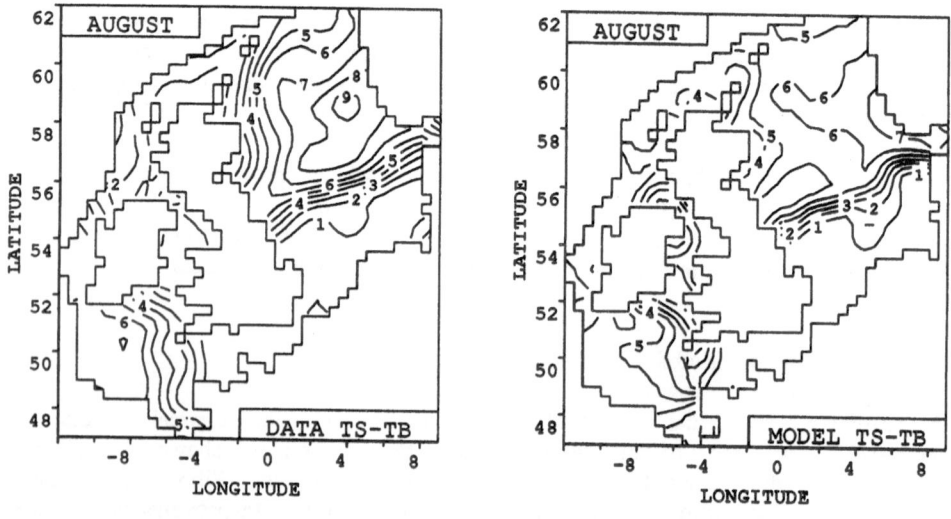

Figure 4.   Stratification (°C) in August from the databank and model.

run the model. The Meteorological Office provided mean monthly values of wind speed, dry bulb temperature, humidity, cloud cover and atmospheric pressure for each 1°×1° square of latitude and longitude. Following James (1977), the annual variation of these parameters was assumed to be sinusoidal and a best fit curve of the form: $P = a_1 + a_2 \cdot \sin(\omega t + a_3)$ was determined for each 1°×1° square where ω is the angular frequency

$(2\pi/365 \text{ days}^{-1})$. The radiation data, Qs, were not available for each $1° \times 1°$ cell. Consequently the variation of Qs with latitude was determined from 10 coastal and island sites around the UK by regression. The variation of Qs with longitude was found to be insignificant. Hydrographic data, which were tabulated for each grid point, consisted of mean water depths taken from Admiralty Charts and depth averaged M_2 tidal stream amplitudes (U) extracted from the model results of Pingree and Griffiths (1978).

An example of the model predictions for a grid point in the Celtic Sea (at about 7°W, 51°N) is shown in Figure 2. The left portion of the figure shows the meteorological forcing, represented by the incoming solar radiation, air temperature and wind speed over a period of 4 years. (Time is measured in days from March 1 which is near the start of the annual heating cycle.) The radiation peaks at about day 110, June 20, which is close to the minimum in wind speed. In contrast, maximum air temperature lags the radiation peak by about one month. The right portion of the figure shows the surface and bottom temperatures, Ts and TB. Starting from an arbitrary value of 5 °C, the model settles down to an annual cycle after about 3 years. Maximum surface temperatures are predicted to occur towards the end of August when stratification is at its strongest. Figure 2 shows clearly the onset of stratification in the spring and its breakdown during the autumn, with vertically-mixed water being present throughout the winter months.

To obtain a quantitative assessment of the accuracy of the model results, a databank consisting of sea surface and sea bed temperatures, thermocline depths and heat contents was created using 45,000 XBT casts obtained from MIAS (Marine Information and Advisory Service, Wormley) and the Hydrographic Office, Taunton (Elliott and Clarke, 1990b). Databank files of similar format to the model output were produced by averaging the data within each grid square by month. In certain areas, such as the Irish Sea, Norwegian coastal waters and off the coast of Scotland, data were noticably sparse (on average less than one observation per box). However, contour maps of temperature obtained from the databank were in good agreement with other results in all areas except for the shelf break region (200 m contour line).

Observed summer temperatures are shown in Figure 3; in the North Sea the surface values range from about 17 °C in the south to 13 °C in the north. The North Sea bottom temperatures vary from 17 °C in the south to 7 °C in the north and a strong temperature gradient is evident in the central North Sea. This sea bed frontal boundary (1 °C per 50 km) is stronger than the corresponding surface temperature gradient and reflects the manner in which depths increase with latitude. Most of the shelf is thermally stratified in summer, with the exception of the English Channel and southern North Sea where strong tidal currents prevent the formation of a thermocline.

Vertical temperature contrasts, (Ts - TB), presented in Figure 4 show good agreement between the observations and predictions over most regions of the shelf with the exception of the northern North Sea. In this region the observed summer stratification reaches 9 °C whereas the predicted values are around 6 °C. Bottom temperatures agree reasonably well between the

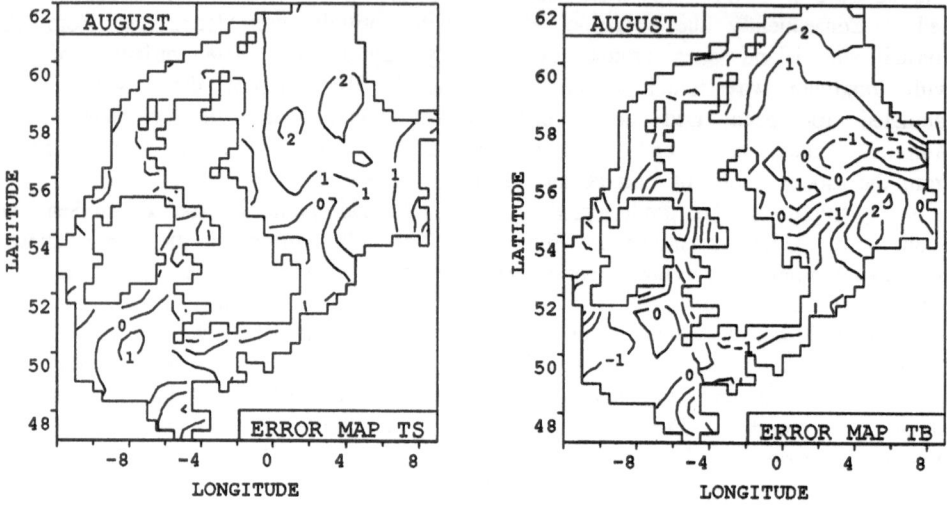

Figure 5.   Error maps (°C) for summer predictions.

predictions and the data, the reduced stratification in the model being a result of the model underestimating the surface temperature in this part of the North Sea during summer.   This error could not be reduced by varying the mixing efficiencies and is probably related to the effects of salinity stratification which limits the depth to which wind mixing can penetrate thus producing a shallower mixed-layer of elevated temperature. Consequently, the predicted thermocline is too deep in this region and this reduces the surface temperature by about 3 °C.   Note that the position of the 1 °C stratification contour is well reproduced by the model as a result of tuning the wind mixing efficiency, and that the model reproduces the Irish Sea fronts (near the Isle of Man and in Cardigan Bay) which are not so clearly revealed by the database.

The errors in the model simulations can be quantified by contouring the difference in temperature between the database and model results.   Figure 5 shows the errors obtained from a simulation of summer conditions in which the effects of salinity stratification are included in the model.   The bottom temperatures are generally accurate to within 1 °C except in the central portion of the North Sea.   However, this is the location at which there is a pronounced north-south temperature gradient (Fig. 3) and consequently a small error in the position of the front can result in a significant temperature differences between the model and the observations. Surface temperatures are underestimated by about 2 °C in the northern North Sea to the west of Norway and by up to 1 °C in the central part of the Celtic Sea.   In general, however, there is an agreement of better than 1 °C between the model and the observations, especially in the vicinity of the frontal zones.

The model results can provide an insight into the seasonal advance and retreat of the frontal zones. The simulations showed that the seasonal stratification does not develop until around May 15 when weak stratification appears near the shelf edge at the southern limit of the Celtic Sea. The frontal boundaries then advance rapidly and by June 1 most of the northern North Sea and the entire Celtic Sea are stratified. By June 15 the fronts have reached their summer positions. During the rest of the summer months the vertical stratification increases in intensity and the horizontal temperature contrasts strengthen across the fronts but their positions remain relatively static. It is not possible to define a speed of frontal advance during the spring because the stratification develops almost simultaneously in the zones of weak tidal mixing. In contrast, when the fronts start to retreat towards the shelf break at about August 15 they move steadily with a speed of about 7 km/day until the breakdown of stratification which occurs around November 15.

The climatic model was used to predict the thermal structure of the shelf seas following an anomalous season, and to estimate the persistence of such events. For example, following an exceptionally cold winter does stratification develop later or earlier in the year than normal, and does this affect the temperature of the water during the following summer months? Anomalous seasons were simulated by allowing the model to spin up under normal forcing and inserting the perturbed forcing into the fourth year before returning to normal forcing during year five. Calculations were made for both summer mixed and stratified locations, and the effects of unusually hot and cold summers and winters quantified. The results suggested that feedback, in the form of a sensible heat flux driven by the air-sea temperature difference, will act to restore the heat content and temperature structure to normal following an anomalous heating period. However, regions that stratify in summer can display temperatures that are lower than normal following a hot summer, and vice versa. This arises through the reduced wind speed, which is experienced during a hot summer, limiting the depth of the mixed-layer. This produces a shallower mixed-layer than normal into which the incoming radiation is absorbed so that elevated surface temperatures are produced. The higher temperature then increases the stability of the thermocline, further restricting the penetration of the wind mixing, and drives a sensible heat flux from the sea to the atmosphere. This, paradoxically, results in the net heat flux being lower than normal so that the heat content of the water column is reduced. Consequently, when overturning occurs in the autumn, the mixed water is cooler than normal - and this cool water forms the bottom water during the following heating season. Thus water temperatures will be lower, and vertical stratification stronger, during the following year. A similar process acts to produce warmer winter temperatures than normal following a cool summer. For both events, the time scale of persistence is of the order of 9-12 months.

The two-layer model of the thermal structure can be applied to the NW European shelf seas to predict the seasonal stratification and the advance and retreat of frontal features. Such predictions are of value because the observational coverage of temperatures, particularly bottom values, in the

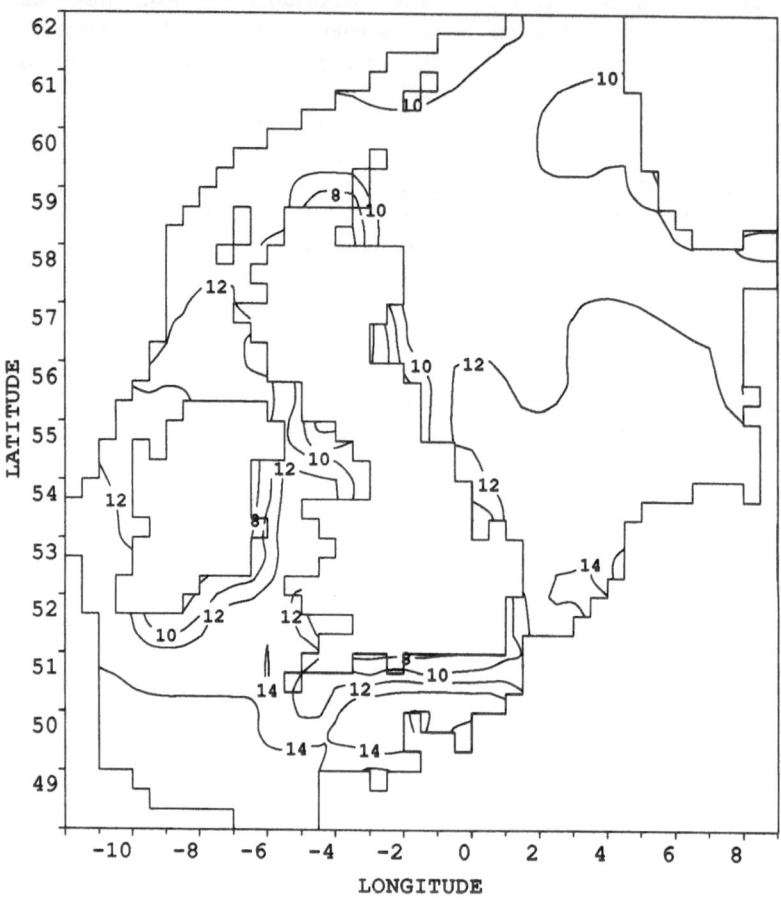

Figure 6.   Surface temperatures (°C) computed by the closure model using the BMO surface flux predictions, results shown are for 0300 November 2 1989.

shelf seas is relatively sparse and the frequency of cloud cover prevents satellites from providing detailed spatial and temporal surface coverage. The numerical scheme, which has similarities to the method discussed by Stigebrandt (1985), is computationally efficient since it uses energy arguments to calculate the entrainment between the surface and bottom layers. The results from the climatic model and the database parameters have been combined into a software package called MERMAID (Marine EnviRonmental Model And Integrated Database) which provides a graphical display of the advance and retreat of frontal features (Elliott, 1990).

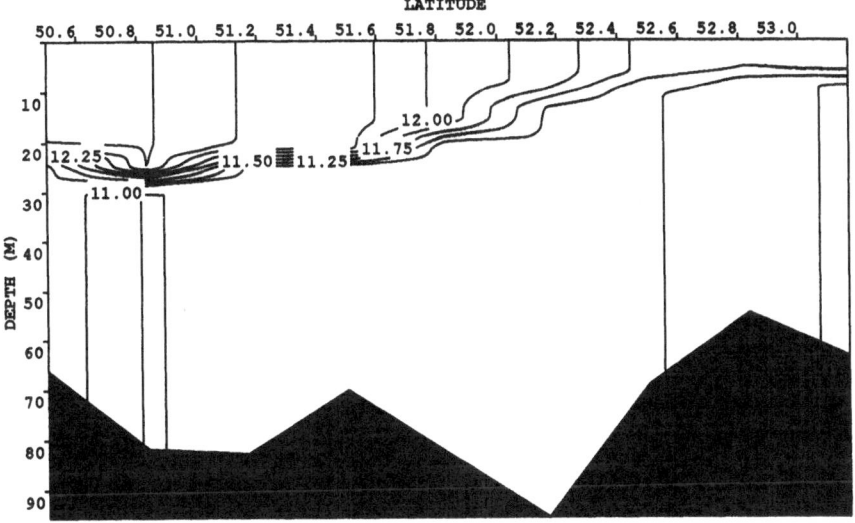

Figure 7.   Initial temperature field (°C) predicted for 0000 June 7 1984 by the two-layer model.

3.   The Closure Model

The two-layer model was developed so that only a modest amount of computation would be required to reproduce the seasonal heating cycle. However, observations of the vertical structure of temperature in the Celtic Sea (Wang et al., 1990) suggest that the two-layer approximation may rarely be valid.   Data collected by a thermistor chain extending from the surface to the bottom in 90 m of water showed that, although wind-mixing events could produce a surface mixed-layer 20 m thick, surface heating rapidly caused the mixed-layer to warm so that an approximately linear stratification was produced between the surface and mid-depth.   Even at times soon after a wind event, the thermocline was diffuse being at least 20 m thick.   Limitations of the climatic model, therefore, are likely to lie not in the representation of the vertical mixing via an energy argument but more in the unrealistic assumption that the structure is strictly two-layered with an abrupt thermocline.

In view of the weakness of the two-layer approximation in some areas of the shelf, the work has been extended by using a depth-resolving closure scheme (Chen et al., 1988) to provide operational forecasts of the temperature structure of the shelf seas.   The vertical structure of the temperature field is determined by solving

$$\frac{\partial T}{\partial t} = \frac{\partial}{\partial z}(K_T \frac{\partial T}{\partial z})$$

494

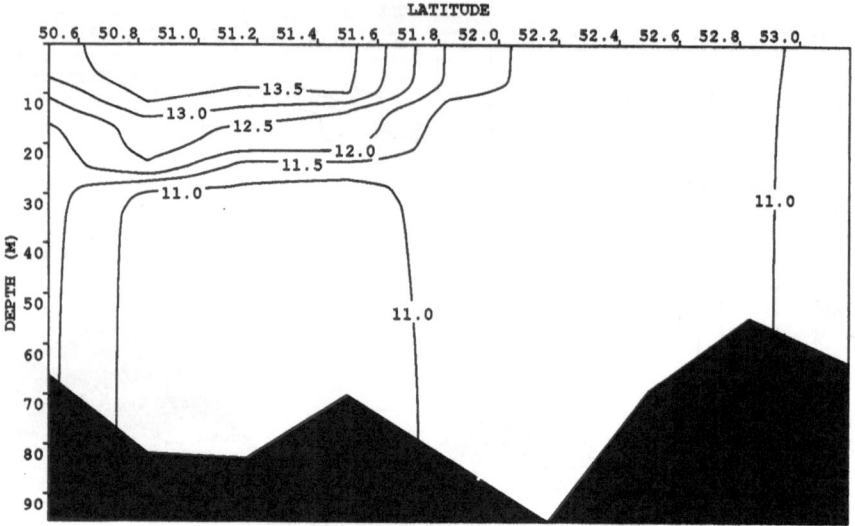

LATITUDE

Figure 8.   Initial temperature field (°C) for 1200 June 12 1984 used by the closure model.

where K_T is the vertical exchange coefficient for the thermal energy (eddy diffusivity). Following Johns and Li (1990), we propose that

$$K_T = S_H \, \ell \, E^{1/2}$$

where S_H is a stability function related to the flux Richardson number, ℓ is a vertical mixing length scale and E is the turbulent energy obtained from

$$\frac{\partial E}{\partial t} = K_M[(\frac{\partial u}{\partial z})^2 + (\frac{\partial v}{\partial z})^2] + \frac{g}{\rho_0} K_T \frac{\partial \rho}{\partial z} - \frac{c^{3/4} E^{3/2}}{\ell} + \frac{\partial}{\partial z}(K_E \frac{\partial E}{\partial z})$$

Here c is a constant, u and v are the Reynolds-averaged components of velocity; K_M, K_E are the exchange coefficients for momentum and turbulent energy, and ρ is the density of sea-water.   This formulation follows closely that of Mellor and Durbin (1975).

The velocity components, u and v, are obtained from conventional momentum equations of the form:

$$\frac{\partial u}{\partial t} - fv = \frac{\partial}{\partial z}(K_M \frac{\partial u}{\partial z}) + P_X$$

$$\frac{\partial v}{\partial t} + fu = \frac{\partial}{\partial z}(K_M \frac{\partial v}{\partial z}) + P_Y$$

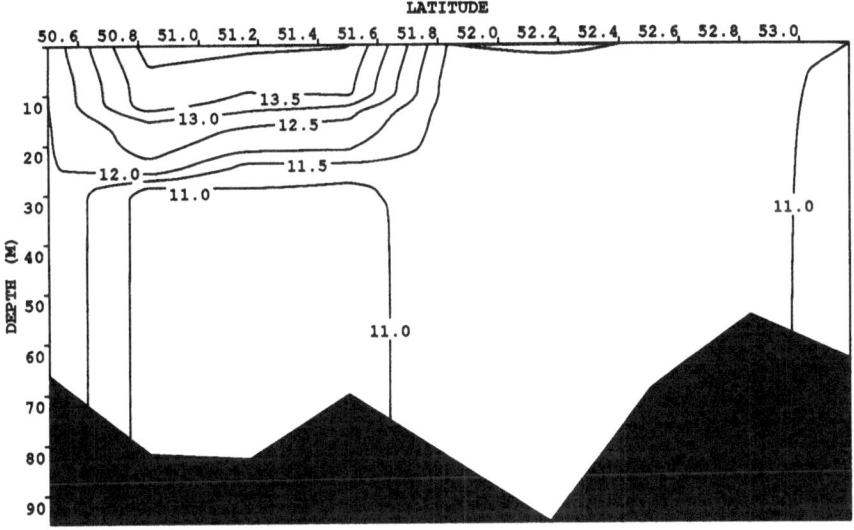

Figure 9.   Predicted temperatures (°C) for 1200 June 14 1984 obtained by the closure model.

The terms Px, Py represent the horizontal pressure gradients associated with the tidal currents and have the form:

$$Px = -\omega Hu \cdot \sin(\omega t + Gu)$$

$$Py = -\omega Hv \cdot \sin(\omega t + Gv)$$

Realistic tidal currents are reproduced in the model by the inclusion of pressure forces due to the M_2 and S_2 constituents, these two constituents accounting for over 95% of the observed flux of tidal energy in the Celtic and Irish Seas (Robinson, 1979).   At each location there were 20 vertical levels with increased resolution near the surface.

4.   Predicting the Thermal and Acoustic Structure

The climatic and closure models have been combined to form an operational mixed-layer model of the shelf seas.   The climatic model was run to cyclic steady state and the results from each grid point were stored at the start of the annual heating cycle (March 1).   From this start point the model can then be stepped forward efficiently to any required date during the year since no spin-up period is required.   This climatic forecast is then supplied as the initial temperature field to the depth-resolving closure scheme.   The closure model is used to advance the prediction on time scales of hours to days using predicted atmospheric winds and heat fluxes as the forcing. Two versions of this scheme have been developed: the first covers the entire shelf region and will be driven by meteorological fluxes supplied by the British Meteorological Office (BMO), the second version will provide local forecasts to a ship at sea and will use meteorological fluxes supplied to a workstation by an operator. The global shelf model will be interfaced with the output from the BMO 15-level fine-resolution operational forecasting model which will provide surface winds and heat fluxes. Using archived BMO forecasts, we have developed a spatial interpolation scheme to map the atmospheric predictions onto the grid of the closure scheme so that the vertical structure of the sea temperatures can be predicted over time scales of 12-24 hours. The BMO model produces predictions for 3 hour intervals twice a day, and these data are interpolated to provide hourly flux values for the closure scheme. It is anticipated that operational forecasts of sea temperature will be validated by ship observations and remote sensing so that corrected predictions can be used as the initial conditions for subsequent forecasts. However, if ground-truth data are lacking, the climatic model can be used to initialize the forecast. Figure 6 shows an example of a hindcast of sea surface temperatures produced using the BMO meteorological model fluxes.

Local predictions can be made at sea using the workstation version of the model. This uses the climatic model to initialize the temperature field, and then makes a local forecast using flux values entered by an operator. Numerical tests have shown that realistic stratification can be simulated if the closure scheme is allowed to integrate the initial two-layer profile for a period of ≈ 5 days before stepping forward the operational forecast. For example, suppose that an operator wishes to run the forecast model at 0000 on August 1 to predict conditions on August 2. A data file would be created containing the expected meteorological fluxes during August 1-2 and these would be supplied to the closure model. This would first call the climatic model to supply two-layer predictions for 1200 hours on July 27. The closure scheme would then use these initial conditions with the meteorological fluxes for 0000 on August 1 (held steady) and step forward to provide an initial depth-resolving profile for 0000 on August 1. This profile would then be stepped forward to provide the operational forecast using the time varying meteorological fluxes supplied by the operator. In practice, the operator would define target positions within the model grid so that forecasts would only be made over relatively small regions of the shelf.

To demonstrate the use of the closure model, Figure 7 shows a two-layer temperature structure predicted for 0000 on June 7, 1984. The section runs N-S at longitude 5° 45′ W, and extends from the Celtic Sea into the Irish Sea; observations (Pingree and Griffiths, 1978) have shown a surface front at a latitude of about 51° 40′ N in this region during the summer. Figure 7 shows a gradual decrease of temperature with latitude but no abrupt surface front as suggested by the measurements. However, when the closure scheme is applied to this initial temperature distribution and the more realistic tidal mixing is integrated with fixed meteorological fluxes for a

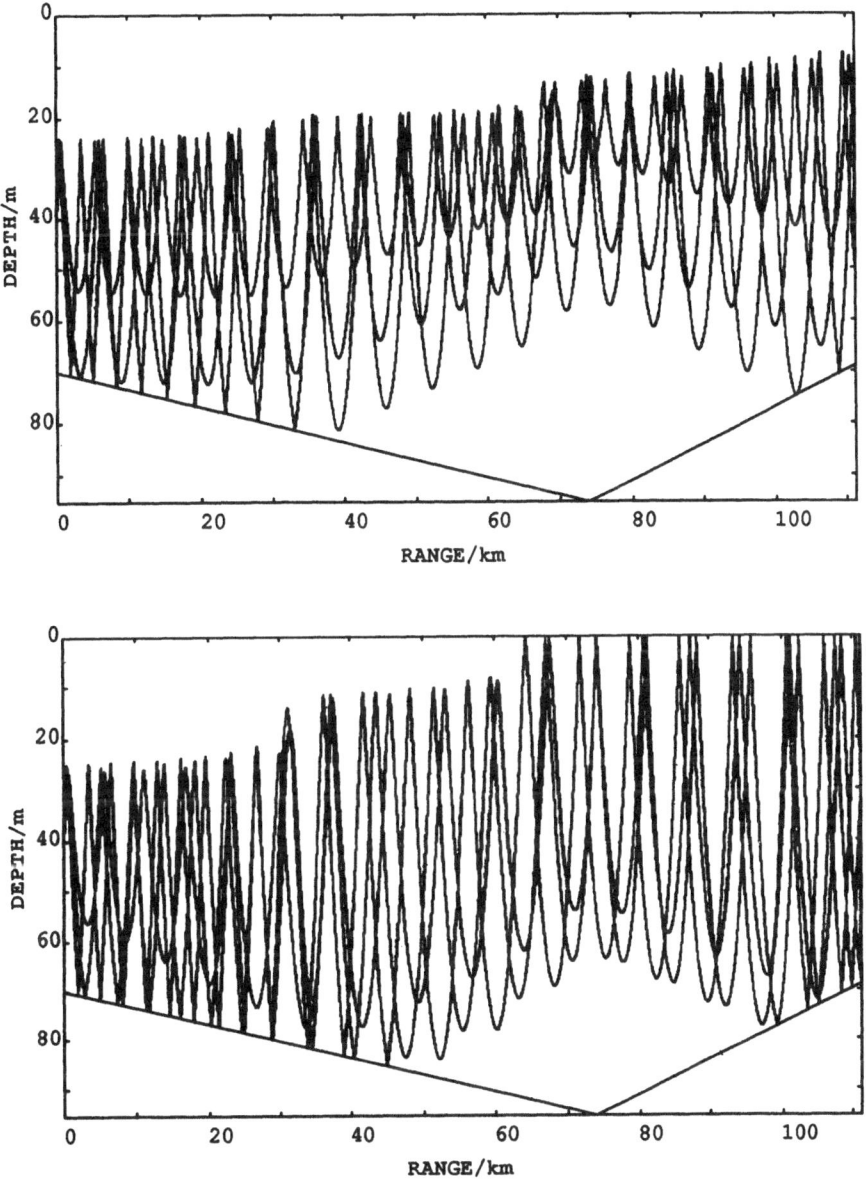

Figure 10. Ray paths computed in the frontal region using the GRASS acoustic model, upper: using the two-layer model temperatures, lower: using the closure model temperatures.

period of 5 days, then the temperature distribution presented in Figure 8 is obtained. This shows a realistic representation of the Celtic Sea front, and if this is stepped forward over a 2 day interval using time varying surface fluxes then the distribution shown in Figure 9 is obtained. This represents the forecast for 1200 on June 14 that would be made at 1200 on June 12, and it shows realistic stratification and a warming of the surface waters due to the summer heating cycle.

This predicted temperature field can now be interfaced with an acoustic model. For demonstration purposes, we have used the GRASS ray tracing model (Cornyn, 1973a,b); however, Harrison (1989) has pointed out that acoustic transmission on the shelf is likely to be dominated by bottom effects and that normal mode models (such as SNAP; Jensen and Ferla, 1979) may be more appropriate in shallow water regions. Figure 10 presents ray tracing calculations for the frontal region (between the latitudes of 51.5 °N and 52.4 °N) shown in Figures 7-8. The upper part of the figure shows the ray paths calculated for the two-layer temperature model (Figure 7) while the lower part shows those calculated from the closure scheme temperature field (Figure 8). In both calculations the acoustic source was set beneath the thermocline at a depth of 40 m. A common feature of both results is the trapping of acoustic energy beneath the thermocline in the stratified region, however the near-surface ray paths are different in the well-mixed water. Thus accurate models of the thermal structure are essential if realistic acoustic predictions are to be made near the frontal zones.

5. Discussion

An operational forecasting scheme for the prediction of the acoustic structure of the UK shelf seas will require the use of meteorological forecasts, database and climate models for initialization, a closure scheme for temperature prediction, and an acoustic model appropriate to shallow water conditions. Special consideration will need to be given to the derivation of methods for initializing the temperature field, both through the use of satellite and in-situ measurements. In contrast to the relatively standard techniques used in the predictive models, data assimilation in marine forecasting (Robinson et al., 1989) is a new and rapidly developing field.

Our present scheme uses a two-layer climatic model that was tuned against a database derived from 45,000 XBT observations. In practice, this means that the climatic model is being used as a sophisticated interpolation device to provide forecasts in those regions of the shelf where observations are sparse. An operational model will require the climatic field to be merged with real-time temperature observations to produce an initial forecast field. This assimilation process will benefit from additional studies of the climate of the shelf seas, especially with regard to periods of anomalous meteorological forcing. There is a need to collect temperature data in the shelf seas, using instrumented moorings that span the water column, so that the accuracy of both the thermal and acoustic predictive schemes can be evaluated.

Acknowledgements

This study was supported by the Procurement Executive (MOD) through the Admiralty Research Establishment, Portland.

References

Bowers, D.G. (1984) A two-layer model of the seasonal thermocline and its application to the Celtic Sea. UCES Report U84-4, University College of North Wales, 65 pp.

Chen, D., Horrigan, S.G. and Wang, D-P. (1988) The late summer vertical nutrient mixing in Long Island Sound. *J. Marine Res.*, **46**, 753-770.

Cornyn, J.J. (1973a) GRASS: a digital-computer ray-tracing and transmission-loss-prediction system. Volume I—overall description. Nav. Res. Lab., Report No. 7621.

Cornyn, J.J. (1973b) GRASS: a digital-computer ray-tracing and transmission-loss-prediction system. Volume II—user's manual. Nav. Res. Lab., Report No. 7642.

Elliott, A.J. (1990) MERMAID - A sea temperature model and database system. *The Oilman* (in press).

Elliott, A.J. and Clarke, T. (1990a) Seasonal stratification in the NW European shelf seas. *Cont. Shelf Res.* (submitted).

Elliott, A.J. and Clarke, T. (1990b) Monthly distributions of surface and bottom temperatures in the NW European shelf seas. *Cont. Shelf Res.* (submitted).

Flather, R.A. (1976) A tidal model of the North-West European continental shelf. *Memoires Societe Royale des Sciences de Liege*, **6**, 141-164.

Harrison, C.H. (1989) Ocean propagation models, *Appl. Acoustics*, **27**, 163-201.

James, I.D. (1977) A model of the annual cycle of temperature in a frontal region of the Celtic Sea. *Estuar. Coast. Mar. Sci.*, **5**, 339-353.

Jensen, F.B. and Ferla, M.C. (1979) SNAP: The SACLANTCEN normal-mode acoustic propagation model. SACLANT Memo SM-121.

Johns, B. and Li, Z. (1990) A comparison of wind-driven upwelling processes derived from two- and three-dimensional models of a coastal ocean. *Dynam. Atmos. Oceans* (submitted).

Mellor, G.L. and Durbin, P.A. (1975) The stucture and dynamics of the ocean surface mixed-layer. *J. Phys. Oceanogr.*, **5**, 718-728.

Pingree, R.D. and Griffiths, D.K. (1978) Tidal fronts in the shelf seas around the British Isles. *J. Geophys. Res.*, **83**, 4615-4622.

Robinson, Allan R., Spall, Michael A., Walstad, Leonard J and Leslie, Wayne G. (1989) Data assimilation and dynamical interpolation in Gulfcast experiments. *Dynam. Atmos. Oceans*, **13**, 301-316.

Robinson, I.S. (1979) The tidal dynamics of the Irish and Celitc Seas. Geophys. *J.R. astr. Soc.*, **56**, 159-197.

Simpson, J.H. and Bowers, D.G. (1981) Models of stratification and frontal movement in shelf seas. *Deep-Sea Res.,* **28**, 727-738.

Simpson, J.H. and Bowers, D.G. (1984) The role of tidal stirring in controlling the seasonal heat cycle in shelf seas. *Annls. Geophys.,* **2**, 411-416.

Simpson, J.H. and Hunter, J.R. (1974) Fronts in the Irish Sea. *Nature,* **250**, 404-406.

Simpson, J.H., Allen, C.M. and Morris, N.C.G. (1978) Fronts on the continental shelf. *J. Geophys. Res.,* **83**, 4607-4614.

Stigebrandt, A. (1985) A model of the seasonal pycnocline in rotating systems with application to the Baltic proper. *J. Phys. Oceanogr.,* **15**, 1392-1404.

Wang, D-P., Chen, D. and Sherwin, T.J. (1990) Coupling between mixing and advection in shallow sea fronts. *Cont. Shelf Res.,* **10**, 123-136.

THE USE OF COUPLED OCEAN-ACOUSTIC MODELS IN THE DESIGN OF NAVAL FORECAST SYSTEMS

A.D.HEATHERSHAW[1], C.E.MOONEY[1] and S.J.MASKELL[2]

1 *Admiralty Research Establishment, Portland, Dorset, DT5 2JS, UK.*
2 *University of Exeter, Exeter, Devon, EX4 4QF, UK.*

ABSTRACT. The use of a 3-D numerical ocean model to simulate mesoscale frontal and eddy environments for the input to range dependent acoustic models is described. The technique has been used to investigate the sensitivity of acoustic predictions to changes in the environment and in ocean model parameters. The ocean model is used to simulate an idealised environment in which a geostrophically balanced frontal jet is perturbed baroclinically to generate eddy features having realistic spatial and temporal scales. The model may also be initialised with discrete eddy features to study their interaction with the front. The results of the acoustic calculations not only confirm the importance of mesoscale ocean variability on sound propagation characteristics, but also show that acoustic predictions may be sensitive to the choice of ocean model parameter, in particular, in primitive equation models, the lateral eddy viscosity coefficient.

1. Introduction

Supercomputers and remote sensing of the oceans by satellite now bring the goal of global eddy resolving ocean prediction models within our reach (Semtner, 1988a,b). While there is a pressing need for such models to study climatic change, in particular the effects of global warming, the scales that can be predicted by such models now overlap those that are of interest to naval oceanographers. In particular, low frequency sonar involves detection over ranges that are comparable with the mesoscale variability of the ocean due to fronts and eddies. In both the USA (eg Clancy, 1987) and UK (eg Heathershaw et al, 1990a), navy-operational ocean forecast systems are being developed that will predict the ocean environment for input to range dependent acoustic models. However, it is by no means clear that such models will adequately resolve the mesoscale variability for acoustic purposes. Not only do we need to understand the sensitivity of the acoustic predictions to changes in the model generated environments, but also their sensitivity to ocean model parameters. The latter may include horizontal and vertical resolution but more importantly will include parameters that reflect some uncertainty in the ocean model physics. In particular, eddy viscosity and diffusion coefficients are used to describe processes that occur on scales that are too small to be resolved by numerical model grids. The values of these coefficients are more often chosen with computational requirements in mind than for their physical representativeness. How then do we address this problem? One solution is to run ocean and acoustic models together, the so-called coupled ocean-acoustic modelling approach.

2. Coupled Ocean-Acoustic Model Development

During the 1970s and early 1980s much was done to study the effects of ocean fronts and eddies, on sound propagation, using generalised models of these features. For example,

J. Potter and A. Warn-Varnas (eds.), Ocean Variability & Acoustic Propagation, 501–516.
© *Her Majesty's Stationary Office, London 1991.*

502

Henrick et al (1977, 1980) developed an analytical model of mesoscale ocean eddies which made it possible to relate acoustic properties to eddy characteristics, including currents. This model was subsequently used by Baer (1981), in conjunction with a 3-D acoustic model, to study horizontal and vertical refraction effects due to eddies. Similarly, Rousseau et al (1982) used an idealised model of an ocean front to study its effect on short range acoustic transmissions. A primary limitation of these studies has been their use of analytical models to describe isolated features and their inability to evolve ocean environments realistically in both space and time. This has had to await the development of sophisticated numerical model codes and powerful computers to achieve the desired resolution.

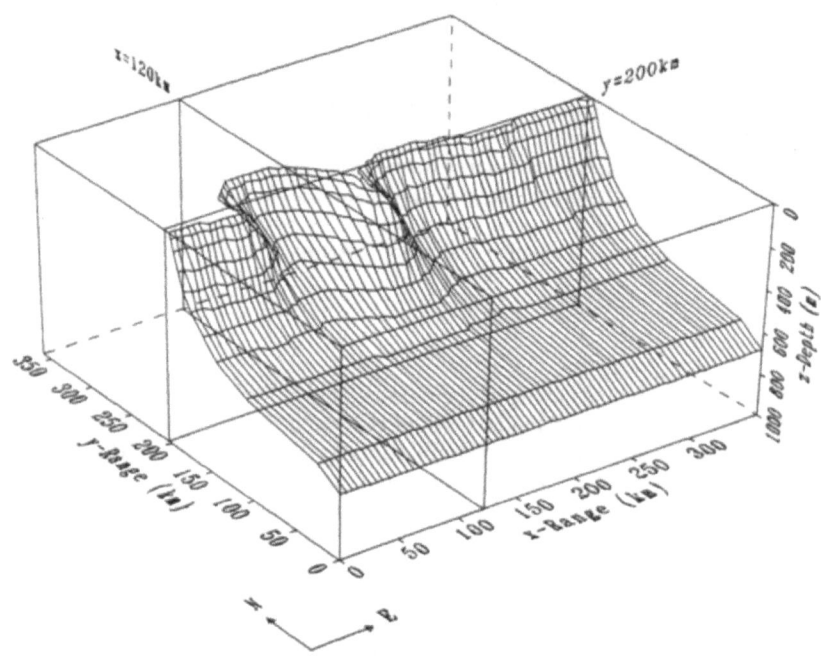

Figure 1. 3-D view of idealised frontal model showing position of 6 °C isothermal surface at 8 days following initialisation with baroclinic step perturbation. Acoustic calculations were performed on the sections at x = 120 km and y = 200 km.

1-D numerical models of the upper ocean have been used in coupled ocean-acoustic model investigations of surface duct propagation (eg McManus, 1985, Fourniol, 1987). Such models, which are less demanding of computer resources than 3-D models, have formed the basis of most early naval ocean forecast systems, for example the US Navy's TOPS (Clancy and Pollack, 1983) and the Royal Navy's NEAT MLM (Heathershaw et al, 1990a). Coupled ocean-acoustic model studies of the upper ocean have provided valuable insights into the way in which sound propagates in this region and Porter et al (1989), using a number of 1-D models, have shown how differences in predicted near-surface temperature gradients, can have a significant effect on sound propagation characteristics.

To study the sensitivity of the acoustics to changes in the mesoscale environment requires a different approach. Specifically, 3-D numerical ocean models are required, in which the time varying fields of temperature and salinity can be properly described for input to range dependent acoustic models. Research along these lines is fairly new; for example Botseas and Seigmann (1989) and Melberg et al (1990) describe the coupling, or interfacing, of the implicit finite difference acoustic model, IFD (Lee and Botseas, 1982), with the Harvard Open Ocean Model (Robinson and Walstad, 1987). While these studies have demonstrated that coupled ocean-acoustic models may be used to study the temporal variability of sound propagation through modelled frontal regions, they did not specifically address the sensitivity of acoustic predictions to changes in the ocean model parameters.

In this paper we describe the development of a coupled ocean-acoustic modelling technique that has been designed to address just this problem. A 3-D numerical ocean model has been used to simulate eddy features at an ocean front and to provide synoptic estimates of the sound speed field, throughout the frontal region, for input to range dependent acoustic models. The ocean model was set up with an idealised frontal feature and simulations are described in which the model has been initialised with a baroclinic step perturbation at the front, and by the introduction of isolated eddy features. The latter, according to their location and mutual vorticity influence on the front, result in a variety of interactions all giving realistic mesoscale environments for acoustic purposes. The approach described here is unique in that we are able to study the sensitivity of the acoustic predictions to the changes in the simulated environments brought about by different ocean model parameter settings. In particular, we have taken advantage of the coupling that is now possible between ocean models and acoustic models to study the effects of changes in spatial resolution and in the horizontal and vertical eddy viscosity coefficients in the ocean model. This work is of particular relevance to the development of naval ocean forecast systems where it is necessary to achieve a proper understanding of the inherent limitations of ocean prediction models in providing simulated environmental data for acoustic studies.

3. The Models

In this section we briefly review the ocean model and the acoustic models that have been used in this study.

3.1 THE OCEAN MODEL

The ocean model employed in this study is an adaptation of the Cox (1984) model. It is configured so as to represent an idealised frontal system in a 3-D, rectangular, flat bottomed ocean domain, with temperature contrasts and physics appropriate to the Iceland-Faeroes front. The set-up is shown in Figure 1 which is a 3-D view of the front at 8 days following an initial baroclinic disturbance (see later).

Following Cox, the equations for the conservation of heat, salt and momentum are written, with Boussinesq, hydrostatic and rigid lid approximations, in spherical coordinates. For simplicity in the following sections, Cartesian coordinates are used to describe the relevant terms. The equations (see Annex) are solved in finite difference form on an Arakawa B-grid with a leap-frog time stepping scheme. The Coriolis term is treated semi-implicitly so as not to resolve inertial oscillations in time. The frictional terms $A_H\nabla^2 q$, where q is the velocity, are lagged by one timestep for numerical stability. A forward timestep is taken every 20 timesteps to avoid a computational mode arising from the leap-frog scheme and the barotropic stream function is solved by successive over relaxation.

The Cox model is used in preference to other models, eg quasi-geostrophic models (Robinson and Walstad, 1987 and Marshall and Nurser, 1986) and isopycnic coordinate models (Bleck and Boudra, 1986), because it is a robust and well understood model, and is widely used in the ocean modelling community in the UK.

The model is set up with 15 levels in the vertical with $\Delta z = 25$ m in the top two levels and $\Delta z = 75$ m in the remaining 13 levels, giving a total depth of 1025 m. There are 72 range increments in both the x and y directions, with $\Delta x = \Delta y = 5$ km, giving overall dimensions of 360 x 360 km. For the bulk of the simulations the horizontal and vertical eddy viscosity coefficients, A_H and A_V (see Annex), are set to 10^7 and 1.0 cm^2 s^{-1} respectively and the horizontal and vertical eddy diffusion coefficients, K_H and K_V, are 10^5 and 1.0 cm^2 s^{-1}. During the acoustic investigations, however, these values were varied within prescribed limits so as to be representative of ocean model studies elsewhere. All simulations were performed with a time step Δt of 360s.

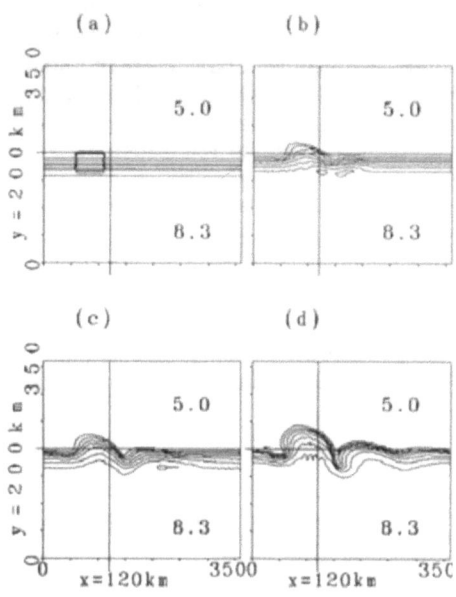

Figure 2. Simulated near-surface ($z = -12.5$ m) temperature fields at (a) 0, (b) 2, (c) 4 and (d) 8 days following initialisation with a baroclinic step perturbation (Figure 2a).

The model was set up with an east-west front having a temperature difference across the front of 3.3 °C. This is an idealised two-layer representation of an ocean front, which for the Iceland-Faeroes Front corresponds to warm North Atlantic water overlying colder Norwegian Sea water, the warmer water increasing asymptotically in depth away from the front, to a mean depth of 500 m. Boundary conditions in the E-W direction are cyclic and may be either closed or open in the N-S direction. In the latter case, open boundary conditions are applied using the method due to Stevens (1988); since the flow in these experiments is essentially zonal (from west to east), the results of simulations with the N-S boundaries closed are not significantly different from those with the boundaries open. When integrated forward in time the model will evolve its own current field although, for the purposes of this study, an initial current field was prescribed by converting the initial temperature distribution into density, assuming constant salinity, and then integrating this in the thermal wind equation.

To generate frontal eddies, a baroclinic step perturbation was introduced at the front by increasing the temperatures in the top 300m of the water column, along a 55 km section of the front, by 1 °C. The model was then integrated forward in time to give eddy like features at the front (Figure 2). Similar features may also be generated by applying a barotropic perturbation (Heathershaw et al, 1988) or a sinusoidal perturbation of arbitrary wavelength and amplitude. Alternatively, discrete eddy features can be introduced into the modelled domain and their subsequent interaction with the front studied. In this case, following Kielmann and Kase (1987) and Smith and Davies (1989), this is achieved by introducing a temperature anomaly of the form:-

$$\Delta T = \frac{A}{\cosh^2(z/D)} \exp\left[-\frac{\{ (x - x_c)^2 + (y - y_c)^2 \}}{2L^2} \right] \tag{1}$$

where A is the amplitude of the eddy temperature perturbation (°C), D is a vertical depth scale (typically 200m), (x_c, y_c) is the horizontal position of the eddy centre, and L is an e-folding width scale. The latter is of the order of the first internal Rossby radius of deformation, R_i, given by:

$$R_i = \frac{1}{f} \left[\frac{(\rho_2 - \rho_1) \, g \, h}{\rho_2} \right]^{1/2} \tag{2}$$

ρ_1 and ρ_2 being the densities of the upper and lower layers in the frontal simulation, g is gravity, h the depth of the upper layer, and f the Coriolis parameter. R_i is approximately 15 km and L has been set equal to R_i.

The acoustic investigations described in this paper are concentrated along two sections, one through the front at x =120 km and the other parallel to it at y = 200 km. Eddy features may be placed anywhere in the modelled domain but in Figure 3 we demonstrate this particular initialisation technique by placing a   cyclonic cold core eddy just to the south of the front and to one side of the section at x = 120 km.

It should be noted that, in the context of this study, the ocean model is used as a process model. There is no external forcing, ie no wind stress at the surface, and no thermodynamically active surface layer. For the purposes of the acoustic investigations described here, the model has been used purely as a 'generator' of mesoscale environmental data. Atmospherically forced models providing comparable simulations are being investigated elsewhere. Further details of the ocean model are given in the Annex and in Heathershaw et al (1988,1990b).

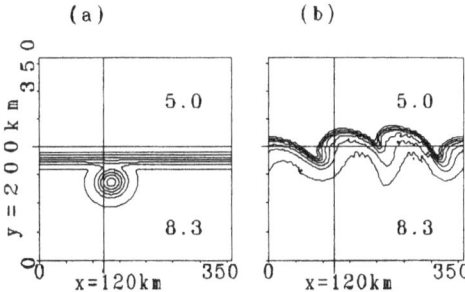

Figure 3. Illustration of discrete eddy initialisation technique showing (left) initial near surface (z = -12.5 m) temperature field, and the situation (right) with the model integrated to 16 days.

3.2 THE ACOUSTIC MODELS

To determine the sensitivity of acoustic predictions to changes in the modelled environments and in the ocean model parameters, the outputs from the eddy resolving model described previously have been used as input to two acoustic models. The models were chosen to span a range of frequencies and represent different features of the acoustic fields. The ray theory model, GRASS, has been used to obtain results representative of a high frequency situation (1 kHz), and the wave theory model, PAREQ, was used to investigate low frequency (150 Hz) propagation characteristics.

GRASS (Germinating Ray-Acoustic Simulation System) is a range dependent ray theory model which can generate ray trace diagrams and frequency dependent propagation loss curves. The primary use of this model has been to assess the influence that mesoscale variability within the ocean has upon sound propagation paths and to study the dependence of acoustic predictions upon ocean model parameter settings. Further details of GRASS may be found in Cornyn (1973) and Harrison (1989).

Using Chien and Millero's (1977) equation, sound speed values were calculated from the model generated temperature fields, with salinity assumed constant, and smooth profiles then fitted to these values using a cubic spline to give continuous first and second order derivatives (see Cornyn, 1973). A total of 72 profiles, at range increments of 5 km, were therefore available for ray tracing with linear interpolation between profiles being used to obtain estimates at intermediate range steps.

For ray tracing 11 rays were used. These were launched at $1°$ intervals for $±5°$ about the horizontal. Propagation loss was calculated with the ray density increased to obtain convergent solutions for the resulting intensities. This was found to occur with 48 rays per degree, for a range of angles $±15°$ about the horizontal, giving a total of 1440 rays.

To determine propagation loss characteristics using GRASS, a phase independent intensity calculation was performed with a frequency dependent volume attenuation coefficient given by Thorp's (1965) equation. At the ocean surface rays were assumed to be specularly reflected, without attenuation, while, for the majority of the investigations, the ocean bottom was assumed to be fully absorbing, so as to isolate the acoustic effects of the in-water changes. Experiments were also carried out in which range dependent bottom loss was incorporated at the lower boundary of the ocean model. In all these tests an omni-directional sound source has been assumed with the rays confined between the angular limits given above. The sound source was placed at a depth of 100 m and the receiver at depths of 100 and 250 m respectively.

To determine the low frequency sound propagation characteristics associated with mesoscale features, we have used the wave theory model PAREQ (Tappert, 1977, Jensen and Krol, 1975, Harrison, 1989). In this case we have performed calculations for a frequency of 150 Hz and with the same source and receiver depth combinations as the GRASS experiments. For these calculations it was necessary to simulate the effects of sound interacting with the ocean bottom and, so as to partially isolate the effects of the in-water changes, PAREQ was set up with a high loss ocean bottom consisting of a 500 m layer of sediment of density 1500 kg m^{-3}, with an attenuation factor of 0.3 dB per wavelength and a seawater/sediment sound speed ratio of 1.003. The sound speed gradient within the sediment layer was 1.3 s^{-1}.

4. Results

As previously mentioned, the majority of this work has been undertaken with the eddy viscosity and diffusion parameters, A_H, A_V, K_H and K_V (equations A1-A2 and A5), set to 10^7, 1, 10^5 and 1 cm^2 s^{-1} respectively. To study the effect that altering the horizontal eddy viscosity has upon acoustic propagation, A_H was varied in the range 0.1 to 2 x 10^7 cm^{2}s^{-1} (see later), but with A_V, K_H and K_V remaining constant.

The results of the frontal simulations are shown in Figures 1-2. These indicate a wave-like disturbance spreading towards the east which, after 8 days, has the characteristic backward-breaking wave shape seen in other numerical simulations of this type (eg Dippner, 1990), and which is also seen in satellite images of the Iceland-Faroes front region (eg Scott and McDowall, 1990) and many other frontal zones. The eddy structure of the front becomes more apparent over longer integrations (see Heathershaw et al, 1990b). Figure 3 is an example of the modelled near surface temperature and current fields following initialisation of the model with a discrete eddy feature (equation 1).

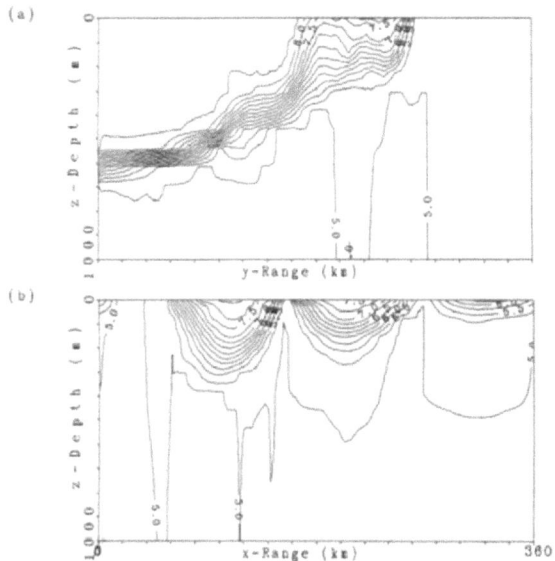

Figure 4. Temperature sections (a) at x = 120 km through the front and (b) at y = 200 km along the front (see Figure 1) for 8 days and with an A_H value of 10^7 cm^2 s^{-1}.

Temperature sections through and along the front, without the discrete eddy, are shown in Figure 4. The classic two-layer structure of the front, which, for the purposes of these experiments, is set up to correspond to warm North Atlantic water overlying colder Norwegian Sea water, is illustrated in Figure 4a. Hallock (1985) has shown that, at depth, the thermal structure of the Iceland-Faeroes front is modified by the Iceland-Faeroes ridge, giving rise to a cold water overflow region. All of the following experiments are carried out assuming a flat bottomed ocean domain so that topographically related features such as this cannot be represented in these model simulations.

Little vertical structure is evident within the two layers because none was present in the initial profiles. Measurements from the Iceland-Faroes front region (eg Scott and McDowall, 1990) reveal considerable small scale variability which cannot be reproduced in the model simulations described here. The acoustic significance of this detail is being investigated elsewhere, although, to first order we would expect these effects, which are mostly due to internal waves and other small scale structure, to give rise to perturbations in the sound velocity field that are an order of magnitude lower than those which occur as a result of mesoscale variability.

Figure 5 illustrates the results of ray tracing along the section at y = 200 km, with the sound source placed at a depth of 100 m and positioned at the western extremity of this section. The effects of mesoscale features on sound propagation paths are clearly evident with sound energy deflected downwards by as much as 300 m, as a result of the simulated eddy features shown in Figure 4b. In two dimensions these features may be considered as warm core eddies comprising acoustically faster water at their centres and giving rise to sound speed changes of about 10 m s^{-1}. Further examples of ray tracing through simulated frontal environments are given in Heathershaw et al (1988, 1990b).

Coupled ocean-acoustic models, of the type described here, also enable the temporal variations, in the sound propagation characteristics of the front, to be studied (see Heathershaw et al, 1988). For the baroclinic perturbation case we have found that the through-front propagation characteristics are little altered in the early stages of fronto-genesis. For periods out to 8 days the only observed change was a gradual broadening of convergence zones (CZ) which appear in the cold water to the north of the front when the sound source is to the south. This is associated with a weakening of the cross-frontal gradients, along this particular section (x = 120 km), as the front evolves. At other locations these gradients may be strengthened giving rise to different propagation characteristics. Few other generalisations are possible at this stage regarding CZs because we have assumed a fully absorbing bottom for rays striking the seabed.

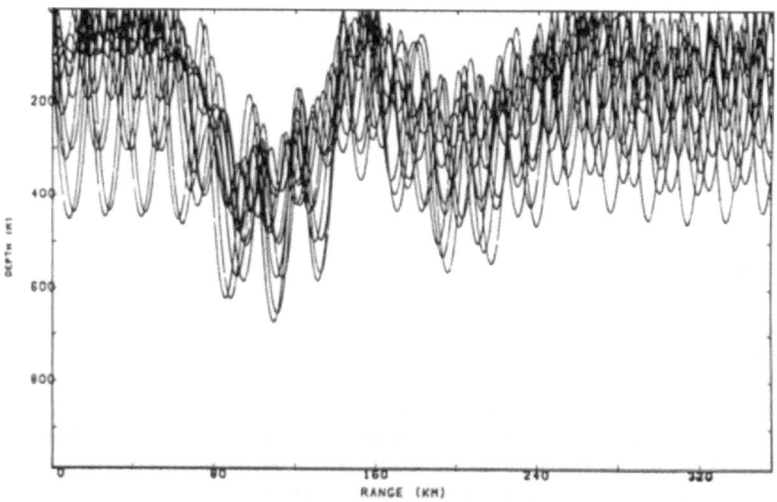

Figure 5. Ray trace for section along front at y = 200 km (Figure 4b) with the sound source at a depth of 100 m at the western extremity of the section (see Figure 1).

Conversely, sound propagation characteristics along the front (y = 200 km), show a great deal of variability as the front evolves from its initial linear state. These changes are not surprising as only relatively small displacements of the front will be required to induce large changes in the along-front temperature and sound velocity fields. Observed alterations in propagation paths will depend critically on where these sections are taken in relation to the front. In general, perturbations in the sound velocity field and the sound propagation characteristics will be greatest when the displacement due to a mesoscale feature is normal to the sound propagation path.

Corresponding to the temperature sections in Figure 4, propagation loss curves were calculated using GRASS. These results are shown in Figure 6. A frequency of 1 kHz was used and source/receiver depth combinations of 100/100 m and 100/250 m, the latter to demonstrate the differences which result when the receiver is located below the general level of mesoscale activity. For both sections, range independent propagation loss curves were calculated by assuming uniform conditions down range of the first profile and with a source receiver depth combination of 100/100 m.

With the model integrated to 8 days the most striking feature of the through-front propagation results (Figure 6) is the 15 dB average propagation loss increase, which occurs, with the source at a depth of 100 m, as sound travels from the warm side of the front to the cold. Beyond this fluctuations of about ±10 dB occur as a result of the CZ behaviour mentioned previously. With the receiver at 250 m, Figure 6 shows that the frontal effect is less pronounced but still equivalent to about 5 dB.

For the section along the front, at y = 200 km, Figure 6 shows a 20 dB increase in propagation loss, relative to the range independent case, associated with the eddy-like features shown in Figures 2 and 4. With the receiver at 250m the effect of the eddies is less pronounced although, for the largest feature (Figure 4b), the effect is still discernible.

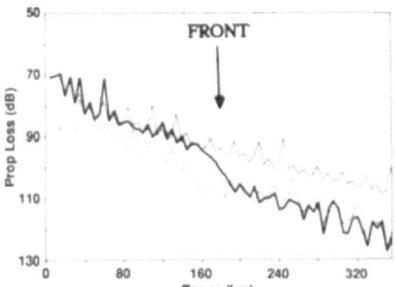 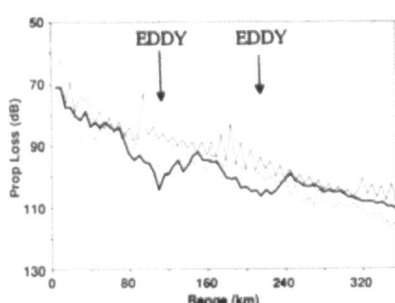

Figure 6. GRASS propagation loss curves for sections through (left) and along (right) front. Results are shown for a frequency of 1000 Hz and for sound source/receiver depths of 100/100 m (thick line) and 100/250 m (broken line). The range independent case (thin line) is shown for a source/receiver depth combination of 100/100 m only.

5. Discussion

The results of this study have demonstrated the ability of ocean models to provide high resolution environmental data sets, with simulated mesoscale variability, for input to range dependent acoustic models. The versatility of this approach may be clearly demonstrated. In particular it is possible to perform ocean-acoustic simulations, under well controlled conditions, varying both the ocean and acoustic model parameters (eg eddy viscosity, sound frequency and source depth) within experimentally and operationally useful ranges, and introducing spatial and temporal variability into the problem in a way that would be difficult to achieve with measurements alone. In addition, the environmental data sets are synoptic, overcoming the difficulty that has been experienced in measurements at sea where mesoscale features may evolve on time scales comparable to those required to survey them. Measurements at sea can also be costly and time consuming and, while observations are important in improving our understanding of basic physical processes in the ocean, and their parameterisation in ocean models, computer simulations provide a cost effective means of studying mesoscale acoustic variability.

510

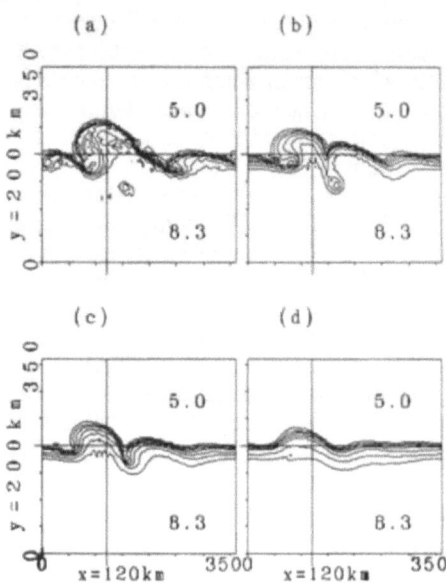

Figure 7 Near surface temperature fields at 8 days for A_H values of (a) 0.1 x 10^7, (b) 0.5 x 10^7, (c) 1 x 10^7 and (d) 2 x 10^7 cm^2 s^{-1}.

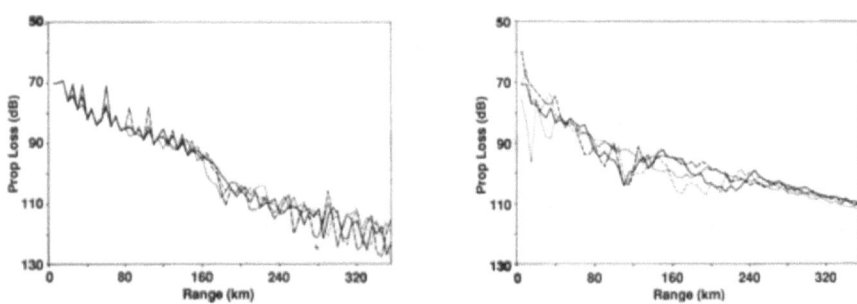

Figure 8. Composite diagram showing variations in propagation loss on sections through (left) and along (right) front at 8 days and corresponding to horizontal eddy viscosity values in the range 0.1 to 2 x 10^7 (see Figure 7).

In utilising these coupled techniques it is essential that we have conviction in the ocean model results. Forecast models (eg Melberg et al, 1990) are initialised and constrained by oceanographic and other data, eg satellite observations of frontal meanders and eddies. On the other hand, verification of process-model results relies on comparisons with theoretical predictions which give maximum growth rates for preferred wavelengths, according to mesoscale frontal dynamics.

For the situation described here, tests were performed by applying a small amplitude sinusoidal perturbation at the front and comparing the observed rates of growth of features having different wavelengths, with those that are predicted by linear theory (Kilworth et al, 1984). The latter predict maximum growth rates to be at a wavelength given by $2\pi R_i$, where R_i is the internal Rossby radius (equation 2). In these studies, with h = 500 m, ρ_2 = 1027.92 kg m^{-3} and ρ_1 = 1027.10 kg m^{-3}, we obtain $R_i \approx$ 15 km, corresponding to a wavelength $2\pi R_i$ = 94 km. The observed growth rates agreed well with those obtained by Kilworth et al, all non-zero growth rates lying within 11% of their results. Further details of these tests are given in Wood (1988).

Comparisons were also made with the observed wavelengths of the features generated by the baroclinic perturbation technique described previously. These tests showed that with the model integrated to 8 days, beyond which the frontal dynamics becomes highly non-linear, the observed wavelengths of frontal features lay within the range 90 to 120 km and, therefore, close to the figure of 94 km given by linear theory.

While these tests have confirmed that realistic mesoscale simulations may be possible with numerical ocean models, they do not provide evidence that the simulated environments will be suitable for acoustic purposes. In particular it is possible to perform simulations that are dynamically correct, and which give realistic spatial and temporal scales for the observed features, but which may contain significant differences in detail. In large measure these differences will be due to our choice of ocean model parameters, in particular those that describe sub-gridscale mixing processes too small to be resolved by the numerical model grids. It is important for naval ocean forecast system design, that we know precisely how significant these differences are acoustically.

To address this problem we have performed numerical simulations in which the horizontal eddy viscosity parameter, A_H, was varied in the range 0.1 to 2 x 10^7 cm^2 s^{-1}. A_H is approximately two orders of magnitude larger than the next largest eddy coefficient, the horizontal eddy diffusion parameter, which is of the order 10^5 cm^2 s^{-1}; the vertical eddy coefficients are both of order 1.0 cm^2 s^{-1}. In these experiments the model was initialised with the baroclinic step perturbation described previously.

Figure 7 compares the results for each value of A_H, after integrating the model for 8 days. It illustrates that, with K_H, A_V and K_V constant, the lowest value of 0.1 x 10^7 cm^2 s^{-1} gives a 'noisy' simulation whereas the largest value, 2 x 10^7 cm^2 s^{-1}, gives results in which the growth rate is reduced although the spatial scales are similar. A similar study was carried out, this time varying K_H in the range 0.1 to 2 x 10^5 cm^2 s^{-1}, and this showed (Heathershaw et al, 1990b) that there was little discernible difference in the modelled fields, after 8 days.

The effect of the choice of A_H in equations (A1) and (A2), can be seen if we consider the effect of the eddy viscosity term in the simplified momentum equation; $\partial u/\partial t$ = $A_H (\partial^2 u)/(\partial x^2)$, where u is the horizontal velocity in the x direction. By inspection, this gives a relationship between time (T) and length (L) scales of the form $T = L^2/A_H$, where L is now a length scale associated with the size of a feature and T a time for its decay as a result of momentum diffusion. Therefore, for an ocean eddy, of diameter $2R_i$ (about 30 km), with $A_H = 10^7$ cm^2 s^{-1} we obtain a characteristic 'spin-down' time of approximately 16 days. This is longer than the time scales of interest in operational ocean forecast models and suggests that with suitable data assimilation schemes it will be possible to predict this level of detail in mesoscale simulations. The question remains, however, as to the acoustic significance of the detailed changes that are shown (Figure 7) in the ocean model output. To assess this problem we have performed propagation loss calculations on the sections already discussed, both through and along the front, for a range of A_H values.

The results of these tests, carried out using GRASS, are illustrated in Figure 8, and suggest that, for sound that is propagating through the front (Figure 8a), there will be an uncertainty of ±5 dB which is associated with the in-water differences due to the different eddy viscosity values. For sound propagating parallel to the front, however, the variations are of the order of ±10 dB and are particularly noticeable at short ranges, ie out to about

40 km. These differences, which are apparent in ray diagrams (see Heathershaw et al, 1990b), will depend on the location of the sound source in relation to the ocean feature, but more importantly from the evidence presented here will depend on eddy viscosity in the ocean model.

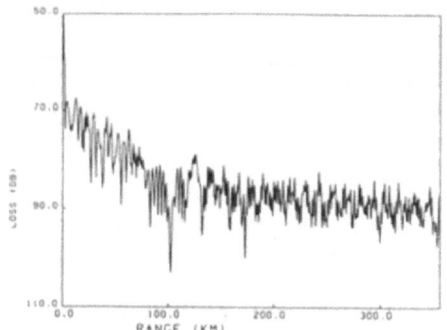 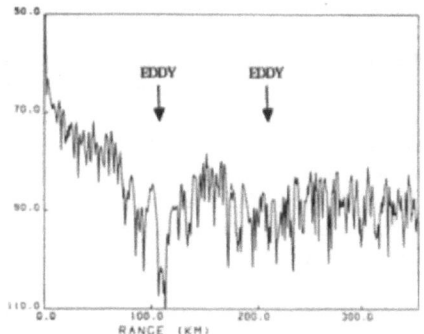

Figure 9. PAREQ low frequency (150 Hz) propagation loss curves for sections along the front at 2 days (left) and 8 days (right) with $A_H = 10^7$ cm^2 s^{-1}. Results are shown for a high bottom loss case.

Similar comparisons have yet to be carried out with PAREQ. However, the early indications are that despite the lower frequencies, in-water effects are still important and that the acoustic predictions may still be sensitive to ocean model parameters. Figure 9 shows propagation loss curves calculated with PAREQ for the section at y = 200 km along the front and through the eddy like features shown in Figure 4b, but for the model integrated to 2 days and 8 days. These results clearly indicate the temporal variability in propagation loss due to the changes occurring in the simulated mesoscale environment, even at low frequencies.

Further work is in progress to exploit the coupling that is now possible between ocean models and acoustic models, to investigate the sensitivity of acoustic predictions to other ocean model parameters, in particular vertical and horizontal resolution, and to study the effect of changes in eddy initialisation parameters such as eddy size and temperature. Work is also in hand to investigate the effects of various frontal and eddy initialisations in relation to ridge and shelf topography in the ocean model, and with a version of the model that incorporates real coastlines and bottom topography, corresponding to the Iceland-Faeroes front region.

These studies aim to provide a better understanding of the optimum setting of ocean model parameters in naval ocean forecast systems and of the limitations inherent in data assimilation from observations and feature modelling.

6. Conclusions

Studies with a 3-D eddy resolving ocean model being used to provide simulated mesoscale environments for input to range dependent acoustic models, have demonstrated the versatility of the coupled ocean-acoustic model technique in providing insights into the sensitivity of acoustic predictions to changes in the modelled environments and in the model parameters. We have found that at high frequencies (of order 1 k Hz) acoustic predictions will be sensitive to ocean model parameters, in particular the horizontal eddy viscosity coefficient. At low frequencies (of order 150 Hz) we find that the acoustic predictions are still sensitive to in-water changes and that consequently they may be influenced by uncertainty in ocean model parameter settings. However, at low frequencies attenuation in the bottom sediments becomes more important although this aspect may also be studied in the coupled ocean-acoustic modelling approach described here by including suitable bottom loss parameterisations.

7. Acknowledgements

We are grateful to former and present colleagues in the Ocean Science Division at ARE Portland, for the assistance in the preparation of this paper. In particular, Bill Cooper was instrumental in providing many of the early simulations with the ocean model, and Roger Hillman has provided invaluable guidance with the use of acoustic models. Davie Barr, of the Computer Support Group at ARE, is also thanked for his unstinting efforts in overcoming computing problems.

8. References

Baer, R. N. (1981). 'Propagation through a three-dimensional eddy including the effects on an array', J. Acoust. Soc. Am., 69, 70-75.

Bleck, R. and Boudra, D. B. (1986). 'Wind driven spin-up in eddy-resolving ocean models formulated in isopycnic and isobaric coordinates', J. Geophys. Res., 91C, 7611-7621.

Botseas, G. and Seigmann, W. L. (1989). 'IFD: Interfaced with Harvard Open Ocean Model forecasts', Naval Underwater Systems Center, Connecticut, Technical Report 8367.

Chien, C. T. and Millero, F. J. (1977). 'Speed of sound in sea water at high pressures', J. Acoust. Soc. Am., 62, 1129.

Clancy, M. and Pollack, K. (1983). 'A real-time synoptic ocean thermal analysis/forecast system', Prog. Oceanog., 12, 383-424.

Clancy M. (1987). 'Real-time applied oceanography at the Navy's Global Center', Mar. Tech. Soc. Journal, 21, 33-46.

Cornyn, J. (1973). 'GRASS: A digital-computer ray-tracing and transmission loss prediction system. Vol 1 - Overall description', Naval Research Laboratory Report, 7621.

Cox, M. D. (1984). 'A primitive equation, 3-dimensional model of the ocean', GFDL Ocean Group Tech Report No 1, Princeton University.

Dippner, J. W. (1990). 'Eddy-resolving modelling with dynamically active tracers', Cont. Shelf Res., 10, 87-101.

Fourniol, J. M. (1987). 'Coupled acoustic and ocean thermodynamic model', Masters Thesis, Naval Postgraduate School, Monterey, California, USA.

Hallock, R. (1985). 'Variability of frontal structure in the southern Norwegian Sea', J. Phys. Oceanog., 15, 1245-1253.

Harrison, C. H. (1989). 'Ocean propagation models', App. Acoust., 27, 163-201.

Heathershaw, A. D., Maskell, S. J., Cooper W. St J. and Hillman, R. C. (1988). 'Studies of sound propagation through a front using an eddy resolving ocean model', Admiralty Research Establishment Tech Memo (UJO) 88144.

Heathershaw, A. D., Gething, M. R. and Foreman S. J. (1990a). 'Ocean forecast models for ASW', Proceedings, Underwater Defence Technology Conference, London, 123-128.

Heathershaw, A. D., Stretch, C. E. and Maskell, S. J. (1990b). 'Coupled ocean-acoustic model studies of sound propagation through an ocean front', J. Acoust. Soc. Am., (submitted).

Henrick, R. F., Siegmann, W. L. and Jacobson, M. J. (1977). 'General analysis of ocean eddy effects for sound transmission applications', J. Acoust. Soc. Am., 62, 860-870.

Henrick, R. F., Jacobson, M. J. and Siegman, W. L. (1980). 'General effects of currents and sound speed variations on short range acoustic transmission in cyclonic eddies', J. Acoust. Soc. Am., 67, 121-134.

Jensen, F.B. and Kroll, H.R. (1975). 'The use of the parabolic equation method in sound propagation modelling', Report No. SM-72, SACLANT ASW Centre, La Spezia, Italy.

Kielmann, J. and Kase, R. H. (1987). 'Numerical modelling of meander and eddy formation in the Azores Current frontal zone', J. Phys. Oceanog., 17, 529-541.

Killworth, P. D., Paldor, N. and Stern, M. E. (1984). 'Wave propagation and growth on a surface front in a two-layer geostrophic current', J. Mar. Res., 42, 761-785.

Lee, D. and Botseas. G. (1982). 'IFD: An Implicit Finite Difference computer model for solving the parabolic equation', Naval Underwater Systems Center, Connecticut, Technical Rept. 6659.

Marshall, J. C. and Nurser, A. J. G. (1986). 'Steady free circulation in a stratified quasi-gesostrophic ocean', J. Phys. Oceanog., 16, 1799-1813.

Melberg, L. E., Robinson, A. R. and Botseas, G. (1990). 'Modeled time variability of acoustic propagation through a Gulf Stream meander and eddies', J. Acoust. Soc. Am., 87, 1044-1054.

McManus J. J. (1985). 'Coupled mixed layer-acoustic model', Master's Thesis, Naval Postgraduate School, Monterey, California, USA.

Porter, M., Piacsek, S., Henderson, L. and Jensen, F. (1989). 'Acoustic impact of upper ocean models', unpublished manuscript, SACLANT Undersea Research Centre.

Robinson, A. R. and Walstad, L. J. (1987). 'The Harvard Open Ocean Model: Calibration and application to dynamical process, forecasting and data assimilation studies', Applied Numerical Mathematics, 3, 89-131.

Rousseau, T. H., Siegmann, W. L. and Jacobson, M. J. (1982). 'Acoustic propagation through a model of shallow fronts in the deep ocean', J. Acoust. Soc. Am., 72, 924-936.

Scott, J. C. and McDowall, A. L. (1990). 'Cross-frontal jets near Iceland: In-water, satellite infra-red and GEOSAT altimeter data', J. Geophys. Res., to appear.

Semtner, A. and Chervin, R. (1988a). 'Breakthroughs in ocean and climate modelling made possible by supercomputers of today and tomorrow', Proceedings of Supercomputer '88 Conference, Vol II, Science and Applications, 230-239.

Semtner, A. and Chervin, R. (1988b). 'A simulation of the global ocean circulation with resolved eddies', J. Geophys. Res., 93, 15502-15522.

Smith, D. C. and Davies, G. P. (1989). 'A numerical study of eddy interaction with an ocean jet', J. Phys. Oceanog., 19, 103-133.

Stevens, D. P. (1990). 'On open boundary conditions for three dimensional primitive equation ocean circulation models', Geophys. Astro. Fluid Dyn., 51, 103-133.

Tappert, F.D. (1977). 'The parabolic approximation method', In: Wave Propagation and Underwater Acoustics, Ed. J.B. Keller and J.S. Papadakis, Springer-Verlag, Berlin, 224-287

Thorp, W. H. (1965). 'Deep-ocean sound attenuation in the sub- and low- kilocycle-per-second region', J. Acoust. Soc. Am., 38, 648-654.

Wood, R. A. (1988). 'Instability of oceanic fronts', Ph.D Thesis, Exeter University, UK.

Annex The Ocean Model

For simplicity in examining the ocean model physics in relation to the acoustics, the description of the Cox (1984) model given here is in Cartesian coordinate form.

Prediction of the time dependent current fields in the ocean model is carried out using the Navier Stokes equations for fluid flow on a rotating earth, with three further basic assumptions. Firstly, the Boussinesq approximation is made, in which density differences are neglected except in the bouyancy terms, the validity of the approximation being due to the relatively small variations in density. Secondly, the equation for the vertical component of motion is reduced to the hydrostatic assumption by the neglect of locally vertical components of acceleration and of the Coriolis effect. The third approximation is that a turbulent viscosity hypothesis is made, in which stress exerted at scales of motion too small to be resolved by the computational grid are represented by enhanced molecular mixing by the use of eddy coefficients.

With these three approximations, the two equations for the horizontal components of velocity (u,v) are:

$$\frac{\partial u}{\partial t} + (q.\nabla)u - fv = -\frac{1}{\rho_0}\frac{\partial p}{\partial x} + A_H\nabla^2 u + A_V\frac{\partial^2 u}{\partial z^2} \tag{A1}$$

and

$$\frac{\partial v}{\partial t} + (q.\nabla)v + fu = -\frac{1}{\rho_0}\frac{\partial p}{\partial y} + A_H\nabla^2 v + A_V\frac{\partial^2 v}{\partial z^2}, \tag{A2}$$

where x is measured west-east, y is measured south-north and z is locally vertically upwards. The velocity $q = (u,v)$ and $(q.\nabla)u = u\partial u/\partial x + v\partial v/\partial y$. The parameter f governs the Coriolis effect, and $f = 2\Omega \sin\lambda$, where Ω is the angular rotation rate of the earth about its axis, and λ is the latitude. The density is held constant at ρ_0 in these equations, and p is pressure. The operator ∇^2 represents $\partial^2/\partial x^2 + \partial^2/\partial y^2$. A_H, A_V are, respectively, the horizontal and vertical eddy viscosity coefficients described above.

The hydrostatic equation is:

$$\frac{\partial p}{\partial z} = -\rho\, g, \tag{A3}$$

where g is gravity and ρ is density.

A fourth approximation is made; that water is incompressible, giving the continuity equation:

$$\frac{\partial u}{\partial x} + \frac{\partial v}{\partial y} + \frac{\partial w}{\partial z} = 0. \tag{A4}$$

To obtain an equation for the density, ρ, the assumption is made that the temperature, T, is governed by a transport equation where the sub-gridscale transfers of heat energy are describable using horizontal and vertical (K_H and K_V) eddy diffusion coefficients. The equation for the temperature is then:

$$\frac{\partial T}{\partial t} + (q.\nabla)T = K_H\nabla^2 T + K_V\frac{\partial^2 T}{\partial z^2}, \tag{A5}$$

Density ρ is related to T through an elementary equation of state:

$$\rho = \rho_0(1 - \alpha T), \tag{A6}$$

where α is a constant (the thermal expansion coefficient).

It is also possible to have an equation of state similar to (A5) but for salinity, S, instead of for T, and then replace (A6) by a more elaborate, empirically based equation of state; $\rho = \rho(S,T)$.

Equations (A1) - (A6) form a closed set of equations and may be solved numerically to obtain solutions for u, v, w, T, p and ρ. Further details of the method of solution of these equations, in spherical coordinates, are given in Cox (1984).

THE ENVIRONMENTAL ACOUSTIC TACTICAL SUPPORT SYSTEM: LOW-FREQUENCY MESOSCALE OCEAN FEATURE ENVIRONMENTAL ACOUSTIC RESULTS

G. A. Kerr, King D. B., Cloy G. P., Gomes B. R., Bucca P. J.
Naval Oceanographic and Atmospheric Research Laboratory*
Stennis Space Center, MS 39529-5004 USA

ABSTRACT. The Environmental Acoustic Tactical Support (EATS) System is a developmental tool created for test and evaluation of environmental acoustic software specifically related to anti-submarine warfare (ASW). The design of the EATS System is modular, which allows easier interchange of various ocean acoustic modules. A personal computer version of the EATS System will be discussed in conjunction with EATS System analyses of recent oceanographic exercise data and a modeled Gulf Stream regional oceanographic environment. Use of the EATS System led to the early discovery of a mesoscale eddy that occupied one exercise area. The EATS System model of the eddy caused convergence zone range differences of up to 6.7 nautical miles (nmi) at the fourth convergence zone, and up to 9.5-dB transmission loss differences for a high-frequency source over values that would have been calculated using standard U.S. Navy climatologies. Later exercise use of the EATS System led to the discovery of a temperature gradient change across a weak front. This discovery resulted in a 0.5-nmi change in the range to the first convergence zone. Multiple radial acoustic model runs made in a modeled Gulf Stream environment were used to investigate the combined effect of the front and associated eddies. When convergence zones coincided with warm-(cold-) water features, the turning depth sound energy at the next down-track convergence zone was determined to increase (decrease). In addition, ranges between convergence zones increased (decreased) down-track of convergence zones coinciding with warm (cold) oceanographic eddies.

1. Introduction

The Environmental Acoustic Tactical Support (EATS) System is a modular set of computer programs developed under the sponsorship of the U.S. Navy's Advanced Underwater Acoustic Modeling Program. This program was initiated to develop acoustic modeling support for future ASW acoustic systems. The EATS System was designed to be used as a development tool in a test and evaluation mode. Particular computer system attributes were not employed to achieve across-system flexibility. ANSII Standard FORTRAN 77 was used throughout the system. Inexpensive, commercially available software and libraries were used for all system graphics. All graphics in this paper were initially produced by the EATS System.

---

*Formerly the Naval Ocean Research and Development Activity.

J. Potter and A. Warn-Varnas (eds.), Ocean Variability & Acoustic Propagation, 517–525.
© 1991 *Kluwer Academic Publishers.*

This paper briefly describes the EATS System and its use during two oceanographic experiments. During these experiments the EATS System was found to be a useful tool in the determination of convergence zone ranges for simulated platform placement purposes. In addition to these experimental results, acoustic results from an oceanographic simulation in the Gulf Stream region of the North Atlantic Ocean are discussed.

2. System Description

The EATS System Version PC1.0 contains three major modules: an oceanographic data module, an acoustic model module, and an acoustic parameter module. A fourth oceanographic feature model module is being developed to provide better first-guess fields for the oceanographic data module's optimal interpolation (OI) routines. All modules interface with a common data base of environmental information stored in direct access files. Access to these files is accomplished through single-record sequential files, which contain the information necessary to unlock and access the direct access files.

The oceanographic data module was designed to create three-dimensional oceanographic fields from in situ measurements of temperature, salinity, and sound speed through OI techniques. First-guess oceanographic climatological data files currently available for OI include the Naval Oceanographic Office's (NAVOCEANO) Generalized Digital Environmental Model (GDEM) seasonal half degree resolution data, the Fleet Numerical Oceanography Center's (FNOC) Optimal Thermal Interpolation System predictions, and a data base constructed from any previous analysis. Various submodules within the oceanographic data module allow the editing of in situ data; the automated and unautomated OI of raw temperature, salinity, and sound speed fields; the merge of deeper climatologies to the resulting shallower OI fields; the calculation of various oceanographic acoustic parameters; and the testing of the resulting oceanographic synoptic fields. The variable resolution gridded synoptic oceanographic fields created in this module are made available to the other EATS System modules via updated direct-access files. Formatted synoptic ocean files are also created for transport to other systems.

The acoustic model module presently contains the Navy Standard Split-Step Parabolic Equation (PE) Model Version 3.1, the Automated Signal Excess Prediction System Transmission Loss (ASTRAL) Model Version 3.0, and the SACLANT Normal Mode Acoustic Propagation (SNAP) model. Interface to the climatological, synoptic oceanographic, and ocean model data bases is handled through a menu-driven system. This system creates a batch file, which runs the acoustic model selected and produces graphical results: these graphics include a full-field PE transmission loss plot, transmission loss versus range plots for up to three source receiver combinations, and a down-track environmental plot. The PE model's input data file is reformatted for ASTRAL and SNAP use.

The acoustic parameter module can be invoked to display graphics of environmental acoustic parameters of interest to the acoustic tactician. Such parameters as the range to the first convergence zone, depth excess, layer depth, and launch bottom grazing angle are some of the parameters available from this menu-driven module. The future oceanographic feature model module will enable ocean feature (i.e., front and eddy) locations to be used in creating synoptic three-dimensional temperature, salinity, and sound speed fields. NAVOCEANO's three-dimensional sound velocity grid package will be used in the first version of this module.

3. EATS Exercise Use

The first at-sea use of the EATS System took place during an oceanographic exercise held in the Norwegian Sea during the summer of 1988. The oceanographic environment at that time was considered complex. The center of the exercise area was occupied by an eddy colder than the surrounding waters at the surface and warmer than the surrounding waters at depth. The EATS System allowed this phenomenon to be detected after the first aircraft bathythermograph (BT) survey of the exercise area. Although the BT sample was widespread, the eddy was clearly indicated both at the surface and at depth. The eddy's early detection led to a revision in the exercise sampling plan. Over the duration of the exercise, the eddy remained nearly fixed. By the end of the exercise enough samples were being taken over periods of several days to justify a high-resolution (1/8°) OI. Figure 1.0 is a temperature cross section through the eddy from 67.58°N to 68.25°N along 4.0°W. This particular section was derived from the OI of a high-resolution set of BTs taken through the feature. Notice the upwelling of cool water near the surface and the core of warmer water located near 550 m.

Figure 2.0 shows 500-m temperature contours, which resulted from a high-resolution (1/8°) OI of data taken between August 23 and 29. The track, superimposed over the temperature contours, was used to test the acoustic significance of this oceanographic phenomenon. A second, smaller eddy was indicated, approximately halfway between points F and Z. Sound speed gradient changes between 200 m and 400 m at the eddy locations were readily apparent from plots of the along-track sound speed profiles.

In an effort to determine the acoustic effect of these eddies, a number of PE transmission loss model runs were made along the track indicated in Figure 2.0. The first model run was made between locations A and Z using both the GDEM climatological sound speed data set and the EATS synoptic sound speed data set. Transmission loss versus range plots for a high-frequency source for both cases are given in Figure 3.0. The table within this figure describes the absolute transmission loss

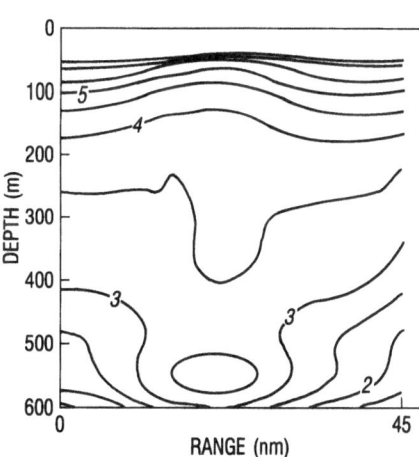

Figure 1.0. A temperature cross section through the deep, warm mesoscale feature which occupied the Norwegian Sea exercise area from 67.50°N to 68.25°N along 4.00°W. Contour values are given in degrees Celsius.

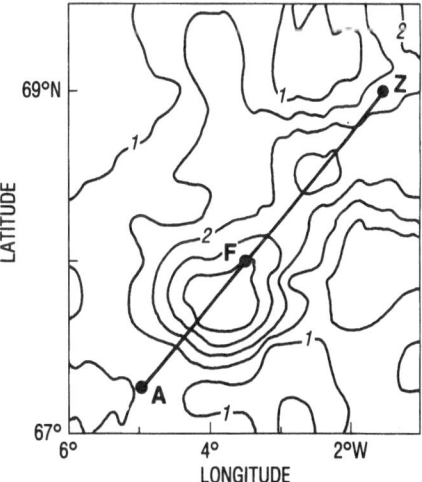

Figure 2.0. A 500-m temperature contour derived from an OI analysis of the Norwegian Sea exercise data of 23-29 August 1988. Contours in degrees Celsius. The track A, F, Z superimposed on the contours was used for acoustic model runs.

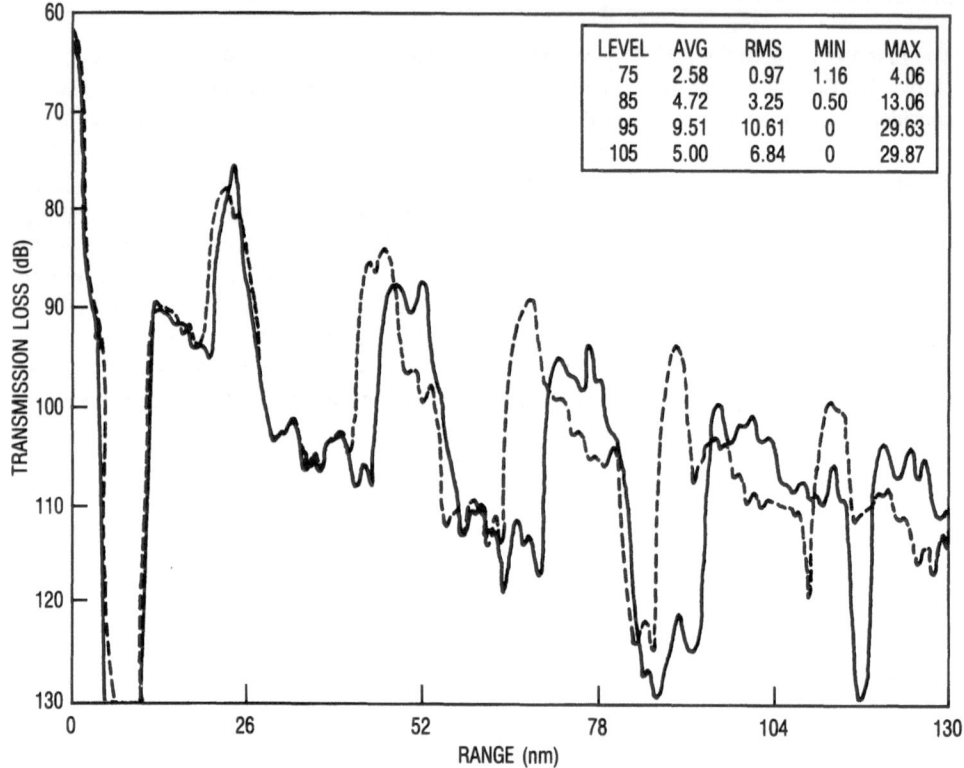

LEVEL	AVG	RMS	MIN	MAX
75	2.58	0.97	1.16	4.06
85	4.72	3.25	0.50	13.06
95	9.51	10.61	0	29.63
105	5.00	6.84	0	29.87

Figure 3.0. Transmission (dB) loss versus range from location A to location Z (Figure 2) using a GDEM (– –) climatology and an EATS synoptic environment (—).

difference statistics for several loss levels. The average transmission loss difference ranged from 2.6 dB to 9.5 dB. As indicated in Figure 3.0, the GDEM climatology resulted in lower transmission loss at convergence zone ranges than the EATS synoptic environment, except at the first convergence zone. Convergence zone range differences varied from 1.5 nautical miles (nmi) at the first convergence zone to 6.7 nmi at the fourth convergence zone; the EATS environment produced the shorter ranges.

Another set of similar high-frequency PE transmission loss model runs were made between location F, near the core of the largest eddy, and location Z along the same track. In this case, range to the first and second convergence zones differed by 3.5 nmi and 8.7 nmi, respectively. The synoptic environment produced the larger ranges. Transmission loss differences near the convergence zones varied from 0.4 dB at the first zone to 6.0 dB at the second.

An additional case in which the first along-track GDEM profile (location F) was replaced by the synoptic profile was tested to simulate a single BT drop near the eddy. In this case there was no difference in the range to the first convergence zone between the the two scenarios. However, the range to the second convergence differed by 1.4 nmi, with the totally synoptic environment produced the larger range. At the 92-dB level, the width of the second convergence zone for the modified GDEM and full synoptic oceanographic fields were 4.8 nmi and 8.2 nmi, respectively. The average transmission loss differences for this case ranged from 0.2 dB to 3.7 dB. In a similar analysis along A-Z, the difference in the convergence zone ranges were approximately 5 nmi for the second and third convergence zones.

The third at-sea utilization of the EATS System took place during an oceanographic exercise held in the Sargasso Sea during the summer of 1989. Acoustically and oceanographically the area was expected to be invariant, except for slight depth excess variations induced by the weak Sargasso front over which the temperature near the surface was expected to vary by about 2°C.

A comparison of sound speed profiles obtained from the GDEM climatology, an FNOC prediction, and an in situ CTD indicated that climatological data bases may contain sound speed magnitude and more serious sound speed gradient errors at both deep and shallow depths.

The location of the Sargasso front is identified in Figure 4.0 by the ocean area between the 22.0° and 23.2° isotherms. The most significant acoustic effect of the oceanographic environment was the temperature—hence, sound speed—gradient change that took place across the front. Figure 5.0 shows the temperature difference between 200 m and 400 m in the area. Several transmission loss runs were made to determine the acoustic significance of the temperature gradient change across the front. A track (see Figure 5.0) through the most variable gradient region was selected. Another track (see Figure 5.0) was selected to represent the least variable gradient region of the front. Figure 6.0 presents the transmission loss curves along the two tracks using ASTRAL for a high-frequency source. The transmission loss for the high-variability case is about 3.75 dB larger than the low-variability case. The leading and trailing edges of the CZ at the 90-dB level are about 0.5 nmi larger for the high variability track.

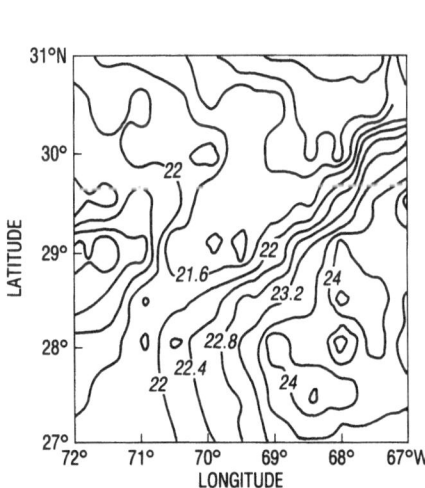

Figure 4.0. A Sargasso Sea exercise 75-m temperature contour derived from an AXBT sample of 1 and 2 August 1989.

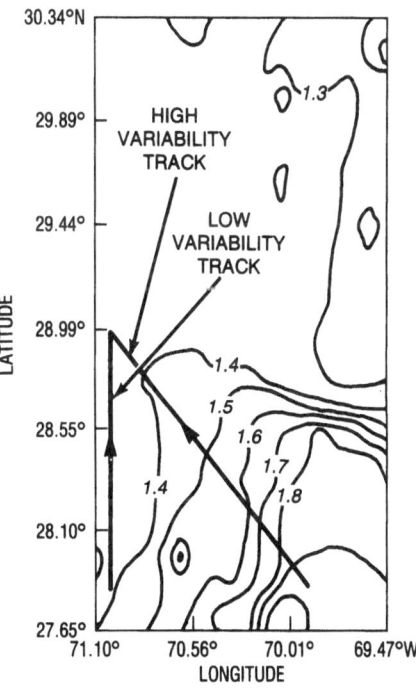

Figure 5.0. The temperature difference between 200 m and 400 m determined from raw data entering the Sargasso Sea exercise OI analysis of 22 August 1989. Temperature differences are given in degrees Celsius.

522

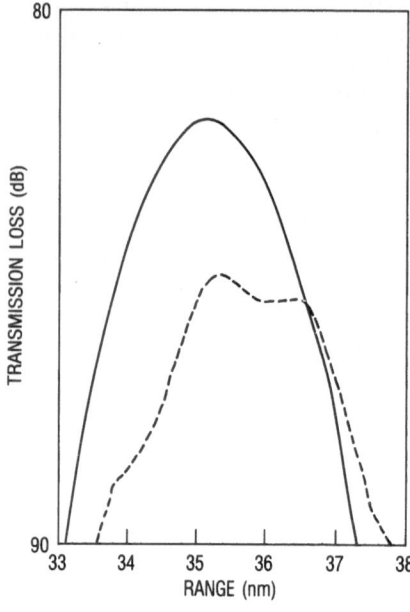

Figure 6.0. Transmission loss near the first CZ through the low temperature gradient region (—) and high temperature gradient region (---) for a high frequency source.

4. Gulf Stream Simulation

A preliminary version of the U.S. Navy's Tactical Environmental Support System's (TESS) Gulf Stream oceanographic feature model was obtained during 1987 for possible EATS System integration. Similar but more efficient capabilities available from NAVOCEANO will be implemented within the EATS System by the end of FY 1990.

The preliminary TESS ocean model was used in conjunction with the Harvard Quasi Geostrophic Gulf Stream Model to produce a realistic synoptic ocean environment for August 26, 1986. The Harvard model predicted temperature field, at a depth of 100 m, was used to determine the location of the Gulf Stream and associated eddies. This information was passed to the TESS Gulf Stream feature model, which created the final three-dimensional temperature field. An old version of the Integrated Carrier ASW Prediction System's water-mass-based climatological data base was used to supply salinity values for each TESS temperature profile.

The ASTRAL model was used to produce transmission loss along 180 radials emanating from various locations in the three-dimensional synoptic oceanographic field. All model runs discussed here used a source and receiver located at 100 m and a frequency of 500 Hz.

With the source in the warm-core eddy located near shore (location 1), the transmission loss pattern (Figure 8.0) indicates primarily bottom bounce propagation. The mode of propagation changes dramatically (Figure 9.0) to pure convergence zone propagation if the source is moved just outside this eddy (location 2). The reduction of depth excess caused by the presence of the warm-core eddy (the bottom depth was slightly greater at location 1) was sufficient to cause the change in propagation mode.

Figure 8.0 also indicates approximately 10-dB transmission loss differences between locations behind the warm-core eddy to the east of the source and adjacent regions, and south of the Gulf Stream. In an effort to determine the cause of these differences, Parabolic Equation (PE) model runs were made along the 96.0° radial (Figure 9.0), which ran through the warm-core eddy to the east of the source and along the 140.0° radial (Figure 10.0), which ran through the lower transmission loss

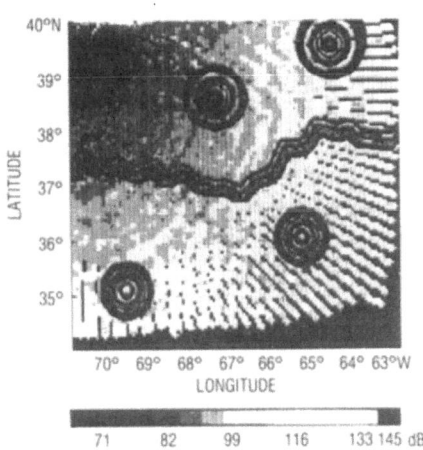

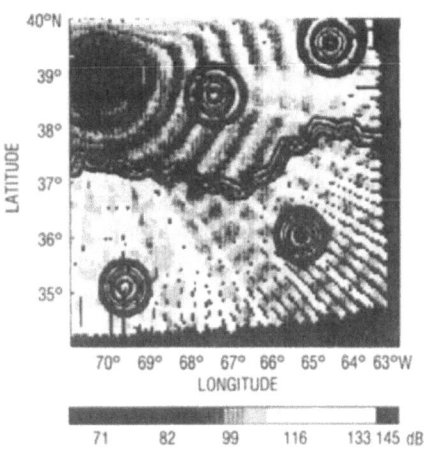

Figure 7.0. Transmission loss versus range at location 1 (center of a warm core eddy).

Figure 8.0. Transmission loss versus range at location 2 (just to the east of a warm core eddy).

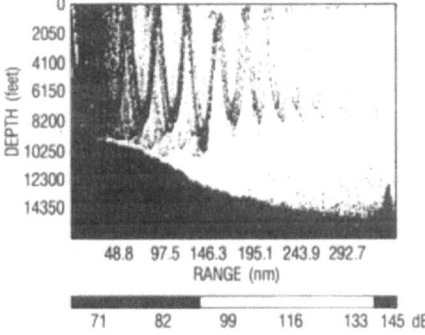

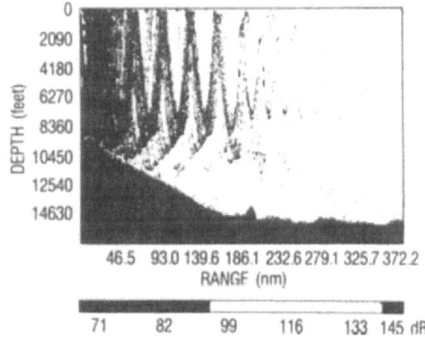

Figure 9.0. Transmission loss versus range at location 2 (Figure 8.0) along the 96.0° radial.

Figure 10.0. Transmission loss versus range from location 2 (Figure 8.0) along the 140.0° radial.

area between the two cold-core eddies located to the south of the Gulf Stream. Along the 140.0° radial (Figure 10.0), high sound energy reaches into the shallower part of the water column at all convergence zone locations. However, along the 96.0° radial (Figure 9.0) high energy at the fourth convergence zone does not approach the near surface. It is believed that the deeper turning depths at this convergence zone are due to the sound speed gradient change caused by the presence of the warm-core eddy. Along a 180° radial (Figure 8.0) the warm Gulf Stream front had the same effect. When the near-surface N-1 convergence zone is close to a warmer water-mass-induced sound speed gradient, which is larger near the surface, it appears that the turning depth of the near-surface and near-bottom Nth convergence zones deepen. One would expect that the degree of deepening would be proportional to the gradient through which the high sound energy travels. This probably explains the banded variation in transmission loss through the Gulf Stream. Here the Nth near-surface convergence zones both hit and miss the high gradient portion of the front. The high transmission loss bands appear to coincide with those paths where the Nth near-surface convergence zone coincides with the near-surface high gradient portion of the front, while the lower transmission loss areas appear to coincide with tracks where the Nth near-bottom convergence zone coincides with the lower gradient portion of the front.

At a location approximately midway between two warm-core eddies, an increase in the convergence zone range was indicated behind the eastern-most eddy. Similar sets of model runs made with the source located in and near the Gulf Stream revealed a decrease in convergence zone range and a decrease in transmission loss behind the cold-core eddy to the southeast of the source, respectively. The decrease in transmission loss and convergence zone range are believed to be due to the opposite effect of that encountered behind warm-core eddies. Here the colder water sound speed gradient forces the sound energy into shallower-than-normal depths, and the reduced sound speed magnitude decreases convergence zone distances.

5. Summary

The EATS System's oceanographic module has been used in numerous at sea exercises to rapidly obtain synoptic views of the water column sound speed structure. This information has, in turn, been used in conjunction with the EATS System's acoustic model module to relate the ocean's structure to acoustic propagation. As a result exercise platforms have been repositioned to sample oceanographic phenomena. A gain in acoustic advantage based on the ocean environment has also been demonstrated. Such an on-board capability has led to rapid dissemination of oceanographic acoustic information to the exercise scientific party both during the exercise and at concluding exercise briefs.

The EATS System modules use common direct-access files, whose structure is contained in one-record sequential files. Through the use of these common files, the EATS-produced synoptic environments and climatological data are passed to the acoustic models resident in the acoustic model module. All acoustic model information is first formatted for PE model input then reformatted, if required, for either ASTRAL or SNAP input.

In both the exercise and Gulf Stream simulation examples given, the dominant oceanographic acoustic factor affecting the propagation of sound in the sea has been the change in sound speed gradient induced by oceanographic phenomena. During the first exercise, held in the Norwegian Sea, a cold, near-surface, warm, sub-subsurface eddy caused significant range to convergence zone and transmission loss magnitude differences in the area due primarily to the change in sound speed gradient near the deep, warm core. During the second exercise, held in the Sargasso Sea, a slight variation in the sound speed gradient across a portion of the Sargasso front caused a correspondingly small variation in the range to the first convergence zone.

In the Gulf Stream simulation, convergence zones coincident with warm-core eddies and the front caused an increase in down-track convergence zone ranges and down-track deepening of the sound energy field. A corresponding increase also occurred in transmission loss near the surface. Behind cold-core eddies a decrease in convergence zone range and transmission loss can be attributed to the opposite effect of the colder water. The amount of time the high sound energy spends in the high oceanographic gradient appears to influence the magnitude of convergence zone distances and widths. In addition, in cold-water marginal depth excess areas, the presence of warm-core eddies can dramatically change the structure of sound energy propagation from bottom bounce to convergence zone modes of propagation.

6. References

Kerr, G. A. and B. R. Gomes (1989). *Environmental/Acoustic Results During the Critical Sea Test (CST) I Exercise Using the Environmental Acoustic Tactical Support (EATS) System.* Naval Ocean Research and Development Activity, Stennis Space Center, MS, NORDA TN 440, May.

Kerr, G. A. (1989). *Environmental Acoustic Resulting During the Critical Sea Test III Exercise Using the Environmental Acoustic Tactical Support System.* Naval Oceanographic and Atmospheric Research Laboratory, Stennis Space Center, MS, NOARL TN 1, December.

Kerr, G. A., G. P. Cloy, M. A. Hebert, and M. R. Wooten (1990). *Environmental Acoustic Tactical Support Oceanographic Data Module Users Guide.* Naval Oceanographic and Atmospheric Research Laboratory, Stennis Space Center, MS, NOARL TN 7, January.

This document was approved for public release; NOARL Contribution Number PR90:045:222.

ENVIRONMENTAL FOCUSING AND SOURCE LOCALIZATION IN THE OCEAN

M.D. Collins and W.A. Kuperman
Naval Research Laboratory
Washington, DC 20375
USA

ABSTRACT. Accurately determining the sound-speed distribution in large oceanographically-active regions of the ocean is very difficult. Since mismatch can severely degrade matched-field processing (MFP), this task might seem to be essential for MFP. However, it is possible to determine the sound speed and the source locations in a single process that we refer to as focalization. This approach should be much more efficient than attempting to determine the sound speed separately because the maximum requirement for source localization is information about the sound speed along the paths between the sources and the receiving array to the resolution appropriate for the frequencies of the sources. We present simulations that suggest that focalization might be feasible and that the sound-speed distribution that brings the sources into focus is not unique. We discuss the implementation of focalization on a large scale with methods such as simulated annealing and adiabatic normal modes.

1. INTRODUCTION

Propagation models are often used to estimate the position of acoustic sources in the ocean. The conventional wisdom in ocean acoustics is that accurate oceanographic information is required a priori for this inversion problem. In this paper, we demonstrate that it is possible to estimate source location with limited a priori oceanographic information.

Matched-field processing (MFP) has been an active area of investigation in underwater acoustics for more than ten years [1,2] (see [3] for an explanation of MFP). Matched-field processors are inverse methods that estimate some or all of the coordinates of acoustic sources by constructing forward solutions of the wave equation for many test source

527

J. Potter and A. Warn-Varnas (eds.), Ocean Variability & Acoustic Propagation, 527–538.

locations (the replica fields). An ambiguity function is defined in terms of the data and the replicas, and the peaks of the ambiguity function correspond to the estimates of the source locations. MFP can be computationally intensive in large regions of the ocean because it may be necessary to construct thousands of replicas.

The method of adiabatic normal modes [3] has been applied to develop a large-scale numerical laboratory for simulating MFP. The adiabatic normal mode model is very efficient for computing solutions of the wave equation and is usually very accurate in deep-water environments. By precomputing and storing the normal modes and eigenvalues for large regions of the ocean, the wide-area rapid acoustic prediction (WRAP) model can compute the Green's function for thousands of source and receiver locations very rapidly [4]. The WRAP model can handle three-dimensional MFP problems while simulating the noise field at the receiving array by computing the contributions of thousands of noise sources distributed near the ocean surface [5].

Standard MFP methods require knowledge of the sound-speed distribution and other acoustic parameters of the ocean. If this a priori information is not known accurately, a situation that is not unlikely in large oceanographically-active regions of the ocean, MFP can fail even if the signal-to-noise ratio is high. One approach to reducing this mismatch problem is tomography [6], which involves remote sensing of the sound-speed distribution in the ocean through extensive measurement and inversion computation. Since tomography probably determines more information than is required for MFP, one is motivated to consider other approaches for reducing the problem of mismatch.

An alternative approach is to bring the sources into focus by adjusting the acoustic parameters of the ocean. This process, which we refer to as focalization, has the primary goal of determining source location and possibly the secondary goal of determining sound speed. The sound speed is determined at most along the propagation paths between the sources and the array to the appropriate resolution for the frequency, source distribution, and array configuration involved. Focalization is more difficult than standard MFP methods because there may be a large number of parameters to vary (as opposed to two or three parameters for MFP).

An approach that should be useful for focalization is simulated annealing [7,8]. This Monte Carlo optimization technique, which has only recently been developed, is very efficient for finding the global minimum of a function of many variables. Simulated annealing is analogous to slowly cooling a pure liquid substance to form a perfect crystal (the lowest energy state of the system). A random perturbation is chosen at each iteration of the simulated annealing Markov process. The perturbation is accepted automatically as the new parameter state if the energy is reduced. To allow escape from local minima, the perturbation is accepted according to a Boltzmann probability distribution if the energy is increased.

The cooling process is controlled by the distribution of the perturbations and by an artificial parameter, the temperature, which is decreased slightly after each iteration. In

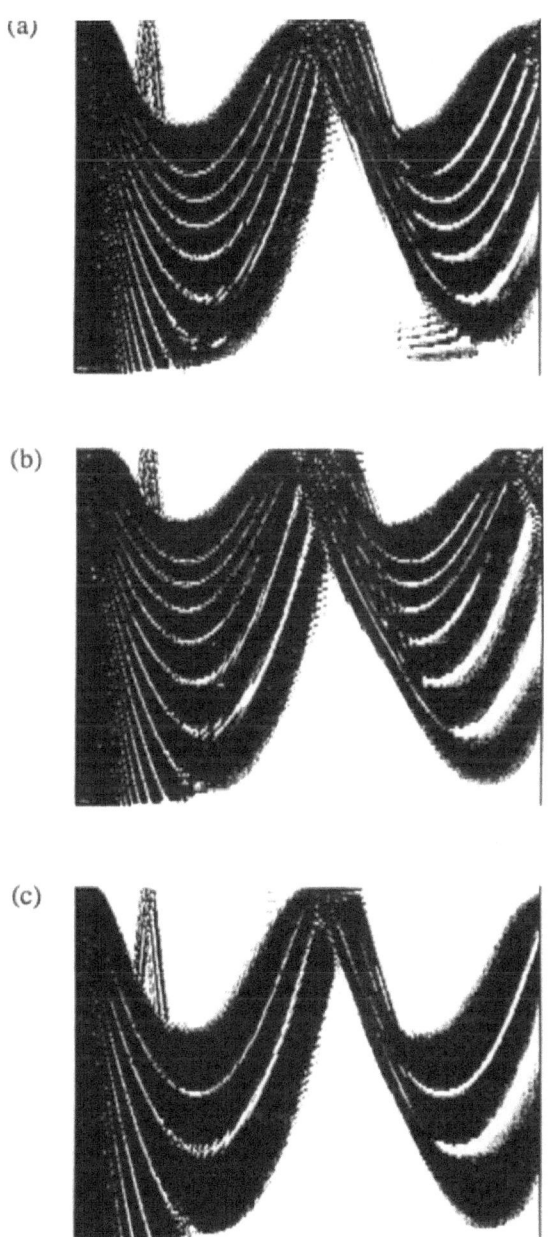

Figure 1. Transmission loss over 100 km for a 50 Hz point source in a deep ocean with Munk's sound-speed profile: (a) 1000 m channel depth, 400 m source depth; (b) 800 m channel depth, 400 m source depth; and (c) 1000 m channel depth, 200 m source depth. The loss levels range from 80 dB (black) to 100 dB (white).

general, simulated annealing converges if the temperature decreases inverse logarithmically [9]. The fast simulated annealing algorithm [10] converges with an inverse-linear cooling schedule if the perturbations are selected from a nonuniform distribution that occasionally allows large perturbations. In this paper, we apply a modified version of the fast simulated annealing algorithm described in [11] to perform a preliminary investigation into extending the WRAP model to perform focalization.

We apply an efficient ray-tracing method [12] to investigate the feasibility of focalization in the ocean. The acoustic field due to a point source near the ocean surface can be represented by tracing rays in the directions of the Lloyd's mirror beams. An efficient method for localization involves back propagating the beams received at an array. The rays focus near the ocean surface at the source range for the correct sound-speed distribution. If the sound speed is unknown, it is adjusted until the rays focus. We present a simulation involving uncertainties in Munk's sound-speed profile [13] to demonstrate that focalization can correctly determine both sound speed and source location in the deep ocean.

A wave-theoretic analog of this approach involves adiabatically back propagating the phase of each normal mode received at an array and adjusting the sound speed until the phases focus. We present a simulation in shallow water to demonstrate that wave-theoretic formulations of focalization can correctly determine source location and sound speed. Figure 1 contains acoustic point-source fields in deep oceans with Munk's sound-speed profile. We observe that the acoustic field is much more sensitive to the source location than to the sound-speed profile. This suggests that source location is more important than sound speed in focalization and that the sound-speed distribution that brings the phases into focus may not be unique. We demonstrate this with a simulation.

2. FEASIBILITY IN THE DEEP OCEAN

We work in cylindrical coordinates with the range r being the horizontal distance from the receiving array and the depth z being the distance below the ocean surface. For range-dependent problems in the deep ocean, accurate solutions of the wave equation require extensive computation times. Since three-dimensional MFP typically requires several thousand replica fields, an efficient method for solving the wave equation approximately would serve well in an attempt to study the nature of the problem of source localization in the presence of variable mesoscale features.

For high-frequency sources (>100 Hz) near the ocean surface, it is possible to save up to several orders of magnitude of computation time by using a qualitative solution obtained with ray tracing. The nearfield resembles the Lloyd's mirror beams due to a point source in an infinite half space. At long ranges, the Lloyd's mirror beams are bent by refraction and repeatedly bounce off the ocean surface at the convergence zones. This type of behavior is evident in the acoustic fields appearing in Figure 1. A qualitative representation of the field

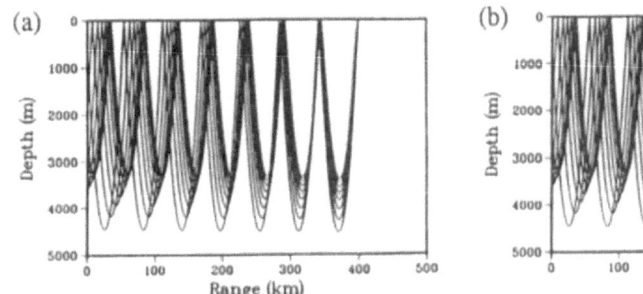
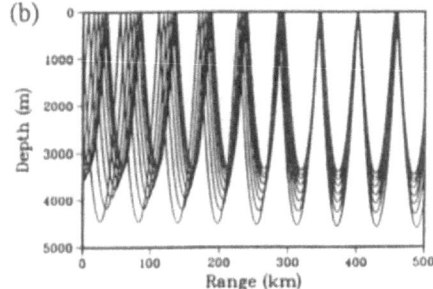

Figure 2. The rays of Example A that correspond to the Lloyd's mirror beams for a source at 400 km range from a vertical array: (a) the rays incident on the array; and (b) the back-propagated rays.

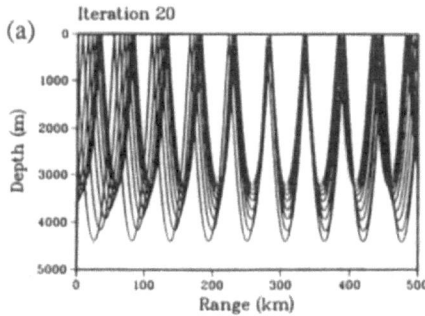
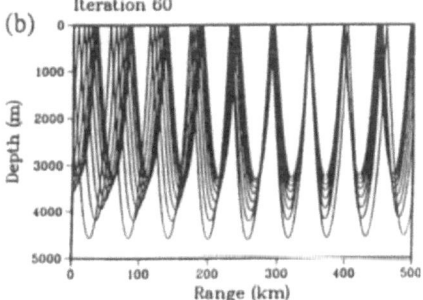
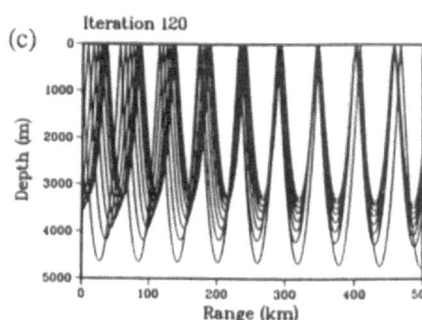
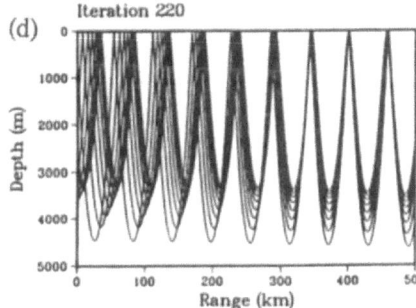

Figure 3. The back-propagated rays for the test sound-speed distributions of Example A: (a) the rays are far out of focus after 20 iterations; (b) and (c) the rays begin to focus after 60 and 120 iterations; and (d) the rays are in focus after 220 iterations.

532

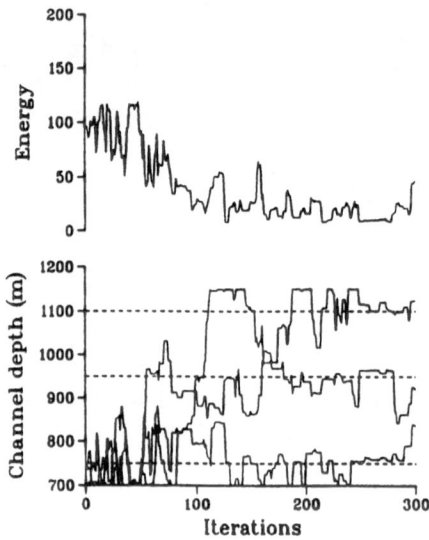

Figure 4. The energy and the channel depth estimates for Example A.

is obtained by tracing rays in the directions corresponding to the Lloyd's mirror beams from the point on the ocean surface directly above the source.

This ray representation of the field is used to estimate source location with the method of back propagation. The beams incident on a vertical array of hydrophones are represented by rays that originate at the centers of the beams and propagate in the directions of the beams. If the sound-speed distribution is known, the rays are traced away from the array, and the rays nearly focus at the ocean surface at the source range (the rays do not focus perfectly because of refraction near the ocean surface). The depth of the source is determined by the Lloyd's mirror beam pattern.

For the mismatch case, the sound-speed distribution is adjusted until the rays focus at a point near the ocean surface. With this approach, it is necessary to trace rays for each test sound-speed distribution. The cost function E is defined as

$$E = \min_r \sum_{m=1}^{M} z_m(r)^2 , \tag{1}$$

where the ray $z_m(r)$ corresponds to the m-th beam and M is the number of beams. The focalization process involves searching the sound-speed parameter space for the minimum of E. At each iteration of the simulated annealing process, the estimate of the source range is the range at which the sum in Eq. 1 is minimized.

We illustrate focalization in the deep ocean with Example A. The sound speed is defined in terms of the range-dependent Munk sound-speed profile for which the channel depth [13] varies linearly between the unknown values of 750 m at r = 0, 950 m at r = 250 km, and 1100 m at r = 500 km. The incident rays and the back-propagated rays, which correspond to the incident rays between the source at r = 400 km and the array, appear in Figure 2. The back-propagated rays appear in Figure 3 at various stages of the simulated annealing process. The rays are initially far out of focus. After 220 iterations of the simulated annealing algorithm, the rays are nearly focused at the ocean surface at the source range. The energy and the channel depth estimates appear in Figure 4. The Markov process converges to the correct sound speed parameters. This example demonstrates that focalization can determine source location and sound speed in the presence of variable mesoscale oceanographic features.

3. FEASIBILITY WITH WAVE MODELS

To investigate the possibility of performing focalization with adiabatic normal modes, we perform MFP in mode space [14]. We adjust the sound speed and back propagate the phase integrals,

$$\psi_n(r) = \int_0^r k_n(s)\,ds\,, \tag{2}$$

until they match the measured phases θ_n, where k_n is the n-th eigenvalue (see [3] for details of the adiabatic approximation). This approach is a generalization of the method of [15] to range-dependent problems. The cost function E is defined as

$$E = \min_r \sum_{n=2}^N \Delta[\theta_n - \theta_{n-1} - \psi_n(r) + \psi_{n-1}(r)]\,, \tag{3}$$

$$\Delta(x) \equiv \min([x]_{2\pi}, [-x]_{2\pi})\,, \tag{4}$$

where N is the number of modes used for focalization and $[x]_{2\pi} \equiv x(\text{modulo } 2\pi)$. At each iteration, the estimate for the source range is the range at which the minimum of the sum in Eq. 3 occurs.

To illustrate the performance of wave-theoretic focalization, we consider two examples involving a 50 Hz source in a shallow-water environment in which the sound speed in the water column is a linear function of depth at each range. The sound speed at the ocean bottom z = 500 m is taken to be 1500 m/s at all ranges. The sound speed at the ocean surface is taken to be piecewise linear and is defined in terms of the values at the nodes (the

534

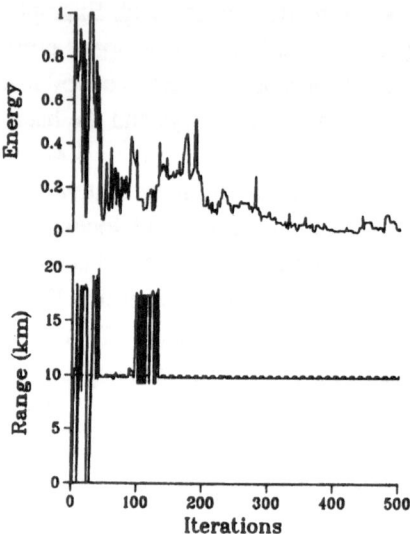

Figure 5. The energy and the source range estimates for Example B.

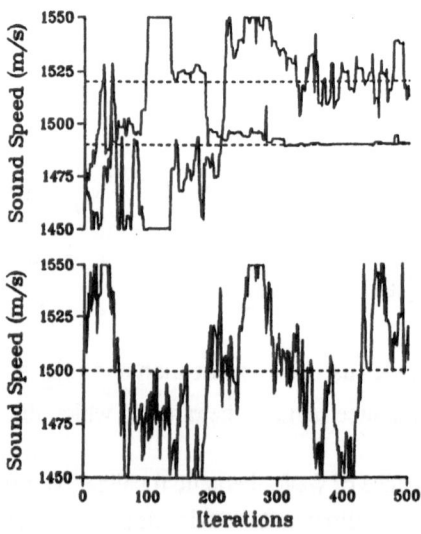

Figure 6. The sound speed estimates for Example B. The sound speed estimates converge for the points between the source and receiver (top) but not for the other point (bottom).

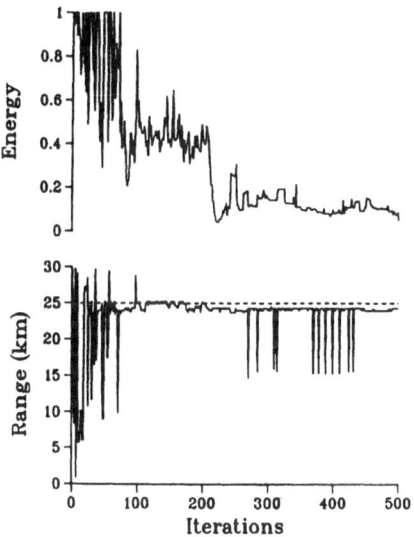

Figure 7. The energy and the source range estimates for Example C.

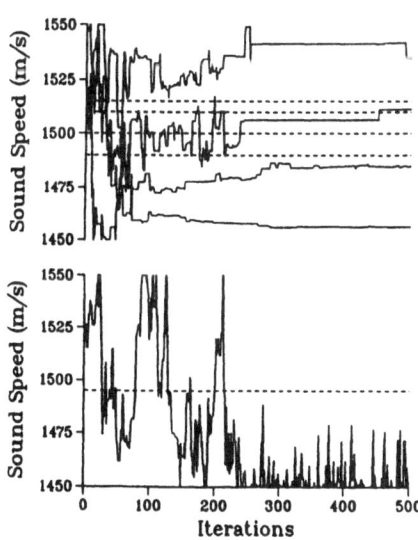

Figure 8. The sound speed estimates for Example C. Although the source is localized, the sound speed estimates do not converge.

endpoints of the linear segments). In the lossless sediment, the sound speed is 1700 m/s and the density is 1.5 g/cm³.

For Example B, the source is at r = 10 km, and the nodal sound-speed values are 1470 m/s at r = 0, 1490 m/s at r = 20/3 km, 1520 m/s at r = 40/3 km, and 1500 m/s at r = 20 km. The nodal value at the array location is assumed to be known. Only two of the three unknown nodal values lie between the source and the array. We perform focalization using the first eight trapped modes. The simulated annealing Markov processes appear in Figures 5 and 6.

The algorithm locks onto the correct source range after about 50 iterations. An ambiguity associated with a sidelobe near 17 km appears for a short time near 100 iterations. The two nodal values between the source and the array are recovered. The fact that the source range is recovered much earlier that the sound speed in the Markov process is consistent with the discussion in the Introduction relating to the results appearing in Figure 1. The third nodal value does not settle down. We conclude that combined tomography and source localization is possible with wave-theoretic focalization.

For Example C, the source is at 25 km, and the nodal sound-speed values are 1520 m/s at r = 0, 1515 m/s at r = 6 km, 1510 m/s at r = 12 km, 1500 m/s at r = 18 km, 1490 m/s at r = 24 km, and 1495 m/s at r = 30 km. The nodal value at the array location is assumed to be known. Only four of the five unknown nodal values lie between the source and the array. We perform focalization using the first twelve trapped modes. The simulated annealing Markov processes appear in Figures 7 and 8.

The algorithm locks onto the correct source range after about 100 iterations and remains within about 1 km of the correct source range with the exception of several encounters with a sidelobe near 15 km. The fact that the algorithm does not recover the correct sound speed is consistent with the discussion in the Introduction relating to the results appearing in Figure 1. We conclude that focalization can be successful (i.e. determine the correct source location) even when the algorithm does not converge to the correct ocean parameters.

4. LARGE-SCALE IMPLEMENTATION

In this section, we discuss approaches that might be useful for implementing focalization into the WRAP model for multiple source location with adiabatic normal modes in a noisy ocean. Some of the essential tools for this problem might include: (a) an efficient parameterization of the environment; (b) an efficient algorithm for searching the parameter space; (c) an efficient model for constructing replica fields; (d) an effective cost function; (e) an oceanography model; and (f) an ambient noise model.

Item (a) is very important because both ambiguity and computational expense increase with the dimension of the parameter space. Empirical orthogonal functions (EOF's) [16] seem to be a good candidate for this role. These functions, which are derived from representative data, are known to be very efficient for approximating data. For the focalization problem, EOF's might be useful for representing the sound speed as a function of depth or the eigenvalues as a function of index. For each of these applications, the coefficients of the EOF's would depend on the two horizontal position coordinates.

Although there might be more efficient methods for the relatively low-dimensional examples we considered, simulated annealing is an attractive choice for Item (b). Based on our experience with the WRAP model, we consider adiabatic normal modes as the most sensible choice for Item (c). Item (d) is perhaps the central issue of conventional frequency-domain beamforming. Deciding on a cost function usually involves compromises between resolution and stability. A frequency-domain beamformer has been developed that seems to achieve both of these to a high degree [17].

Item (e) might be useful for constructing representative sound speed profiles from which one could construct EOF's. Furthermore, it might be useful to implement focalization and oceanography as a single process with some of the oceanographic parameters as unknowns. Item (f) would be useful for problems involving correlated noise. With some knowledge of the noise correlation at the array, it would be possible to treat the correlated component as signal and effectively increase the signal-to-noise ratio.

5. CONCLUSION

Focalization appears to be an efficient method for dealing with the problem of mismatch. Depending on the amount of data available and the number of unknown parameters, successful focalization might or might not determine the correct sound-speed parameters. If the ultimate goal is to locate the sources rather than to determine the sound speed, however, ambiguities in the sound speed are tolerable. The presence of distributed multiple sources might actually help focalization by improving the uniqueness of the sound-speed distribution. Two of the most important tools for implementing large-scale focalization appear to be simulated annealing and adiabatic normal modes.

[1] H.P. Bucker, "Use of calculated fields and matched-field detection to locate sound sources in shallow water," J. Acoust. Soc. Am. 59, 368-373 (1976).

[2] A.B. Baggeroer, W.A. Kuperman, and H. Schmidt, "Matched field processing: Source localization in correlated noise as an optimum parameter estimation problem," J. Acoust. Soc. Am. 83, 571-587 (1988).

[3] W.A. Kuperman and J.S. Perkins, "Three-dimensional oceanography and acoustics", this proceedings.

[4] W.A. Kuperman, M.B. Porter, J.S. Perkins, and R.B. Evans, "Rapid computation of acoustic fields in three-dimensional ocean environments," submitted to J. Acoust. Soc. Am.

[5] W.A. Kuperman and F. Ingenito, "Spatial correlation of surface generated noise in a stratified ocean," J. Acoust. Soc. Am. 67, 1988-1996 (1980).

[6] W.H. Munk, "Ocean acoustic tomography: A scheme for large scale monitoring," Deep-Sea Res. 55, 220-226 (1979).

[7] N. Metropolis, A.W. Rosenbluth, M.N. Rosenbluth, A.H. Teller, and E. Teller, "Equations of state calculations by fast computing machines," J. Chem. Phys. 21, 1087-1091 (1953).

[8] S. Kirkpatrick, C.D. Gellatt, and M.P. Vecchi, "Optimization by simulated annealing", IBM Thomas J. Watson Research Center, Yorktown Heights, New York (1982).

[9] S. Geman and D. Geman, "Stochastic Relaxation, Gibbs Distributions, and the Bayesian Restoration of Images," IEEE Trans. Patt. Anal. Mach. Intel. 6, 721-741 (1984).

[10] H. Szu and R. Hartley, "Fast Simulated Annealing," Phys. Let. 122, 157-162 (1987).

[11] W.A. Kuperman, M.D. Collins, J.S. Perkins, and N.R. Davis, "Optimal time-domain beamforming with simulated annealing including application of a priori information", accepted for publication in J. Acoust. Soc. Am.

[12] M.D. Collins, "A nearfield asymptotic analysis for underwater acoustics," J. Acoust. Soc. Am. 85, 1107-1114 (1989).

[13] W.H. Munk, "Sound channel in an exponentially stratified ocean with applications to SOFAR", J. Acoust. Soc. Am. 55, 220-226 (1974).

[14] T.C. Yang, "A method of range and depth estimation by modal decomposition," J. Acoust. Soc. Am. 82, 1736-1745 (1987).

[15] E.C. Shang, "An efficient high-resolution method for source localization processing in mode space," J. Acoust. Soc. Am. 86, 1960-1964 (1989).

[16] R.E. Davis, "Predictability of sea surface temperature and sea level pressure anomalies over the North Pacific Ocean," J. Phys. Ocean. 6, 249-266 (1976).

[17] H. Schmidt, A.B. Baggeroer, W.A. Kuperman, and E.K. Scheer, "Environmentally tolerant beamforming for high resolution matched field processing: Deterministic mismatch," accepted for publication in J. Acoust. Soc. Am.

REFRACTION OF ACOUSTIC MODES IN VERY LONG-RANGE TRANSMISSIONS

WALTER H. MUNK
Scripps Institution of Oceanography
University of California, San Diego
La Jolla, CA 92093

ABSTRACT. We consider the role of wave refraction in long-range acoustic transmissions. Refraction by a variable depth of the sea floor and a variable depth of the sound axis lead to distinctly different results.

Introduction

An experiment has been proposed (Munk & Forbes, 1989) to study acoustic transmission over very long ranges (order 10^4 km). The purpose is to test the feasibility of measuring global ocean warming by acoustic transmissions. Atmosphere-ocean models are consistent with a warming at the sound channel of order 0.005°C per year *at the present time*. The month-to-month mesoscale variability at the sound channel is of order 1°C. It would take several centuries to detect the warming trend over and above mesoscale noise from local measurements.

However the situation is greatly improved by taking advantage of the efficient transmission of low-frequency sound to obtain spacial averages of temperature. The coherence scale of mesoscale variability is 100 km, giving $n - 100$ independent samples of mesoscale noise in a 10^4 km transmission. Travel time perturbations are proportional to temperature perturbation averaged over the entire transmission paths and are thus reduced by a factor of $n^{-\frac{1}{2}}$ relative to a uniform signal. It should then be possible to detect a global warming trend in decades rather than centuries. More precise estimates indicate that detection should be possible on a 95% probability level from *one* decade of measurements.

There are many problems with this estimate. The inherent ocean variability on a larger than meso-scale is probably the limiting factor. Further, "global" warming is by no means uniform, but can be expected to be structured on a gyre scale.

This is not the place to discuss these problems in any detail. Rather we wish to emphasize certain acoustic properties of ocean transmissions which can be expected to be crucial in a 10^4 km ranges, but have not been studied since they are of much less importance at shorter ranges.

Modal Phase Velocity

Here we consider *adiabatic* changes in the phase velocity of acoustic modes, that is, changes sufficiently gradual so that the scattering of energy from mode to mode and from frequency to frequency can be ignored. (In ray language this is equivalent to preserving *action*.) Changes in phase velocity lead to horizontal *refraction* of the ray paths.

J. Potter and A. Warn-Varnas (eds.), Ocean Variability & Acoustic Propagation, 539–543.
© 1991 *Kluwer Academic Publishers.*

540

 Figure 1 illustrates two models giving rise to wave refraction: (i) a coastal wedge with rising sea bottom but a fixed sound axis; and (ii) a situation found at high (northern and southern) latitudes with a rising sound axis but fixed sea bottom. These models are illustrative of a wide class of problems. Transmission across any major ocean current system is associated with a marked transition in axial depth (but not necessarily an outcropping of the axis). Upwelling in shallow water gives rise to a change in axial depth over a shoaling bottom, thus combining the two models.

The two models can be treated with the same formalism. Dashen and Munk (in preparation) have treated a shoaling bottom for a straight coast, Munk and Zachariasen (in preparation) the shoaling bottom around circular islands, and seamounts and Munk (in preparation) has treated the case of the axial wedge. For analytical convenience the sound profile $C(z)$ was represented by a bi-linear $S^2(z)$ profile ($S = 1/C$ is the sound slowness) with an unrealistic kink at the axis. Solutions are then in the form of Airy functions.

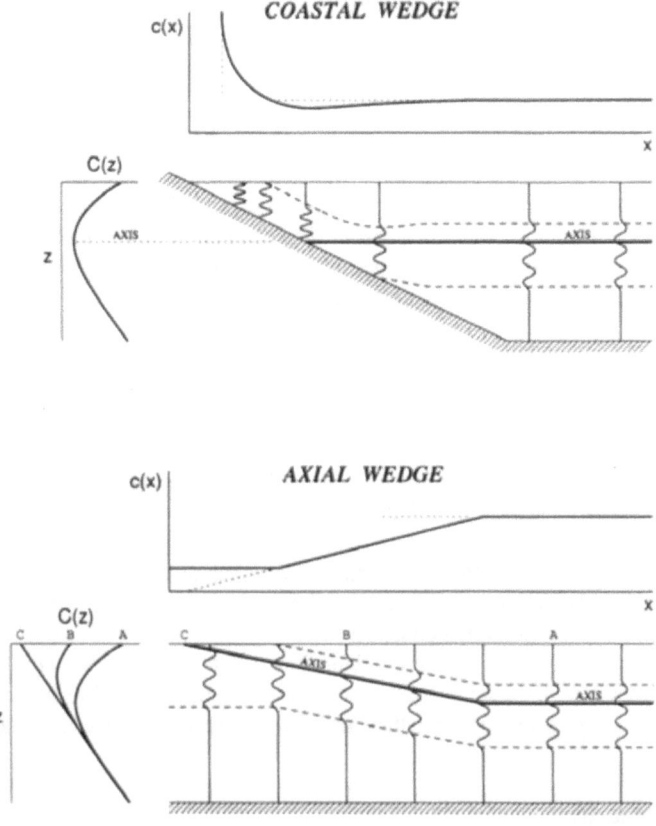

Figure 1. Schematic representation of the mode function for a variable depth of bottom (upper panel) and variable depth of sound axis (lower panel). The acoustic energy is essentially confined to the depth interval between the WKBJ turning points (dashed lines).

Even though the formalism in the two cases is very similar, the resultant phase velocities $c(x)$ are very different. In the case of a coastal wedge the phase velocity reaches a slight minimum where the lower turning point intersects the sloping bottom (see Fig. 1), then rises sharply with decreasing depth, and becomes infinite at a point where the "squeezed" vertical wave function has a vertical wave length equal to the scalar wave length of the transmitted sound. In ray language, the minimum corresponds to the transition from a purely refracted R/R path to a bottom reflected R/BR path. The shoreward traveling ray is steepened at each successive collision with the sloping bottom until it becomes vertical at the depth of infinite phase velocity.

For the axial wedge the phase velocity decreases monotonically with decreasing axial depth, as represented schematically by the three straight lines in Fig. 1. The offshore kink occurs where the axis starts to shoal. The nearshore kink occurs where the upper turning point hits the surface, or, in ray language, where the R/R transmission changes to R/SR.

Coastal Refraction

From our experience with surface and internal waves, we are accustomed to see phase velocity diminish with decreasing depth. Call this the "normal" situation. For the coastal wedge the situation is then normal up to the point of minimum phase velocity. In the x, y-plane the wave rays are deflected towards shore, as if the incoming acoustic modes were *attracted* by the island. Shoreward of the minimum point the horizontal rays are *repelled* by the coast. This is illustrated in Figure 2 for the case of circular islands. Waves passing at a great distance from the island are slightly turned towards the island. Waves on a course too near the island center are turned sharply, they are refractively repelled. I believe this is a more likely mechanism to account for the observed reflections from distant islands of explosively generated sound waves than some kind of specular reflections from steep underwater slopes.

The Axial Wedge

Figure 3 shows the case of an axial wedge associated with the antarctic Circumpolar Current. This is the "normal" situation, with rays attracted towards the South Pole. The effect is quite dramatic. Rays that would have had a clear geodesic path from Perth to Bermuda in the absence of lateral refraction now collide with Brazil. Rays that barely miss the Cape of Good Hope do not make it to Bermuda. The situation has been discussed by Munk, O'Reilly and Reid (1988) in connection with an acoustic transmission experiment from Perth, Australia to Bermuda.

Chromatic Aberration

The dependence of phase velocity on depth differs from mode to mode, and is also a function of wave frequency. Refractive bending in an axial wedge is most pronounced for low modes of high frequency. Figure 3 illustrates the extent to which the refracted paths might differ, from the axially refracted low modes of high frequency, to the unrefracted geodesics for high modes of low frequency. The interpretation of long-range transmissions calls for a separation of incoming modes and frequencies, since these have traveled along quite distinct ocean paths. Separation can be achieved with a vertical array.

542

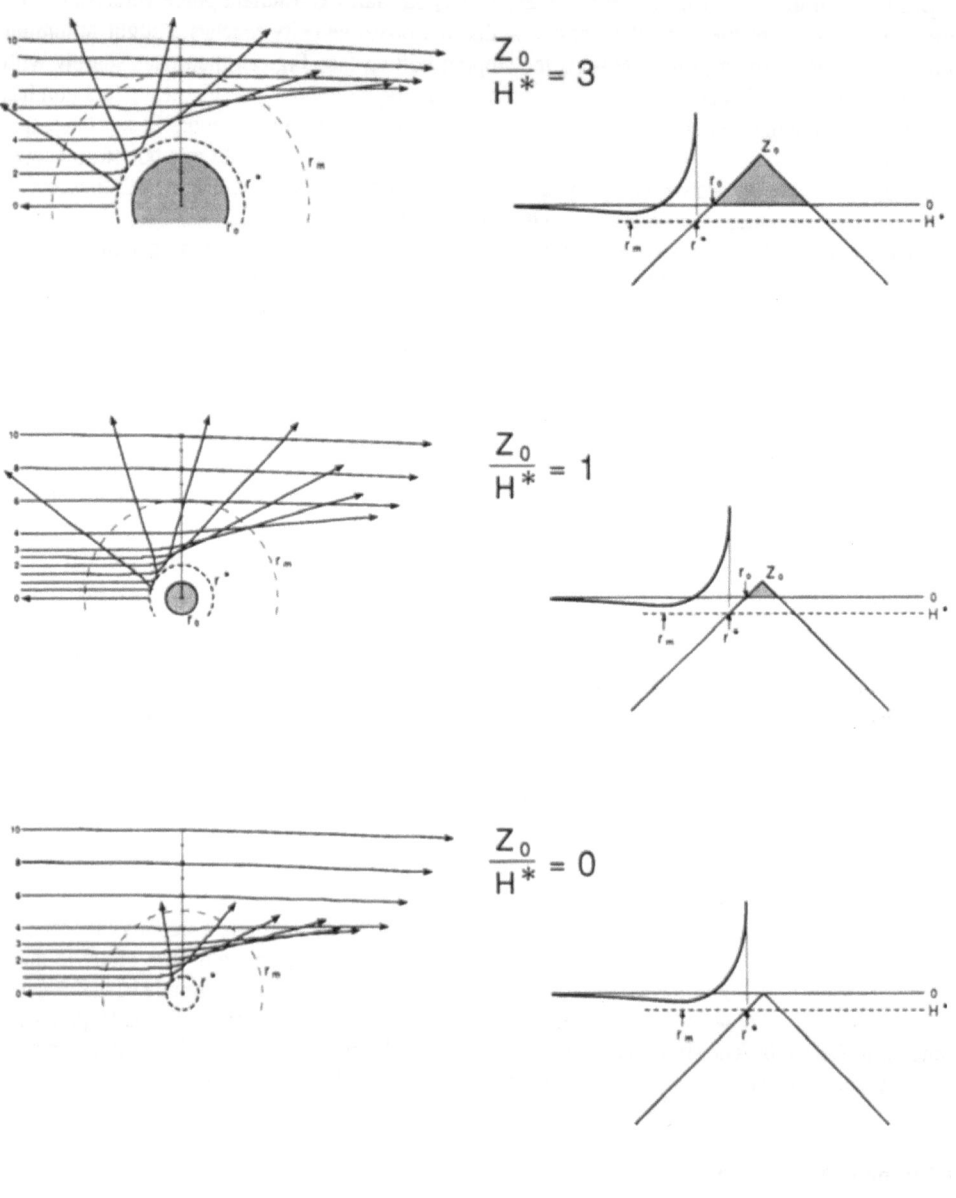

Figure 2. Refraction of sound waves by circular islands. Rays passing at a distance are refracted towards the island; rays moving towards the vicinity of the island center are repelled. Left figures give the top views, right figures show sections through the island center with the phase velocity profile indicated. The three panels are drawn for different island diameters (from Munk and Zachariasen, in preparation).

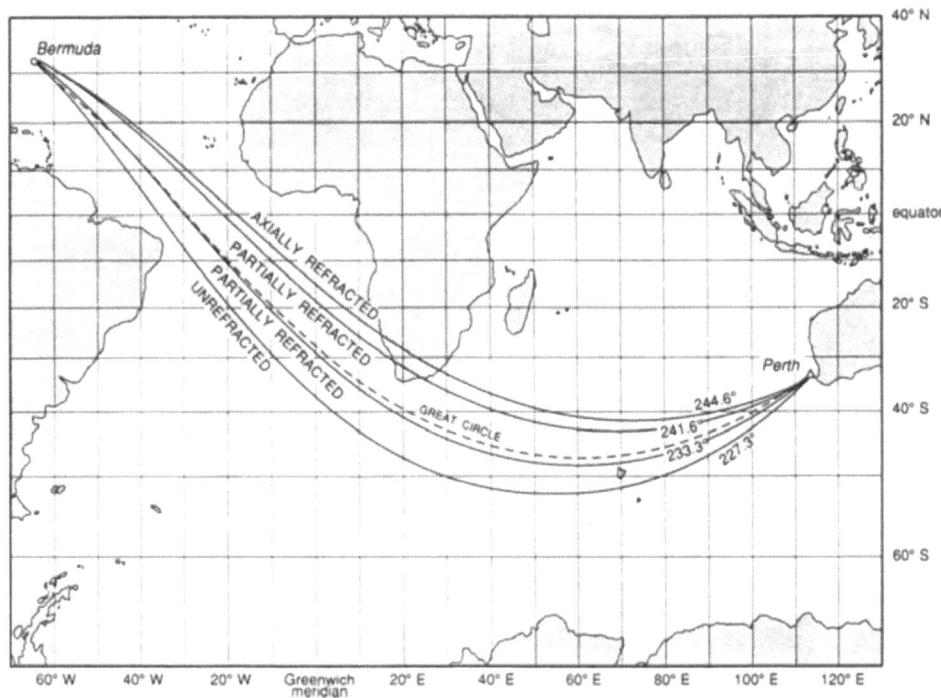

Figure 3. Refracted rays from Perth to Bermuda. Rays are strongly refracted in the antarctic Circumpolar Current associated with a shoaling of the sound axis from 40° south to 60° south. The southernmost rays correspond to high modes of low frequency; the northernmost rays to low modes of high frequency. The latter intersect the Cape of Good Hope and do not reach Bermuda.

Acknowledgement: I am indebted to the Office of Naval research for having supported our work on long-range acoustic transmissions.

References

Munk, W. H., W. C. O'Reilly and J. L. Reid (1988). Australia-Bermuda sound transmission experiment (1960) revisited. *J. Phys. Oceanogr.*, **18**:1876–1898.

Munk, W. H. and A. M. G. Forbes (1989). Global Ocean Warming: An Acoustic Measure? *J. Phys. Oceanogr.*, **19**:1765–1778

Dashen, R. and W. H. Munk. A model of ocean noise. *J. Acoust. Soc. Am.* (In preparation 1990).

Munk, W. H. and F. Zachariasen. Refraction of Sound by Islands and Seamounts. *J. Phys. Oceanogr.* (In preparation 1990).

Munk, W. H. Refraction of Sound Waves at Polar Latitude. (In preparation 1990).

ENVIRONMENTAL SENSITIVITY STUDIES WITH AN INTERFACED OCEAN - ACOUSTICS SYSTEM

Allan R. Robinson and Scott M. Glenn
Harvard University
Cambridge, MA 02138

William L. Siegmann
Rensselaer Polytechnic Institute
Troy, NY 12180

Ding Lee and George Botseas
Naval Underwater Systems Center
New London, CT 06320

ABSTRACT. An examination of environmental sensitivities of acoustic propagation loss predictions is performed using a coupled system of ocean and acoustic models. Three-dimensional time series of environmental fields are obtained from output of the Harvard Open Ocean Model and are interfaced with acoustic propagation models FOR3D and IFD. Sensitivity of the propagation loss to ocean model horizontal and vertical resolution is investigated, along with sensitivity to methodology for determining sound speed fields from the dynamical forecast model. Resolution recommendations and research issues are formulated based on preliminary propagation loss comparisons for acoustic frequencies up to 100 Hz and ranges up to 200 km.

1. Introduction

Coupled environmental and acoustic prediction systems have been described by Robinson (1987), Botseas et al. (1989), Lee et al. (1989), Chiu and Ehret (1990), Mellberg et al. (1990), Newhall et al. (1990), and Siegmann et al. (1990). In these systems, three-dimensional (3-D) time series of environmental fields have been generated from output of the Harvard Open Ocean Model, and most often interfaced with finite-difference parabolic approximation acoustic models FOR3D or IFD. Recent improvements to the environmental fields have been made in order to increase their accuracy and resolution for acoustics applications. Coupled with the accuracy and capabilities of the propagation models, the fields offer new and exciting opportunities for examinations of two types of environmental sensitivities of acoustic predictions. The first is resolution sensitivity, describing behavior of acoustic predictions to horizontal and vertical resolution of the dynamical forecast model and the coupling interface. The second is feature sensitivity, indicating the acoustic variability due to the occurrence and location of mesoscale features in the forecast domain and the methodology for determining environmental fields from the dynamical forecast model. This study concentrates on a systematic evaluation of acoustic propagation loss sensitivities, leading to preliminary resolution recommendations and formulation of significant research issues.

J. Potter and A. Warn-Varnas (eds.), Ocean Variability & Acoustic Propagation, 545–559.
© 1991 *Kluwer Academic Publishers.*

546

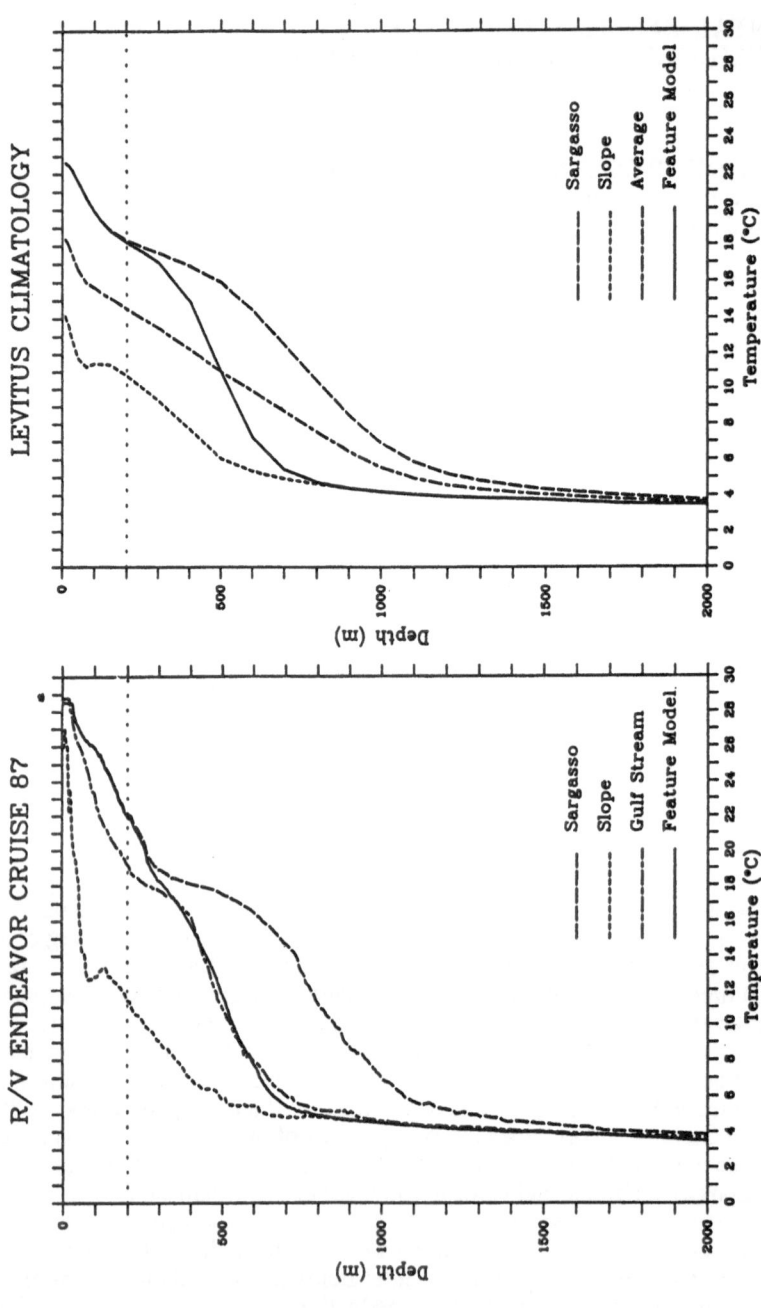

Figure 1. Temperature profiles to construct 3–D environmental fields from Harvard Open Ocean Model forecasts. Left: case 1, typical Sargasso, Slope, Gulf Stream profiles measured by R/V Endeavor. Gulf Stream profile (solid curve) obtained using feature model to interpolate measured Sargasso and Slope profiles. (Feature model does not account for surface boundary layer effects above dotted line.) Right: case 2, Sargasso and Slope profiles from Levitus climatology. Gulf Stream profiles obtained by averaging Sargasso and Slope profiles (dash-dot curve), and by using feature model to interpolate between Sargasso and Slope profiles (solid curve).

2. Environmental Prediction System

The environmental prediction system is the Harvard University Gulfcast (Robinson et al. 1989a), an implementation of the Ocean Descriptive Predictive System (Robinson and Leslie 1985) that uses dynamical and statistical models, an ocean observation network, and a data assimilation scheme to generate optimal forecast of the 3-D environmental fields. Initial forecasts are updated with new data to generate new estimates of the Gulf Stream and ring feature locations. Gulf Stream and ring feature models (Robinson et al. 1988) placed in the new locations are used to define the initial velocity structure for the dynamical forecast model, the Quasi-Geostrophic Harvard Open Ocean Model (Robinson and Walstad 1987). The forecast model is routinely used to generate 7-day forecasts in real time (Glenn et al. 1987), and has been shown capable of hindcasting synoptic dynamical events such as ring formations or absorptions over 2-3 week time periods (Robinson et al. 1988, 1989b).

2.1 ENVIRONMENTAL FIELDS

The time series of 3-D environmental fields (temperature, salinity, density, sound speed, etc.) are generated from the dynamical model forecasts by the methods presented in Glenn and Robinson (1990), which are briefly reviewed here. The dynamical model generates daily forecasts of the 3-D streamfunction field. At any model level, the streamfunction has a sharp gradient across the Gulf Stream, but is approximately constant in the Slope water to the north and in the Sargasso Sea to the south. Characteristic Slope, Sargasso, and Gulf Stream axis temperature and salinity values at each internal model level (plus top and bottom) are assigned to characteristic values of the Slope, Sargasso, and Gulf Stream axis streamfunction. Intermediate streamfunction values were assigned temperatures and salinities by interpolation between the characteristic values. The methodology has been shown to reproduce the observed frontal structure of the Gulf Stream with a minimum number of input temperature and salinity profiles (Glenn and Robinson 1990). The temperature and salinity fields then are used to calculate the water density and sound speed fields using the standard relations found in Fofonoff and Millard (1983).

This coupling sensitivity study involves two cases that use the 7-day dynamical forecast running from 6-13 May 1987. For case 1, the environmental fields were generated from the forecast streamfunction using typical summer temperature and salinity profiles as measured by the R/V Endeavor on Cruise 87. Plots of the R/V Endeavor Slope, Sargasso, and Gulf Stream temperature profiles are found in Fig. 1. Also shown is a plot of a synthetic Gulf Stream temperature profile obtained using a feature model to interpolate between the Slope and Sargasso profiles. The feature model adequately reproduces the characteristics of the temperature profile in the deep thermocline, but does not include surface boundary layer effects which are usually confined to the upper 200 m.

For case 2, the environmental fields were generated using the Levitus climatology (NOAA, 1982) for the Slope and Sargasso, also shown in Fig. 1. Since the Gulf Stream axis temperature profile cannot be recovered from the climatology, case 2A uses a simple average and case 2B uses the Gulf Stream feature model to obtain synthetic Gulf Stream temperature profiles from the Levitus Slope and Sargasso.

3. Acoustic Prediction System

An acoustic prediction model which permits treatment of 3-D propagation effects is FOR3D, an implementation of a wide vertical angle parabolic approximation with narrow

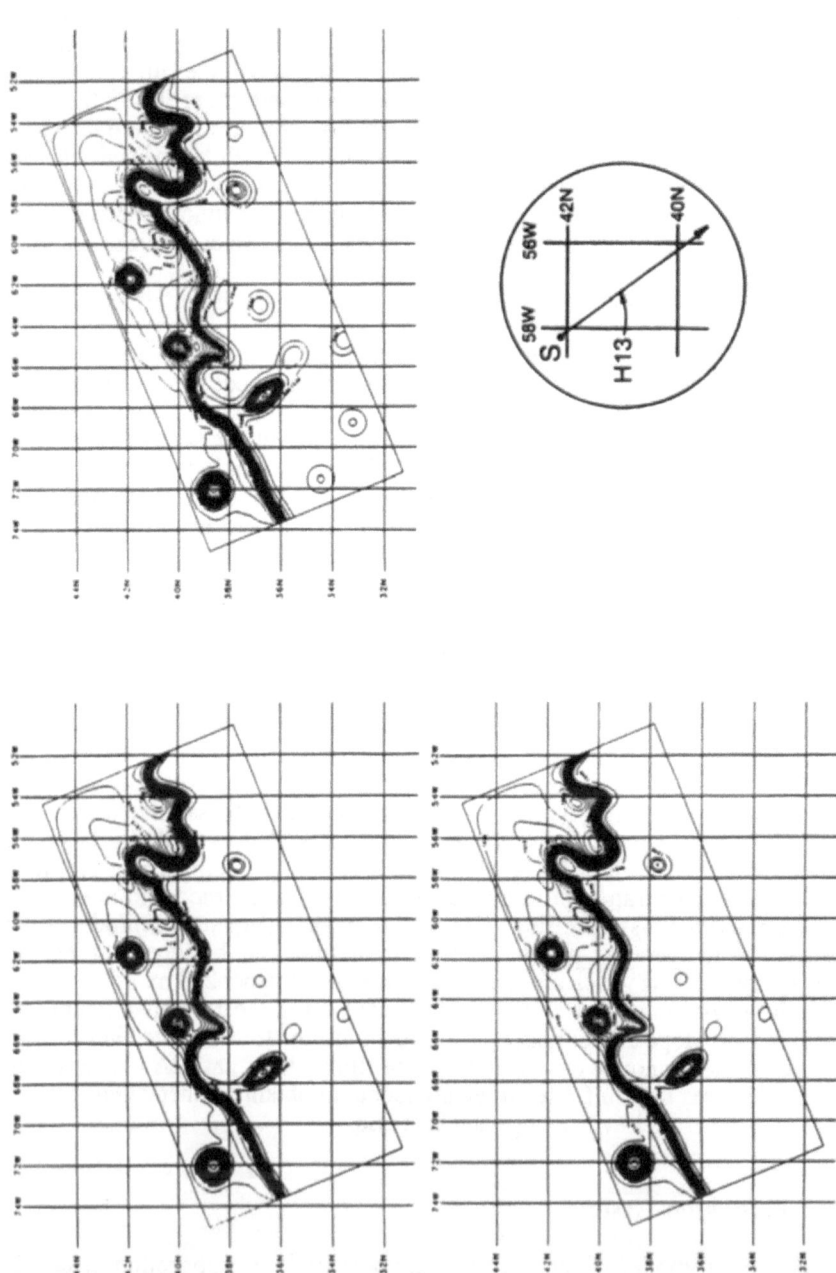

Figure 2. Level curves of sound speed from Harvard University Gulfcast, 13 May 1987 (forecast day 7). Contour interval 2 m/s. Upper left: case 1A (9 internal levels, 15 km horizontal resolution), depth 50 m. Lower left: case 1B (10 km horizontal resolution), depth 50 m. Upper right: case 1C (12 internal levels), depth 100 m. Lower right: location of propagation path H13.

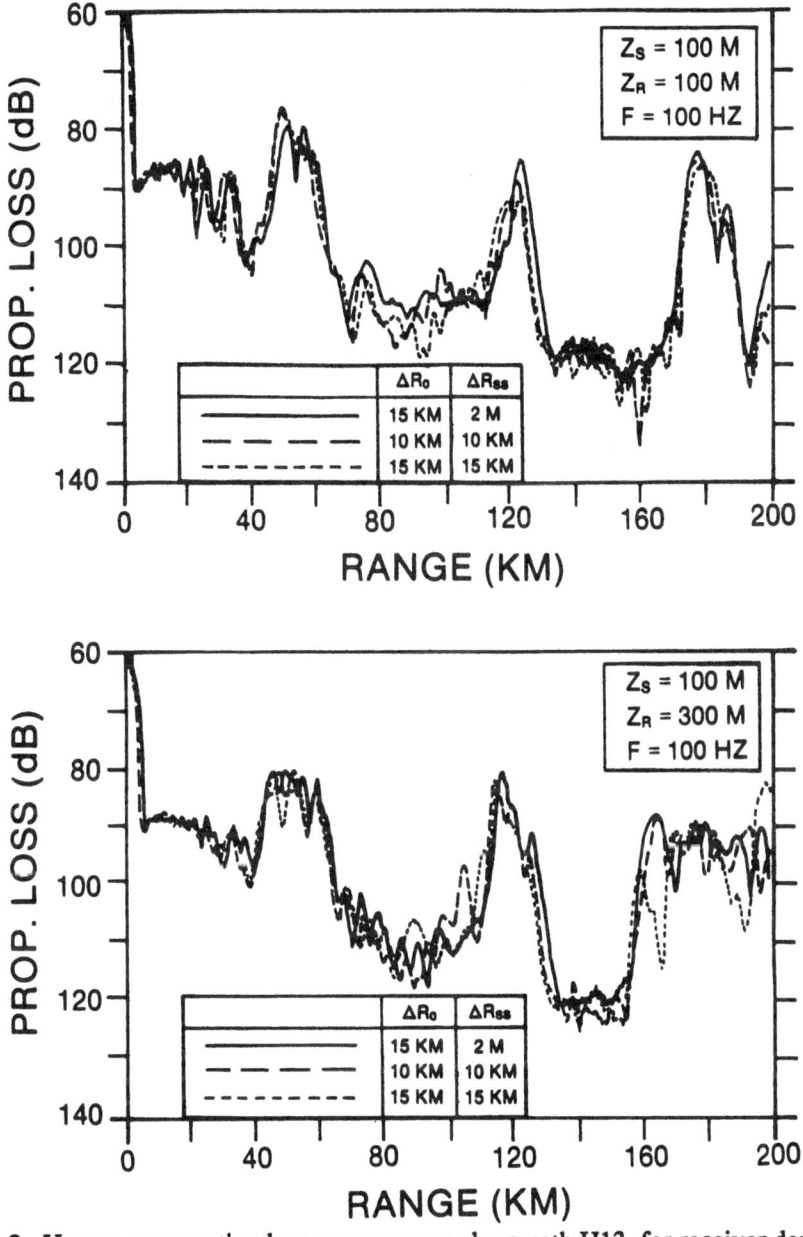

Figure 3. Upper: propagation loss versus range along path H13, for receiver depth 100 m, source depth 100 m, frequency 100 Hz. Lower: same, receiver depth 300 m. Solid and short-dashed curves from case 1A, long-dashed curves from case 1B. Solid curves are accurate (interpolated sound speed profiles accepted into acoustic model every $\Delta R_{SS} = 2$ m). Prop. loss differences: insignificant for ΔR_{SS} less than several km; only visible for larger ΔR_{SS} (dashed curves).

angle cross-range coupling fully incorporated (Lee et al. 1988, 1990b, Botseas et al. 1987). This algorithm is designed to maintain high accuracy and unconditional numerical stability, in addition to attaining capabilities for treating complex acoustic environments including bottom layering, density variations, and cross-range variations. Accuracy tests and verifications of its capabilities have been performed (Siegmann et al. 1988, 1990b, Lee et al. 1990a). The algorithm is a successor to the 2-D parabolic approximation propagation model IFD (Lee and Botseas, 1982) and its wide angle extension, for which performance characteristics are well documented (Lee and McDaniel 1988).

Couplings between the Harvard Open Ocean Model and both of the models FOR3D and IFD have been performed, and descriptions of the necessary interpolations and linkages have been provided (Botseas et al. 1989, Siegmann et al. 1990).

3.1 ACOUSTIC PROPAGATION SCENARIOS

All the propagation loss calculations to be described have an acoustic source positioned at latitude 42.09 N, longitude 58.20 W at depth 100 m, on May 13 of the aforementioned dynamical forecast. The propagation direction of interest has true bearing 145 deg, and this path is referred to as H13 and sketched in Fig. 2. When 3-D calculations were performed, path H13 was the center line of the wedge domain. This region of the ocean has generally flat bottom topography but considerable mesoscale feature activity, as is apparent from the sound speed contours in Fig. 2. Other examples of propagation paths have been and are being investigated.

Particular features of the bottom model were selected as in Siegmann et al. (1990), where other features and parameter settings necessary for generation of the acoustic computations are specified. We remark that to ensure accuracy, small computational range steps of 2 m were usually employed, as well as vertical steps of 5 m (at 25 Hz) and 2 m (at 100 Hz). Smaller step sizes were used for convergence and accuracy checks. An averaging window length of 2 km is used in plotting propagation loss values.

4. Sensitivity Results

Case 1 environmental fields, as described in Sec. 2.1, were generated in three versions for the sensitivity studies described next. The baseline case 1A uses 9 vertical levels in the ocean volume (plus surface and bottom) and 15 km horizontal grid spacing. Case 1B differs in having 10 km horizontal grid spacing, and case 1C differs from the baseline in having 12 vertical levels. Horizontal contours of the sound speed are shown for these three cases in Fig. 2. Figure 3 shows representative propagation loss curves from cases 1A and 1B for a frequency of 100 Hz. The parameter ΔR_{ss} measures the range interval in which new (usually interpolated) sound speed profiles are accepted into the propagation model. The solid curves correspond to a new profile at each range step and are verifiably accurate. Curves (not shown) for ΔR_{ss} relatively large, up to about 5 km, are negligibly different from the solid curves. Only when ΔR_{ss} becomes comparable to horizontal grid size ΔR_0 do propagation loss differences become evident. Figure 4 illustrates comparisons between propagation loss curves from cases 1A and 1C for a frequency of 25 Hz. No significant differences are apparent, as is consistent with earlier simulations at frequencies near 50 Hz employing degradations of data (Maurais et al. 1990, Chiu and Ehret 1990). These curves at 100 Hz occasionally show small differences that imply corresponding differences in acoustic wavefields. Specifically, the convergence zone (CZ) widths differ by up to about 5 km, and wavefield magnitudes at CZs differ by up to about a factor of two (i.e. 6 dB in

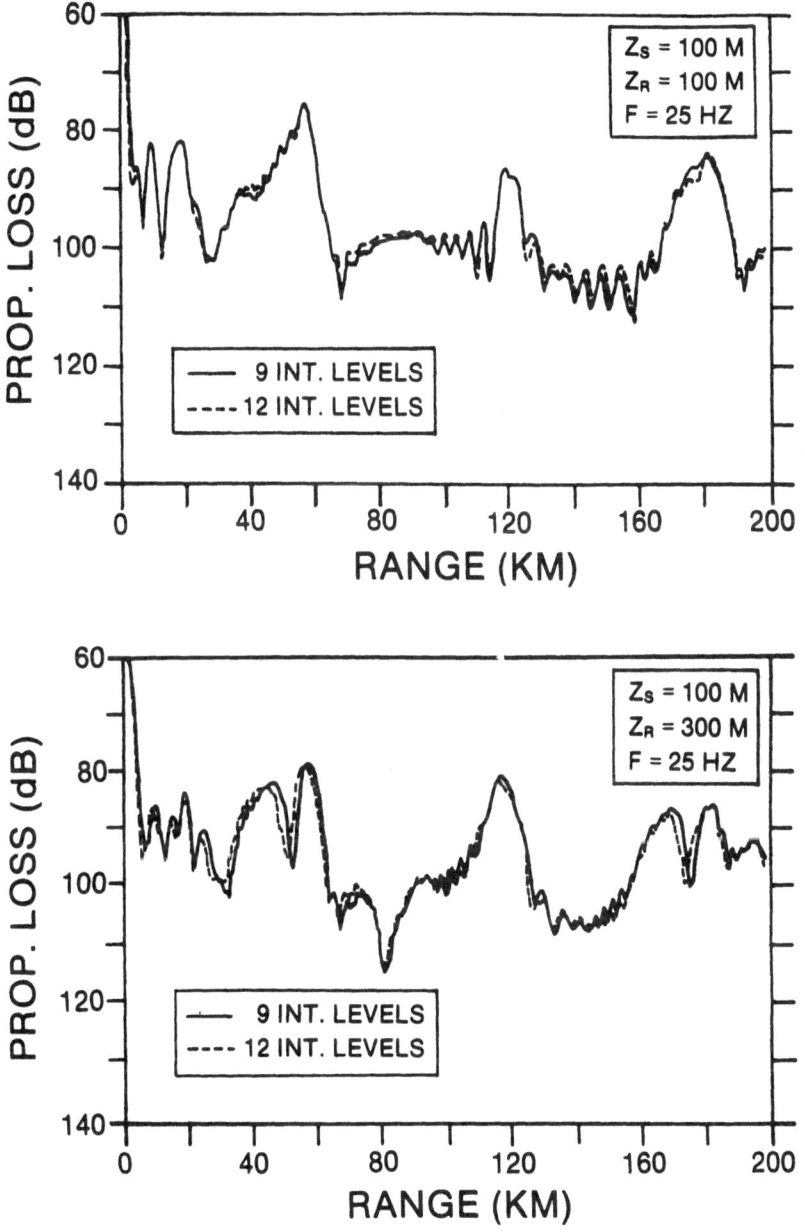

Figure 4. Upper: propagation loss versus range along path H13, for receiver depth 100 m, source depth 100 m, frequency 25 Hz. Lower: same, receiver depth 300 m. Solid curves from case 1A, dashed curves from case 1C. No significant prop. loss differences occur. (At frequency 100 Hz, occasional small differences occur in convergence zone levels and widths.)

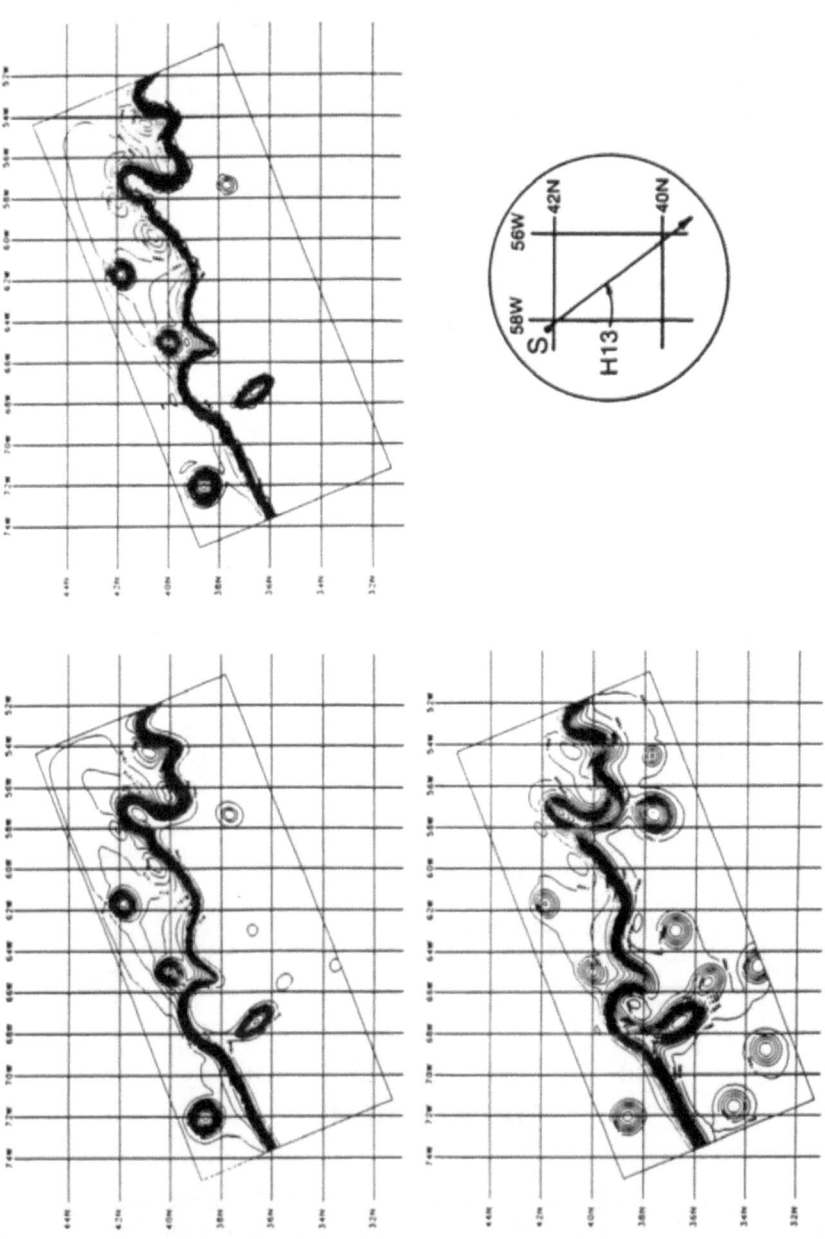

Figure 5. Level curves of sound speed from Harvard University Gulfcast, 13 May 1987 (forecast day 7). Contour interval 2 m/s, depth 50 m. Upper left: case 1A (sound speed generated from 3 Endeavor profiles). Lower left: case 2A (2 climatological profiles). Upper right: case 2B (2 climatological profiles used with feature model). Feature positions in case 2A displaced significantly from others. Lower right: location of propagation path H13.

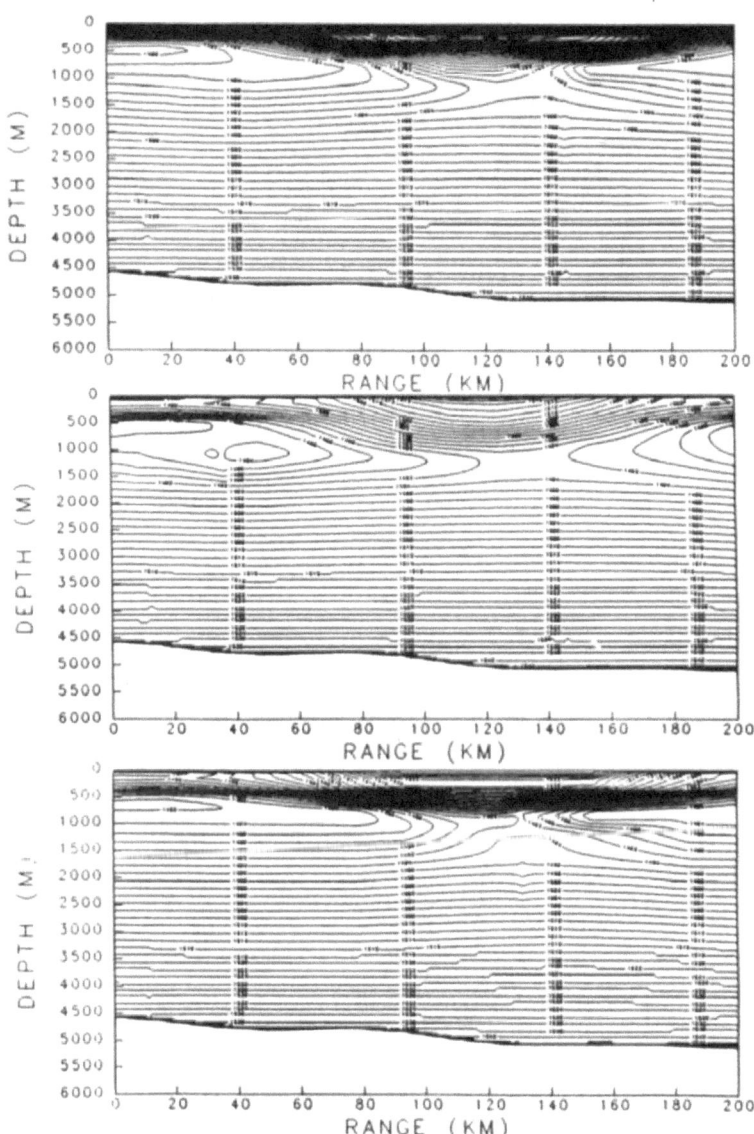

Figure 6. Sound speed contours in vertical plane of propagation, path H13. Contour interval 2 m/s. Upper: case 1A. Middle: case 2A. Lower: case 2B. Patterns significantly different above 1000 m, nearly identical below 1500 m.

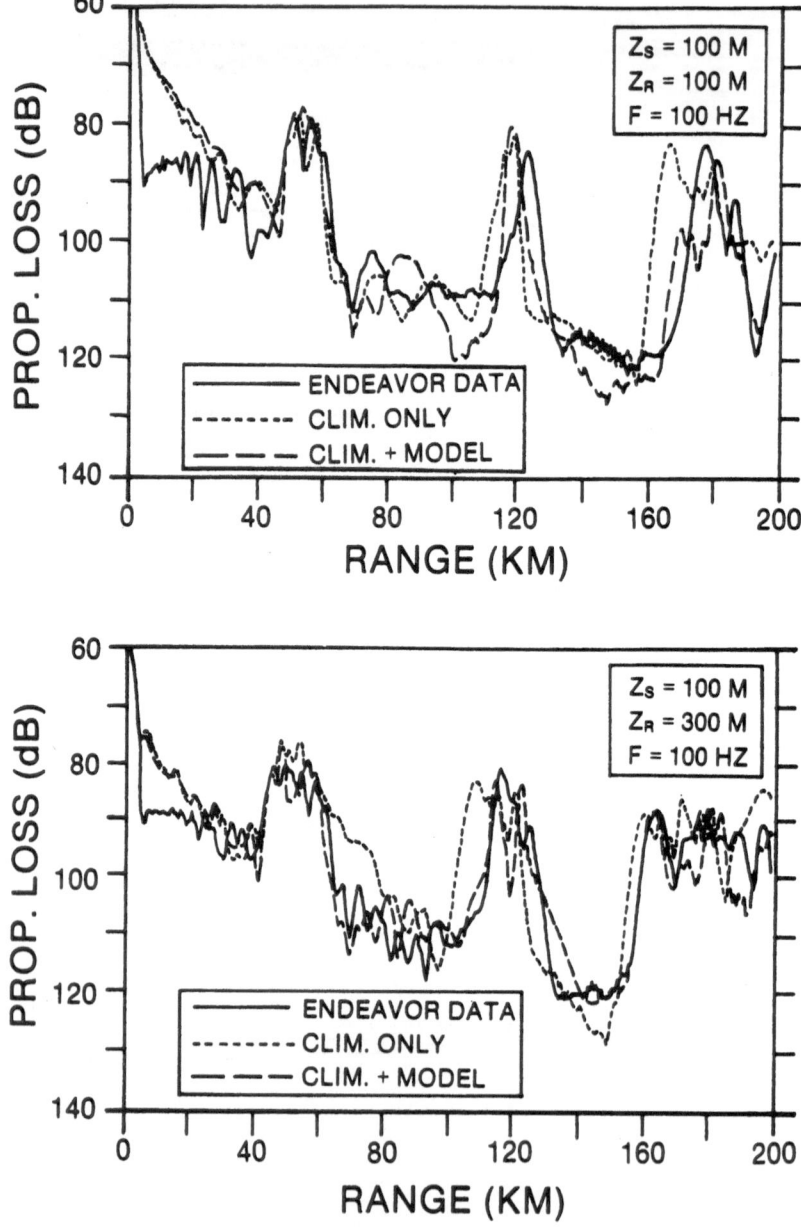

Figure 7. Upper: propagation loss versus range along path H13, for receiver depth 100 m, source depth 100 m, frequency 100 Hz. Lower: same, receiver depth 300 m. Solid curves from case 1A, short-dashed curves from case 2A, long-dashed curves from case 2B. Significant prop. loss differences occur. Closest agreement typically between cases 1A and 2B.

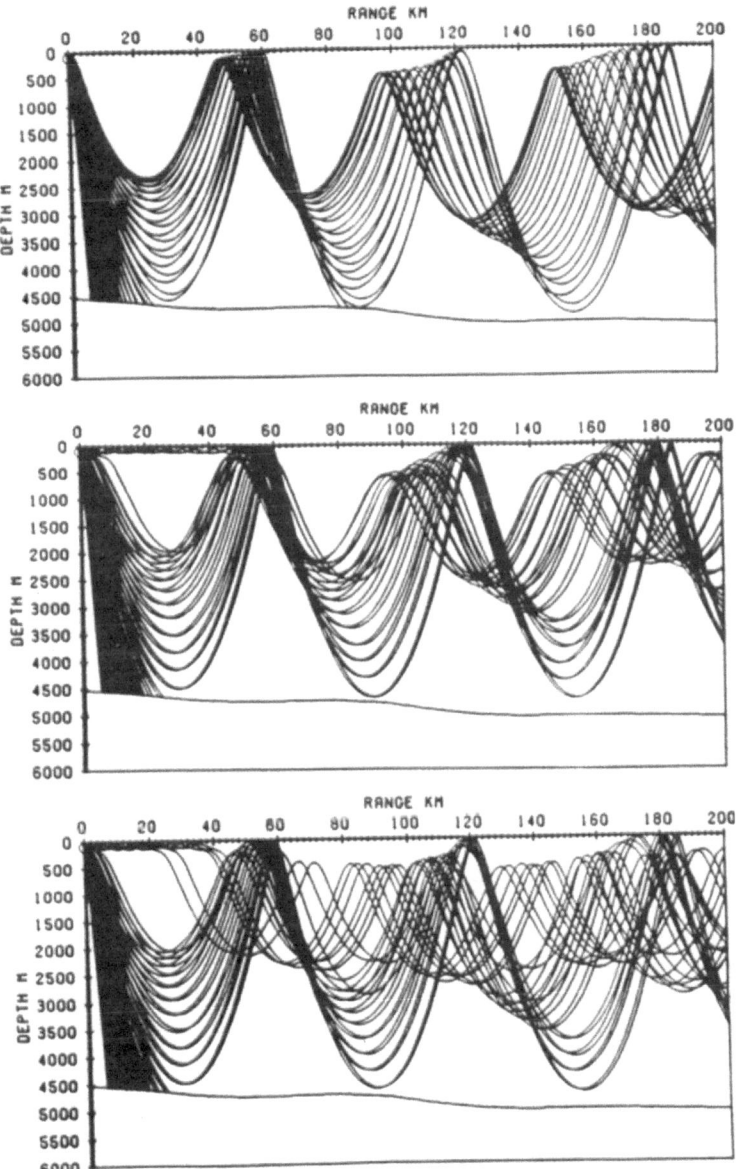

Figure 8. Ray paths for source depth 100 m. Initially 80 rays traced, those with more than 1 bottom or 10 surface interactions suppressed. Upper: case 1A. Middle: case 2A. Lower: case 2B. Ray trace features visible in prop. loss curves such as Figure 7.

propagation loss). Essentially no differences were observed between 3-D and 2-D propagation computations in this region of nearly uniform bottom topography.

Horizontal sound speed contours for cases 2A and 2B are shown in Fig. 5, along with the baseline case 1A. Frontal gradient positions and intensities for the Gulf Stream and rings in case 2A, which uses only climatological profiles in sound speed generation, are significantly different from those in case 1A, which employs typical measured profiles, and from those in case 2B, which uses the feature model along with climatology. Figure 6 shows sound speed contours in the vertical plane of propagation path H13. Propagation loss curves for the three cases appear in Fig. 7, where substantial differences are evident for a frequency of 100 Hz. Significant differences between curves (not shown) can also occur at 25 Hz. Typically the closest agreement between the patterns happens for cases 1A and 2B. Ray paths which further emphasize the differences between the three cases are shown in Fig. 8. Some general features of the propagation loss curves are interpretable in terms of the ray paths, as for example the behavior for a receiver depth of 100 m and ranges less than 40 km in Fig. 7 and the ducting evident in the bottom two traces in Fig. 8.

5.   Conclusions

A coupled environmental and acoustic prediction system is being exercised for evaluation of acoustic propagation sensitivities to the dynamical forecast model configuration, interfacing parameters, and methodology for generating 3-D environmental fields. Preliminary results of the systematic study are presented here for two cases. Case 1 tests sensitivity to horizontal and vertical resolutions of the dynamical forecast model and the acoustic interface. Case 2 tests the quality and quantity of background environmental data required to generate sufficiently accurate sound speed fields for acoustic propagation predictions.

The case 1 horizontal resolution studies indicate that typical eddy-resolving dynamical model grid resolution of 10-15 km is adequate for acoustic propagation calculations at the frequencies considered (25 Hz and 100 Hz) and for ocean regions with small bottom variations. The acoustic propagation sound speed input interval should be somewhat finer, with a resolution less than about 5-7 km at these frequencies. Case 1 vertical resolution studies indicate that 9-12 internal model levels (plus top and bottom) is adequate for the dynamical model at the lower frequency. The number of vertical model levels affects the locations and averaged strengths of CZs, for example, by less than 3 km and 3 dB (i.e. wavefield magnitude differences of up to a factor of 1.4), respectively. The number of dynamical model levels was starting to affect results at 100 Hz, with differences in the strengths and locations of some CZs, within 200 km range, of up to about twice the values given for 25 Hz. Additional testing with propagation in the presence of bottom variability and at higher frequencies is necessary.

The case 2 acoustic propagation sensitivity to the temperature profiles used to generate the environmental fields was much more pronounced. For example, differences of 12 dB and 20 km in the strengths and locations of CZs were observed. The sensitivity was observed to increase with increasing frequency. At both frequencies, propagation loss results using fields prepared with the Levitus Slope and Sargasso temperatures and the simple average for the Gulf Stream (case 2A) differed the greatest, emphasizing the need to properly represent the structure of the Gulf Stream front. Propagation loss results using environmental fields generated with Levitus Slope and Sargasso temperatures and the feature model for the Gulf Stream (case 2B) were closer in structure to results using fields prepared with the R/V Endeavor typical summer profiles (case 1A) at the lower frequency, but differences tended to increase with frequency. This emphasizes the need to replace climatological background temperature profiles with typical synoptic temperature profiles for accurate acoustic propagation predictions, with growing importance as frequency is increased.

Based on these preliminary results of the systematic sensitivity study of the coupled system, it appears that accuracy of acoustic predictions will benefit most from improved representations of the sound speed field from the dynamical forecast model. Possible forecast system improvements include: (1) the addition of a coupled Surface Boundary Layer model for better predictions of the near-surface temperature field; (2) the use of Primitive Equation dynamical models to more accurately model the flow around topography and to extend forecasts to longer time periods; (3) the use of an objective data assimilation scheme to improve the accuracy of the fields for initialization of each forecast; and (4) the use of feature models for interpolation functions throughout the forecast domain to better represent the observed structure of the Gulf Stream and ring temperature fronts.

Acknowledgements

This research was performed under the sponsorship of the Office of Naval Research Contract N00014-90-J-1593 to Harvard University; Office of Naval Research Contract No. N00014-86-K-0129 to Rensselaer Polytechnic Institute; Office of Naval Research Contract No. N00014-90-WX-24019 to Naval Underwater Systems Center; and NUSC Independent Research Project A65025. Cray computer time was provided through a special initiative of the Associate Technical Director for Research and Technology (NUSC Code 10) to support innovative technology applications. The ray trace model was provided by Dr. H. Weinberg.

References

Botseas, G., Lee, D., and King, D. (1987) 'FOR3D: A Computer Model for Solving the LSS Three-Dimensional Wide Angle Wave Equation', NUSC Tech. Rep. 7943.

Botseas, G., Lee, D., and Siegmann, W.L. (1989) 'IFD: Interfaced with Harvard Open Ocean Model Forecasts', NUSC Tech. Rep. 8367.

Chiu, C.S. and Ehert, L. L. (1990) 'Computations of Sound Propagation in a Three-Dimensionally Varying Ocean: A Coupled Mode Approach', D. Lee, A. Cakmak, and R. Vichnevetsky (eds.), Proc. Second IMACS Symposium on Computational Acoustics, to appear.

Fofonoff, N.P. and Millard, R. C., Jr., (1983) 'Algorithms for Computation of Fundamental Properties of Seawater', UNESCO Tech. Papers in Marine Science.

Glenn, S. M., Robinson A. R., and Spall, M. A. (1987) 'Recent Results from the Harvard Gulf Stream Forecasting Program,' Oceanographic Monthly Summary 7, 3, 12-13.

Glenn, S. M. and Robinson, A. R. (1990) 'Nowcasting and Forecasting of Oceanic Dynamic and Acoustic Fields', D. Lee, A. Cakmak, and R. Vichnevetsky (eds.), Proc. Second IMACS Symposium on Computational Acoustics, to appear.

Lee, D. and Botseas, G. (1982) 'IFD: An Implicit Finite-Difference Computer Model for Solving the Parabolic Equation', NUSC Tech. Rep. 6659.

Lee, D. and McDaniel, S. T. (1988) Ocean Acoustic Propagation by Finite Difference Methods, Pergamon, Oxford.

Lee, D., Saad, Y., and Schultz, M. H. (1988) 'An Efficient Method for Solving the Three-Dimensional Wide Angle Equation', D. Lee, R. Sternberg, and M. Schultz (eds.), Computational Acoustics: Wave Propagation, North-Holland, Amsterdam, pp. 75-90.

Lee, D., Botseas, G., Siegmann, W. L., and Robinson, A. R. (1989) 'Numerical Computations of Acoustic Propagation Through Three-Dimensional Ocean Eddies', in W. F. Ames (ed.), IMACS Annals on Computing and Applied Mathematics, J. C. Baltzer, Basel, pp. 317-321.

Lee, D., Schultz, M. H., and Saad, Y. (1990a) 'A 3-Dimensional Wide Angle Wave Equation with Vertical Density Variations', D. Lee, A. Cakmak, and R. Vichnevetsky (eds.), Proc. Second IMACS Symposium on Computational Acoustics, to appear.

Lee, D., Botseas, G., and Siegmann, W. L. (1990b) 'Examination of Three-Dimensional Effects Using a Propagation Model with Azimuth-Coupling Capability (FOR3D)', submitted.

Maurais, D., Siegmann, W. L., Jacobson, M. J., and Glenn, S. M. (1990) 'Acoustic Intensity Dependence on Vertical Resolution and Interpolation of Sound Speed', in preparation.

Mellberg, L. E., Robinson, A. R., and Botseas, G. (1990) 'Modeled Time Variability of Acoustic Propagation Through a Gulf Stream Meander and Eddies', J. Acoust. Soc. Am. 87, 1044-1054.

Newhall, A. E., Lynch, J. F., Chiu, C. S., and Daugherty, R. (1990) 'Improvements in Three-Dimensional Raytracing Codes for Underwater Acoustics', D. Lee, A. Cakmak, and R. Vichnevetsky (eds.), Proc. Second IMACS Symposium on Computational Acoustics, to appear.

NOAA (1982) Climatological Atlas of the World Ocean, Professional Paper No. 13, Rockville, Maryland.

Robinson, A. R. and Leslie, W. G. (1985) 'Estimation and Prediction of Oceanic Fields', Prog. Oceanogr. 14, 485-510.

Robinson, A. R. (1987) 'Dynamical Forecasting of Mesoscale Fronts and Eddies for Acoustical Applications', J. Acoust. Soc. Am. Suppl. 1, S90.

Robinson, A. R. and Walstad, L. J. (1987) 'The Harvard Open Ocean Model: Calibration and Application to Dynamical Process Forecasting and Data Assimilation Studies', J. Appl. Numer. Math. 3, 89-131.

Robinson, A. R., Spall, M. A., and Pinardi, N. (1988) 'Gulf Stream Simulations and the Dynamics of Ring and Meander Processes', J. Phys. Oceanogr. 18, 1811-1853.

Robinson, A. R., Spall, M. A., Walstad, L. J., and Leslie, W. G. (1989a) 'Data Assimilation and Dynamical Interpolation in GULFCAST Experiments', Dynamics of Atmospheres and Oceans, 13, 301-316.

Robinson, A. R., Glenn, S. M., Spall, M. A., Walstad, L. J., Gardner, G. M., and Leslie, W. G. (1989b) 'Forecasting Gulf Stream Meanders and Rings', EOS, The Oceanography Report 70, 1464-1473.

Siegmann, W. L., Lee, D., and Botseas, G. (1988) 'Finite Difference Computations of Three-Dimensional Sound Propagation', D. Lee, R. Sternberg, and M. Schultz (eds.), Computational Acoustics: Wave Propagation, North-Holland, Amsterdam, pp. 91-109.

Siegmann, W. L., Jacobson, M. J., Lee, D., Botseas, G., Robinson, A. R., and Glenn, S. M. (1990a) 'Interfacing Mesoscale Ocean Prediction and Parabolic Acoustic Propagation Models Including Bottom Topography', D. Lee, A. Cakmak, and R. Vichnevetsky (eds.), Proc. Second IMACS Symposium on Computational Acoustics, to appear.

Siegmann, W. L., Lee, D., and Botseas, G. (1990b) 'Analytic Solutions for Testing Accuracy and Azimuthal Coupling in Three-Dimensional Acoustic Propagation', D. Lee, A. Cakmak, and R. Vichnevetsky (eds.), Proc. Second IMACS Symposium on Computational Acoustics, to appear.

SIMULATING TEMPERATURE, SALINITY AND CURRENTS IN THE OCEAN

Kim David Saunders & David B. King

Naval Oceanographic and Atmospheric Research Laboratory
John C. Stennis Space Center, Mississippi 39529-5004
United States of America

ABSTRACT. Effective use of scarce resources requires a coherent and rational approach to oceanic system development. Included in this approach is the simulation of the performance of the proposed system, be it oceanographic or acoustic, under simulated realistic environmental conditions. Over the past year and a half, we have been working on methods for simulating the physical oceanic environment. The work began with a simulation of the internal wave field in the ocean, based on one formulation of the Garrett-Munk model. Internal tides were then included and the work extended to frontal features. We present a summary of methods for simulating the environment, with particular emphasis on the methods we have employed. These methods include Fourier transforms, wave function models, direct solution of the equations of motion, extended generalized stochastic subdivision, and impulse response models. The simulations have been successfully coupled to acoustic transmission loss models and have produced results consistent with observations. The results of this coupling applied to temporal and spatial variation are presented.

1. Introduction

About a year and a half ago, a program to provide realistic simulations of the ocean environment was begun at the Naval Oceanographic and Atmospheric Research Laboratory (NORDA as it was then known). The initial simulations concentrated on the internal wave field as this is one of the most energetic components in the oceanic current spectrum. Our goal was to use the simplest possible algorithm to obtain realistic current, temperature and salinity fields in representative regions of the deep ocean and to apply these simulations to problems of interest to the U.S. Navy. This paper reports our methodology and application to one acoustic problem.

1.1. DEFINITION OF OCEANIC SIMULATION MODELING

Simulation modeling may be defined as the process by which a realistic field of oceanographic variables is generated. The field can include any number of variables defined for any given range of space and time coordinates. If the simulation is a discrete simulation, the resolution of the dependent variables must also be specified. In this paper, we will be concerned only with discrete simulations that may be computer generated.

1.2. NEED FOR SIMULATION MODELS

There are a variety of reasons one might require direct simulations of the oceanic environment. Some fields of interest requiring such simulations are systems design, algorithm performance, survey design and basic research. The common aspect of these different fields is the need for complete (in some sense) fields of oceanic properties to test some hypothesis or provide insight.

Simulation techniques provide an attractive alternative to direct observations as they can be used to generate large scale sets of synoptic "data" over specified ranges at defined resolutions. This statement should not be taken to imply that simulations can ever replace direct observations,

J. Potter and A. Warn-Varnas (eds.), Ocean Variability & Acoustic Propagation, 561–577.
© 1991 *Kluwer Academic Publishers.*

rather, they supplement them. If the simulations are to be useful, they must produce output that is representative of the actual oceanic variables. Therefore, the formulation of the algorithms that comprise the simulation models must be based on our present knowledge of the ocean. In fact, one of the uses of simulation models is to uncover or emphasize gaps in our knowledge of the ocean.

2. Types of Simulation Methods

There are a wide variety of methods that can be used to produce oceanic simulations. Some of these include Fourier series/ transform methods, modal representations, numerical integration of the equations of motion with random forcing, extended generalized stochastic subdivision, interpolation from data bases, AI methods, signal processing/filters with random input, semi-analytical methods and various combinations of all these. In order to lay the basis for the current work at the Naval Oceanographic and Atmospheric Research Laboratory, a short outline of the methods, their strengths and weaknesses follows.

2.1. FOURIER METHODS

Perhaps the simplest method of simulating a field in space and time is to employ Fourier transforms to go from a spectral representation to a specific realization. This method has been used by many investigators to obtain realizations of fields in the upper ocean, e.g. Areté Associates (Dugan, 1989) have used this method for computing fields of temperature and salinity; Macaskill and Ewart (1984) have used this technique in simulating propagation of acoustic waves in a random field of sound velocity and Saunders and Knauer (1987) have used the method to simulate the upper ocean shear field for modeling the response of dropped instruments.

The method may be summarized as follows. Let $S_{xx}(k)$ be a given spectrum of a scalar variable $x(r)$ in an n dimensional space. The spectrum is defined as:

$$S_{xx}(k) \, \delta(k-k') = \langle X(k)X^*(k') \rangle, \tag{2.1}$$

$$X(k) = \frac{1}{2\pi} \int_{-\infty}^{\infty} x(r) \, e^{-ikr} \, dr \, , \tag{2.2}$$

$$= \mathscr{F}_k \{ x(r) \} \, . \tag{2.3}$$

We can approximate a single realization of $x(r)$ by using the following relation:

$$x(r) = \mathscr{F}_k^{-1} \{ S_{xx}^{1/2}(k) \, e^{i\varphi(k)} \} \, , \tag{2.4}$$

$$\langle \varphi(k)\varphi(k+l) \rangle = \delta(|l|) \, . \tag{2.5}$$

The Fourier transform method is useful in situations where the statistics of the field do not vary with **r**. In the oceanic case, this can often be assumed to hold if the independent variable is time or a horizontal coordinate in the deep ocean well away from frontal regions. The statistics generally are not invariant in the vertical direction or near frontal boundaries. In these latter situations, other methods are better employed.

2.2. MODAL REPRESENTATIONS

For certain processes, such as internal waves, the spatial non-uniformity of the statistics can be bypassed by expanding the fields of motion in terms of a set of eigenfunctions. For the internal wave field, which we will use as an example, this permits the use of arbitrary profiles of the Brunt-Väisälä frequency, N(z), to define the basic state. This method has been used by Dynamics Technology (Borchardt, 1985, Botto, et al., 1985, Cox, 1986, Berge, 1987, Huxtable, 1988, and Borchardt, 1989) in a number of programs to computer vertical profiles of horizontal current or shear.

Under certain conditions, such as small mean vertical shear, the equations of motion may be reduced to a simple set of a linear partial differential equations. For the internal waves occurring in the frequency band between the inertial frequency and the Brunt-Väisälä frequency, the equations of motion reduce to:

$$\left\{ \left(\frac{\partial^2}{\partial t^2} + f^2 \right) \left[\nabla_h^2 + \frac{1}{\rho_0} \frac{\partial}{\partial z} \left(\rho_0 \frac{\partial}{\partial z} \right) \right] + (N^2 - f^2)\nabla_h^2 \right\} \ w = 0 \tag{2.6}$$

with the boundary conditions that $w = 0$ at $z = 0, -H$. If N is only a function of z, and H is constant in x and y, then the equation is separable, allowing plane waves of the form:

$$w = W(z) \ e^{i (kx - \omega t)} \ , \tag{2.7}$$

which gives an ordinary differential equation for W, which, after a few approximations, becomes

$$W'' + k^2 \frac{N^2(z) - \omega^2}{\omega^2 - f^2} W = 0 \ , \quad W = 0 \ at \ z = \left\{ \begin{array}{l} 0 \\ -H \end{array} \right. \tag{2.8}$$

where $k = |\mathbf{k}|$.

This is an eigenvalue problem, as non-trivial solutions exist only for certain values of k_n , n = 1,2,3.... It is easy to prove that the corresponding eigenfunctions, W_n, form a complete basis (under certain conditions on N(z) and if complex eigenvalues are permitted) and thus, the field of internal waves can be expressed as a linear combination of the eigenfunctions in the form:

$$w(x,y,z,t) = \sum_{n=0}^{\infty} \int_f^{N_{max}} \beta_n(\omega)W_n(z) \ e^{i \{ k_n [x \cos\theta + y \sin\theta] + \omega t \}} \ \sqrt{d\omega} \ . \tag{2.9}$$

The problem is to specify the coefficients β_n . In order to specify these parameters, we require a knowledge of how the variance in the simulated field is distributed between the various modes and over frequency. The details of this specification will be discussed in detail in section 3 for a specific application.

The modal, or eigenfunction, method is valuable when the underlying physics are known and a complete set of eigenfunctions can be found. In the case of internal waves, this allows the construction of a kinematically and dynamically consistent field of motion with arbitrary stratification. It is equivalent to the Fourier transform method for constant $N(z)$, but is more general.

2.3. NUMERICAL INTEGRATION

Simulations can also be obtained by integrating the equations of motion, using random forcing in time and space at the boundaries (or forcing derived from observations, it desired). This method has been used to investigate internal wave dynamics using spectral integration codes (Shen and Holloway, 1986, and Winters and D'Asaro, 1989) and by direct integration using finite difference formulations, such as the Lax-Wendroff scheme ((Rubenstein, 1989). This technique has been used by a wide variety of investigators to model the large scale features in the ocean. A good review of the state of the art is provided by Holland (1989).

This method is adaptable to simulation modeling by directly performing the integrations or by using the results of such models in a reduced form in deriving simulation algorithms. It has the following advantages and disadvantages.
1. Advantages:
   - Energy is naturally distributed by the equations of motion.
   - Nonlinear balances are resolved directly.
   - Mean buoyancy and shear fields are naturally included.
   - Horizontal variations in depth, mean current and buoyancy pose no problem.
2. Disadvantages:
   - Computational time can be very large.
   - Choice of horizontal boundary conditions pose problems.
   - Three dimensional, time dependent simulations are presently very expensive.
   - Subgrid scales cannot be simulated.

2.4. EXTENDED GENERALIZED STOCHASTIC SUBDIVISION (EGSS)

EGSS (Saunders, 1989) is based on generalized stochastic subdivision (Lewis, 1987) that in turn was based on fractal interpolation (see Barnsley, 1988, chapter 6 for an introduction).

Generalized stochastic subdivision is a method for generating or interpolating a time (or multidimensional independent variable) series that has a specific correlation function. It is a local method that is externally consistent, that is, existing data can be incorporated easily and consistently. The extension of generalized stochastic subdivision, EGSS, consists of the method by which non-stationary data sets may be produced for multiple series with given cross correlation functions.

2.5. INTERPOLATION FROM DATA BASES

One of the oldest methods for simulating oceanic variable is interpolation from data bases. This process is usually not thought of as a simulation, but that is what it is. Any type of interpolation (or worse, extrapolation) is an attempt to fill in data where none exist. Various types of both subjective and objective analysis have been developed, but these methods cannot be given adequate treatment here. The common feature of these methods is that specific algorithms, sometimes based on known properties of the fields to be contoured are employed but no details on the data between the observations are known. In all cases of interpolation, the interpolated data are not the same as the actual oceanic fields, thus they are a simulation of the true fields.

Interpolated fields are extremely useful, and we may regard them as a type of simulation. Large scale climatological data bases such as Levitus's (1982) and GDEM (Teague, et al., 1990) provide a first order simulation of the world ocean on basin scales and a basis on which to construct finer level (subgrid) simulations. The primary drawback of the large scale data bases is that they are very coarse in time (typical time scales on the order of a month) and space (typical length scales on the order of degrees.) A local data base such as dynamic GDEM (Teague, et al. 1990) can be used to provide initial conditions for other types of simulation models in ocean frontal regions or to provide initial conditions for numerical integrations in the manner of Käse, et al. (1989).

2.6. ARTIFICIAL INTELLIGENCE TECHNIQUES

The use of artificial intelligence (AI), in particular, expert systems methods for ocean simulation or prediction is relatively new. Much of the work to date has been directed at using these techniques to assist in analyzing satellite imagery (Lybanon, et al., 1986) and for predicting the path of major ocean fronts such as the Gulf Stream (Lybanon, 1988). The expert system approach consists of a set of rules that describe a model of the process to be simulated. These rules are implemented in a specialized programming language and operate on an initial state of the field of variables to produce a predicted state at a later time, based on the set of rules. In the initial tests for Gulf Stream prediction, the method has shown definite promise and could probably be applied to other types of oceanic simulations.

2.7. SIGNAL PROCESSING / FILTER METHODS

The filter (or signal processing) method of simulation is closely related to the Fourier transform method. The basic idea is that the process to be simulated is modeled as a filter (either linear or nonlinear) and a random (white noise) signal is input to the filter. The simulated variable is the output. For linear, time independent filters, this can be shown to be equivalent to the Fourier transform method. The advantage lies in the ability to construct recursive, nonlinear filters that produce data with strongly non-Gaussian statistics. (A good introduction to such filter systems in given in Priestly, 1988). We have used some of these techniques to produce time series that possess nearly lognormal statistics.

2.8. SEMI-ANALYTICAL METHODS

These methods are partly ad hoc in that the physical processes on which the algorithms are based are not well understood. They are also partly based on well understood principles that can be applied in a simple manner. Although some of the assumptions are not justified on physical grounds, they produce a realistic simulation that is useful for acoustic model and other system testing. The advantage of using such a technique lies in the ease and small cost of computing the field.

3. Simulating the Internal Wave Field

For the first level simulation model, we chose to simulate the internal wave field as motions in the internal wave band account for a major portion of the kinetic energy in the upper ocean with energy levels on the order of 4000 J/m². These energy densities are of the same order as those found in major ocean current systems (Ierley, 1990).

3.1. MODAL REPRESENTATION OF THE INTERNAL WAVE FIELD

3.1.1. Assumptions and Approximations

Our guiding principle in constructing the simulation was to strive for maximum simplicity while producing results that were representative of the internal wave field. Another principle was to construct the computer programs in such a manner that they would be computationally efficient.
  Our assumptions were:
  1. Linearity,
  2. No mean shear,
  3. Boussinesq fluid,
  4. Flat (or slowly varying) bottom,
  5. Horizontally homogeneous (or slowly varying) density field.
We placed no especial restrictions on the form the Brunt-Väisälä frequency profiles in the vertical.

3.1.2. The Internal Wave Equation

We start with the linear Boussinesq form for the equations of motion:

$$u_t - fv = -p_x , \tag{3.1.a}$$

$$v_t + fu = -p_y , \tag{3.1.b}$$

$$w_t = -p_z - g\frac{\rho}{\rho_0} , \tag{3.1c}$$

$$\rho_t + w\frac{d\bar\rho}{dz} = 0 , \tag{3.1.d}$$

$$\zeta_t = w . \tag{3.1.e}$$

If we represent the dependent variables by a horizontally propagating plane wave:

$$(u, v, w, \rho, \zeta) = (U, V, W, R, Z) e^{i(k_x x + k_y y + \omega t)} , \tag{3.2}$$

we obtain

$$U - \frac{f k_x + i \omega k_y}{\omega k^2} W_z ,$$ (3.3.a)

$$V - \frac{f k_y + i \omega k_x}{\omega k^2} W_z ,$$ (3.3.b)

$$|U|^2 + |V|^2 - \left[\frac{1 + \dfrac{f^2}{\omega^2}}{k^2} \right] |W_z|^2 .$$ (3.3.c)

3.1.3. Normal Mode Expansion

If the equations 3.1.a - 3.1.d and 3.2 are combined, an ordinary differential equation for the vertical component of velocity is obtained

$$\frac{d^2 W}{dz^2} + k^2 \frac{N^2(z) - \omega^2}{\omega^2 - f^2} W - 0 ,$$ (3.4.a)

along with the boundary conditions of zero vertical velocity at the surface and bottom

$$W - 0 \ at \ z - \begin{cases} 0 \\ -H \end{cases} ,$$ (3.4.b)

comprise an eigenvalue problem where the k's assume a discrete set of values, k_n, $n = 1, 2,$ . . . , and the W's form (under certain conditions) a complete set of eigenfunctions, W_n. As an example, for the Brunt-Väisälä frequency profile, typical of summer Sargasso Sea conditions, shown in figure 1, the first few modes at 2 cph are shown in figure 2.

3.1.4. Stochastic Representation

We wish to express a representation of the internal wave field as a linear combination of the eigenmodes, distributed over all permitted frequencies, that is, from the inertial frequency to the maximum of the Brunt-Väisälä frequency. We approximate the integral in equation 2.9 by a sum over a finite set of frequencies and further assume that the phase of each component is uncorrelated with the phase of any other component. In addition, for each frequency and mode, we assume that there are a number of wavetrains approaching from different, random directions. The expression for the vertical velocity is therefore given by

$$w - \sum_{l-1}^{N_{dir}} \sum_{j-1}^{N_{modes}} \sum_{\substack{\omega \in \{\omega_n\}, \\ n-1 \ldots N_j}} \beta_j(\omega) W_j(z ; \omega) \cdot$$

$$e^{i[\omega t + k_j(x \cos \theta_y(\omega) + y \sin \theta_y(\omega) + \varphi_y(\omega)]} \sqrt{\frac{\Delta \omega}{N_{dir}}} ,$$ (3.5)

where $x = r \cos \alpha$, $y = r \sin \alpha$, $r =$ the range and $\alpha =$ the direction of the section. In order to determine the weights β_n, we will make use of the Garrett-Munk spectral model of the internal wave field.

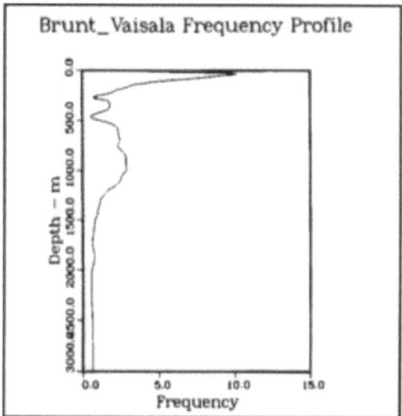

Figure 1. Brunt-Väisälä frequency profile typical of summer Sargasso Sea conditions.

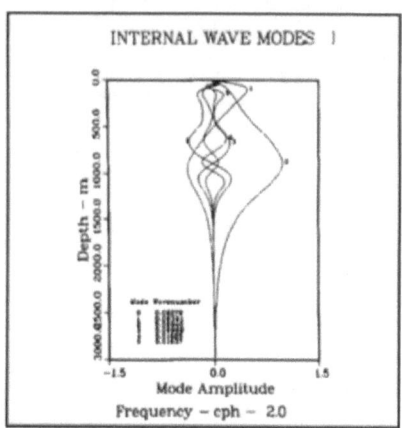

Figure 2. First few modes corresponding to Figure 1.

3.2. CONSISTENCY IN GARRETT-MUNK MODEL

3.2.1. Overview of the GM Model

The Garrett-Munk internal wave model actually refers to a series of models that have been developed and improved over the past 20 years (Garrett and Munk, 1972, 1975, Cairns, et al., 1975, Desaubies, 1976 and Munk, 1978). We use the Munk (1978) formulation as our starting point. The Garrett-Munk (GM) models all assume a very smooth, exponential form for the Brunt-Väisälä frequency profile, as well as linearity and no mean shear. The smooth profile allows the use of a WKBJ approximation for the wave functions and by assuming the so-called "Tijuana boundary conditions" (Garrett and Munk, 1972), they are able to make the transition from a discrete eigenvalue spectrum to a continuous one. A form for the distribution of energy over frequency and wavenumber is chosen that is consistent with observations and the results combined into model spectra for horizontal kinetic energy, vertical kinetic energy and potential energy (or equivalently, a displacement spectrum).

3.2.2. Spectral Form of GM 78

The Garrett-Munk spectra are given as

$$F_u(\omega, j) = b^2 N_0 N(z) \left[\frac{\omega^2 + f^2}{\omega^2}\right] E(\omega, j), \qquad (3.6.a)$$

$$F_\zeta(\omega, j) = b^2 N_0 N^{-1}(z) \left[\frac{\omega^2 - f^2}{\omega^2}\right] E(\omega, j), \qquad (3.6.b)$$

$$E(\omega, j) = E \cdot B(\omega) \cdot H(j), \qquad (3.6.c)$$

$$B(\omega) = \frac{2f}{\pi \omega \sqrt{\omega^2 - f^2}}, \qquad (3.6.d)$$

and

$$H(j) = \frac{(j^2 + j_*^2)^{-1}}{\sum_{j=1}^{\infty}(j^2 + j_*^2)^{-1}}. \qquad (3.6.e)$$

The mode scale number, j_* is usually chosen equal to 3. The energy parameter, E, is a non-dimensional number = 6.5×10^{-5}. The distribution of kinetic and potential energy as a function a frequency is shown in figure 3, and the distribution of energy given by equation 3.6.e is shown in figure 4.

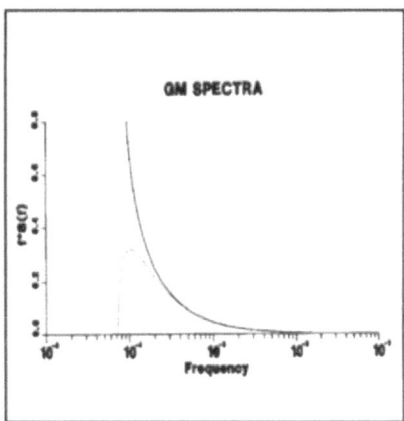

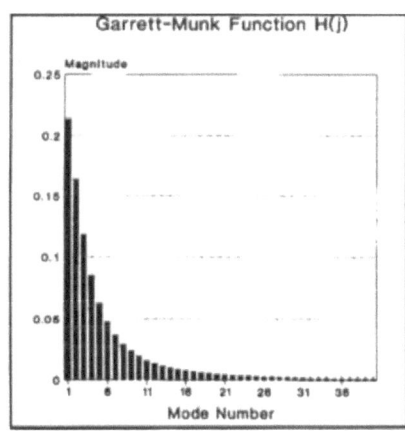

Figure 3. Energy distribution as a function of frequency in the Garrett-Munk model. The solid line represents F_u and the dotted line, F_ζ.

Figure 4. Modal distribution of energy in the Garrett-Munk model.

3.2.3. Inconsistency in the Spectra

The problem we encountered in trying to use the GM formulation for reconstructing a representative field of internal waves was the problem of getting a stationary spectrum in the vertical wavenumber, horizontal wavenumber and frequency that could be inverted. From the form of GM '78 equations (3.6.a-3.6.e), we can see that a vertical dependence has been imposed on the variance spectra at each mode and frequency. If one thinks of the modes as being defined by the eigen-wavefunctions, these have a number of nodes equal to the ordinal number of the eigenfunction. Clearly, at a node, there can be no contribution to the variance spectrum, hence to specify a modal, frequency and depth dependent spectrum is self inconsistent. Another way of looking at this problem is to note that the relation between the wavenumber and frequency as given in GM '78 only holds where the scales of variation of the internal wave modes are small compared to the scale of the variations in the Brunt-Väisälä frequency profile. As the major amount of energy in the spectra are concentrated in the low modes and at low frequencies, this assumption does not hold very well. We therefore chose, as the simplest approach, to keep the form of the GM '78 spectra in the frequency and modal distribution of the wave energy and to expand the wave field in terms of the numerically computed eigen-wavefunctions, based on the linear internal wave equation, 3.4.a with boundary conditions 3.4.b.

3.3. MODIFICATION FOR GENERAL BRUNT-VÄISÄLÄ FREQUENCY PROFILES

3.3.1. Normalization of the Horizontal Kinetic Energy

From equations 3.4.a and 3.4.b, we can immediately get the orthogonality condition for the internal wave modes

$$\int_{-H}^{0} \frac{N^2 (z) - \omega^2}{\omega^2 - f^2} \ W_i \ W_j \ dz \ - \ \delta_{ij} \ , \tag{3.7}$$

where δ_{ij} is the Kronecker delta function. We also can obtain from the same equations, the relation

$$\int_{-H}^{0} \left[\frac{dW_j}{dz} \right]^2 \ dz \ - \ k_j^2 \ . \tag{3.8}$$

3.3.2. Rescaling the Modal Contributions

If we scale the horizontal modal components of the wave field as

$$(\ \hat{U}_j \ , \ \hat{V}_j \ , \ \hat{W}_j \ ) \ - \ \beta_j^2 \ (\ U_j \ , \ V_j \ , \ W_j \ ) \ , \tag{3.9}$$

we can equate the horizontal kinetic energy spectrum at each mode with

$$F_u (\ \omega, j \ ) \ - \ \tfrac{1}{2} [\ \hat{U}_j \ \hat{U}_j^* + \hat{V}_j \ \hat{V}_j^* \ ] \ - \ \tfrac{1}{2} \ \frac{1 \ + \ f^2/\omega^2}{k_j^2} \left| \frac{dW_j}{dz} \right|^2 \ . \tag{3.10}$$

If we integrate the spectra over depth, we obtain

$$HKE \ = \ \int_{-H}^{0} F_u \ (\ \omega, j \ ) \ dz \ = \ \tfrac{1}{2} \ \beta_j^2 \left[1 \ + \ \frac{f^2}{\omega^2} \right] \ , \tag{3.11}$$

and using the relation from the Garrett-Munk spectrum, equation 3.6.a, we obtain

$$HKE \ = \ b^2 \left[1 + \frac{f^2}{\omega^2} \right] E \ (\ \omega, j \ ) \ N_0 \int_{-H}^{0} N \ (\ z \ ) \ dz \ . \tag{3.12}$$

We then obtain a relation for the β_j,

$$\beta_j^2 \ = \ 2 \ b^2 \ N_0 \ \overline{N} \ H \ E \ (\ \omega, j \ ) \ , \tag{3.13}$$

$$\overline{N} \ = \ \frac{1}{H} \int_{-H}^{0} N \ (z) \ dz \ . \tag{3.14}$$

3.3.3. Consistency

We have obtained the weights for the various modes by assuming a consistent form for the horizontal kinetic energy. We now need to see how this weighting affects the potential energy and vertical kinetic energy spectrum. If we expand w(x,y,z,t) for a constant x and y, and take the auto correlation, we get, following Kinsman (1965),

$$< w_j \ (\ t, \ z \ ) \ w_j \ (\ t', \ z \ ) > \ =$$

$$= \ \tfrac{1}{2} \int_{0}^{\infty} \beta_j^2 \ (\omega) \ W_j^2 \ (\ z \ ; \ \omega \ ) \ \cos [\ \omega \ (\ t \ - \ t' \ ) \ ] \ d\omega \ . \tag{3.15}$$

From this relation, we immediately see that the spectrum of the vertical velocity fluctuations of the j_th mode at a given level (denoted S_w to distinguish it from the Garrett-Munk spectrum, F_w) is given by

$$S_w \ (\ \omega, z, j \ ) \ = \ \beta_j^2 \ W_j^2 \ . \tag{3.16}$$

It is immediately obvious that this is different from the form of the GM vertical velocity and displacement spectra. The GM relations relating the vertical spectra to the horizontal spectra no longer apply and therefore, the spectrum of the horizontal velocities must be written in terms of the derivatives of the vertical eigenmodes:

$$S_u \ (\omega \ , z \ j \ ) \ = \ \frac{\beta_j^2}{k_j^2} \left[1 \ + \ \frac{f^2}{\omega^2} \right] \left| \frac{dW_j}{dz} \right|^2 \ . \tag{3.17}$$

3.4. IMPLEMENTATION

The vertical velocity component of the internal wave field were modeled according to the modal decomposition (3.5). The displacement and horizontal velocity fields were modeled in similar manner, with the appropriate wave functions and frequency weights. The choice of frequencies to be used in the summation were chosen logarithmically spaced and the form of the summand changed to reflect that change in variable. The model was coded and run on both a VAX system running under the VMS operating system and on a CONVEX system running under the UNIX operating system.

4. Application to Acoustic Transmission Loss Modeling

Acoustic models offer a very cost efficient method of isolating and quantifying effects of oceanographic variation on acoustic processes. The validity of conclusions drawn from using acoustic models to study ocean variation is closely related to the accuracy and resolution of the environmental data that is supplied to the acoustic model. Unfortunately, highly resolved environmental data is often not available. It is too expensive in both time and resources to acquire field oceanographic data at the resolution that would be required for analysis of "fine-scale" oceanographic features.

Simulation of oceanographic features holds the promise of being able to provide the necessary oceanographic data that will allow quantization of the acoustic effects of such features as internal waves. This section contains a discussion of a numerical experiment that uses a simulation of the sound speed structure with internal waves as input to a transmission loss model. The purpose of this exercise was not to quantify the effect on the acoustics that might be expected from internal waves but rather to demonstrate the coupling of an ocean simulation model to an acoustic model.

A Parabolic Equation (PE) model was used as the transmission loss model for this effort. The parabolic approximation was first applied to radio wave propagation in the troposphere in the mid 1940's [Leontovich and Fock, 1946]. The first use of the parabolic approximation to underwater acoustics came about by the use of the split-step Fourier transform [Hardin and Tappert, 1973] which allowed for acceptable computational times. This approach was implemented by Brock [Brock, 1978]. PE and its underlying physics has been well documented in the literature over the years. If the reader is interested in more detail, an excellent source is a review article by Tappert [Tappert, 1977]

The split-step PE model is based on the parabolic approximation to the reduced wave equation. Far field solutions can be expressed in the form

$$p(r, z) \sim U(r, z) \, r^{-1/2} \, e^{i \, k_0 \, r} , \tag{4.1}$$

where k_0 is ω/c. Substitution of the above form for the pressure into the wave equation, with suitable approximations, gives the result

$$\frac{\partial^2 U}{\partial z^2} + 2 i k_0 \frac{\partial U}{\partial r} + k_0^2 [n^2(r, z) - 1] U = 0 , \tag{4.2}$$

which is valid provided $k_0 r >> 1$ and

These two approximations essentially transform a boundary value problem into an initial value problem, thus allowing a marching solution. The final form of the solution, using the split-step Fourier approach, gives the following:

573

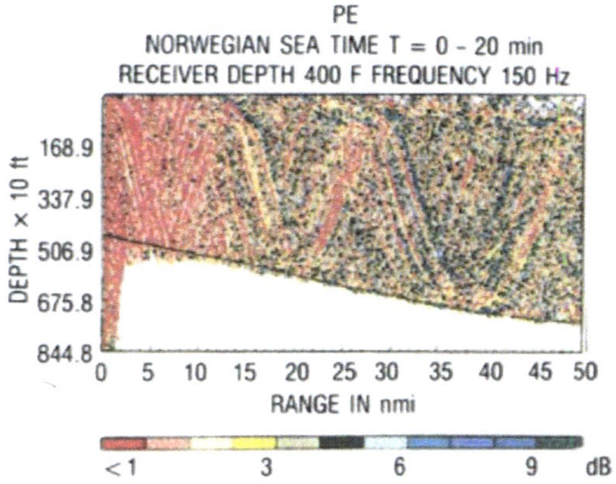

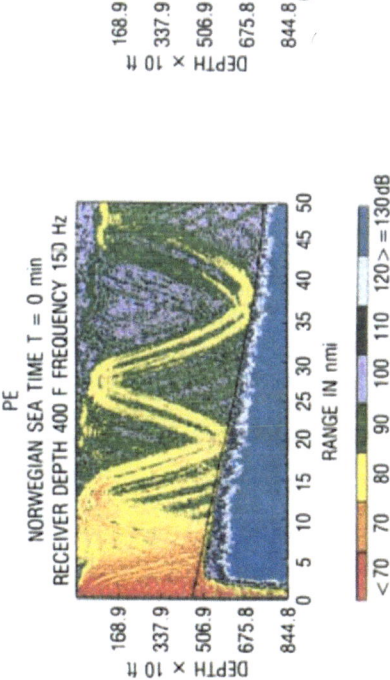

Figure 5c. Pixel by pixel difference of the two plots in dB.

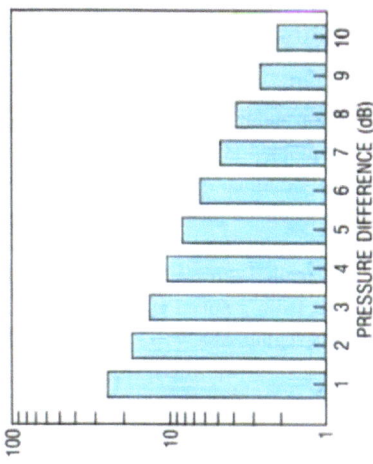

Figure 5a. Transmission loss (in dB) as a function of range and depth at time = 0 minutes. The vertical axis is in units of feet and the horizontal range in nautical miles.

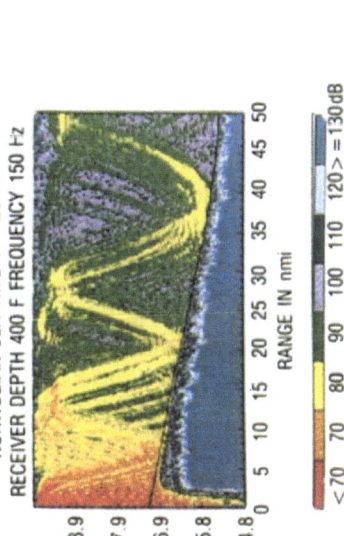

Figure 5d. Histogram of the differences in dB vs. the percentage of the acoustic field.

Figure 5b. Same as Figure 5a, but at time = 20 minutes.

$$\left| \frac{\dfrac{\partial^2 U}{\partial r^2}}{2 \, i \, k_0 \, \dfrac{\partial U}{\partial r}} \right| \; < \; 1 \; . \tag{4.3}$$

$$U(\, r + \Delta r \,) \; - \; e^{i \frac{k_0}{2} \, \Delta r \, (n^2 - 1)} \; F_k^{-1} \left\{ \, e^{i \frac{k^2}{2} \, \Delta \, \frac{r}{k_0}} \, F_k \, [\; U(\, r, z \,) \,] \, \right\} \; . \tag{4.4}$$

This equation is the basis of the split-step PE that was used in this demonstration. There are a few things about this equation that should be pointed out. First, the environmental information is fully contained in the index of refraction function, $n(r,z)$. Second, it should be noticed that the solution is obtained for all depths, z. This capability will be used in the analysis that follows. Last, the environment should be finely discretized in both depth and range. To start the calculation, the solution must be initialized at some range (say $r = 0$) by using an initial value of $U(r=0,z)$ (called the source function). The solution is then marched out in range by successive transforms from z space to k space and back to z space.

The Norwegian Sea was selected as the ocean area for this demonstration. The PE model described above was merged with the ocean simulation model (discussed in section 3) which supplied the sound speed structure data required by PE. The bottom topography was extracted from a gridded bathymetry data base, and a bottom composition description needed to complete the environmental inputs was obtained by using a geoacoustic description typical of the Norwegian Sea.

Since the effects of an internal wave on the sound speed structure is temporal in nature, two temporally separated representations of the sound speed field were generated and used as inputs to PE. The acoustic results of these two sound speed field representations are shown in figures 5 a and b.

Figure 5a shows the transmission loss for all ranges and depths at time $t = 0$ minutes. The vertical axis is depth given in feet, and the horizontal axis is the range in nautical miles. The color bar at the bottom of the plot is the key to the different transmission loss values measured in dB, that are used in the plot. The bathymetry is shown on the plot as a thin black line. A fixed point depth of 400 ft and a frequency of 150 hz were selected for this demonstration.

Figure 5b is the transmission field plot for time $t = 20$ minutes. Subtle differences in the transmission loss for the different time snapshots can be picked out by eye.

However, a more quantitative result can be obtained by differencing the two acoustic fields. Figure 5c shows a pixel by pixel difference plot of the two fields in dB space (the $t = 0$ minutes field minus the $t = 20$ minute field). Because of the logarithmic form of transmission loss, a 2 dB difference at 80 dB is more important than a 2 dB error at 100 dB. To partially compensate for that, no differences were taken if the transmission loss was greater than 105 dB. These points are represented by the white areas. The color values have been changed to correspond to the difference in dB. All differences greater than 10 dB were collected in the last color bin. As expected the areas of largest differences show up in the region where the intensity is the greatest. For that reason the refracted paths can be seen in the difference plot.

Figure 5d is a histogram of the differences in dB vs the percentage of the acoustic field (e.g., 24.75% of the two fields are within 1 dB of each other).

One further test was run with these two environments. The frequency was changed to 5 hz and the fields produced by the PE model were in very good agreement (99.99% of the fields were within 1 Db of each other). This supports the contention that acoustic variation caused by internal waves will become more pronounced as the frequency increases.

5. Conclusions

We have developed a method for simulating the internal wave environment of the upper ocean for a general stratification. The simulated fields of temperature and salinity produced by our model have been used to derive time dependent fields of sound velocity that, in turn, have provided the input data in a parabolic equation transmission loss model. From the acoustic model, we have obtained variations in transmission loss on the order of 6 - 9 Db at ranges on the order of 50 nautical miles, though the variations in the total field have indicated by about 1/4 of the field has not changed by more than 1 Db. This type of environmental simulation is relatively simple to construct and efficient to generate and will probably play a large rôle in future acoustic simulations.

ACKNOWLEDGEMENTS

We would like to thank Drs. A. Green, M. Briscoe and D. Rubenstein for their help and support in this work. This work was supported by the Office of Naval Technology, program element 602435N, CDR. L. Bounds, program manager.

REFERENCES

Barnsley, M. (1988) Fractals Everywhere. Academic Press, Inc., New York.

Berge, D. (1987) Program VREAL2D for Correlating Velocity Profiles. Dynamics Technology Memo DTM-8721E-16_DAB.

Borchardt, S. (1985) Garrett-Munk Parameters. Dynamics Technology Memo DTM-8506C-05-SRB.

Borchardt, S. (1987) Personal Communication.

Botto, D., D. Berge, R. Cox (1985) Internal Wave Formalism. Dynamics Technology Memo DTM-8412B-33-DJB/ Revised.

Brock H. (1978) The AESD Parabolic Equation Model, NORDA Technical Report 12, January 1978.

Cox, R. (1986) Realization of Ambient Iws at the Surface. Dynamics Technology Memo DTM-8607I-26-RWC.

Dugan, J. (1989) Personal Communication.

Fock V. (1965) Electromagnetic Diffraction and Propagation Problems, Pergamon Press, New York. [Chapters 11, 12, and 13].

Hardin R. and F. Tappert (1973) Applications of the Split-step Fourier Method to the Numerical Solution of Non-linear and Variable Coefficient Wave Equations, SIAM Rev 15,423

Holland, W.R. (1989) Experiences with Various Parameterizations of Sub-Grid Scale Dissipation and Diffusion in Numerical Models of Ocean Circulation. Proc. 'Aha Huliko'a Hawaiian Winter Workshop, University of Hawaii at Manoa, Jan. 17-20, 1989, P. Müller and D. Henderson, Editors.

Huxtable, B. (1988) VREAL - Derivation of the Formulas for the Velocity Cross Spectral Density. Dynamics Technology Memo DTM-8748B-113-BDH.

Ierley, G. (1990) Boundary Layers in the General Ocean Circulation, Ann. Rev. Fluid Mech. 22,111-140.

Käse, R.H., A. Beckmann and H. Hinrichsen. (1989) Observational Evidence of Salt Lens Formation in the Iberian Basin. J. Geophys. Res. 94(C4), 4905-4912.

Kinsman, B. (1965) Wind Waves. Prentice Hall, Inc., Englewood Cliffs, New Jersey.

Leontovich M. and Fock V. (1946) Zh. Eks. Teor. Fiz. 16, 557-573 [J. Phys USSR 10(1946)], 13-24; Also See Fock 1965.
Lewis, J.P. (1987) Generalized Stochastic Subdivision. ACM Trans. on Graphics. 6(3), 167-190.

Levitus, S. (1982) Climatological Atlas of the World Ocean. NOAA Prof. Paper 12, 173 pp.

Lybanon, M., J.D. McKendrick, R.E. Blake, J.R.B. Cockett, and M.G. Thomason (1986) A Prototype Knowledge-Based System to the Oceanographic Image Analyst. SPIE 635, 203-206.

Lybanon, M. (1988) Oceanographic Expert System Functional Description. NORDA Technical Note 368. NORDA, Stennis Space Center, MS.

Macaskill, C. and T.E. Ewart (1984) Computer Simulation of Two-Dimensional Random Wave Propagation. IMA J. Appl. Math. 33,1-15.

Priestley, M.B. (1988) Non-Linear and Non-Stationary Time Series Analysis. Academic Press, New York.

Rubenstein, D. (1989) Personal Communication.

Saunders, K.D. (1989) Ocean Simulation Modeling. Proc. EAS Workshop on SONAR Simulation. NORDA, Stennis Space Center. May, 1989.

Saunders, K.D. and L. Knauer. (1987) XDP Mechanics. Proc. Oceans '87. Halifax, N.S.

Shen, C.Y. and G. Holloway (1986) A Numerical Study of the Frequency and Energetics of Nonlinear Internal Gravity Waves. J. Geophys. Res. 91(C1), 953-974.

Tappert F. (1977) The Parabolic Approximation Method: in Lecture Notes in Physics No. 70, Wave Propagation in Underwater Acoustics, Springer-Verlag, New York.

Teague, W.J., M.J. Carron and P. Hogan. (1989) A Comparison of the GDEM and Levitus Climatologies. (in press).

Winters, K.B. and E.A. D'Asaro (1989) Two Dimensional Instability of Finite Amplitude Internal Gravity Wave Packets Near a Critical Layer. J. Geophys. Res. 94(C9), 12709-12720.

A NUMERICAL INVESTIGATION OF SEMI-DIURNAL FLUCTUATIONS IN ACOUSTIC INTENSITY AT A SHELF EDGE

T. J. SHERWIN

Unit for Coastal and Estuarine Studies
University College of North Wales
Marine Science Laboratories
Menai Bridge, Gwynedd LL59 5EY
United Kingdom

ABSTRACT. Numerical techniques have been used to investigate how the strength of an acoustic signal from a fixed source located at the head of a shelf slope varies as the sound velocity field modulates in response to an internal tide. A numerical model simulated a large internal tide at a shelf edge in which typical internal elevations were > 100 m near the head of the shelf but of order 5-15 m in the upper and lower sound channels. An acoustic ray tracing model was then used to investigate the variations in the acoustic field for sources at a frequency of 1000 Hz located at depths of 10 m and 100 m. The acoustic intensity varied over a tidal cycle by more than 5 dB in places and was dependent on the relationships between source location, bottom topography and bottom loss characteristics. For a totally absorbing bottom variations in excess of 15 dB occurred.

1. Introduction

Synoptic iso-pycnal (and iso-velocity) surfaces are rarely horizontal at the edge of an ocean so range independent acoustic propagation assumptions are strictly invalid there. The list of processes that contribute to variability of the density field at an ocean-shelf boundary, includes (roughly in order of increasing frequency) slope currents, upwelling, eddies, continental shelf waves, Kelvin and edge waves, inertial motions, internal tides and other internal waves (Huthnance, 1981). Most of these processes have time scales of the order of days upwards, but internal waves (a term that includes internal tides) typically have periods of order half a day and less. The significance of such waves to short period modulations in the ocean acoustic field is not only recognized (e.g. Kirby, 1988) but, following recent developments in internal wave modelling techniques, is also open to numerical examination.

This paper examines the variations in acoustic propagation loss at a shelf edge and determines the order of magnitude and spatial scale of such variations as an acoustic field modulates in response to a large internal tide.

J. Potter and A. Warn-Varnas (eds.), Ocean Variability & Acoustic Propagation, 579–592.
© 1991 *Kluwer Academic Publishers.*

2. Internal Tide Simulation

2.1. BACKGROUND

It is generally accepted that internal tides are created by the interaction of the barotropic tide with stratification in the presence of rapidly changing topography. Two formative papers by Baines (1973 and 1982) demonstrated the generation process analytically using simple topography and density fields. Since then Craig (1987 and 1988) has modelled generation over uneven topography with gentle slopes and New (1988) has modelled steep linear slopes. More recently, Cushman-Roisin et al. (1989) have used a finite difference technique to model internal waves in fjords. The numerical approach favoured here however - first proposed by Chuang and Wang (1981) - has been shown by Sherwin and Taylor (1990a and 1990b) to be able to model linear internal tide generation for any reasonable combination of topography or density field. The model is called SPRITE (Software for Producing Realistic Internal Tidal Energy) and is described briefly below.

2.2. A DESCRIPTION OF SPRITE

SPRITE is a 2-D vertical slice linear f-plane numerical internal wave model which computes the amplitude and phase of a baroclinic stream function, ψ, at a particular wave frequency (ω, in our case M_2 or $2\pi/12\cdot42$ h^{-1}). The horizontal x-axis of the modelled region runs normal to the shelf edge and the vertical z-axis is positive upwards. There is no variation along the y-axis normal to the plane of the model (i.e. $\partial/\partial y = 0$). Density (ρ) can vary over the whole model domain and horizontal density gradients are assumed to be in geostrophic balance with the vertical shear in a mean current <v> flowing along the y-axis. (In this application, however, we assume $\partial\rho/\partial x = 0$ and <v> = 0). The internal tide is forced by barotropic tidal currents flowing parallel to the x-axis with a volume flux amplitude (per unit y) of Q and internal tide energy generated within the model radiates out of either end, onto the shelf and into the ocean.

With $\partial\rho/\partial x = 0$ the program solves the hyperbolic equation

$$\frac{\partial^2\psi}{\partial x^2} - \lambda^2 \frac{\partial^2\psi}{\partial z^2} = R \tag{1}$$

where $\psi = \Psi(x,z)\exp(i\omega t)$ is a complex stream function defined such that the horizontal and vertical baroclinic velocities are

$$u = -\frac{\partial\psi}{\partial z} \quad \text{and} \quad w = \frac{\partial\psi}{\partial x} \tag{2}$$

respectively. Energy flows along characteristics which have a slope, λ, given by

$$\lambda^2 = (\omega^2-f^2)/(N^2-\omega^2) \tag{3}$$

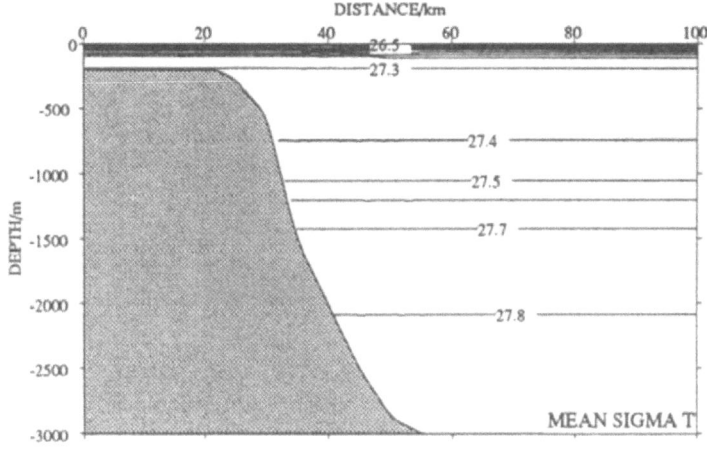

Figure 1. Topography and σ_t (intervals of 0.1 kg m^{-3}) used by SPRITE.

where f is the local Coriolis frequency (here $2\pi\sin(55°)/12$ h^{-1}) and

$$N^2 = -(g/\rho)\,\frac{\partial\rho}{\partial z} \qquad (4)$$

is the square of the local buoyancy frequency.

The solution is driven by the forcing function,

$$R = \frac{N^2}{(N^2 - \omega^2)}\, z\,\frac{\partial^2(h^{-1})}{\partial x^2}\, Q \qquad (5)$$

where h is the total depth of water at x.   Equation (1) can be transformed to sigma co-ordinates ($\Sigma = z/h$, $X = x$) and the problem solved using finite differences and a vectorized Gaussian elimination technique.

In order to make the solution matrix well conditioned in the presence of steep topography ($\partial h/\partial x > |\lambda|$), a linear damping coefficient is included by replacing ω with $\omega + i/\tau$ where τ is an e-fold decay time.   The stream function, ψ, is set to zero at the upper and lower boundaries and the ocean surface is assumed rigid.

2.3. THE SIMULATED INTERNAL TIDE

SPRITE was applied to a 100 km section over topography which varied from 200 m deep on the shelf, down a steep slope to 3000 m in the ocean (Fig. 1). A temperature/salinity profile observed at 56° N, 10·5° W in the Rockall Trough in August 1987 was used to determine the density profile - the raw data were smoothed by a Hanning filter with a 30 m vertical width to 270 m and by a running mean with 210 m vertical width below that.   The amplitude of a generated internal tide increases with the size of the barotropic tide

582

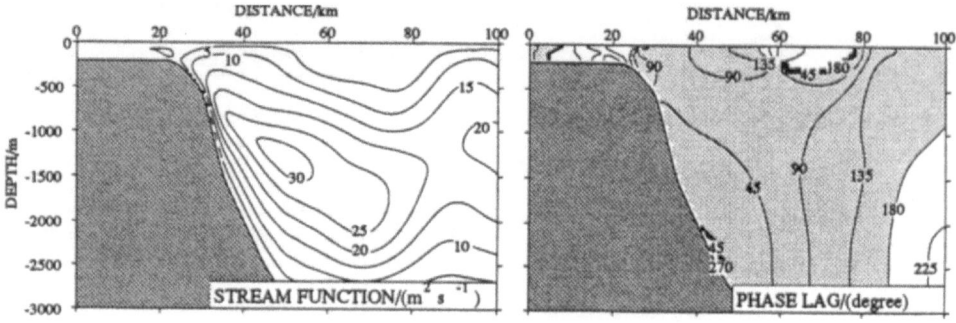

Figure 2. Stream function and phase lag of the internal tide solution. Shaded region lags the onshore barotropic tidal current.

and the steepness of the slope, and both the topography and Q (amplitude 50 m^2 s^{-1}, equivalent to a speed of 25 cm s^{-1} on the shelf) were chosen so that the simulation approximates to average (but not extreme) conditions in the SW approaches to the UK, where very large internal tides exist (Baines, 1982).

A horizontal grid spacing of 1 km was used with 51 levels in the vertical sigma grid giving a resolution of 4 m on the shelf and 60 m in the ocean. The decay time was 2 days and the results were smoothed, in the horizontal direction only, using a Hanning filter. The resulting residuals (LHS-RHS in Eq. (1)) were sufficiently small to give confidence in the matrix inversion process.

The results from SPRITE are summarized in Fig. 2 which shows the amplitude of the stream function and associated phase lag, relative to the onshore barotropic flux, over the solution domain. Most of the activity is concentrated near the top of the slope where maximum internal elevation amplitudes of over 100 m and horizontal velocity amplitudes of about 20 cm s^{-1} are computed near the sea-bed. In the upper and lower sound channels however (at about 100 m and 1600 m depth), internal elevation amplitudes were much smaller, from 5 to 15 m.

However, our interest concerns the variation in the σ_t (and hence sound velocity) profile over a tidal cycle. Figure 3 shows the variation in the level of the iso-pycnal surfaces near the top of the slope at four phases of the tidal cycle. The density at phase ωT was computed as

$$\rho(T) = \rho_0 - \frac{\partial \rho}{\partial z} \eta \cos(\omega T - \gamma) \tag{6}$$

where ρ_0 is the mean density at depth z and γ is the local phase lag associated with η, the sum of the baroclinic and barotropic elevations at z (from continuity, the barotropic tide contributes to the elevation at all depths over a sloping sea-bed). Iso-pycnals near the head of the slope reach a maximum elevation between the times of maximum barotropic flood and high water (HW, 3·1 h later), when the σ_t contour labelled 27·3 advances

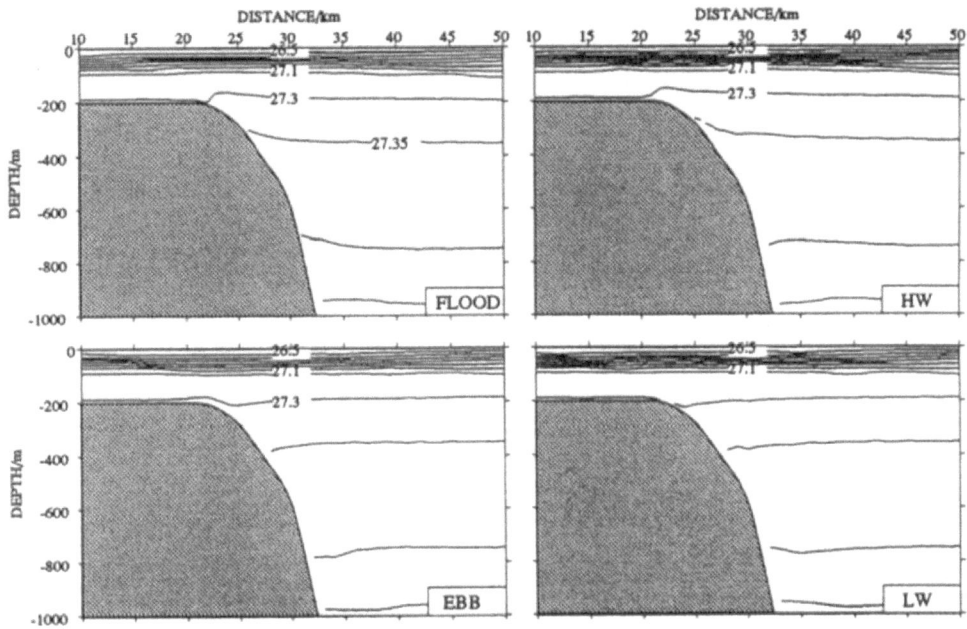

Figure 3. σ_t surface displacement at phases of the barotropic tide. Intervals are 0·1 kg m^-3 with the addition of contours at 27·35 and 27·45.

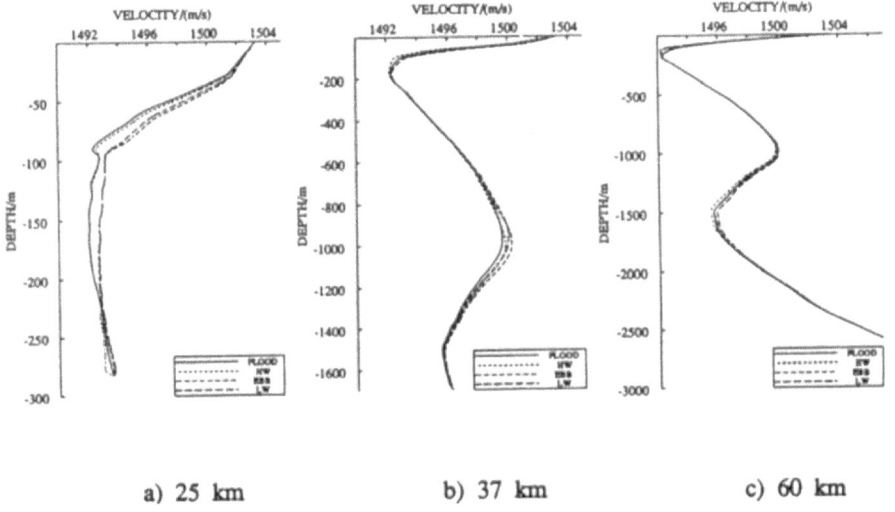

a) 25 km b) 37 km c) 60 km

Figure 4. Variation in sound velocity profiles over a tidal cycle at different locations over the slope. Note the different vertical scales.

towards the coast in a relatively steep wave with a phase speed of order 10 cm s⁻¹. At the other extreme, maximum depression of this contour occurs between maximum ebb and low water (LW). Contours in the deeper water (e.g. 27·4) lead those higher up by a few hours.

Sound velocity profiles were calculated from formulae due to Leroy (1969) using temperature and salinity profiles derived from the σ_t data (Fig. 4). At 25 km, on the edge of the shelf, the internal tide is mainly mode 1 with a phase that gives flood and HW (and ebb and LW) similar profiles. Minimum sound velocities near HW are about 0·7 m s⁻¹ less than near LW. At 37 km however, the phase of the internal elevation causes the sound velocity channel at 150 m to have a minimum at HW and a maximum at LW, whilst, by contrast, at 1000 km the maximum sound velocity is encountered on the ebb with a minimum on the flood. Thus the phase of the sound velocity profiles on the ocean side (at both 37 and 60 km) shifts with depth making it necessary to examine propagation loss at several stages of the barotropic tide in order to obtain a complete picture of the possible variations in acoustic propagation loss over a tidal cycle.

3. Modelling Acoustic Propagation Loss

3.1. A DESCRIPTION OF GRASS

GRASS (Cornyn, 1973) is an established acoustic propagation loss model which uses ray theory to solve the 3-D scalar wave equation

$$\frac{1}{c^2} \frac{\partial^2 \Phi}{\partial t^2} - \left[\frac{\partial^2 \Phi}{\partial x^2} + \frac{\partial^2 \Phi}{\partial y^2} + \frac{\partial^2 \Phi}{\partial z^2} \right] = 0 \tag{7}$$

where Φ is a field variable, for example acoustic pressure, and c is the velocity of sound in seawater. Ray theory assumes that the amplitude and phase of a sound wave with a particular frequency are slowly varying functions of space, from which it can be shown (e.g. Apel, 1987) that lines of equal phase (rays) satisfy

$$\left(\frac{\partial W}{\partial x} \right)^2 + \left(\frac{\partial W}{\partial y} \right)^2 + \left(\frac{\partial W}{\partial z} \right)^2 = \frac{c_0^2}{c^2} \tag{8}$$

where W is a phase function (with units of length) and c_0 is a constant, or reference, velocity.

If, for simplicity, it is assumed that c is horizontally invariant then it follows that ray paths can be constructed from Snell's law:

$$\frac{c_0}{\sin \alpha_0} = \frac{c}{\sin \alpha} \tag{9}$$

Here c_0 and α_0 are the sound velocity and incident angle (i.e. relative to the vertical) of a ray at its start position and c and α apply to some other depth.

GRASS uses these concepts to fire a fan of rays from a specified source position across a horizontal range. By tracing the paths of individual rays over the solution domain and computing their density at fixed locations (receivers), a 2-D picture of acoustic propagation loss is produced. A particular advantage of GRASS in the present application is that it can handle topography and sound velocity fields that vary with range. Disadvantages are that it does not accommodate combinations of bottom reflected and bottom refracted rays (Harrison, 1989) and that it does not apply in regions where rapid change of amplitude occur, for example, near caustics and at physical boundaries. Ray tracing also requires that the rate of change of sound velocity be small relative to the wavelength and, for this and other reasons, it is normally restricted to frequencies greater than about 1000 Hz (Cornyn, 1973).

Transmission losses between source and receiver, L, occur for various reasons and are incorporated as

$$L = A+B+S+G \quad dB \ km^{-1} \tag{10}$$

where A is the attenuation loss due to absorption (computed from Thorp's equation, see Cornyn, 1973), B is the loss encountered by bottom reflections, S is the loss encountered by surface reflections (assumed nil here) and G is the loss due to geometrical spreading. Bottom losses (when used) were computed in the present exercise using the standard values given in Table 1.

TABLE 1. Sea-bed propagation loss as a function of grazing angle

Grazing angle (degree)	0	10	20	30	40	50	60	70	80	90
Loss (dB)	5·5	7·0	8·0	9·0	10·0	10·5	11·0	11·5	11·5	11·0

All propagation loss plots were computed using a 30° ray fan, centred on the horizontal, with a density of 48 rays/degree giving a total of 1440 rays. Receivers were set on a 1 km × 25 m x × z grid and the intensities computed using a random phase summation technique. For presentation purposes, the results have been smoothed using a 2-D filter in which the 4 neighbouring vertical and horizontal points were weighted with 1/8 and the original point with 1/2. Sensitivity tests were conducted using i) 30 rays/degree and a 30° fan and ii) 30 rays/degree and a 48° fan. There was no significant difference between case i) and case ii), suggesting that the results are insensitive to fan width. Some differences were noticed between case i) and the values presented here, suggesting that the results may be sensitive to ray density, but they were less than those encountered for the different situations discussed below.

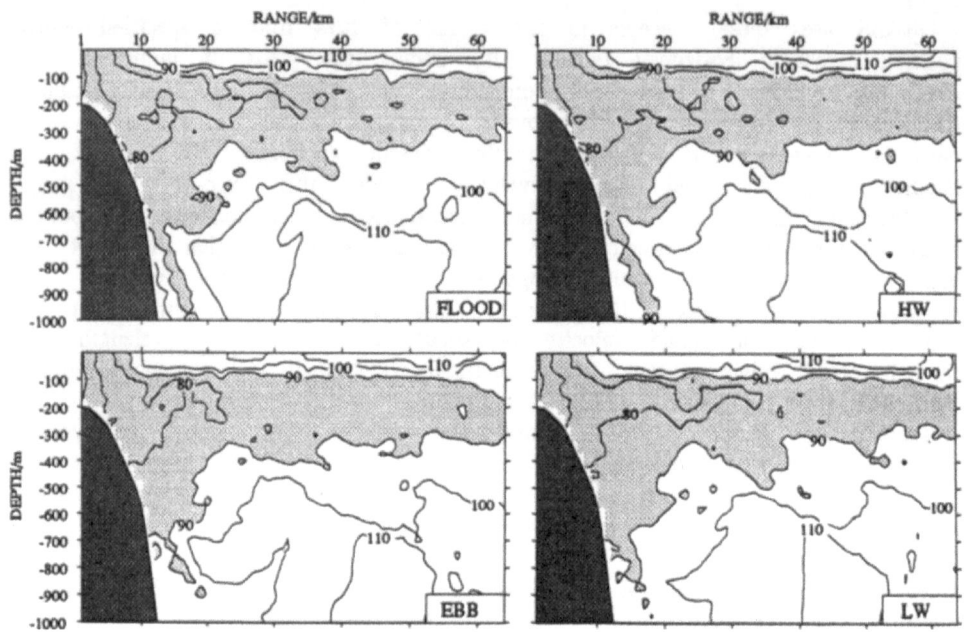

Figure 5.   Acoustic propagation loss (dB) for a source sited at 20 km from
the SPRITE origin and 100 m deep. Shaded region has a loss of < 90 dB.

3.2.   SOURCE AT 20 KM AND 100 M DEPTH

In the first trial, the source was located at mid depth near the edge of the
shelf in 200 m of water.   (In Fig. 5 and following, the x-axis (range) has
its origin at the source rather than the origin of SPRITE).   Most of the
sound from this source radiates out along the surface sound channel beneath
the quiet layer found in the top 100 m beyond 10 km although some sound also
propagates down the slope.   Evidently there are differences in the spatial
distribution of propagation loss over the tidal cycle.   For example, the
80 dB contour increases its range from 23 km on the ebb to 34 km at LW, and
the 90 dB contour penetrates to 1000 m on the flood but only to about 850 m
on the ebb and at LW.   In Fig. 6, however, it appears that the differences
between the tidal phases (in the upper 500 m out to 40 km) are generally
less than 5 dB although at some places pockets greater than 10 dB occur -
mainly on the upper and lower edges of the sound channel between 50 and
100 m and between 200 and 300 m.   Typically these pockets are of order 10 km
apart and are small, having dimensions that are similar to the smoothed grid
spacing of the receivers (2 km by 50 m), and are thus probably caused by the
formation and removal of local caustics.

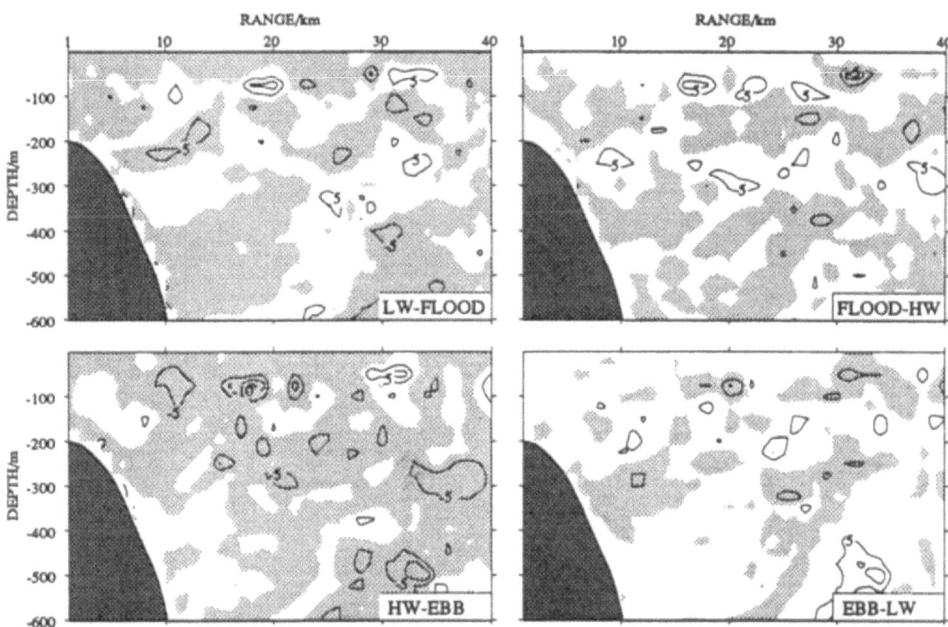

Figure 6. Changes in acoustic intensity over a tidal cycle near the head of the slope for the source pictured in Fig. 5. Contour intervals are 5 dB with the 0 dB contour omitted. Shaded regions get louder as the phase changes between the two times indicated in the figure captions (for example, the top left hand figure shows the change from LW to flood).

3.3. SOURCE AT 20 KM AND 10 M DEPTH

In the second trial the source was again located at the edge of the shelf, but this time near the surface. By comparison with the deeper source, most of the energy is directed down the slope (Fig. 7) where it is refracted back towards the surface, reaching a minimum depth at 30 km before propagating down again. As before, there appear to be significant variations in the solution over the tidal cycle with, for example, the 90 dB contour having a maximum depth penetration on the flood (> 1000 m) and a minimum on the ebb (800 m). There appears to be less variability for this source between phases of the tidal cycle, at least in the upper 400 m (Fig. 8). Overall, the variations appear to be more orderly than before, and long ribbons of constant change are apparent - for example between HW and ebb, the intensity increases by up to 5 dB along a line extending from the surface at 10 km to a depth of about 400 m at 20 km, where it merges into a group of ribbons running upwards to reach the surface at 30 km. These ribbons are probably caused by the change in the path of a group of relatively high energy rays. In deep water, between 400 and 600 m, they can be seen to be associated with larger changes of over 10 dB (see e.g. LW to flood).

588

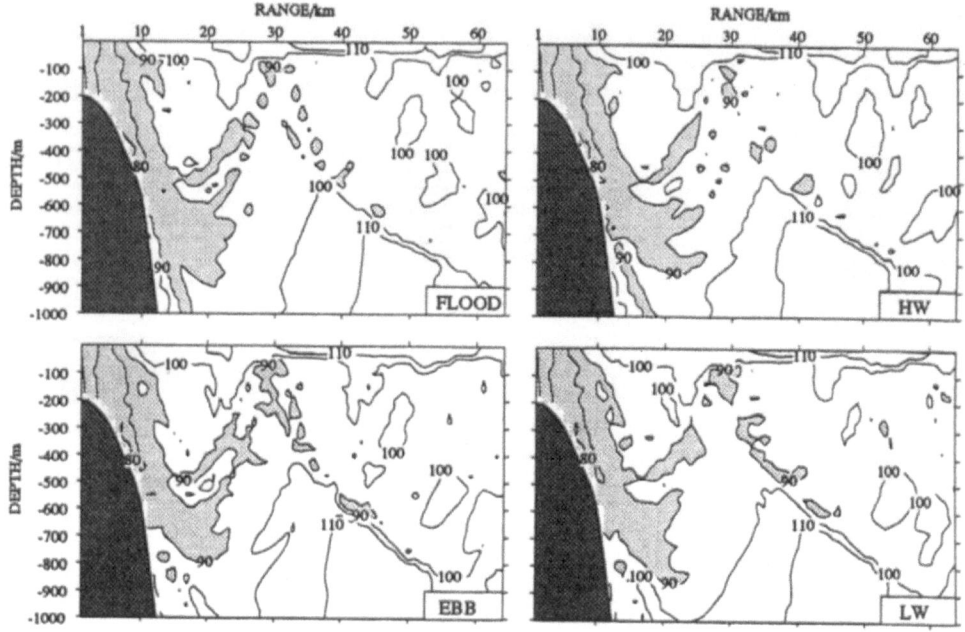

Figure 7.   As Fig. 5, but with a source at 10 m depth.

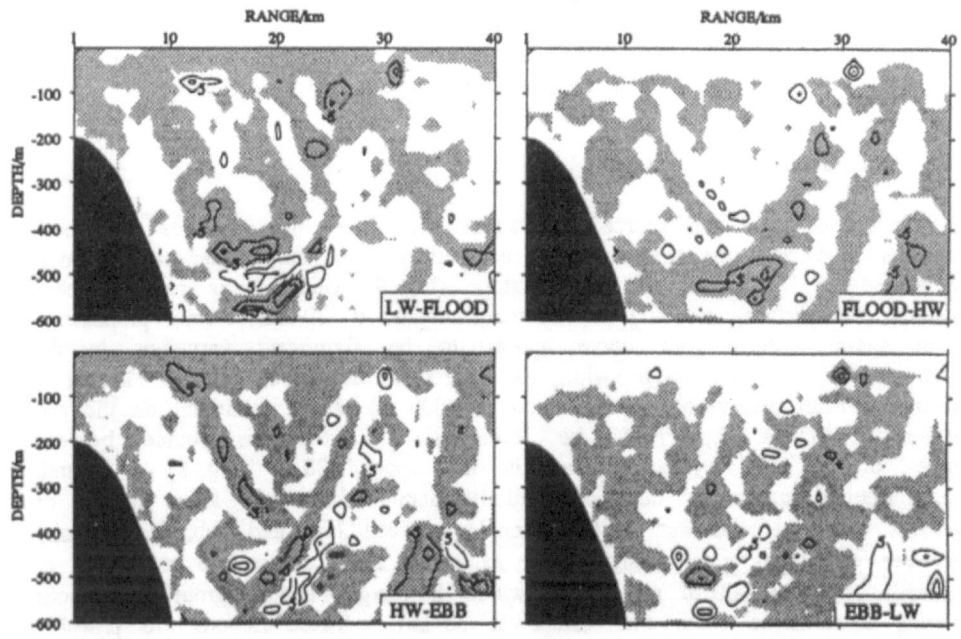

Figure 8.   As Fig. 6, but with a source at 10 m depth.

3.4. BOTTOM LOSS DEPENDENCE FOR A SOURCE AT 25 KM

It is apparent from the above trials that much of the sound energy propagates close to the sea-bed. To examine the sensitivity of the results to bottom absorption, tests were conducted with a source located at a depth of 100 m over the top of the slope (in 280 m of water). In the first case (Fig. 9a) the standard propagation losses of Table 1 were used, whilst in the second (Fig. 9b) a totally absorbing sea-bed was assumed. In both examples more sound is directed into the ocean than for the two sources at 20 km, probably because less energy hits (and is absorbed by) the sea-bed.

For standard sediments, there was virtually no difference between the flood and ebb tide conditions. However, for the totally absorbing sediments, in a region between 5 and 10 km from the source and at a depth range of about 100 to 200 m, the propagation loss was between 10 and 20 dB greater on the ebb than on the flood. This increased loss is due to a significant deepening of the quiet surface layer - as can be seen in Fig. 4a, the top of the sound channel is typically 10 m deeper on the ebb than on the flood.

4.   Discussion

These investigations have shown how a large internal tide modulates acoustic signals emanating from a fixed source over a tidal cycle. No fixed pattern or typical length scale for the variability has emerged, although for most of the solution domain the change in acoustic propagation loss was less than 5 dB. In certain cases, however, this variation exceeded 10 dB and for a totally absorbing bottom with a source near the head of the slope it was locally greater than 15 dB.

It is apparent that the variations are sensitive to the relationship between source position, bottom topography and bottom loss characteristics, as well as the phase of the tide; it would therefore be useful to conduct a more complete investigation over a wider range of conditions than considered here. Before doing so, however, one should question the suitability of GRASS for such work since, as indicated earlier, it does not handle combinations of bottom reflected and refracted waves adequately. It may be better to use a parabolic equation model which can cope not only with horizontal variations in sound velocity and topography but also lower frequencies, the complexities of the sea-bed and even current shear (Harrison, 1989).

In the present exercise we have considered barotropic tidal forcing at a fixed amplitude ($Q = 50$ m^2 s^{-1}), approximating to M$_2$ tides in the Celtic Sea or spring tides on the Malin Shelf, which is in some sense typical of the NW European shelf edge. Although the response of SPRITE is linear with respect to Q, the response of GRASS is inevitably non-linear so it is probably unwise to extrapolate the present findings too freely to other values of Q. However, since most of the gradients in the acoustic intensity field are small compared with any reasonable internal elevation amplitudes

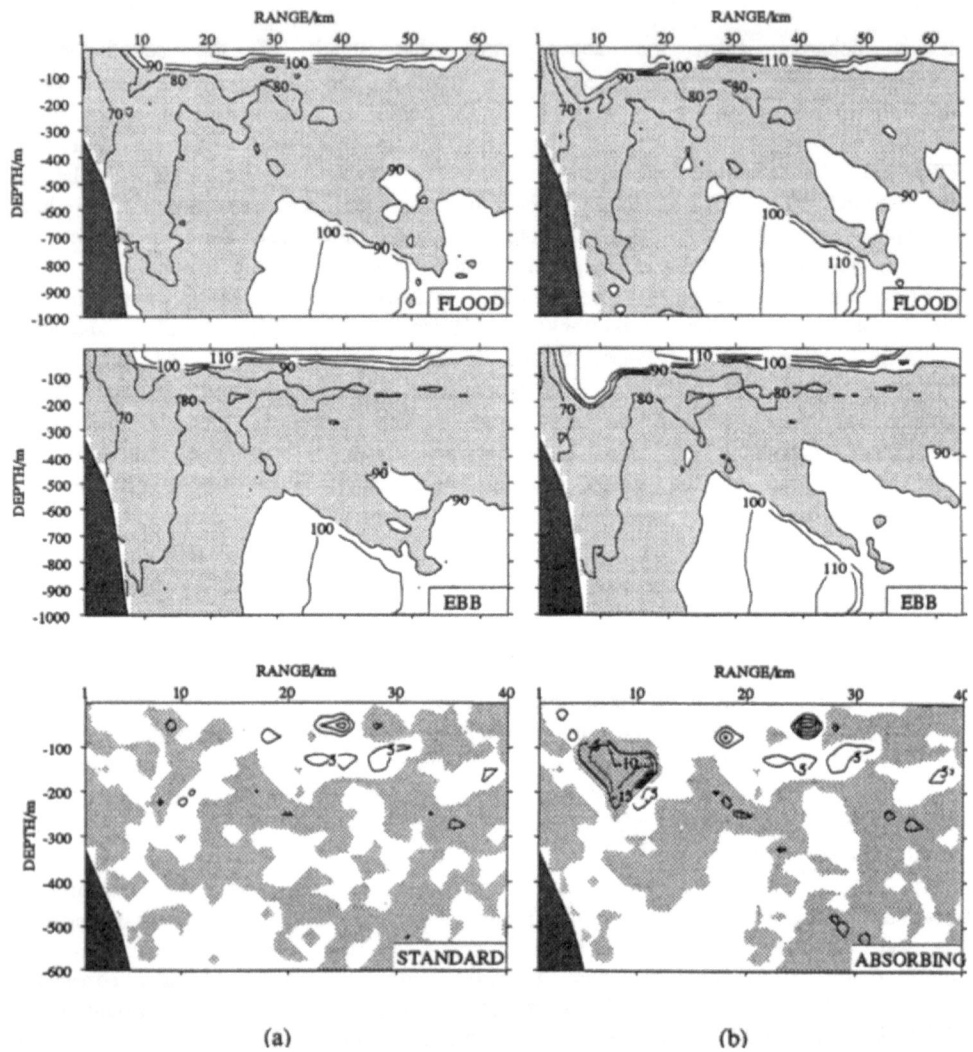

Figure 9. Propagation losses and changes in acoustic intensity between ebb and flood for a source located 25 km from the origin of SPRITE and 100 m deep. Bottom sediments a) have standard bottom loss (Table 1) and b) are totally absorbing.

(the main exceptions being at the base of the quiet surface layer and possibly at the head of the slope), variations may scale linearly over much of the solution domain.

The investigations would probably be improved by using a more realistic density field - in particular large internal tides, such as in the Celtic Sea, probably create significant internal mixing at the head of a slope (e.g. New, 1988). Such mixing will obviously alter the sound velocity profiles and reduce the surface velocity gradient in a sensitive part of the models. It is also possible to improve the internal tide predictions from SPRITE, by using a more sophisticated parametrization of decay.

This has been an essentially qualitative study to examine how internal tides modify an acoustic field, and has demonstrated that the main effect in the vicinity of a shelf edge is to oscillate the positions of the acoustic beams over a tidal cycle. The question of whether such modulations have tactical significance is beyond our scope. However, it is suggested that the scale of these modulations is too small for acoustic intensity measurements to be used as an oceanographic tool for monitoring internal tides at present.

Acknowledgements

This study was supported by the Procurement Executive (MOD) through the Admiralty Research Establishment, Portland.

References

Apel, J.R. (1987) *Principles of Ocean Physics*, Academic Press, London, pp 631 (Chap.7).

Baines, P.G. (1973) The generation of internal tides by flat-bump topography. *Deep-Sea Res.*, **20**, 179-205.

Baines, P.G. (1982) On internal tide generation models. *Deep-Sea Res.*, **29**, 307-338.

Chuang, W.-S. and D.-P. Wang, (1981) Effects of density front on the generation and propagation of internal tides. *J. Phys. Oceanogr.*, **11**, 1357-1374.

Cornyn, J.J. (1973) GRASS: a digital-computer ray-tracing and transmission-loss-prediction system. Volume I—overall description. Nav. Res. Lab. Report No. 7621, pp 67.

Craig, P.D. (1987) Solutions for internal tidal generation over coastal topography. *J. Mar. Res.*, **45**, 83-105.

Craig, P.D. (1988) A numerical model study of internal tides on the Australian Northwest Shelf. *J. Mar. Res.*, **46**, 59-76.

Cushman-Roisin, B., V. Tverberg and E.G. Pavia, (1989) Resonance of internal waves in fjords: A finite-difference model. *J. Mar. Res.*, **47**, 547-567.

Harrison, C.H. (1989) Ocean propagation models. *Appl. Acoustics*, **27**, 163-201.

Huthnance, J.M. (1981) Waves and currents near the continental shelf edge. *Prog. Oceanog.*, **10**, 193-226.

Kirby, G.J. (1988) Some range dependent effects in sonar performance. *Acoustics Bull.*, 12-14.

Leroy, C.C. (1969) Development of simple equations for accurate and more realistic calculation of the speed of sound in seawater. *J. Acoustic Soc. Am.*, **46**, 216-226.

New, A.L. (1988) Internal tidal mixing in the Bay of Biscay. *Deep-Sea Res.*, **35**, 691-709.

Sherwin, T.J. and N.K. Taylor (1990a) The application of a finite difference model of internal tide generation to the NW European Shelf. *Deut. Hydrog. Zeit.* [in press].

Sherwin, T.J. and N.K. Taylor (1990b) Numerical investigations of linear internal tide generation in the Rockall Trough. *Deep-Sea Res.* [in press].

Summary of Session 4

Heathershaw and Kuperman

This session provided an interesting mix of papers ranging from those dealing with fundamental aspects of acoustic propagation, to those dealing with the application of both ocean and acoustic models in operational forecast and tactical support systems. In the following review the papers have been grouped by topic rather than in the order in which they were presented.

The concept of acoustic propagation as an ordered sequence of localised events was emphasized in the paper by Steinberger and McCoy (given by McCoy). The marching of an acoustic field and the use of phase space representations of the field were both concepts that had led to improvements in our understanding of acoustic propagation in the presence of ocean variability, and in the development of computational algorithms. This approach only requires a description of that portion of the environment that effects a localised event, at the 'time' of the occurrence of that event.

A number of papers dealt with the problem of simulating the ocean environment for input to acoustic models and for other purposes. The use of data bases and dynamical models to simulate conditions in the shallow waters around the British Isles was described by Elliott. This work used a database or observations to provide initial conditions, and a fully depth resolving turbulence closure model to estimate vertical profile conditions at shorter timescales. The ocean model physics represented a balance between atmospheric forcing and mixing at the sea bed by tidal currents. The output from these simulations had been used as input to GRASS, which showed significant differences in the acoustic fields than when using forecast conditions.

The acoustic effects of internal tides generated at the shelf-edge were described by Sherwin. A sigma coordinate linear internal tide model was used to simulate the internal tide on a 2-D section across the shelf edge and these data were then input to GRASS to examine changes over the tidal cycle. Robinson described the Harvard Open Ocean Model (HOOM) which provides environmental simulations for input to acoustic models. Simulations of the mesoscale variability associated with Gulf Stream rings and meanders, employing feature models of eddies and jets, had been used as input to the 3-D acoustic model (FOR3D) to study the sensitivity of the acoustic predictions to ocean model and feature model parameters, in particular spatial resolution.

The sensitivity of acoustic predictions to changes in simulated mesoscale environments and ocean model parameters was described by Heathershaw, using a 3-D dynamical primitive equation ocean model and ray and wave theory models (GRASS and PAREQ). Realistic simulations of an idealised frontal environment were described. Significant variations in the predicted propagation loss are possible as a result of variations in lateral eddy viscosity in the ocean model. Saunders described the use of dynamical, stochastic, spectral semi-analytic and empirical models – along with databases and Artificial Intelligence and Experts Systems – to provide ocean simulations over a range of space and time scales. Specific examples dealt with simulations of internal waves and tides, and fronts, and the coupling together of models of these features with acoustic models.

J. Potter and A. Warn-Varnas (eds.), Ocean Variability & Acoustic Propagation, 593–595.
© 1991 *Kluwer Academic Publishers.*

The prediction of low frequency mesoscale acoustic variability in tactical support systems was described by Kerr. The EATS system has been developed as a tool for testing and evaluating environmental acoustic software specifically related to anti-submarine warfare. The acoustic models available on EATS are the US Navy's standard range dependent models, PE and ASTRAL. Examples were shown of the Gulf Stream and eddy simulations using multiradial acoustic transmission loss calculations.

Two contrasting aspects of ocean-acoustics were presented in the papers by Collins and Kuperman, and by Munk. Collins described how new methods in non-linear optimization theory (viz. 'simulated annealing') and recent advances in acoustic modelling now made source localization in a mesoscale environment possible. For this purpose the ocean might be regarded as a lens which could be focused to localise the sound source, even in the presence of noise. Results of a deep-ocean matched-field processing simulation were described in which the Munk canonical sound speed profile had been used to give a range dependent environment although the use of dynamical ocean models to provide estimates of the environment was also hinted at. Munk described an ambitious international experiment to use acoustics to measure decadal variations in ocean temperature in response to global warming. A pilot experiment is planned for 1991 in which a sound source will be deployed at Herd Island in the Southern Ocean, with listening stations throughout the Atlantic, Indian and Pacific Oceans. The propagation range, of $O(10^4)$ km, enables statistical averaging of the 'noise' associated with mesoscale variability, hopefully revealing global warming trends.

Session 4 Discussion

A number of topics emerged during the discussion following session 4. Weller thought that a significant point to emerge from the presentations was the use that could be made of acoustics to image the ocean. Farmer said that he was planning experiments to do this that would employ towed arrays and Ewart claimed that it was possible to image even a cubic metre of ocean. The question of using natural sound in the ocean to do this was raised and Farmer reported that he had studied the noise produced by breaking internal waves. Pinkel suggested that acoustics might be used to study heat uptake in the upper ocean and Robinson reminded us that the inverted echosounder had been widely employed in oceanographic studies; such an instrument when used in conjunction with feature models and EOFs would provide a useful technique for probing ocean structure. Munk added that acoustic tomography could do the heat content problem as well. Ewart and Kuperman both remarked that ambient noise from storms was likely to help in matched field processing.

Kuperman suggested that the study of ocean currents and acoustics might benefit from inputs from people working on other complex systems. Osborne felt that this was unlikely because the ocean-acoustic problem was too complex. Systems people in general were not looking at problems as complex as those addressed in this meeting. In this context Briscoe referred to some work by Holloway in which it had been found that an ocean with an assumed random distribution of eddies, and realistic topography, was able to give realistic estimates of bottom currents.

Pollard asked if from what had been heard at this meeting whether oceanographers should go on pushing sound through their ocean models. Ewart wanted to encourage the oceanographers to 'walk across the hall' and talk to the acousticians. Robinson remarked that there should not be uneducated collaboration between the two groups – collaboration needs to be a two way exchange of ideas. Flatté observed that the best kind of collaboration or merging together of the two fields occurs within the individual while Frisk felt that ocean models and acoustic models should have equivalent levels of sophistication (who's are the most sophisticated at the moment though?). Kuperman asked if there were any simple problems. Briscoe suggested that we study the effects of rain falling on the ocean (was this for ambient noise or to study the effect on sea state?). In this context Farmer remarked that there was a lot of signal and that both passive and active techniques might be employed. Ewart touched on the educational issues and wondered whether we needed to do more to educate young scientists in both areas. Munk felt that the two sides might be brought together through the establishment of a civilian ocean-acoustic observatory and drew an analogy with cosmology and the discovery of radio stars as the result of research on radars. It was suggested by Flatté that oceanographers perhaps need to ask more sophisticated questions of the acousticians. Pinkel thought that perhaps a fundamental point to emerge from the meeting was that for years people had used a spectral approach to describe ocean variability but that it might now be possible to describe ocean features with feature models. Others suggested that something better than the Fourier transform was needed. This was failing to describe the non-linear nature of the problem. The general consensus of opinion was that it would be another 10–20 years before such techniques were available and that in the meantime we should continue to gather over-sampled data sets while at the same time carrying out the acoustics experiments.

LIST OF PARTICIPANTS

Hubert Bouxin
Delegation Generale pour l'Armement
DCAN Toulon GERDSM
Le Brusc sixfours les plages 83140
France

Alan Brandt
Office of Naval Research
800 North Quincy St.
Arlington, VA 22217-5000
USA

Melbourne Briscoe
Office of Naval Research, Code 124
800 North Quincy St.
Arlington, VA 22217-5000
USA

Michael Broadhead
Naval Ocean and Atmospheric
Research Laboratory
Stennis Space Center
Bay St. Louis, MS.
USA

Michael Collins
Naval Research Laboratory
4555 Overlook Avenue
Washington DC 20375-5000
USA

Jean-Marc Darras
Société AERO
3 Avenue de l'Opera
Paris 75001
France

Lewis Dozier
Science Applications International Corp.
1710 Goodridge Drive
PO Box 1303, McLean VA 22102
USA

Brian Dushaw
Scripps Institute of Oceanography
University of California
La Jolla CA 92093
USA

Alan Elliott
University College of North Wales
Menai Bridge
Anglesey, LL59 5EY
UK

Dale Ellis
Defence Research Establishment Atlantic
9 Grove Street, P.O. Box 1012
Dartmouth, N.S. B2Y 3Z7
Canada

Terry Ewart
APL, Univ. Washington
1013 NE 40th St.
Seattle, WA 98105-6698
USA

David Farmer
Institute of Ocean Sciences
P.O. Box 6000, 9860 West Saanich Road
Sidney, BC V8L 4B2
Canada

Robert Field
Naval Ocean and Atmospheric
Research Laboratory
Stennis Space Center
MS 39529 5004
USA

Stan Flatté
University of California
Santa Cruz
CA 95064
USA

George Frisk
Woods Hole Oceanographic Institution
Bigelow 211
Woods Hole, MA 02543
USA

Tony Heathershaw
Admiralty Research Establishment
Ocean Science Division
Southwell, Portland, Dorset DT5 2JS
UK

598

George Heburn
Naval Ocean and Atmospheric
Research Laboratory
Stennis Space Center, code 323
MS 39529 5004
USA

Tom Hopkins
SACLANT Undersea Research Centre
Viale San Bartolomeo, 400
La Spezia 19026
Italy

Bruce Howe
APL, Univ. Washington
1013 NE 40th St.
Seattle, WA 98105
USA

Pam Jackson
Naval Ocean and Atmospheric
Research Laboratory
Stennis Space Center
Bay St. Louis, MS.
USA

Larry Jendro
Office of Naval Research (Europe)
223, Old Marylebone Road
London NW1 5TH
UK

Finn Jensen
SACLANT Undersea Research Centre
Viale San Bartolomeo, 400
La Spezia 19026
Italy

George Kerr
Naval Ocean and Atmospheric
Research Laboratory
Stennis Space Center
MS 39529-5004
USA

Bill Kuperman
Naval Research Laboratory
4555 Overlook Avenue
Washington DC 20375-5000
USA

Robert Laval
Societé AERO
3 Avenue de l'Opera
Paris 75001
France

K. Maroihi
GeoHydrodynamics & Environ. Res.
University of Liege
B5 Sart Tilman, B-4000 Liege
Belgium

John McCoy
The Catholic University of America
School of Engineering
Washington DC 20064
USA

Ed McDonald
Naval Research Laboratory, code 5105
4555 Overlook Avenue
Washington DC 20375-5000
USA

Robert Mellen
The Kildare corporation
95 Trumbull Street, Suite D
New London, CT 06320
USA

Claire Mooney
Admiralty Research Establishment
Ocean Science Division
Southwell, Portland, Dorset DT5 2JS
UK

Walter Munk
Scripps Institute of Oceanography
University of California
La Jolla CA 92093-0225
USA

Alfred Osborne
Istituto di Cosmo-Geofisica
Corso fiume, 4
10133 Torino
Italy

Hank Perkins
SACLANT Undersea Research Centre
Viale San Bartolomeo, 400
La Spezia 19026
Italy

Joseph Phillips
Institute for Geophysics, Univ. Texas
8701 Mopac Boulevard
Austin, TX 78759-8345
USA

Robert Pinkel
Scripps Institute of Oceanography
University of California
La Jolla CA 92093 Mail code A-013
USA

Raymond Pollard
Insitute of Oceanographic Sciences
Deacon Laboratory, Wormley
Godalming, Surrey GU8 5UB
UK

John Potter
SACLANT Undersea Research Centre
Viale San Bartolomeo, 400
La Spezia 19026
Italy

John Preston
SACLANT Undersea Research Centre
Viale San Bartolomeo, 400
La Spezia 19026
Italy

Allan Robinson
Harvard University
29 Oxford Street
Cambridge, MA 02138
USA

David Rubenstein
Science Applications International Corp.
1710 Goodridge Drive
PO Box 1303, McLean VA 22102
USA

Roger Samelson
Woods Hole Oceanographic Institution
Woods Hole, MA 02543
USA

Kim Saunders
Naval Ocean and Atmospheric
Research Laboratory
Stennis Space Center
MS 39529 5004
USA

Hans Schneider
Forschungsanstalt der Bundeswehr
für Wasserschall und Geophysik
2300 Kiel 14
Klausdorfer Weg 2-24
FRG

John Scott
Admiralty Research Establishment
Ocean Science Division
Southwell Portland Dorset DT5 2JS
UK

Jürgen Sellschopp
Forschungsanstalt der Bundeswehr
für Wasserschall und Geophysik
2300 Kiel 14
Klausdorfer Weg 2-24
FRG

Toby Sherwin
University College of North Wales
Menai Bridge
Anglesey, LL59 5EY
UK

Kevin Smith
AMP, RSMAS
4600 Rickenbacker causeway
Miami, Florida 33149-1098
USA

Rolf Thiele
Forschungsanstalt der Bundeswehr
für Wasserschall und Geophysik
2300 Kiel 14
Klausdorfer Weg 2-24
FRG

Dong-Ping Wang
Marine Sciences Research Center
SUNY, Stony Brook, NY 11794-5000
NY
USA

Alex Warn-Varnas
SACLANT Undersea Research Centre
Viale San Bartolomeo, 400
La Spezia 19026
Italy

Chick Weinberg
Naval Underwater Systems Center
New London
CT 06320
USA

Robert Weller
Woods Hole Oceanographic Institution
Clark 204a
Woods Hole, MA 02543
USA

602

Munk, W.H. : *261, 539.*

n

New, A.L. : *417.*

o

Osborne, A.R. : *407.*

p

Pearson, C.R. : *265.*
Perkins, J.S. : *433.*
Phillips, J.D. : *199.*
Pinkel, R. : *103.*
Pollard, R.T. : *417.*
Potter, J.R. : *481.*
Prior, M. : *359.*

r

Rajan, S.D. : *69.*
Ramp, S.R. : *343.*
Reynolds, S.A. : *23.*
Robinson, A.R. : *545.*
Rovner, G. : *161.*
Rubenstein, D. : *215.*

s

Samelson, R.M. : *463.*
Saunders, K.D. : *561.*
Schneider, H.G. : *283, 323.*
Scott, J.C. : *449, 481.*
Sellschopp, J. : *229, 293.*
Sherman, J.T. : *103.*
Sherwin, T.J. : *579.*
Siegmann, W.L. : *545.*
Smith, K.B. : *139.*
Spindel, R.C. : *81.*
Stergiopoulos, S. : *237.*

t

Tappert, F.D. : *139.*
Thiele, R. : *313.*

W

Wang, D. : *251.*
Weller, R.A. : *463.*
Worcester, P.F. : *81, 119.*

SUBJECT INDEX

a

AATE: *23.*
acoustic
  backscatter: *173.*
  image: *433.*
  impedance interface: *199.*
  intensity: *103, 579.*
  mechanisms: *343.*
  modelling: *87, 359, 517.*
  modes: *433, 539.*
  normal mode: *119.*
  prediction: *327.*
  propagation in shallow water: *69.*
  propagation loss: *545.*
  propagation: *41.*
  ray paths: *103.*
  spectrum: *215.*
  structure: *485.*
  transmission loss models: *561.*
  travel time: *81.*
  variability: *215.*
adiabatic
  changes: *539.*
  mode theory: *433.*
  normal modes: *527.*
air–sea interaction: *391.*
airgun pulses: *199.*
altimetry: *327.*
ambient noise: *433.*
ambiguity function: *433, 527.*
angular spectrum: *23.*
anisotropic medium: *237.*
anisotropy: *41.*
anomalous
  diffusion: *407.*
  seasons: *485.*
Antarctic circumpolar current: *539.*
Arctic
  transmission loss data: *265.*
  water: *359.*
assimilation techniques: *327.*
ASTRAL model: *265.*
ATHENA: *327.*
Atlantic
  inflow experiment: *391.*
  intermediate water (AIW): *87.*
  Mediterranean outflow: *375.*
  water: *359.*

autocorrelation functions: *391.*
AXBT: *57.*

b

Baltic Sea: *229, 283, 313.*
bandlimited impulse response: *57.*
baroclinic perturbation: *501.*
barotropic
  electromagnetic and pressure
    experiment: *119.*
  perturbation: *501.*
Bartlett: *433.*
batfish: *313.*
bathythermograph: *57.*
beamforming: *81.*
BEMPEX: *119.*
Biot theory: *69.*
Born approximation: *173.*
bottom
  loss: *501.*
  sediments: *501.*
  topography: *501.*
Bragg scatter: *173.*
breaking waves: *173.*
broadband acoustic source: *81.*
Brunt–Väisälä: *0.*
bubble plumes: *173.*
buoyancy frequency: *23.*

c

California current system: *343.*
Celtic Sea front: *485.*
channel leakage: *293.*
chaos: *139.*
chromatic aberration: *539.*
circulation: *119.*
climatic model: *485.*
closure model: *485.*
Clyde Sea: *485.*
coastal
  environment: *41.*
  oceans: *251.*
  zone color scanner: *343.*
Cobb Seamount: *103.*
cold core eddies: *501.*

..